BUSINESS/SCIENCE/TECHNOLOGY DIVISION
CHICAGO PUBLIC LIBRARY
400 SOUTH STATE STREET
CHICAGO, IL 60605

REVIEWS in MINERALOGY and GEOCHEMISTRY

VOLUME 57   2005

MICRO- AND MESOPOROUS MINERAL PHASES

EDITORS:

Giovanni Ferraris   Università di Torino
                          Torino, Italy

Stefano Merlino   Università di Pisa
                          Pisa, Italy

FRONT COVER: Perspective view of the crystal structure of montregianite seen along [101]. Channels delimited by eight-membered rings of silicon tetrahedra form (010) layers that alternate with continuous layers of sodium–yttrium octahedra. Potassium cations and water molecules within the channels are shown as blue and red spheres, respectively. Atomic coordinates from Ghose et al. (1987) Am Mineral 72:365-374. The photoluminescence spectrum and image of synthetic montregianite-(Tb) excited by X-rays (8.050 keV) are also depicted. See chapter 6 for details.

BACK COVER: *(top)* Mesopores in the hollow center of chrysotile fibers (average diameter close to 7 nm); see chapter 12 for details. HRTEM image by courtesy of C. Viti. *(bottom)* Association of two microporous titanosilicate minerals: cauliflower like penkvilksite (2 × 1 cm) on radiating spherical aggregates of pink zorite within a cavity of the peralkaline Yubileynaya pegmatite (Mt. Karnasurt, Lovozero, Kola Peninsula, Russia). See chapters 1, 2, 4, 5 and 6 for details. From the collection of Igor V. Pekov; photo by Natalia A. Pekova.

Series Editor: **Jodi J. Rosso**

MINERALOGICAL SOCIETY of AMERICA
GEOCHEMICAL SOCIETY
ACCADEMIA NAZIONALE dei LINCEI

COPYRIGHT 2005

MINERALOGICAL SOCIETY OF AMERICA

The appearance of the code at the bottom of the first page of each chapter in this volume indicates the copyright owner's consent that copies of the article can be made for personal use or internal use or for the personal use or internal use of specific clients, provided the original publication is cited. The consent is given on the condition, however, that the copier pay the stated per-copy fee through the Copyright Clearance Center, Inc. for copying beyond that permitted by Sections 107 or 108 of the U.S. Copyright Law. This consent does not extend to other types of copying for general distribution, for advertising or promotional purposes, for creating new collective works, or for resale. For permission to reprint entire articles in these cases and the like, consult the Administrator of the Mineralogical Society of America as to the royalty due to the Society.

REVIEWS IN MINERALOGY
AND GEOCHEMISTRY

(Formerly: REVIEWS IN MINERALOGY)

ISSN 1529-6466

Volume 57

Micro- and Mesoporous Mineral Phases

ISBN 093995069-3

Additional copies of this volume as well as others in this series may be obtained at moderate cost from:

THE MINERALOGICAL SOCIETY OF AMERICA
1015 EIGHTEENTH STREET, NW, SUITE 601
WASHINGTON, DC 20036 U.S.A.
WWW.MINSOCAM.ORG

MICRO- AND MESOPOROUS MINERAL PHASES

57   *Reviews in Mineralogy and Geochemistry*   **57**

FROM THE SERIES EDITOR

This volume was jointly published by the Mineralogical Society of America (MSA) and the Accademia Nazionale dei Lincei (ANL). The chapters were invited contributions to the December 6-7, 2004 meeting on Micro- and Mesoporous Mineral Phases held in Rome, Italy. A volume of extended abstracts was published in 2004 in occasion of the Rome meeting ("Micro- and Mesoporous Mineral Phases," Accad Lincei, Rome; ISBN 88-86409-41-9). It reports the abstracts of 73 contributions including the 12 invited ones that are the content of this RiMG volume.

Errata (if any) can be found at the MSA website *www.minsocam.org*.

Jodi J. Rosso, Series Editor
West Richland, Washington
May 2005

PREFACE

In Materials Science, investigations aiming to prepare new types of molecular sieves (porous materials) have opened a productive field of research inspired by the crystal structures of minerals. These new molecular sieves are distinct from zeolites in that they have different kinds of polyhedra that build up their structures. Of particular interest are the new molecular sieves characterized by a mixed "octahedral"-tetrahedral framework (heteropolyhedral frameworks), instead of a purely tetrahedral framework as in zeolites.

Heteropolyhedral compounds have been extensively studied since the early 1990's, with particular attention having been focused on titanosilicates, such as ETS-4 (synthetic analog of the mineral zorite) and ETS-10. However, titanosilicates are not the only representatives of novel microporous mineral phases. The search for "octahedral"-tetrahedral silicates was extended to metals other than titanium, for instance, the zirconosilicates with the preparation of synthetic counterparts of the minerals gaidonnayite, petarasite and umbite. Many microporous heteropolyhedral compounds containing metals such as Nb, V, Sn, Ca and lanthanides, have been reported and a wide number of distinct structural types (e.g., rhodesite-delhayelite and tobermorite) have been synthesized and structurally characterized. Moreover, the potential applications of these novel materials have been evaluated, particularly in the areas of catalysis, separation of molecular species, ion exchange and optical and magnetic properties.

A comprehensive review of the mineralogical, structural, chemical and crystal-chemical studies carried on natural phases may be extremely useful to inspire and favor investigations on analogs or related synthetic materials. A similar synergy between mineralogists and materials scientists already occurred in the "classical" case of zeolites, in which the wide and deep structural and crystal-chemical knowledge accumulated in the study of the natural phases was extraordinarily useful to the chemists who are active in the field of molecular sieves. In particular, the structural investigation of the natural phases may be extremely rewarding and helpful in orienting the work of synthesis and in understanding the nature of the synthetic products, for the following reasons:

- Whereas rarely the crystalline synthetic products are suitable for single-crystal structural investigations, the natural counterparts are often well crystallized.
- Crystallization in nature occurs from chemical systems characterized by a wide compositional range, thus producing compounds with a very rich and variable crystal chemistry, which may provide precious information, suggesting possible substituting elements and addressing the synthetic work in a very productive way.

The present volume follows a meeting on "Micro- and mesoporous mineral phases" (Rome, December 6-7, 2004) that was jointly organized by the Accademia Nazionale dei Lincei (ANL) and the International Union of Crystallography (IUCr) via its Commission on Inorganic and Mineral Structures (CIMS). The meeting was convened by Fausto Calderazzo, Giovanni Ferraris, Stefano Merlino and Annibale Mottana and financially supported by several other organizations representing both Mineralogy (e.g., the International Mineralogical Association and the European Mineralogical Union) and Crystallography (e.g., the European Crystallographic Association and the Italian Association of Crystallography). To participants, ANL staff, organizations, and, in general, all involved persons, our sincere acknowledgments; in particular, we are grateful to Annibale Mottana who was able to convince the ANL Academicians to schedule and support the meeting.

This volume of the RiMG series highlights the present knowledge on micro- and mesoporous mineral phases, with focus on their crystal-chemical aspects, occurrence and porous activity in nature and experiments. As zeolites are the matter of numerous *ad hoc* meetings and books—including two volumes in this series—they do not specifically appear in the present volume. The phases of the sodalite and cancrinite-davyne groups, which mineralogists consider distinct from zeolites, are instead considered (in the order, chapter 7 by W. Depmeier and part of chapter 8 by E. Bonaccorsi and S. Merlino, respectively).

The first two chapters of the volume cover general aspects of porous materials. This includes the application of the IUPAC nomenclature developed for ordered porous materials to non-zeolite mineral phases (L.B. McCusker, chapter 1) and the extension to heteropolyhedral structures of a topological description by using nodes representing the coordination polyhedra (S.V. Krivovichev, chapter 2). Chapters from 3 to 7 are dedicated to various groups of heteropolyhedral porous structures for which the authors emphasize some of the more general aspects according to their research specialization. G. Ferraris and A. Gula (chapter 3) put the emphasis on the modular aspects of well-known porous phases (such as sepiolite, palygorskite and rhodesite-related structures) as well as on heterophyllosilicates that may be not strictly porous phases (according to the definition given in chapter 1) but could be the starting basis for pillared materials. The porous mineral phases typical of hyperalkaline rocks (such as eudialytes and labuntsovites) are discussed by N.V. Chukanov and I.V. Pekov under their crystal-chemical (chapter 4) and minerogenetic (chapter 5) aspects showing the role of ion exchange during the geological evolution from primary to later phases, with experimental cation exchange data also being reported. J. Rocha and Z. Lin (chapter 6) emphasize how research on the synthesis of octahedral-pentahedral-tetrahedral framework silicates has been inspired and motivated by the many examples of such materials provided by nature; synthesis, structure and possible technological applications of a wide number of these materials are also described. Following chapters 7 and 8—which besides the cancrinite-davyne group, presents the crystallographic features of the minerals in the tobermorite and gyrolite groups—M. Pasero (chapter 9) illustrates the topological and polysomatic aspects of the "tunnel oxides," a historical name applied to porous oxides related to MnO_2, and reviews their main technological applications. The next two chapters (10 and 11) draw attention to "unexpected" porous materials like apatite and sulfides. T.J. White and his team (chapter 10) convincingly show that the apatite structure type displays porous properties, some of which are already exploited. Chapter 10 also contains two appendices that report crystal and synthesis data for hundreds of synthetic apatites, a number that demonstrates how wide the interest is for this class of compounds. E. Makovicky (chapter 11) analyzes the structures of natural and synthetic sulfides and selenides showing that, even if experimental work proving porous activity is practically still missing, several structure types display promising channels. Chapter 12, by M. Mellini, is the only one dedicated to mesoporous mineral phases—which are crystalline compounds with pores wider than 2 nm. Examples discussed are carbon nanotubes, fullerenes—which occur also in nature—chrysotile, opal and, moving from channels to cages, clathrates.

We highly appreciate the hard work and patience of series editor Jodi J. Rosso and are grateful to the colleagues who accepted to contribute to the meeting and this volume, both as authors and reviewers in the process of cross refereeing. We hope that the topics presented in this volume will contribute to establish further links between the Minerals Science and the wider Materials Science.

May 2005

Giovanni Ferraris   *Stefano Merlino*
Torino, Italy   Pisa, Italy

TABLE OF CONTENTS

1   IUPAC Nomenclature for Ordered Microporous and Mesoporous Materials and its Application to Non-zeolite Microporous Mineral Phases

Lynne B. McCusker

INTRODUCTION	1
Concept	1
GENERAL DEFINITIONS	2
The host	2
The pores	4
CRYSTAL CHEMICAL FORMULA	5
Chemical composition of the guest species	6
Chemical composition of the host	6
Structure of the host	7
Structure of the pores	7
Symmetry	7
IZA code	7
EXAMPLES	8
Hollandite	8
Synthetic hollandite	8
Lovozerite	9
Labuntsovite	10
Tobermorite	11
Tetrahedrite	13
Rhodesite	14
CONCLUSIONS	15
REFERENCES	15

2   Topology of Microporous Structures

Sergey Krivovichev

INTRODUCTION	17
BASIC CONCEPTS AND TOOLS	17
Nodal description of complex structural architectures	17
Polyhedra	18
Tilings	20
Regular and quasiregular nets	20
HETEROPOLYHEDRAL FRAMEWORKS: CLASSIFICATION PRINCIPLES	22
FRAMEWORKS BASED UPON FUNDAMENTAL BUILDING BLOCKS (FBBs)	26
Some definitions	26
Leucophosphite-type frameworks	26
Frameworks with oxocentered tetrahedral cores	27
Pharmacosiderite-related frameworks	28

Micro- & Mesoporous Mineral Phases – Table of Contents

FRAMEWORKS BASED UPON POLYHEDRAL UNITS 29
    Example 1: minerals of the labuntsovite group...29
    Example 2: shcherbakovite-batisite series 30
    Combinatorial topology of polyhedral units 31
    Topological complexity of polyhedral units: petarasite net 33
FRAMEWORKS BASED UPON 1D UNITS..33
    Fundamental chains as bases of complex frameworks..33
    Frameworks with parallel and non-parallel orientations of fundamental chains 35
    Zorite and ETS-4 35
    Benitoite net as based upon arrangement of polyhedral units and tubular units.........37
    Tubular units, their topology, symmetry and classification..38
FRAMEWORKS BASED UPON 2D UNITS..40
    Octahedral-tetrahedral frameworks in silicates with 2D silicate anions 40
    Komarovite..49
    Umbite-related frameworks..49
    Terskite 50
    The use of 2D nets to recognize structural relationships..50
    2D nets and polytypism: the case of penkvilksite 52
    Framework uranyl phosphates and arsenates 52
FRAMEWORKS BASED UPON BROKEN SILICATE FRAMEWORKS 53
DYNAMIC TOPOLOGY OF FRAMEWORKS:
    CELLULAR AUTOMATA MODELING 53
    Introductory remarks 53
    Pentlandite-djerfisherite-bartonite frameworks...55
    Cellular automata and frameworks of cubes 57
    Transition rules of cellular automata and chemical requirements...............................58
CONCLUDING REMARKS: THE ART OF STRUCTURE DESCRIPTION 59
ACKNOWLEDGMENTS 59
REFERENCES 59

3   Polysomatic Aspects of Microporous Minerals – Heterophyllosilicates, Palysepioles and Rhodesite-Related Structures

Giovanni Ferraris & Angela Gula

INTRODUCTION 69
HETEROPHYLLOSILICATES 70
    Polysomatism in the heterophyllosilicates 71
    The bafertisite series...76
    Structural aspects in the bafertisite series...79
    Porous features in heterophyllosilicates 83
    Leaching and solid state transformations 84
    Pillared derivatives? 88
PALYSEPIOLES..88
    Sepiolite, palygorskite and related structures...88

Merotypes and plesiotypes of palysepioles 90
Microporous features of palysepioles 91
Some porous structures related to palysepioles 92
SEIDITE-(Ce) AND RELATED STRUCTURES 92
Modeling the structure of seidite-(Ce) 93
Microporous properties 94
Modular aspects 95
Structures related to the rhodesite series 97
CONCLUSIONS 99
ACKNOWLEDGMENTS 99
REFERENCES 99

4 Heterosilicates with Tetrahedral-Octahedral Frameworks: Mineralogical and Crystal-Chemical Aspects

Nikita V. Chukanov & Igor V. Pekov

INTRODUCTION 105
GENERAL CHARACTERIZATION AND CLASSIFICATION OF MICROPOROUS
HETEROSILICATE MINERALS 105
GENERAL CRYSTAL-CHEMICAL CHARACTERISTICS 106
Force parameters 106
T:M ratios 107
Framework density 108
CRYSTAL STRUCTURES OF THE MOST IMPORTANT MHM 109
Minerals with lowest framework densities 109
Labuntsovite-group minerals 121
Other important MHM 124
Factors influencing the crystal-chemical diversity of MHM 128
PROPERTIES AND SOME GENETIC IMPLICATIONS 129
Ion exchange, cation leaching and gas sorption 129
Other properties 132
Interaction of MHM with organic substance in nature 134
CONCLUSIONS 136
ACKNOWLEDGMENTS 136
REFERENCES 136

5   Microporous Framework Silicate Minerals with Rare and Transition Elements: Minerogenetic Aspects

Igor V. Pekov & Nikita V. Chukanov

INTRODUCTION 145
DISTRIBUTION AND GENETIC FEATURES 146
    Occurrences and genetic types 146
    Crystal chemical aspects of minerogenesis 151
EVOLUTION OF PARAGENESES: STRUCTURAL ASPECTS 154
    Evolution series 154
    Dual character of MHM evolution series 155
    Role of epitaxy 156
SOLID-STATE TRANSFORMATIONS WITH THE PRESERVATION OF THE
    FRAMEWORK 159
    Decationization and hydration 159
    Natural ion exchange 162
MHM AS GEOCHEMICAL AND MINEROGENETIC INDEX MINERALS 163
    MHM as selectors and concentrators of litophile rare elements 163
    The index role of decationized MHM 165
    The minerogenetic role of ion exchange processes 165
    The catalytic role of MHM in nature 166
CONCLUSIONS 166
ACKNOWLEDGMENTS 167
REFERENCES 168

6   Microporous Mixed Octahedral-Pentahedral-Tetrahedral Framework Silicates

João Rocha & Zhi Lin

INTRODUCTION 173
SYNTHESIS 174
STRUCTURE 176
    Titanosilicates 176
    Zirconosilicates 179
    Vanadosilicates and niobosilicates 181
    Stannosilicates and indosilicates 182
    Rare-earth silicates 184
PROPERTIES AND APPLICATIONS 186
    Catalysis and sorption 186
    Cation exchange 190
    Optical and magnetic properties 190
CONCLUSIONS AND OUTLOOK 193
ACKNOWLEDGMENTS 193
REFERENCES 193

7   The Sodalite Family – A Simple but Versatile Framework Structure

Wulf Depmeier

ABSTRACT 203
INTRODUCTORY REMARKS 203
    A functional chemical formula 203
    Poroates, zeoates, clathrates 204
    Basic structural features and chemistry of sodalite-type structures 205
SODALITES AS FUNCTIONAL MATERIALS 209
    Sodalites as a border-line case? 209
    Sodalites as clathrates; matrix isolation 211
    Sodalites as zeoates; ion exchange 213
CRYSTALLOGRAPHY – STRUCTURAL DETAILS 213
    Basic crystallography of sodalite 213
    The topology of the sodalite framework type 215
    The classical sodalite tilt system 221
    Tetragonal tetrahedron distortion 226
    Twisting and cage cation ordering 228
    Cage anion ordering and shearing 229
    T site ordering, Loewenstein's rule 229
STRUCTURAL DEPENDENCE ON SUBSTITUTION, TEMPERATURE AND
    PRESSURE 232
    Phase transitions and modulated structures 232
    Thermal expansion 233
    High pressure studies 234
CONCLUSION 235
ACKNOWLEDGMENT 235
REFERENCES 235

8   Modular Microporous Minerals: Cancrinite-Davyne Group and C-S-H Phases

Elena Bonaccorsi & Stefano Merlino

CANCRINITE-DAVYNE GROUP 241
    Structural aspects 241
    Genesis of natural compounds 252
    Thermal behavior 252
    Synthetic cancrinites 253
    Ionic exchanges 256
C-S-H COMPOUNDS 257
NATURAL AND SYNTHETIC COMPOUNDS OF THE TOBERMORITE GROUP 258
    Historical outlook 258
    OD character of the phases in the tobermorite group 259

    Genesis and parageneses of the natural compounds 261
    Synthetic counterparts 262
    General structural aspects..263
    11 Å phases: clinotobermorite and tobermorite 11 Å..265
    9 Å phases: "clinotobermorite 9 Å" and riversideite 267
    14 Å phase: plombierite 268
    Thermal behavior 269
    Technological properties 270
NATURAL AND SYNTHETIC COMPOUNDS OF THE GYROLITE GROUP 271
    Natural phases: occurrences and composition...272
    Synthetic phases: preparation and composition 274
STRUCTURAL ASPECTS..275
    Thermal behavior 279
    Related mineral phases..280
    Ambiguities in the layer stacking and polytypism 282
ACKNOWLEDGMENTS 282
REFERENCES 283

9   A Short Outline of the Tunnel Oxides

Marco Pasero

INTRODUCTION 291
BASIC STRUCTURAL FEATURES..291
    1 × 1 tunnels 292
    1 × 2 tunnels 292
    1 × 3 tunnels 292
    2 × 2 tunnels 293
    2 × 3 tunnels 295
    2 × 4 and 2 × 5 tunnels 295
    3 × 3 tunnels 295
    3 × 4 tunnels 296
STRUCTURAL DETAILS..296
    Octahedral distortion 296
    Extra-framework positions 296
POLYSOMATIC RELATIONSHIPS AMONG COMPOUNDS WITHIN THE
    PYROLUSITE-"BUSERITE" FAMILY 299
    Polysomes with infinite ribbons 300
    True 1 × 2 polysome 301
APPLICATIONS 301
ACKNOWLEDGMENTS 303
REFERENCES 303

10   Apatite – An Adaptive Framework Structure

Tim White, Cristiano Ferraris,
Jean Kim & Srinivasan Madhavi

INTRODUCTION 307
DESCRIPTIVE CRYSTALLOGRAPHY..308
    Stuffed alloy representation...308
    Close-packed metalloid units 313
    Derivation from triangular anion nets 314
CRYSTAL CHEMICAL SYSTEMATICS 316
    X anion ordering..317
    Correlation of cell parameters and ionic radii 318
    Classification by metaprism twist angle (φ) 318
    Prediction of cell parameters 322
    Symmetry and flexibility 324
MICROPOROSITY IN NATURAL APATITES 327
    Occurrence..327
    Spinodal decomposition 329
MICROPOROSITY IN SYNTHETIC APATITES 331
    Cell constant discontinuities and miscibility gaps 333
    Nanodomain intergrowths at disequibrium 334
    Influence of pressure 336
ION EXCHANGE PROPERTIES 337
APATITE TECHNOLOGIES 341
    Remediation of radioactive wastes 342
    Catalysis 342
    Fuel cell electrolytes...342
    Permeable reaction barriers 343
CONCLUSION...344
ACKNOWLEDGMENTS 344
REFERENCES 344
APPENDIX A – APATITE LATTICE PARAMETERS 355
APPENDIX B – APATITE SYNTHESIS METHODS..373

11   Micro- and Mesoporous Sulfide and Selenide Structures

Emil Makovicky

INTRODUCTION 403
SULFOSALTS: GENERAL FEATURES..403
    The family of cetineites 404
    Halogen sulfides of Bi with anions in the channels...406
    KBi_3S_5 and related channel structures 407
    Hutchinsonite merotypes..410
    Rod-based sulfosalts...413

Galkhaite: a cage-like sulphide	417
SULFIDES AND SELENIDES: GENERAL FEATURES	**418**
Djerfisherite and bartonite: derivatives of pentlandite	418
Microporous sulfides with octahedral walls	420
Cage-like arsenide and antimonide structures	422
Crookesite and the related thallium-copper chalcogenides	423
Structures based on supertetrahedra	428
EPILOGUE	**430**
ACKNOWLEDGMENTS	**431**
REFERENCES	**431**

12 Micro- and Mesoporous Carbon Forms, Chrysotile, and Clathrates

Marcello Mellini

INTRODUCTION	**435**
Microporosity, mesoporosity and macroporosity	435
CARBON FORMS	**438**
Anthracite	438
Carbon nanotubes	438
Fullerenes	439
SERPENTINE	**439**
Chrysotile mesopores	439
Pore-dependent properties	440
The origin of pores in chrysotile	442
More water than space?	442
Serpentine mesopores: from gas carriers to nanowires templates	443
CLATHRATES	**443**
Enclosure compounds	443
Energy resource and/or geological hazard	444
CONCLUSIONS	**445**
ACKNOWLEDGMENTS	**445**
REFERENCES	**445**

IUPAC Nomenclature for Ordered Microporous and Mesoporous Materials and its Application to Non-zeolite Microporous Mineral Phases

Lynne B. McCusker

Laboratory of Crystallography
ETH - Hönggerberg
CH-8092 Zürich, Switzerland
lynne.mccusker@mat.ethz.ch

INTRODUCTION

A few years ago, the IUPAC recognized a need in the area of ordered microporous and mesoporous materials for a system of terms, whose definitions are generally accepted. A subcommission was formed to address this problem, and eventually a set of recommendations was published (McCusker et al. 2001). These recommendations, which include both basic nomenclature and a standardized crystal chemical formula, are based on common usage and on a systematic classification scheme recently published by Liebau (2003).

Zeolites and zeolite-like materials, with their 3-dimensional, 4-connected inorganic frameworks, served as the basis for this terminology, because they constitute the largest group of ordered microporous materials and because a nomenclature already existed for them (Barrer 1979). However, the new nomenclature was developed with the idea of encompassing all ordered, microporous and mesoporous materials with inorganic hosts, including those with non-zeolitic chemical compositions, with non-tetrahedral building units and/or with host structures that do not extend in three dimensions. The only restrictions imposed were that the pores must be ordered, accessible, and have free diameters of less than 50 nm (generally accepted range for microporous and mesoporous materials).

In most cases, the atoms of the host (and therefore the voids) are arranged periodically with long-range order and produce sharp maxima in a diffraction experiment. That is, the materials are crystalline. However, there are some materials for which the host displays only short-range order (i.e., is amorphous with respect to diffraction experiments), but the pores are of uniform size with long-range order and produce diffraction maxima at d-values reflecting the pore-to-pore distance. That is, the pore structure is "crystalline." The following recommendations apply to any material in which the arrangement of the pores within the inorganic host is highly ordered. They do not apply to materials such as pillared clays, where the pores are disordered (see Schoonheydt et al. 1999), to the large class of microporous layered compounds with continuous 2-dimensional voids (unless the layers themselves contain ordered pores), or to materials with partly or fully organic hosts. To demonstrate how this nomenclature can be applied to non-zeolitic materials with ordered micro- or mesopores, some microporous phases of mineralogical interest have been examined.

Concept

In principle, any material can be considered to consist of atoms linked by chemical bonds and the voids between these linked atoms. In an ordered microporous or mesoporous material, the voids between the linked atoms have a free volume larger than that of a sphere with a 0.25 nm diameter, and they are arranged in an ordered manner. For subsequent discussion, the linked atoms are called the host and the voids the pores. The pores can be empty or they can be occupied by guest species. According to IUPAC recommendations (Rouquérol et al. 1994), pores with free diameters of less than 2 nm are called micropores, and those in the range of 2 to 50 nm mesopores.

The terminology defined by IUPAC and reproduced in the following sections describes the pertinent characteristics of the host and the pores. Because these materials are often used as catalysts or molecular sieves, features controlling the diffusion of guest species and the space restrictions for reaction intermediates are considered to be particularly relevant. A set of descriptors have been selected for inclusion in a standardized crystal chemical formula that allows the user to highlight specific aspects of a structure. That is, the formula can be abbreviated or expanded as needed.

GENERAL DEFINITIONS

In this section, the structures of the mineral hollandite (Bystroem and Bystroem 1950; $Mn_{6.952}Fe_{0.64}Al_{0.264}Si_{0.016}Ba_{0.47}K_{0.328}Pb_{0.112}Na_{0.110}O_{16}$; $I4/m$; $a = 9.80$ Å, $c = 2.86$ Å) and a synthetic analog (Franchon et al. 1987; $Ba_{1.2}Ti_{6.8}Mg_{1.2}O_{16}$; $I2/m$, $a = 10.227$ Å, $b = 14.907$ Å, $c = 9.964$ Å, $\beta = 90.77°$) will be used to illustrate the terms as they are defined.

The host

Topology. The topology of a host structure describes the connectivity of its host atoms without reference to chemical composition or observed symmetry. The topologies of hollandite, with Mn, Fe, Al and Si in the octahedral sites, and the synthetic analog, with Ti and Mg in these sites, are identical. That topology is shown in Figure 1.

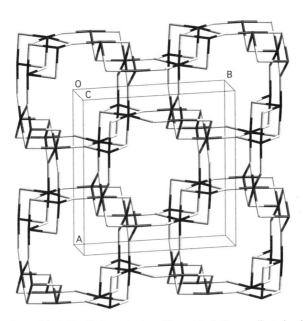

Figure 1. The topology of the hollandite host structure. The octahedrally coordinated cations are shown in black and the bridging oxygen atoms in gray.

Topological symmetry. The highest possible symmetry for a host structure is the symmetry of its topology (topological symmetry). Although the symmetry of a particular material can be as high as the topological symmetry, it is often a subgroup thereof. Whatever the observed symmetry, however, the number of framework atoms in the unit cell will be an integer multiple of the number in the topological unit cell. Distortions of the host structure due to the chemical composition of the host and/or to the presence of guest species in the pores are common. The symmetry of the mineral hollandite described by Bystroem and Bystroem (1950) is *I*4/*m* and that is also the topological symmetry. However, the synthetic analog has the symmetry *I*2/*m* and a unit cell five times larger than that of the topology.

Host dimensionality. A host structure can extend in zero (finite), one (chain), two (layer) or three (framework) dimensions. Most known microporous and mesoporous materials (including hollandite) have three-dimensional host structures, but it is also possible for a lower dimensional host to have an ordered pore system. For example, the mineral rhodesite (Hesse et al. 1992) can be considered to have a 2-dimensional silicate host structure with an ordered arrangement of channels within the layers (Fig. 2).

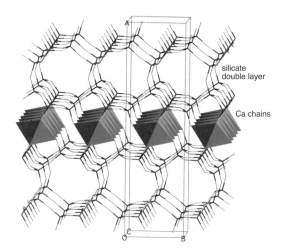

Figure 2. The 2-dimensional silicate host structure of rhodesite showing an ordered arrangement of channels within the layers. The silicate double layers are linked via chains of Ca octahedra.

Basic building units. The host structure can be constructed by linking basic building units (usually coordination polyhedra sharing corners, edges or faces). In the case of zeolite structures, these basic building units (**BBU**) are tetrahedra, where the central atom ($_{ce}$**H**) is typically Si or Al, and the peripheral atoms ($_{pe}$**H**) are O. In the vast majority of microporous and mesoporous materials with inorganic hosts, the central atoms are cations and the peripheral atoms anions. In hollandite, the central atom is (primarily) Mn, the peripheral atom O, and the basic building units are MnO$_6$ octahedra. As can be seen from the polyhedral drawing in Figure 3, these share both edges and corners with neighboring octahedra.

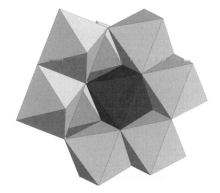

Figure 3. The basic building unit of the hollandite host structure is an octahedron (MnO$_6$), which shares corners and edges with neighboring octahedra.

Composite building units. It is sometimes useful to combine **BBU**'s to construct a larger composite building unit (**CBU**) that is characteristic of the topology. For example, rings, chains or polyhedral building units, composed of a finite or infinite number of **BBU**'s can be chosen. Rings are described using the notation n-ring, where n is the number of $_{ce}\text{H}$ atoms in the ring, and polyhedral building units using the notation $[n_i^{m_i}]$, where m is the number of n-rings defining the polyhedron and $\sum m_i$ is the total number of faces. For hollandite, the twisted 4-ring unit $[3^44^2]$ (Fig. 4a) or the 8-ring (Fig. 4b) might be useful **CBU**'s.

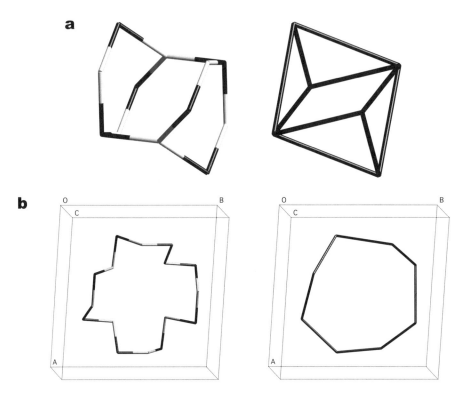

Figure 4. Two composite building units that are characteristic of the hollandite host structure: (a) the twisted 4-ring $[3^44^2]$ unit with and without the bridging oxygens shown, and (b) the 8-ring with and without the bridging oxygens.

The pores

Windows. The n-rings defining the faces of a polyhedral pore are called windows.

Cages and cavities. A polyhedral pore whose windows are too narrow to be penetrated by guest species larger than H_2O is called a cage. For oxides ($_{pe}\text{H} = \text{O}$), the limiting ring size is considered to be $n = 6$. For example, the $[4^66^8]$ polyhedron of sodalite is a cage. A polyhedral pore, which has at least one face defined by a ring large enough to be penetrated by guest species, but which is not infinitely extended (i.e., not a channel), is called a cavity. For example, the $[4^{18}6^412^4]$ polyhedron in the zeolite faujasite is a cavity. Hollandite does not have cages or cavities.

Channels. A pore that is infinitely extended in one dimension and is large enough to allow guest species to diffuse along its length is called a channel. Channels can intersect to form 2- or 3-dimensional channel systems. The 8-ring channel system along the [001] direction in hollandite is 1-dimensional (Fig. 5). It can be described in terms of the repeating unit along the channel as $[3^{16}8^{2/2}]$. Note that the delimiting 8-rings are halved, because only ½ of each terminating 8-ring "belongs" to a single repeat.

Effective channel width. The effective width of a channel is a fundamental characteristic of a microporous or mesoporous material that describes the accessibility of the pore system to guest species (i.e., the "bottleneck"). It is generally defined in terms of either the smallest *n*-ring (topological description) or the smallest free aperture (metrical description) along the dimension of infinite extension. The 1-dimensional channel system in hollandite has 8-ring pore openings (topological description) with a free diameter of ca 0.19 nm (metrical description). The free diameter takes into account the size of the $_{pe}H$ atoms (e.g., an ionic radius for oxygen of 0.135 nm has been assumed, so 0.27 nm has been subtracted from 0.46 nm, the O···O distance defining the pore diameter). For distorted rings, the metrical description is more useful for assessing how accessible the pore system is. It should also be borne in mind that channels that are not straight (e.g., sinusoidal) will impose additional restrictions on accessibility. The pore width along a channel, topologically as well as metrically, can be constant or it may undulate (e.g., if the channel consists of a series of cavities).

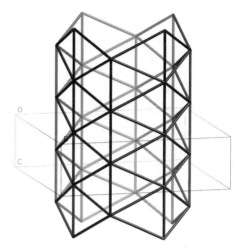

Figure 5. The 1-dimensional channel along the [001] direction in hollandite, $[3^{16}8^{2/2}]$. Oxygen atoms have been omitted for clarity.

Pore dimensionality. The number of dimensions in which the pore has infinite extension is the pore dimensionality (D^p). For cages, $D^p = 0$, and for channels, $D^p = 1$. For systems of interconnected channels, $D^p = 2$ or 3. The dimensionality of the pore system, D^p, will always be less than or equal to the dimensionality of the host, D^h. Hollandite has a 1-dimensional pore system along the [001] direction.

CRYSTAL CHEMICAL FORMULA

The properties of microporous and mesoporous materials are controlled principally by their chemical composition and their structures, so a crystal chemical formula that describe these features in a standardized manner has been developed. It is consistent with the recommendations of the International Union of Crystallography for the nomenclature of inorganic structure types (Lima-de-Faria et al. 1990), with IUPAC rules (Leigh 1990), and with common usage, but is flexible enough to allow the researcher to emphasize only those features that are relevant to the discussion.

The crystal chemical formula contains up to 6 terms in the following order:

|*guest composition*| [*host composition*] $_h${*host structure*} $_p${*pore structure*} (*Sym*) – **IZA**

The first two terms describe the chemical composition of the guest species (between boldface vertical bars) and the host (between boldface square brackets), respectively. The next two terms (between boldface curly brackets) contain information about the structure of the host and the pores, respectively. The fifth term (between boldface round brackets) gives the (actually observed) symmetry of the material. If the host structure of the material belongs to a zeolite framework type, the sixth term (preceded by a dash) is the IZA code in boldface type.

Each of these terms has a prescribed content, which can be reduced or expanded in accordance with the detail appropriate for a specific problem. In the following sections, each term is considered in turn.

Chemical composition of the guest species

The chemical formulae for the various guest species present in the pores of the material are placed inside boldface vertical bars and arranged in the order

cations (**A**) – *anions* (**X**) – *neutral molecules* (**M**)

to yield the general term

$$|A_a X_x M_m|_n$$

This term should reflect the chemical composition of the guest species for the complete unit cell. The subscript n is the number of topological unit cells within the unit cell of the material. In hollandite, the guest species are Ba^{2+}, K^+, Pb^{2+} and Na^+, and this term is

$$| Ba^{2+}_{0.47} K^+_{0.328} Pb^{2+}_{0.112} Na^+_{0.110} |$$

For the synthetic analog, only Ba^{2+} cations are present, but the unit cell is five times larger than the topological one, so the term would be

$$| Ba^{2+}_{1.2} |_5$$

Chemical composition of the host

The symbols for the atoms constituting the host are placed inside boldface square brackets and are arranged in the order

interstitial species ($_i$**A**, $_i$**X** or $_i$**M**) – *central host atoms* ($_{ce}$**H**) – *peripheral host atoms* ($_{pe}$**H**)

An interstitial species is a non-exchangeable cation, anion or molecule located in a void with a free diameter of less than 0.25 nm. The element symbols can be complemented in the usual way to indicate oxidation state V (Roman numeral superscript) and coordination number CN (Arabic number superscript within square brackets). The linkedness L and connectedness s of the central host atoms, which indicate the number (s) of neighboring **BBU**'s that share corners, edges or faces ($L = 1$, 2 or 3, respectively) with the $_{ce}$H under consideration (Lima-de-Faria et al. 1990), can also be specified within Japanese brackets (⌈ ⌋), if desired. The general form of the term can be written as

$$[_iA_a^{V\,[CN]}\,_iX_x^{V\,[CN]}\,_iM_m\,_{ce}H_c^{V\,[CN]\,\lceil L;s \rfloor}\,_{pe}H_p^{V\,[CN]}]_n$$

From this, the basic building unit(s) (**BBU**) of the host, [$_{ce}$H($_{pe}$H)$_n$], can be deduced. As for the guest species, this term should reflect the chemical composition of the host for a complete unit cell and be expressed as multiples (n) of the topological unit cell, if it is known. For hollandite this term would be

$$[\,(Mn_{6.766}^{IV}Mn_{0.186}^{III}Fe_{0.64}^{III}Al_{0.264}^{III}Si_{0.016}^{IV})^{[6]\lceil 1;4/2;4 \rfloor}\,O_{16}^{[3]}\,]$$

indicating that Mn is present in two oxidation states, that all cations are 6-coordinate, share corners with four neighboring octahedra and edges with four additional ones, and that all oxygen atoms bond to three central host cations.

For the synthetic hollandite this term would be

$$[\,(Ti_{6.8}^{IV}Mg_{1.2}^{II})^{[6]\lceil 1;4/2;4 \rfloor}\,O_{16}^{[3]}\,]_5$$

indicating that the Ti and Mg cations have the oxidation states IV and II, respectively, and that the unit cell is five times larger than the topological one.

Structure of the host

The parameters describing the host structure are placed inside boldface curly brackets preceded by a subscript h and are arranged in the order

dimensionality of the host (D^h) – *composite building unit* (**CBU**)

to yield the general term

$$_h\{\textbf{\textit{D}}^h \textbf{ CBU}\}$$

The composite building unit(s) can be chosen to highlight those aspects of the host structure that are relevant to the discussion. For hollandite, the host structure could be given as

$$_h\{\ 3\ [3^4 4^2]\ \}$$

indicating that the framework is 3-dimensional and contains the unit shown in Figure 4a.

Structure of the pores

The parameters describing the pore system are placed inside boldface curly brackets preceded by a subscript p and are arranged in the order

dimensionality of the pore system $\textbf{\textit{D}}^p$ – *shape of the pore* $[n_i^{m_i}]$ –
direction of the channel $[uvw]$ – *effective channel width* ($W_{(\mathrm{eff})}^{channel}$)

to yield the general term

$$_p\{\ \textbf{\textit{D}}^p\ [n_i^{m_i}]\ [uvw]\ (W_{(\mathrm{eff})}^{channel})\}$$

The term $[uvw]$ can be replaced with $<uvw>$ to indicate that all crystallographically equivalent directions are involved. If more than one pore system is present, the descriptions are separated by a slash (/). For hollandite, this term could be given as

$$_p\{\ 1\ [3^{16} 8^{2/2}]\ [001]\ (8\text{-ring})\ \}$$

indicating that it has a 1-dimensional channel system with an 8-ring pore opening along the c-axis. The channel itself is lined with 3-rings (Fig. 5).

Symmetry

Materials with the same host topology do not necessarily have the same symmetry. If this aspect of the structure is pertinent, it can be included as the fifth term of the crystal chemical formula enclosed in boldface round brackets. Following the recommendations of the International Union of Crystallography (Guinier et al. 1984) and the International Mineralogical Association (Nickel and Mandarino 1987), this information can be given in form of the crystal system (C for cubic, H for hexagonal, T for trigonal, R for rhombohedral, Q for tetragonal, O for orthorhombic, M for monoclinic and A for triclinic) or the space group symbol. For hollandite, this would be ($I4/m$) and for the synthetic analog, as ($I2/m$).

IZA code

For zeolites and zeolite-like materials, full descriptions of the confirmed zeolite framework types are given in the *Atlas of Zeolite Framework Types* (Baerlocher et al. 2001). The three-letter code alone conveys a wealth of information about the host structure and the pore system, so when relevant, it should be appended to the crystal chemical formula following a dash. For non-zeolites, such as hollandite, this term is not applicable.

EXAMPLES

In the following sections, a few examples taken from the realm of microporous minerals are used to illustrate how the IUPAC nomenclature and crystal chemical formula can be applied to these materials.

Hollandite

Putting together all of the terms discussed in the previous section, the following crystal chemical formula for the hollandite example emerges

$$| Ba^{2+}_{0.47} K^+_{0.328} Pb^{2+}_{0.112} Na^+_{0.110} | [(Mn^{IV}_{6.766} Mn^{III}_{0.186} Fe^{III}_{0.64} Al^{III}_{0.264} Si^{IV}_{0.016})^{[6\lceil 1;4/2;4\rfloor]} O_{16}^{[3]}]_h\{3 [3^4 4^2]\}_p\{1 [3^{16} 8^{2/2}] [001] \text{(8-ring)}\} (I4/m)$$

However, it is not necessary to include all of this information in a single formula. If the discussion only requires the chemical composition, the formula can be shortened to

$$| Ba_{0.47}K_{0.328}Pb_{0.112}Na_{0.110} | [Mn_{6.952}Fe_{0.64}Al_{0.264}Si_{0.016}O_{16}]$$

If the nature of the pore system and not the stoichiometry is of interest, it can be written without the exact chemical composition as

$$| Ba - K - Pb - Na | [Mn - Fe - Al - Si - O]_p\{ 1 [001] \text{(8-ring, 0.19)} \}$$

If the connectivity of the host atoms is of particular interest, it might be given as

$$| Ba - K - Pb - Na | [(Mn, Fe, Al, Si)_{7.87}^{[6\lceil 1;4/2;4\rfloor]} O_{16}^{[3]}]$$

If a comparison of structures with similar chemical composition but different CBU's were being made, it might be appropriate to use the formula

$$| Ba - K - Pb - Na | [Mn - Fe - Al - Si - O]_h\{ 3 [3^4 4^2] \}$$

highlighting the $[3^4 4^2]$ unit.

Synthetic hollandite

For the synthetic analog of hollandite, the more complete formula would be

$$| Ba^{2+}_{1.2} |_5 [(Ti^{IV}_{6.8} Mg^{II}_{1.2})^{[6\lceil 1;4/2;4\rfloor]} O_{16}^{[3]}]_5 {}_h\{3 [3^4 4^2]\} {}_p\{1 [3^{16} 8^{2/2}] [010] \text{(8-ring)}\} (I2/m)$$

and the shorter formula giving just the chemical composition of the material and indicating that the observed unit cell is 5 times larger than the topological unit cell

$$| Ba_{1.2} |_5 [Ti_{6.8} Mg_{1.2} O_{16}]_5$$

Here it might also be appropriate to include the space group to emphasize the fact that the symmetry is reduced from tetragonal to monoclinic

$$| Ba_{1.2} |_5 [Ti_{6.8}Mg_{1.2}O_{16}]_5 (I2/m)$$

The effective pore width of the 8-ring in this synthetic analog is larger than in the mineral and is not quite circular, and this could be indicated using

$$| Ba - | [Ti - Mg - O]_p\{ 1 [010] \text{(8-ring, 0.24, 0.25)} \}$$

The formula showing the connectivity of the host atoms (and thereby this material's relationship to the mineral) would be written

$$| Ba - | [(Ti, Mg)_8^{[6\lceil 1;4/2;4\rfloor]} O_{16}^{[3]}]$$

Finally, the presence of a $[3^4 4^2]$ CBU could be indicated with

$$| Ba - | [Ti - Mg - O]_h\{ 3 [3^4 4^2] \}$$

Lovozerite

A projection of the the zirconium silicate layer in the the mineral lovozerite (Yamnova et al. 2001; $Na_2CaZr[Si_6O_{12}(OH,O)_6]\cdot H_2O$; $R3$; $a = 10.18$ Å, $c = 13.13$ Å) is shown in Figure 6. These highly puckered layers of Si_6O_{18} 6-rings linked via Zr atoms are stacked in an ABAB sequence, creating the cage shown in Figure 7. An expanded crystal chemical formula might be given as

$$| Ca^{2+} Na^+_2 H_2O | [Zr^{IV[6]\lceil 1;6\rfloor} Si_6^{IV[4]\lceil 1;3\rfloor} O_{12}^{[2]} O_4^{[1]} (OH)_2^{[1]}]\ _h\{3\ [6^48^3]\}\ _p\{3\ (8\text{-ring})\}\ (R3)$$

From this formula, it is apparent that the Zr atoms of the host are 6-coordinate and are corner-linked to six neighboring polyhedra. The Si atoms, on the other hand are 4-coordinate and are corner-linked to only three neighboring polyhedra. In fact, the Zr atoms are connected to six Si via bridging oxygens and the Si atoms to one Zr and two Si, but this detailed information cannot be gleaned from the crystal chemical formula. Two thirds of the oxygens bridge between two central host atoms and one third are terminal. The host structure is 3-dimensional,

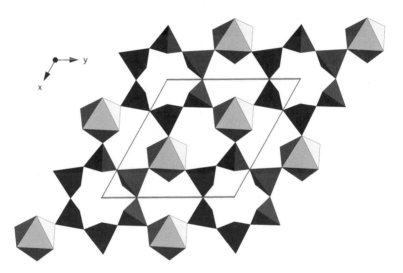

Figure 6. Projection of the zirconium silicate layer in the mineral lovozerite with highly puckered silicate 6-rings linked via Zr octahedra. Dark tetrahedra are SiO_4 units and light octahedra ZrO_6 units.

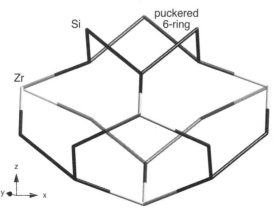

Figure 7. The $[6^48^3]$ cavity characteristic of the lovozerite structure (dark: Zr; light: Si; oxygens have been omitted for clarity). The 3-dimensional channel system with 8-ring pore openings passes through these cavities.

and contains [6⁴8³] cavities (see Fig. 7). The pore system, which goes through these cavities is also 3-dimensional and has 8-ring pore openings. One of the features of the host structure is that the Si atoms form 6-rings, which are linked by the Zr octahedra to form the layer shown in Figure 6. This could be indicated with

$$| CaNa_2H_2O | [ZrSi_6O_{16}(OH)_2]_h\{3 \text{ (6-ring)} \}$$

However, it should be noted that the formula does not distinguish between the two different polyhedra of the host, so the 6-ring here could refer to either the Si_6O_6 or the $Si_4Zr_2O_6$ 6-rings. The 8-rings defining the pore system are quite distorted, so if the pore system is of particular interest, it might be better to use the metrical description rather than the topological one

$$| CaNa_2H_2O | [ZrSi_6O_{16}(OH)_2]_p\{3 \text{ (0.19, 0.44)}\}$$

The topological symmetry of the lovozerite host is $R\bar{3}$, but no subscripts are necessary for the chemical composition terms, because the unit cell volume does not change with the reduction to $R3$ symmetry.

Labuntsovite

A projection of the structure of the mineral nenadkevichite (Perrault et al. 1973; Labuntsovite group; $(Na_{3.76}K_{0.24}Ca_{0.11}Mn_{0.03})(Nb_{2.76}Ti_{1.18})(O_{2.80}OH_{1.20})(Si_8O_{24})\cdot 8H_2O$; *Pbam*; $a = 7.41$ Å, $b = 14.20$ Å, $c = 7.15$ Å) is shown in Figure 8. It is readily apparent that the host structure has 8-ring pores along the [001] direction. The crystal chemical formula could be written as

$$|Na^+_{3.76} K^+_{0.24} Ca^{2+}_{0.11} Mn^{2+}_{0.03} (H_2O)_8 | [(Nb_{2.76}^V Ti_{1.18}^{IV})^{[6]\lceil 1;6 \rfloor} Si_8^{IV[4]\lceil 1;4\rfloor} O_{26.8}^{[2]} (OH)_{1.2}^{[2]}]$$
$$_h\{3 \text{ (4-ring)}\} _p\{1 [3^4 4^2 6^6 8^{2/2}] [001] \text{ (8-ring)}\} (Pbam)$$

The one-dimensional 8-ring channels along the [001] direction contain a variety of cations and water molecules. The host consists of Nb and Ti octahedra and Si tetrahedra, all of which are fully corner-connected. As in the case of lovozerite, the formula does not indicate which polyhedra are connected to which. The space group is different from the topological one (*Cmmm*), but the unit cell has the same size.

Figure 8. Projection of the structure of the mineral nenadkevichite showing the chains of (Nb,Ti) corner-sharing octahedra (light) linked to one another via silicate (dark tetrahedra) 4-rings (perpendicular to *x*-axis) to form 8-ring channels along the [001] direction. The latter contain cations (dark spheres) and water molecules (light spheres).

A simple chemical composition could be given with

| Na$_{3.76}$K$_{0.24}$Ca$_{0.11}$Mn$_{0.03}$(H$_2$O)$_8$ | [(Nb$_{2.76}$Ti$_{1.18}$)Si$_8$O$_{26.8}$(OH)$_{1.2}$]

and the nature of the channel system (Fig. 9) by

| Na - K - Ca - Mn - (H$_2$O) - | [(Nb, Ti) - Si - O - OH -] $_p$\{1 [3^{4}4^{2}6^{6}8$^{2/2}$] [001] (8-ring)\}

the presence of Si$_4$O$_{12}$ 4-rings by

| Na - K - Ca - Mn - (H$_2$O) - | [(Nb, Ti) - Si - O - OH -] $_h$\{3 (4-ring)\}

and the presence of corner-linked chains of (Nb,Ti)O$_6$ octahedra by

| Na - K - Ca - Mn - (H$_2$O) - | [(Nb, Ti) - Si - O - OH -] $_h$\{3 ([100] *uB*, 2, 1, 1)\}

The chain descriptor above means that the chains extend in the [100] direction, that they are unbranched, have a periodicity of two, a multiplicity of one and a dimensionality of one. Here again, which of the two types of central host atom forms the chain cannot be discerned from the formula.

An idealized host structure might be given as just

[(Nb$_3^V$Ti^{IV})$^{[6]⌊1;6⌋}$ Si$_8^{IV[4]⌊1;4⌋}$ O$_{28}^{[2]}$]$^{5-}$

and the effect of Ti substitution on the charge of the host could then be underlined with

[(Nb^{V}Ti$_3^{IV}$)$^{[6]⌊1;6⌋}$ Si$_8^{IV[4]⌊1;4⌋}$ O$_{28}^{[2]}$]$^{7-}$

Chukanov et al. (2002) have classified the monoclinic labuntsovite-group minerals using the formula A$_4$B$_4$C$_4$D$_2$M$_8$(T$_4$O$_{12}$)$_4$(OH,O)$_8$·nH$_2$O where A, B, C and D represent different guest sites in the structure, M is Nb and/or Ti, and T is Si. This could be rewritten as

| A$_2$B$_2$C$_2$D(H$_2$O)$_y$ |$_n$ [M$_4$T$_8$O$_{28-x}$(OH)$_x$]$_n$ (SG)

where n is the volume of the unit cell compared with that of the topological one (2 ≤ n ≤ 4, for all known monoclinic labuntsovite group minerals) and SG is the space group.

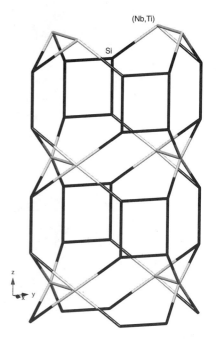

Figure 9. The 1-dimensional [3^{4}4^{2}6^{6}8$^{2/2}$] channel along the [001] direction that is characteristic of the labuntsovite group of minerals.

Tobermorite

The tobermorite 11 Å structure (Merlino et al 1999; Ca$_4$(Si$_6$O$_{15}$)(OH)$_2$(H$_2$O)$_5$; *Bm*; $a = 6.735$ Å, $b = 7.385$ Å, $c = 22.487$ Å, 123.25°) consists of silicate double chains running parallel to the *b*-axis connected by corrugated layers of edge-sharing 7-coordinate Ca ions (Fig. 10). The crystal chemical formula of tobermorite from Wessels mine could be written as

| (H$_2$O)$_6$ | [Ca$_4^{II\,[7]⌊1,2;11⌋}$Ca$_4^{II\,[7]⌊1,2;12⌋}$Si$_8^{IV[4]⌊1,2;7⌋}$Si$_4^{IV[4]⌊1;4⌋}$O$_{16}^{[4]}$O$_{14}^{[2]}$(OH)$_4^{[2]}$(H$_2$O)$_4^{[1]}$]
$_h$\{3 (*uB*, 3, 2, 1)\} $_p$\{2 ⊥[001] (8-ring)\} (*Bm*)

indicating that the only guest species are water molecules, that the host structure consists of 7-coordinate Ca and 4-coordinate Si atoms and, that the Ca are 11- and 12-connected and the Si 4- and 7-connected via shared corners and edges (Fig. 11). The framework structure

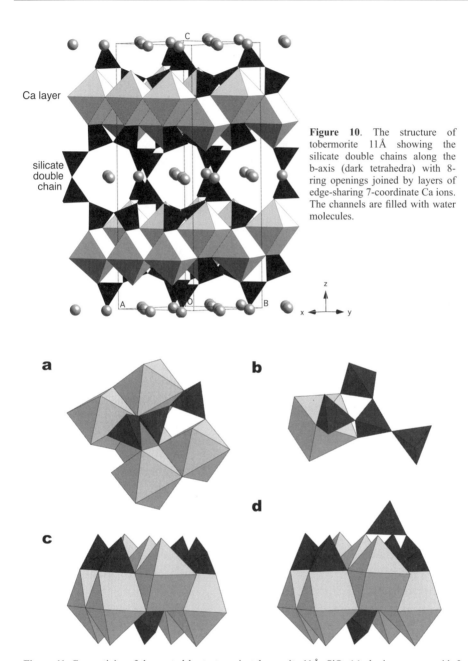

Figure 10. The structure of tobermorite 11Å showing the silicate double chains along the b-axis (dark tetrahedra) with 8-ring openings joined by layers of edge-sharing 7-coordinate Ca ions. The channels are filled with water molecules.

Figure 11. Connectivity of the central host atoms in tobermorite 11Å. SiO_4 (a) sharing corners with 2 tetrahedra and 4 Ca polyhedra and an edge with another Ca polyhedron and (b) sharing corners with three tetrahedra and one Ca polyhedron. CaO_7 (c) sharing corners with 4 tetrahedra and edges with 6 Ca polyhedra and one tetrahedron and (d) sharing corners with 5 tetrahedra and edges with 6 Ca polyhedra and one tetrahedron.

is 3-dimensional and the 8-ring channel system 2-dimensional, running perpendicular to the [001] direction. The silicate double chain (condensation of two "dreiereinfachketten" of the wollastonite-type) is given in the $_h\{\ \}$ term in the notation described by Liebau (2003) as an unbranched chain (uB) with a periodicity of 3 (number of tetrahedra per repeat period of the chain), a multiplicity of 2 (double chain) and a dimensionality of 1 (a chain, not a sheet or a framework).

A simpler version could be given as

$$| (H_2O)_6 | [Ca_8Si_{12}O_{30}(OH)_4(H_2O)_4]\ _p\{2\ \bot[001]\ (8\text{-ring})\}$$

Up to 1/6 of the Si sites can be occupied by Al. In this case, the negative charge is compensated by H^+ ions, and the formula becomes

$$| (H_2O)_6 | [Ca_8Al_2Si_{10}O_{28}(OH)_6(H_2O)_4]\ _p\{2\ \bot[001]\ (8\text{-ring})\}$$

In Ca-rich tobermorite 11 Å (Merlino et al. 2001), up to two Ca^{2+} ions are located in the pores as guest species, making the crystal chemical formula

$$| Ca_2(H_2O)_6 | [Ca_8^{[7]}\ Si_{12}^{[4]}\ O_{16}^{[4]}\ O_{18}^{[2]}\ (H_2O)_4^{[1]}]$$

The fact that Ca^{2+} ions are present not only as host but also as guest species in this mineral is immediately apparent in the above formula. It is believed that these guest Ca^{2+} ions have a profound effect on the thermal behavior of the mineral.

Tetrahedrite

In 1997, the synthesis of a stoichiometric analog of the copper antimony sulfide mineral tetrahedrite was reported (Pfitzner et al. 1997; $Cu_{12}Sb_4S_{13}$; $I\bar{4}3m$; $a = 10.3293$ Å). It has a host structure composed of $Cu_{12}S_{11}$ [3^84^6] cages (Fig. 12), each containing an Sb species. The crystal chemical formula could be written as

$$| Sb_8 | [Cu_{12}^{I,II\ \lceil 3\rceil\lceil 1;8\rfloor}\ Cu_{12}^{I,II\ \lceil 4\rceil\lceil 1;8\rfloor}\ S_{24}^{[3]}\ S_2^{[6]}]\ _h\{3\ [3^8]\}\ _p\{0\ [3^84^6]\}\ (I\bar{4}3m)$$

From this, it can be deduced that Sb is the guest species, that the host is composed of two types of Cu atoms (one 3-coordinate and the other 4-coordinate), that both types share corners with eight neighboring Cu atoms, that most sulfur atoms bond to 3 Cu, but that a few bond to six, that the host is 3-dimensional and contains octahedra of Cu atoms, that the pore structure is zero dimensional with cuboctahedral cages (Fig. 12); and that the symmetry is $I\bar{4}3m$.

The naturally occurring tetrahedrites are not stoichiometric. For example, Makovicky and Skinner (1979) described two tetrahedrites with the chemical compositions $Cu_{12.3}Sb_4S_{13}$ and $Cu_{13.8}Sb_4S_{13}$, where the tetrahedrally coordinated Cu site is not fully occupied, and the missing Cu could not be located crystallographically and is assumed to be mobile (i.e., a guest species). Curiously, the Cu-rich phase proved to have even less Cu in the host

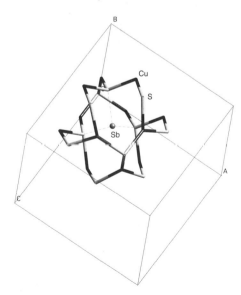

Figure 12. The zero dimensional $Cu_{12}S_{11}$ [3^84^6] cage in tetrahedrite (dark: Cu; light: S) with an Sb guest species coordinated to three S atoms.

$Cu^{[4]}$ site than the Cu-poor phase. The crystal chemical formulas are

$$| Sb_8Cu_{1.8} | [Cu_{12}^{I,II\,[3]} (Cu_{10.8} \square_{1.2})^{I,II\,[4]} S_{24}^{[3]} S_2^{[6]}] {}_h\{ 3\,[3^8] \} {}_p\{ 0\,[3^84^6] \} \; (I\bar{4}3m)$$

for the Cu-poor phase, and

$$| Sb_8Cu_{7.6} | [Cu_{12}^{I,II\,[3]} (Cu_8 \square_4)^{I,II\,[4]} S_{24}^{[3]} S_2^{[6]}] {}_h\{ 3\,[3^8] \} {}_p\{ 0\,[3^84^6] \} \; (I\bar{4}3m)$$

for the Cu-rich phase.

Rhodesite

The structure of rhodesite (Hesse et al 1992; $HK_{1-x}Na_{x+2y}Ca_{2-y}Si_8O_{19} \cdot 6\text{-}z(H_2O)$; *Pmam*; $a = 23.416$ Å, $b = 6.555$ Å, $c = 7.050$ Å) can be viewed as having a 3-dimensional calcium silicate host structure with potassium and water in the pores (Figs. 2 and 13), or it can be viewed as having a 2-dimensional double layer silicate host structure with Ca ions as "additional" host species. In the first case, the crystal chemical formula (for $x = 0$, $y = 0$ and $z = 4$) would be

$$| K_2(H_2O)_4 | [Ca_4^{II\,[6]\lceil 1;4/2;2\rfloor} Si_8^{IV\,[4]\lceil 1;4\rfloor} Si_8^{IV\,[4]\lceil 1;5\rfloor} O_{28}^{[2]} (OH)_2^{[2]} O_8^{[3]}]$$
$${}_h\{ 3\,([001]\,\boldsymbol{uB}, 1, 1, 1) \} {}_p\{ 2 \perp [100]\,(8\text{-ring}) \} \; (\textit{Pmam})$$

indicating that (1) the guest species are K^+ and water molecules, (2) the host structure consists of octahedrally coordinated Ca and tetrahedrally coordinated Si atoms, (3) that the Ca^{2+} ions share corners with 4 other ${}_{ce}H$ atoms (Si in this case) and edges with two others (Ca), (4) half of the Si atoms share corners with 4 other ${}_{ce}H$ atoms while the other half share corners

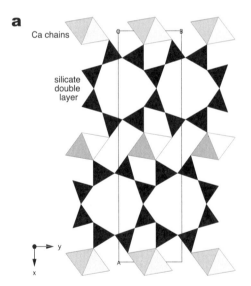

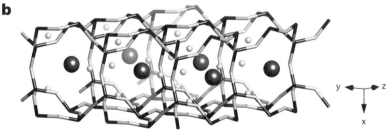

Figure 13. (a) Projection of the host structure of rhodesite showing the silicate double layers linked via Ca octahedra. (b) the silicate double layer alone showing the K^+ (dark spheres) and H_2O (light spheres) guest species in the 2-dimensional 8-ring channel system.

with five, (5) the host structure is 3-dimensional with a single chain of periodicity 1 (chain of Ca octahedra) along the [001] direction, (6) there is a 2-dimensional 8-ring channel system perpendicular to the [100] direction, and (7) the space group is *Pmam*.

The alternative suggested by Liebau (2003) would be

| K$_2$(H$_2$O)$_4$ | | [$_{add}$Ca$_4^{II\,[6]}$ Si$_{12}^{IV\,[4]\lceil 1;3\rfloor}$ Si$_4^{IV\,[4]\lceil 1;4\rfloor}$ O$_{28}^{[2]}$ (OH)$_2^{[2]}$ O$_8^{[3]}$] $_h${ 2 }
$_p${2 ⊥[100] (8-ring) } (*Pmam*)

where Ca^{2+} ions are not considered to be $_{ce}$H atoms but "additional" host atoms and the Si connectivities are changed accordingly. In this case, the host structure is 2-dimensional. The presence of [8^6] cavities in this layer could be indicated by

| K$_2$(H$_2$O)$_4$ | | [$_{add}$Ca$_4^{II\,[6]}$ Si$_{12}^{IV\,[4]\lceil 1;3\rfloor}$ Si$_4^{IV\,[4]\lceil 1;4\rfloor}$ O$_{28}^{[2]}$ (OH)$_2^{[2]}$ O$_8^{[3]}$] $_h${ 2 }
$_p${2 [8^6] } (*Pmam*)

CONCLUSIONS

The IUPAC nomenclature for microporous and mesoporous materials can be applied in a relatively straightforward manner to non-zeolite mineral phases. Although not all details of a structure are immediately apparent from the crystal chemical formula, the most relevant attributes usually are. Features such as (1) the chemical composition of both the host structure and the guest species, (2) the composite building unit(s) characteristic of the host structure, (3) the nature of the pore system, and (4) the symmetry of the material can be indicated. Similarities and differences between related minerals (e.g., those within a group) can be seen easily if their crystal chemical formulas are compared. Furthermore, the formula can be expanded or contracted to suit the user's requirements, so only those terms relevant to a particular discussion need be given. Perhaps the most important difference in the application of the nomenclature to non-zeolite versus zeolite microporous phases lies in the availability of the **IZA** code for zeolites and their analogs. If a similar system describing the host structures of microporous minerals could be devised, the information content of the crystal chemical formula could be significantly increased.

REFERENCES

Baerlocher Ch, Meier WM, Olson DH (2001) Atlas of Zeolite Framework Types. 5th ed. Elsevier, Amsterdam

Barrer RM (1979) Chemical nomenclature and formulation of compositions of synthetic and natural zeolites. Pure Appl Chem 51:1091-1100

Bystroem A, Bystroem AM (1950) The crystal structure of hollandite, the related manganese oxide minerals and α-MnO$_2$. Acta Crystallogr 3:146-154

Chukanov NV, Pekov IV, Khomyakov AP (2002) Recommended nomenclature for labuntsovite-group minerals. Eur J Mineral 14:163-173

Franchon E, Vicat J, Hodeau J-L, Wolfers P, Qui DT, Strobel P (1987) Commensurate ordering and domains in the Ba$_{1.2}$Ti$_{6.8}$Mg$_{1.2}$O$_{16}$ hollandite. Acta Crystallogr B43:440-448

Guinier A, Bokii GB, Boll-Dornberger K, Cowley JM, Durovic S, Jagodzinski H, Krishna P, de Wolff PM, Zvyagin BB, Cox DE, Goodman P, Hahn Th, Kuchitsu K, Abrahams SC (1984) Nomenclature of polytype structures - report of the International Union of Crystallography ad hoc committee on the nomenclature of disordered, modulated and polytype structures. Acta Crystallogr A40:399-404

Hesse K-F, Liebau F, Merlino S (1992) Crystal structure of rhodesite, HK$_{1-x}$Na$_{x+2y}$Ca$_{2-y}$(1B,3,2 infinity(2))[Si$_8$O$_{19}$]·(6-z)H$_2$O, from 3 localities and its relation to other silicates with dreier double layers. Z Kristallogr 199:25-48

Leigh GJ (ed) (1990) IUPAC Nomenclature of Inorganic Chemistry (Recommendations 1990). Blackwell, Oxford UK

Liebau F (2003) Ordered microporous and mesoporous materials with inorganic hosts: definitions of terms, formula notation and systematic classification. Microporous Mesoporous Mater 58:15-72

Lima-de-Faria J, Hellner E, Liebau F, Makovicky E, Parthé E (1990) Nomenclature of inorganic structure types. Report of the International Union of Crystallography Commission on Crystallographic Nomenclature Subcommittee on the Nomenclature of Inorganic Structure Types. Acta Crystallogr A46:1-11

Makovicky E, Skinner BJ (1979) Crystal structures of the exsolution products $Cu_{12.3}Sb_4S_{13}$ and $Cu_{13.8}Sb_4S_{13}$ of unsubstituted synthetic tetrahedrite. Can Mineral 17:619-634

McCusker LB, Liebau F, Engelhardt G (2001) Nomenclature of structural and compositional characteristics of ordered microporous and mesoporous materials with inorganic hosts. Pure Appl Chem 73:381-394

Merlino S, Bonaccorsi E, Armbruster T (1999) Tobermorites: their real structure and order-disorder (OD) character. Am Mineral 84:1613-1621

Merlino S, Bonaccorsi E, Armbruster T (2001) The real structure of tobermorite 11Å: normal and anomalous forms, OD character and polytypic modification. Eur J Mineral 13:577-590

Nickel EH and Mandarino JA (1987) Procedures involving the IMA Commission on New Minerals and Mineral Names, and guidelines on mineral nomenclature. Am Mineral 72:1031-1042

Perrault PG, Boucher C, Vicat J (1973) Structure Cristalline du Nenadkevichite $(Na,K)_{2-x}(Nb,Ti)(O,OH)Si_2O_6 \cdot 2H_2O$. Acta Crystallogr B29:1432-1438

Pfitzner A, Evain M, Petricek V (1997) $Cu_{12}Sb_4S_{13}$: a temperature-dependent structure investigation. Acta Crystallogr B53:337-345

Rouquérol J, Avnir D, Fairbridge CW, Everett DH, Haynes JH, Pericone N, Ramsay JDF, Sing KSW, Unger KK (1994) Recommendations for the characterization of porous solids. Pure Appl Chem 66:1739-1758

Schoonheydt RA, Pinnavaia T, Lagaly G, Gangas N (1999) Pillared clays and pillared layered solids. Pure Appl Chem 71:2367-2371

Yamnova NA, Egorov-Tismenko YK, Pekov, IV (2001) Refined crystal structure of lovozerite $Na_2CaZr[Si_6O_{12}(OH,O)_6] \cdot H_2O$. Crystallogr Rep 46:937-941

Topology of Microporous Structures

Sergey Krivovichev

Department of Crystallography
St. Petersburg State University
199034 St. Petersburg, Russia
sergey.krivovichev@uibk.ac.at

INTRODUCTION

Microporous crystalline compounds have attracted considerable attention because of their applications in various areas of technology, including catalysis, radioactive waste management, gas separation, adsorption, ion-exchange, etc. Until recently, the major part of research in this field was focused on zeolites and related materials based upon open three-dimensional (3D) frameworks of linked tetrahedra. However, within the last 10–15 years, the quest for new microporous materials led researchers to explore other possibilities, e.g., framework materials containing non-tetrahedral cations that play a crucial role in a framework construction. As for zeolites, this research direction renewed interest in mineral phases as materials formed in the beautiful laboratories of Nature.

The shift of interest from zeolites to materials with non-tetrahedral cations opened a wide range of new possibilities for synthesis and exploration. The chemical range of possible compositions was significantly extended and, from the structural point of view, this chemical extension resulted in endless topological opportunities of framework construction and almost every issue of a journal that deals with structures of minerals and inorganic materials provides new information on framework structures. In this situation, there is a critical need of structure description and classification, from both chemical and topological viewpoints.

The aim of this chapter is to provide a review of possible approaches that can be used for topological and geometrical analysis of microporous structures. For obvious reasons, we do not consider zeolites and related tetrahedral materials here; their detailed topological descriptions can be found in (Smith 1988, 2000; Baerlocher et al. 2001). Most attention will be focused on octahedral-tetrahedral frameworks as important for both mineralogy and material sciences.

BASIC CONCEPTS AND TOOLS

Nodal description of complex structural architectures

Complex structures of inorganic compounds are usually interpreted in terms of coordination polyhedra of cations, i.e., polyhedra centered by cations with anions at their vertices. The most common coordination polyhedra are tetrahedra and octahedra. Three-dimensional (3D) structures of linked tetrahedra and octahedra are sometimes very difficult to describe. To reduce the complexity of the structures, one has to used simplified approaches such as *nodal* (or *graphical*) representations. Within this approach, each node corresponds to either an octahedron (black node) or a tetrahedron (white node). Nodes are connected if the polyhedra share at least one common vertex (= anion), and the number of connecting lines (called *edges*) corresponds to the number of anions common to the polyhedra. Using this approach, each octahedral-tetrahedral unit is associated with a graph with black and white vertices that represents a topology of polyhedral linkage. Nodal representation is well-developed for tetrahedral frameworks (Smith 1988, 2000). For heteropolyhedral units, it was suggested by Moore (1970b) and Hawthorne (1983). The nodal representation of heteropolyhedral units is a useful tool in simplifying, classifying and relating structures of inorganic oxysalts (Hawthorne et al. 2000; Krivovichev and Burns 2003d; Krivovichev 2004a,c).

As nodal representation involves description of structural topology in terms of graphs, there are a number of special tools and concepts taken from the graph theory and related areas of topology (theory of polyhedra, tilings, etc.). Below we provide a short list of such concepts as adopted for crystal chemical purposes by Smith (2000) and O'Keeffe and Hyde (1996).

Vertex valency. A *valency* of a vertex in a graph is the number of edges incident upon it. In the case of octahedral-tetrahedral structures, maximal valence of a white vertex (that corresponds to a tetrahedron) is 4, whereas the maximal valence of a black vertex (symbolizing an octahedron) is 6.

Coordination sequence of a vertex in a graph is a sequence of numbers N_k of vertices at a *topological distance* of k edges. For instance, Figure 1 shows two 2D graphs that describe topologies of heteropolyhedral sheets in a large number of inorganic compounds (Krivovichev and Burns 2003d; Krivovichev 2004a). The graphs have the same black:white ratio of 1:2 and the same vertex valences. Moreover, they have the same number of 4- and 8-membered rings. However, the graphs are different by their coordination sequences of black vertices. In Figure 1, vertices at different topological distances (i.e., separated from a black vertex by k edges) are separated. For both graphs, $N_1 = 5$, $N_2 = 5$, $N_3 = 15$. However, $N_4 = 11$ for the graph in Figure 1a and $N_4 = 10$ for the graph in Figure 1b. Coordination sequences of 3D nets corresponding to octahedral-tetrahedral frameworks in zircono- and titano-silicates were calculated by Ilyushin and Blatov (2002).

Polyhedra

Some 3D graphs can be described as assembled from polyhedral units of different shape and topology. In this regard, the following topological concepts and equations are relevant.

Face symbol is the number of polygonal faces comprising the polyhedron, each face denoted by the number of their vertices; for instance, face symbol for a cube is 4^6 [it has six (superscript) square faces]; face symbol for a hexagonal prism is $6^2 4^6$.

Vertex symbol is the number of edges meeting at vertices; vertex symbol for a cube is 3^8 as it has eight (superscript) 3-valent vertices; vertex symbol for a hexagonal prism is 3^{12}.

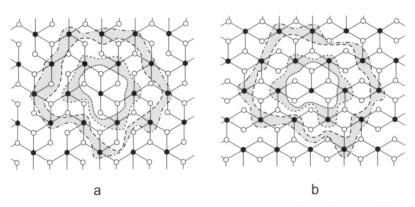

Figure 1. Coordination sequences. Two black-and-white graphs with black:white ratio of 1:2. Four coordination spheres for a black vertex are selected.

Schläfli symbol of a polyhedron provides list of numbers n of N-gonal faces meeting at the same vertex. Schläfli symbols of a tetrahedron is 3^3 (three triangular faces meeting at a vertex), whereas Schläfli symbol of a cube is 4^3 (three square faces meeting at a vertex).

Euler equation. For a polyhedron with v vertices, e edges and f faces, the following equation takes place: $v + f - e = 2$. This equation was derived (in more general context) by the Swiss mathematician Leonhard Euler.

Steiniz theorem (in its simple formulation) states that a net of edges and vertices of a 3D polyhedron can be projected onto 2D plane without intersections of edges. The resulting projection is a ***Schlegel projection*** or ***Schlegel diagram***. Schlegel diagrams were first used in crystal chemistry by Moore (1970a) to compare geometries of coordination polyhedra in basic iron phosphates. Later, they were recognized as an important tool for study of structural topologies and geometries by Hoppe and Köhler (1988), Dvoncova and Lii (1993), and Krivovichev (1997). Nowadays they are actively used in discussions of isomerization and topological transformations of carbon clusters, fullerenes and metallofullerenes (Sokolov 1992; Chiu et al. 2000).

Connectivity diagram is based upon a Schlegel diagram. It is a convenient tool for describing the topology of polyhedral linkages and for its comparison with polyhedral distortions. Figure 2a shows the local environment about WO_6 octahedron in the structures of $A_2[(UO_2)W_2O_8]$ (A = Na, Ag) (Krivovichev and Burns 2003a). The connectivity diagram for the octahedron is shown in Figure 2c. In the diagram, heavy black and double lines represent polyhedral edges shared between the equivalent (in the chemical sense) and non-equivalent polyhedra, respectively. Circles at vertices indicate that the vertex is shared between the respective polyhedron and that adjacent to it, without regard to the chemical specificity of the polyhedron with which it shares the vertex. The unfilled and filled circles are used to indicate

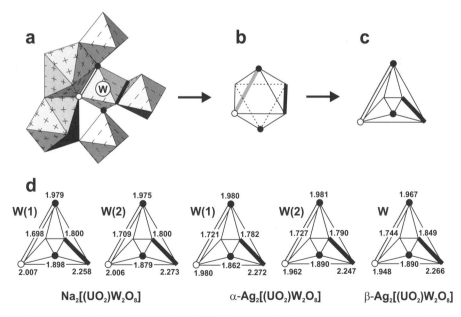

Figure 2. (a) Local environment about the WO_6 octahedra in the $[(UO_2)W_2O_8]$ sheet in the structures of $Na_2[(UO_2)W_2O_8]$ and α- and β-$Ag_2[(UO_2)W_2O_8]$; (b) projection of the octahedron along its threefold symmetry axis with shared elements marked (see text for details); (c) connectivity diagram of the octahedron; (d) connectivity diagrams for WO_6 octahedra (c) with W-O distances given.

that the vertex is shared with the equivalent or non-equivalent (in the chemical sense) polyhedron, respectively. Figure 2d depicts connectivity diagrams for WO_6 octahedra in three uranyl tungstates with the same structural topology, with the W-O distances provided near the vertices. Analysis of the diagrams shows that the W-O bond lengths for the vertices of the same topological type are identical within 0.06 Å. For recent applications of connectivity diagrams to the discussion of structures of minerals, see (Krivovichev et al. 1997; Krivovichev and Burns 2000).

Extended face symbol can be introduced for description of polyhedra with two types of vertices (e.g., black and white). It provides a sequence of black (**b**) and white (**w**) vertices of a polygonal face taken in a cyclic order. For example, the polygonal face shown in Figure 3a has the sequence **bbwwbbww** or $b^2w^2b^2w^2$. The extended face symbol for the cage shown in Figure 3b is $(b^2w^2b^2w^2)^2(bw^2bw^2)^4(w^4)^2(b^2w)^4$.

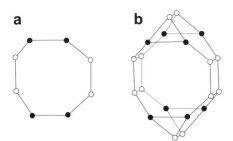

Figure 3. Polygonal face with an extended symbol **bbwwbbww** or $b^2w^2b^2w^2$ that gives sequence of black and white vertices (a) and polyhedral unit with the extended face symbol $(b^2w^2b^2w^2)^2(bw^2bw^2)^4(w^4)^2(b^2w)^4$.

Tilings

Tiling of 3D Euclidean space is a countable number of 3D bodies (*tiles*) that cover space without gaps and overlaps. Here we shall consider only tilings with polyhedral tiles. Tiling is called *face-to-face* if adjacent polyhedral tiles either share vertices, whole edges, whole faces or have no points in common. If a tiling has a finite number *n* of topologically and geometrically different tiles, it is called *n-hedral*. If *n* = 1, tiling is *isohedral*; if *n* = 2, tiling is *dihedral*; if *n* = 3, tiling is *trihedral*, etc.

A special class of isohedral face-to-face tiling of 3D space is that in which all tiles are in the same orientation. Such tiles are called *parallelohedra*. For 3D Euclidean space, there are exactly five types of parallelohedra derived in 1885 by Fedorov (1971): cube, hexagonal prism, truncated octahedron, rhombic dodecahedron and elongated rhombic dodecahedron. Tiles of isohedral face-to-face tiling of 3D space with not necessarily parallel orientations of tiles are called *stereohedra*. The complete list of stereohedra for a 3D Euclidean space is still unknown. Even the maximal number of faces for a 3D stereohedron is unknown [minimal theoretical boundary is 226 (Krivovichev 1999), but this number is certainly too high]. Engel (1981) discovered four Dirichlet-Voronoi stereohedra with 38 faces but it is not known whether this number is a maximum.

Regular and quasiregular nets

There are several important high-symmetrical 3D nets that represent underlying topologies of a number of inorganic and metal-organic frameworks. Such nets were recently considered by Delgado-Friedrichs et al. (2003). These authors consider 3D nets with the following properties: (i) all vertices are of the same type; (ii) convex hull of coordination figure of a vertex in a net is a regular polygon or a regular polyhedron (i.e., one of five Platonic solids); (iii) site symmetry of a vertex in the net is at least the rotation symmetry of that regular polygon or polyhedron. According to Delgado-Friedrichs et al. (2003), nets that fulfill these three conditions are *regular*. There are exactly five types of regular nets that correspond to the following regular polygons and polyhedra: equilateral triangle, square, tetrahedron, octahedron and cube (Fig. 4). Following Delgado-Friedrichs et al. (2003), we denote regular nets as **pcu, bcu, srs, nbo** and **dia**. The **pcu** and **bcu** nets correspond to the primitive **cu**bic lattice and **b**ody-**c**entered **cu**bic lattice, respectively. The **srs**, **nbo**, and **dia** nets correspond to structural topologies of SrS, NbO, and diamond, respectively. The **fcu** net shown in Figure 4 is *quasiregular* and corresponds to the **f**ace-**c**entered **cu**bic lattice.

The regular and quasiregular nets themselves are relatively dense. However, distances between the vertices of these nets can be considerably increased by topological operations identified by O'Keeffe et al. (2000) as *decoration, augmentation* and *expansion*.

Decoration is replacement of a vertex by a group of vertices.

Augmentation is a special case of decoration and involves replacement of vertices of *n*-connected net by a group of *n* vertices.

Expansion is increasing of the length of an edge by a group of edges. This process is very common for metal-organic frameworks (O'Keeffe et al. 2000).

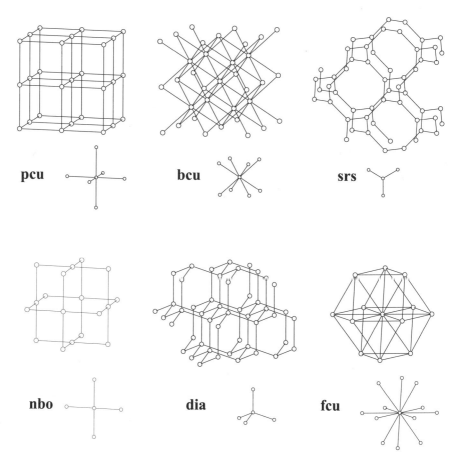

Figure 4. Regular and quasiregular nets.

An excellent example of transformation of a net is the transition from the **dia** net to a range of cristobalite-related tetrahedral structures. Replacing of vertices of the **dia** net by Si atoms and insertion of O atoms into middle points of edges results in the structure of cristobalite, SiO_2. Tetrahedral framework in cristobalite can thus be considered as an expanded **dia** net (here expansion corresponds to insertion of O atoms). Replacement of a single tetrahedron (T1) by a group of four tetrahedra (also called a *supertetrahedron* T2, Fig. 5) results in a metal sulfide framework found in a number of sulfides, including $[Ge_4S_{10}][(CH_3)_4N]_4$ (Yaghi et al. 1994) and $[C_7H_{13}N][Mn_{0.25}Ge_{1.75}S_4](H_2O)$ (Cahill et al. 1998a). Increasing dimensions of supertetrahedron from T2 to T3 produces metal sulfide frameworks in $[(CH_3)_2NH_2]_6[In_{10}S_{18}]$ (Cahill et al. 1998b) and ASU-34, $In_{10}S_{18}[C_6H_{12}NH_2]_6[C_6H_{12}NH](H_2O)_5$ (Li et al. 1999a). Note that the structure of $[(CH_3)_2NH_2]_6[In_{10}S_{18}]$ contains two interpenetrating cristobalite-like frameworks (and thus is similar to the structure of cuprite, Cu_2O, that contains two interpenetrating frameworks of oxocentered OCu_4 tetrahedra), the structure of ASU-34 consists of interpenetrating supertetrahedral framework with organic molecules in the framework cavities (and, in this sense, is similar to the structure of melanothallite, Cu_2OCl_2; Krivovichev et al. 2002). Replacement of a vertex of the **dia** net by a T4 supertetrahedron results in a metal-sulfide framework in the structure of $[C_{10}H_{28}N_4]_{2.5}[Cd_4In_{16}S_{33}](H_2O)_{20}$ (Li et al. 2001).

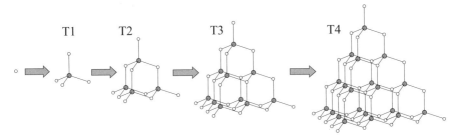

Figure 5. Replacement of a single vertex in the **dia** regular net by a tetrahedron (T1) and supertetrahedra of increasing size (T2-T4) generates cristobalite-like and supertetrahedral diamondoid frameworks.

Schindler et al. (1999) considered many examples of correspondence (in mathematical terminology, homeomorphisms) between simple 3D nets and complex inorganic frameworks. They suggest to call complex structures which can be obtained by transformations of simple 3D nets as *metastructures*. The transformations considered by Schindler et al. (1999) are in fact those recognized by O'Keeffe et al. (2000) as decoration, augmentation and expansion.

Regular and quasiregular nets are those of the simplest 3D nets. These nets dominate topologies of metal-organic solids with extremely large pores as discussed by O'Keeffe et al. (2000), Eddaoudi et al. (2001). However, they are not so common in structures of microporous mineral phases owing to the high complexity of their polyhedral arrangements.

HETEROPOLYHEDRAL FRAMEWORKS: CLASSIFICATION PRINCIPLES

Within a nodal approach to complex heteropolyhedral frameworks, each 3D framework is symbolized by a 3D graph with n different vertices, where n is the number of topologically and geometrically different coordination polyhedra. In the simple case of $n = 2$, we have vertices of two types only: black and white. The simplest example is an octahedral-tetrahedral framework that corresponds to a 3D net with black and white vertices symbolizing octahedra and tetrahedra, respectively.

In each black-and-white 3D net associated with an octahedral-tetrahedral framework we can subdivide a connected subnet of black vertices (that corresponds to an octahedral substructure within the framework) and a connected subnet of white vertices (= tetrahedral substructure). For instance, if there are no edges connecting black vertices, the subnet of black vertices is a single vertex and octahedral substructure is a single octahedron. If octahedra are linked into a chain, corresponding black vertices form a chain.

Dimensionality (D) of a graph is defined as the number of dimensions in which it has an infinite extension. Thus, for a chain and a sheet, D is equal to 1 and 2, respectively. For a 3D net corresponding to an octahedral-tetrahedral framework, we define D_{nt} as the dimensionality of an octahedral (or, more generally, non-tetrahedral) substructure (black subgraph) and D_{tetr} as the dimensionality of a tetrahedral substructure (white subgraph). Thus, we can classify 3D nets according to the pairs of values (D_{nt}; D_{tetr}). The classification can be represented as a 4 × 4 table where rows correspond to particular values of D_{nt} and columns to particular values of D_{tetr}. Table 1 presents an example of classification of minerals and inorganic compounds with mixed frameworks according to the (D_{nt}; D_{tetr}) pairs.

The most common groups of frameworks are those that have ($D_{tetr} = 0$; $D_{nt} = 0$) and ($D_{tetr} = 1$; $D_{nt} = 0$). The lists of materials corresponding to these classification criteria are given in Tables 2 and 3, respectively. Minerals and compounds that have ($D_{tetr} = 2$; $D_{nt} = 0$) will be discussed later.

We note that Table 1 provides only very general and formal classification of mixed frameworks. For detailed topological systematics, one has to find an appropriate description of each 3D net that provides its simple construction and allows its comparison with other nets. We found it useful to subdivide frameworks given in Table 1 into five major groups: (i) frameworks based upon fundamental building blocks (FBBs); (ii) frameworks assembled from polyhedral units; (iii) frameworks based upon 1D units (fundamental chains or tubular units), (iv) frameworks based upon 2D units, and (v) frameworks consisting of broken tetrahedral subframework. This classification is simple and of convenient description.

Table 1. Classification of materials based upon frameworks of tetrahedra and non-tetrahedrally coordinated high-valent metal cations.

D_{nt} \ D_{tetr}	0	1	2	3
0	wadeite, benitoite, kostylevite, petarasite, keldyshite	hilairite, umbite, terskite, vlasovite, elpidite, gaidonnayite, zektzerite	lemoynite, armstrongite, $Cs_2(ZrSi_6O_{15})$, $Na_{1.8}Ca_{1.1}{}^{VI}Si({}^{IV}Si_5O_{14})$, penkvilksite	—
1	belkovite, labuntsovite-group minerals	batisite-shcherbakovite, zorite, ETS-4	USH-8 = $[(CH_3)_4N]$ $[(C_5H_5NH)_{0.8}$ $((CH_3)_3NH)_{0.2}]$ $(UO_2)_2[Si_9O_{19}]F_4$	ETS-10
2	komarovite	—	—	—
3	—	—	—	—

Table 2. Minerals and inorganic compounds containing frameworks of isolated (or finite clusters of) tetrahedra and isolated (or finite clusters of) metal polyhedra ($D_{tetr} = 0$; $D_{nt} = 0$).

Framework type*	T unit	Mineral/ name	Chemical formula	Channel directions	Ref.
WAD	3MR	wadeite	$K_2Zr(Si_3O_9)$	[100],[010]	1,2
		catapleite	$Na_2ZrSi_3O_9(H_2O)_2$		3
		FDZG-2	$(NH_4)_2ZrGe_3O_9$		4
		-	$Cs_2Zr(Si_3O_9)$		5
		-	$K_2Sn(Si_3O_9)$		6
		-	$K_2Si^{VI}Si_3^{IV}O_9$		7
BEN	3MR	benitoite	$BaTiSi_3O_9$	[100],[001]	8,9
		bazirite	$BaZrSi_3O_9$		10
		pabstite	$BaSnSi_3O_9$		10,11
		-	$BaSi^{VI}Si_3^{IV}O_9$		12
		-	$KCaP_3O_9$		13
KST	6MR	kostylevite	$K_4Zr_2Si_6O_{18}(H_2O)_2$	[100]	14
		UND-1	$Na_{2.7}K_{5.3}Ti_4Si_{12}O_{36}(H_2O)_4$		15
		AM-7	$K_{1.5}Na_{0.5}SnSi_3O_9(H_2O)$		16
		AV-8	$Na_{0.2}K_{1.8}ZrSi_3O_9(H_2O)_n$		17
		-	$K_2Pb^{4+}Si_3O_9(H_2O)$		18
PET	6MR	petarasite	$Na_5Zr_2Si_6O_{18}(Cl,OH)(H_2O)_2$	[100],[001]	19
		AV-3	$Na_5Zr_2Si_6O_{18}(Cl,OH)(H_2O)_2$		20,21
ILI	6MR	diversilite-Ce	$(Ba,K,Na)_{11-12}(Ce,La,Th,Nd)_4$ $(Ti,Nb,Fe^{3+})_6(Si_6O_{18})_4(OH)_6 \cdot 6H_2O$	[100]	22
KEL	Si_2O_7	keldyshite	$NaZr(Si_2O_6OH)$	[010]	23
		parakeldyshite	$Na_2Zr(Si_2O_7)$		24
		khibinskite	$K_2Zr(Si_2O_7)$		25
		-	$Na_2Si^{VI}(Si^{IV}_2O_7)$		26
CZA	4MR	-	$Ca_2Zr(Si_4O_{12})$	[010]	27
GIT	Si_2O_7	gittinsite	$CaZr(Si_2O_7)$	[001]	28
SZA	Si_2O_7	-	$SrZr(Si_2O_7)$	[101]	29
SZB	Si_2O_7	-	$Sr_7Zr(Si_2O_7)_3$	[100],[010], [001]	30

* for most framework types, abbreviations introduced by Ilyushin and Blatov (2002)

References: (1) Henshaw 1955; (2) Blinov et al. 1977; (3) Ilyushin et al. 1981c; (4) Liu et al. 2003a; (5) Balmer et al. 2001; (6) Rudenko et al. 1983; (7) Swanson and Prewitt 1983; (8) Zachariasen 1930; (9) Fischer 1969; (10) Hawthorne 1987; (11) Choisnet et al. 1972; (12) Finger et al. 1995; (13) Sandstroem and Bostroem 2004; (14) Ilyushin et al. 1981a; (15) Liu et al. 1997; (16) Lin et al. 2000; (17) Ferreira et al. 2001a; (18) Pertierra et al. 2001; (19) Ghose et al. 1980; (20) Rocha et al. 1998; (21) Lin et al. 1999a; (22) Krivovichev et al. 2003b; (23) Khalilov et al. 1978; (24) Voronkov et al. 1970; (25) Chernov et al. 1970; (26) Fleet and Henderson 1995; (27) Colin et al. 1993; (28) Roelofsen-Ahl and Peterson 1989; (29) Huntelaar et al. 1994; (30) Plaisier et al. 1994.

Table 3. Minerals and inorganic compounds containing frameworks of chains of tetrahedra and isolated (or finite clusters of) metal polyhedra ($D_{tetr} = 1$; $D_{nt} = 0$).

Framework type	Mineral/ name	Chemical formula	Channel directions	Ref.
HIL	hilairite	$Na_2ZrSi_3O_9(H_2O)_3$	[001]	1
	komkovite	$BaZrSi_3O_9(H_2O)_{2.43}$		2
	calciohilairite	$CaZrSi_3O_9(H_2O)_3$		3
	pyatenkoite-(Y)	$Na_5YTiSi_6O_{18} \cdot 6H_2O$		4
	sazykinaite-(Y)	$Na_5YZrSi_6O_{18} \cdot 6H_2O$		5,6
	Ti-AV-11	$K_2TiSi_3O_9$		7
	Sn-AV-11	$K_2SnSi_3O_9$		8
UMB	umbite	$K_2Zr(Si_3O_9)(H_2O)$	[001]	9,10
	AM-2	$K_2Zr(Si_3O_9)(H_2O)$		11-13
	—	$K_2Ti(Si_3O_9)(H_2O)$		14,15
	AV-6	$K_2Sn(Si_3O_9)(H_2O)$		16,17
	AV-12	$K_2Hf(Si_3O_9)(H_2O)$		18
	—	$K_2Zr(Ge_3O_9)(H_2O)$		19
	ASU-25	$[C_3H_{12}N_2]ZrGe_3O_9$		19
ASU-26	ASU-26	$[C_2H_{10}N_2]ZrGe_3O_9$	[010]	19
LEM	lemoynite	$Na_2CaZr_2(Si_{10}O_{26})(H_2O)$	[101]	20,21
	natrolemoynite	$Na_2H_2Zr_2(Si_{10}O_{26})(H_2O)$		22
	altisite	$Na_3K_6Ti_2(Al_2Si_8O_{26})Cl_3$		23
TER	terskite	$Na_4Zr(H_4Si_6O_{18})$	[001]	24
VLS	vlasovite	$Na_2ZrSi_4O_{11}$	[001]	25-29
	—	$Na_2ZrSi_4O_{11}$		30
ELP	elpidite	$Na_2ZrSi_6O_{15}(H_2O)_3$	[100],[010]	31-33
	—	$Na_2ZrSi_6O_{15}(H_2O)_3$		12
GDN	gaydonnayite	$Na_2Zr(Si_3O_9)(H_2O)_2$	[010],[001]	34
	AV-4	$Na_2Zr(Si_3O_9)(H_2O)_2$		12,16
	georgechaoite	$NaKZr(Si_3O_9)(H_2O)_2$		35
	—	$(Ca,Na,K)_{2-x}ZrSi_3O_9 \cdot nH_2O$		36
ZEK	zektzerite	$NaLiZrSi_6O_{15}$	[100]	37
AV-10	AV-10	$Na_2SnSi_3O_9 \cdot 2H_2O$	[010]	38

References: (1) Ilyushin et al. 1981d; (2) Sokolova et al. 1991; (3) Pushcharovskii et al. 2002; (4) Rastsvetaeva and Khomyakov 1996; (5) Khomyakov et al. 1993; (6) Rastsvetaeva and Khomyakov 1992; (7) Lin et al. 2003; (8) Ferreira et al. 2004; (9) Ilyushin et al. 1981b; (10) Ilyushin 1993; (11) Lin et al. 1997; (12) Jale et al. 1999; (13) Poojary et al. 1997; (14) Dadachov and Le Bail 1997; (15) Bortun et al. 2000; (16) Lin et al. 1999b; (17) Pertierra et al. 2002; (18) Lin and Rocha 2002; (19) Plevert et al. 2003; (20) Blinov et al. 1974; (21) le Page and Perrault 1976; (22) McDonald and Chao 2001; (23) Ferraris et al. 1995; (24) Pudovkina and Chernitsova 1991; (25) Voronkov and Pyatenko 1961; (26) Pyatenko and Voronkov 1961; (27) Fleet and Cann 1967; (28) Voronkov et al. 1974; (29) Gobechiya et al. 2003; (30) Vitins et al. 1996; (31) Neronova and Belov 1963, 1964; (32) Cannillo et al. 1973; (33) Sapozhnikov and Kashaev 1978, 1980; (34) Chao 1985; (35) Ghose and Thakur 1985; (36) Belyayevskaya et al. 1991; (37) Ghose and Wan 1978; (38) Ferreira et al. 2001b

FRAMEWORKS BASED UPON FUNDAMENTAL BUILDING BLOCKS (FBBs)

Some definitions

As it was already mentioned, in inorganic crystal chemistry and structural mineralogy, a preferred approach to describe a crystal structure is to do it in terms of coordination polyhedra of cations. A coordination polyhedron can be considered as a *basic building unit* (BBU) following the IUPAC nomenclature (McCusker et al. 2001) or a *structural subunit* if one follows terminology recommended by the International Union of Crystallography (Lima-de-Faria et al. 1990). Sometimes, crystal structure can be effectively described as based upon clusters of coordination polyhedra usually called *fundamental building blocks* (FBBs) (Hawthorne 1994). An alternative term is *a composite building unit* (CBU) which is used as a second level of structural hierarchy (the first is a BBU which is a single polyhedron) (McCusker et al. 2001). Description of inorganic structures in terms of FBBs is especially justified if the same FBB occurs in many related structures. This indicates possible role of FBBs as clusters pre-existing in a crystallization media (Férey 2001). FBBs (or CBUs) are effectively used for description of Mo phosphates (Haushalter and Mundi 1992) and a large family of open-framework Fe, Al and Ga phosphates with Fe, Al and Ga in non-tetrahedral coordinations (materials known as ULM-n and MIL-n; see Férey 1995, 1998, 2001; Riou-Cavellec et al. 1999 for reviews).

An interesting feature of frameworks based upon FBBs is that FBBs are frequently arranged according to positions of nodes of simple regular nets discussed above. Below we consider some examples of frameworks based upon FBBs of various shapes and compositions.

Leucophosphite-type frameworks

Figure 6a,b shows a cluster consisting of four octahedra and six tetrahedra. The octahedra share edges and corners to form octahedral tetramers surrounded by tetrahedra sharing their corners with octahedra. This "butterfly-shaped" cluster occurs as a FBB in a number of framework phosphates (Férey 1995, 1998, 2001). Figure 6c shows the structure of leucophosphite, $K_2[Fe_4(OH)_2(H_2O)_2(PO_4)_4](H_2O)_2$ (Moore 1972). In this structure, "butterfly-shaped" FBBs have composition $[Fe_4(OH)_2(H_2O)_2(PO_4)_6]^{8-}$ and are linked into framework by sharing corners of PO_4 tetrahedra with FeO_6 octahedra of adjacent clusters. The arrangement of FBBs in leucophosphite is schematically shown in Figure 6d. As it is clearly seen, its topology corresponds to the **bcu** regular net (Fig. 4). It is noteworthy that the leucophosphite structure type based upon the highly symmetrical regular net is common among minerals and inorganic compounds. It has been observed for minerals chemically analogous of leucophosphite: tinsleyite $K_2[Al_4(OH)_2(H_2O)_2(PO_4)_4](H_2O)_2$ (Dick 1999) and spheniscidite $(NH_4)_2[Fe_4(OH)_2$

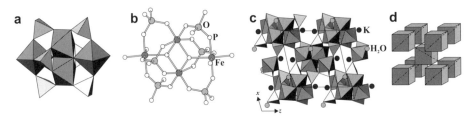

Figure 6. "Butterfly-shaped" cluster (FBB) consisting of four octahedra and six tetrahedra shown in polyhedral (a) and ball-and-stick (b) aspects. The structure of leucophosphite, $K_2[Fe_4(OH)_2(H_2O)_2(PO_4)_4](H_2O)_2$ (c) and the scheme demonstrating arrangement of FBBs in leucophosphite (d).

$(H_2O)_2(PO_4)_4](H_2O)_2$ (Yakubovich and Dadachov 1992). It is also common for alkali metal Mo phosphates of general formula $A_n[Mo_2O_2(PO_4)_2](H_2O)_m$ ($n = 1–2$; $m = 0–1$) (King et al. 1991a; Guesdon et al. 1993; Leclaire et al. 1994).

Frameworks with oxocentered tetrahedral cores

Figure 7 shows the structure of $(C_2H_{10}N_2)_2[Fe_4O(PO_4)_4](H_2O)$ (de Bord et al. 1997; Song et al. 2003) which is based upon pentahedral-tetrahedral cluster consisting of four FeO_5 trigonal bipyramids and four PO_4 tetrahedra (Fig. 7a, b). It is noteworthy that four FeO_5 polyhedra share the same O atom that can be considered as being at the center of a tetrahedron formed by four Fe atoms (Fig. 7c). The $[Fe_4O(PO_4)_4]$ FBBs are linked by sharing O corners with adjacent clusters (Fig. 7d). Framework topology corresponds to the **bcu** regular net.

Harrison et al. (1996, 2000) reported a series of microporous materials with general formula $M_3[Zn_4O(XO_4)_3](H_2O)_n$ (M = Na, K, Rb, Cs; X = P, As; n = 3–6). Structures of these materials are based upon clusters shown in Figure 8a,b. The core of the $[Zn_4O(XO_4)_6]$

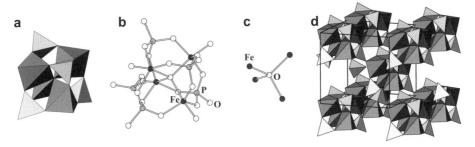

Figure 7. Pentahedral-tetrahedral cluster consisting of four FeO_5 trigonal bipyramids and four PO_4 tetrahedra shown in polyhedral (a) and ball-and-stick (b) aspects, and the Fe_4O oxocentered tetrahedron (c) as a core of the cluster. Arrangement of the $[Fe_4O(PO_4)_4]$ clusters in the structure of $(C_2H_{10}N_2)_2[Fe_4O(PO_4)_4](H_2O)$ (d).

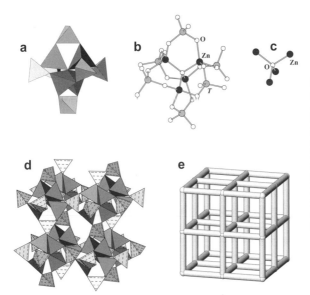

Figure 8. The $[Zn_4O(XO_4)_6]$ cluster from the structures of $M_3[Zn_4O(XO_4)_3](H_2O)_n$ (M = Na, K, Rb, Cs; X = P, As; n = 3-6) shown in polyhedral (a) and ball-and-stick (b) aspects, and the Zn_4O oxocentered tetrahedron (c) as a core of the cluster. The $[Zn_4O(XO_4)_6]$ clusters are linked through common XO_4 groups to form a 3D framework (d) with topology of the **pcu** regular net (e).

cluster is an oxocentered OZn$_4$ tetrahedron similar to the OFe$_4$ tetrahedron in (C$_2$H$_{10}$N$_2$)$_2$ [Fe$_4$O(PO$_4$)$_4$](H$_2$O) (de Bord et al. 1997) (Fig. 8c). However, in contrast to the latter structure, Zn atoms in M_3[Zn$_4$O(XO$_4$)$_3$](H$_2$O)$_n$ are tetrahedrally coordinated. The [Zn$_4$O(XO$_4$)$_6$] clusters are linked through common XO$_4$ groups to form a 3D framework with topology of the **pcu** regular net (Fig. 8d,e). Note that the function of the XO$_4$ groups in the organization of the [Zn$_4$O(XO$_4$)$_3$] is to link adjacent OZn$_4$ tetrahedra into a 3D network. Replacement of XO$_4$ groups by 1,4-benzenedicarboxylate (BDC) groups, C$_8$H$_4$O$_4$, leads to the formation of an *expanded* framework material Zn$_4$O(BDC)$_3$ with unusually large pores (Li et al. 1999b; Zaworotko 1999).

Pharmacosiderite-related frameworks

Minerals of the pharmacosiderite group have general formula A_x[M_4(OH)$_4$(AsO$_4$)$_3$](H$_2$O)$_y$, where A = K, Na, Ba; x = 0.5–1; M = Al, Fe^{3+}, y = 6–7 (Strunz and Nickel 2001). Structure of pharmacosiderite (A = K, M = Fe^{3+}) was reported by Zemann (1948) and refined by Buerger et al. (1967). It is based upon octahedral-tetrahedral framework that consists of octahedral tetramers M_4(O,OH)$_{16}$ linked via isolated AsO$_4$ tetrahedra. The FBB of this structure is shown in Figure 9a,c. It is interesting to note that core of this FBB is a cubic M_4(OH)$_4^{8+}$ polycation consisting of four M^{3+} cations and four OH$^-$ groups. The FBBs are linked into 3D framework of the **pcu** regular net topology (Fig. 9d). Zemann (1959) pointed out that pharmacosiderite is isostructural with a family of open-framework germanates with general formula A_xH$_y$[Ge$_7$O$_{16}$](H$_2$O)$_z$ (A = NH$_4$, Na, K, Rb, Cs; $x + y = 4$; $z = 0$–4) (Nowotny and Wittman 1954; Sturua et al. 1978; Bialek and Gramlich 1992; Roberts et al. 1995; Roberts and Fitch 1996). In germanates, framework contains Ge atoms in both tetrahedral

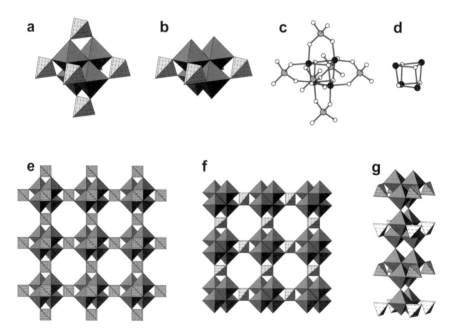

Figure 9. Octahedral-tetrahedral clusters in pharmacosiderite-related frameworks (a, b, c) and M_4(OH)$_4^{8+}$ polycation consisting of four M^{3+} cations and four OH$^-$ groups at the core of these clusters (d). The structures of the minerals of the pharmacosiderite group are based upon 3D framework of the **pcu** regular net topology (e). The The [M_4(O,OH)$_4$(XO$_4$)$_4$] FBBs (b) in the structures of A_2[Ti$_2$O$_3$(SiO$_4$)](H$_2$O)$_n$ (A = Na, H) are linked into 3D framework (e) consisting of chains shown in (f).

and octahedral coordinations; thus, its formula can be written as $[^{VI}Ge_4O_4(^{IV}GeO_4)_3]$. The pharmacosiderite-type framework was also observed in the structures of $A_3[Mo_4O_4(PO_4)_3]$ (A = NH_4, Cs) (Haushalter 1987; King et al. 1991b). In 1990, Chapman and Roe prepared a number of titanosilicate analogs of pharmacosiderite, including Cs, Rb and exchanged protonated phases. Harrison et al. (1995) reported the structure of $Cs_3H[Ti_4O_4(SiO_4)_3](H_2O)_4$. Behrens et al. (1996), Behrens and Clearfield (1997) and Dadachov and Harrison (1997) provided data on preparation, structures and properties of $A_3H[Ti_4O_4(SiO_4)_3](H_2O)_n$ (A = H, Na, K, Cs). Structures and ion-exchanged properties of $HA_3[M_4O_4(XO_4)_3](H_2O)_4$ (A = K, Rb, Cs; M = Ti, Ge; X = Si, Ge) were reported by Behrens et al. (1998). These compounds were considered as perspective materials for the selective removal of Cs and Sr from wastewater solutions.

In pharmacosiderite framework with the $M:X$ = 4:3, each $M_4(O,OH)_4$ cluster is surrounded by six XO_4 tetrahedra thus forming $[M_4(O,OH)_4(XO_4)_6]$ FBB. Removal of two opposing tetrahedra from this FBB results in another FBB with composition $[M_4(O,OH)_4(XO_4)_4]$ (Fig. 9b). This FBB serves as a basis for open frameworks in the structures of phosphovanadylite, $(Ba,K,Ca,Na)_x[(V,Al)_4P_2(O,OH)_{16}](H_2O)_{12}$ (Medrano et al. 1998), and $[(CH_3)_4N]_{1.3}(H_3O)_{0.7}[Mo_4O_8(PO_4)_2](H_2O)_2$ (Haushalter et al. 1989). However, removal of two XO_4-links results in a change of framework topology. Instead of the **pcu** regular net topology of pharmacosiderite $M:X$ = 4:3 framework, the $M:X$ = 2:1 framework topology in phosphovanadylite is that of the **nbo** regular net. Note that this topology is more open in terms of its framework density calculated as the number of metal atoms per 1000 Å^3.

The $[M_4(O,OH)_4(XO_4)_4]$ FBBs (Fig. 9b) may also link through non-shared corners of their $M(O,OH)_6$ octahedra (Fig. 9e,f). This situation is realized in the structures of $A_2[Ti_2O_3(SiO_4)](H_2O)_n$ (A = Na, H) (Poojary et al. 1994; Clearfield et al. 2000). In these structures, octahedral tetramers are linked to form chains running along the c axis. Linkage of the chains along the a and b axes is of the same type as in pharmacosiderite 4:3 frameworks. More details about structures and properties of open-framework pharmacosiderite titanosilicates may be found in Clearfield (2001).

FRAMEWORKS BASED UPON POLYHEDRAL UNITS

Description of frameworks in terms of polyhedral units is widely used in crystal chemistry of zeolites and related tetrahedral structures (Smith 1988, 2000; Baur and Fischer 2000, 2002; Liebau 2003). However, it is rarely employed in descriptions of more complex frameworks, e.g., those based upon linked octahedra and tetrahedra. Below we provide some examples of description of titanosilicate octahedral-tetrahedral frameworks in terms of assemblies of polyhedral units.

Example 1: minerals of the labuntsovite group

Labuntsovite-group minerals have recently attracted considerable attention of mineralogists owing to the high variability of their structure and chemical composition (Chukanov et al. 2002, 2003; Armbruster et al. 2004, etc.). The basis of their structures is an octahedral-tetrahedral framework shown in Figure 10a. The $M\phi_6$ octahedra (M = Ti, Nb; ϕ = O, OH) share corners to produce single chains that are interlinked by Si_4O_{12} four-membered silicate rings. The resulting framework has channels oriented perpendicular to the octahedral chains and occupied by low-valent cations and H_2O molecules. Note that the framework shown in Figure 10a has D_{nt} = 1 (octahedra are linked to form a chain) and D_{tetr} = 0 (silicate tetrahedra form 4-membered rings (4MRs)). Nodal representation of the labuntsovite-type framework is shown in Figure 10b. It can be considered to consist of two types of polyhedral units shown in Figure 10c,d. Face and vertex symbols of the unit shown in Figure 10c are

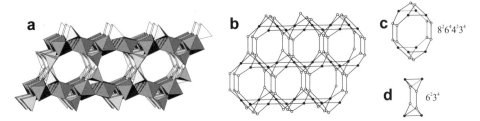

Figure 10. Octahedral-tetrahedral framework in the structures of labuntsovite-group minerals (a) and its nodal representation. The labuntsovite 3D net consists of two types of polyhedral units with extended face symbols $(b^2w^2b^2w^2)^2(b^2wb^2w)^4(w^4)^2(b^2w)^4$ (c) and $(bw^2bw^2)^2(b^2w)^4$ (d).

$8^26^44^23^4$ and 3^{20}, respectively. Face and vertex symbols of the unit shown in Figure 10d are 6^23^4 and 3^8, respectively. Extended face symbols (see above) of the polyhedral units shown in Figure 10c,d are $(b^2w^2b^2w^2)^2(b^2wb^2w)^4(w^4)^2(b^2w)^4$ and $(bw^2bw^2)^2(b^2w)^4$, respectively. The $8^26^44^23^4$ units share 8-membered faces to form tunnels.

Owing to the fact that the labuntsovite-type 3D net can be assembled from polyhedral units of two types, it is possible to construct a *dihedral* tiling of 3D space that corresponds to this net. This tiling is demonstrated in Figure 11. It consists of two polyhedra that are not topologically equivalent to the polyhedral units shown in Figure 10 and can be obtained from the latter by addition of edges linking two black vertices (oriented vertically in Fig. 10). The polyhedra of the tiling are arranged in the following way. First, polyhedra of the type shown in Figure 11b are arranged into layers by sharing faces on their sides (Fig. 11c). These layers have polyhedral hollows (Fig. 11d) that are perfectly suited for the polyhedra of the type shown in Figure 11a. Whole tiling is obtained by assembling layers of larger tiles one under another and by filling gaps with the smaller tiles (Fig. 11e).

Example 2: shcherbakovite-batisite series

Minerals of the shcherbakovite-batisite series have the general formula $NaA_2(Ti,Nb)_2O_2[Si_4O_{12}]$, where A = K, Ba for shcherbakovite and Ba, K for batisite. The two minerals are isotypic with structures based upon chains of corner-linked MO_6 octahedra (M = Ti, Nb) and chains of SiO_4 tetrahedra (Fig. 12a) (Nikitin and Belov 1962; Schmahl and Tillmans 1987; Rastsvetaeva et al. 1997; Uvarova et al. 2003; Krivovichev et al. 2004a). The

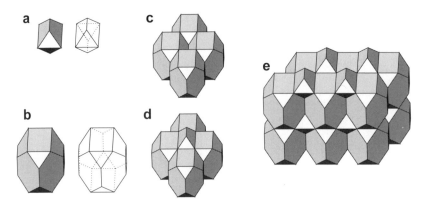

Figure 11. Dihedral tiling of 3D space that corresponds to the labuntsovite net. See text for details.

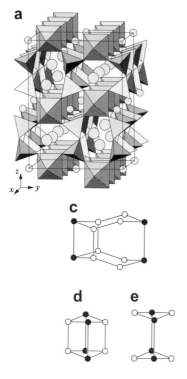
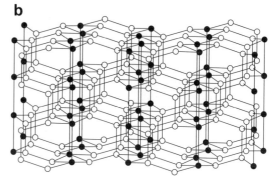

Figure 12. Octahedral-tetrahedral framework in minerals of the shcherbakovite-batisite series (a) and its nodal representation (b). The 3D black-and-white net can be constructed by assembling three different polyhedral units characterized by the following face symbols: 6^44^2 (c), 4^6 (d) and 6^24^2 (e).

nodal representation of the shcherbakovite-batisite framework is shown in Figure 12b. This 3D black-and-white net can be constructed by assembling three different polyhedral units characterized by the following face symbols: 6^44^2 (Fig. 12c), 4^6 (Fig. 12d) and 6^24^2 (Fig. 12e). Note that the 6^44^2 unit contains 6-membered faces of two types: **bw²bw²** and **b²w⁴**. The 4^6 and 6^24^2 units are closely related and can be obtained one from another by adding (deleting) two edges linking white vertices.

Using simple topological and geometrical operations, the 6^44^2 unit can be transformed into a hexagonal prism, whereas the 4^6 and 6^24^2 units can be transformed into square prisms. Thus the 3D net shown in Figure 12b can be considered as a modified version of the tiling of 3D space into hexagonal and square prisms shown in Figure 13.

Combinatorial topology of polyhedral units

A detailed consideration of polyhedral units that occur in different structures reveals some interesting topological relationships. About half of the units can be produced from some simple convex polyhedra by a series of topological operations including deleting edges, inserting new vertices and edges and merging vertices. As an example, we consider a hexagonal prism (Fig. 14). Its face and vertex symbols are 6^24^6 and 3^{12}, respectively. Let us number vertical edges of the prism (edges that link two hexagonal bases) in a cyclic order (Fig. 14a). If two vertical edges of the prism are deleted, the result is a *two-edges-deleted* hexagonal prism. If two edges are not of the same square face, there are two ways to delete them: (i) to delete edges 1 and 4 (Fig. 14b) and (ii) to delete edges 1 and 3 (Fig. 14c). The results are polyhedral units shown in Figure 14d,e. Note that they have the same face and vertex symbols: 6^44^2 and 2^43^8, respectively. As the vertices of 3D black-and-white nets can be either black or white, there are many ways to colour vertices of these two units and all can be easily

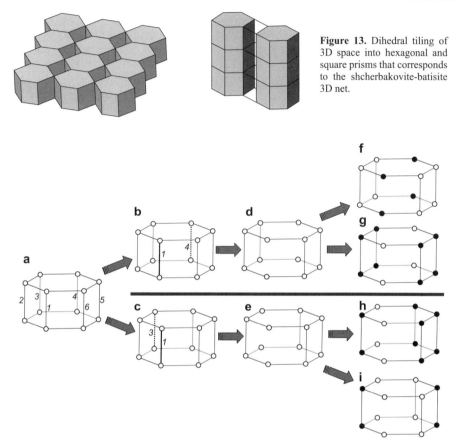

Figure 13. Dihedral tiling of 3D space into hexagonal and square prisms that corresponds to the shcherbakovite-batisite 3D net.

Figure 14. Topological evolution of hexagonal prism into polyhedral units that occur in 3D nets corresponding to octahedral-tetrahedral frameworks. Note that the paths shown include only deleting of edges and color changes for the vertices. See text for details.

enumerated. However, we consider only those cages that are realized in the real structures. The units shown in Figure 14f,g are hexagonal prisms with 1,4-deleted edges. However, their black-and-white coloring is different: vertices that are black in one cage are white in the other and *vice versa*. The units shown in Figure 14h,i are 1,3-edge-deleted hexagonal prisms with different coloring of their vertices. From the four units considered, the first (Fig. 14f) occurs in the structure of zektzerite, $NaLiZrSi_6O_{15}$ (Ghose and Wan 1978), the second and third (Fig. 14g,h) in $K_4Nb_8O_{14}(PO_4)_4(SiO_4)$ (Leclaire et al. 1992), and the fourth (Fig. 14i) is the 6^44^2 unit from the structures of minerals of the batisite-shcherbakovite series (see above).

Another type of topological operations used to relate different polyhedral units is inserting a vertex into an edge (*stellation* according to Smith 2000) or inserting an edge into an edge (adding *a handle*). Inserting an edge into the vertical edge of the hexagonal prism results in the unit shown in Figure 15b which is observed in zektzerite (Ghose and Wan 1978). The unit shown in Figure 15c (observed again in zektzerite) can be produced from a hexagonal prism by inserting three vertices, deleting two edges and *contracting an edge*. The latter procedure corresponds to deleting an edge and merging vertices incident to this edge. The unit shown in Figure 15d is derived from a hexagonal prism by deleting four edges, inserting a vertex into

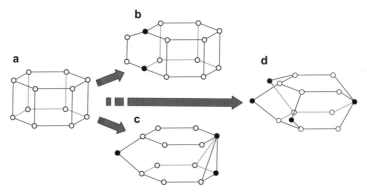

Figure 15. Topological evolution of hexagonal prism into polyhedral units that occur in 3D nets corresponding to octahedral-tetrahedral frameworks. The paths shown include stellation, adding a handle, contracting an edge, addition of edges. See text for details.

an edge and addition of two diagonal edges. This can be found in the octahedral-tetrahedral framework of the high-pressure phase $Na_{1.8}Ca_{1.1}{}^{VI}Si({}^{IV}Si_5O_{14})$ (Gasparik et al. 1995). It is noteworthy that similar topological relationships between different polyhedral units have been observed for tetrahedral frameworks by Smith (1988, 2000).

Topological complexity of polyhedral units: petarasite net

It is important to note that not all polyhedral units can be considered as a result of topological transformations of simple convex polyhedra. As an example of topological complexity, we consider a polyhedral unit with 22 faces observed in the structure of petarasite, $Na_5Zr_2Si_6O_{18}(Cl,OH)\cdot 2H_2O$ (Ghose et al. 1980) (Fig. 16). This unit is noteworthy because (i) it has a maximal number of faces observed in zircono-, niobo- and titanosilicates, (ii) it can be realized as a convex polyhedron, (iii) it is quite large ($6.6 \times 9.6 \times 14.4$ Å^3). Note that this polyhedral unit has 4-valent vertices which is impossible for polyhedral units in zeolites and other tetrahedral frameworks (Smith 2000).

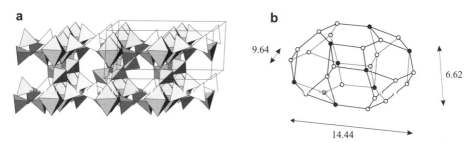

Figure 16. Octahedral-tetrahedral framework in the structure of petarasite, $Na_5Zr_2Si_6O_{18}(Cl,OH)$ $2H_2O$ (a) and its topologically complex polyhedral unit with 22 faces (b).

FRAMEWORKS BASED UPON 1D UNITS

Fundamental chains as bases of complex frameworks

Liebau (1985) suggested that the description of silicate sheets and frameworks can be based upon simple silicate chains (*fundamental* chains) and formulated special rules to subdivide these chains in 2D and 3D silicate anions. A similar approach can be applied to heteropolyhedral frameworks as well. For example, the titanosilicate framework observed in the labuntsovite-group minerals can be described as based upon chains of corner-sharing octahedra interlinked by silicate 4MRs. Octahedral-tetrahedral framework in shcherbakovite-batisite series consists of linked octahedral and tetrahedral chains. The choice of one structure description or another is again dictated by the goal, simplicity and convenience.

The concept of fundamental chain is especially convenient in the case when structure contains no obvious 0D or 2D units. As an example, we consider a family of chiral open-framework uranyl molybdates based upon corner-linked UO_7 pentagonal bipyramids and MoO_4 tetrahedra. The general chemical formula of these compounds can be written as $A_r[(UO_2)_n(MoO_4)_m(H_2O)_p](H_2O)_q$, where A is either inorganic or organic (e.g., protonated amine) cation. There are three topological types of chiral open-framework uranyl molybdates that are characterized by different U:Mo = n:m ratios of 6:7 (Tabachenko et al. 1984; Krivovichev et al. 2005a,b), 5:7 (Krivovichev et al. 2003a) and 4:5 (Krivovichev et al. 2005c). The structure of the only known 5:7 compound, $(NH_4)_4[(UO_2)_5(MoO_4)_7](H_2O)$, consists of a three-dimensional framework of composition $[(UO_2)_5(MoO_4)_7]^{4-}$ (Fig. 17a). The framework contains a three-dimensional system of channels. The largest channel is parallel to [001] and has the dimensions 7.5 × 7.5 Å, which result in a crystallographic free diameter (effective pore width) of 4.8 × 4.8 Å (based on an oxygen radius of 1.35 Å). Smaller channels run parallel to [100], [110], [010], [$\bar{1}$10], [1$\bar{1}$0] and [$\bar{1}\bar{1}$0] and have dimensions 5.2 × 6.3 Å (giving an effective

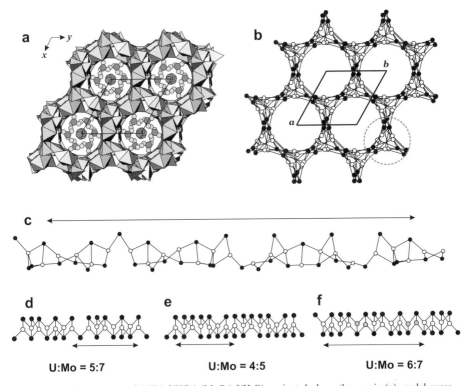

Figure 17. The structure of $(NH_4)_4[(UO_2)_5(MoO_4)_7](H_2O)$ projected along the c-axis (a), nodal representation of its $[(UO_2)_5(MoO_4)_7]$ framework (b), nodal representation of its fundamental chain (c), and graphs isomorphous to nodal representations of fundamental chains of chiral uranyl molybdate frameworks with the U:Mo ratio of 5:7, 4:5 and 6:7 (d, e and f, respectively).

pore width of 2.5 × 3.6 Å). Four symmetrically unique NH_4^+ cations and H_2O molecules are located in the framework channels. The $[(UO_2)_5(MoO_4)_7]^{4-}$ framework is unusually complex. Its nodal representation is shown in Figure 17b. Each node corresponds to a UO_7 bipyramid (black) or a MoO_4 tetrahedron (white). All black vertices are 5-connected and all white vertices are either 3- or 4-connected. Using the nodal representation, the uranyl molybdate framework in the structure of $(NH_4)_4[(UO_2)_5(MoO_4)_7](H_2O)$ can be described in terms of fundamental chains. The nodal representation of the fundamental chain corresponding to the uranyl molybdate framework is shown in Figure 17c. The chain is a sequence of 3- and 4-connected MoO_4 tetrahedra (white vertices) linked through one, two or three UO_7 pentagonal bipyramids (black vertices). The graph shown in Figure 17c can be further reduced to the simplified isomorphic graph shown in Figure 17d. This reduction preserves all topological linkages between the nodes. Note that the graph shown in Figure 17d is periodic and its identity unit includes seven white vertices, whereas, in the real structure, the identity period of the fundamental chain includes 21 white vertices. Thus the topological structure of the fundamental chain is simpler than its geometrical realization.

Distinction of a fundamental chain permits the comparison of the 5:7 uranyl molybdate framework in $(NH_4)_4[(UO_2)_5(MoO_4)_7](H_2O)$ with the 6:7 and 4:5 chiral uranyl molybdate frameworks. The two latter frameworks can also be described as based upon fundamental chains of UO_7 and MoO_4 polyhedra (Krivovichev et al. 2005a,b,c). The reduced black-and-white graphs of these chains are shown in Figure 17e,f for 6:7 and 4:5 frameworks, respectively. Detailed examination of these graphs demonstrated that they cannot be transformed one into another without significant topological reconstruction. Thus, the fundamental chains that form bases for the chiral uranyl molybdate frameworks with U:Mo = 5:7, 4:5 and 6:7 are topologically different, though closely related.

Frameworks with parallel and non-parallel orientations of fundamental chains

We note that, in the uranyl molybdate frameworks discussed above, all fundamental chains are parallel to each other. Of course, non-parallel orientations of fundamental chains are also possible. Figure 18 shows the complex octahedral-tetrahedral framework observed by Lii and Huang (1997) in the structure of $[H_3N(CH_2)_3NH_3]_2[Fe_4(OH)_3(HPO_4)_2(PO_4)_3] \cdot xH_2O$ (Fig. 18a). This structure is based upon "butterfly-type" octahedral-tetrahedral cluster shown in Figure 6a,b. In the structure, these clusters are linked to form chains oriented parallel to [100] and [010] (Fig. 18b). The chains are cross-linked to form a framework as schematically shown in Figure 18c.

Another example of structure with non-parallel orientation of fundamental chains is that of $[(UO_2)_3(PO_4)O(OH)(H_2O)_2](H_2O)$ recently reported by Burns et al. (2004). In this structure, fundamental chains are ribbons of edge-sharing UO_8 hexagonal bipyramids, UO_7 pentagonal bipyramids and PO_4 tetrahedra (Fig. 19a). The ribbons are cross-linked to form an open framework with cavities filled by H_2O molecules (Fig. 19b). It is interesting that topologically identical ribbons form continuous sheets in the structures of the phosphuranylite-group (Burns et al. 1996).

Zorite and ETS-4

Zorite, $Na_6[Ti(Ti,Nb)_4(Si_6O_{17})_2(OH)_5](H_2O)_{10.5}$ (Mer'kov et al. 1973; Sandomirskii and Belov 1979), and its synthetic counterpart, ETS-4 (Philippou and Anderson 1996; Cruciani et al. 1998; Braunbarth et al. 2000; Kuznicki et al. 2001; Nair et al. 2001), received much attention within recent years because of their applications in catalysis, gas separation, optoelectronics, ion-exchange, etc. The structure of zorite was first reported by Sandomirskii and Belov (1979). It is based upon a complex octahedral-tetrahedral framework that consists

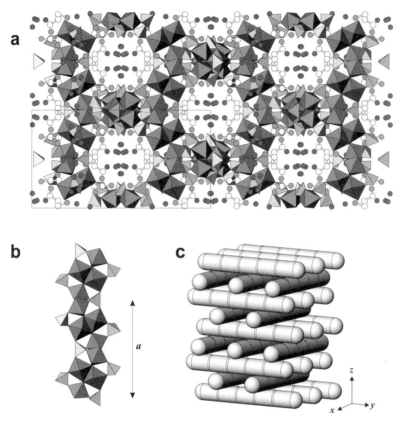

Figure 18. The structure of $(C_3H_{12}N_2)[Fe_4(OH)(H_2O)_2(PO_4)_5](H_2O)_4$ projected along the a-axis (a), its fundamental chain consisting of "butterfly-type" octahedral-tetrahedral clusters (b), and schematic representation of linkage of the chains in the framework (c).

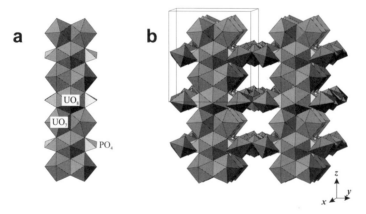

Figure 19. Ribbons of edge-sharing UO_8 hexagonal bipyramids, UO_7 pentagonal bipyramids and PO_4 tetrahedra (a) as fundamental chains in the structure of $[(UO_2)_3(PO_4)O(OH)(H_2O)_2](H_2O)$ (b).

of chains of corner-sharing octahedra cross-linked by double chains of disordered silicate tetrahedra (Fig. 20a). The octahedral chains run along [010] (Fig. 20c), whereas tetrahedral chains run along [001] (Fig. 20b). The tetrahedral chains are further linked by additional half-occupied Ti octahedra (or pentahedra, see discussion in Braunbarth et al. 2000). As a result, the 3D network corresponding to the zorite framework contains large polyhedral voids in the form of octagonal prisms (Fig. 20d). These voids are arranged into columns so that large tunnels are formed along [001] (Fig. 20e).

Benitoite net as based upon arrangement of polyhedral units and tubular units

Figure 21a shows the structure of benitoite, $BaTiSi_3O_9$, projected along the c axis (Zachariasen 1930; Fischer 1969). The benitoite structure type is common for many minerals and inorganic compounds, including bazirite, $BaZrSi_3O_9$, pabstite, $BaSnSi_3O_9$ (Hawthorne 1987; Choisnet et al. 1972), $BaSi^{VI}Si_3^{IV}O_9$ (Finger et al. 1985), $KCaP_3O_9$ (Sandstroem and Bostroem 2004), etc. The structure represents a framework of isolated MO_6 octahedra and Si_3O_9 silicate rings. Nodal representation of the framework is shown in Figure 21b. Analysis of its topology allows to subdivide two polyhedral units, 4^3 and 6^32^2, shown in Figs. 21c and d, respectively. However, the benitoite 3D net cannot be constructed solely by these units as it contains the 1D tubular unit shown in Figure 21e. This unit cannot be unequivocally separated into polyhedral units and thus should be considered as an independent building unit of the benitoite net.

The tubular unit shown in Figure 21e can be considered as a black-and-white graph on the surface of a cylinder. It consists of two types of rings: 6-membered rings **bw²bw²** and 4-membered rings **bwbw**. The tubular unit can be constructed using a procedure which is used to describe the topology of carbon nanotubes and which is known as *folding and gluing* (Kirby 1997). First, one constructs a tape-like black-and-white graph that is a fragment of a tiling of a 2D plane. Equivalent points on the sides of the tape are identified by letters *a*, *b*, *c*, *d*.... In order to get the tubular unit, the tape is folded and opposite sides are glued by joining

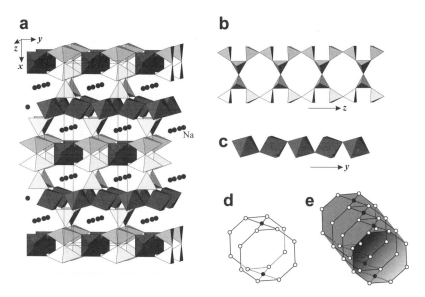

Figure 20. The structure of zorite, $Na_6[Ti(Ti,Nb)_4(Si_6O_{17})_2(OH)_5](H_2O)_{10.5}$ (a) is based upon chains of corner-sharing octahedra (c) cross-linked by double chains of disordered silicate tetrahedra (b). The 3D network corresponding to the zorite framework contains large polyhedral voids in the form of octangular prism (d). These voids are arranged into columns so that large tunnels along [001] are formed (e).

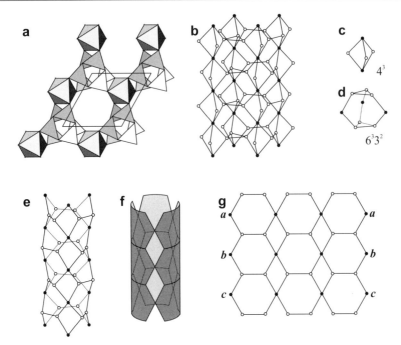

Figure 21. Octahedral-tetrahedral framework in the structure of benitoite, BaTiSi$_3$O$_9$ projected along the *c* axis (a) and its nodal representation (b). The benitoite net can be represented as consisting of polyhedral voids of two types (c and d) and tubular units (e, f). The tubular unit can be constructed by folding and gluing the tape-like black-and-white graph shown in (g). See text for details.

equivalent points to make a cylinder. The idealized model for a tubular unit in benitoite and its *prototape* are shown in Figure 21f,g.

Tubular units, their topology, symmetry and classification

Figure 22a shows the octahedral-tetrahedral framework observed in the structure of vlasovite, Na$_2$ZrSi$_4$O$_{11}$ (Voronkov and Pyatenko 1961; Voronkov et al. 1974; Vitins et al. 1996). This framework consists of chains of SiO$_4$ tetrahedra interlinked by ZrO$_6$ octahedra. The corresponding 3D net (Fig. 22b) is based upon tubular units (Fig. 22c) that can be constructed by folding and gluing the tape of regular hexagons as it is shown in Figure 22d,e. The tubular unit is *achiral* as it has four mirror planes: three parallel and one perpendicular to the extension of the unit.

A different kind of tubular unit is observed in the octahedral-tetrahedral framework of hilairite type Na$_2$ZrSi$_3$O$_9$(H$_2$O)$_3$ (Ilyushin et al. 1981d) (Fig. 23). This framework is built by corner sharing of isolated octahedra ZrO$_6$ and silicate tetrahedra of the 6-er (or *sechster*) single Si$_6$O$_{18}$ chains. The hilairite net is assembled from tubular units composed of 8- and 3-membered rings. Unfolding of the tubular unit provides the tape shown in Figure 23d. To make the tubular unit, one has to fold the tape and to join the points identified by the same letters. In contrast to benitoite and vlasovite, equivalent points are not opposite to each other but, instead, are in diagonal orientation. Their joining produces a *chiral* unit that has a *helical* structure. It is noteworthy that the structures of the hilairite-group minerals (sazykinaite-(Y), hilairite, calciohilairite, komkovite and pyatenkoite-(Y)) have the same space group *R*32 that contains only rotational symmetry elements.

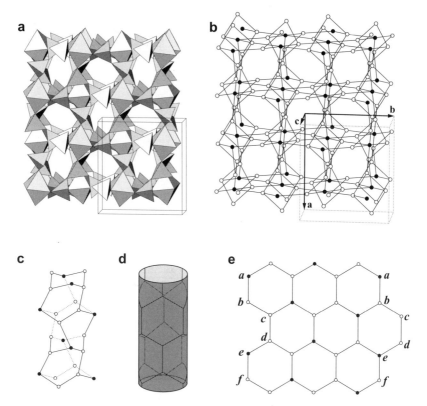

Figure 22. Octahedral-tetrahedral framework in the structure of vlasovite, $Na_2ZrSi_4O_{11}$ (a) and its nodal representation (b). The 3D net (b) is based upon tubular units (c,d) that can be constructed by folding and gluing the tape of regular hexagons (e). Note that the tubular unit is achiral.

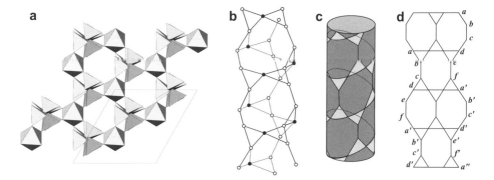

Figure 23. Octahedral-tetrahedral framework in the structure of hilairite $Na_2ZrSi_3O_9(H_2O)_3$ (a) consists of tubular units (b,c) composed from 8- and 3-membered rings. Unfolding of the tubular unit provides the tape shown in (d). Note that the tubular unit is chiral.

The concept of tubular units is useful not only from the point of structure description. It helps to recognize the internal structure of framework channels in terms of its symmetry. For example, the 3D nets corresponding to the chiral 4:5 and 5:7 uranyl molybdate frameworks discussed above can also be described in terms of tubular units (Fig. 24). In both cases, tubular units have a chiral helical structure that is in agreement with the symmetry of the channels in the real structures ($6_5 2 2$ for 4:5 and 6_5 for 5:7 frameworks).

Thus, tubular units can be classified according to: (i) topology of their prototapes; (ii) their symmetry. The first feature allows to classify tubular units from the topological viewpoint and can be useful to identify possible windows and pores. The second helps to recognize such important property of a framework channel as its chirality. Structures with chiral channels may be useful for a wide range of applications, including speculations about the "origin of life."

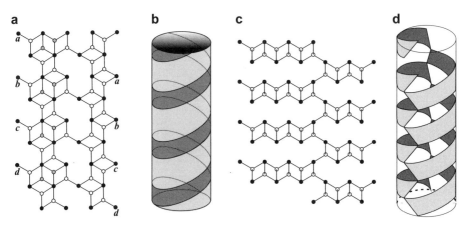

Figure 24. Chiral tubular units in the structures of the U:Mo = 4:5 and 5:7 uranyl molybdate frameworks can be described in terms of tubular units (b and d, respectively). The tubular units can be obtained by folding and gluing tapes shown in (a) and (c), respectively.

FRAMEWORKS BASED UPON 2D UNITS

Octahedral-tetrahedral frameworks in silicates with 2D silicate anions

It is natural to describe octahedral-tetrahedral frameworks with $D_{tetr} = 2$ as based upon 2D silicate anions interlinked by arrays of MO_6 octahedra. Obviously, topological classification of such structures can be made using topological properties of silicate sheets as classification criteria (Table 4).

Figure 25a shows the structure of cavansite, $Ca(VO)(Si_4O_{10})(H_2O)_4$ (Evans 1973; Rinaldi et al. 1975; Solov'ev et al. 1993). It consists of the $[VO(Si_4O_{10})]$ framework based upon Si_4O_{10} silicate sheets linked through isolated VO_5 tetragonal pyramids. Figure 25b shows the octahedral-tetrahedral framework present in the structure of KNAURSI [= $KNa_3(UO_2)_2$ $(Si_4O_{10})_2(H_2O)_4$], the phase that formed during vapor hydration of an actinide-bearing borosilicate waste glass (Burns et al. 2000). The KNAURSI framework is based upon Si_4O_{10} silicate sheets interlinked via UO_6 distorted octahedra. The silicate sheets in the structures of cavansite and KNAURSI are shown in Figure 25c,e, with their idealized versions shown in Figure 25d,f. Obviously, overall topology of the sheets is identical: they consist of 4- and 8-membered rings of tetrahedra present in the 1:1 ratio. Thus, we can designate the sheet topology using its ring symbol as $8^1 4^1$.

Table 4. Minerals and inorganic compounds containing frameworks built upon tetrahedral sheets linked via isolated metal polyhedra or finite clusters of polyhedra ($D_{tetr} = 2$; $D_{nt} = 0$).

Net symbol	Sheet isomer	Mineral/name	Chemical formula	Ref.
6^1	$(u^3d^3)(d^3u^3)$	VSH-3Rb	$Rb_2(VO)_2(Si_6O_{15})(H_2O)_{1.6}$	1
		VSH-3K	$K_2(VO)_2(Si_6O_{15})(H_2O)_{1.6}$	1
	$(u^2d^2)(d^2u^2)$	VSH-1K	$K_2(VO)(Si_4O_{10})(H_2O)$	2
		VSH-14Na	$Na_2(VO)(Si_4O_{10})(H_2O)_{1.4}$	1
	$(u^4d^4)(d^3u^4d)(u^2d^4u^2)$ $(du^4d^3)(d^4u^4)(u^3d^4u)$ $(d^2u^4d^2)(ud^4u^3)$	pentagonite	$Ca(VO)(Si_4O_{10})(H_2O)_4$	3
4^18^1	$(u^2du)(d^3u)$	KNAURSI	$KNa_3(UO_2)_2(Si_4O_{10})_2(H_2O)_4$	4
		NAURSI	$Na_4(UO_2)_2(Si_4O_{10})_2(H_2O)_4$	5
		USH-1	$Na_4(UO_2)_2(Si_4O_{10})_2(H_2O)_4$	6
	$(du^2d)(du^2d)$	cavansite	$Ca(VO)(Si_4O_{10})(H_2O)_4$	7-9
		FDZG-1	$(C_4N_2H_{12})[ZrGe_4O_{10}F_2]$	10
		VSH-13Na	$Na_2(VO)(Si_4O_{10})(H_2O)_3$	1
	$(du^3)(d^3u)(ud^3)(u^3d)$	VSH-12Cs	$Cs_2(VO)(Si_4O_{10})(H_2O)_x$	1
		VSH-12LiX	$Li_2(VO)(Si_4O_{10})(H_2O)_x$	1
	$(ud^3)(d^2ud)$	montregianite	$Na_2[Y(Si_8O_{19})](H_2O)_5$	11
		AV-1	$Na_2[Y(Si_8O_{19})](H_2O)_5$	12
		AV-5	$Na_2[Ce(Si_8O_{19})](H_2O)_5$	13
		rhodesite	$HKCa_2(Si_8O_{19})(H_2O)_5$	14
		delhayelite	$Na_3K_7Ca_5(Al_2Si_{14}O_{38})F_4Cl_2$	15
		macdonaldite	$BaCa_4H_2(Si_{16}O_{38})(H_2O)_{10.4}$	16
$4^16^18^1$	$(u^3)(d^3)$	davanite	$K_2Ti(Si_6O_{15})$	17
		dalyite	$K_2Zr(Si_6O_{15})$	18
		armstrongite	$CaZr[Si_6O_{15}](H_2O)_3$	19,20
	$(ud^2)(d^3)(du^2)(u^3)$	—	α-$K_2Ti(Si_6O_{15})$	21
		—	α-$K_3NdSi_6O_{15}(H_2O)_2$	22
	$(du^2)(ud^2)(ud^2)(du^2)$	sazhinite-(Ce)	$Na_2Ce[Si_6O_{14}(OH)](H_2O)_{1.5}$	23
		—	β-$K_3NdSi_6O_{15}$	24
5^28^1	—	—	$Cs_2Zr(Si_6O_{15})$	25
		SNL-A	$A_2Ti(Si_6O_{15})$ A = K, Rb, Cs	26,27
		—	$Cs_2Ti(Si_6O_{15})$	28
$4^15^26^18^2$	—	—	$Na_3Nd(Si_6O_{15})(H_2O)_2$	29,30
$4^66^28^312^1$	—	VSH-11RbNa	$(Rb,Na)_2(VO)(Si_4O_{10})(H_2O)_x$	1
$4^36^212^1$	—	VSH-4Cs	$Cs_2(VO)(Si_4O_{10})(H_2O)_{2.7}$	1
		VSH-4Rb	$Rb_2(VO)(Si_4O_{10})(H_2O)_3$	1
		VSH-6CsK	$(Cs,K)_2(VO)(Si_4O_{10})(H_2O)_3$	1
		VSH-6Rb	$Rb_2(VO)(Si_4O_{10})(H_2O)_3$	1
		VSH-9CsNa	$CsNa(VO)(Si_4O_{10})(H_2O)_4$	1
4^16^1	—	VSH-2Cs	$Cs_2(VO)(Si_6O_{14})(H_2O)_3$	2

References: (1) Wang et al. 2002a; (2) Wang et al. 2001; (3) Evans 1973; (4) Burns et al. 2000; (5) Li and Burns 2001; (6) Wang et al. 2002b; (7) Evans 1973; (8) Rinaldi et al. 1975; (9) Solov'ev et al. 1993; (10) Liu et al. 2003b; (11) Ghose et al. 1987; (12) Rocha et al. 1997; (13) Rocha et al. 2000; (14) Hesse et al. 1992; (15) Cannillo et al. 1970; (16) Cannillo et al. 1968; (17) Gebert et al. 1983; (18) Fleet 1965; (19) Kashaev and Sapozhnikov 1978; (20) Kabalov et al. 2000; (21) Zou and Dadachov 1999; (22) Haile and Wuensch 2000a; (23) Shumyatskaya et al. 1980; (24) Haile and Wuensch 2000b; (25) Jolicart et al. 1996; (26) Nyman et al. 2000; (27) Nyman et al. 2001; (28) Grey et al. 1997; (29) Karpov et al. 1977; (30) Haile et al. 1997.

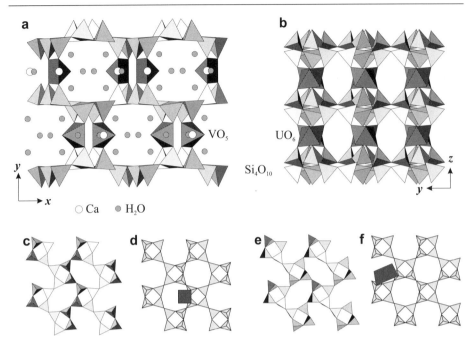

Figure 25. Structure of cavansite, Ca(VO)(Si$_4$O$_{10}$)(H$_2$O)$_4$ (a) and KNAURSI, KNa$_3$(UO$_2$)$_2$(Si$_4$O$_{10}$)$_2$(H$_2$O)$_4$ (b) are based upon Si$_4$O$_{10}$ silicate sheets shown in (c,d) and (e,f), respectively. The sheets consist of 4- and 8-membered rings of tetrahedra and can designated using their ring symbol as 8^14^1. The square in (d) and distorted rectangle in (f) indicate modes of linkage of 2D silicate anion to an adjacent sheet via interlayer octahedron.

Inspection of the idealized versions of the Si$_4$O$_{10}$ silicate sheets shown in Figure 25d,f demonstrates that, despite identical ring symbols, the two sheets are topologically different: one cannot be transformed to the other without topological reconstructions. The point is that non-shared corners of tetrahedra have different patterns of "up" and "down" orientations relative to the plane of the sheet. Thus, the two sheets should be considered as *geometrical isomers*. The concept of geometrical isomerism was introduced into inorganic crystal chemistry by Moore (1975) for description of heteropolyhedral sheets in phosphate minerals (see also Krivovichev (2004a)). A detailed search of octahedral-tetrahedral frameworks based upon Si$_4$O$_{10}$ sheets with the 8^14^1 ring symbol allowed to identify four different geometrical isomers shown in Figure 26. A 2D graph of the Si$_4$O$_{10}$ sheet in KNAURSI corresponding to the 8^14^1 topology is given in Figure 26e with the letters **u** and **d** written near the white nodes. The **u** and **d** designations indicate that the silicate tetrahedra symbolized by the white nodes have their non-shared corners oriented up or down relative to the plane of the sheet, respectively. With the graph oriented as in Figure 26e, the sequences of tetrahedral orientations along the horizontal line may be written in rows. Thus, the first complete row of the **u** and **d** symbols near the white nodes may be written as …**uuduuuduu**… with the (**uudu**) repeat. The repeat of the second row can be written as …**dddudddud**… with the repeat (**dddu**). The third row again has the (**uudu**) repeat. Thus, one may write **u** and **d** symbols for a given row in a tabular form (it is important to maintain vertical order of the rows: the symbol of the next row should be exactly below the corresponding symbol of the preceding row). As the graph shown in Figure 26e is thought to be infinite in the horizontal and vertical directions, the corresponding table of the **u** and **d** symbols is also infinite within the plane of the figure. In this symbolic table, one can choose a rectangular domain

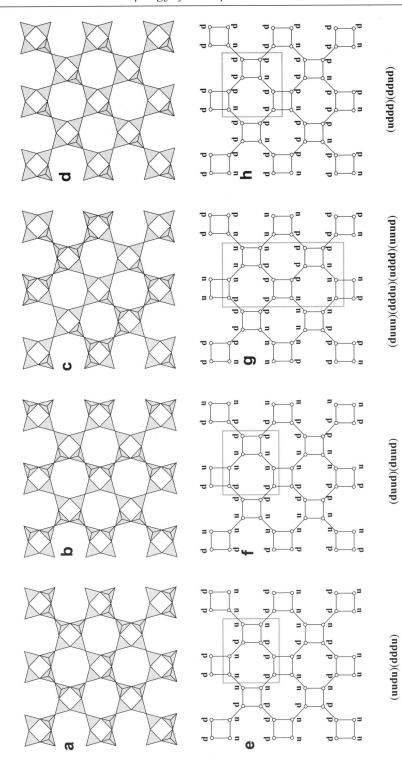

Figure 26. Geometrical isomers of the Si_4O_{10} sheets with the 8^14^1 topology, their graphs and orientation matrices. See text for details.

(elementary unit) that may be used to reproduce the entire table by independent translations along the horizontal and vertical directions. This elementary unit consists of four columns and two rows and may be used to construct the entire table by vertical and horizontal translations. We call this unit *an orientation matrix of tetrahedra*. Obviously, this matrix can also be written in a row as (**uudu**)(**dddu**) implying that the first four symbols in brackets correspond to the first row and the second four symbols in brackets correspond to the second row of the matrix. The orientation matrix can be used as a code for the reconstruction of a whole topology and for the purpose of distinction and classification of different sheet topologies. The concept of orientation matrix (but not the **u** and **d** symbols themselves) was first introduced by Krivovichev and Burns (2003b,c) for investigations of geometrical isomers in uranyl chromates. The four geometrical isomers of the 8^14^1 sheets shown in Figure 26 have different orientation matrices. The list of respective minerals and compounds is given in Table 4.

It should be noted that the 8^14^1 sheet with the orientation matrix (**uddd**)(**ddud**) (Fig. 26 d,h) is not observed as a single silicate sheet. Instead, it occurs as a half of double silicate sheet Si_8O_{19} in the structures of monteragianite, $Na_2[Y(Si_8O_{19})](H_2O)_5$ (Ghose et al. 1987), and its synthetic analogs (Rocha et al. 1997; Rocha et al. 2000).

It is noteworthy that the mixed frameworks based upon silicate tetrahedral sheets dominate structural topologies in natural and synthetic vanadium silicates. Recently, Wang et al. (2002a) reported a number of alkali metal vanadium silicates, VSH-n (see Table 5 for the list of abbreviations of microporous materials), based upon sheets of different topology.

Table 5. Some abbreviations used for microporous compounds.

Abbreviation	Complete name	Place of Origin
AM-n	*A*veiro-*M*anchester-n	Department of Chemistry, University of Aveiro, Aveiro, Portugal *and* Department of Chemistry, University of Manchester Institute of Science and Technology, Manchester, U.K.
ASU-n	*A*rizona *S*tate *U*niversity-n	Department of Chemistry, Arizona State University, Tempe, Arizona, U.S.A.
AV-n	*Av*eiro Microporous Solid-n	Department of Chemistry, University of Aveiro, Aveiro, Portugal
FDZG-n	*F*u*d*an *Z*irconi*g*ermanate-n	Department of Chemistry, Fudan University, Shanghai, China
MIL-n	*M*aterials of *I*nstitute *L*avoisier	Institut Lavoisier, Université du Versailles-St.Quentin, Versailles, France
SNL-N	*S*andia *N*ational *L*aboratory-N	Sandia National Laboratory, Albuquerque, New Mexico, U.S.A.
ULM-n	*U*niversity *L*e *M*ans-n	Laboratoire des Fluorures, Université du Maine, Le Mans, France
UND-n	*U*niversity of *N*otre *D*ame-n	Department of Chemistry and Biochemistry, University of Notre Dame, Notre Dame, Indiana, U.S.A.
USH-nA	*U*ranium *S*ilicate *H*ouston-n	Department of Chemistry, University of Houston, Houston, Texas, U.S.A.
VSH-nA	*V*anadium *S*ilicate *H*ouston-nA (n = framework type; A = non-framework cation)	Department of Chemistry, University of Houston, Houston, Texas, U.S.A.

Figure 27 shows three isomers of the Si_2O_5 silicate tetrahedral sheets of the mica-like topology. These sheets are composed solely of 6-membered rings of tetrahedra and therefore have the ring symbol 6^1. The isomers can be distinguished by their orientation matrices. The list of respective minerals and compounds is given in Table 4.

The structures of armstrongite, $CaZr[Si_6O_{15}](H_2O)_3$ (Kashaev and Sapozhnikov 1978; Kabalov et al. 2000), davanite $K_2Ti(Si_6O_{15})$ (Gebert et al. 1983), and dalyite, $K_2Zr(Si_6O_{15})$ (Fleet 1965), contain octahedral-tetrahedral frameworks based upon silicate sheets consisting of 4-, 6- and 8-membered rings of tetrahedra (ring symbol $4^16^18^1$). Figure 28a shows three geometrical isomers of the $4^16^18^1$ sheets. It is interesting that the structure of sazhinite-(Ce), $Na_2Ce[Si_6O_{14}(OH)](H_2O)_{1.5}$ (Shumyatskaya et al. 1980), is based upon the $4^16^18^1$ sheet isomer different from that observed in armstrongite, owing, probably, to the protonation of one of the O atoms at non-shared corners of silicate tetrahedra.

Figure 29 shows models of two silicate tetrahedral sheets that contain 5-membered rings of tetrahedra. The 5^28^1 sheets (Fig. 29a,c) are observed in the structures of the compounds of general formula $A_2M(Si_6O_{15})$ (A = K, Rb, Cs; M = Ti, Zr). It is noteworthy that the Cs framework titanosilicates have been extensively investigated owing to their application as ceramic waste materials. It was found that there are two modifications of $Cs_2Ti(Si_6O_{15})$: the Cc modification reported by Nyman et al. (2001) and the $C2/c$ modification prepared by Grey et al. (1997). Both modifications are based upon silicate sheets of the same type linked into the framework by isolated TiO_6 octahedra. However, the structures are different in mutual

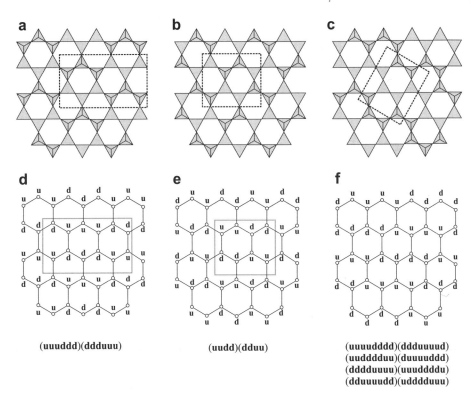

Figure 27. Geometrical isomers of the Si_2O_5 sheets with the 6^1 topology, their graphs and orientation matrices. See text for details.

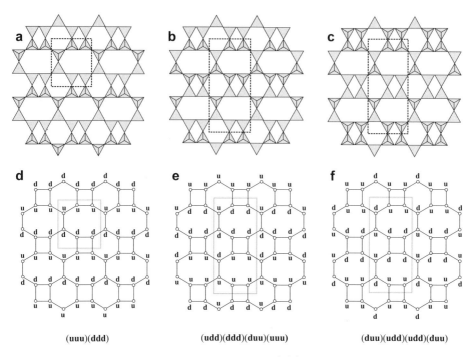

Figure 28. Geometrical isomers of the Si_6O_{15} sheets with the $4^16^18^1$ topology, their graphs and orientation matrices. See text for details.

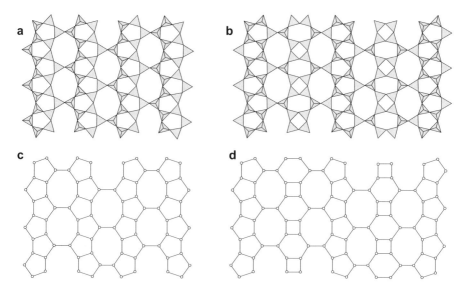

Figure 29. Silicate tetrahedral sheets that contain 5-membered rings of tetrahedra: the 5^28^1 sheet (a,c) in the structures of $A_2M(Si_6O_{15})$ (A = K, Rb, Cs; M = Ti, Zr) and the $4^15^26^18^2$ sheet (b,d) in the structure of $Na_3Nd(Si_6O_{15})(H_2O)_2$.

orientations of adjacent silicate sheets. In the $C2/c$ modification, chains of 5-membered rings of the adjacent sheets are parallel, whereas, in the structure of the Cc modification, these chains in the adjacent sheets are approximately perpendicular to each other. Thus, the octahedral-tetrahedral frameworks in Cc- and $C2/c$-$Cs_2Ti(Si_6O_{15})$ are topologically different, though closely related.

Figure 30 shows models of two porous silicate sheets that contain 12-membered rings. These sheets serve as a base of very open heteropolyhedral frameworks observed in the structures of VSH-11 (the $4^66^28^312^1$ sheet) and VSH-4, -6 and -9 (the $4^36^212^1$ sheet) reported by Wang et al. (2002a).

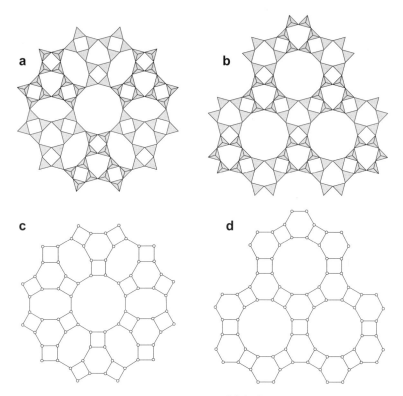

Figure 30. Porous silicate sheets that contain 12MRs: the $4^66^28^312^1$ sheet from the structure of VSH-11 (a) and the $4^36^212^1$ sheet from the structures of VSH-4, -6 and -9 (b).

The number of tetrahedra in a ring is not limited by 12. Figure 31a shows the structure of a new natural phase, $K_3NaCaY_2[Si_{12}O_{30}](H_2O)_4$, recently found by Viktor Yakovenchuk (Geological Institute of the Russian Academy of Sciences, Apatity, Kola peninsula) in hydrothermal veins of the Khibiny alkaline massif. The structure (Krivovichev et al. unpublished) represents an octahedral-tetrahedral framework consisting of $Si_{12}O_{30}$ sheets linked by isolated YO_6 octahedra. The $Si_{12}O_{30}$ sheet (Fig. 31b,c,d) contains 4-, 6- and 14-membered rings—its ring symbol is $14^16^14^3$.

All the structures based upon 2D silicate anions ($D_{tetr} = 2$) described above have $D_{nt} = 0$, i.e., their non-tetrahedral cations either form isolated octahedra or pentahedra or form finite clusters of polyhedra. Figure 32a demonstrates an example of structure for which $D_{tetr} = 2$ and

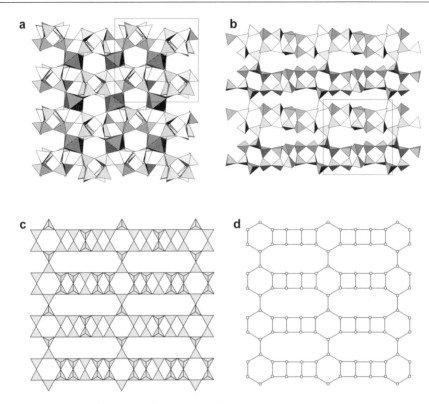

Figure 31. Octahedral-tetrahedral framework in the structure of a new natural phase, $K_3NaCaY_2[Si_{12}O_{30}](H_2O)_4$ (a), its silicate sheet (b) and its ideal (c) and nodal (d) representations.

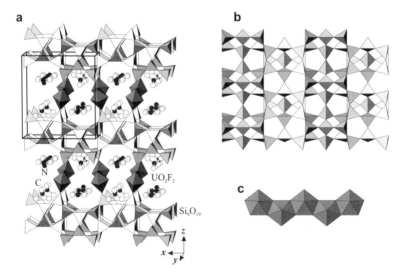

Figure 32. The structure of USH-8, $[(CH_3)_4N][(C_5H_5NH)_{0.8}((CH_3)_3NH)_{0.2}](UO_2)_2[Si_9O_{19}]F_4$ (a), is based upon heteropolyhedral framework consisting of Si_9O_{19} double sheets of silicate tetrahedra (b) and chains of edge-sharing UO_3F_4 pentagonal bipyramids (c).

$D_{nt} = 1$. This structure was recently reported by Wang et al. (2002c) for USH-8, [(CH$_3$)$_4$N] [(C$_5$H$_5$NH)$_{0.8}$((CH$_3$)$_3$NH)$_{0.2}$](UO$_2$)$_2$[Si$_9$O$_{19}$]F$_4$. It is based upon complex Si$_9$O$_{19}$ double sheets of silicate tetrahedra (Fig. 32b) [Wang et al. (2002c) pointed out that sheets of this topology can be found in the structure of zeolite ferrierite, Na$_{1.5}$Mg$_2$Si$_{30.5}$Al$_{5.5}$O$_{72}$ (H$_2$O)$_{18}$ (Vaughan 1966)] and chains of edge-sharing UO$_3$F$_4$ pentagonal bipyramids (Fig. 32c).

Komarovite

The recently determined structure of komarovite, (Na,K)$_{5.5}$(Ca,REE)(Nb,Ti)$_6$[Si$_4$O$_{12}$]O$_{14}$F$_2$(H$_2$O)$_4$ (Balić-Žunić et al. 2002), provides an example of octahedral-tetrahedral system based upon octahedral sheets and 4-membered rings of silicate tetrahedra (Fig. 33a). The octahedral sheet is shown in Figure 33b. It consists of corner-sharing (Nb,Ti)O$_6$ octahedra that form chains running along [100] and [001]. An alternative description can be done in terms of labuntsovite modules interlinked by chains of corner-linked octahedra.

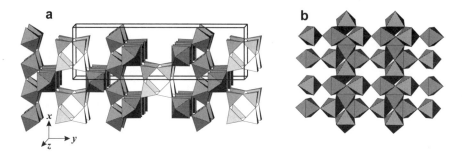

Figure 33. Octahedral-tetrahedral framework in the structure of komarovite, (Na,K)$_{5.5}$(Ca,REE)(Nb,Ti)$_6$[Si$_4$O$_{12}$]O$_{14}$F$_2$(H$_2$O)$_4$ (a) and its octahedral sheet consisting of chains of corner-sharing (Nb,Ti)O$_6$ octahedra (b).

Umbite-related frameworks

The structure of umbite, K$_2$Zr(Si$_3$O$_9$)(H$_2$O) (Ilyushin et al. 1981b, 1993), and its synthetic analog AM-2 (Lin et al. 1997) is based upon framework of Si$_3$O$_9$ chains of corner-sharing SiO$_4$ tetrahedra and isolated ZrO$_6$ octahedra. The Ti analog of umbite, K$_2$Ti(Si$_3$O$_9$)(H$_2$O) (Dadachov and Le Bail 1997; Bortun et al. 2000) possesses interesting ion-exchange properties. Recently, Plevert et al. (2003) reported a series of Zr germanates based upon umbite-related octahedral-tetrahedral frameworks. Despite the fact that the structure is based upon chains of corner-linked tetrahedra, the umbite-related frameworks are better described as based upon sheets of octahedra and tetrahedra shown in Figure 34a,c (Plevert et al. 2003). The sheets are geometrical isomers that differ by the orientation of one of the tetrahedra relative to the plane of the sheet. Nodal representations of the two sheets are shown in Figure 34e,f. The structures of umbite itself and ASU-25, [C$_3$H$_{12}$N$_2$]ZrGe$_3$O$_9$, are based upon sheets with alternative **u** and **d** orientations of tetrahedra (Fig. 34g). The structure of ASU-26, [C$_2$H$_{10}$N$_2$]ZrGe$_3$O$_9$, is based upon sheets with one kind of orientation of tetrahedra (this accounts for the non-centrosymmetric space group, Pn, of this material) (Fig. 34h). In the structure of ASU-24, [C$_6$H$_{18}$N$_2$][C$_6$H$_{17}$N$_2$]$_2$[Zr$_3$Ge$_6$O$_{18}$(OH$_2$,F)$_4$F$_2$](H$_2$O)$_2$, sheets with alternate **u** and **d** orientations of tetrahedra are interlinked by additional Zr(O,F)$_6$ octahedra, thus forming a pillared layered structure (Fig. 34i). As it was pointed out by Plevert et al. (2003), the ASU-24 material has lowest framework density of any oxide material, i.e., 8.48 metal atoms/nm^3.

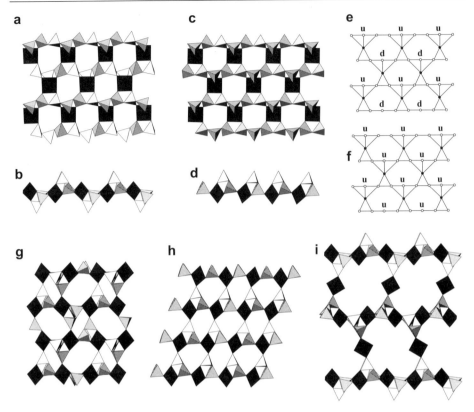

Figure 34. The umbite-related octahedral-tetrahedral frameworks can be described as based upon sheets of octahedra and tetrahedra (top view: a and c; side view: b and d, respectively), nodal representations of which are given in (e) and (f), respectively. Linkage of the (a) type produces frameworks in the structures of umbite and ASU-25 (g), whereas linkage of the sheets of the (c) type results in formation of the ASU-26 framework (h). Octahedral-tetrahedral framework in ASU-24 consists of the type (a) sheets interlinked by additional $Zr(O,F)_6$ octahedra, thus forming pillared layered structure (i).

Terskite

Another example of octahedral-tetrahedral framework with $D_{tetr} = 1$ and $D_{nt} = 0$ which is better described on the basis of 2D sheets is that observed in the structure of terskite, $Na_4Zr(H_4Si_6O_{18})$ (Pudovkina and Chernitsova 1991) (Fig. 35). The basis for the 3D net in terskite (Fig. 35b) is a corrugated 2D net parallel to (100) and consisting of hexagons $\mathbf{bw^2bw^2}$ and pentagons $\mathbf{bw^4}$ (Fig. 35c). This 2D net is a dihedral tiling of a 2D plane. The 2D nets are linked into the 3D net via additional white vertex located between the nets.

The use of 2D nets to recognize structural relationships

Elpidite and $Ca_2ZrSi_4O_{12}$. The structures of elpidite, $Na_2ZrSi_6O_{15}(H_2O)_3$ (Neronova and Belov 1963,1964; Cannillo et al. 1973; Sapozhnikov and Kashaev 1978), and $Ca_2ZrSi_4O_{12}$ (Colin et al. 1993) can easily be related to each other if described as based upon 2D nets (Fig. 36). The 2D net in $Ca_2ZrSi_4O_{12}$ (Fig. 36d) can be obtained from that in elpidite (Fig. 36c) by deleting pairs of white vertices and all edges incident upon those vertices. Thus the $ZrSi_4O_{12}$ framework in $Ca_2ZrSi_4O_{12}$ can be obtained from the $ZrSi_6O_{15}$ framework in elpidite by extraction of two silicate tetrahedra and rearrangement of bonds between the sheets.

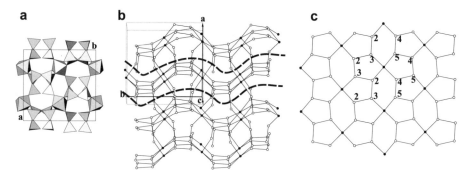

Figure 35. Octahedral-tetrahedral framework in the structure of terskite, $Na_4Zr(H_4Si_6O_{18})$ (a). Its 3D net (b) is based upon corrugated 2D nets parallel to (100) and consisting of hexagons bw^2bw^2 and pentagons bw^4 (c). The 2D nets are linked into the 3D net via additional white vertex located between the nets.

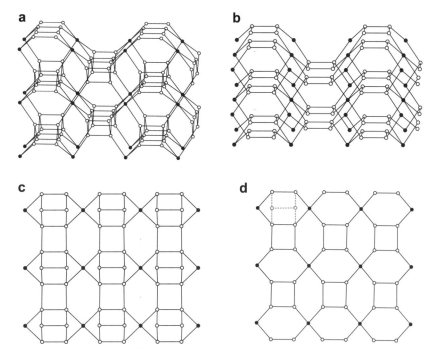

Figure 36. 3D nets corresponding to the octahedral-tetrahedral frameworks in the structures of elpidite, $Na_2ZrSi_6O_{15}(H_2O)_3$ (a), and $Ca_2ZrSi_4O_{12}$ (b) can be described as based upon 2D nets (c and d, respectively).

Gittinsite and $SrZrSi_2O_7$. The analysis of 3D nets in gittinsite, $CaZrSi_2O_7$ (Roelofsen-Ahl and Peterson 1989) and $SrZrSi_2O_7$ (Huntelaar et al. 1994) reveals that both nets are based upon the same 2D net shown in Figure 37a. It consists of heptagons $bwbw^2bw$ and triangles bw^2. The structures differ in the mode the 2D nets are linked to each other. In gittinsite, the adjacent 2D nets have the same orientation of triangles (Fig. 37b), whereas, in $SrZrSi_2O_7$, the adjacent 2D nets are related to each other by rotation by 180° around the axis vertical to the plane of the nets (the triangles in the adjacent nets have opposite orientation; Fig. 37c).

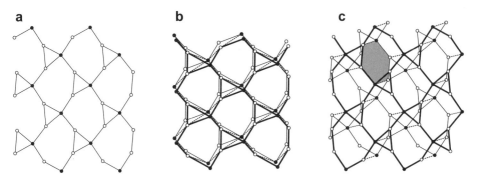

Figure 37. The 2D net consisting of heptagons **bwbw²bw** and triangles **bw²** (a) is a basis for the 3D nets in gittinsite, CaZrSi$_2$O$_7$ (b) and SrZrSi$_2$O$_7$ (c). In gittinsite, the adjacent 2D nets have the same orientation of triangles (b), whereas, in SrZrSi$_2$O$_7$, the adjacent 2D nets are related to each other by rotation by 180° around axis vertical to the plane of the nets (c).

2D nets and polytypism: the case of penkvilksite

Existence of dense 2D nets within 3D nets raises a possibility of polytypism if there exists enough flexibility in the links of the 2D nets. Example are the structures of penkvilksite-2O, penkvilksite-2M Na$_4$(Ti,Zr)(Si$_8$O$_{22}$)(H$_2$O)$_4$ (Merlino et al. 1994) and tumchaite Na$_4$(Zr,Ti)(Si$_8$O$_{22}$)(H$_2$O)$_4$ (Subbotin et al. 2000) which are based on the same type of 2D net shown in Figure 38a. This net is not planar and is constructed from hexagons **bw²bw²**, squares **bw³** and triangles **bw²**. The monoclinic and orthorhombic polytypes are different in the stacking order of the 2D nets within the 3D nets (for OD description see Merlino et al. 1994).

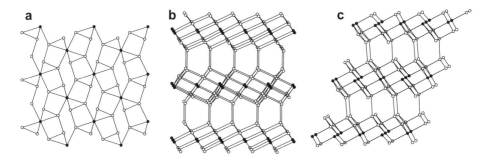

Figure 38. 2D net shown in (a) is a basis for 3D nets in the structures of penkvilksite-2O (b) and penkvilksite-2M (c). The nets shown in (b) and (c) are different in the stacking order of the 2D nets.

Framework uranyl phosphates and arsenates

Locock and Burns (2002a,b; 2003a,b) recently reported a family of framework uranyl phosphates and arsenates based upon dense sheets of UO$_7$ pentagonal bipyramids and TO$_4$ tetrahedra (T = P, As) (Fig. 39a). The sheets have composition [UO$_2$(TO$_4$)] and correspond to the uranophane anion-topology as discussed by Burns et al. (1996). The [UO$_2$(TO$_4$)] sheets are linked into 3D framework by additional U(O,H$_2$O)$_7$ pentagonal bipyramids locating in the interlayer (Fig. 39b). Each TO$_4$ tetrahedron within the sheet has three shared and one non-shared corners. The latter feature gives rise to the "up-and-down" isomeric variations of the [UO$_2$(TO$_4$)] sheets. Locock and Burns (2003b) subdivided five orientational isomers of the [UO$_2$(TO$_4$)] sheets with uranophane anion-topology.

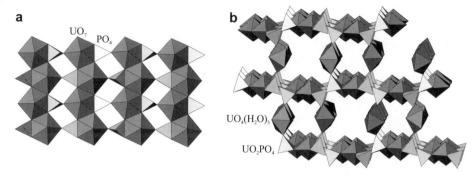

Figure 39. Dense sheets of UO_7 pentagonal bipyramids and TO_4 tetrahedra (T = P, As) (a) serve as bases for a family of framework uranyl phosphates and arsenates (b).

FRAMEWORKS BASED UPON BROKEN SILICATE FRAMEWORKS

Zeolite-type tetrahedral frameworks are not considered in this review. However, there is a specific class of zeolites that contain *broken* or *interrupted* frameworks, i.e., frameworks with less-than-four-connected tetrahedra. The interruption of a framework is usually associated with insertion of additional non-tetrahedral cation that act as a framework-forming component along with tetrahedrally coordinated cations (Si^{4+}, Al^{3+}, P^{5+}, etc.). An example is thornasite, $Na_{12}Th_3(Si_8O_{19})_4(H_2O)_{18}$ (Li et al. 2000), that is based upon interrupted Si_8O_{19} silicate framework with ThO_8 square antiprisms located at the points of framework interruptions.

The titanosilicate molecular sieve ETS-10, $Na_2(TiSi_5O_{13})$ (Anderson et al. 1994; Wang and Jacobson 1999), is not considered as a zeolite. However, its structure is based upon interrupted tetrahedral framework containing chains of corner-sharing TiO_6 octahedra in the framework channels (Fig. 40a). The 3D black-and-white net corresponding to the octahedral-tetrahedral framework in ETS-10 is shown in Figure 40b. Topological structure of the interrupted silicate framework can be described as based upon 12^35^2 polyhedral units shown in Figure 40c. These units are arranged into columns to form large channels parallel to [100] and [010] (Fig. 40d). In the real structure, half of the channels are statistically occupied by chains of corner-sharing octahedra, whereas another half is free.

DYNAMIC TOPOLOGY OF FRAMEWORKS: CELLULAR AUTOMATA MODELING

Introductory remarks

The topological models of frameworks described above refer to static structures. The framework is thought to be perfect, periodic and infinite. However, real crystals are never infinite. Moreover, they grow and change in space and time. To model topological growth of 3D frameworks, one can use the theory of cellular automata (Krivovichev 2004b). This theory also provides an alternative approach to description of complex structural architectures. Another methods of description of 3D organizations of frameworks can be found in Ilyushin (2003) and Draznieks et al. (2000).

Cellular automata are discrete deterministic dynamical systems first introduced by von Neumann (1951) as models for self-reproductive biological systems. A version of a two-dimensional cellular automaton is known as a Life game invented by the mathematician John Conway. Application of cellular automata to modeling physical laws was considered in

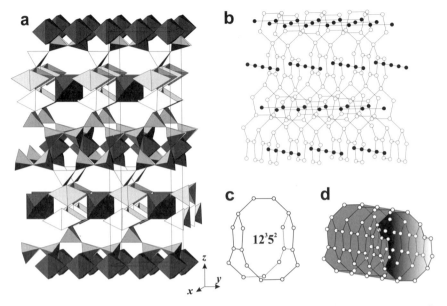

Figure 40. Octahedral-tetrahedral framework in the structure of ETS-10 (a) and its nodal representation (b). The silicate substructure of the 3D net can be described as based upon 12^35^2 polyhedral units (c). These units are arranged into columns to form large channels parallel to [100] and [010] (d). In the real structure, half of the channels are statistically occupied by chains of corner-sharing octahedra, whereas another half is empty.

details by Toffoli and Margolus (1987). A recent interest to cellular automata is due to Stephen Wolfram and his book "A New Kind of Science" (Wolfram 2002), where he pointed out that cellular automata can be considered as universal models for physical, biological, sociological and other systems (for critical comments see Krantz 2002 and Gray 2003). A comprehensive account of theory of cellular automata may be found in (Ilachinski 2001).

A simple example of a cellular automaton is shown in Figure 41a. It consists of a line of square cells. Each cell takes on one of two possible states: black (1) or white (0). Each horizontal line of cells corresponds to a single step in time, t. At each step in time, each cell updates its current state according to a transition rule taking into account the states of cells in its neighborhood. The transition rule for the automaton shown in Figure 41a is depicted in Figure 41b. This rule makes a cell white whenever both of its immediate neighbors were white on the step before. The initial condition ($t = 0$) is one black cell. The result of work of this automaton is a periodic checkerboard-like pattern of black and white cells.

In general, a cellular automaton consists of a n-dimensional *lattice* of cells that take on one of a finite number of possible states. Dynamics of automaton is defined by local interactions of each cell with cells in its local neighborhood. At each step in time, each cell updates its state according to a *transition rule* (Ilachinski 2001).

Wolfram (2002) classified cellular automata into four basic classes according to their behavior: (1) all initial conditions lead to the same uniform final state; (2) the final state is periodic or nested pattern; (3) the behavior is more complicated than in classes 1 and 2 and is in many respects random, though some small-scale structures are present at all steps; (4) the behavior is very complicated and involves a complex mixture of order and randomness. Since crystal structures are periodic, in the following we shall use only cellular automata of class 2.

One important class of cellular automata is *mobile automata* described by Wolfram (2002). They are similar to usual cellular automata but differ in that they have active cells and the rules that specify how active cells should move from one step to the next. The example of a mobile automaton is shown in Figure 42. Active cells in this automaton are marked with a black circle. Transition rules that are enough to describe evolution of this automaton from a single grey cell are defined in Figure 42b. The result of work of the mobile automaton is shown in Figure 42a. It is a pattern of white and grey cells with *pores* consisting of five white cells each.

Pentlandite-djerfisherite-bartonite frameworks

Figure 43a shows a M_8S_{14} cluster that occurs in a number of transition metal sulfides (M = Fe, Cu, Ni, Co). The cluster consists of eight M atoms tetrahedrally coordinated by four S atoms. The MS_4 tetrahedra share edges to form the cubic M_8S_{14} cluster shown in Figure 43b. The M_8S_{14} clusters were found as separate entities in several metal-organic compounds (Christou et al. 1982, 1985) and as FBBs of complex frameworks in a number of metal sulfides of the pentlandite-djerfisherite-bartonite series (Table 6). In order to describe topologies of metal sulfide frameworks based upon M_8S_{14} clusters, we represent one cluster as a cube (Fig. 43c) and linkage of two adjacent clusters (Fig. 43d) as a linkage of two cubes via a common edge (Fig. 43e). Using this approach, frameworks in compounds of the pentlandite-djerfisherite-bartonite series can be described as arrangements of layers consisting of cubes (Fig. 44).

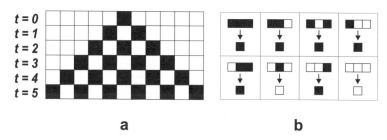

Figure 41. Cellular automaton (a) with transition rule (b) that makes a cell white whenever both of its immediate neighbors were white on the step before.

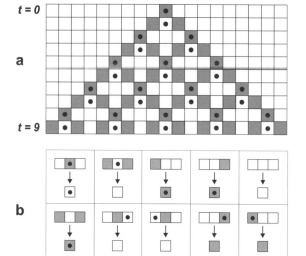

Figure 42. Mobile automaton (a) with active cells marked by black circles and its transition rule (b).

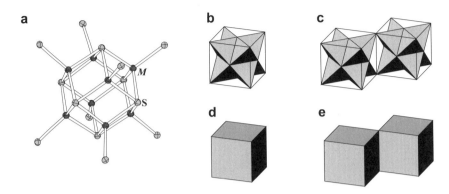

Figure 43. The M_8S_{14} cluster (M = Fe, Cu, Ni, Co) that occurs in transition metal sulfides of the pentlandite-djerfisherite-bartonite series (a) and its representation as a group of eight MS_4 tetrahedra (b) and as a cube (d). Linkage of two clusters (c) can be represented as linkage of two cubes via an edge (e).

Table 6. Crystallographic data for metal sulfide minerals based upon M_8S_{14} clusters.

Mineral	Formula	Framework	Space group	Ref.
pentlandite	$(Fe,Ni)_9S_8$	$[(Fe,Ni)_8S_8]$	$Fm\overline{3}m$	1
cobalt pentlandite	Co_9S_8	$[Co_8S_8]$	$Fm\overline{3}m$	2
argentopentlandite	$(Fe,Ni)_8AgS_8$	$[(Fe,Ni)_8S_8]$	$Fm\overline{3}m$	3
djerfisherite	$K_6Na(Fe,Cu)_{24}S_{26}Cl$	$[(Fe,Cu)_{24}S_{26}]$	$Fm\overline{3}m$	4
owensite	$(Ba,Pb)_6(Cu,Fe,Ni)_{25}S_{27}$	$[(Cu,Fe,Ni)_{24}S_{26}]$	$Fm\overline{3}m$	5
thalfenisite	$Tl_6(Fe,Ni,Cu)_{25}S_{26}Cl$	$[(Fe,Ni,Cu)_{24}S_{26}]$	$Fm\overline{3}m$	6
bartonite	$K_6Fe_{24}S_{26}(S,Cl)$	$[Fe_{24}S_{26}]$	$I4/mmm$	7
chlorbartonite	$K_6Fe_{24}S_{26}(Cl,S)$	$[Fe_{24}S_{26}]$	$I4/mmm$	8

References: (1) Rajamani and Prewitt 1973; (2) Rajamani and Prewitt 1975; (3) Hall and Stewart 1973; (4) Dmitrieva and Ilyukhin 1975; (5) Szymański 1995; (6) Rudashevskii et al. 1977; (7) Evans and Clark 1981; (8) Yakovenchuk et al. 2003.

In the pentlandite structure type, layers are composed of cubes arranged in a checker-board fashion (Fig. 44a). Two adjacent layers are shifted relative to each other by one cube. The structure of djerfisherite and related compounds are based upon layers of two types: one identical to the layers in pentlandite and another one that consists of isolated cubes only (Fig. 44b). The layers alternate and two layers of the same type are translationally equivalent. Metal sulfide framework in the structure of bartonite contains the same types of layers as observed in djerfisherite. The layers of different types alternate; however, their stacking sequence is different. Whereas the pentlandite-like layers are in translationally equivalent positions, the layers of isolated cubes are in two positions shifted relative to each other by one cube (Fig. 44c). Figure 44d shows the construction of a framework topologically related to those shown in Figure 44a–c. This framework contains only layers of isolated cubes; each cube shares four corners with cubes from the layer below and four corners with cubes from the layer above. Framework of this type has not been observed in metal sulfides. It can be found in the structures of $[(C_2H_{10}N_2)_4(H_2O)_2][Co_{7.12}Al_{0.88}P_8O_{32}]$ (ACO zeolite topology; Feng et al. 1997), $(C_2H_{10}N_2)_2[Fe_4O(PO_4)_4](H_2O)$, leucophosphite, $K_2[Fe_4(OH)_2(H_2O)_2(PO_4)_4](H_2O)_2$ (see above

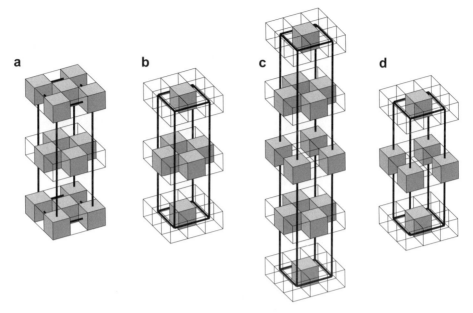

Figure 44. Expanded description of pentlandite (a), djerfisherite (b), bartonite (c) and zeolite ACO (d) frameworks in terms of layers of cubes that symbolize cubic FBBs (see text for details).

section about frameworks based upon FBBs). We note that the framework of cubes shown in Figure 44d corresponds to the **bcu** regular net described by Delgado-Friedrichs et al. (2003).

Cellular automata and frameworks of cubes

The 3D frameworks of cubes shown in Figure 44 can be obtained as a product of evolution of two-dimensional cellular automata. In contrast to the patterns shown in Figures 41 and 42, these frameworks should be constructed by successive addition of layers of cells that are updated at each step in time according to a transition rule.

The cellular automata corresponding to the framework shown in Figure 44 and the configurations resulting after twelve steps in their evolution are depicted in Figure 45. The automata are based upon cubic lattices of either black or white (empty) cubes. Initial condition for all automata is a single black cell.

Cellular automaton for the pentlandite framework is very simple (Fig. 45a). Its transition rule states that if the cell is black, all its immediate (non-diagonal) neighbors in the next layer are black. Cellular automaton corresponding to the framework shown in Figure 44d is also simple (Fig. 45d). It has the following transition rule: if the cell is black, all its diagonal neighbors in the next layer are black.

Construction of the djerfisherite and bartonite frameworks (Fig. 44a,b) requires mobile automata (Fig. 45b,c). Transition rules for these automata are similar. In both cases, if a white cell is active, it is black and active in the next layer. The rules for a black active cell differ. In both automata, it is white in the next step and its immediate neighbors are black. However, the djerfisherite automaton (Fig. 45b) generates five active white cells, whereas the chlorbartonite automaton generates only four (Fig. 45c). The difference in the transition rules results in different framework topologies.

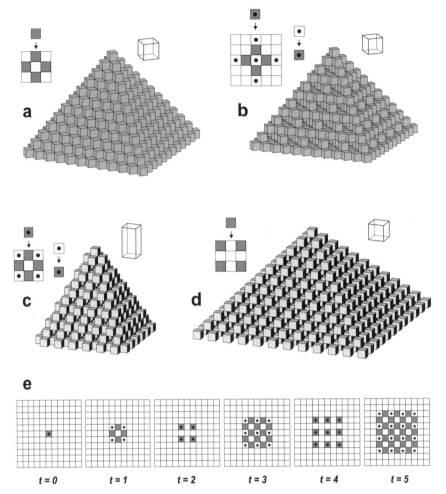

Figure 45. Cellular automata that correspond to the frameworks of cubes shown in Figure 44: cellular automata for pentlandite (a), djerfisherite (b), bartonite (c) and zeolite ACO (d). Pictures show configurations resulting after twelve steps in the evolution of the automata and their transition rules. Picture (e) gives successive steps in evolution of the two-dimensional automaton that results in the bartonite-type framework of cubes (c).

Transition rules of cellular automata and chemical requirements

The metal sulfide frameworks shown in Figure 44a–c can be generated by the successive addition of cubic Fe_8S_{14} clusters using different transition rules. What controls these rules, at least in chemically related systems such as metal sulfides based upon M_8S_{14} clusters? A partial answer to this question may be derived from the chemical composition of respective compounds. The metal sulfide framework in the minerals of the pentlandite group contains relatively small cubic cavities occupied by transition metal atoms M in octahedral coordinations (Fig. 46a). Both djerfisherite and bartonite frameworks contain large cavities occupied by the octahedral A_6X moieties (A = K, Tl, Pb, Ba; X = S, Cl) (Fig. 46b,c). In addition to the large cavities, there are additional small cubic cavities in djerfisherite (as in pentlandite) occupied by Na (or by other relatively small cations such as Li or Mg). In chlorbartonite framework, only large

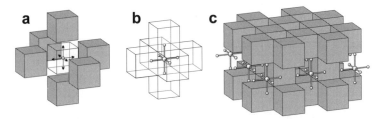

Figure 46. Octahedrally coordinated M atoms in pentlandite and Na atoms in djerfisherite are in a cubic cavity that is created by six M_8S_{14} clusters (shown as cubes) (a); $\mathbf{A}_6\mathbf{X}$ complex in djerfisherite and bartonite is at the center of a cross-like cavity consisting of seven cubes (b); location of the $\mathbf{A}_6\mathbf{X}$ complexes within framework of cubic clusters in bartonite (c).

cavities are present. Thus, the presence of potassium favors "modification" of transition rules of cellular automaton in order to accommodate K_6X complexes into the framework cavities.

CONCLUDING REMARKS:
THE ART OF STRUCTURE DESCRIPTION

Geometrical and topological approaches to the description of 3D frameworks used in this chapter (graph theory, nets, tilings, polyhedra, cellular automata) are useful tools in recognizing complex structural architectures and their relationships to each other. Choice of one or another approach depends upon the preferences of a researcher. Sometimes, one has to emphasize certain aspects of a given structure such as pore dimensions, widths of windows, chirality of channel surfaces, etc. In this case, the approach is dictated by the specific goals of investigation. In other cases, criteria are simplicity, clearness and (finally) beauty of the model. The models described here may be considered as a kind of Platonic ideas that underlie organization of natural matter. However, one still has to remember that models remain to be models and that "…there are more things in heaven and earth, Horatio, than are dreamt of in your philosophy."

ACKNOWLEDGMENTS

Thanks are due to G. Ferraris, L.B. McCusker and N.V. Chukanov for their comments, and to Andrew Locock and Thomas Armbruster for discussions of many questions considered here. Partial financial support was provided by the Alexander von Humboldt Foundation in the form of equipment donation.

REFERENCES

Anderson MW, Terasaki O, Ohsuna T, Philippou A, MacKay SP, Ferreira A, Rocha J, Lidin S (1994) Structure of the microporous titanosilicate ETS-10. Nature 367:347-351

Armbruster T, Krivovichev SV, Weber T, Gnos E, Organova NN, Yakovenchuk VN (2004) Origin of diffuse superstructure reflections in labuntsovite-group minerals. Am Mineral 89:1655-1666

Baerlocher Ch, Meier WM, Olson DH (2001) Atlas of Zeolite Framework Types, Elsevier: Amsterdam London New York Oxford Paris Shannon Tokyo

Balić-Žunić T, Petersen OV, Bernhardt HJ, Micheelsen HI (2002) The crystal structure and mineralogical description of a Na-dominant komarovite from the Ilimaussaq alkaline complex, South Greenland. N Jb Miner Mh 2002:497-514

Balmer ML, Su Y, Xu H, Bitten E, McCready D, Navrotsky A (2001) Synthesis, structure determination, and aqueous durability of $Cs_2ZrSi_3O_9$. J Am Ceram Soc 84:153-160

Baur W, Fischer RX (2000) Zeolite Structure Codes from ABW to CZP. Landolt-Bernstein: Numerical Data and Functional Relationships in Science and Technology – New Series. Vol 14, subvolume B. Springer, Berlin

Baur W, Fischer RX (2002) Zeolite-Type Crystal Structures and Their Chemistry. Framework Type Codes DAC to LOV. Landolt-Bernstein: Numerical Data and Functional Relationships in Science and Technology – New Series. Vol 14, subvolume C. Springer, Berlin

Behrens EA, Clearfield A (1997) Titanium silicates, $M_3HTi_4O_4(SiO_4)_3·4H_2O$ (M = Na^+, K^+), with three-dimensional tunnel structures for the selective removal of strontium and cesium from wastewater solutions. Microporous Mater 11:65-75

Behrens EA, Poojary DM, Clearfield A (1996) Syntheses, crystal structures, and ion-exchange properties of porous titanosilicates, $HM_3Ti_4O_4(SiO_4)_3·4(H_2O)$ (M = H^+, K^+, Cs^+), structural analogues of the mineral pharmacosiderite. Chem Mater 8:1236-1244

Behrens EA, Poojary DM, Clearfield A (1998) Syntheses, X-ray powder structures, and preliminary ion-exchange properties of germanium-substituted titanosilicate pharmacosiderites: $HM_3(AO)_4(BO_4)_3·4(H_2O)$ (M = K, Rb, Cs; A = Ti, Ge; B = Si, Ge). Chem Mater 10:959-967

Belyayevskaya GP, Borutskii BYe, Marsii IM, Vlasova YeV, Sivtsov AV, Golovanova TI, Vishnev AI (1991) Potassiium-calcium gaidonnayite, $(Ca,Na,K)_{2-x}ZrSi_3O_9·nH_2O$, a new mineral variety from the Khibiny block. Dokl Akad Sci USSR, Earth Sci 321A:111-116

Bialek R, Gramlich V (1992) The superstructure of $K_3HGe_7O_{16}·4H_2O$. Z Kristallogr 198:67-77

Blinov VA, Shumyatskaya NG, Voronkov AA, Ilyukhin VV, Belov NV (1977) Refinement of the crystal structure of wadeite $K_2Zr(Si_3O_9)$ and its relation to derivative structural types. Sov Phys Crystallogr 22:31-35

Blinov VA, Voronkov AA, Ilyukhin VV, Belov NV (1975) Crystal structure of lemoynite: a new type of mixed framework. Sov Phys Dokl 19(7):397-398

Bortun AI, Bortun LN, Poojary DM, Xiang O, Clearfield A (2000) Synthesis, characterization, and ion exchange behaviour of a framework potassium titanium trisilicate $K_2TiSi_3O_9·H_2O$ and its protonated phases. Chem Mater 12:294-305

Braunbarth C, Hillhouse HW, Nair S, Tsapatis M, Burton A, Lobo RF, Jacubinas RM, Kuznicki SM (2000) Structure of strontium ion-exchanged ETS-4 microporous molecular sieves. Chem Mater 12:1857-1865

Buerger MJ, Dollase WA, Garaycochea-Wittke I (1967) The structure and composition of the mineral pharmacosiderite. Z Kristallogr 125:92-108

Burns PC, Alexopoulos CM, Hotchkiss PJ, Locock AJ (2004) An unprecedented uranyl phosphate framework in the structure of $[(UO_2)_3(PO_4)O(OH)(H_2O)_2](H_2O)$. Inorg Chem 43:1816-1818

Burns PC, Miller ML, Ewing RC (1996) U^{6+} minerals and inorganic phases: a comparison and hierarchy of structures. Can Mineral 34:845-880

Burns PC, Olson RA, Finch RJ, Hanchar JM, Thibault Y (2000) $KNa_3(UO_2)_2(Si_4O_{10})_2(H_2O)_4$, a new compound formed during vapor hydration of an actinide-bearing borosilicate waste glass. J Nucl Mater 278:290-300

Cahill CL, Ko Y, Hanson JC, Tan K, Parise J (1998a) Structure of microporous QUI-MnGS-1 and in situ studies of its formation using time-resolved synchrotron X-ray powder diffraction. Chem Mater 10:1453-1458

Cahill CL, Ko Y, Parise J (1998b) A novel 3-dimensional open framework sulfide based upon the $[In_{10}S_{20}]^{10-}$ supertetrahedron: DMA-InS-SB1. Chem Mater 10:19-21

Cannillo E, Rossi G, Ungaretti L (1970) The crystal structure of delhayelite. Rend Soc Ital Mineral Petrol 26:63-75

Cannillo E, Rossi G, Ungaretti L (1973) The crystal structure of elpidite. Am Mineral 58:106-109

Cannillo E, Rossi G, Ungaretti L, Carobbi SG (1968) The crystal structure of macdonaldite. Atti della Accad Naz Lincei, Scienze Fis Mat Natur Rend, Ser 8, 45:399-414

Chao GY (1985) The crystal structure of gaidonnayite $Na_2Zr(Si_3O_9)(H_2O)_2$. Can Mineral 23:11-15

Chapman DM, Roe AL (1990) Synthesis, characterization and crystal chemistry of microporous titanium-silicate materials. Zeolites 10:730-737

Chernov AN, Maksimov BA, Ilyukhin VV, Belov NV (1971) Crystalline structure of monoclinic modification of K, Zr-diorthosilicate ($K_2ZrSi_2O_7$). Sov Phys Dokl 15(8):711-713

Chiu YN, Xiao J, Merritt CD, Liu K, Huang WX, Ganelin PV, Li NN (2000) Special geminals and Schlegel diagrams of molecular structures of fullerenes and metallofullerenes. J Mol Struct 530:67-83

Choisnet J, Deschanvres A, Raveau B (1972) Sur de nouveaux germanates et silicates de type benitoite. J Solid State Chem 4:209-218

Christou G, Hagen KS, Bashkin JK, Holm RH (1985) The clusters $[Co_8S_6(SPh)_8]^{4-,5-}$: preparations, properties, and structural relationship of near-cubic $Co_8(\mu_4-S)_6$ cores to the clusters in synthetic pentlandite. Inorg Chem 24:1010-1018

Christou G, Hagen KS, Holm RH (1982) Synthesis, structure, and properties of $[Co_8S_6(SC_6H_5)_8]^{4-}$, containing an octanuclear Co_8S_6 rhombic dodecahedron related to that of cobalt pentlandite. J Am Chem Soc 104: 1744-1745

Chukanov NV, Pekov IV, Khomyakov AP (2002) Recommended nomenclature for labuntsovite group minerals. Eur J Mineral 14:165-173

Chukanov NV, Pekov IV, Zadov AE, Voloshin AV, Subbotin VV, Sorokhtina NV, Rastsvetaeva RK, Krivovichev SV (2003) Minerals of the Labuntsovite Group. Moscow, Nauka (in Russian)

Clearfield A (2001) Structure and ion exchange properties of tunnel type titanium silicates. Solid State Sci 3: 103–112

Clearfield A, Bortun LN, Bortun AI (2000) Alkali metal ion exchange by the framework titanium silicate $M_2Ti_2O_3SiO_4 \cdot nH_2O$ (M = H, Na). React Funct Polymers 43:85–95

Colin S, Dupre B, Venturini G, Malaman B, Gleitzer C (1993) Crystal structure and infrared spectrum of the cyclosilicate $Ca_2ZrSi_4O_{12}$. J Solid State Chem 102:242-249

Cruciani G, De Luca P, Nastro A, Pattison P (1998) Rietveld refinement of the zorite strucyture of ETS-4 molecular sieves. Microporous Mesoporous Mater 21:143-153

Dadachov MS, Harrison WTA (1997) Synthesis and crystal structure of $Na_4[(TiO)_4(SiO_4)_3] \cdot 6H_2O$, a rhombohedrally distorted sodium titanium silicate pharmacosiderite analogue. J Solid State Chem 134: 409-415

Dadachov MS, Le Bail A (1997) Structure of zeolitic $K_2TiSi_3O_9 \cdot H_2O$ determined *ab initio* from powder diffraction data. Eur J Solid State Inorg Chem 34:381-390

de Bord JRD, Reiff WM, Warren CJ, Haushalter RC, Zubieta J (1997) A 3-D organically templated mixed valence (Fe^{2+}/Fe^{3+}) iron phosphate with oxide-centered Fe4O(PO4)4 cubes: hydrothermal synthesis, crystal structure, magnetic susceptibility, and Mössbauer spectroscopy of $[H_3NCH_2CH_2NH_3]_2$ $[Fe_4O(PO_4)_4] \cdot H_2O$. Chem Mater 9:1994-1998

Delgado-Friedrichs O, O'Keeffe M, Yaghi OM (2003) Regular and quasiregular nets. Acta Crystallogr A59: 22-27

Dick S (1999) Über die Struktur von synthetischem Tinsleyit $K(Al_2(PO_4)_2OH)(H_2O)) \cdot (H_2O)$. Z Naturforsch B54:1385-1390

Dmitrieva MT, Ilyukhin VV (1975) Crystal structure of djerfisherite. Sov Phys Dokl 20(7):469-470

Draznieks CM, Newsam JM, Gorman AM, Freeman CM, Férey G (2000) De novo prediction of inorganic structures developed through automated assembly of secondary building units. Angew Chem Int Ed 39: 2270-2275

Dvoncova E, Lii KH (1993) Hydrothermal synthesis and structural characterization of mixed-valence iron phosphates: $AFe_5(PO_4)_5(OH) H_2O$ (A = Ca, Sr). Inorg Chem 32:4368-4372

Eddaoudi M, Moler DB, Li H, Chen B, Reineke T, O'Keeffe M, Yaghi OM (2001) Modular chemistry: secondary building units as a basis for the design of highly porous and robust metal-organic carboxylate frameworks. Acc Chem Res 34:319-330

Engel P (1981) Über Wirkungsbereihsteilungen von kubischer Symmetrie. Z Kristallogr 154:199-215

Evans HT Jr (1973) The crystal structures of cavansite and pentagonite. Am Mineral 58:412-424

Evans HT Jr, Clark JR (1981) Crystal structure of bartonite, a potassium iron sulfide, and its relationship to pentlandite and djerfisherite. Am Mineral 66:376-384

Fedorov ES (1971) Symmetry of Crystals. ACA Monograph number 7, American Crystallographic Assoc., New York

Feng P, Bu X, Stucky GD (1997) Hydrothermal syntheses and structural characterization of zeolite analog compounds based on cobalt phosphate. Nature 388:735-741

Férey G (1995) Oxyfluorinated microporous compounds ULM *n*: chemical parameters, structures and a proposed mechanism for their molecular tectonics. J Fluor Chem 72:187-193

Férey G (1998) The new microporous compounds and their design. C R Acad Sci Ser IIc 1:1-13

Férey G (2001) Microporous solids: from organically templated inorganic skeletons to hybrid frameworks. Ecumenism in chemistry. Chem Mater 13:3084-3098

Ferraris G, Ivaldi G, Khomyakov AP (1995) Altisite $Na_3K_6Ti_2(Al_2Si_8O_{26})Cl_3$, a new hyperalkaline aluminosilicate from Kola Peninsula (Russia) related to lemoynite: crystal structure and thermal evolution. Eur J Mineral 7:537-546

Ferreira A, Lin Z, Rocha J, Morais CM, Lopes M, Fernandez C (2001b) *Ab initio* structure determination of a small-pore framework sodium stannosilicate. Inorg Chem 40:3330-3335

Ferreira A, Lin Z, Soares MR, Rocha J (2004) Synthesis and *ab initio* structure determination from powder diffraction data of $K_4Sn_2Si_6O_{18}$. Mater Sci Forum 443-444:329-332

Ferreira P, Ferreira A, Rocha J, Soares MR (2001a) Synthesis and structural characterization of zirconium silicates. Chem Mater 13:355-363

Finger LW, Hazen RM, Fursenko BA (1995) Refinement of the crystal structure of $BaSi_4O_9$ in the benitoite form. J Phys Chem Solids 56:1389-1393

Fischer K (1969) Verfeinerung der Kristallstruktur von Benitoit BaTi(Si$_3$O$_9$). Z Kristallogr 129:222-243
Fleet ME, Henderson GS (1995) Sodium trisilicate: a new high-pressure silicate structure (Na$_2$Si(Si$_2$O$_7$)). Phys Chem Mineral 22:383-386
Fleet SG (1965) The crystal structure of dalyite. Z Kristallogr 121:349-368
Fleet SG, Cann JR (1967) Vlasovite; a second occurrence and a triclinic to monoclinic inversion. Mineral Mag 36:233-241
Gasparik T, Parise JB, Eiben BA, Hriljac JA (1995) Stability and structure of a new high pressure silicate, Na$_{1.8}$Ca$_{1.1}$Si$_6$O$_{14}$. Am Mineral 80:1269-1276
Gebert W, Medenbach O, Floerke OW (1983) Darstellung und Kristallographie von K$_2$TiSi$_6$O$_{15}$ isotyp mit Dalyit K$_2$ZrSi$_6$O$_{15}$. Tscherm Mineral Petrogr Mitt 31:69-79
Ghose S, Sen Gupta PK, Campana CF (1987) Symmetry and crystal structure of montregianite Na$_4$K$_2$Y$_2$Si$_{16}$O$_{38}$(H$_2$O)$_{10}$, a double-sheet silicate with zeolitic properties. Am Mineral 72:365-374
Ghose S, Thakur P (1985) The crystal structure of georgechaoite NaKZr(Si$_3$O$_9$)(H$_2$O)$_2$. Can Mineral 23:5-10
Ghose S, Wan C (1978) Zektzerite, NaLiZrSi$_6$O$_{15}$: a silicate with six-tetrahedral-repeat double chains. Am Mineral 63:304-310
Ghose S, Wan C, Chao GY (1980) Petarasite, Na$_5$Zr$_2$Si$_6$O$_{18}$(Cl,OH)·2H$_2$O, a zeolite-type zirconosilicate. Can Mineral 18:503-509
Gobechiya ER, Pekov IV, Pushcharovskii DY, Ferraris G, Gula A, Zubkova NV, Chukanov NV (2003) New data on vlasovite: Refinement of the crystal structure and the radiation damage of the crystal during the X-ray diffraction experiment. Crystallogr Rep 48:750-754
Gray L (2003) A mathematician looks at Wolfram's New Kind of Science. Not Am Math Soc 50:200-211
Grey IE, Roth RS, Balmer ML (1997) The crystal structure of Cs$_2$TiSi$_6$O$_{15}$. J Solid State Chem 131:38-42
Guesdon A, Borel MM, Leclaire A, Grandin A, Raveau B (1993) A series of mixed-valent molybdenum monophosphates, isotypic with leucophosphite represented by CsMo$_2$P$_2$O$_{10}$ and K$_{1.5}$Mo$_2$P$_2$O$_{10}$·H$_2$O. Z Anorg Allg Chem 619:1841-1849
Haile SM, Wuensch BJ (2000a) Structure, phase transitions and ionic conductivity of K$_3$NdSi$_6$O$_{15}$(H$_2$O)$_x$. I. α-K$_3$NdSi$_6$O$_{15}$(H$_2$O)$_2$ and its polymorphs. Acta Crystallogr B56:335-348
Haile SM, Wuensch BJ (2000b) Structure, phase transitions and ionic conductivity of K$_3$NdSi$_6$O$_{15}$(H$_2$O)$_x$. II. Structure of β-K$_3$NdSi$_6$O$_{15}$. Acta Crystallogr B56:349-362
Haile SM, Wuensch BJ, Laudise RA, Maier J (1997) Structure of Na$_3$NdSi$_6$O$_{15}$·2(H$_2$O) - a layered silicate with paths for possible fast-ion conduction. Acta Crystallogr B53:7-17
Hall SR, Stewart JM (1973) Crystal structure of argentian pentlandite (Fe,Ni)$_8$AgS$_8$, compared with the refined structure of pentlandite (Fe,Ni)$_9$S$_8$. Can Mineral 12:169-177
Harrison WTA, Broach RW, Bedard RA, Gier TE, Bu XH, Stucky GD (1996) Synthesis and characterization of a new family of thermally stable open-framework zincophosphate/arsenate phases: M$_3$Zn$_4$O(XO$_4$)$_3$·n(H$_2$O). Crystal structures of Rb$_3$Zn$_4$O(PO$_4$)$_3$·3.5(H$_2$O), K$_3$Zn$_4$O(AsO$_4$)$_3$·4(H$_2$O) and Na$_3$Zn$_4$O(PO$_4$)$_3$·6(H$_2$O). Chem Mater 8:691-700
Harrison WTA, Gier TE, Stucky GD (1995) Single-crystal structure of Cs$_3$HTi$_4$O$_4$(SiO$_4$)$_3$·4(H$_2$O), a titanosilicate pharmacosiderite analog. Zeolites 5:408-412
Harrison WTA, Phillips MLF, Bu XH (2000) Synthesis and single-crystal structure of Cs$_3$Zn$_4$O(AsO$_4$)$_3$·4(H$_2$O), an open-framework zinc arsenate. Microporous Mesoporous Mater 39:359-365
Haushalter RC (1987) (Mo$_4$O$_4$)$^{6+}$ cubes in Cs$_3$Mo$_4$P$_3$O$_{16}$. J Chem Soc Chem Commun 1987:1566-1568
Haushalter RC, Mundi LA (1992) Reduced molybdenum phosphates: octahedral-tetrahedral framework solids with tunnels, cages, and micropores. Chem Mater 4:31-48
Haushalter RC, Strohmaier KG, Lai FW (1989) Structure of a three-dimensional microporous molybdenum phosphate with large cavities. Science 246:1289-1291
Hawthorne FC (1983) Graphical enumeration of polyhedral clusters. Acta Crystallogr A39:724-736
Hawthorne FC (1987) The crystal chemistry of the benitioite group minerals and structural relations in (Si$_3$O$_9$) ring structures. N Jb Mineral Mh 1987:16-30
Hawthorne FC (1994) Structural aspects of oxide and oxysalt crystals. Acta Crystallogr B50:481-510
Hawthorne FC, Krivovichev SV, Burns PC (2000) Crystal chemistry of sulfate minerals. Rev Mineral Geochem 40:1-112
Henshaw DE (1955) The structure of wadeite. Mineral Mag 30:585-595
Hesse KF, Liebau F, Merlino S (1992) Crystal structure of rhodesite, HK$_{1-x}$Na$_{x+2y}$Ca$_{2-y}${lB,3,2$^2_\infty$}(Si$_8$O$_{19}$)(6−z)H$_2$O, from three localities and its relation to other silicates with dreier double layers. Z Kristallogr 199:25-48
Hoppe R, Köhler J (1988) SCHLEGEL projections and SCHLEGEL diagrams - new ways to describe and discuss solid state compounds. Z Kristallogr 183:77-111
Huntelaar ME, Cordfunke EHP, van Vlaanderen P, Ijdo DJW (1994) SrZr(Si$_2$O$_7$). Acta Crystallogr C50:988-991

Ilachinski A (2001) Cellular Automata: a Discrete Universe, New Jersey, London, Singapore, Honkong: World Scientific
Ilyushin GD (1993) New data on the crystal structure of umbite $K_2ZrSi_3O_9 \cdot H_2O$. Izv Akademii Nauk SSSR Neorg Mater 29:971-975
Ilyushin GD (2003) Modeling of Self-Organization Processes in Crystal-Forming Systems. Moscow, URSS, p 376 (in Russian)
Ilyushin GD, Blatov VA (2002) Crystal chemistry of zirconosilicates and their analogs: topological classification of MT-frameworks and suprapolyhedral invariants. Acta Crystallogr B58:198-218
Ilyushin GD, Khomyakov AP, Shumyatskaya NG, Voronkov AA, Nevsky NN, Ilyukhin VV, Belov NV (1981a) Crystal structure of a new natural zirconium silicate $K_4Zr_2Si_6O_{18} \cdot 2(H_2O)$. Sov Phys Dokl 26(2):118-120
Ilyushin GD, Pudovkina ZV, Voronkov AA, Khomyakov AP, Ilyukhin VV, Pyatenko YuA (1981b) The crystal structure of the a natural modification of $K_2ZrSi_3O_9 \cdot H_2O$. Sov Phys Dokl 26(3):257-258
Ilyushin GD, Voronkov AA, Ilyukhin VV, Nevsky NN, Belov NV (1981c) Crystal structure of natural monoclinic catapleiite $Na_2ZrSi_3O_9 \cdot 2(H_2O)$. Sov Phys Dokl 26(9):808-810
Ilyushin GD, Voronkov AA, Nevsky NN, Ilyukhin VV, Belov NV (1981d) Crystal structure of hilairite $Na_2ZrSi_3O_9 \cdot 3(H_2O)$. Sov Phys Dokl 26(10):916-917
Jale SR, Ojo A, Fitch FR (1999) Synthesis of microporous zirconosilicates containing ZrO_6 octahedra and SiO_4 tetrahedra. Chem Commun 1999:411-412
Jolicart G, le Blanc M, Morel B, Dehaudt P, Dubois S (1996) Hydrothermal synthesis and determination of $Cs_2ZrSi_6O_{15}$. Eur J Solid State Inorg Chem 33:647-657
Kabalov YK, Zubkova NV, Pushcharovsky DY, Schneider J, Sapozhnikov AN (2000) Powder Rietveld refinement of armstrongite, $CaZr[Si_6O_{15}] \cdot 3H_2O$. Z Kristallogr 215:757-761
Karpov OG, Pushcharovskii DYu, Pobedimskaya EA, Burshtein IF, Belov NV (1977) The crystal structure of the rare-earth silicate $NaNdSi_6O_{13}(OH)_2 \cdot n(H_2O)$. Sov Phys Dokl 22(9):464-466
Kashaev AA, Sapozhnikov AN (1978) Crystal structure of armstrongite. Sov Phys Crystallogr 23:539-545
Khalilov AD, Khomyakov AP, Makhmudov SA (1978) Crystal structure of the keldyshite $NaZr[Si_2O_6OH]$. Sov Phys Dokl 23(1):8-10
Khomyakov AP, Nechelyustov GN, Rastsvetaeva RK (1993) Sazykinaite-(Y) $Na_5YZrSi_6O_{18} \cdot 6H_2O$ - a new mineral. Zap Vseross Mineral Obshch 122(5):76-82 (in Russian)
King HE Jr, Mundi LA, Strohmaier KG, Haushalter RC (1991a) A synchrotron single crystal X-ray structure determination of a small crystal: Mo-Mo double bonds in the 3-D microporous molybdenumphosphate $NH_4(Mo_2P_2O_{10}) \cdot H_2O$. J Solid State Chem 92:1-7
King HE Jr, Mundi LA, Strohmaier KG, Haushalter RC (1991b) A synchrotron single crystal X-ray structure determination of $(NH_4)_3Mo_4P_3O_{16}$: a microporous molybdenum phosphate with $Mo_4O_4^{6+}$ cubes. J Solid State Chem 92:154-158
Kirby EC (1997) Recent work on toroidal and other exotic fulleren structures. In: From Chemical Topology to Three-Dimensional Geometry. Balaban AT (ed) Plenum New York, p. 263-296
Krantz SG (2002) A New Kind of Science, by Stephen Wolfram, book review. Bull Am Math Soc 40:143-150
Krivovichev SV (1997) Use of the Schlegel diagrams for description and classification of mineral crystal structures. Zap Vseross Mineral Obshch 126(2):37-46 (in Russian)
Krivovichev SV (1999) Theory of regular systems of points and partition of space. II. Upper bound for the number of faces in stereohedra. Crystallogr Rep 44:349-355
Krivovichev SV (2004a) Combinatorial topology of inorganic oxysalts: 0-, 1- and 2-dimensional units with corner-sharing between coordination polyhedra. Crystallogr Rev 10:185-232.
Krivovichev SV (2004b) Crystal structures and cellular automata. Acta Crystallogr A60:257-262
Krivovichev SV (2004c) Topological and geometrical isomerism in minerals and inorganic compounds with laueite-type heteropolyhedral sheets. N Jb Mineral Mh 2004:209-220
Krivovichev SV, Armbruster T, Chernyshov DYu, Burns PC, Nazarchuk EV, Depmeier W (2005a) Chiral open-framework uranyl molybdates. 3. Synthesis, structure and the $C222_1 \rightarrow P2_12_12_1$ low-temperature phase transition of $[C_6H_{16}N]_2[(UO_2)_6(MoO_4)_7(H_2O)_2](H_2O)_2$. Microporous Mesoporous Mater 78:225-234
Krivovichev SV, Burns PC (2000) Crystal chemistry of uranyl molybdates. I. The structure and formula of umohoite. Can Mineral 38:717-726
Krivovichev SV, Burns PC (2003a) A novel rigid uranyl tungstate sheet in the structures of $Na_2[(UO_2)W_2O_8]$ and α- and β-$Ag_2[(UO_2)W_2O_8]$. Solid State Sci 5:373-381
Krivovichev SV, Burns PC (2003b) Geometrical isomerism in uranyl chromates I. Crystal structures of $(UO_2)(CrO_4)(H_2O)_2$, $[(UO_2)(CrO_4)(H_2O)_2](H_2O)$ and $[(UO_2)(CrO_4)(H_2O)_2]_4(H_2O)_9$. Z Kristallogr 218: 568-574
Krivovichev SV, Burns PC (2003c) Geometrical isomerism in uranyl chromates II. Crystal structures of $Mg_2[(UO_2)_3(CrO_4)_5](H_2O)_{17}$ and $Ca_2[(UO_2)_3(CrO_4)_5](H_2O)_{19}$. Z Kristallogr 218:683-690

Krivovichev SV, Burns PC (2003d) Combinatorial topology of uranyl molybdate sheets: syntheses and crystal structures of $(C_6H_{14}N_2)_3[(UO_2)_5(MoO_4)_8](H_2O)_4$ and $(C_2H_{10}N_2)[(UO_2)(MoO_4)_2]$. J Solid State Chem 170: 106-117

Krivovichev SV, Burns PC, Armbruster T, Nazarchuk EV, Depmeier W (2005b) Chiral open-framework uranyl molybdates. 2. Flexibility of the U:Mo = 6:7 frameworks: syntheses and crystal structures of $(UO_2)_{0.82}$ $[C_8H_{20}N]_{0.36}[(UO_2)_6(MoO_4)_7(H_2O)_2](H_2O)_n$ and $[C_6H_{14}N_2][(UO_2)_6(MoO_4)_7(H_2O)_2](H_2O)_m$. Microporous Mesoporous Mater 78:217-224

Krivovichev SV, Cahill CL, Burns PC (2003a) New uranyl molybdate open framework in the structure of $(NH_4)_4[(UO_2)_5(MoO_4)_7](H_2O)$. Inorg Chem 42:2459-2464

Krivovichev SV, Cahill CL, Nazarchuk EV, Burns PC, Armbruster T, Depmeier W (2005c) Chiral open-framework uranyl molybdates. 1. Topological diversity: synthesis and crystal structure of $[(C_2H_5)_2NH_2]_2$ $[(UO_2)_4(MoO_4)_5(H_2O)](H_2O)$. Microporous Mesoporous Mater 78:209-215

Krivovichev SV, Filatov SK, Burns PC (2002) Cuprite-like framework of OCu_4 tetrahedra in the crystal structure of synthetic melanothallite, Cu_2OCl_2, and its negative thermal expansion. Can Mineral 40: 1185-1190

Krivovichev SV, Filatov SK, Semenova TF (1997) On the systematics of polyions of linked polyhedra. Z Kristallogr 212:411-417

Krivovichev SV, Yakovenchuk VN, Armbruster T, Pakhomovskiy YaA, Depmeier W (2003b) Crystal structure of Ti analogue of ilimaussite-(Ce), $(Ba,K,Na)_{11-12}(Ce,La,Th,Nd)_4(Ti,Nb,Fe^{3+})_6(Si_6O_{18})_4(OH)_6\cdot 6H_2O$: revision of structural model and structural formula. Z Kristallogr 218:392-396

Krivovichev SV, Yakovenchuk VN, Pakhomovsky YaA (2004a) Topology and symmetry of titanosilicate framework in the crystal structure of shcherbakovite, $Na(K,Ba)_2(Ti,Nb)_2O_2[Si_4O_{12}]$. Zap Vseross Mineral Obshch 133(3):55-63 (in Russian)

Kuznicki SM, Bell VA, Nair S, Hillhouse HW, Jacubinas, Braunbarth CM, Toby BH, Tsapatis M (2001) A titanosilicate molecular sieve with adjustable pores for size-selective adsorption of molecules. Nature 412:720-724

le Page Y, Perrault G (1976) Structure cristalline de la lemoynite, $(Na,K)_2CaZr_2Si_{10}O_{26}\cdot 5-6H_2O$. Can Mineral 14:132-138

Leclaire A, Borel MM, Chardon J, Grandin A, Raveau B (1992) A niobium silicophosphate belonging to the niobium phosphate bronze series: $K_4Nb_8P_4SiO_{34}$. Acta Crystallogr C48:1744-1747

Leclaire A, Borel MM, Grandin A, Raveau B (1994) The mixed valent molybdenum monophosphate $RbMo_2P_2O_{10}\cdot(1-x)H_2O$: an intersecting tunnel structure isotypic with leucophosphite. J Solid State Chem 108:177-183

Li H, Eddaoudi M, Laine A, O'Keeffe M, Yaghi OM (1999a) Noninterpenetrating indium sulfide supertetrahedral cristobalite framework. J Am Chem Soc 121:6096-6097

Li H, Eddaoudi M, O'Keeffe M, Yaghi OM (1999b) Design and synthesis of an exceptionally stable and highly porous metal-organic framework. Nature 402:276-279

Li H, Kim J, Groy TL, O'Keeffe M, Yaghi OM (2001) 20 Å $Cd_4In_{16}S_{35}^{14-}$ supertetrahedral T4 clusters as building units in decorated cristobalite frameworks. J Am Chem Soc 123:4867-4868

Li Y, Burns PC (2001) The structures of two sodium uranyl compounds relevant to nuclear waste disposal. J Nucl Mater 299:219-226

Li Y, Krivovichev SV, Burns PC (2000) The crystal structure of thornasite: a novel interrupted silicate framework. Am Mineral 85:1521-1525

Liebau F (1985) Structural Chemistry of Silicates: Structure, Bonding, and Classification. Springer-Verlag, Berlin

Liebau F (2003) Ordered microporous and mesoporous materials with inorganic hosts: definitions of terms, formula notation, and systematic classification. Microporous Mesoporous Mater 58:15-72

Lii KH, Huang YF (1997) Large tunnels in the hydrothermally synthesized open-framework iron phosphate $[H_3N(CH_2)_3NH_3]_2[Fe_4(OH)_3(HPO_4)_2(PO_4)_3]\cdot xH_2O$. Chem Commun 1997:839-840

Lima-de-Faria J, Hellner E, Liebau F, Makovicky E, Parthé E (1990) Nomenclature of inorganic structure types: report of the IUCr commission on crystallographic nomenclature subcommittee on the nomenclature of inorganic structure types. Acta Crystallogr A46:1-11

Lin Z, Ferreira A, Rocha J (2003) Synthesis and structural characterization of novel tin and titanium potassium silicates $K_4M_2Si_6O_{18}$. J Solid State Chem 175:258-263

Lin Z, Rocha J (2002) The first example of a small-pore framework hafnium silicate. Stud Surf Sci Catal 142A: 319-325

Lin Z, Rocha J, Brandao P, Ferreira A, Esculcas AP, Pedrosa de Jesus JD, Philippou A, Anderson MW (1997) Synthesis and structural characterization of microporous umbite, penkvilksite, and other titanosilicates. J Phys Chem B101:7114-7120

Lin Z, Rocha J, Ferreira P, Thursfield A, Agger JR, Anderson MW (1999a) Synthesis and structural characterization of microporous framework zirconium silicates. J Phys Chem B103:957-963

Lin Z, Rocha J, Pedrosa de Jesus JD, Ferreira A (2000) Synthesis and structure of a novel microporous framework stannosilicate. J Mater Chem 10:1353-1356

Lin Z, Rocha J, Valente A (1999b) Synthesis and characterisation of a framework microporous stannosilicate. Chem Commun 1999:2489-2490

Liu XS, Shang MY, Thomas JK (1997) Synthesis and structure of a novel microporous titanosilicate (UND-1) with a chemical composition of $Na_{2.7}K_{5.3}Ti_4Si_{12}O_{36} \cdot 4H_2O$. Microporous Mater 10:273-281

Liu Z, Weng L, Chen Z, Zhao D (2003a) $(NH_4)_2ZrGe_3O_9$: a new microporous zirconogermanate. Acta Crystallogr C59:i29-i31

Liu ZC, Weng LH, Zhou YM, Chen ZX, Zhao DY (2003b) Synthesis of a new organically templated zeolite-like zirconogermanate $(C_4N_2H_{12})[ZrGe_4O_{10}F_2]$ with cavansite topology. J Mater Chem 13:308-311

Locock AJ, Burns PC (2002a) Crystal structures of three framework alkali metal uranyl phosphate hydrates. J Solid State Chem 167:226–236

Locock AJ, Burns PC (2002b) The crystal structure of triuranyl diphosphate tetrahydrate. J Solid State Chem 163:275-280

Locock AJ, Burns PC (2003a) Structures and synthesis of framework Rb and Cs uranyl arsenates and their relationships with their phosphate analogues. J Solid State Chem 175:372–379

Locock AJ, Burns PC (2003b) Structures and syntheses of framework triuranyl diarsenate hydrates. J Solid State Chem 176:18–26

McCusker LB, Liebau F, Engelhardt G (2001) Nomenclature of structural and compositional characteristics of ordered microporous and mesoporous materials and inorganic hosts. Pure Appl Chem 73:381-394

McDonald AM, Chao GY (2001) Natrolemoynite, a new hydrated sodium zirconosilicate from Mont Saint-Hilaire, Quebec: description and structure determination. Can Mineral 39:1295-1306

Medrano MD, Evans HTjr, Wenk HR, Piper DZ (1998) Phosphovanadylite: a new vanadium phosphate mineral with a zeolite-type structure. Am Mineral 83:889-895

Mer'kov AN, Bussen IV, Goiko EA, Kul'chitskaya EA, Men'shikov YuP, Nedorezova AP (1973) Raite and zorite, new minerals from the Lovozero tundra. Zap Vseross Mineral Obshch 102:54-62 (in Russian)

Merlino S, Pasero M, Artioli G, Khomyakov AP (1994) Penkvilksite, a new kind of silicate structure: OD character, X-ray single-crystal ($1M$), and powder Rietveld ($2O$) refinements of two MDO polytypes. Am Mineral 79:1185-1193

Moore PB (1970a) Crystal chemistry of the basic iron phosphates. Am Mineral 55:135-169

Moore PB (1970b) Structural hierarchies among minerals containing octahedrally coordinating oxygen. I. Stereoisomerism among corner-sharing octahedral and tetrahedral chains. N Jb Miner Mh 1970:163-173

Moore PB (1972) Octahedral tetramer in the crystal structure of leucophosphite, $K_2[Fe^{3+}_4(OH)_2(H_2O)_2(PO_4)_4](H_2O)_2$. Am Mineral 57:397-410

Moore PB (1975) Laueite, pseudolaueite, stewartite and metavauxite: a study in combinatorial polymorphism. N Jb Miner Abh 1975:148-159

Nair S, Jeong HK, Chandrasekaran A, Braunbarth C, Tsapatis M, Kuznicki SM (2001) Synthesis and structure determination of ETS-4 single crystals. Chem Mater 13:4247-4254

Neronova NN, Belov NV (1963) Crystal structure of elpidite $Na_2Zr[Si_6O_{15}] \cdot 3H_2O$, and dimorphism of $[Si_6O_{15}]$ dimetasilicate radicals. Dokl Akad Sci USSR, Earth Sci 150:115-118

Neronova NN, Belov NV (1964) Crystal structure of elpidite, $Na_2Zr(Si_6O_{15}) \cdot 3H_2O$. Sov Phys Crystallogr 9: 700-705

Nikitin AV, Belov NV (1962) Crystal structure of batisite $Na_2BaTi_2Si_4O_{14} = Na_2BaTi_2O_2[Si_4O_{12}]$. Dokl Akad Sci USSR, Earth Sci 146:142-143

Nowotny H, Wittmann A (1954) Zeolitische Alkaligermanate. Monatsh Chem 85:558-574

Nyman M, Bonhomme F, Maxwell RS, Nenoff TM (2001) First Rb silicotitanate phase and its K-structural analogue: new members of the SNL-A family (Cc-A2TiSi6O15; A = K, Rb, Cs). Chem Mater 13:4603-4611

Nyman M, Bonhomme F, Teter DM, Maxwell RS, Gu BX, Wang LM, Ewing RC, Nenoff TM (2000) Integrated experimental and computational methods for structure determination and characterization of a new, highly stable cesium silicotitanate phase, $Cs_2TiSi_6O_{15}$ (SNL-A). Chem Mater 12:3449-3458

O'Keeffe M, Eddaoudi M, Li H, Reineke T, Yaghi OM (2000) Frameworks for extended solids: geometrical design principles. J Solid State Chem 152:3-20

O'Keeffe M, Hyde BG (1996) Crystal Structures. I. Patterns and Symmetry. Mineralogical Society of America, Washington D.C.

Pertierra P, Salvado M, Garcia-Granda S, Bortun AI, Khainakov SA, Garcia JR (2002) Hydrothermal synthesis and structural characterization of framework microporous mixed tin-zirconium silicates with the structure of umbite. Inorg Chem Commun 5:824-828

Pertierra P, Salvado MA, Garcia-Granda S, Garcia JR, Bortun AI, Bortun LN, Clearfield A (2001) Synthesis and structural study of $K_2PbSi_3O_9 \cdot H_2O$ with the structure of kostylevite. Mater Res Bull 36:717-725

Philippou A, Anderson MW (1996) Structural investigation of ETS-4. Zeolites 16:98-107

Plaisier JR, Huntelaar ME, de Graaff RAG, Ijdo DJW (1994) A contribution to the understanding of phase equilibria (structure of $Sr_7Zr(Si_2O_7)_3$). Mater Res Bull 29:701-707

Plevert J, Sanchez-Smith R, Gentz TM, Li H, Groy TL, Yaghi OM, O'Keeffe M (2003) Synthesis and characterization of zirconogermanates. Inorg Chem 42:5954-5959

Poojary DM, Bortun AI, Bortun LN, Clearfield A (1997) Syntheses and X-ray powder structures of $K_2(ZrSi_3O_9) \cdot (H_2O)$ and its ion-exchanged phases with Na and Cs. Inorg Chem 36:3072-3079

Poojary DM, Cahill RA, Clearfield A (1994) Synthesis, crystal structures and ion-exchange properties of a novel porous titanosilicate. Chem Mater 6:2364-2368

Pudovkina ZV, Chernitsova NM (1991) Crystal structure of terskite $Na_4Zr[H_4Si_6O_{18}]$. Sov Phys Dokl 36(3): 201-203

Pushcharovskii DY, Pekov IV, Pasero M, Gobechiya ER, Merlino S, Zubkova NV (2002) Crystal structure of cation-deficient calciohilairite and possible mechanisms of decationization in mixed-framework minerals. Crystallogr Rep 47:748-752

Pyatenko YuA, Voronkov AA (1961) Vlasovite, a zirconium silicate with a new type of silicon-oxygen radical. Dokl Akad Sci USSR, Earth Sci 141:1294-1298

Rajamani V, Prewitt CT (1973) Crystal chemistry of natural pentlandites. Can Mineral 12:178-187

Rajamani V, Prewitt CT (1975) Refinement of the structure of Co_9S_8. Can Mineral 13:75-78.

Rastsvetaeva RK, Khomyakov AP (1992) Crystal structure of a rare earth analog of hilairite. Sov Phys Crystallogr 37:845-847

Rastsvetaeva RK, Khomyakov AP (1996) Pyatenkoite-(Y) $Na_5YTiSi_6O_{18} \cdot 6H_2O$ a new mineral of the hilairite group: Crystal structure. Doklady Chemistry 351:283-286

Rastsvetaeva RK, Pushcharovskii DYu, Konev AA, Evsyunin VG (1997) Crystal structure of K-containing batisite. Crystallogr Rep 42:770-773

Rinaldi R, Pluth JJ, Smith JV (1975) Crystal structure of cavansite dehydrated at 220°C. Acta Crystallogr B31: 1598-1602

Riou-Cavellec M, Riou D, Férey G (1999) Magnetic iron phosphates with an open framework. Inorg Chim Acta 291:317-325

Roberts MA, Fitch AN (1996) The crystal structures of hydrated and partially dehydrated $M_3HGe_7O_{16} \cdot n(H_2O)$ (M = K, Rb, Cs) determined from powder diffraction data using synchrotron radiation. Z Kristallogr 211: 378-387

Roberts MA, Fitch AN, Chadwick AV (1995) The crystal structures of $(NH_4)_3HGe_7O_{16} \cdot n(H_2O)$ and $Li_{4-x}H_xGe_7O_{16} \cdot n(H_2O)$ determined from powder diffraction data using synchrotron radiation. J Phys Chem Solids 56:1353-1358

Rocha J, Ferreira P, Carlos LD, Ferreira A (2000) The first microporous framework cerium silicate. Angew Chem Int Ed 39:3276-3279

Rocha J, Ferreira P, Lin Z, Agger JR, Anderson MW (1998) Synthesis and characterization of a microporous zirconium silicate with the structure of petarasite. Chem Commun 1998:1269-1270

Rocha J, Ferreira P, Lin Z, Brandaõ P, Ferreira A, Pedrosa de Jesus JD (1997) The first synthetic microporous yttrium silicate containing framework sodium atoms. Chem Commun 1997:2103-2104

Roelofsen-Ahl JN, Peterson RC (1989) Gittinsite: a modification of the thortveitite structure. Can Mineral 27: 703-708

Rudashevskii NS, Mintkenov GA, Karpenkov AM, Shishkin NN (1977) Silver-containing pentlandite – the independent mineral species argentopentlandite. Zap Vses Mineral Obshch 106:688-691 (in Russian)

Rudenko VN, Rozhdestvenskaya IV, Nekrasov IYa, Dadze TP (1983) Crystal structure of Sn-wadeite. Mineral Zh 5:70-72

Sandomirskii PA, Belov NV (1979) The OD structure of zorite. Sov Phys Crystallogr 24:686-693

Sandstroem M, Bostroem D (2004) Calcium potassium cyclo-triphosphate. Acta Crystallogr E60:i15-i17

Sapozhnikov AN, Kashaev AA (1978) Features of the crystal structure of calcium-containing elpidite. Sov Phys Crystallogr 23:24-27

Sapozhnikov AN, Kashaev AA (1980) The crystal structure of calcined Ca-containing elpidite. Sov Phys Crystallogr 25:357-359

Schindler M, Hawthorne FC, Baur W (1999) Metastructures: homeomorphisms between complex inorganic structures and three-dimensional nets. Acta Crystallogr B55:811-829

Schmahl WW, Tillmanns E (1987) Isomorphic substitutions, straight Si-O-Si geometry, and disorder of tetrahedral tilting in batisite, $(Ba,K)(K,Na)Na(Ti,Fe,Nb,Zr)Si_4O_{14}$. N Jb Mineral Mh 1987:107-118

Shumyatskaya NG, Voronkov AA, Pyatenko YuA (1980) Sazhinite, $Na_2Ce(Si_6O_{14}(OH)) \cdot nH_2O$: a new representative of the dalyite family in crystal chemistry. Sov Phys Crystallogr 25:419-423

Smith JV (1988) Topochemistry of zeolites and related materials. 1. Topology and geometry. Chem Rev 88: 149-182

Smith JV (2000) Tetrahedral Frameworks of Zeolites, Clathrates and Related Materials. Landolt-Bernstein: Numerical Data and Functional Relationships in Science and Technology. Vol 14, subvolume A, Springer, Berlin

Sokolov VI (1992) Buckminsterfullerene as a ligand of variable hapticity. Symmetry analysis of possible combinations of face pairs and its consequence for chiral stereochemistry of C_{60}. Dokl Akad Nauk 326: 647-649 (in Russian)

Sokolova EV, Arakcheeva AV, Voloshin AV (1991) Crystal structure of komkovite. Sov Phys Dokl 36(10): 666-668

Solov'ev MV, Rastsvetaeva RK, Pushcharovskii DYu (1993) Refined crystal structure of cavansite. Crystallogr Rep 38:274-275

Song Y, Zavalij PY, Chernova NA, Suzuki M, Whittingham SM (2003) Comparison of one-, two-, and three-dimensional iron phosphates containing ethylenediamine. J Solid State Chem 175:63–71

Strunz H, Nickel EH (2001) Strunz Mineralogical Tables. Chemical-Structural Mineral Classification System. E. Schweizerbart'sche Verlagsbuchhandlung, Stuttgart

Sturua GI, Belokoneva EL, Simonov MA, Belov NV (1978) Crystal structure of Ge zeolite $K(H_3O)_3[Ge_7O_{16}] = KH_3[Ge_7O_{16}]\cdot 3H_2O$. Sov Phys Dokl 23(10):703-704

Subbotin VV, Merlino S, Pushcharovskii DYu, Pakhomovskii YaA, Ferro O, Bogdanova AV, Voloshin AV, Sorokhtina NV, Zubkova NV (2000) Tumchaite $Na_2(Zr,Sn)Si_4O_{11}\cdot 2(H_2O)$ - a new mineral from carbonatites of the Vuoriyarvi alkali-ultrabasic massif, Murmansk region, Russia. Am Mineral 85:1516-1520

Swanson DK, Prewitt CT (1983) The crystal structure of potassium silicate $K_2Si^{VI}Si_3^{IV}O_9$. Am Mineral 68: 581-585

Szymański JT (1995) The crystal structure of owensite, $(Ba,Pb)_6(Cu,Fe,Ni)_{25}S_{27}$, a new member of the djerfisherite group. Can Mineral 33:671-677

Tabachenko VV, Kovba LM, Serezhkin VN (1984) Crystal structures of $Mg(UO_2)_6(MoO_4)_7(H_2O)_{18}$ and $Sr(UO_2)_6(MoO_4)_7(H_2O)_{15}$. Koord Khim 10:558-562

Toffoli T, Margolus N (1987) Cellular Automata Machines: a New Environment for Modeling. MIT Press, Boston

Uvarova YuA, Sokolova EV, Hawthorne FC, Liferovich RP, Mitchell RH (2003) The crystal chemistry of shcherbakovite from the Khibina Massif, Kola Peninsula, Russia. Can Mineral 41:1193-1201

Vaughan PA (1966) The crystal structure of the zeolite ferrierite. Acta Crystallogr 21:983-990

Vitins G, Grins J, Hoerlin T (1996) Synthesis and structural and ionic conductivity studies of $Na_2ZrSi_4O_{11}$. Solid State Ionics 86-88:119-124

von Neumann J (1951) Cerebral mechanisms in behaviour. *In*: The Hixon Symposium, Jeffress LA (ed), Wiley, New York, p 1-32

Voronkov AA, Pyatenko YuA (1961) The crystal structure of vlasovite. Sov Phys Crystallogr 6:755-760

Voronkov AA, Shumyatskaya NG, Pyatenko YuA (1970) Crystal structure of a new natural modification of $Na_2Zr(Si_2O_7)$. Zh Strukt Khim 11:932-933

Voronkov AA, Zhdanova TA, Pyatenko YuA (1974) Refinement of the structure of vlasovite $Na_2ZrSi_4O_{11}$ and some characteristics of the composition and structure of the zirconosilicates. Sov Phys Crystallogr 19: 152-156

Wang X, Huang J, Jacobson AJ (2002c) $[(CH_3)_4N][(C_5H_5NH)_{0.8}((CH_3)_3NH)_{0.2}]U_2Si_9O_{23}F_4$ (USH-8): an organically templated open-framework uranium silicate. J Am Chem Soc 124:15190-15191

Wang X, Huang J, Liu L, Jacobson AJ (2002b) The novel open-framework uranium silicates $Na_2(UO_2)(Si_4O_{10})\cdot 2.1H_2O$ (USH-1) and $RbNa(UO_2)(Si_2O_6)\cdot H_2O$ (USH-3). J Mater Chem 11:406-410

Wang X, Jacobson AJ (1999) Crystal structure of the microporous titanosilicate ETS-10 refined from single crystal X-ray diffraction data. Chem Commun 1999:973-974

Wang X, Liu L, Jacobson AJ (2001) The novel open-framework vanadium silicates $K_2(VO)(Si_4O_{10})\cdot H_2O$ (VSH-1) and $Cs_2(VO)(Si_6O_{14})\cdot 3H_2O$ (VSH-2). Angew Chem Int Ed 40:2174-2176

Wang X, Liu L, Jacobson AJ (2002a) Open-framework and microporous vanadium silicates. J Am Chem Soc 124:7812-7820

Wolfram S (2002) A New Kind of Science. Wolfram Media, Inc., Champaign

Yaghi OM, Sun Z, Richardson DA, Groy TL (1994) Directed transformation of molecules to solids: synthesis of a microporous sulfide from molecular germanium sulfide cages. J Am Chem Soc 116:807-808

Yakovenchuk VN, Pakhomovsky YaA, Men'shikov YuP, Ivanyuk GYu, Krivovichev SV, Burns PC (2003): Chlorbartonite, $K_6Fe_{24}S_{26}(Cl,S)$, a new mineral from a hydrothermal vein in the Khibina massif, Kola peninsula, Russia: description and crystal structure. Can Mineral 41:503-511

Yakubovich OV, Dadachov MS (1992) Synthesis and crystal structure of ammonium analog of leucophosphite, $NH_4\{Fe_2(PO_4)_2(OH)(H_2O)\}(H_2O)$. Sov Phys Crystallogr 37:757-760

Zachariasen WH (1930) The crystal structure of benitoite $BaTiSi_3O_9$. Z Kristallogr 74:139-146

Zaworotko MJ (1999) Open season for solid frameworks. Nature 402:242-243

Zemann J (1948) Formel und Strukturtyp des Pharmakosiderits. Tscherm Mineral Petrogr Mitt 1:1-13
Zemann J (1959) Isotypie zwischen Pharmakosiderit und zeolithischen Germanaten. Acta Crystallogr 12:252
Zou X, Dadachov MS (1999) A new mixed framework compound with corrugated $[Si_6O_{15}]_{\infty\infty}$ layers: $K_2TiSi_6O_{15}$. J Solid State Chem 156:135-142

Polysomatic Aspects of Microporous Minerals – Heterophyllosilicates, Palysepioles and Rhodesite-Related Structures

Giovanni Ferraris[1,2] and Angela Gula[2]

[1]*Dipartimento di Scienze Mineralogiche e Petrologiche*
Università di Torino
10125 Torino, Italy

[2]*Istituto di Geoscienze e Georisorse*
Consiglio Nazionale delle Ricerche
10125 Torino, Italy

giovanni.ferraris@unito.it   angela.gula@uito.it

INTRODUCTION

Several structural families reported in this book can be described by using concepts of modular crystallography (Thompson 1978; Veblen 1991; Merlino 1997; Ferraris et al. 2004). This chapter presents three groups of microporous minerals emphasizing the modular aspects of their crystal structures and the role that modularity plays in correlating different structures as well as structure and properties, namely aspects aimed at an engineering of microporous materials (cf. Rocha and Lin 2005).

The description of a crystal structure as an edifice consisting of complex building modules that occur also in other structures implicitly leads to identify features that are common to a group of compounds. This kind of group can often be expressed as a series of structures that are collinear in composition and cell parameters, information that may be crucial to model unknown structures related to the series, as illustrated by some examples in this chapter.

Biopyriboles (Fig. 1) represent a first and now classical example of modular structures established by Thompson (1978). He showed that the structures of micas, pyroxenes and amphiboles share, according to different ratios, the same modules of mica (M) and pyroxene (P) and are members of a polysomatic series M_mP_p. The ideal chemical composition and cell parameters of the members of the series are linear functions of the ratio m/p. The classification of biopyriboles as members of a polysomatic series, a type of series belonging to the wider category of the homologous series (cf. Ferraris et al. 2004), and the consequent discovery of the multiple-chain-width biopyriboles jimthompsonite and chesterite (Veblen and Buseck 1979) dramatically proved the predictive power of these series in terms of structure characterization and modeling. The modeling of carlosturanite

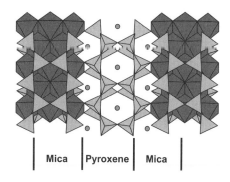

Figure 1. Projection along [100] of the crystal structure of an amphibole showing its slicing as modules of mica and pyroxene.

(Mellini et al. 1985) and of other modular structures reviewed by Veblen (1991) represents some of the earlier successes of polysomatism. These successful examples opened a prolific route as shown by comprehensive reviews in Merlino (1997) and Ferraris et al. (2004). Among other purposes, some microporous structures described in this chapter further clarify the mechanism that is behind the definition of polysomatic and related series.

All the structures described in this chapter contain silicate tetrahedra, which are grouped in sheets and/or strips, and condensed or isolated octahedra. Thus, these structures belong to the larger category of the heterosilicates described in this volume by Chukanov and Pekov (2005) and Rocha and Lin (2005).

The porous properties of the structures here presented are well documented and technologically applied in same cases (e.g., palygorskite and sepiolite); they are instead to some extent speculative in other cases (e.g., members of the heterophyllosilicate series). In the latter case, the structures are discussed with the aim of attracting the attention of those materials scientists who are looking for novel potentially useful porous structures. In particular, the structural parallelism between the 2:1 layer silicates, like micas, and members of the heterophyllosilicate polysomatic series leads to speculate on the possible use of some of these compounds as starting material to produce pillared porous structures analogous to the technologically important pillared clays (cf. Corma 1997). A main problem to be solved on this route is the synthesis of the mineral analogues, not only to have simple and defined chemical compositions, but also because most heterophyllosilicates are quite rare in nature (cf. Khomyakov 1995). Exceptions are astrophyllite, lamprophyllite and, to a minor extent, lomonosovite and murmanite. In the Khibina and Lovozero massifs, lamprophyllite reaches 1–4% as component of lujavrites and is a common accessory mineral in khibinites; lomonosovite and murmanite can be locally considered rock-forming minerals in some formations of the Lovozero massif (N.V. Chukanov, personal communication).

HETEROPHYLLOSILICATES

Octahedral O close-packing sheets with full (brucite-type) or partial (e.g., gibbsite-, spinel-, and corundum-type) occupancy of the octahedral sites are recurrent in modular structures. In particular, coupling of a tetrahedral silicate T sheet with a dioctahedral (gibbsite) or trioctahedral (brucite) O sheet constitutes the building blocks of TO (or 1:1) and TOT (or 2:1) layer silicates (or phyllosilicates; Fig. 2). For example, the 2:1 layer silicates can be grouped in an A_nB_m merotype series[1] where the TOT module (A) is the fixed building module and B is an interlayer variable module. Talc is representative of $n = 1$ and $m = 0$; micas (B = alkaline or alkaline-earth cation), chlorites (B = octahedral sheet), and smectites (B = alkaline or alkaline-earth cation, H_2O, □) are well known members of the series with $n = 1$ and $m = 1$.

Strips (modules) of TOT layers occur in several silicate structures: biopyriboles (Fig. 1), heterophyllosicates and palysepioles (see below for the definition of the two latter groups). This section reports the description of a wide family of layer titanosilicates (heterophyllosilicates) that bear features very close to the TOT-based phyllosilicates from which they can be formally derived. In Russian literature, heterophyllosilicates are often referred to as titanosilicate micas and, more generally, as amphoterosilicates (cf. Khomyakov 1995 and Chukanov and Pekov

---

[1] According to Makovicky (1997), in a merotype series, whereas one building module is kept constant, a second (third, etc.) module is peculiar of each member. A series is said to be plesiotype when all members share modules that, however, may still slightly differ in chemistry and configuration: see the mero-plesiotype series of bafertisite and rhodesite described later in this chapter and other examples in Ferraris et al. (2004).

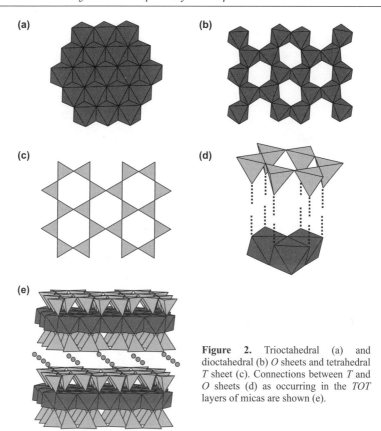

Figure 2. Trioctahedral (a) and dioctahedral (b) O sheets and tetrahedral T sheet (c). Connections between T and O sheets (d) as occurring in the TOT layers of micas are shown (e).

2005). This term alludes to the fact that the cation Ti^{4+} participates[2], together with Si, in building a mixed tetrahedral/octahedral framework (or just layers), which acts as a complex anion. In its turn, this complex anion hosts in its cavities a variety of cations, including the octahedral ones already present in the anionic part; thus, these cations are said to be "amphoteric."

Polysomatism in the heterophyllosilicates

The characterization of nafertisite, a rare titanosilicate first reported from the Khibina hyperalkaline massif (Kola Peninsula, Russia; Khomyakov et al. 1995) and later from the Igaliko nepheline syenite complex (South Greenland; Petersen et al. 1999)[3], prompted Ferraris et al. (1996, 1997) to correlate a group of titanium silicates whose structures are based on TOT-like layers and to introduce the term *heterophyllosilicate*. This correlation has been established via the definition of the polysomatic series of the heterophyllosilicates. In the members of this series, a row of Ti polyhedra (or substituting cations, see footnote 2) periodically substitutes a row of disilicate tetrahedra (silicate diorthogroups) in the T tetrahedral sheet that is typical of the layer silicates; the octahedral O sheet is instead maintained (Fig. 3). HOH layers are thus

---

[2] Ti^{4+} is the main central cation in these polyhedra, but often it is partially substituted by Nb^{5+}, Zr^{4+} and Fe^{3+} (Tables 1 and 2). Sometimes the latter cations are dominating on Ti, like Zr in seidozerite, Fe in orthoericssonite and Nb in vuonnemite. In the text, for short, only Ti is indicated as centering the relevant polyhedra.

[3] According to N.V. Chukanov (pers. comm.), the IR spectra of samples from the two localities quoted in the text show some differences likely related to cation ordering and other details of the crystal structure.

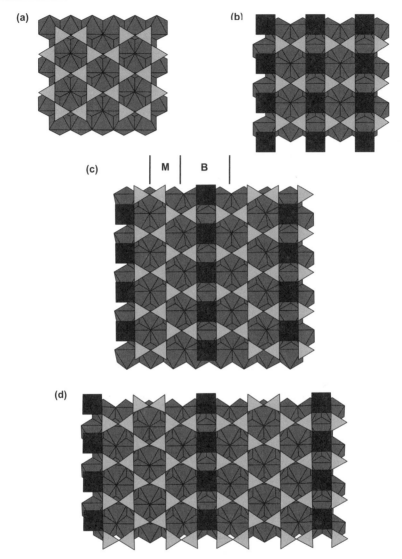

Figure 3. *TOT* layer of a phyllosilicate (a) as transformed, by periodic substitution of rows of octahedra (dark grey) for tetrahedra, to *HOH* heterophyllosilicate layers typical of bafertisite (b), astrophyllite (c) and nafertisite (d). *B* bafertisite-type and *M* mica-type modules are shown in (c).

obtained where *H* stands for *hetero* to indicate the presence of rows of 5- or 6-coordinated Ti in a sheet corresponding to the *T* sheet of the layer silicates. The length of the O⋯O edges (~ 2.7 Å) of the substituting polyhedra is very close to that of the same edges in the substituted tetrahedra, thus the insertion of the polyhedra in a *T* sheet does not produce strain.

As summarized by Ferraris (1997), depending on the periodicity of the Ti substitution and ignoring some minor topological features, three types of *HOH* layers (Fig. 3) are known so far. The slice of *HOH* layer containing rows of Ti polyhedra in its *H* sheet is conventionally called bafertisite-type module and has composition $B = I_2Y_4[X_2(O)_4Si_4O_{14}](O,OH)_2$; this module may be intercalated with a mica-type module $M = IY_3[Si_4O_{10}](O,OH)_2$. In the two formulae, I and Y represent interlayer and octahedral cations, respectively; X corresponds to Ti and its vicariant elements as stated in footnote 2.

$(HOH)_B$ bafertisite-type layer. In this layer, a bafertisite-type module alone is periodically repeated (Fig. 3). The heterophyllosilicates based on the resulting $(HOH)_B$ layer are some tens and are described in a subsequent section as the mero-plesiotype bafertisite series. We shall see, however, that in this series the so-called bafertisite-type layer does actually correspond to two topologically different modules. Anyway, if it is not necessary to specify, the undifferentiated layer is indicated as $(HOH)_B$. As discussed in the section on the rhodesite group of structures, a H bafertisite-type sheet occurs also in the structure of jonesite (Krivovichev and Armbruster 2004).

$(HOH)_A$ astrophyllite-type layer. Relative to the $(HOH)_B$ layer, in a $(HOH)_A$ astrophyllite-type layer a one-chain-wide mica-like module M is present between two bafertisite-type modules (Fig. 3). As recently discussed by Piilonen et al. (2003a,b), the known heterophyllosilicates based on a $(HOH)_A$ layer form a complex isomorphous series (Table 1) and mineral species are defined by the chemical nature of I, Y and X cations in the formula given above; some polytypes are known. As mentioned above, astrophyllite is one of the few non-rare heterophyllosilicates. Except in magnesium astrophyllite (Shi et al. 1998), in these structures the layers are locked together by sharing an oxygen atom between two Ti (X cation) octahedra belonging to adjacent layers (Fig. 4). Thus, [100] channels are realized with windows characterized by a small (~ 1.5 Å) minimum effective width[4]. Chelishchev (1972, 1973) has proved that under supercritical conditions (400–600°C; pressure of 1000 kg/cm^2) astrophyllite exchanges K with Na, Rb and Cs; anyway, this group of heterophyllosilicates will no longer be considered in this chapter.

Eveslogite. $\{(Ca,K,Na,Sr,Ba)_{48}[(Ti,Nb,Fe,Mn)_{12}(OH)_{12}Si_{48}O_{144}](F,OH,Cl)_{14}$; $P2/m$, $a = 14.069$, $b = 24.937$, $c = 44.31$ Å, $\gamma = 95.02°\}$ is a titanosilicate recently described from Khibina massif (Men'shikov et al. 2003). It shows an ab base 6-times larger than that of astrophyllite and 3-times larger than that of magnesium astrophyllite (Table 1). Its c parameter corresponds to the thickness of four $(HOH)_A$ astrophyllite-type layers. On this basis, it was suggested (cf. Ferraris et al. 2004) that the crystal structure of eveslogite is based on a [001] stacking of four $(HOH)_A$ layers. That supports a fitting of the eveslogite chemical data to an astrophyllite-like chemical formula. However, according to IR data (N.V. Chukanov, personal communication) the crystal structure of eveslogite is likely close to that of yuksporite [$(Sr,Ba)_2$ $K_4(Ca,Na)_{14}(\square,Mn,Fe)\{(Ti,Nb)_4(O,OH)_4[Si_6O_{17}]_2[Si_2O_7]_3\}(H_2O,OH)_n$; $a = 7.126$, $b = 24.913$, $c = 17.075$ Å, $\beta = 101.89°$; $P2_1/m$], a mineral from Khibina massif to which Men'shikov et al. (2003) tentatively assigned a layer structure, together with eveslogite. Recently, Krivovichev et al. (2004) have instead shown that yuksporite has a microporous structure based on [100] channels delimited by Si tetrahedra and Ti octahedra.

$(HOH)_N$ nafertisite-type layer. Relative to the $(HOH)_B$ layer, in a $(HOH)_N$ nafertisite-type layer two one-chain-wide mica-like modules M are present between two bafertisite-type modules [Fig. 3; alternatively, one can say that a second M module is intercalated in the $(HOH)_A$ astrophyllite-type layer].

Nafertisite $\{nfr$; $(Na,K,\square)_4(Fe^{2+},Fe^{3+},\square)_{10}[Ti_2O_3Si_{12}O_{34}](O,OH)_6$; $A2/m$, $a = 5.353$, $b = 16.176$, $c = 21.95$ Å, $\beta = 94.6°\}$ is the only compound known to be based on a $(HOH)_N$ layer. No single crystals suitable for X-ray diffraction analysis are available; a model of the

---

[4] The minimum effective width is calculated as the minimum O⋯O distance across the interlayer minus the ionic diameter of O^{2-} (2.7 Å; McCusker et al. 2003).

Table 1. Members of the astrophyllite group. Space groups are shown in agreement with the following choice of axes: $a \sim 5.4$ and $b \sim 11.9$ Å (or multiples) in the *HOH* layer; $c \sim 11.7$ Å (or multiple) outside the layer.

Name	Chemical Formula	Space group
Niobokupletskite	$K_2Na\{(Mn,Zn,Fe)_7[(Nb,Zr,Ti)_2O_3Si_8O_{24}](O,OH,F)_4\}$	$P\bar{1}$
Kupletskite-1A	$K_2Na\{(Mn,Fe^{2+})_7[(Ti,Nb)_2(O_2F)Si_8O_{24}](OH)_4\}$	$P\bar{1}$
Kupletskite-$Ma2b2c$	$K_2Na\{(Mn,Fe^{2+})_7[Ti_2(O_2F)Si_8O_{24}](OH)_4\}$	$C2/c$
Kupletskite-(Cs) [c]	$(Cs,K)_2Na\{(Mn,Fe,Li)_7[(Ti,Nb)_2(O_2F)Si_8O_{24}](OH)_4\}$	?
Astrophyllite	$K_2Na\{(Fe^{2+},Mn)_7[Ti_2(O_2F)Si_8O_{26}](OH)_4\}$	$A\bar{1}$
Magnesium astrophyllite [a]	$K_2Na\{[Na(Fe,Mn)_4Mg_2][Ti_2O_2Si_8O_{24}](OH)_4\}$	$C2$
Niobophyllite	$K_2Na\{(Fe^{2+},Mn)_7[(Nb,Ti)_2(F,O)_3Si_8O_{24}](OH)_4\}$	$P\bar{1}$
Zircophyllite [c]	$K_2(Na,Ca)\{(Mn,Fe^{2+})_7[(Zr,Nb)_2(O_2F)Si_8O_{24}](OH)_4\}$	?
Fe-dominant zircophyllite [b, c]	$K_2(Na,Ca)\{(Fe^{2+},Mn)_7[(Zr,Nb)_2(O_2F)Si_8O_{24}](OH)_4\}$	?
"Hydroastrophyllite" [b, c]	$(H_3O,K)_2Ca\{(Fe^{2+},Mn)_{5-6}[Ti_2(O_2F)Si_8O_{24}](OH)_4\}$	?

[a] Ti is 5 coordinated (see Fig. 4b).
[b] Not approved as mineral species.
[c] Structure unknown.

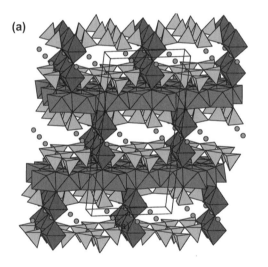

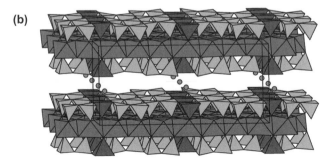

Figure 4. Structure of astrophyllite (a) and magnesium astrophyllite (b) seen along [100]. The connection of two adjacent *HOH* layers by sharing a corner between two Ti (X cation) polyhedra (dark grey) occurs in the majority of minerals reported in Table 1 and in some of Table 2: [100] channels are thus formed (a). In (b), Ti has coordination number 5 according to a tetragonal pyramid (dark grey). Circles represent interlayer cations.

crystal structure of nafertisite was obtained (Ferraris et al. 1996) by comparing its chemical composition and cell parameters with those of:

- bafertisite {*bft*; Ba$_2$(Fe,Mn)$_4$[Ti$_2$O$_4$Si$_4$O$_{14}$](O,OH)$_2$; $P2_1/m$, a = 5.36, b = 6.80, c = 10.98 Å, β = 94°; Pen and Shen (1963); see a discussion in the section 'Pseudosymmetries'};
- astrophyllite {*ast*; (K,Na)$_3$(Fe,Mn)$_7$[Ti$_2$O$_3$Si$_8$O$_{24}$](O,OH)$_4$; $P\bar{1}$, a = 5.36, b = 11.63, c = 11.76 Å, α = 112.1, β = 103.1, γ = 94.6°; P, instead of the A cell of Woodrow (1967), is here adopted}.

The following observations allowed to build a model of the structure of nafertisite by inserting a further M mica-type module in the *HOH* layer of astrophyllite (Fig. 5).

a. The difference in composition between astrophyllite and nafertisite is about $(I,\square)(Y,\square)_3$[Si$_4$O$_{10}$](OH,O)$_2$; it is comparable to the composition of an M mica-type module and corresponds to half the difference between the compositions of nafertisite and bafertisite.

b. Bafertisite, astrophyllite and nafertisite have a common value of $a \sim 5.4$ Å, which matches the a value of mica.

c. The value of the differences ($b_\text{ast} - b_\text{bft}$) ~ $1/2(b_\text{nfr} - b_\text{bft})$ ~ 4.7 Å corresponds to the value of $b/2$ in mica; $(d_{002})_\text{nfr}$ = 10.94 Å matches the thickness of one *HOH* layer in bafertisite and astrophyllite.

In the structure of nafertisite, the layers are locked together via sharing an oxygen atom between two Ti octahedra that belong to adjacent layers (Fig. 5). Thus, the interlayer space is subdivided in [100] channels with a small effective width like that mentioned for astrophyllite.

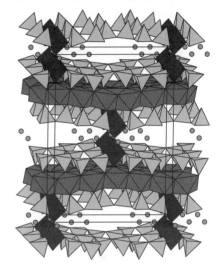

Figure 5. View along [100] of the crystal structure of nafertisite. Pairs of Ti polyhedra (dark grey) share a corner across the interlayer space, thus forming [100] channels. Circles as in Figure 4.

The heterophyllosilicate polysomatic series. Following the description given above, bafertisite, astrophyllite and nafertisite are members of a B_mM_n polysomatic series with general formula $I_{?+n}Y_{4+3n}[X_?(\text{O'})_{?+n}\text{Si}_{4+4n}\text{O}_{14+10n}](\text{O"})_{?+?n}$. In this formula, atoms belonging, even in part, to the H sheet are shown in square brackets. I represents large (alkali) interlayer cations and Y octahedral cations; O' (bonded to X) and O" (belonging to the octahedral sheet only) can be an oxygen atom, OH, F or H$_2$O; the 14+10n oxygen atoms are bonded to Si. The X cation is 5- or 6-coordinated according to polyhedra that share one corner with the octahedral sheet and four corners with four Si tetrahedra of the T sheet. The value of p (0, 1, 2) depends on the configuration around X. In case of octahedral coordination, the sixth corner can be (i) unshared (p = 2), (ii) shared with an octahedron of the adjacent layer (p = 1) or (iii) with an interlayer anion (p = 0); p = 0 holds also when (iv) an edge is shared between two octahedra or (v) the coordination number of X is 5 (see below figures with examples). In the cases (ii) and (iv) a "layered" framework structure is actually formed.

The heterophyllosilicates have also been described by using differently defined modules (Christiansen et al. 1999), a possibility which is not rare in modular crystallography.

Heterophyllosilicates vs. porosity. Layer structures like heterophyllosilicates cannot be included in the definition of "ordered microporous and mesoporous materials" adopted by IUPAC (McCusker et al. 2003; McCusker 2005). That because, even if wide enough to allow exchange of at least water molecules, in these structures the interlayer space extends in two dimensions, namely microporosity does not develop according to specific ordered spaces corresponding to channels, cages and cavities.

Actually, as mentioned above, some heterophyllosilicates show the interlayer space subdivided in channels by bridges established via polyhedra that share corners, thus locking together the layers; however, the minimum effective width of these channels is below or at the bottom scale of porosity. Subdivision of the interlayer space is observed also in several so-called modulated phyllosilicates (cf. Guggenheim and Eggleton 1988), namely those silicates where a periodic perturbation to the basic layer structure occurs (see Section on Palysepioles below).

Both the heterophyllosicates with locked layers and the mentioned modulated phyllosilicates actually show a framework structure, although usually they are considered "layered structures" because condensation of polyhedra in layers is an outstanding structural feature. In particular, the heterophyllosilicates with locked layers can be considered a transition structural type between true layer titanosilicates and framework titanosilicates extensively described in this volume by Chukanov and Pekov (2005) and Pekov and Chukanov (2005).

The main interest to consider heterophyllosilicates here is however related to a hypothesis of using their layers as starting modules to build pillared materials, another kind of porous structures defined (Schoonheydt et al. 1999) separately from the ordered porous structures mentioned above.

The bafertisite series

The $(HOH)_B$ bafertisite-type layer is the most versatile of the three heterophyllosilicate layers defined above, being able to sandwich a large variety of interlayer contents. The number of crystal structures containing this layer is comparable to that known for structures containing the *TOT* phyllosilicate layer. However, the latter appears in important rock-forming minerals, like micas and clay minerals, while the titanosilicates we are dealing with occur only in rare hyperalkaline rocks (Table 2; Khomyakov 1995).

Two main groups of $(HOH)_B$-bearing compounds are known: the götzenite group, where Ca is at the centre of the hetero-octahedra and is not of interest in this chapter (cf. Christiansen and Rønsbo 2000; Ferraris et al. 2004), and the complex bafertisite series (Ferraris et al. 1997) where the $(HOH)_B$ layer alternates with a large variety of interlayer contents [cf. Egorov-Tismenko and Sokolova (1990) and Egorov-Tismenko (1998) for earlier crystal-chemical analyses of part of the series and quotation of former Russian papers on layer titanosilicates]. The members of this series are related by merotypy to the B_1M_0 member (i.e., bafertisite) of the heterophyllosilicate polysomatic series, in the sense (cf. footnote 1) that whereas the $(HOH)_B$ layer appears in all members, the interlayer content is characteristic of each member. For the members of the bafertisite series the general formula given above becomes $A_2\{Y_4[X_2(O')_{2+p}Si_4O_{14}](O'')_2\}W$; the interlayer content labeled I in the general formula is here split into A (alkaline and alkaline-earth large cations) and W (H_2O molecules and complex anions). In this formula:

1. $[X_2(O')_{2+p}Si_4O_{14}]^{n-}$ is a complex anion representing the heterophyllosilicate H sheet alone;
2. $\{Y_4[X_2(O')_{2+p}Si_4O_{14}](O'')_2\}^{m-}$ is a larger complex anion representing the whole $(HOH)_B$ layer; it is within braces that leave outside the interlayer contents A and W (Table 2).

Table 2. Members of the mero-plesiotype bafertisite series in increasing order of the cell parameter (t, Å) (either c or a) stacking the layers. The content of the heteropolyhedral H sheet is shown in square brackets, and that of the HOH layer is within braces; the composition of the interlayer is shown outside the braces.

Name	Chemical formula	t^m (Å)	Space group	Reference[1]
Murmanite [a]	$(Na,\square)_2\{(Na,Ti)_4[Ti_2(O,H_2O)_2Si_4O_{14}](OH,F)_2\}\cdot 2H_2O$	11.70	$P\bar{1}$	Németh et al. 2005
Bafertisite [a]	$Ba_2\{(Fe,Mn)_4[Ti_2O_2(O,OH)_2Si_4O_{14}](O,OH)_2\}$	11.73	Cm	Guan et al. 1963
Hejtmanite [a]	$Ba_2\{(Mn,Fe)_4[Ti_2(O,OH)_4Si_4O_{14}](OH,F)_2\}$	11.77	Cm	Rastsvetaeva et al. 1991
Epistolite [b]	$(Na,\square)_2\{(Na,Ti)_4[Nb_2(O,H_2O)_4Si_4O_{14}](OH,F)_2\}\cdot 2H_2O$	12.14	$P\bar{1}$	Németh et al. 2005
Vuonnemite [b]	$Na_8\{(Na,Ti)_4[Nb_2O_2Si_4O_{14}](O,OH,F)_2\}(PO_4)_2$	14.45	$P\bar{1}$	Ercit et al. 1998
Lomonosovite [a, n]	$Na_8\{(Na,Ti)_4[Ti_2O_2Si_4O_{14}](O,OH)_2\}(PO_4)_2$	14.50	$P\bar{1}$	Belov et al. 1978
Synthetic vanadate [a]	$Na_8\{(Na,Ti)_4[Ti_2O_2Si_4O_{14}]O_2\}(VO_4)_2$	14.75	$P1$	Massa et al. 2000
Yoshimuraite [a, c]	$Ba_4\{Mn_4[Ti_2O_2Si_4O_{14}](OH)_2\}(PO_4)_2$	14.75	$P\bar{1}$	McDonald et al. 2000
Innelite [b, c]	$(Ba,K)_2Ba_2\{(Na,Ca,Ti)_4[Ti_2O_2Si_4O_{14}]O_2\}(SO_4)_2$	14.76	$P1$	Chernov et al. 1971
Bussenite [a]	$Ba_4Na_2\{(Na,Fe,Mn)_2[Ti_2O_2Si_4O_{14}](OH)_2\}(CO_3)_2F_2\cdot 2H_2O$	16.25	$P\bar{1}$	Zhou et al. 2002
Seidozerite [b, g, i]	$Na_2\{(Na,Mn,Ti)_4[(Zr,Ti)_2O_2Si_4O_{14}]F_2\}$	18.20	$P2/c$	Pushcharovsky et al. 2002
Lamprophyllite [b, c]	$(Sr,Ba)_2\{(Na,Ti)_4[Ti_2O_2Si_4O_{14}](OH,F)_2\}$	19.49	$C2/m$	Rastsvetaeva et al. 1990
Nabalamprophyllite [b, c]	$Ba(Na,Ba)\{(Na,Ti,)_4[Ti_2O_2Si_4O_{14}](OH,F)_2\}$	19.74	$P2/m$	Rastsvetaeva and Chukanov 1999
Barytolamprophyllite [b, c]	$(Ba,Na)_2\{(Na,Ti)_4[Ti_2O_2Si_4O_{14}](OH,F)_2\}$	19.83	$C2/m$	Pen et al. 1984
Orthoericssonite [b, c]	$Ba_2\{Mn_4[Fe_2O_2Si_4O_{14}](OH)_2\}$	20.23	$Pnmn$	Matsubara 1980
Quadruphite [a]	$Na_{13}Ca\{(Ti,Na,Mg)_4[Ti_2O_2Si_4O_{14}]O_2\}(PO_4)_4F_2$	20.36	$P\bar{1}$	Sokolova and Hawthorne 2001

Continued on following page

Table 2 continued from previous page.

Name	Chemical formula	t^m (Å)	Space group	Reference [1]
Ericssonite [e]	$Ba_2\{Mn_4[Fe_2O_2Si_4O_{14}](OH)_2\}$	20.42	$C2/m$	Moore 1971
Surkhobite [a,i]	$(Ca,Na,Ba,K)_2\{(Fe,Mn)_4[Ti_2O_2Si_4O_{14}](F,O,OH)_3\}$	20.79	$C2$	Rozenberg et al. 2003
Jinshajiangite [e,p]	$(Na,Ca)(Ba,K)\{(Fe,Mn)_4[(Ti,Nb)_2O_3Si_4O_{14}](F,O)_2\}$	20.82	$C2/m?$	Hong and Fu 1982
Perraultite [a,i]	$(Na,Ca)(Ba,K)\{(Mn,Fe)_4[(Ti,Nb)_2O_3Si_4O_{14}](OH,F)_2\}$	20.84	$C2$	Yamnova et al. 1998
Delindeite [b]	$Ba_2\{(Na,Ti,\square)_4[Ti_2(O,OH)_4Si_4O_{14}](H_2O,OH)_2\}$	21.51	$A2/m$	Ferraris et al. 2001b
Polyphite [a]	$Na_{14}(Ca,Mn,Mg)_5\{(Ti,Mn,Mg)_4[Ti_2O_2Si_4O_{14}]F_2\}(PO_4)_6F_4$	26.56	$P\bar{1}$	Sokolova et al. 1987
M55C [e]	$Na_8\{(Na,Ti)_4[Ti_2O_2Si_4O_{14}](O,F)_2\}(PO_4)_2$?	28.1	$P112_1/m$	Németh 2004
Shkatulkalite [e]	$\{(Na,Mn,Ca,\square)_4[(Nb,Ti)_2(H_2O)_2Si_4O_{14}](OH, H_2O,F)_2\}\cdot 2(H_2O,\square)$	31.1	$P2/m$	Men'shikov et al. 1996
M55A [e]	$\{(Na,Na,\square)_5\{(Na_3Ti)[(Ti,Nb)_2O_2Si_4O_{14}](F,OH,O)_2\}(PO_4)_2\cdot 4H_2O$	38.11	$P112_1/m$	Németh 2004[f], Khomyakov 1995
Sobolevite [a]	$Na_{12}CaMg\{(Ti,Na,Mg)_4 [Ti_2O_2Si_4O_{14}]O_2\}(PO_4)_4F_2$	40.62	$P1$	Sokolova et al. 1988
M55B [e]	$\{(Na,K,Ba,\square)_4[(Ti,Nb)_2(H_2O,OH)_2Si_4O_{14}](F,H_2O,\square)_2\}$	43.01	$P112_1/m$	Németh 2004
Bornemanite [a,i]	$BaNa_3\{(Na,Ti,Mn)_4[(Ti,Nb)_2O_2Si_4O_{14}](F,OH)_2\}PO_4$	47.95	Ib	Ferraris et al. 2001a
M73 [a,d]	$(Ba,Na)_2\{Na,Ti,Mn)_4[(Ti,Nb)_2(OH)_3Si_4O_{14}](OH,O,F)_2\}\cdot 3H_2O$	48.02	$A2/m$	Németh 2004[f], Khomyakov 1995
M72 [a,d,h]	$BaNa\{(Na,Ti)_4[(Ti,Nb)_2(OH,O)_2Si_4O_{14}](OH,F)_2\}\cdot 3H_2O$	50.94	Im	Németh 2004[f], Khomyakov 1995

[a] $(HOH)_B$ layer. - [b] $(HOH)_V$ layer. - [c] X cation in coordination 5. - [d] X cation in coordination 5 and 6. - [e] Structure unknown; the inclusion of the mineral in this table is based mainly on chemical and crystal data. - [f] Crystallographic data. - [g] Grenmarite (Bellezza et al. 2004) is a new species that differs from seidozerite for having Zr dominating also in one site of the O sheet. Actually, seidozerite is included in this Table more for historical [see the first introduction of this series by Egorov-Tismenko and Sokolova (1990)] than structural reasons; in fact, sharing of edges between adjacent HOH layers and the consequent lack of a real interlayer space suggest the inclusion of seidozerite in the götzenite-rosenbuschite-seidozerite family (cf. Christiansen et al. 2003 and Bellezza et al. 2004). - [h] Approved as mineral species: IMA No 2003-044 (Burke and Ferraris 2004). - [i] Two octahedra of X cations share either a corner or an edge. - [j] Reference to the most recent paper describing the structure, when known, otherwise to the paper describing the species. [m] The cells given by authors have been converted to reduced cells, if the case. [n] "Betalomonosovite" is a discredited mineral species with composition close to that of lomonosovite; its crystal structure ($a = 5.326$, $b = 14.184$, $c = 14.47$ Å, $\alpha = 102.2$, $\beta = 95.5$, $\gamma = 90.17°$; P-1, Rastsvetaeva 1998). [p] IR spectra (N. V. Chukanov, personal communication), cell parameters and composition suggest that jinshajingite has the same structure of perraultite thus corresponding to the Fe dominant of the latter mineral.

Structural aspects in the bafertisite series

In the known members of the bafertisite series (Table 2), both the H sheet and the whole $(HOH)_B$ layer are negatively charged, thus the more or less complex interlayer content acts as a cation. In some cases, e.g., epistolite, murmanite, M72 and M73, the charge of the layer is very weak, an aspect discussed later. Fixing as c the cell parameter outside the $(HOH)_B$ layer, all the heterophyllosilicates listed in Table 2 are characterized by similar values of $a \sim 5.4$ Å and $b \sim 7.1$ Å (or multiples), i.e., of the periodicities within the layer.

The HOH layer. In terms of merotypy and plesiotypy (see footnote 1), the heterophyllosilicates of Table 2 form a mero-plesiotype series (Ferraris et al. 2001b). This series is merotype because the *HOH* module is constantly present in the crystal structure of all members, whereas a second module, namely the interlayer content, is peculiar to each member. At the same time, the series has a plesiotype character because chemical nature and coordination number of the X and Y cations and the linkage between the H and O sheets may be modified. In fact, as discussed by Sokolova and Hawthorne (2004) (cf. also Christiansen et al. 1999), the two H sheets sandwiching an O sheet are either facing each other via the same type of polyhedra (i.e., heteropolyhedra face each other) or showing a relative shift (i.e., a heteropolyhedron faces a tetrahedron), thus realizing two different topologies. The two kinds of *HOH* layer occurring in vuonnemite, $(HOH)_V$, and in bafertisite, $(HOH)_B$, are here considered typical examples of the two topologies (Fig. 6). Besides, the coordination

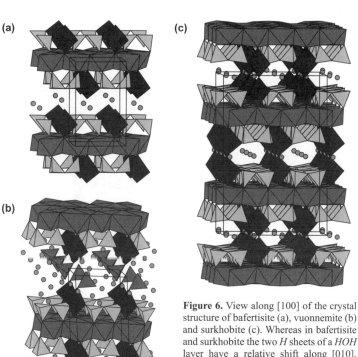

Figure 6. View along [100] of the crystal structure of bafertisite (a), vuonnemite (b) and surkhobite (c). Whereas in bafertisite and surkhobite the two H sheets of a *HOH* layer have a relative shift along [010], in vuonnemite this shift is absent. When a discrimination between the two types of topology is necessary, in the text the corresponding *HOH* layers are labeled $(HOH)_B$ (bafertisite type) and $(HOH)_V$ (vuonnemite type), respectively. Circles as in Figure 4; the dark-grey tetrahedra in the interlayer represent PO_4 groups.

number of the X cations can be either 5 (square pyramid) or 6 (octahedron); the two types of coordination occur even in the same structure (Figs. 7 and 8). Note that the configuration with coordination 5 is a way to decrease the total negative charge of the *HOH* layer; the same result can be obtained via a higher charge of the X cation (e.g., Nb^{5+} instead of Ti^{4+}) or presence of high-charge Y cations (e.g., Fe^{3+} and Ti^{4+}).

The $(HOH)_B$ layer is more common (see footnotes a and b in Table 2). Nèmeth et al. (2005) inferred that the different topology observed for the *HOH* layer in the pair epistolite, $(HOH)_V$, and murmanite, $(HOH)_B$, in spite of their similar chemical composition (Table 2), is likely related to the Na/Ti ratio in the O sheet and to the different charge of the X cation (either Nb^{5+} of Ti^{4+}). The following reasons, that likely are valid for other members of Table 2, were given. An oxygen atom shared between H and O sheets is bonded to four cations: three belonging to the O sheet and one to the H sheet. Because of bond-valence balance, even an O^{2-} anion cannot be bonded at the same time to four high-charge cations like Si^{4+} and Ti^{4+} or Nb^{5+}; consequently, constraints are derived not only to the composition, but also to the topology of the *HOH* layer. Thus, to reach a suitable bond-valence balance, Na octahedra share edges between them in epistolite but not in murmanite and a different connectivity between the H and O sheets is established.

The interlayer. As does the *TOT* layer in the 2:1 phyllosilicates, in the heterophyllosilicates two adjacent *HOH* modules delimit an interlayer space that contains (Table 2) either a single cation or a complex composition which may even correspond to that of a known mineral, like nacaphite, $Na_2Ca[PO_4]F$ (Sokolova et al. 1989a) occurring in quadruphite, polyphite

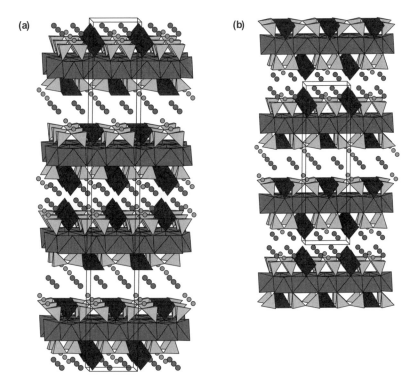

Figure 7. View along [100] of the crystal structure of two M73 polytypes. Smaller and larger circles represent cations and H_2O molecules, respectively. Note that the X cation shows either octahedral or pyramidal coordination (dark grey polyhedra).

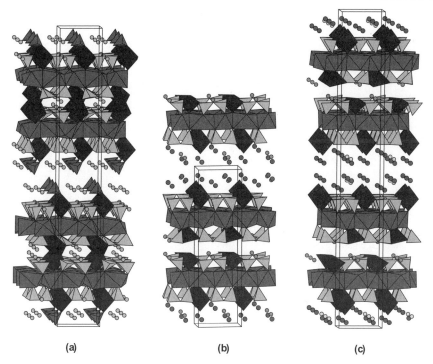

Figure 8. View along [100] of the crystal structure of bornemanite (a) and two polytypes of M72 (b and c). Circles and X cation as in Figure 7; dark-grey tetrahedra as in Figure 6.

(Khomyakov et al. 1992) and sobolevite (Sokolova et al. 1988). Besides nacaphite, other stoichiometric compositions may correspond to interlayer contents: $BaSO_4$ in innelite (Chernov et al. 1971); Na_3PO_4 in lomonosovite (Belov et al. 1978), vuonnemite (Ercit et al. 1998) and bornemanite (Ferraris et al. 2001a); Na_3VO_4 in $Na_8\{(NaTi)_2[Ti_2O_2Si_4O_{14}]O_2\}(VO_4)_2$ (Massa et al. 2000).

The interlayer content determines the value of the c parameter, which necessarily increases with the complexity of the sandwiched module (Table 2). On its own, the thickness of the *HOH* layer is about 10 Å from apex to apex of the X coordination polyhedra belonging to two facing H sheets. In the structures where the interlayer is simple, the c parameter is close to $n \times 10$ Å, with n indicating the number of *HOH* layers in the cell. Typical separations between the bases of the Si tetrahedra belonging to two facing layers are 4.2 Å in bafertisite, about 8 Å in lomonosovite and vuonnemite, 13.5 Å in quadruphite and sobolevite. However, in the minerals with the larger separation, the interlayer, besides cations, contains also complex anions even if, on the whole, behaves as a complex cation.

In some cases, more than one type of interlayer occurs in the same member of the bafertisite series. That happens in bornemanite (Fig. 8; $a = 5.498$, $b = 7.120$, $c = 47.95$ Å, $\gamma = 88.4°$, $I11b$; Ferraris et al. 2001a) and the two related minerals labeled M72 and M73 in Khomyakov (1995). By transmission electron microscopy (TEM), Nèmeth (2004) and Nèmeth et al. (2004a) have found two polytypes for both the latter two minerals and modeled their structures (Figs. 7 and 8). The following crystal data have been obtained by electron diffraction and refined from powder diffraction data:

M72: $a = 5.552$, $b = 7.179$, $c = 25.47$ Å, $\gamma = 91.10°$, $P11m$; $a = 5.552$, $b = 7.179$, $c = 50.94$ Å, $\gamma = 91.10°$, $I11m$.

M73: $a = 5.40$, $b = 6.88$, $c = 24.01$ Å, $\beta = 91.3°$, $P2/m$; $a = 5.40$, $b = 6.88$, $c = 48.02$ Å, $\beta = 91.3°$, $A2/m$.

The structure of bornemanite can be described as a [001] stack of $(HOH)_B$ layers: a lomonosovite-like content alternates with a seidozerite-like one[5]. A similar situation with two different interlayer contents occurs in the structures of M72 and M73 that shall be further discussed later.

In some members [bornemanite (in part; Fig. 8), perraultite, surkhobite (Fig. 6), and likely jinshajiangite (see footnote p in Table 2)] of the bafertisite series the layers are locked together by sharing an oxygen atom between two X octahedra belonging to adjacent layers. Thus, even if the HOH layer remains an outstanding feature of the structure, these members of the series with locked layers become framework heterosilicates and the interlayer space is divided in [100] channels with a small window like those observed in astrophyllite and nafertisite. The height of these windows corresponds to the separation between the bases of the Si tetrahedra mentioned above. Likely the "framework" members form in a later stage during the evolution of the hyperalkaline massifs; in fact, Pekov and Chukanov (2005) describe epitactic crystals with composition jinshajiangite-perraultite overgrowing true layer members with composition bafertisite-hejtmanite. These authors note that in the hyperalkaline complexes early pegmatitic titanosilicates show mainly chain- and layer-based structures, which are replaced by framework phases at a later hydrothermal stage.

Pseudosymmetries. Most of the compounds in Table 2 are either monoclinic or triclinic. As noted by Ferraris and Nèmeth (2003) and further discussed by Ferraris et al. (2004), often the values of the third periodicity and β angle are such that $c\sin(\beta - 90) \sim a/n$ ($n = 3, 4, ...$) holds. This relation implies that a $[uvw]$ row with periodicity $c_o = nc\sin\beta$ and normal to the ab plane does exist. The supercell with parameters a, b and c_o is (pseudo)orthorhombic if $\alpha = 90°$ (monoclinic members) and (pseudo)monoclinic (angle $\alpha_m \neq 90°$) in the triclinic members with $\gamma \sim 90°$.

The occurrence of supercells favors twinning (Ferraris et al. 2004); besides, if the same supercell is shared by different members of the series, phenomena of syntactic intergrowth can occur, as reported by Nèmeth et al. (2005) for vuonnemite + epistolite + shkatulkalite and lomonosovite + murmanite. Note that, as described by Khomyakov (1995) and Pekov and Chukanov (2005), epitactic overgrowths are widespread among the minerals of Table 2, being favored by the common periodicities of their HOH layers.

Syntaxy and likely twinning are a main source of problems in solving and, at least, properly refining the structures of members of the bafertisite series, as discussed by Nèmeth et al. (2005) for epistolite and murmanite. In particular, these authors within a matrix of epistolite ($a = 5.455$, $b = 7.16$, $c = 12.14$ Å, $\alpha = 104.01$, $\beta = 95.89$, $\gamma = 90.03°$) have observed syntaxy of murmanite ($a = 5.387$, $b = 7.079$, $c = 11.74$ Å, $\alpha = 93.80$, $\beta = 97.93$, $\gamma = 90.00°$) and shkatulkalite ($a = 5.468$, $b = 7.18$, $c = 31.1$ Å, $\beta = 94.0°$). The following closely related supercells favor the syntaxy. Epistolite: $a = 5.455$, $b = 7.160$, $c = 93.728$ Å, $\alpha = 88.75$, $\beta = 90.57$, $\gamma = 90.03°$; murmanite: $a = 5.387$, $b = 7.079$, $c = 92.843$ Å, $\alpha = 89.47$, $\beta = 91.35$, $\gamma = 90.01°$; shkatulkalite: $a = 5.468$, $b = 7.18$, $c = 93.079$ Å, $\alpha = 90$, $\beta = 90.64$, $\gamma = 90°$.

An interesting case of intergrowth has been reported by Rastsvetaeva et al. (1991) for a sample of hejtmanite (Table 2), a species defined later by Vrána et al. (1992) with cell parameters

---

[5] Ferraris et al. (2001b), followed by Ferraris et al. (2004), called seidozerite-like the module of bornemanite with a seidozerite-like interlayer content. Note, instead, that in this module the topology of the HOH layer is that of $(HOH)_B$ and not of $(HOH)_V$ as in seidozerite (see also footnote 8).

(a and c exchanged): a = 10.698, b = 13.768, c = 11.748 Å, β = 112.27°. Rastsvetaeva et al. (1991) suggested that Sokolova et al. (1989b) failed in properly solving the structure of hejtmanite because the "single crystals" consisted of a syntaxy between two phases differing in their structures mainly for the position of Ba and the doubling of the a and b parameters. To justify the syntaxy Rastsvetaeva et al. (1991) used the following non-reduced cells: Phase (I) $P2_1/m$, a = 5.361, b = 6.906, c = 12.556 Å, β = 119.8° (the corresponding reduced cell is obtained by the transformation 100/010/101: a = 5.361, b = 6.906, c = 10.931 Å, β = 94.61°); Phase (II) Cm, a = 10.723, b = 13.812, c = 12.563 Å, β = 119.9° [the corresponding reduced cell used by Vrána et al. (1992) is obtained by the transformation 100/0$\bar{1}$0/$\bar{1}$0$\bar{1}$].

The same type of syntaxy described for hejtmanite likely occurs also in bafertisite, the Fe-equivalent of hejtmanite, whose crystal structure has been approximately solved both in space group $P2_1/m$ (a = 5.36, b = 6.80, c = 10.98 Å, β = 94°; Pen and Shen 1963) and Cm (a = 10.60, b = 13.64, c = 12.47 Å, β = 119.5°; the reduced cell obtained by $\bar{1}$00/0$\bar{1}$0/101 is a = 10.60, b = 13.64, c = 11.73 Å, β = 112.3°; Guan et al. 1963). On the basis of a single-crystal diffraction pattern Yang et al. (1999) proposed that the cells of bafertisite $P2_1/m$ (Pen and Shen 1963) and Cm (Guan et al. 1963) correspond to a sub and true cell, respectively. The intergrowth of two phases proposed by Rastsvetaeva et al. (1991) is supported by the impossibility of properly refining the crystal structures of bafertisite and hejtmanite even in the supposed "true" cell (space group Cm). In fact, according to the hypothesis of the intergrowth, a proper refinement is hindered in this cell because a part of the measured reflections are contributed by different intergrown individuals.

Porous features in heterophyllosilicates

As shown above, the crystal structures of the members of the bafertisite mero-plesiotype series are based on the alternating stacking of a *HOH* layer with an interlayer module, a situation to be compared with that occurring in the 2:1 phyllosilicates. The swelling capacity of the layer silicates, normally via adsorption of H_2O molecules, depends mainly from the negative charge of the *TOT* layer. If the charge is zero (talc, pyrophyllite) or higher than about 0.6 (vermiculites, micas), swelling is not possible or can be obtained only by special procedures; smectites (e.g., montmorillonite, saponite), which have a low (0.25–0.60) charge, typically swell. The capacity to swell of the smectites is the property that promotes the use of these clay minerals to prepare pillared clays (cf. Cool et al. 2002).

Pillaring. According to the IUPAC nomenclature (Schoonheydt et al. 1999), "pillaring is the process by which a layered compound is transformed into a thermally stable micro- and/or mesoporous material with retention of the layer structure. A pillared derivative is distinguished from an ordinary intercalate by virtue of intracrystalline porosity made possible by the lateral separation of the intercalated guest." A pillared material differs from an ordered porous material by having disordered channels. In fact, whereas pillared materials have periodically stacked layers, let say along [001], and their diffraction patterns show sharp 00*l* maxima, the interlayer pillared content is not stacked coherently.

The process of pillaring of a layered material includes the following steps (Fig. 9) whose details are not yet completely understood (Cool et al. 2002): swelling in a polar solvent (usually water); substitution of the original interlayer cations by bulky (inorganic/organic) cations (pillaring agent); washing (further chemical reactions occur at this step); calcination (pillars are formed).

Pillaring materials (e.g., pillared clays) show the following basic characteristics.

- The interlayer spacing increases from the original value of ~ 5 Å to at least ~15 Å.
- The layers do not collapse under calcination.

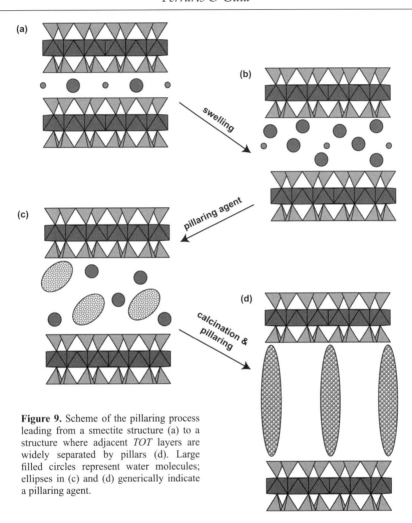

Figure 9. Scheme of the pillaring process leading from a smectite structure (a) to a structure where adjacent *TOT* layers are widely separated by pillars (d). Large filled circles represent water molecules; ellipses in (c) and (d) generically indicate a pillaring agent.

- The pillaring agent (inorganic/organic) is laterally spaced.
- The interlayer space is porous: at least N_2 molecules can go through, namely the pore has a window of about 3.2 Å.
- Not necessarily, pores and pillars are ordered; in fact, in a pillared material usually only the 00*l* reflections are sharp.

Leaching and solid state transformations[6]

As documented in this volume by Chukanov and Pekov (2005) and Pekov and Chukanov (2005), leaching of cations in natural conditions is quite common for the microporous

---

[6] Note that the wording "solid-state transformation" widely used in this chapter does not exclude the intervention of local dissolution and re-crystallization processes. "Solid state transformation" intends to emphasize that primary and secondary phases differ mainly for the interlayer contents in a way that, at least formally, is amenable to leaching/exchange processes. On the other side, the real nature of the processes of mineral substitutions in "solid state" at atomic level is currently matter of debate (cf. Putnis 2002).

framework heterosilicates that often occur with the heterophyllosilicates. As we shall see in this section, the phenomenon is documented also for the latter minerals. The mentioned authors affirm that ion exchange with the circulating solutions, during the post-crystallization steps in the course of the evolution of the hyperalkaline massifs, is the easiest way to adequate the stability of the crystal structure to the changing geological conditions (see also the Section "Transformation Minerals" below).

In a layered structure like those of the heterophyllosilicates of Table 2, deficiency of interlayer cations, in particular Na^+, can be expected and is not uncommon even in micas[7]. Such deficiency not necessarily is due to leaching; it can be there since the original crystallization process. A later loss of cations (decationization) by leaching must involve a re-adjustment in the structure and requires some exchange capacity with the surrounding medium. Whereas in phyllosilicates the non-tetrahedral cations of the *TOT* layer are tightly locked within the octahedral *O* sheet and typical leachable cations, like Na^+, are normally absent there (but see the recently described mica shirokshinite containing octahedral Na; Pekov et al. 2003), in the heterophyllosilicates the presence of Na^+ in the *O* sheet is common. Because of its large ionic radius, the introduction of this cation leads usually to some deformation of the *O* sheet, a feature that may favor leaching as exemplified below by delindeite.

Delindeite. The structure of delindeite (Fig. 10) offers an example of a solid-state modification involving leaching of Na from the *O* sheet. Ferraris et al. (2001b) proved that an Na deficiency affects sites in the *O* sheet of the *HOH* layer in delindeite (Table 2; $a = 5.327$, $b = 6.856$, $c = 21.51$ Å, $\beta = 93.80°$, $A2/m$), a quite unusual situation for a layer structure where the more weakly bonded cations reside in the interlayer. Unlike most compounds of Table 2, where the *O* sheet is occupied by typical octahedral cations as Fe, Mn and Mg, two independent Na sites occur in the *O* sheet of delindeite. One Na atom incorporates in its large and disordered coordination sphere also a bridging Si-O-Si oxygen atom that, in heterophyllosilicates, normally belongs to the *H* sheet only. In other words, in delindeite a basal oxygen atom of a silicon tetrahedron is captured within the large coordination sphere of Na. Thus, the Na^+ cation "sees" directly the interlayer space through a window of the *H* sheet and some exchange with the surrounding medium becomes feasible. A statistical distribution of the captured oxygen on two independent positions avoids the strain of an O···O edge shared between *H* and *O* sheets.

A similar coordination to a bridging oxygen atom belonging to the *H* sheet has been reported in the structure of seidozerite (Pushcharovsky et al. 2002)[8]

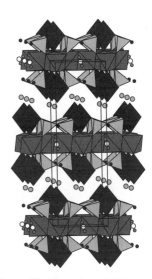

Figure 10. View along [100] of the crystal structure of delindeite. The black circles represent the disordered bridging oxygen atom that is within the coordination sphere of a Na^+ cation (blank circles) belonging to the *O* sheet (see text). Grey circles represent interlayer cations.

---

[7] A murmanite-related phase that is almost devoid of interlayer content has been found in Lovozero massif by N.V. Chukanov (personal communication).

[8] The inclusion of a bridging oxygen atom within the coordination sphere of Na belonging to the *O* sheet introduces a further peculiarity in the *HOH* layer of seidozerite. Consequently, contrary to what reported in previous literature, in this chapter statements like "seidozerite derivatives" or "seidozerite-like *HOH* layer" for the heterophyllosilicates of Table 2 are avoided (see also footnotes 5 in the text and *g* in Table 2).]

and can be compared with situations found in members of the rhodesite group described below. Na$^+$ vacancies in the O sheet require a charge balance that is provided by the O^{2-} → OH substitution, like in steenstrupine-(Ce) (Makovicky and Karup-Møller 1981).

Transformation minerals. There are consistent evidences of solid-state transformation from one to another member of Table 2 via leaching/substitution of the interlayer composition. In some cases, swelling processes become evident by comparing the structures of the parent and daughter phases (see below). As summarized by Khomyakov (1995) on the basis of his previous work (cf. also Chukanov and Pekov 2005 and Pekov and Chukanov 2005), an active interaction with water is a characteristic of many highly alkaline titanosilicates like the heterophyllosilicates of Table 2. Several stage-by-stage replacements of anhydrous titanosilicates by their hydrated and decationated analogues are described in the quoted literature, including the following two reactions that are of interest in this chapter: lomonosovite + H$_2$O → murmanite + Na$_3$PO$_4$; vuonnemite + H$_2$O → epistolite + Na$_3$PO$_4$ (+shkatulkalite?).

Often the transformation is topotactic as witnessed by the occurrence of oriented intergrowths and pseudomorphs after the parent phase. Based on electron and X-ray diffraction studies, intergrowths of epistolite and shkatulkalite (transformation phases) with murmanite (primary phase) are reported by Nèmeth et al. (2005); Sokolova and Hawthorne (2001) report intergrowths of lomonosovite (primary phase) with the transformation phases quadruphite, polyphite and sobolevite, all containing PO$_4$ groups that share a corner with the octahedron of the X cation (Table 2). Pekov and Chukanov (2005) report the following pseudomorphs of framework heterosilicates after members of the bafertisite series: labuntsovite-group minerals after lomonosovite, vuonnemite, astrophyllite and lamprophyllite; zorite and Na-komarovite after vuonnemite; sitinatikite and narsarsukite after lomonosovite; vinogradovite and kukisvumite after lamprophyllite. According to Khomyakov (1995), the new phases, that have been called transformation minerals, can be formed only by transforming primary minerals that instead crystallize from melts or fluids.

Khomyakov (1995) wrote that for some phases the transformation is active at atmospheric conditions ("Under natural atmospheric conditions anhydrous peralkaline titanosilicates undergo a spontaneous transition to their low-alkaline hydrogen-bearing analogues. This transition may be described as a reaction between the sodium-supersaturated solid phase and the atmospheric water."). Other phases, as it happens for the mentioned transformations of lomonosovite and vuonnemite, "which are stable under atmospheric conditions, are readily hydrated in the epithermal and hypergene processes."

Following Khomyakov's suggestion (cf. Khomyakov 1995), in Russian literature, the often-observed preservation of structure modules through the transformation from primary to secondary phases (usually via a topotactic reaction) is known as *inheritance principle*; the minerals involved in the steps of a transformation are said to form an *evolutionary series*, a concept useful in understanding the mineral associations and the geological evolution of the hyperalkaline formations (cf. Pekov and Chukanov 2005). For example, in the Ilímaussaq peralkaline massif the secondary phases epistolite and murmanite have been first discovered; only later, following the hypothesis on transformations reported above, the corresponding primary phases vuonnemite and lomonosovite have been identified.

Transformation minerals after vuonnemite and lomonosovite. Vuonnemite (Fig. 6) and lomonosovite (Fig. 11; Table 2) are primary (parent) minerals that, as written above, by leaching of their PO$_4$ groups and hydration transform into secondary (daughter) phases (Figs. 11 and 12). The solid-state nature of these transformations (but see footnote 6) is supported by abundant findings of pseudomorphs of the secondary phases after the primary ones and a wide occurrence of oriented intergrowths between them, as recently proved

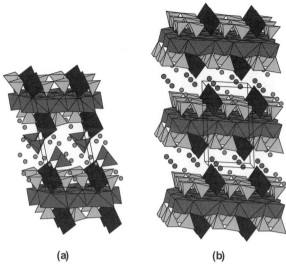

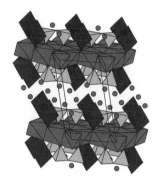

(a) (b)

Figure 11. View along [100] of the crystal structure of the primary phase lomonosovite (a) that is transformed into the secondary phase murmanite (b) via leaching of PO$_4$ groups (dark grey tetrahedra) and hydration. Circles as in Figure 7.

Figure 12. View along [100] of the crystal structure of epistolite, a secondary phases derived from vuonnemite (Fig. 6b) via leaching of PO$_4$ groups and hydration (large grey circles).

by electron and single-crystal X-ray diffraction studies (Nèmeth et al. 2005). Note that the occurrence of intergrowths involving primary and secondary phases and, if the case, other phases is considered a main source of difficulties in properly solving and refining the corresponding structures [cf. vuonnemite + epistolite + shkatulkalite in Nèmeth et al. (2005)]. Because of the loosing of PO$_4$, the width of the interlayer is smaller in the secondary phases, but the swelling contribution of hydration is evident.

Transformation minerals after bornemanite. Khomyakov (1995) reported that the mineral labeled M72 is pseudomorph after bornemanite (Table 2). Khomyakov (2004) showed that bornemanite exchange Na$_3$(PO$_4$) for H$_2$O when treated with boiling water.

As mentioned above, by electron (SAED) and X-ray (powder) diffraction, Nèmeth et al. (2004a) have shown that M72 consists of two polytypes and modeled their structures (Fig. 8). The swelling action of the transformation is evident in the separation of the two X octahedra, which in bornemanite share an edge, and by the overall increase of the c parameter (Table 2) in spite of the fact that PO$_4$ groups are leached.

A similar transformation is observed for M73 (Nèmeth 2004); the parent phase of M73 is not exactly identified but likely is a bornemanite-type phase. Again, two polytypes have been identified and their structure modeled. The two polytypes of M73 differ from those of M72 mainly for a different arrangement of the two interlayer contents (Fig. 7). The X cation is present with coordination numbers 5 and 6 both in M72 and M73. Interesting to note that, contrary to M72, M73 shows the widest interlayer between two H sheets with X in coordination 5.

Note that in some cases the coordination number of the X cation decreases from six in the parent phase to five in the daughter phase (Figs. 8, 12). That may be related either to the removal of PO$_4$ groups that share an oxygen atom with the X cation, or to the separation of X octahedra that were locked together in the primary phase.

Pillared derivatives?

We have mentioned that the *TOT* layers of smectites, which bear an intermediate negative charge, are the best starting material to prepare pillared clays because of their easy swelling, a property observed in nature and exploited in laboratory. Above it has been shown that processes able to modify the interlayer content of the heterophyllosilicates via solid-state reactions are active in nature. In particular, as recently discussed by Ferraris (2004) and Nèmeth et al. (2004b), the secondary phases that are formed via the following transformation seem to be suitable candidates for investigating swelling properties in view of preparing pillared materials based on *HOH* layers: vuonnemite → epistolite (plus shkatulkalite?); lomonosovite → murmanite plus an unidentified phase reported by Semenov et al. (1962); bornemanite → M72; bornemanite-type → M73.

So far, there is no record in the literature of the use of heterophyllosilicates for obtaining pillaring porous materials. Even the knowledge on the exchange capacity of these minerals leading to secondary phases is acquired mainly from field observations, being experimental investigation still scarce (but see Chelishchev 1972).

Reasons of the limited experimental investigation can be a limited knowledge of the heterophyllosilicates among materials scientists, and difficulties in synthesizing these compounds; a step, this one, which for sure is needed because the minerals listed in Table 2 are generally rare and show a complex chemical composition. The only *ab initio* synthetic bafertisite-type compound reported in the literature is an anhydrous vanadate, $Na_8\{(Na,Ti)_4[Ti_2O_2Si_4O_{14}]O_2\}(VO_4)_2$, that corresponds to the phosphate lomonosovite and has been prepared by crystallization from a melt (Massa et al. 2000). It can be mentioned also that by heating a mixture of natural lamprophyllite and nepheline at about 900°C, Zaitsev et al. (2004) have obtained new formed lamprophyllite with a composition different from that of the starting sample. However, by analogy with the 2:1 phyllosilicate smectites, one may conclude that the hydrated phases of heterophyllosilicates look as the best candidates to attempt swelling in these titanosilicates.

According to Khomyakov (1995) (cf. Pekov and Chukanov 2005), the secondary hydrated phases of heterophyllosilicates are likely to be obtained only via solid-state transformation, for sure an obstacle towards synthesis. An alternative is finding conditions to swell directly primary anhydrous phases as it has been possible with micas (cf. Cool et al. 2002). Recently, Ustinov and Ul'yanov (1999) have obtained murmanite by hydrothermal transformation of lomonosovite at 100°C. Khomyakov (2004) obtained murmanite and epistolite from, in the order, lomonosovite and vuonnemite treated with boiling water; as mentioned above, under the same conditions he showed also that "beta-lomonosovite" and bornemanite exchange $Na_3(PO_4)$ and Na for H_2O.

Some inspiration on the synthesis of the secondary hydrated phases of heterophyllosilicates might be found in the variety of methods utilized for the synthesis of smectites (cf. Güven 1988).

PALYSEPIOLES

Sepiolite, palygorskite and related structures

Sepiolite, ideally $Mg_8[Si_{12}O_{30}](OH)_4 \cdot 12H_2O$, and palygorskite (also known as attapulgite), ideally $Mg_5[Si_8O_{20}](OH)_2 \cdot 8H_2O$, are important clay minerals. Their crystal structures are known only through powder diffraction data. Sepiolite is orthorhombic, *Pncn*, with cell parameters $a = 13.40$, $b = 26.80$, $c = 5.28$ Å (Brauner and Preisinger 1956). For palygorskite two polytypes are known (Artioli and Galli 1994); Chiari et al. (2003) report the following crystal data for the two polytypes: $C2/m$, a = 13.337, b = 17.879, c = 5.264 Å, β = 105.27°; $Pbmn$, a = 12.672, b = 17.875, c = 5.236 Å.

The structures of sepiolite and palygorskite are based on a framework of chessboard connected TOT ribbons. These ribbons develop along [001] and correspond to cuts with different width of a phyllosilicate 2:1 layer; they delimit [001] channels (Fig. 13). In the [010] direction, the $(TOT)_S$ ribbon of sepiolite is one chain wider than that, $(TOT)_P$, of palygorskite. This feature requires for sepiolite a b value about 9 Å longer than that of palygorskite, i.e., about 4.5 Å per added T chain. The minimum effective width of the channel windows in palygorskite and sepiolite is the same (about 4.5 Å). The frameworks of sepiolite and palygorskite can be seen also as modulated TOT silicate layers showing a waving T sheet and a discontinuous O sheet (Guggenheim and Eggleton 1988).

By refinement carried out with neutron-diffraction data collected on a deuterated powder sample of palygorskite, Giustetto and Chiari (2004) showed that only highly disordered zeolitic H$_2$O is present in the channels of the structure. The same evidence is obtained by ^{1}H NMR investigation on dehydrated and rehydrated palygorskite (Kuang et al. 2004). Based on a critical review of compositional data available in literature, Galan and Carretero (1999) proved that whereas sepiolite is a true trioctahedral mineral, palygorskite is intermediate between di- and trioctahedral composition because some trivalent cations (e.g., Al) always occur in the octahedral sites. The intermediate di- trioctahedral character of palygorskite is confirmed by IR spectroscopy (Chahi et al. 2002). A recent IR investigation (García Romero et al. 2004) reveals absence of trioctahedral Mg in the Mg-richest (3.11 apfu) sample of palygorskite so far known.

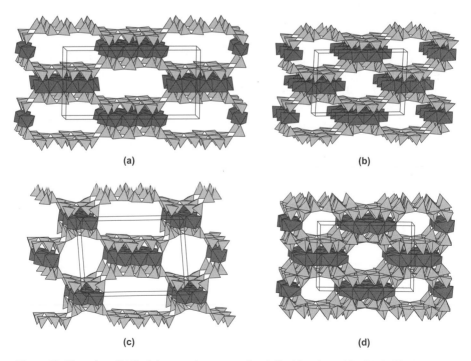

Figure 13. View along [001] of the crystal structure of sepiolite (a), palygorskite (b), kalifersite (c) and intersilite (d); a vertical axis. The content of the [001] channels is not drawn.

Kalifersite – a hybrid between sepiolite and palygorskite. Kalifersite, $(K,Na)_5(Fe^{3+})_7$ $[Si_{20}O_{50}](OH)_6 \cdot 12H_2O$, is a rare [001] fibrous silicate found in a hydrothermally altered pegmatite at Mt. Kukisvumchorr (Khibina alkaline massif, Kola Peninsula, Russia). It is triclinic ($a = 14.86$, $b = 20.54$, $c = 5.29$ Å, $\alpha = 95.6$, $\beta = 92.3$, $\gamma = 94.4°$, $P\bar{1}$); Ferraris et al. (1998) succeeded to model its crystal structure (Fig. 13) after realizing a modular relationship with sepiolite and palygorskite as follows.

(i) Kalifersite, sepiolite and palygorskite have close values of their a and c parameters; the latter corresponds to the fibrous direction of these silicates and to the periodicity of a pyroxene chain.

(ii) The b value of kalifersite is intermediate between that of palygorskite and sepiolite.

(iii) The $[Si_{20}O_{50}](OH)_6$ silicate anion of kalifersite corresponds to the sum of those of sepiolite, $[Si_{12}O_{30}](OH)_4$, and palygorskite, $[Si_8O_{20}](OH)_2$.

(iv) Martin-Vivaldi and Linares-Gonzales (1962) have interpreted X-ray powder diffraction patterns intermediate between those of palygorskite and sepiolite as random intergrowths of $(TOT)_S$ and $(TOT)_P$ ribbons.

Taking into account the above chemical and crystallographic aspects, a structure model for kalifersite based on a 1:1 chessboard arrangement of $(TOT)_P$ and $(TOT)_S$ [001] ribbons and the filling of the channels with alkalis and water molecules was obtained (Fig. 13). Due to the longer a parameter, the minimum effective widths (~ 5.4 and ~ 6.2 Å) of the two types of channels in kalifersite are larger than the corresponding ones in palygorskite and sepiolite (~ 4.5 Å).

The polysomatic series of palysepioles. Palygorskite (P), and sepiolite (S) are the end members of the *palysepiole* (*paly*gorskite + *sepiol*ite) polysomatic series P_pS_s defined by Ferraris et al. (1998); kalifersite is the P_1S_1 member. Falcondoite (Springer 1976) and loughlinite (Fahey et al. 1960) differ from sepiolite only in the composition of the octahedral part; the same situation is valid for yofortierite (Perrault et al. 1975) and tuperssuatsiaite (Cámara et al. 2002) in comparison to palygorskite. Chukanova et al. (2002) have described an orthorhombic unnamed mineral from Mt. Flora (Lovozero massif) with composition $(Fe,Mn)_4Si_6O_{15}(OH)_2 \cdot nH_2O$ and cell parameters $a = 13.53$, $b = 26.70$, $c = 5.13$ Å; it corresponds to a Fe-dominant sepiolite.

Merotypes and plesiotypes of palysepioles

Raite – a structure based on a palygorskite framework. The titanosilicate raite, $Na_3Mn_3Ti_{0.25}[Si_8O_{20}](OH)_2 \cdot 10H_2O$, is a rare silicate found in the Yubileinaya pegmatite at Mt. Karnasurt (Lovozero alkaline massif, Kola Peninsula, Russia). Raite is monoclinic ($C2/m$, $a = 15.1$, $b = 17.6$, $c = 5.290$ Å, $\beta = 100.5°$; Pushcharovsky et al. 1999) and its crystal structure consists of a palygorskite-like framework, but the channel content differs substantially from that of palygorskite both in chemistry and structure. Practically, in raite the O sheet is not interrupted even if a part of the octahedra that reside in the channels are 3/4 vacant and only 1/4 occupied by Ti. Thus, raite and palygorskite share only the TOT building module and are therefore in merotypic relationship.

Intersilite. This rare titanosilicate occurs at Mt. Alluaiv (Lovozero alkaline massif, Kola Peninsula, Russia) and has chemical composition $(Na,K)Mn(Ti,Nb)Na_5(O,OH)(OH)_2[Si_{10}O_{23}(O,OH)_2] \cdot 4H_2O$. In its structure ($a = 13.033$, $b = 18.717$, $c = 12.264$ Å, $\beta = 99.62°$, $I2/m$; Yamnova et al. 1996), sepiolite-like ribbons partially overlap along [010] because of tetrahedral inversions within the same ribbon (Fig. 13). Due to the substantial modification of the sepiolite framework, intersilite is in plesiotype relationship with the palysepiole polysomatic series. The minimum effective width of the channels is ~ 3.5 Å.

Microporous features of palysepioles

Modern chemical (Galan and Carretero 1999) and structural (Giustetto and Chiari 2004) analyses prove that only H_2O molecules reside in the structural channels of the clay minerals sepiolite and palygorskite. The recent finding of kalifersite and raite shows that in nature, under appropriate conditions, the channels of sepiolite- and palygorskite-like frameworks can be almost fully filled with M^{n+} cations (n =1 to 4), including Ti; in these structures the "octahedral" O sheet is no longer discontinuous. The occurrence of Ti in raite realizes a bridge between palysepioles and titanosilicates.

In nature, interactions at nanoscale between the clay minerals palygorskite and sepiolite and the surrounding medium are well known, even if not always it is clear if they are related to adsorption via the wide free surface of the particles, to absorption in the interlayer or to both mechanisms (cf. Velde 1992). This is the case, for example, of the adsorption of humic acid by palygorskite and sepiolite (Singer and Huang 1989) and the curious association of quinone pigments with Eocene sepiolite known as quincyte (Louis et al. 1968; Prowse et al. 1991).

Pre-tech era applications. As summarized by Chiari et al. (2003) in the introduction of their paper on the interaction between the host palygorskite framework and the guest indigo, Maya Blue is a characteristic pigment produced by the Mayas around the VIII century AD; its color ranges from a bright turquoise to a dark greenish blue. It has been shown that the pigment contains palygorskite (in some cases, minor sepiolite) and its color is due to the presence of indigo that the Mayas obtained from a vegetable. Experimentally, it has been shown that the simple technique required to prepare the pigment, namely mixing palygorskite and *Indigofera suffruticosa* then filtering after heating at about 100°C, was accessible to the Mayas.

Sepiolite has been used for hundreds of years in southern Spain for the purification of wine (Galan and Ferrero 1982).

Modern technological applications. The exploitation of the capacity of palygorskite and sepiolite to absorb organic molecules already known to Mayas is a wide practice in modern technology. However, not always the structural channels are the only sites active in sorption, surface mechanisms being active (cf. Shariatmadari et al. 1999). To the list of applications reviewed by Jones and Galan (1988) and Galan (1996), only some quotations of recent results are here added. These include: sorption of heavy metals from industrial waste water (cf. Garcia Sanchez et al. 1999); removal of dyes from tannery waste waters (cf. Espantaléon et al. 2003); stabilization of dyes and pesticides (cf. Casal et al. 2001; Rytwo et al. 2002); template synthesis of carbon nanofibers (Fernandez-Saavedra et al. 2004); preparation of Ni and Pd catalysts (Anderson and Galan-Fereres 1999; Corma et al. 2004).

Future directions of research. No technological uses are known for the other palysepioles and related structures mentioned above. Actually, all of them are rare minerals and have been discovered only recently. To test their properties it would be necessary to have some quantity of synthetic material. Presumably, the synthesis is feasible by hydrothermal method as done for sepiolite (Mizutani et al. 1991).

Actually, only kalifersite is based on a guest framework substantially differing from those of both palygorskite and sepiolite. In fact, kalifersite incorporates in its structure the two types of channels that occur singly in sepiolite and palygorskite; thus, in principle, the composite silicate framework of kalifersite could display at the same time the absorption features of the two clay minerals. Other structures described in this section incorporate the high-charge cation Ti^{4+}, an element well known for its catalytic activity. Likely, palysepiole frameworks based on ribbons wider than those occurring in sepiolite could be synthesized, thus realizing pores with at least one larger dimension of the window. In fact, the bridges that connect the segments of inverted tetrahedra limit the second dimension of the window.

Some porous structures related to palysepioles

Ferraris et al. (2004) pointed out that the crystal structure of the lithium silicate silinaite (NaLiSi$_2$O$_5$·2H$_2$O; C2/c, a = 14.383, b = 8.334, c = 5.061 Å, β = 96.6°; Grice 1991) shows a chessboard arrangement of narrow *TOT*-like ribbons comparable to those of palysepioles such that ten-membered channels filled by Na and H$_2$O are formed (Fig. 14). Whereas in palygorskite and sepiolite the inversion of the tetrahedral sheet is every four and six tetrahedra, respectively, in silinaite the same inversion is every two tetrahedra. Besides, the *O* part of the silinaite *TOT*-like ribbon consists of Li tetrahedra instead of octahedra as in palysepioles and, in general, in the phyllosilicates. The ten-membered channels are delimited by eight Si tetrahedra and two Li tetrahedra. The minimum effective width of these channels is about 5.5 Å.

Ten-membered channels (Fig. 14) occur also in lintisite [Na$_3$LiTi$_2$(Si$_2$O$_6$)$_2$O$_2$·2H$_2$O; Merlino et al. 1990], a mineral mentioned in this volume by Chukanov and Pekov (2005) together with other titanosilicates, which Ferraris et al. (2004) reported as members of a merotype series based on three structural modules that are typical of the following phases: silinaite, lorenzenite, Na$_4$Ti$_4$(Si$_2$O$_6$)$_2$O$_2$, and a hypothetical *Z* zeolite, (Na,K)Si$_3$AlO$_8$·2H$_2$O. Whereas the channels of silinaite consist of eight Si tetrahedra and two Li tetrahedra, those of lintisite have four Si tetrahedra replaced by Ti octahedra. The minimum effective width of the channels in lintisite is slightly smaller than in silinaite.

SEIDITE-(Ce) AND RELATED STRUCTURES

The silicate double layer with eight-membered channels that occurs in the crystal structure of the titanosilicate seidite-(Ce) (Fig. 15) is present also in a group of well known microporous structures reported in this volume by Rocha and Lin (2005). In this chapter we refer to the frameworks of these heterosilicates as the rhodesite-type structure; this structure type is compared with the structure of seidite-(Ce) and their modular aspects are emphasized (cf. Ferraris et al. 2004).

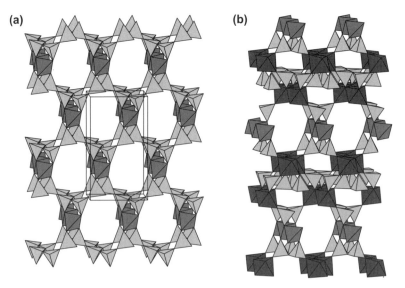

Figure 14. View of the crystal structure of silinaite, along [100] (a), and lintisite along [001] (b; *a* vertical axis). The dark grey tetrahedra are centered by Li. The content of the channels is not shown.

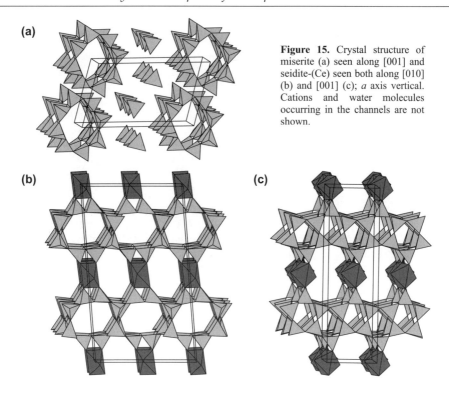

Figure 15. Crystal structure of miserite (a) seen along [001] and seidite-(Ce) seen both along [010] (b) and [001] (c); *a* axis vertical. Cations and water molecules occurring in the channels are not shown.

Modeling the structure of seidite-(Ce)

Seidite-(Ce), $Na_4(Ce,Sr)_2\{Ti(OH)_2(Si_8O_{18})\}(O,OH,F)_4 \cdot 5H_2O$, is a titanosilicate found in the Yubileynaya pegmatite at Mt. Karnasurt (Lovozero alkaline massif, Kola Peninsula, Russia) and first described by Khomyakov et al. (1998). Because of its disordered [010] fibrous morphology, single crystals suitable for X-ray crystallography are not available and its crystal structure (a = 24.61, b = 7.23, c = 14.53 Å, β = 94.6°; $C2/c$; Fig. 15) was modeled by Ferraris et al. (2003) on the basis of electron (SAED) and X-ray powder diffraction data. Even if finally the structure of seidite-(Ce) turned out to share its silicate module with rhodesite $\{K_2Ca_4[Si_8O_{18}(OH)]_2 \cdot 12H_2O$; a = 23.416, b = 6.555, c = 7.050 Å, $Pmam$; Hesse et al. 1992$\}$ and related structures (see below), Ferraris et al. (2003) have got the key for its modeling after comparison with the structure of miserite (Scott 1976). The latter structure shows microporous features (Fig. 15), being about 2.8 Å the minimum effective width of its isolated eight-membered channels.

Ferraris et al. (2003) noted that the cell parameters of miserite (*m*) [$KCa_5(Si_2O_7)$ $(Si_6O_{15})(OH)F$; a = 10.100, b = 16.014, c = 7.377 Å, α = 96.41, β = 111.15, γ = 76.57°; $P\bar{1}$] and seidite-(Ce) (*s*) are related as follows: $a_s \cong 2a_m$, $c_s \cong b_m$, $b_s \cong c_m$. Furthermore, the oblique (010) lattice net of miserite can be described as based on a centered rectangular cell with parameters very close to those of the centered (001) net of seidite-(Ce). It was also noted that formally the same $(Si_8O_{22})^{12-}$ silicate anion occurs in both minerals; instead, the number of non-silicate cations is 6 in miserite and 7 in seidite-(Ce), the chemically most important difference being the presence of the high-charge Ti^{4+} cation in the latter mineral. By rotation of the isolated Si_2O_7 groups that alternate with [001] eight-membered silicate channels in the structure of miserite, the (100) silicate layer consisting of interconnected channels typical of seidite-(Ce) is obtained

(Fig. 15). In their turn, the (100) silicate layers are connected by isolated Ti octahedra, thus obtaining a three-dimensional mixed tetrahedral/octahedral framework.

The (100) layer of the eight-membered channels corresponds to the double silicate layer that occurs in the microporous mineral rhodesite (Hesse et al. 1992) and related compounds. A xonotlite-like [$Ca_6Si_6O_{17}(OH)_2$; $P2/a$, a = 17.032, b = 7.363, c = 7.012 Å, β = 90.36°; Hejny and Armbruster 2001] chain, i.e., a double wollastonite-like chain, delimits the (010) front windows of the double layer. As discussed below, both the upper and lower (100) walls of the silicate double layer consist of an apophyllite-type [$KCa_4(Si_8O_{20})F \cdot 8H_2O$; $P4/mnc$, a = 8.965, b = 8.965, c = 15.768 Å; Colville et al. 1971] net that includes eight- and four-membered rings. Being the upper (100) wall staggered by $b/2$ relative to the lower wall, the window rings of the two walls are not aligned, and the [100] direction does not show pores. A second type of eight-membered channels crosses the structure of seidite-(Ce) along [010]; in these channels, two out of eight polyhedra forming the window are Ti octahedra.

A further double system of channels extends along [001] (Fig. 15). One set of channels of this system consists of ten-membered pores with two Ti octahedra as ring members; the second set corresponds to five-membered channels delimited by tetrahedra only. Each set forms alternating (100) layers. The front windows of the five-membered channels correspond to the (001) wall of the [010] eight-membered silicate channels.

The wollastonite chains that develop along [010] and [001] determine the periodicity of both sets of channels. Note that the [001] five-membered rings of seidite-(Ce) correspond to the eight-membered [010] rings in rhodesite (Fig. 16), the different cross-section being related to the $b/2$ stagger of the (100) apophyllite-type sheets mentioned above.

Microporous properties

The first experimental evidence of microporous features in seidite-(Ce) was the high value of its measured density, 3.21 g/cm³ as determined by Clerici liquid [an aqueous solution of $CH_2(COO)_2Tl_2 \cdot HCOOTl$] vs. the calculated value of 2.75 g/cm³. This high value was due to absorption of Tl, as confirmed by an electron microprobe analysis of grains of the mineral

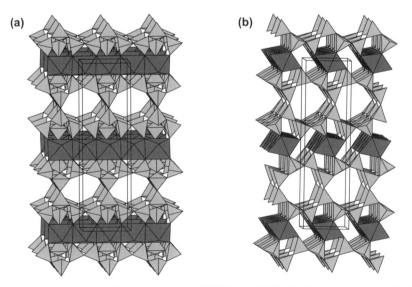

Figure 16. Crystal structure of rhodesite seen along [010] (a) and [001] (b). Cations and water molecules occurring in the channels are not shown; a axis vertical.

that had been kept in aqueous solution of Tl^+, K^+, Rb^+, Cs^+ and Ba^{2+} at room temperature for periods of 6–8 weeks: Na was substituted by the shown cations (Khomyakov et al. 1998).

The presence of channels in the structure and weak bonding of guest Na to the host framework clearly support the ion-exchange properties of seidite-(Ce). Presumably, all the extra-framework ions and H_2O groups are guests that can be removed from the host structure through the described system of pores. The minimum and maximum effective widths of the two types of eight-membered channels described above are about 3 and 5 Å, respectively; the same widths are about 1 and 6.5 Å in the ten-membered channels, the bottleneck corresponding to two facing Ti octahedra (Fig. 15).

Modular aspects

The (100) layer consisting of eight-membered channels that occur in seidite-(Ce) (Fig. 15) is well known as a double layer of corner-sharing Si tetrahedra, and is present in the structures listed in Table 3 (cf. Rocha and Lin 2005). In all these compounds, the silicate double layer alternates with a sheet of cations that mainly show an octahedral coordination.

Variety of tetrahedral and "octahedral" sheets. The tetrahedral sheets forming the double layers are comparable with the so-called apophyllite sheet, which consists of a net of four-membered rings connected to form eight-membered rings. However, taking into account the orientation of the tetrahedra in the four-membered rings, three types of tetrahedral sheets can be distinguished (Fig. 17): all tetrahedra pointing either upwards or downwards (apophyllite); three tetrahedra pointing upwards and one downwards [all the compounds of Table 3 except seidite-(Ce)]; two tetrahedra pointing upwards and two downwards [seidite-(Ce)]. Cavansite [$Ca(VO)Si_4O_{10}\cdot 4H_2O$; *Pcmn*, $a = 9.792$, $b = 13.644$, $c = 9.629$ Å; Evans 1973] has a seidite-(Ce)-like tetrahedral sheet but the two tetrahedra pointing in the same direction share a corner instead of occupying opposite sides of a four-membered ring.

Five types of "octahedral" sheets occur in the compounds of Table 3 (Fig. 18).

- Isolated Ti octahedra in seidite-(Ce) (Fig. 15).
- Chains of edge-sharing Ca octahedra in rhodesite (Figs. 16, 18a), macdonaldite and hydrodelhayelite. This sheet corresponds to a trioctahedral sheet where one out of two chains of octahedra is missing.

Table 3. Members of the rhodesite mero-plesiotype series.

Name	Chemical formula	a, b, c (Å), β (°)	S.G.	Ref.
Seidite-(Ce)	$Na_4(Ce,Sr)_2\{Ti(OH)_2(Si_8O_{18})\}$ $(O,OH,F)_4\cdot 5H_2O$	24.61, 7.23, 14.53, 94.6	$C2/c$	(1)
Rhodesite	$K_2Ca_4[Si_8O_{18}(OH)]_2\cdot 12H_2O$	23.416, 6.555, 7.050	*Pmam*	(2)
Macdonaldite	$BaCa_4[Si_8O_{18}(OH)]_2\cdot 10H_2O$	14.081, 13.109, 23.560	*Cmcm*	(3)
Delhayelite	$K_7Na_3Ca_5[Si_7AlO_{19}]_2F_4Cl_2$	24.86, 7.07, 6.53	*Pmmn*	(4)
Hydrodelhayelite	$K_2Ca_4[Si_7AlO_{17}(OH)_2]_2\cdot 6H_2O$	6.648, 23.846, 7.073	$Pnm2_1$	(5)
Monteregianite-(Y)	$K_2Na_4Y_2[Si_8O_{19}]_2\cdot 10H_2O$ *	9.512, 23.956, 9.617, 93.85	$P2_1/n$	(6)
AV-9	$K_2Na_4Eu_2[Si_8O_{19}]_2\cdot 10H_2O$ **	23.973, 14.040, 6.567, 90.35	$C2/m$	(7)

* $K_2Na_4Ce_2[Si_8O_{19}]_2\cdot 10H_2O$ (AV-5 in Rocha et al. 2000) is isostructural with monteregianite-(Y).
** Isostructural compounds with Tb (Ananias et al. 2001), Er (Ananias et al. 2004) and Nd (Rocha et al. 2004) substituting Eu are known.
References: (1) Ferraris et al. 2003; (2) Hesse et al. 1992; (3) Cannillo et al. 1968; (4) Cannillo et al. 1970; (5) Ragimov et al. 1980; (6) Ghose et al. 1987; (7) Ananias et al. 2001

- Chains of alternating Y and Na octahedra connected by pairs of further Na octahedra, to form a trioctahedral sheet with vacancies, in monteregianite-(Y) and its isostructural compound with Ce substituting Y (Fig. 18b).
- A trioctahedral-like sheet in delhayelite (Fig. 18c), where edge-sharing chains like those of rhodesite alternate with chains of eight-coordinated Na-polyhedra.
- A modification of the sheet of monteregianite-(Y) in the AV-9 compounds (Fig. 18d), where the coordination number of the inter-chain Na-polyhedra is 9. This type of sheet requires that two H_2O molecules per formula unit, which in other structures of the group reside in the channels, move to the coordination sphere of Na within the "octahedral" sheet.

The Na atoms that belong to the "octahedral" sheet and have coordination number higher than six also bind to bridging oxygen atoms of the silicate double layer. Thus, these Na atoms are in contact with the channels and leaching can be favored, as it has been described above for the heterophyllosilicate delindeite.

The rhodesite mero-plesiotype series. According to the categorization of modular series given by Makovicky (1997) (cf. Ferraris et al. 2004), the compounds of Table 3 form a mero-plesiotype series (cf. footnote 1). In fact, nearly the same double silicate layer occurs in all members and alternates with a variable module (merotype aspect). The series is also plesiotypic because, as seen above, the silicate double layer may show a different ratio between upwards and downwards pointing tetrahedra. This aspect is related to the number, charge and coordination number of the cations in the interlayer sheet and to the Si/Al ratio within the layer. Most of the compounds in Table 3 have a T:O ratio of 8:19 (T = Si, Al), corresponding to a 3:1 ratio between the tetrahedra sharing three and four corners, respectively. In seidite-(Ce), the ratio between the two types of tetrahedra is 1:1, consequently T:O = 8:18; the different connectivity of the tetrahedra is related to the relative shift between the two apophyllite-type sheets forming the double silicate layer. As mentioned above, depending on this shift, the lateral view of the double layer shows either five- (Fig. 15) or eight-membered rings (Fig. 16).

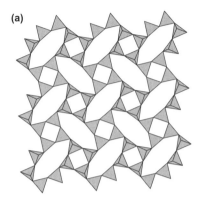

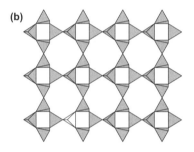

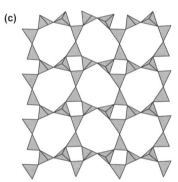

Figure 17. Tetrahedral sheets in apophyllite (a), rhodesite (b) and seidite-(Ce) (c).

Further modular interpretation. The porous structures of the rhodesite series and those related to this series (see next section) can be described as based on *OTT* layers and compared with the 1:1 (*TO*) and 2:1 (*TOT*) layer silicates and the heterophyllosilicates. The two facing

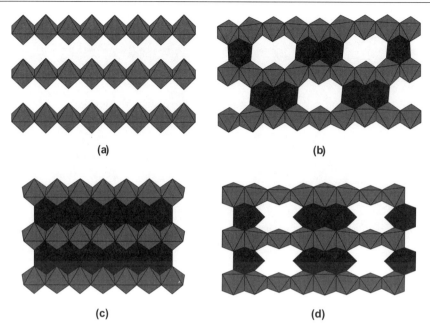

Figure 18. "Octahedral" sheets in rhodesite (a), monteregianite-(Y) (b), delhayelaite (c) and AV-9 (d). The Na polyhedra (dark grey) joining parallel rows of edge-sharing octahedra (light grey) have coordination number 6, 8, and 9 in (b), (c) and (d), respectively.

T sheets, which present two types of rings instead of one only as in the 1:1 and 2:1 silicates, are joined by sharing apical corners of tetrahedra to form the porous TT layer. The formal O sheet is by far more variable than in the layer silicates and connects two adjacent TT sheets; thus, a three-dimensional framework is formed without weaker bonds between the formal OTT layers.

Structures related to the rhodesite series

Reyerite and fedorite. These two minerals belong to the group of gyrolite, $Ca_{16}NaSi_{23}$ $AlO_{60}(OH)_8 \cdot 14H_2O$, which is extensively described in this volume by Bonaccorsi and Merlino (2005), and are members of a merotype series (cf. Ferraris et al. 2004). A double tetrahedral layer consisting of eight-membered rings and comparable to that described for the rhodesite group occurs in reyerite, $Ca_{14}(Na,K)_2Si_{22}Al_2O_{58}(OH)_8 \cdot 6H_2O$, and fedorite, $[K_2(Ca_5Na_2)$ $Si_{16}O_{38}(OH,F)_2 \cdot H_2O$, (Fig. 19). The minimum effective channel width is about 3 Å in both structures. In fedorite, the double layer alternates with a trioctahedral sheet of Ca octahedra, thus reinforcing similarity with the rhodesite group. However, the two equivalent silicate sheets forming the double layer do not display the apophyllite-type but consist of two types of six-membered rings; one type is quite deformed.

A heterophyllosilicate H sheet in jonesite. The partially disordered crystal structure of jonesite $\{Ba_2(K,Na)[Ti_2(Si_5Al)O_{18}(H_2O)] \cdot (H_2O)_n;$ $a = 8.694$, $b = 25.918$, $c = 8.694$ Å, $\beta = 104.73°$, $P2_1/m$; Krivovichev and Armbruster 2004} shows a double layer forming [100] eight-membered channels (Fig. 20) similar to those described for the rhodesite series. However, the (010) sheet delimiting the double layer has the same chemical composition and topology of the H sheet typical of the HOH bafertisite-type layer (Fig. 3): some of the silicate tetrahedra are substituted by Ti octahedra. The minimum effective width of the pores is about 3.3 Å.

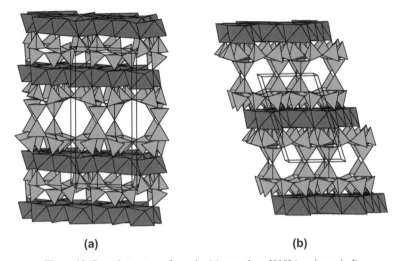

Figure 19. Crystal structure of reyerite (a) seen along [010] (*c* axis vertical) and fedorite seen along [100] (b).

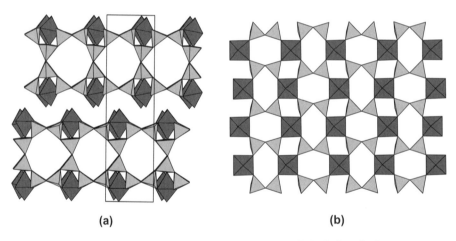

Figure 20. Jonesite: (010) double layer seen along [100] (a); (010) projection of the bafertisite-type *H* sheet (b) delimiting the double layer.

Analogues of seidite-(Ce). Ermolaeva et al. (2005) describe a new phase from the Lovozero and Khibina massifs that has chemical composition (Th,REE)(Ti,Nb)(Si,P)$O_6 \cdot n$H$_2$O and is supposed to be an analog of seidite-(Ce) with Th > REE. This phase occurs with bitumen-like substances whose formation from light hydrocarbons (like methane) is likely catalyzed by the presence of microporous heterosilicates. At the same time, the transformation of metal organic complexes occurring in the bitumen-like substances leads to the formation of microphases of Ti- and Nb-silicates of Th, Ca, Ln, Y, Na and K (cf. Chukanov and Pekov 2005). On the basis of the chemical formula, the latter authors suppose that also the amorphous mineral umbozerite, Na$_3$Sr$_4$Th(Mn,Zn,Fe)Si$_8$O$_{24}$(OH) (Es'kova et al. 1974), is an analog of seidite-(Ce).

CONCLUSIONS

Modularity (Ferraris et al. 2004) is a common crystallographic feature of the microporous structures presented in this and other chapters of this volume. In itself, this aspect is not important for the investigation of microporous properties but, as in general holds for modular structures, can be crucial in the process of modeling the structures of new microporous phases related to others already known. Thus, modularity may be a tool useful to solve difficult structures.

Modularity can also inspire paths towards the synthesis of tailored microporous compounds, for example by playing on the type and stacking of modules, according to well established patterns widely documented in modular materials. Instructive under this aspect, are the members of the mero-plesiotype series for which it is possible to play on a variable module and on the specific configurations of the modules to introduce useful properties, as described in this volume by Rocha and Lin (2005) for compounds of the rhodesite series.

In principle, as already Nature does, the *HOH* layer of the heterophyllosilicates can be exploited to build materials with a variety of interlayer contents. Besides, following some indications of swelling observed in the field, presumably pillared materials can be prepared by using *HOH* layers; the presence of Ti polyhedra both in the *H* and *O* sheets might be useful for applications in heterogeneous catalysis. However, whereas the microporous structures described in the sections on palysepioles and rhodesite series are well known to materials scientists and several of them are promising for technological applications, the porous properties occurring in some heterophyllosilicates are active in nature, but still unknown and unexploited in the world of technology.

ACKNOWLEDGMENTS

The authors are indebted with N.V. Chukanov for communicating unpublished IR data and constructively reading the manuscript and to Peter Nèmeth for useful discussions on aspects of the heterophyllosilicates. Research supported by MIUR (Rome) as the FIRB project "Properties and technological applications of minerals and their synthetic analogues" and the PRIN project "Microstructural and modular aspects in minerals: analyses and applications."

REFERENCES

Ananias D, Ferreira A, Rocha J, Ferreira P, Rainho JP, Morais C, Carlos LD (2001) Novel microporous europium and terbium silicates. J Am Chem Soc 123:5735-5742
Ananias D, Rainho JP, Ferreira A, Rocha J, Carlos LD (2004) The first examples of X-ray phosphors, and C-band infrared emitters based on microporous lanthanide silicates. J Alloys Compd 374:219-222
Anderson JA, Galan-Fereres M (1999) Precursor-support interactions in the preparation of sepiolite-supported Ni and Pd catalysts. Clay Minerals 34:57-66
Artioli G, Galli E (1994) The crystal structures of orthorhombic and monoclinic palygorskite. Mater Sci Forum 166-169:647-652
Bellezza M, Franzini M, Larsen AO, Merlino S, Perchiazzi N (2004) Grenmarite, a new member of the götzenite-seidozerite-rosenbuschite group from the Langesundsfjord district, Norway: definition and crystal structure. Eur J Mineral 16:971-978
Belov NV, Gavrilova GS, Solov'eva LP, Khalilov AD (1978) The refined structure of lomonosovite. Sov Phys Dokl 22:422-424
Bonaccorsi E, Merlino S (2005) Modular microporous minerals: cancrinite-davyne group and C-S-H phases. Rev Mineral Geochem 57:241-290
Brauner K, Preisinger A (1956) Struktur und Entstehung des Sepioliths. Tsch Mineral Petrogr Mitt 6:120-140
Burke EAJ, Ferraris G (2004) New minerals and nomenclature modifications approved in 2003 by the Commission on New Minerals and Mineral Names, International Mineralogical Association. Can Mineral 42:905-913

Camara F, Garvie LAJ, Devouard B, Groy TL, Buseck PR (2002) The structure of Mn-rich tuperssuatsiaite: a palygorskite-related mineral. Am Mineral 87:1458-1463

Cannillo E, Rossi G, Ungaretti L (1970) The crystal structure of delhayelite. Rend Soc Ital Miner Petrol 26: 63-75

Cannillo E, Rossi G, Ungaretti L, Carobbi SG (1968) The crystal structure of macdonaldite. Atti Accad Naz Lincei, Classe Sci Fisiche 45:399-414

Casal B, Merino J, Serratosa JM, Ruiz Hitzky E (2001) Sepiolite-based materials for the photo- and thermal-stabilization of pesticides. Appl Clay Sci 18:245-254

Chahi A, Petit S, Decarreau A (2002) Infrared evidence of dioctahedral-trioctahedral site occupancy in palygorskite. Clays Clay Mineral 50:306-313

Chelishchev NF (1972) Ion-exchange properties of astrophyllites under supercritical conditions. Geokhimiya 7:856-860 (in Russian)

Chelishchev NF (1973) Ion exchange properties of minerals. Nauka, Moscow (in Russian)

Chernov AN, Ilyukhin VV, Maksimov BA, Belov NV (1971) Crystal structure of innelite, $Na_2Ba_3(Ba,K,Mn)(Ca,Na)Ti(TiO_2)_2[Si_2O_7]_2(SO_4)_2$. Sov Phys Crystallogr 16:87-92

Chiari G, Giustetto R, Ricchiardi G (2003) Crystal structure refinements of palygorskite and Maya Blue from molecular modelling and powder synchrotron diffraction. Eur J Mineral 15:21-33

Christiansen CC, Johnsen O, Makovicky E (2003) Crystal chemistry of the rosenbuschite group. Can Mineral 41:1203-1224

Christiansen CC, Makovicky E, Johnsen ON (1999) Homology and typism in heterophyllosilicates: an alternative approach. N Jb Mineral Abh 175:153-189

Christiansen CC, Rønsbo JG (2000) On the structural relationship between götzenite and rinkite. N Jb Mineral Mh 2000:496-506

Chukanov NV, Pekov IV (2005) Heterosilicates with tetrahedral-octahedral frameworks: mineralogical and crystal-chemical aspects. Rev Mineral Geochem 57:105-144

Chukanova VN, Pekov IV, Chukanov NV, Zadov AE (2002) Iron-rich analogue of sepiolite and the conditions of its formation in the contact aureole of the Lovozero alkaline massif. Geochem Intern 40:1225-1229

Colville AA, Anderson CP, Black PM (1971) Refinement of the crystal structure of apophyllite: I. X-ray diffraction and physical properties. Am Mineral 56:1222-1233

Cool P, Vansant EF, Poncelet G, Schoonheydt RA (2002) Layered structures and pillared layered structures. In: Handbook of porous solids. Vol 1. Schüth F, Sing KSW, Weitkamp J (eds) Wiley-VCH, p 1251-1310

Corma A (1997) From microporous to mesoporous molecular sieve materials and use in catalysis. Chem Rev 97:2373-2419

Corma A, Garcia H, Leyva A, Primo A (2004) Alkali-exchanged sepiolites containing palladium as bifunctional (basic sites and noble metal) catalysts for the Heck and Suzuki reactions. Appl Catal A 257:77-83

Egorov-Tismenko YuK (1998) Polysomatic series of seidozerite-nacaphite minerals of titanosilicate mica analogues. Crystallogr Rep 43:306-312

Egorov-Tismenko YuK, Sokolova EV (1990) Structural mineralogy of the homologic series seidozerite-nacaphite. Mineral Zh 12(4):40-49 (in Russian)

Ercit TS, Cooper MA, Hawthorne FC (1998) The crystal structure of vuonnemite, $Na_{11}Nb_2(Si_2O_7)(PO_4)_2O_3(F,OH)$, a phosphate-bearing sorosilicate of the lomonosovite group. Can Mineral 36:1311-1320

Ermolaeva VN, Chukanov NV, Pekov IV, Nekrasov AN, Sokolov SV, Kogarko LN (2005) On the role of organic substances in transport and concentration of thorium and other rare elements in alkaline pegmatites of Lovozero and Khibiny peralkaline massifs. Geochem Intern (in press)

Es'kova YeM, Semenov YeI, Khomyakov AP, Mer'kov AN, Lebedeva SI, Dubakina LS (1974) Umbozerite, a new mineral. Dokl Akad Sci USSR, Earth Sci 216:124-126

Espantaleón AG, Nieto JA, Fernandez M, Marsal A (2003) Use of activated clays in the removal of dyes and surfactants from tannery waste waters. Appl Clay Sci 24:105-110

Evans HT Jr (1973) The crystal structures of cavansite and pentagonite. Am Mineral 58:412-424

Fahey JJ, Ross, M, Axelrod JM (1960) Loughlinite, a new hydrous sodium magnesium silicate. Am Mineral 45:270-281

Fernandez-Saavedra R, Aranda P, Ruiz-Hitzky E (2004) Templated synthesis of carbon nanofibers from polyacrylonitrile using sepiolite. Adv Functional Mater 14:77-82

Ferraris G (1997) Polysomatism as a tool for correlating properties and structure. EMU Notes Mineral 1: 275-295

Ferraris G (2004) Are the HOH layers of the heterophyllosilicates suitable to prepare pillared materials? Acta Crystallogr A60:s53

Ferraris G, Belluso E, Gula A, Khomyakov, AP, Soboleva SV (2003) The crystal structure of seidite-(Ce), $Na_4(Ce,Sr)_2\{Ti(OH)_2(Si_8O_{18})\}(O,OH,F)_4 \cdot 5H_2O$, a modular microporous titanosilicate of the rhodesite group. Can Mineral 41:1183-1192

Ferraris G, Belluso E, Gula A, Soboleva SV, Ageeva OA, Borutskii BE (2001a) A structural model of the layer titanosilicate bornemanite based on seidozerite and lomonosovite modules. Can Mineral 39:1665-1673

Ferraris G, Ivaldi G, Khomyakov AP, Soboleva SV, Belluso E, Pavese A (1996) Nafertisite, a layer titanosilicate member of a polysomatic series including mica. Eur J Mineral 8:241-249

Ferraris G, Ivaldi G, Pushcharovsky DYu, Zubkova NV, Pekov IV (2001b) The crystal structure of delindeite, $Ba_2\{(Na,K,\square)_3(Ti,Fe)[Ti_2(O,OH)_4Si_4O_{14}](H_2O,OH)_2\}$, a member of the mero-plesiotype bafertisite series. Can Mineral 39:1307-1316

Ferraris G, Khomyakov AP, Belluso E, Soboleva SV (1997) Polysomatic relationship in some titanosilicates occurring in the hyperagpaitic alkaline rocks of the Kola Peninsula, Russia. Proc 30th Inter Geol Cong 16:17-27

Ferraris G, Khomyakov AP, Belluso E, Soboleva SV (1998) Kalifersite, a new alkaline silicate from Kola Peninsula (Russia) based on a palygorskite-sepiolite polysomatic series. Eur J Mineral 10:865-874

Ferraris G, Makovicky E, Merlino S (2004) Crystallography of Modular Materials. IUCr Monographs in Crystallography, Oxford University Press, Oxford

Ferraris G, Nèmeth P (2003) Pseudo-symmetry, twinning and structural disorder in layer titanosilicates. Abstracts of ECM-21 (Durban), p 41

Galan E (1996) Properties and applications of palygorskite-sepiolite clays. Clay Minerals 31:443-453

Galan E, Carretero I (1999) A new approach to compositional limits for sepiolite and palygorskite. Clays Clay Mineral 47:399-409

Galan E, Ferrero F (1982) Palygorskite-sepiolite clays of Lebrija, Southern Spain. Clays Clay Mineral 30: 191-199

García Romero E, Suárez Barrios M, Bustillo Revuelta MA (2004) Characteristics of a Mg-palygorskite in Miocene rocks, Madrid Basin (Spain). Clays Clay Mineral 52:484-494

Garcia Sanchez A, Alvarez Ayuso E, Jimenez de Blas O (1999) Sorption of heavy metals from industrial waste water by low-cost mineral silicates. Clay Minerals 34:469-477

Ghose S, Sen Gupta PK, Campana CF (1987) Symmetry and crystal structure of monteregianite, $Na_4F_2Y_2Si_{16}O_{38}\cdot 10H_2O$, a double-sheet silicate with zeolitic properties. Am Mineral 72:365-374

Giustetto R, Chiari G (2004) Crystal structure refinement of palygorskite from neutron powder diffraction. Eur J Mineral 16:521-532

Grice JD (1991) The crystal structure of silinaite, $NaLiSi_2O_5\cdot 2H_2O$: a monophyllosilicate. Can Mineral 29: 363-367

Guan YaS, Simonov VI, Belov NV (1963) Crystal structure of bafertisite, $BaFe_2TiO[Si_2O_7](OH)_2$. Dokl Akad Sci 149:123-126 (published 1965)

Guggenheim S, Eggleton RA (1988) Crystal chemistry, classification, and identification of modulated layer silicates. Rev Mineral 19:675-725

Güven N (1988) Smectites. Rev Mineral 19:497-559

Hejny C, Armbruster T (2001) Polytypism in xonotlite $Ca_6Si_6O_{17}(OH)_2$. Z Kristallogr 216:396-408

Hesse KF, Liebau F, Merlino S (1992) Crystal structure of rhodesite, $HK_{1-x}Na_{x+2y}Ca_{2-y}$ $\{lB,3,2^2_\infty\}$ $[Si_8O_{19}]\cdot(6-z)H_2O$, from three localities and its relation to other silicates with dreier double layers. Z Kristallogr 199:25-48

Hong W, Fu P (1982) Jinshajiangite—a new Ba-Mn-Fe-Ti-bearing silicate mineral. Geochemistry (China) 1: 458-464 (in Chinese)

Jones BF, Galan E (1988) Sepiolite and palygorskite. Rev Mineral 19:631-674

Khomyakov AP (1995) Mineralogy of hyperagpaitic alkaline rocks. Clarendon Press, Oxford

Khomyakov AP (2004) Zeolite-like amphoterosilicates of hyperagpaitic rocks and their unique properties. Micro and Mesoporous Mineral Phases (Accad Lincei, Roma), Volume of Abstracts, p 231-234

Khomyakov AP, Ferraris G, Belluso E, Britvin SN, Nechelyustov GN, Soboleva SV (1998) Seidite-(Ce), $Na_4SrCeTiSi_8O_{22}F\cdot 5H_2O$—a new mineral with zeolite properties. Zap Vseross Mineral Obs 127(4):94-100 (in Russian)

Khomyakov AP, Ferraris G, Nechelyustov GN, Ivaldi G, Soboleva SV (1995) Nafertisite $Na_3(Fe^{3+}, Fe^{2+})_6[Ti_2Si_{12}O_{34}](O,OH)_7\cdot 2H_2O$, a new mineral with a new type of band silicate radical. Zap Vseross Mineral Obs 124(6):101-108 (in Russian)

Khomyakov AP, Nechelyustov GN, Sokolova EV, Dorokhova GI (1992) Quadruphite $Na_{14}CaMgTi_4[Si_2O_7]_2[PO_4]_2O_4F_2$ and polyphite $Na_{17}Ca_3Mg(Ti,Mn)_4[Si_2O_7]_2[PO_4]_6O_2F_6$, two new minerals of the lomonosovite group. Zap Vseross Mineral Obs 121(3):105-112 (in Russian)

Krivovichev SV, Armbruster T (2004) The crystal structure of jonesite, $Ba_2(K,Na)[Ti_2(Si_5Al)O_{18}(H_2O)](H_2O)_n$: A first example of titanosilicate with porous double layers. Am Mineral 89:314-318

Krivovichev SV, Yakovenchuk VN, Armbruster T, Döbelin N, Pattison P, Weber HP, Depmeier W (2004) Porous titanosilicate nanorods in the structure of yuksporite, $(Sr,Ba)_2K_4(Ca,Na)_{14}(\square,Mn,Fe)\{(Ti,Nb)_4(O,OH)_4[Si_6O_{17}]_2[Si_2O_7]_3\}(H_2O,OH)_n$, resolved using synchrotron radiation. Am Mineral 89:1561-1565

Kuang W, Facey GA, Detellier C (2004) Dehydration and rehydration of palygorskite and the influence of water on the nanopores. Clays Clay Mineral 52:635-642

Louis M, Guillemin CJ, Goni JC, Ragot JP (1968) Coloration rose-carmin d'une sepiolite eocene, la quincyte, par des pigments organiques. *In:* Advances in Organic Geochemistry 1968. Schenck PA, Havenaar I (eds) Pergamon Press, p 553-566

Makovicky E (1997) Modularity – different types and approaches. EMU Notes Mineral 1:315-344

Makovicky E, Karup-Møller S (1981) Crystalline steenstrupine from Tunugdliarfik in the Ilímaussaq alkaline intrusion, South Greenland. N Jb Mineral Abh 140:300-330

Martin-Vivaldi JL, Linares-Gonzales J (1962) A random intergrowth of sepiolite and attapulgite. Clays Clay Mineral 9:592-602

Massa W, Yakubovich OV, Kireev VV, Mel'nikov OK (2000) Crystal structure of a new vanadate variety in the lomonosovite group: $Na_5Ti_2O_2[Si_2O_7](VO_4)$. Solid State Sci 2:615-623

Matsubara S (1980) The crystal structure of orthoericssonite. Mineral J 10:107-121

McCusker LB (2005) IUPAC nomenclature for ordered microporous and mesoporous materials and its application to non-zeolite microporous mineral phases. Rev Mineral Geochem 57:1-16

McCusker LB, Liebau F, Engelhardt G (2003) Nomenclature of structural and compositional characteristics of ordered microporous and mesoporous materials with inorganic hosts (IUPAC recommendations 2001). Microporous Mesoporous Mater 58:3-13

McDonald AM, Grice JD, Chao GY (2000) The crystal structure of yoshimuraite, a layered Ba-Mn-Ti silicophosphate, with comments of five-coordinated Ti^{4+}. Can Mineral 38:649-656

Mellini M, Ferraris G, Compagnoni R (1985) Carlosturanite: HRTEM evidence of a polysomatic series including serpentine. Am Mineral 70:773-781

Men'shikov YuP, Khomyakov AP, Polezhaeva LI, Rastsvetaeva RK (1996) Shkatulkaite, $Na_{10}MnTi_3Nb_3(Si_2O_7)_6(OH)_2F\cdot 12H_2O$: a new mineral. Zap Vseross Mineral Obs 125(1):120-126 (in Russian)

Men'shikov YuP, Khomyakov AP, Ferraris G, Belluso E, Gula A, Kulchitskaya EA (2003) Eveslogite, $(Ca,K,Na,Sr,Ba)_{24}[(Ti,Nb,Fe,Mn)_6(OH)_6Si_{24}O_{72}](F,OH,Cl)_7$, a new mineral from the Khibina alkaline massif, Kola Peninsula, Russia. Zap Vseross Mineral Obs 132(1):59-67 (in Russian)

Merlino S (ed) (1997) Modular aspects of minerals. EMU Notes in Mineralogy. Vol 1. Budapest, Eötvös University Press

Merlino S, Pasero M, Khomyakov AP (1990) The crystal structure of lintisite, $Na_3LiTi_2(Si_2O_6)_2O_2\cdot 2H_2O$, a new titanosilicate from Lovozero (USSR). Z Kristallogr 193:137-148

Mizutani T, Fukushima Y, Okada A, Kamigaito O (1991) Hydrothermal synthesis of sepiolite. Clay Minerals 26:441-445

Moore PB (1971) Ericssonite and orthoericssonite. Two new members of the lamprophyllite group, from Långban, Sweden. Lithos 4:137-145

Nèmeth P (2004) Characterization of new mineral phases belonging to the heterophyllosilicate series. Doctorate Dissertation, Dipartimento di Scienze Mineralogiche e Petrologiche, Università di Torino

Nèmeth P, Ferraris G, Dódony I, Radnóczi G, Khomyakov AP (2004a) Models of the modular structures of two new heterophyllosilicates related to bornemanite and barytolamprophyllite. Acta Cryst A60:s196

Nèmeth P, Ferraris G, Radnóczi G, Ageeva OA (2005) TEM and X-ray study of syntactic intergrowths of epistolite, murmanite and shkatulkalite. Can Mineral, in press

Nèmeth P, Gula A, Ferraris G (2004b) Towards pillared heterophyllosilicates? Suggestions from nature. Micro- and Mesoporous Mineral Phases (Accad Lincei, Roma), Volume of Abstracts, p 265-267

Pekov IV, Chukanov NV (2005) Microporous framework silicate minerals with rare and transition elements: minerogenetic aspects. Rev Mineral Geochem 57:145-172

Pekov IV, Chukanov NV, Ferraris G, Ivaldi G, Pushcharovsky DYu, Zadov AE (2003) Shirokshinite, $K(NaMg_2)Si_4O_{10}F_2$, a new mica with octahedral Na from Khibiny massif, Kola Peninsula: descriptive data and structural disorder. Eur J Mineral 15:447-454

Pen ZZ, Shen TC (1963) Crystal structure of bafertisite, a new mineral from China. Sci Sin 12:278-280 (in Russian)

Pen ZZ, Zhang J, Shu J (1984) The crystal structure of barytolamprophyllite. Kexue Tongbao 29:237-241

Perrault G, Harvey Y, Pertsowsky R (1975) La yofortierite, un nouveau silicate hydraté de manganèse de St-Hilaire, P.Q. Can Mineral 13:68-74

Petersen OV, Johnsen O, Christiansen CC, Robinson GW, Niedermayr G (1999) Nafertisite - $Na_3Fe_{10}Ti_2Si_{12}(O,OH,F)_{43}$ - from the Nanna Pegmatite, Narsaarsuup Qaava, South Greenland. N Jb Mineral Mh 1999:303-310

Piilonen PC, Lalonde AE, McDonald AM, Gault RA, Larsen AO (2003a) Insights into astrophyllite-group minerals. I. Nomenclature, composition and development of a standardized general formula. Can Mineral 41:1-26

Piilonen PC, McDonald AM, Lalonde AE (2003b) Insights into astrophyllite-group minerals. II. Crystal chemistry. Can Mineral 41:27-54

Prowse WG, Arnot KI, Recka JA, Thomson RH, Maxwell JR (1991) The quincyte pigments—fossil quinones in an eocene clay mineral. Tetrahedron 47:1095-1108

Pushcharovsky DYu, Pasero M, Merlino S, Vladykin NV, Zubkova NV, Gobechiya ER (2002) Crystal structure of zirconium-rich seidozerite. Crystallogr Rep 47:196-200

Pushcharovsky DYu, Pekov IV, Pluth J, Smith J, Ferraris G, Vinogradova SA, Arakcheeva AV, Soboleva SV, Semenov EI (1999) Raite, manganonordite-(Ce) and ferronordite-(Ce) from Lovozero massif: crystal structures and mineral geochemistry. Crystallogr Rep 44:565-574

Putnis A (2002) Mineral replacement reactions: from macroscopic observations to microscopic mechanisms. Mineral Mag 66:689-708

Ragimov Kg, Chiragov MI, Mamedov KS, Dorfman MD (1980) Crystal structure of hydrodelhayelite, $KH_2Ca(Si,Al)_8O_{19} \cdot H_2O$. Dokl Akad Nauk Azerbaid SSR 36:49-51 (in Russian)

Rastsvetaeva RK (1998) Crystal structure of betalomonosovite from the Lovozero region. Sov Phys Crystallogr 31:633-636

Rastsvetaeva RK, Chukanov NV (1999) Crystal structure of a new high-barium analogue of lamprophyllite with a primitive unit cell. Doklady Chemistry 368(4-6):228-231

Rastsvetaeva RK, Sokolova MN Gusev AI (1990) Refined crystal structure of lamprophyllite. Mineral Zh 12(5):25-28 (in Russian)

Rastsvetaeva RK, Tamazyan RA, Sokolova EV, Belakovskii DI (1991) Crystal structures of two modifications of natural Ba,Mn-titanosilicate. Sov Phys Crystallogr 36:186-189

Rocha J, Anderson MW (2000) Microporous titanosilicates and other novel mixed octahedral-tetrahedral framework oxides. E J Inorg Chem 2000:801-818

Rocha J, Carlos LD, Ferreira A, Rainho J, Ananias D, Lin Z (2004) Novel microporous and layered luminescent lanthanide silicates. Mater Sci Forum 455-456:527-531

Rocha J, Lin Z (2005) Microporous mixed octahedral-pentahedral-tetrahedral framework silicates. Rev Mineral Geochem 57:173-202

Rozenberg KA, Rastsvetaeva RK, Verin IA (2003) Crystal structure of surkhobite: new mineral from the family of titanosilicate micas. Crystallogr Rep 48:384-389

Rytwo G, Tropp D, Serban C (2002) Adsorption of diquat, paraquat and methyl green on sepiolite: experimental results and model calculations. Appl Clay Sci 20:273-282

Schoonheydt RA, Pinnavaia T, Lagaly G, Gangas N (1999) Pillared clays and pillared layered solids. Pure Appl Chem 71:2367-2371

Scott JD (1976) Crystal structure of miserite, a Zoltai type 5 structure. Can Mineral 14:515-528

Semenov EI, Organova NI, Kukharchik MV (1962) New data on minerals of the lomonosovite-murmanite group. Sov Phys Crystallogr 6:746-751

Shariatmadari H, Mermut AR, Benke MB (1999) Sorption of selected cationic and neutral organic molecules on palygorskite and sepiolite. Clays Clay Mineral 47:44-53

Shi N, Ma Z, Li G, Yamnova NA, Pushcharovsky DYu (1998) Structure refinement of monoclinic astrophyllite. Acta Crystallogr B54:109-114

Singer A, Huang PM (1989) Adsorption of humic acid by palygorskite and sepiolite. Clay Minerals 24:561-564

Sokolova E, Hawthorne FC (2001) The crystal chemistry of the $[M_3\Phi_{11-14}]$ trimeric structures from hyperagpaitic complexes to saline lakes. Can Mineral 39:1275-1294

Sokolova E, Hawthorne FC (2004) The crystal chemistry of epistolite. Can Mineral 42:797-806

Sokolova EV, Egorov-Tismenko YuK, Khomyakov AP (1987) Special features of the crystal structure of $Na_{14}CaMgTi_4[Si_2O_7]_2[PO_4]_2O_4F_2$ - homologue of sulphohalite and lomonosovite structure types. Mineral Zh 9(3):28 35 (in Russian)

Sokolova EV, Egorov-Tismenko YuK, Khomyakov AP (1988) Crystal structure of sobolevite. Sov Phys Dokl 33:711-714

Sokolova EV, Egorov-Tismenko YuK, Khomyakov AP (1989a) The crystal structure of nacaphite. Sov Phys Dokl 34:9-11

Sokolova EV, Egorov-Tismenko YuK, Pautov LA, Belakovskii DI (1989b) Structure of the natural barium titanosilicate $BaMn_2TiO[Si_2O_7](OH)_2$, a member of the seidozerite-nacaphite homologous series. Zap Vses Mineral Obs 118(4):81-84 (in Russian)

Springer G (1976) Falcondoite, nickel analogue of sepiolite. Can Mineral 14:407-409

Thompson JB Jr (1978) Biopyriboles and polysomatic series. Am Mineral 63:239-249

Ustinov VI, Ul'yanov AA (1999) Intrastructural isotopic investigation of the transformation of minerals with a layered structure. Geochem Intern 37:908-911

Veblen DR (1991) Polysomatism and polysomatic series: A review and applications. Am Mineral 76:801-826

Veblen DR, Buseck PR (1979) Chain-width order and disorder in biopyriboles. Am Mineral 64:687-700

Velde B (1992) Introduction to clay minerals. Chapman & Hall, London

Vrána S, Rieder M, Gunter ME (1992) Hejtmanite, a manganese-dominant analogue of bafertisite, a new mineral. Eur J Mineral 4:35-43
Woodrow PJ (1967) The crystal structure of astrophyllite. Acta Crystallogr 22:673-678
Yamnova NA, Egorov-Tismenko YuK, Khomyakov AP (1996) Crystal structure of a new natural (Na,Mn,Ti)-phyllosilicate. Crystallogr Rep 41:239-244
Yamnova NA, Egorov-Tismenko YuK, Pekov IV (1998) Crystal structure of perraultite from the Coastal Region of the Sea of Azov. Crystallogr Rep 43:401-410
Yang Z, Cressey G, Welch M (1999) Reappraisal of the space group of bafertisite. Powder Diffraction 14: 22-24
Zaitsev VA, Krigman LD, Kogarko LN (2004) Pseudobinary phase diagram lamprophyllite-nepheline. Lithos 73(1-2):s122
Zhou K, Rastsvetaeva RK, Khomyakov AP, Ma Z, Shi N (2002) Crystal structure of new micalike titanosilicate – Bussenite, $Na_2Ba_2Fe^{2+}[TiSi_2O_7][CO_3]O(OH)(H_2O)F$. Crystallogr Rep 47:50-53

Heterosilicates with Tetrahedral-Octahedral Frameworks: Mineralogical and Crystal-Chemical Aspects

Nikita V. Chukanov[1] and Igor V. Pekov[2]

[1]*Institute of Problems of Chemical Physics,*
142432 Chernogolovka, Moscow Region, Russia
chukanov@icp.ac.ru

[2]*Faculty of Geology, Moscow State University*
119899 Moscow, Russia

INTRODUCTION

Steady progress in methods of synthesis and investigation of physical properties and crystal-chemical features of silicate microporous materials with transition elements (so-called zeolite-like amphoterosilicates) took place during the last decade. Materials of this type are promising ion exchangers, sorbents, catalysts or catalyst carriers and, accordingly, can be used in chromatography, catalysis, water purification, etc. Unlike common zeolites being aluminosilicates, frameworks of amphoterosilicates with transition metals are built of both tetrahedral fragments and "strong" cations (Ti, Nb, Zr, Ta, Sn, W, Fe, Mn, Zn, etc., see Chukanov et al. 2004) with coordination numbers 6 or 5. Structural and chemical diversity of synthetic and natural amphoterosilicates is a basis for the wide variety of their properties. On the other hand, the number of different combinations of structure types and compositions of non-zeolite microporous minerals is still larger than the number of different synthetic materials of this type. In addition, whereas minerals are often available as crystals suitable for structural investigations, synthetic materials are often microcrystalline. Thus, natural titanosilicates, niobosilicates and zirconosilicates can be considered as prototypes of new materials with ion exchange, sorptional and catalytic properties (cf. Rocha and Lin 2005). The brightest examples are members of the eudialyte, labuntsovite, lovozerite, and hilairite groups, elpidite, zorite, gaidonnayite, penkvilksite. These minerals were formed in late formations related to alkaline massifs (alkaline pegmatites, hydrothermalites and metasomatites; cf. Pekov and Chukanov 2005).

GENERAL CHARACTERIZATION AND CLASSIFICATION OF MICROPOROUS HETEROSILICATE MINERALS

Among about 4100 known mineral species, several hundreds minerals have open-framework structures but only about one hundred are classified as proper zeolites (Bish and Ming 2001) with frameworks consisting of atoms having only 4-fold coordination. In particular, a large number of zeolite-like microporous heterosilicate minerals (MHM) is known whose frameworks contain transition elements (mainly Ti, Nb and Zr, but also Fe, Mn, Zn, Ta, Sn, W) having 6-fold, rarely 5-fold coordination (see Voronkov et al. 1975). At present more than 100 mineral species that can be considered as framework MHM's are known. Though the number of main heteroframework types known in minerals is only ~30, a wide variety of MHM mineral species and structural varieties is realized in nature. The diversity of minerals having frameworks of the same type is caused by many factors, the most important being composition of framework and extra-framework sites, symmetry, vacancies in the framework, complex mechanisms of isomorphous substitutions involving several atoms. The structural diversity of such minerals leads to a wide variety of their properties.

Framework MHM's are closely related to the large family of heterophyllosilicate minerals (Ferraris and Gula 2005). Most representatives of these two large supergroups contain the same sets of major cations and often occur together in alkaline rocks. Moreover, in crystal structures of some minerals traditionally considered as phyllosilicates or heterophyllosilicates, the layers are connected (e.g., by Si-O-Si or Ti-O-Ti links) to form 3D frameworks with channels containing exchangeable (at least under supercritical conditions) cations. The best known examples of such minerals are members of the astrophyllite group (Piilonen et al. 2003a,b), of the perraultite-surkhobite isomorphous series (Pekov et al. 1999a; Es'kova et al. 2003), of the palygorskite polysomatic series (Ferraris et al. 2001; Ferraris and Gula 2005) etc. In particular, the transition from the layered bafertisite structure to the framework perraultite structure can be represented as a shift of neighboring HOH layers by 1/4 along the [100] and [010] directions. That results in the formation of Ti-O-Ti links between the layers (Yamnova et al. 1998; Rozenberg et al. 2003); see also Chelishchev (1972) for ion exchange in astrophyllite; Grim (1953) and Shariatmadari et al. (1999) for cation sorption in palygorskite and sepiolite. Recently, the crystal structure of jonesite, a titanosilicate with porous double layers has been reported (Krivovichev and Armbruster 2003). Such layered microporous heterosilicates and their framework derivatives are not discussed in this chapter.

GENERAL CRYSTAL-CHEMICAL CHARACTERISTICS

Force parameters

The force properties of the framework cations are determined by the strengths of their chemical bonds with the oxygen atoms. The expression for vibrational potential energy of a unit cell in the harmonic approximation can be written as follows:

$$U = \sum_{i,j=1}^{N} k_{ij} q_i q_j \tag{1}$$

where N is the number of vibrational degrees of freedom per unit cell; q_i and q_j are local coordinates (e.g., lengths of chemical bonds and angles between them). For real crystals, the values of the harmonic force-field parameters k_{ij} with $i = j$ are usually much higher than for $i \neq j$, and therefore the frequency $\nu = |\partial^2 U/\partial q_i^2|$ can be considered as the force characteristic of the i-th bond.

Generally speaking, the frequencies of the absorption bands in the IR spectra of crystalline substances are related to collective vibrations of a large number of atoms within the area of coherent vibrations. However in most cases (precisely, in the absence of strong resonance effects) collective vibrations of each kind involve mainly chemical bonds of a certain type, and, as a result, frequencies of normal vibrations depend on local force characteristics of corresponding bonds (cluster approximation; for details see the papers: Loghinov et al. 1979, Chukanov and Kumpanenko 1988). The frequency ν_i of a spectral band of stretching vibrations involving chemical bonds of a given type is connected with the force characteristic k_{ii} of these bonds according the equation $k_{ii} = 4\pi^2 \mu \nu_i^2$ where μ is the reduced mass of normal vibrations (for MHM the value of μ is usually close to the mass of oxygen atom). It means that the force characteristics of the bonds between different cations and oxygen atoms in MHM can be easily evaluated from IR spectra.

Zeolite properties of a large number of alkaline silicates containing Ti, Nb or Zr are related with the force characteristics k_{ii} of cations. Cations in zeolite-like minerals with mixed frameworks can be arranged in order of decreasing force characteristics (Rastsvetaeva and Chukanov 2002; Chukanov et al. 2004; see Table 1). In the sequence $T \rightarrow M \rightarrow D \rightarrow A$ (the meaning of T, M, D and A is given in Table 1), a lowering of the frequencies of the corresponding stretching vibrations is observed (Rastsvetaeva and Chukanov 2002). T, M, and usually D cations belong to the frameworks of microporous amphotherosilicates, whereas A cations are weekly bonded with the framework and can be involved in ion-exchange processes if the latter are not sterically hindered.

Ions of the transition elements with coordination numbers ≤ 6 and $v_i > 400$ cm^{-1} are characterized by high values of the force parameters k_{ii} and, together with Si, contribute to build the frameworks. The values of k_{ii} related to T, M and D cations determine the flexibility of the framework; in fact, the energy necessary to stretch a bond increases with k_{ii}. Alkaline and alkaline-earth cations have the lowest force parameters, the highest coordination numbers and $v_i < 400$ cm^{-1}; thus, together with water molecules may be disordered on different sites within cavities of the framework. The lowest force characteristics and the largest ionic radii thus characterize exchangeable cations.

Table 1. Properties of cations in amphotherosilicates.

Cation site	Coordination number	Ionic radius, (Å)	Main cations	Trend to isomorphous substitutions	Trend to vacancies	Trend to site splitting	Stretching frequencies, (cm^{-1})
T	4	< 0.5	Si, Al, P, B, Be*	–	–	–	850–1100
M**	6	0.6–0.7	Ti, Nb, Zr, Fe^{3+}, Y, Mn^{3+}, Sn^{4+}, W^{6+}	+	–	–	550–750
D***	6	0.7–1.0	Mn^{2+}, Fe^{2+}, Mg, Zn, Ca	+	+	–	400–500
A	> 6	1.1–1.5	Na, K, Ca, Sr, Ba, Pb^{2+}, (H$_3$O)$^+$	+	+	+	< 400

*For Zn with 4-fold coordination the effective ionic radius is ≈ 0.6 Å.
**High-force-strength octahedral cations with charges > 2
***Medium-force-strength bivalent octahedral cations.

T:M ratios

The structures of mixed frameworks depend on the T:M ratio, where T represents tetrahedral framework cations (mainly Si) and M represents framework cations with coordination numbers greater than 4. In most cases, the Si:M ratio varies from 1.6 to 8.0. Rarely the T:M ratio is less than 1. The lowest values of the Si:M ratio (0.5–0.7) are realized in komarovites, sitinakite and baotite.

In MHM with $3 \leq$ Si:$M \leq 8$, the SiO$_4$ tetrahedra are usually polymerized to form infinite 1D or 2D structures (chains, bands or layers) connected by isolated M octahedra into 3D frameworks. More than one type of chains is present in minerals of hilairite and gaidonnayite groups, umbite, terskite and vlasovite; bands are known only in the elpidite structure. Tetrahedral layers are present in the crystal structures of altisite, lemoynite, penkvilksite, dalyite and seidite-(Ce). In the minerals with $3 \leq$ Si:$M \leq 8$ the prevailing M cation is usually Ti, Zr or Y; natural Nb-silicates with frameworks of this type are unknown.

A lower Si:M ratio leads to a gradual decrease of the polymerization degree of the tetrahedral subsystem. The frameworks of most microporous Ti- and Nb-silicates with $1.6 \leq$ Si:$M \leq 3.6$ consist of Si$_n$O$_{3n}$ rings (n is equal to 3, 4, 6 or 8) or diorthogroups Si$_2$O$_7$ connected by isolated octahedra (kostylevite, catapleiite, parakeldyshite), groups of octahedra (eudialyte- and lovozerite-group minerals, steenstrupine) or chains of octahedra (labuntsovite-group minerals). The presence of a considerable amount of Nb substituting Ti in titanosilicates is common. Nevertheless the number of natural MHM with Nb > Ti is lower than that of Ti dominant microporous titanosilicate minerals. Among the structurally investigated MHM, Nb prevails over Ti only in seven members of the labuntsovite group, in ilímaussite-(Ce) and in Na-komarovite (see Tables 2, 3). Besides, the Nb-silicate mongolite, Ca$_4$Nb$_6$Si$_5$O$_{24}$(OH)$_{10}$·5–6H$_2$O, with unknown structure can be tentatively classified as MHM.

In most structures, there is a limited isomorphous substitution of Zr for Ti as discussed by Pyatenko and Voronkov (1977) and Pyatenko et al. (1999). Complete isomorphism between Ti- and Zr-silicates is known only for the members of the lovozerite group (Pekov et al. 2003b) and is supposed to occur in the eudialyte-group minerals (Ageeva et al. 2002).

Whereas the NbO$_6$ and the TiO$_6$ octahedra show pronounced tendency to condensation, the ZrO$_6$ octahedra in microporous Zr-silicates are always isolated. For this reason, the occurrence of zirconosilicates with Si:Zr < 1 in nature is improbable; the lowest Si:Zr ratio (2) occurs in keldyshite and some related zirconosilicates (Voronkov et al. 1970; Khomyakov et al. 1974b; Khalilov et al. 1978) whose framework consists of Si$_2$O$_7$ diorthogroups connected by isolated ZrO$_6$ octahedra. Frameworks with both octahedral and tetrahedral infinite subsystems are known only for MHM with $1.33 \leq T{:}M \leq 2.4$ and occur in zorite and members of the vinogradovite-lintisite family. The frameworks of MHM with $T{:}M \leq 2$ consist usually of infinite (1D or 2D) octahedral edifices connected by four-membered rings of SiO$_4$ tetrahedra (labuntsovite-group minerals, komarovites, baotite) or by isolated SiO$_4$ tetrahedra (sitinakite).

The Si:M ratios in MHM can be very high. For example, mesoporous molecular sieves Ti-MCM-41 and Zr-MCM-41 with Si:M ratios in the range from 11 to 96 (M = Ti or Zr), have been synthesized by hydrothermal method (Chaudhari et al. 2001). The SiO$_2$ polymorphs with loose frameworks (e.g., melanophlogite) on one side, and microporous niobates, titanates and related oxides with framework-based structures, on the other side, can be considered as limit cases of natural "amphotherosilicates" with Si:$M = \infty$ and Si:M = 0 respectively. For examples see: Chen et al. 1995; Balzer and Langbein 1996; Mao et al. 2000; Teraoka et al. 2000; Zhao et al. 2000; Möller et al. 2002; tunnel oxides reviewed by Pasero (2005).

As it will be illustrated below, the topological diversity of MHM dependent on the $T{:}M$ ratio leads to a variety of structural channels.

Framework density

The wide variety of fine structural features and properties of MHM is connected with many factors including type of framework (which, in turn, depends on the $T{:}M$ ratio), cation ordering, presence of additional structural fragments, isomorphous substitutions and so on. One of the most important characteristics governing ion-exchange and other important properties of microporous materials is their framework density (*FD*). *FD* is defined as the number of framework knots (i.e., cations with high force characteristics) per 1000 Å^3. The *FD* value usually lies in the range from 14 to 22 for zeolites as well as for most microporous minerals with mixed frameworks.

CRYSTAL STRUCTURES OF THE MOST IMPORTANT MHM
Minerals with lowest framework densities

Zeolite-like framework heterosilicates with very loose frameworks (FD < 16) are especially important for technological applications (cf. Rocha and Lin 2005). The titanosilicates ETS-10 (for its crystal structure see Anderson et al. 1995, Wang and Jacobson 1999) and ETS-4 (Pattison et al. 1998), that corresponds to the mineral zorite (Sandomirsky and Belov 1979; see Table 2, Fig. 1), are examples of synthetic materials with low FD. The main crystal-chemical data for heteroframework minerals with low FD are summarized in the Table 2. Actually, only three types of alkaline heterosilicates with FD < 15 are known in nature: zorite, seidite-(Ce) and some members of the steenstrupine-(Ce) – thorosteenstrupine series. Though these minerals are very rare, they occur together in the Yubileynaya pegmatite at Mt. Karnasurt, Lovozero alkaline massif, together with penkvilksite (Bussen et al. 1974) and other rare MHM. Seidite-(Ce) and zorite are known only in the Lovozero massif. For zorite see Rocha and Lin (2005). Seidite-(Ce) (Khomyakov et al 1998; Ferraris et al. 2003; Table 2) has the lowest framework density (14.05 cations per 1000 Å^3) and shows pronounced cation-exchange properties; further data are given by Ferraris and Gula (2005).

Steenstrupine series. Steenstrupine-(Ce) is a typical mineral of hypersodic rocks of the Ilímaussaq and Lovozero massifs. For complexity of crystal structure, diversity of chemical composition and some crystal-chemical features it resembles eudialyte-group minerals. Actually, the name steenstrupine-(Ce) is applied now to a large group of minerals that are badly investigated because of their metamict state. The crystal structure has been investigated only for a Zr-bearing sample from Tunugdliarfik in the Ilímaussaq alkaline massif, South Greenland (Moore and Shen 1983). It is described as alternation of two types of [001] rods. Rod I, $M(1)$ – Na(1) – OH – $M(3)$ – $M(2)$ – □(1) – T – □(1) – $M(2)$ – $M(3)$ – OH – Na(1) is at (0 0 z); $M(1)$ is mainly (Zr,Th), $M(2)$ = Mn$^{2+}_{0.5}$Mn$^{3+}_{0.5}$, $M(3)$ = Fe^{3+}, T is mainly P. Rod II, Na(3) – P – P – Na(3) – Si(2) – Na(2) – REE – Si(1) – Si(1) – REE – Na(2) – Si(2), is at (1/3, 1/6, z). This rod model describes the alternation of different cations, but does reflect neither their coordination, nor the topology of the mixed framework. A traditional polyhedral pattern (Fig. 2) seems clearer: the SiO$_4$ tetrahedra form six-membered rings connected by alternating ZrO$_6$ and MnO$_6$ octahedra thus generating a 3D framework (Fig. 3). The MnO$_6$ and Fe^{3+}O$_6$ octahedra form clusters; the latter share vertices with the PO$_4$ tetrahedra. From the structural data, one can suggest that among extra-framework cations, Na(2) and Na(3) can be easily removed by leaching. Moore and Shen (1983) on the basis of their structure refinement suggested the following simplified formula for steenstrupine-(Ce) from Ilímaussaq: Na$_{14}$Ce$_6$Mn$_2$Fe$^{3+}{}_2$(Zr,Th)(OH)$_2$(PO$_4$)$_6$(Si$_6$O$_{18}$)$_2$·3H$_2$O. However, the general formula Na$_{1-12}$H$_{7-0}$Ca$REE_6$$Me_5$[Si$_6O_{18}$]$_2$[(P,Si)O$_4$]$_6$·$nH_2$O, ($Me$ = Mn, Fe, Th, Zr, Ti, Al, U, n < 15) suggested earlier by Makovicky and Karup-Møller (1981) on the basis of chemical data of a number of samples, better reflects the ranges of chemical compositions (Pekov et al. 1997). In particular, Na is not a necessary component of steenstrupine-type minerals even in highly alkaline parageneses. Th can substitute not only Zr, but also REE; in thorosteenstrupine (Kupriyanova et al. 1962; Pekov et al. 1997), Na$_{0-5}$Ca$_{1-3}$(Th,REE)$_6$(Mn,Fe,Al,Ti)$_{4-5}$[Si$_6$O$_{18}$]$_2$[(Si,P)O$_4$]$_6$(OH,F,O)$_x$·nH$_2$O, thorium prevails over rare-earth elements.

Komarovite group. The crystal structure of a Na-komarovite has been solved by Balić-Žunić et al. (2002) on a sample from Ilímaussaq. (001) pyrochlore-like layers are connected by [Si$_4$O$_{12}$] rings (Fig. 3). In the [100] channels formed between layers and rings, Na, K and water molecules are statistically distributed between two positions. Inside the pyrochlore-like layers the distribution of cations is: Nb1 and Ca in the middle and Nb2 and Na close to layer surfaces.

Table 2. Crystal-chemical data for MHM.

Name, Formula*	Symmetry	a (Å)	b (Å)	c (Å)
Traskite $Ba_{24}(Ca,Sr)(Ti,Fe,Mn)_{16}(Si_{12}O_{36})(Si_2O_7)_6(O,OH)_{30}Cl_6 \cdot 14H_2O$	$P6m2$	17.88	17.88	12.30
Seidite-(Ce) $Na_4SrCeTi(Si_8O_{22})F \cdot 5H_2O$	$C2/c$	24.74	7.19	14.47
Zorite $Na_6Ti_5(Si_6O_{17})_2O(OH)_4 \cdot 11H_2O$	$Cmmm$	23.24	7.24	6.95
Steenstrupine-type minerals** $Na_{1-12}H_{7-0}CaREE_6Me_{5-x}[Si_6O_{18}]_2[(P,Si)O_4]_6 \cdot nH_2O$ (Me = Mn, Fe, Th, Zr, Ti, Al, U; $n < 15$)	? (trigonal)			
Steenstrupine-(Ce) $Na_{14}Ce_6(Mn^{2+}Mn^{3+})Fe^{3+}_2(Zr,Th)(OH)_2(PO_4)(PO_4)_6(Si_6O_{18})_2 \cdot 3H_2O$	$R\text{-}3m$	10.46	10.46	45.48
Thorosteenstrupine $Na_{0-5}Ca_{1-3}(Th,REE)_6(Mn,FeTi)_{4-5}(Si_6O_{18})_2[(Si,P)O_4]_6(OH,O)_x \cdot nH_2O$	Metamict			
Na-komarovite $(Na,K)(Na,Ca)_{6-x}Li_x(Nb,Ti)_6Si_4O_{26}F_2 \cdot 4H_2O$	$Cmmm$	7.31	24.59	7.40
Komarovite** $Ca_{2-x}(Nb,Ti)_6Si_4O26(OH)2 \cdot nH2O$?	(ortho-rhombic)	21.30	14.00	17.19
Sazykinaite-(Y) $Na_5YZrSi_6O_{18} \cdot 6H_2O$	$R32$	10.82	10.82	15.81
Pyatenkoite-(Y) $Na_5YTiSi_6O_{18} \cdot 6H_2O$	$R32$	10.70	10.70	15.73
Hilairite $Na_2ZrSi_3O_9 \cdot 3H_2O$	$R32$	10.56	10.56	15.85
Calciohilairite $CaZrSi_3O_9 \cdot 3H_2O$	$R32$	10.50	10.50	7.97
Komkovite $BaZrSi_3O_9 \cdot 3H_2O$	$R32$	10.53	10.53	15.74
Pentagonite $Ca(VO)Si_4O_{10} \cdot 4H_2O$	$Pcm2_1$	10.39	14.05	8.975
Cavansite $Ca(VO)Si4O10 \cdot 4H2O$	$Pcmn$	9.79	13.64	9.63
Gaidonnayite $Na_2ZrSi_3O_9 \cdot 2H_2O$	$P2_1nb$	11.74	12.97	6.73
Georgechaoite $NaKZrSi_3O_9 \cdot 2H_2O$	$P2_1nb$	11.84	12.94	6.73
Ilímaussite-(Ce) $(Ba,Na)_{10}K_3Na_{4.5}Ce_5(Nb,Ti)_6[Si_{12}O_{36}][Si_9O_{18}(O,OH)_{24}]O_6$	$R32$	10.77	10.77	61.05
Diversilite-(Ce) $(Ba,K,Na,Ca)_{11-12}(REE,Fe,Th)_4(Ti,Nb)_6(Si_6O_{18})_4(OH)_{12} \cdot 4.5H_2O$	$R32$	10.71	10.71	60.07
Altisite $Na_3K_6Ti_2Si_8Al_2O_{26}Cl_3$	$C2/m$	10.36	16.31	9.13
Lemoynite $(Na,K)_2CaZr_2Si_{10}O_{26} \cdot 5\text{-}6H_2O$	$C2/c$	10.37	15.92	18.60
Natrolemoynite $Na_4Zr_2Si_{10}O_{26} \cdot 10H_2O$	$C2/m$	10.515	16.25	9.10
Jagoite $(Pb,Na)_{11}(Fe^{2+},Mg)_4(Si,Al)_{13}(O,OH)_{41}Cl_3$	$P62c$	8.53	8.53	33.33
Sitinakite $Na_2KTi_4(SiO_4)_2O_5OH \cdot 4H_2O$	$P4_2/mcm$	7.82	7.82	12.02
Kostylevite $K_4Zr_2Si_6O_{18} \cdot 2H_2O$	$P2_1/a$	13.17	11.73	6.56
Umbite $K_2ZrSi_3O_9 \cdot H_2O$	$P2_12_12_1$	10.21	13.24	7.17

α, β, γ (°)	FD	T:M	Framework-forming fragments	References
120	11.75	1.5	$Si_{12}O_{36}$ rings, Si_2O_7 diortho groups, TiO_6 octahedra.	Malinovskii et al. 1976
95.2	14.05	8	Layers with eight-membered silicate channels, TiO_6 octahedra	Khomyakov et al. 1998; Ferraris et al. 2003
	14.5	1.2	Nenadkevichite-type chains of TiO_6 octahedra, xonotlite-type bands Si_6O_{17}.	Mer'kov et al. 1973; Sandomirsky and Belov 1979; Pattison et al. 1998
	14.5-16.7 ?		Si_6O_{18} rings, clusters of ZrO_6 and MnO_6 octahedra, PO_4 tetrahedra.	Makovicky and Karup-Møller 1981; Pekov et al. 1997; Chukanova et al. 2004; Moore and Shen 1983
120	16.7	3.6		
				Kupriyanova et al. 1962; Pekov et al. 1997
	15.0	≤0.67	Pyrochlore-like layers, Si_4O_{12} rings	Balić-Žunić et al. 2002
	15.6	0.67		Portnov et al. 1971; Krivokoneva et al. 1979
120	15.0	3	Hilairite group. See description of the crystal structure in the text.	Khomyakov et al. 1993; Rastsvetaeva and Khomyakov 1992
120	15.4	3		Khomyakov et al. 1996; Rastsvetaeva and Khomyakov 1996
120	15.7	3		Chao et al. 1974; Ilyushin et al. 1981a
120	15.8	3		Boggs 1988; Pushcharovsky et al. 2002
120	15.9	3		Voloshin et al. 1990; Voloshin et al. 1991; Sokolova et al. 1991
	15.3	4	Layers consisting of the rings Si_6O_{18} (in pentagonite) or Si_4O_{12} and Si_8O_{24} (in cavansite); VO^{2+} groups.	Howard 1973
	15.6	4		
	15.6	3	Zigzag $(Si_6O_{18})_\infty$ chains; ZrO_6 octahedra	Chao 1985
	15.5	3		Ghose and Thakur 1985
120	15.7	1.9	Layers of CeO_6 trigonal prisms, Layers of NbO_6 or TiO_6 octahedra	Semenov et al. 1968; Ferraris et al. 2004
120	17.1	2.4		Khomyakov et al. 2003b; Krivovichev and Armbruster 2003
105.3	16.1	5	Thick layers consisting of alternating six-membered rings of the SiO_4 tetrahedra; ten-membered rings of the SiO_4 tetrahedra; isolated ZrO_6 or TiO_6 octahedra.	Ferraris et al. 1995
104.6	16.15	5		LePage and Perrault 1976
105.5	16.0	5		McDonald and Chao 2001
120	16.2	3.25	Double (consisting of the rings Si_6O_{18}) and single tetrahedral layers; $(Fe^{2+},Mg)O_6$ octahedra	Mellini and Merlino 1981
	16.3	0.5	Chains of groups $Ti_4O_{12}(O,OH)_4$; isolated SiO_4 tetrahedra	Menshikov et al. 1992; Sokolova et al. 1989
105.3	16.4	3	Si_6O_{18} rings; isolated ZrO_6 octahedra.	Ilyushin et al. 1981a; Khomyakov et al. 1983b
	16.5	3	Si_3O_9 wollastonite-type chains, isolated ZrO_6 octahedra	Khomyakov et al. 1983c; Ilyushin 1993

Table 2 continued. Crystal-chemical data for MHM.

Name, Formula*	Symmetry	a (Å)	b (Å)	c (Å)
Nenadkevichite subgroup $Na_{4-x}(Nb,Ti)_2Si_4O_{12}(O,OH)_2 \cdot 4H_2O$	$Pbam$	≈7.4	≈14.2	≈7.1
Vuoriyarvite subgroup $(K,Na,Ca,Ba,Sr,H_3O)_{3-x}(Nb,Ti)_2Si_4O_{12}(O,OH)_2 \cdot nH_2O$	Cm	≈14.7	≈14.2	≈7.9
Paratsepinite subgroup $(Na,Ba,Ca,K,Sr,H_3O)_{3-x}(Ti,Nb)_2Si_4O_{12}(OH,O)_2 \cdot nH_2O$	$C2/m$	≈14.6	≈14.1	≈15.7
Lemmleinite subgroup $NaK(K,Ba)_{1-x}Ti_2Si_4O_{12}(OH,O)_2 \cdot 2H_2O$	$C2/m$	≈14.3	≈13.8	≈7.75
Karupmøllerite-Ca $Na_2(Ca,\square)(Nb,Ti)_4[Si_4O_{12}]_2(OH,O)_4 \cdot 7H_2O$	$C2/m$	14.64	14.21	7.915
Organovaite subgroup $K_2(Mn,Zn,Fe)(Nb,Ti)_4(Si_4O_{12})_2(O,OH)_4 \cdot nH_2O$	$C2/m$	≈14.5	≈14.0	≈15.7
Kuzmenkoite subgroup $(K,Ba)_2(Mn,Zn,Fe)(Nb,Ti)_4(Si_4O_{12})_2(O,OH)_4 \cdot nH_2O$	$C2/m$ or Cm	≈14.4	≈13.85	≈7.8
Gutkovaite subgroup $Na_{1-x}(Ca,Sr,K)(K,Ba)_2(Mn,Zn,Fe)(Ti,Nb)_4(Si_4O_{12})_2(O,OH)_4 \cdot nH_2O$	Cm	≈14.45	≈13.9	≈7.8
Labuntsovite subgroup $Na_2K_2(Mn,Mg,Fe)_{1-x}Ti_4(Si_4O_{12})_2(O,OH)_4 \cdot 5\text{-}6H_2O$	$C2/m$	≈14.3	≈13.8	≈7.75
Catapleiite $Na_2ZrSi_3O_9 \cdot 2H_2O$	$P6_2/mmc$, $B2/b$	7.40	7.40	10.05
Calcium catapleiite $CaZrSi_3O_9 \cdot 2H_2O$	Hexagonal or $Pbnn$	7.32	7.32	10.15
Wadeite $K_2ZrSi_3O_9$	$P6_3/m$	6.89	6.89	10.17
Bazirite $BaZrSi_3O_9$	$P\bar{6}c2$	6.755	6.755	9.98
Pabstite $Ba(Sn,Ti)Si_3O_9$	$P\bar{6}c2$	6.71	6.71	9.83
Benitoite $BaTiSi_3O_9$	$P\bar{6}c2$	6.64	6.64	9.795
Petarasite $Na_5Zr_2Si_6O_{18}(Cl,OH) \cdot 2H_2O$	$C2/m$	10.80	14.49	6.62
Kapustinite $Na_{5.5}Mn_{0.25}ZrSi_6O_{16}(OH)_2$	$C2/m$	10.69	10.31	7.41
Litvinskite $Na_{3-x}(\square,Na,Mn^{2+})Zr\ Si_6O_{12}(OH,O)_6 \cdot nH_2O$	Cm	10.59	10.22	7.35
Imandrite $Na_{12}Ca_3Fe_2(Si_6O_{18})_2$	$Pnnm$	10.33	10.55	7.43
Zirsinalite $Na_6(Ca,Mn,Fe^{2+})Zr\ Si_6O_{18}$	$R3m$	10.29	10.29	26.31
Koashvite $Na_6(Ca,Mn)(Ti,Fe)\ Si_6O_{18} \cdot H_2O$	$Pbam$	7.32	21.08	10.26
Lovozerite $Na_{2-x}Ca(Zr,Ti)\ Si_6(O,OH)_{18}$	$R3$	10.18	10.18	13.13
Kazakovite $Na_6MnTiSi_6O_{18}$	$R3m$	10.19	10.19	13.07
Tisinalite $Na_3MnTiSi_6O_{15}(OH)_3$	$P\bar{3}$	10.02	10.02	12.88

α, β, γ (°)	FD	T:M	Framework-forming fragments	References
	16.0-16.2	2	Chains of vertex-sharing TiO_6 or NbO_6 octahedra, Si_4O_{12} rings. The chains can be linked additionally by DO_6 octahedra (D = Fe^{2+}, Mn^{2+}, Zn, Mg or Ca).	Chukanov et al. 2002, 2003b,c, 2004 (for all labuntsovite group including 27 mineral species)
≈118	16.3-17.8	1.8-2		
≈118	16.5-18.0	1.8-2		
≈117	17.3-18.4	1.8-2		
117.4	≈17.5	1.6-1.8		
≈118	18.3-18.8	≈1.6		
≈117	18.5	1.6-1.8		
117.5	18.5	1.6-1.8		
≈117	18.4-18.9	1.6-1.8		
120	16.8	3	Si_3O_9 rings, isolated ZrO_6 octahedra	Brunowsky 1936; Ilyushin et al. 1981b
120	17.0	3		Portnov 1964; Merlino et al. 2004
120	19.1	3		Henshaw 1955; Blinov et al. 1977
120	20.3	3		Young et al. 1978
120	20.9	3		Gross et al. 1965
120	21.4	3		Fischer 1969; Hawthorne 1987
113.2	16.8	3	Minerals of the lovozerite group contain six-membered rings $[Si_6O_9(O,OH)_9]$, isolated MO_6 octahedra (M = Zr, Ti or Fe^{3+}) and additional CO_6 octahedra (C = Mn^{2+}, Ca^{2+}, Fe^{2+} or Na^+) sharing faces with MO_6 octahedra.	Ghose et al. 1980
92.4	17.9	6		Pekov et al. 2003b
92.9	17.6	6		Pekov et al. 2000; Yamnova et al. 2001b
	19.5	6		Khomyakov et al. 1979
120	19.9	6		Kapustin et al. 1974a; Pudovkina et al. 1980
	20.2	6		Kapustin et al. 1974b; Chernitsova et al. 1980
120	20.4	6		Gerasimovsky 1940; Ilyukhin and Belov 1960; Yamnova et al. 2001b
120	20.4	6		Khomyakov et al. 1974a; Voronkov et al. 1979
120	21.4	6		Kapustin et al. 1980; Yamnova et al. 2003

Table 2 continued. Crystal-chemical data for MHM.

Name, Formula*	Symmetry	a (Å)	b (Å)	c (Å)
Khibinskite K$_2$**Zr**Si$_2$O$_7$	$C2/m$	19.22	11.10	14.10
Parakeldyshite Na$_2$**Zr**Si$_2$O$_7$	$P\bar{1}$	9.31	5.42	6.66
Keldyshite (Na,H)$_2$**Zr**Si$_2$O$_7$·nH$_2$O	$P\bar{1}$	9.0	5.34	6.96
Tumchaite Na$_2$(**Zr,Sn**)Si$_4$O$_{11}$ ·2H$_2$O	$P2_1/c$	9.14	8.82	7.54
Penkvilksite-1M Na$_4$**Ti**$_2$Si$_8$O$_{22}$ ·5H$_2$O	$P2_1/c$	8.96	8.73	7.39
Penkvilksite-2O Na$_4$**Ti**$_2$Si$_8$O$_{22}$ ·5H$_2$O	$Pnca$	16.37	8.75	7.40
Elpidite Na$_2$**Zr**Si$_6$O$_{15}$ ·3H$_2$O	$Pbcm$	7.14	14.68	14.65
Belkovite Ba$_3$(**Nb,Ti**)$_6$(Si$_2$O$_7$)$_2$O$_{12}$	$P\bar{6}2m$	8.97	8.97	7.80
Strakhovite NaBa$_3$**Mn**$_4$ [(Si$_4$O$_{10}$(OH)$_2$] (Si$_2$O$_7$)O$_2$ (OH)·H$_2$O	$Pmna$	23.42	12.27	7.18
Eakerite Ca$_2$**Sn**(Al$_2$Si$_6$O$_{18}$) (OH)$_2$·2H$_2$O	$P2_1/a$	15.89	7.72	7.44
Kukisvumite Na$_6$Zn**Ti**$_4$(Si$_2$O$_6$)$_4$O$_4$ ·4H$_2$O	$Pccn$	28.89	8.6	5.22
Manganokukisvumite** Na$_6$**Mn**Ti$_4$(Si$_2$O$_6$)$_4$O$_4$ ·4H$_2$O	$Pccn$	29.05	8.61	5.22
Vinogradovite Na$_5$**Ti**$_4$AlSi$_7$O$_{26}$·3H$_2$O	$C2/c$	24.22	8.56	5.35
Paravinogradovite (Na,☐)$_2$(**Ti,Fe^{3+}**)$_4$(Si$_2$O$_6$)$_2$ (Si$_3$AlO$_{10}$)(OH)$_4$·H$_2$O	$C2/c$	24.49	8.66	5.20
Lintisite Na$_3$Li**Ti**$_2$(Si$_2$O$_6$)$_2$O$_2$ ·2H$_2$O	$C2/c$	28.58	8.60	5.22
Dalyite K$_2$**Zr**Si$_6$O$_{15}$	$P\bar{1}$	7.37	7.73	6.91
Davanite** K$_2$**Ti**Si$_6$O$_{15}$	$P\bar{1}$	7.14	7.53	6.93
Armstrongite Ca**Zr**(Si$_6$O$_{15}$)·3H$_2$O	$C2/m$	14.04	14.16	7.81
Sazhinite-(Ce) Na**NaCe**(Si$_6$O$_{14}$OH)·1.5H$_2$O	$Pmm2$	7.50	15.62	7.35
Baotite Ba$_4$(**Ti,Nb**)$_8$Si$_4$O$_{28}$Cl	$I4_1/a$	19.99	19.99	5.91
Batisite BaNaNa**Ti**$_2$Si$_4$O$_{12}$O$_2$	$Imma$	8.10	10.40	13.85
Shcherbakovite KKNa(**Ti,Nb**)$_2$Si$_4$O$_{12}$(O,OH)$_2$	$Imma$	8.16	10.56	3.99

α, β, γ (°)	FD	T:M	Framework-forming fragments	References
116.5	17.8	2	ZrO_6 octahedra; Si_2O_7 diortho groups	Khomyakov et al. 1974b
94.3, 115.3, 89.6	19.8	2		Khomyakov 1977; Sizova et al. 1974
92, 116, 88	20.0	2		Gerasimovsky 1962; Khalilov et al. 1978
113.2	17.9	4	Layers with Si_4O_{11} composition formed by helical Si_6O_{18} chains; $(Ti,Zr,Sn)O_6$ octahedra	Subbotin et al. 2000
112.7	18.8	4		Merlino et al. 1994
	18.9	4		Merlino et al. 1994
	18.2	6	Bands composed of alternating Si_4O_{12} rings Si_2O_7 groups; ZrO_6 octahedra	Neronova and Belov 1963; Cannillo et al. 1973
120	18.4	0.67	Columns of face-sharing $(Nb,Ti)O_6$ octahedra; Si_2O_7 groups	Voloshin et al. 1991
	19.4	1.5	Si_4O_{12} rings, Si_2O_7 diortho groups, MnO_6 octahedra and MnO_5 semioctahedra.	Kalinin et al. 1994
101.3	20.1	8	Si_4O_{12} and $Si_{12}O_{36}$ rings; SnO_6 octahedra	Kossiakoff and Leavens 1976
	20.1	1.6	Pyroxene chains, columns of edge-sharing TiO_6 octahedra, $(Si,Al)_4O_{10}$ chains.	Yakovenchuk et al. 1991; Merlino et al. 2000
	19.9	1.6	In kukisvimite, manganokukisvumite and lintisite – additional columns of edge-sharing ZnO_6, MnO_6 or LiO_6 octahedra	Gault et al. 2004
101	22.0	2	See description of the crystal structure of the vinogradovite crystal-chemical group in the text.	Semenov et al. 1956; Rastsvetaeva and Andrianov 1984; Kalsbeek and Rønsbo 1992
100.2	22.1	2		Khomyakov et al. 2003a
91.0	21.8	1.33		Khomyakov et al. 1990; Merlino et al. 1990
106.2, 111.5, 100.0	20.2	6	Corrugated layers consisting of four-, six- and eight-membered rings of SiO_4 tetrahedra; $(Zr,Ti)O_6$ octahedra; additional linking of the layers via CaO_6 octahedra (in armstrongite) or via CeO_6 octahedra and NaO_5 polyhedra (in sazhinite-(Ce)).	Fleet 1965
103.0, 114.5, 93.9	21.2	6		Luzhnik et al. 1984
109.3	21.8	3		Kabalov et al. 2000
	18.6	3		Es'kova et al. 1974; Shumyatskaya et al. 1980
	20.3	0.5	Four-membered rings of vertex-sharing $(Ti,Nb)O_6$ octahedra; Si_4O_{12}	Pyatenko et al. 1999
	20.6	2	Si_4O_{12} chains, chains of vertex-sharing TiO_6 octahedra	Nikitin and Belov 1962; Rastsvetaeva et al. 1997; Schmahl and Tillmanns 1987
	19.9	2		Uvarova et al. 2003

Table 2 continued. Crystal-chemical data for MHM.

Name, Formula*	Symmetry	a (Å)	b (Å)	c (Å)
Eudialyte group $N_{12}A_3[M(1,1)M(1,2)]_3[M(2,1)M(2,2)M(2,3)]_{3-6}[M(3)M(4)]\ Z_3[Si_{24}O_{72}]$ $OH_{2-6}X_{2-4}$; N =Na, H_3O^+, □; A = Na, K, Sr, REE, Y, Ba, Mn, Ca, H_3O; $M(1,1)$, $M(1,2)$ = Ca, Mn, REE, Na, Sr, Fe; $M(2,1)$ = Fe, Na, Zr, Ta; $M(2,2)$, $M(2,3)$ = Fe, Mn, Zr, Ti, Na, K, Ba, H_3O; $M(3)$, $M(4)$ = Si, Nb, Ti, W, Na; Z = Zr, Ti, Nb; X = Cl, F, H_2O, OH, CO_3, SO_4. (17 minerals; see Table 4)	$R3m$, $R\bar{3}m$ or $R3$	≈14	≈14	≈30 or ≈60
Joaquinite-(Ce) Ba_2Ce_2**Na**Fe^{2+}**Ti**$_2$**Si**$_8O_{26}$(OH)·H_2O	$C2$	10.52	9.69	11.83
Strontiojoaquinite Ba_2Sr_2(**Na,Fe**$^{2+}$)$_2$**Ti**$_2$**Si**$_8O_{24}$(O,OH)$_2$·H_2O	$Pm, P2$ or $P2/m$	10.52	9.76	11.87
Orthojoaquinite-(Ce) Ba_2Ce_2**Na**Fe^{2+}**Ti**$_2$**Si**$_8O_{26}$(OH)·H_2O	$Ccmm, Ccm2_1$ or $Cc2m$	10.49	9.66	22.26
Orthojoaquinite-(La) Ba_2La_2**Na**Fe^{2+}**Ti**$_2$**Si**$_8O_{26}$(OH)·H_2O	$Ccmm$	10.54	9.68	22.345
Byelorussite-(Ce) Ba_2Ce_2**Na****Mn**$^{2+}$**Ti**$_2$**Si**$_8O_{26}$(F,OH)·H_2O	$Ama2$	22.30	10.51	9.67
Strontio-orthojoaquinite Ba_2Sr_2(**Na,Fe**$^{2+}$)$_2$**Ti**$_2$**Si**$_8O_{24}$(O,OH)$_2$·H_2O	$Pcam$ or $Pca2_1$	10.52	9.78	22.39
Bario-orthojoaquinite $(Ba,Sr)_4$**Fe**$^{2+}_2$**Ti**$_2$**Si**$_8O_{26}$·H_2O	$Ccmm, Ccm2_1$ or $Cc2m$	10.48	9.60	22.59
Verplanckite Ba_4(**Mn,Ti,Fe**)$_2$**Si**$_4O_{12}$ (OH,H_2O)$_3Cl_3$	$P6/mmm$	16.40	16.40	7.20
Narsarsukite Na_2(**Ti,Fe**$^{3+}$)**Si**$_4O_{10}$(O,OH,F)	$I4/m$	10.73	10.73	7.95
Naujakasite Na_4**Fe**$^{2+}$(**Al**$_4$**Si**$_8O_{26}$)	$C2/m$	15.04	7.99	10.49
Manganonaujakasite Na_4**Mn**$^{2+}$(**Al**$_4$**Si**$_8O_{26}$)	$C2/m$	15.03	8.00	10.48
Fersmanite $(Na,Ca)_4$**Ca**$_4$(**Ti,Nb**)$_4$(**Si**$_2O_7$)$_2O_8F_3$	$C2/c$	10.18	10.18	20.40
Vlasovite Na_2**Zr****Si**$_4O_{11}$	$C2/c$	11.11	10.11	8.61
Muirite Ba_{10}(**Ca,Mn,Ti**)$_4$**Si**$_8O_{24}$ (OH,Cl,O)$_{12}$·$4H_2O$	$P4/mmm$	14.03	14.03	5.64
Terskite Na_4**Zr**(**Si**$_6O_{16}$) ·$2H_2O$	$Pnc2$	14.12	14.69	7.51
Stokesite Ca_2**Sn**$_2$**Si**$_6O_{18}$·$4H_2O$	$Pnna$	14.465	11.625	5.235

* Framework cations are given in bold face.
**Minerals with unknown crystal structures are included in a crystal-chemical groups taking into account chemical composition and unit cell parameters.

α, β, γ (°)	FD	T:M	Framework-forming fragments	References
120	18-21	1.4-2.0	Si_3O_9 and Si_9O_{27} rings; rings of edge-sharing CaO_6 or alternating CaO_6 and $(Mn,Fe)O_6$ octahedra, isolated ZrO_6 or TiO_6 octahedra, the polyhedra $(Fe,Mn,Ta,Zr...)O_n$ (n=4-6), additional SiO_4 tetrahedra and $(Nb,W,Zr...)O_6$ octahedra	Johnsen and Gault 1997; Johnsen and Grice 1999; Johnsen et al. 1999, 2003; Chukanov et al. 2004 (for all eudialyte-group minerals)
109.7	21.1	2	Sheets formed by Si_4O_{12} rings; sheets containing Ti_2O_{10} pairs of edge-sharing octahedra and REE atoms, $(Mn,Fe,Zn)O_5$ polyhedra, NaO_6 octahedra. See description of the crystal structure of joaquinite-group minerals in the text.	Dowty 1975
109.2	20.8	2		Wise 1982
	21.3	2		Dowty 1975; Wise 1982
	21.1	2		Matsubara et al. 2001
	21.1	2		Shpanov et al. 1989; Zubkova et al. 2004
	20.8	2		Chihara et al. 1974; Wise 1982
	21.1	2		Wise 1982
120	21.5	2	Si_4O_{12} rings, three-membered groups of $(Mn,Ti,Fe)O_6$ octahedra	Kampf et al. 1973
	21.85	4	Bands formed by Si_4O_{12} rings, chains of vertex-sharing $(Ti,Fe)O_6$ octahedra.	Wagner et al. 1991
113.7	22.5	12	Two-layer sheets formed by alternating Si_4O_{12} and Si_6O_{18} rings, FeO_6 and MnO_6 octahedra	Basso et al. 1976
113.5	22.5	12		Khomyakov et al. 2000
97.2	22.9	0.5	A pyrochlore-type octahedral layer, Si_2O_7 groups, ribbons of CaO_6 octahedra	Sokolova et al. 2002
100.7	21.1	4	Chains formed by Si_4O_{12} rings, ZrO_6 octahedra	Tikhonenkova and Kazakova 1961; Voronkov and Pyatenko 1961
	≤21.8	2	Si_8O_{24} rings; pairs of trigonal prisms $(Ca,Mn,Ti)O_6$.	Malinovsky et al. 1976
	18.0	6	Branched $[H_2Si_3O_6]_\infty$ chains; ZrO_6 octahedra	Khomyakov et al. 1983a; Pudovkina and Chernitsova 1991
	18.2	3	The chains Si_6O_{18}; SnO_6 octahedra	Vorma 1963

Table 3. Labuntsovite-group minerals $A_2B_2C_2DM_4(Si_4O_{12})_2(O,OH)_4 \cdot nH_2O$

Name	A	B	C	D	M	Z	Space group	Subgroup
	\multicolumn{5}{c}{Prevailing cations}			
Nenadkevichite	Na (for $A+B+C$)			-	Nb	1	$Pbam$	Nenadkevichite
Korobitsynite	Na (for $A+B+C$)			-	Ti	1	$Pbam$	
Vuoriyarvite-K	K (for $A+B+C$)			□	Nb	2	Cm	Vuoriyarvite
Tsepinite-Na	Na (for $A+B+C$)			□	Ti	2	Cm	
Tsepinite-K	K (for $A+B+C$)			□	Ti	2	Cm	
Tsepinite-Ca	Ca (for $A+B+C$)			□	Ti	2	Cm	
Tsepinite-Sr	Sr (for $A+B+C$)			□	Ti	2	Cm	
Paratsepinite-Ba	Ba (for $A+B+C$)			□	Ti	4	$C2/m$	Paratsepinite
Paratsepinite-Na	Na (for $A+B+C$)			□	Ti	4	$C2/m$	
Lemmleinite-K	Na	K	K	□	Ti	2	$C2/m$	Lemmleinite
Lemmleinite-Ba	Na	K	Ba	□	Ti	2	$C2/m$	
Labuntsovite-Mn	Na	K	□	Mn	Ti	2	$C2/m$	Labuntsovite
Labuntsovite-Mg	Na	K	□	Mg	Ti	2	$C2/m$	
Labuntsovite-Fe	Na	K	□	Fe	Ti	2	$C2/m$	
Paralabuntsovite-Mg	Na	K	□	Mg	Ti	4	$I2/m$	Paralabuntsovite
Karupmøllerite-Ca	□	Na	□	Ca	Nb	2	$C2/m$	Kuzmenkoite
Gjerdingenite-Fe	□	K	□	Fe	Nb	2	$C2/m$	
Gjerdingenite-Mn	□	K	□	Mn	Nb	2	$C2/m$	
Kuzmenkoite-Mn	□	K	□	Mn	Ti	2	$C2/m$ or Cm	
Kuzmenkoite-Zn	□	K	□	Zn	Ti	2	Cm	
Lepkhenelmite-Zn	□	Ba	□	Zn	Ti	2	Cm	
Organovaite-Mn	□	K	□	Mn	Nb	4	$C2/m$	Organovaite
Organovaite-Zn	□	K	□	Zn	Nb	4	$C2/m$	
Parakuzmenkoite-Fe	□	K	□	Fe	Ti	4	$C2/m$	
Gutkovaite-Mn	Ca + □	K	□	Mn	Ti	2	Cm	Gutkovaite
Alsakharovite-Zn	Na + Sr	K	□	Zn	Ti	2	Cm	
Neskevaaraite-Fe	Na + K	K	□	Fe	Ti	2	Cm	

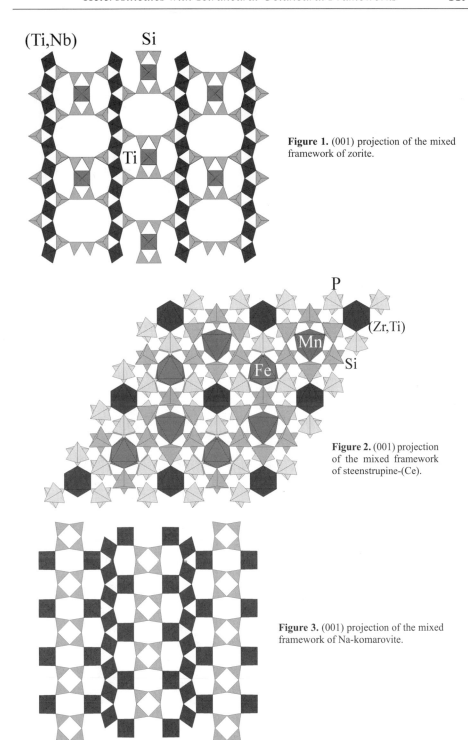

Figure 1. (001) projection of the mixed framework of zorite.

Figure 2. (001) projection of the mixed framework of steenstrupine-(Ce).

Figure 3. (001) projection of the mixed framework of Na-komarovite.

The knowledge of the komarovite structure allows to predict compositional variations. A deficit in cationic valence caused by Ti^{4+} for Nb^{5+} substitution could be compensated by Ca substituting Na. Taking into account also a further lanthanide/Ca substitution, the following general formula for the komarovite-type minerals can be written (Balić-Žunić et al. 2002): $(Na,K)Na_{5+x-y}Ca_{1-2x+y}La_xTi_yNb_{6-y}Si_4O_{26}F_2·4H_2O$. The ranges of the Ti:Nb ratio and content of A cations in minerals of the komarovite group are very wide (Pekov et al. 2004). In particular, samples with Ti > Nb are known whereas the total number of large cations varies from 1.8 up to 7.2 apfu. The most common isomorphous substitutions in the komarovite-type minerals are the same as in pyrochlore-group minerals and involve cations of the pyrochlore-like layer: Nb^{5+} + O^{2-} ↔ Ti^{4+} + OH^-; $2M^+$ ↔ M^{2+} + , where M^+ = Na, K; M^{2+} = Ca, Sr, Ba, Pb; Na ↔ $(H_3O)^+$. For this reason, cation-deficient komarovite-type minerals are considered as possible products of natural leaching and hydration of a Na-dominant protophase (see Azarova et al. 2002).

Hilairite-group minerals. The minerals of this group are represented by the formula $A_{2-5}M_2Si_6O_{18}·6H_2O$ (A = Na, Ca, Ba and minor K, Sr; M_2 = Zr_2, YZr or YTi) and crystallize in the space goup $R32$; they are MHM with very loose frameworks (FD = 15–16; Pekov et al. 2003a). The basis of their structure is a mixed framework formed by infinite [001] screwed Si_3O_9 chains that are linked together by isolated MO_6 octahedra via common vertices (Fig. 4). Each MO_6 octahedron is connected to three silica-oxygen chains. The structure contains large cages and [001] channels occupied by alkaline and alkaline-earth cations and water molecules. Two independent M sites are present: one contains Zr or Ti, the other can be occupied either by Zr, or by Y + *HREE* (Ilyushin et al. 1981c; Sokolova et al. 1991; Rastsvetaeva and Khomyakov 1992, 1996; Pushcharovsky et al. 2002).

Ilímaussite-type minerals. The crystal structure of ilímaussite-(Ce), $(Ba,Na)_{10}K_3Na_{4.5}Ce_5(Nb,Ti)_6[Si_{12}O_{36}][Si_9O_{18}(O,OH)_{24}]O_6$, from the type locality of the Ilímaussaq alkaline complex, South Greenland (Semenov et al. 1968), has been solved and refined to R = 0.11 for 1663 observed X-ray diffractions collected from a poor crystal (Ferraris et al. 2004). The structure of ilímaussite-(Ce) (Fig. 5) consists of three pairs of (001) silicate sheets which are sandwiched: CeO_6 trigonal prisms and Na cations (A layer); NbO_6 octahedra, Ba, and K (O layer); CeO_6 trigonal prisms (A' layer). The layers are stacked according to the sequence $AOA'O$ and the tetrahedral sites in A' are only 50% occupied by Si with consequent disorder of the basal oxygen atoms. For this reason, the number of Si atoms per formula unit with Z = 3 for ilímaussite-(Ce) is 21 instead of 24 as in the case of diversilite-(Ce).

Diversilite-(Ce), the K-rich Ti analogue of ilímaussite-(Ce), was first described as a new mineral with symmetry $R\bar{3}$ containing isolated Si_3O_9 rings and silicate ortho-groups (Khomyakov et al. 2003b; Rastsvetaeva et al. 2003). A reinvestigation of the structure by Krivovichev et al. (2003) on a sample from Mt. Yukspor, Khibiny massif demonstrated that correct space group is $R32$. Like ilímaussite-(Ce), the heteropolyhedral framework of diversilite-(Ce) consists of different types of layers of corner-linked polyhedra. Two non-equivalent, but topologically identical mixed (Si, *REE*) layers are connected by a layer consisting of $(Ti,Nb)O_5(OH)$ octahedra. The SiO_4 tetrahedra share corners to form six-membered rings which are further interlinked via $REEO_6$ trigonal prisms and $(Ti,Nb)(O,OH)_6$ octahedra. Three of five independent Si sites belonging to one of the two non-equivalent mixed layers have refined occupancy factors of 2/3. Three types of extra-framework cations, Ba, K and Na + Ca, occupy different sites with high selectivities. The following formula for ilímaussite-type minerals has been suggested: $(Ba,K,Na,Ca)_{11-12}(REE,Fe,Th)_4(Ti,Nb)_6(Si_6O_{18})_4(OH)_{12}·4.5H_2O$.

Diversilite-(Ce) and ilímaussite-(Ce) are identical by symmetry and have close unit-cell parameters; presumably, a series of ilímaussite-type minerals with 21 ≤ Si ≤ 24 apfu can exist, as well as different polytypes correlated with the chemical composition (Ferraris et al. 2004).

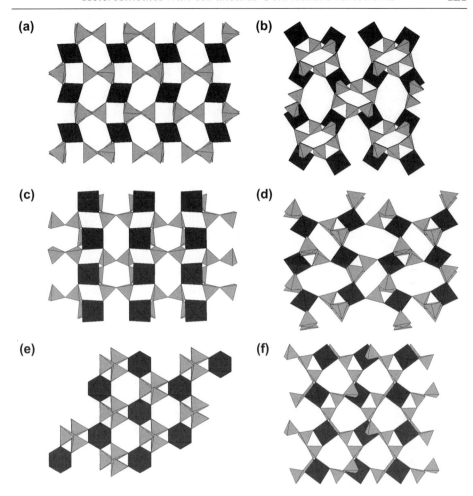

Figure 4. Mixed frameworks of minerals with the framework composition $ZrSi_3O_9$: (100) projection of the crystal structures of catapleiite (a) and kostylevite (b); (001) projection of the crystal structures of petarasite (c), umbite (d), hilairite (e) and gaidonnayite (f).

Labuntsovite-group minerals

Crystal-chemical features of the labuntsovite-group minerals (LGM) have been studied in depth (Chukanov et al. 1999, 2002, 2003b,c, 2004; see also Krivovichev 2005 and McCusker 2005). At present, 27 mineral species belonging to this group are known (Table 3). The LGM are hydrous MHM with frameworks consisting of chains of MO_6 octahedra (M = Ti, Nb), linked by four-membered Si_4O_{12} rings via vertices shared between tetrahedra and octahedra. Channels and cages within the mixed frameworks are occupied by H_2O and extra-framework cations (Na, K, Ca, Sr, Ba).

The LGM are characterized by unusually wide ranges of the main components (wt%): Na_2O 0–14, K_2O 0–15, CaO 0–7, SrO 0–9, BaO 0–17, MgO 0–2, MnO 0–7, FeO 0–5, ZnO 0–7, TiO_2 1–27, Nb_2O_5 0–39, SiO_2 35–46, H_2O 6–17. These variations are connected not only with the compositions of the mineral-forming media, but also with differences in crystal structures of LGM.

Figure 5. (010) projection of the mixed framework of ilímaussite-(Ce) (the upper *REE* layer corresponds to *A'* in the text).

The labuntsovite group includes orthorhombic and monoclinic members. In the orthorhombic members (nenadkevichite-korobitsynite series, see Fig. 6), the [100] chains of corner-sharing MO_6 octahedra have a zigzag planar configuration. The cages contain Na as prevailing cation. In the monoclinic members, the chains of the octahedra are strongly wave-like bent in the (001) plane; in correspondence of the sites where they are closer, the chains can be linked additionally by DO_6 octahedra (D = Fe^{2+}, Mn^{2+}, Zn, Mg or Ca). The presence of vacancies of extra-framework and D cations is typical for the LGM. Three non-equivalent extra-framework cation sites (A, B, and C) occur in the most ordered monoclinic LGM (members of labuntsovite, lemmleinite and paralabuntsovite subgroups); their general formula is: $A_4B_4C_4D_2M_8(Si_4O_{12})_4(OH,O)_8 \cdot nH_2O$; A = Na; B = K, sometimes with minor Na; C = Ba, K. The distance between the sites C and D is short (~ 2.1 Å) and cations cannot occur in both sites at the same time. The following mechanism governs the occupancy of C and D sites:

$$(Fe,Mg,Mn,Zn) + 2H_2O \leftrightarrow \square + 2(K,Ba)$$
$$\quad D \qquad\qquad\qquad C \quad\; D \qquad C$$

The main mechanisms of isomorphous substitutions in LGM involving M cations (i.e., Ti and Nb) are Ti + OH $\leftrightarrow$ Nb + O and Ti + A $\leftrightarrow$ Nb + $\square$. Thus, variations of total amounts of D and M cations are mutually independent; the sum of Ti and Nb contents is constant. Figures 7 and 8 illustrate this conclusion. Similar isomorphic schemes are valid for many other MHM.

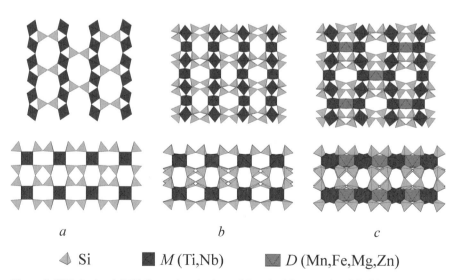

Figure 6. (001) (top) and (100) (bottom) projections of the mixed frameworks of the labuntsovite-group minerals: nenadkevichite type (a), vuoriyarvite type (b) and labuntsovite type (c).

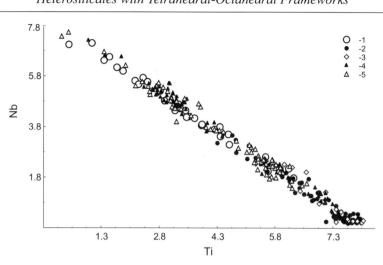

Figure 7. Correlation between Ti and Nb contents (calculated per 16 Si atoms) in labuntsovite-group minerals. 1 – nenadkevichite subgroup; 2 – labuntsovite and paralabuntsovite subgroups; 3 – lemmleinite subgroup; 4 - kuzmenkoite and organovaite subgroups; 5 – vuoriyarvite subgroup.

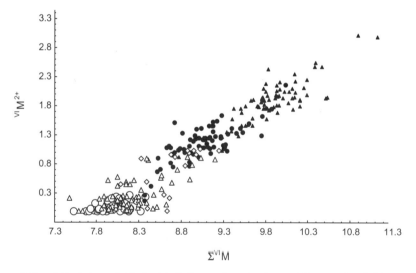

Figure 8. Correlation between total of octahedrally coordinated cations and D-type cations in labuntsovite-group minerals. Symbols as in Figure 7.

Minerals of the labuntsovite subgroup (Table 3) are D-populated (and correspondingly C-vacant) LGM; members of the lemmleinite subgroup are their D-vacant analogues; in paralabuntsovite the c parameter is doubled as a result of ordering in the C and D sites; kuzmenkoite is the A-deficient analog of labuntsovite (it can have split B sites and symmetry lowered to Cm); gjerdingenite-Fe and gjerdingenite-Mn are Nb-dominant analogues of kuzmenkoite-Fe and kuzmenkoite-Mn respectively; karupmøllerite-Ca is an analog of hypothetical gjerdingenite-Ca with Na prevailing in the B site; lepkhenelmite-Zn is a Ba-analog of kuzmenloite-Zn. A specific type of cation ordering is realized in the members of the gutkovaite subgroup. In these minerals the splitting of the A site into two non-equivalent and differently populated positions leads to the loss of the centre of symmetry. In the members of the organovaite subgroup, doubling of the c parameter is due to translational non-equivalency of the adjacent octahedral chains. Paratsepinite-Ba and paratsepinite-Na are D-deficient analogues of organovaite-subgroup minerals. In the structures of paratsepinites, members of the vuoriyarvite subgroup and some members of the kuzmenkoite subgroup, extra-framework cation sites are split into numerous sub-sites having low occupancies and often mixed populations. In paratsepinite-Na, paratsepinite-Ba and lepkhenelmite-Zn the number of such subsites reaches 10.

Isomorphism between LGM belonging to different subgroups is often limited. It is confirmed, in particular, by the presence of sharp interfaces between syntactic or epitactic intergrown pairs of these minerals (labuntsovite-Mn and tsepinite-Na; kuzmenkoite-Mn and lemmleinite-Ba; labuntsovite-Fe and neskevaaraite-Fe; alsakharovite-Zn, kuzmenkoite-Zn and tsepinite-Na; vuoriyarvite-K and korobitsynite, etc.; Chukanov et al. 2003c). Gutkovaite-Mn can occur in close associations with labuntsovite-Mn or with kuzmenkoite-Mn, and in both pairs each phase is well distinct. Continuous isomorphous series are known for LGM within different subgroups and between the members of the labuntsovite and the lemmleinite subgroups. Also, one can-not exclude the possibility of solid solutions between tsepinite and disordered kuzmenkoite in wide ranges of composition.

Other important MHM

Vinogradovite-related minerals. The mixed framework of vinogradovite $Na_5Ti_4Si_7AlO_{26} \cdot 3H_2O$ consists of alternating pyroxene chains, columns of edge-sharing TiO_6 octahedra and Si_4O_{10} chains (Rastsvetaeva and Andrianov 1984; Kalsbeek and Rønsbo 1992; Khomyakov et al. 2003a). Paravinogradovite $(Na,K,\square)_2[(Ti^{4+},Fe^{3+})_4(Si_2O_6)_2(Si_3AlO_{10})(OH)_4] \cdot H_2O$, a new mineral species from the Khibiny alkaline massif (Khomyakov et al. 2003a), is related to vinogradovite both in crystal structure and chemical composition. The mixed framework of paravinogradovite is based on vinogradovite-type tetrahedral chains Si_3AlO_{10} (consisting of the four-membered $[T_4O_{12}]$ rings connected through common vertices), two types of pyroxene-type tetrahedral chains Si_2O_6 and two types of edge-sharing chains of the $(Ti,Fe)O_6$ octahedra. All chains run along [100] and link together sharing vertices of octahedra and tetrahedra. One of two brookite-type octahedral chains is decorated by Na. The remaining Na and K cations are distributed within the [100] channels among five sites with occupancies ≤ 20%.

There is a close relationship between the topology of the frameworks in paravinogradovite and vinogradovite. Ordering of Na and vacancies in paravinogradovite results in two types of octahedral-tetrahedral sheets. In vinogradovite only sheets of one (Na-dominant) type are present.

In kukisvimite, manganokukisvumite and lintisite (minerals structurally related to vinogradovite, Table 2), columns of edge-sharing ZnO_6, MnO_6 or LiO_6 octahedra impart additional rigidity to the framework (Fig. 9).

Umbite. In the framework of this mineral, $K_2ZrSi_3O_9 \cdot H_2O$, $FD = 16.5$ (Fig. 4), Si_3O_9 wollastonite-type chains are linked

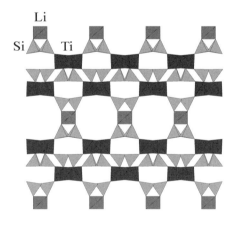

Figure 9. (001) projection of the mixed framework of lintisite.

together by isolated ZrO_6 octahedra; each octahedron is coordinated to six SiO_4 tetrahedra (Ilyushin 1993; Döbelin and Armbruster 2003). [001] channels with elliptical cross-section are formed by rings consisting of four tetrahedra and four octahedra and contain three extra-framework sites, two of which are occupied by K and one by H_2O. A Ti analog of umbite is unknown in nature, though the synthesis of the isostructural solid solutions $K_2(Zr,Ti)Si_3O_9 \cdot H_2O$ including the Ti end member has been reported (Clearfield et al. 1998).

Lovozerite group. In the crystal structures of minerals with the general formula $A_3B_3C_2\{M[Si_6O_9(O,OH)_9]\}$ belonging to this group, six-membered rings $[Si_6O_9(O,OH)_9]$ are linked together by isolated MO_6 octahedra (M = Zr, Ti or Fe^{3+}) via common vertices of tetrahedra and octahedra (Voronkov et al. 1979; Chernitsova et al. 1980; Ghose et al. 1980; Pudovkina et al. 1980; Tamazyan and Malinovsky 1990; Yamnova et al. 2001a,b, 2003; Pekov et al. 2003b, Fig. 10). Free vertices of the tetrahedra and to lower extent common vertices of SiO_4 tetrahedra and the MO_6 octahedra can be occupied by the OH groups. Additional C octahedra share faces with MO_6 octahedra and can contain Mn^{2+}, Ca^{2+}, Fe^{2+}, Na^+ and vacancies. The C site is unstable due to the $C - M$ cation repulsion; thus it is not considered as a part of the framework. The extra-framework sites A and B are occupied mainly by Na, H_2O, or vacant. Kapustinite represents a rare case of the lovozerite-type framework with 30% of vacancies in the M site (Pekov et al. 2003b).

Figure 10. (001) projection of the mixed framework of litvinskite (lovozerite group).

Joaquinite group. This group includes seven monoclinic and orthorhombic minerals with formula $A_6Ti_2Si_8O_{24}(O,OH,F)_{2-3} \cdot 0-1H_2O$; A = Ba, Sr, Na, REE, Fe, Mn (Table 2). For long time joaquinite-(Ce) $NaFe^{2+}Ba_2Ce_2(TiO)_2[Si_4O_{12}]_2(OH) \cdot H_2O$ (space group $C2$) was the only representative of the group with known structure (Dowty 1975). Recently, the crystal structure of its orthorhombic and Mn-dominant analog, byelorussite-(Ce), $NaMnBa_2Ce_2(TiO)_2[Si_4O_{12}]_2(F,OH) \cdot H_2O$, has been solved (Zubkova et al. 2004). In both structures mixed layers consisting of three sheets are present (Fig. 11). The two other sheets are formed by Si_4O_{12} rings; the inner sheet contains Ti_2O_{10} pairs of edge-sharing octahedra. The Ti atoms are shifted from the plane of the sheet leaving space for REE atoms. The layers are linked together via $(Mn,Fe,Zn)O_5$ polyhedra and NaO_6 octahedra to form a heteropolyhedral framework with large cavities containing Ba cations.

Catapleiite type. Catapleiite, $Na_2ZrSi_3O_9 \cdot 2H_2O$, and its calcian analog calcium catapleiite, $CaZrSi_3O_9 \cdot 2H_2O$, (Merlino et al. 2004) are end-members of a continuous solid solution. The mixed

Figure 11. (001) projection of the mixed framework of byelorussite (joaquinite group).

framework of these hexagonal or pseudohexagonal minerals has the same composition $ZrSi_3O_9$ as in umbite, hilairite, gaidonnayite, kostylevite and petarasite, but all these minerals have topologically distinct frameworks (Fig. 4). In the catapleiite-type structures, regular isolated ZrO_6 octahedra and three-membered Si_3O_9 rings are linked together; the large cations (Ca, Na) and water molecules are located in the [100], [110] and [1̄10] channels located in the tetrahedral-octahedral framework. For catapleiite, two close polymorphic modifications are known: $P6_3/mmc$, $a = 7.40$, $c = 10.05$ Å (Brunowsky 1936) and $B2/b$, $a = 23.917$, $b = 20.148$, $c = 7.432$ Å, $\gamma = 147.46°$ (Ilyushin et al. 1981b). The sites occupied by Na atoms and H_2O molecules exchange in the two polymorphs. In calcium catapleiite (*Pbnn*, $a = 7.378$, $b = 12.779$, $c = 10.096$ Å; Merlino et al. 2004), the Ca cations are distributed between two distinct sites with partial occupancy.

Wadeite, bazirite, pabstite and benitoite have catapleiite-type mixed frameworks, but these anhydrous MHM are characterized by higher *FD* values (Table 1). Isomorphous substitutions of Ti and Zr by Sn are a characteristic feature of the wadeite-type minerals (Gross et al. 1965; Rudenko et al. 1983).

Eudialyte-group minerals. Crystal-chemical features and nomenclature of this complex mineral group are described in detail elsewhere (Johnsen and Gault 1997; Johnsen and Grice 1999; Johnsen et al. 2003; Chukanov et al. 2004; Tables 2 and 4). Their framework contains three- and nine-membered rings of SiO_4 tetrahedra and six-membered rings of MO_6 octahedra (M = Ca, Mn, Fe, Na, *REE*; either order or disorder is reported within the rings) linked to each other mainly by cations of transition elements (Zr, Fe, Mn, Ti, Nb, Ta) or Na. The coordination numbers of the linking cations vary between 4 and 6. Additional cations (Si, Nb, W, Ti, Na) with valencies from 1 to 6 and coordination numbers from 4 to 6 can be situated at the center of the nine-membered rings. Different combinations of local situations involving framework and extra-framework cations, cation ordering, and cationic and anionic (Cl, F, OH, CO_3) isomorphous substitutions lead to the unusual variety of the eudialyte-group minerals including variations of their framework density.

Table 4. Members of the eudialyte group approved by CNMMN (after Johnsen et al. 2003)

Mineral Name	Formula
Alluaivite	$Na_{19}(Ca,Mn)_6(Ti,Nb)_5Si_{26}O_{74}Cl \cdot 2H_2O$
Aqualite	$(H_3O)_8(Na,K,Sr)_5Ca_6Zr_3Si_{26}O_{66}(OH)_9Cl$
Carbokentbrooksite	$(Na,\square)_{12}(Na, REE)_3Ca_6Mn_3Zr_3Nb(Si_{25}O_{73})(OH)_3(CO_3) \cdot H_2O$
Eudialyte	$Na_{15}Ca_6Fe_3Zr_3Si(Si_{25}O_{73})(O,OH,H_2O)_3(Cl,OH)_2$
Feklichevite	$Na_{11}Ca_9(Fe^{3+},Fe^{2+})_2Zr_3Nb[Si_{25}O_{73}](OH,H_2O,Cl,O)_5$
Ferrokentbrooksite	$Na_{15}Ca_6Fe_3Zr_3Nb(Si_{25}O_{73})(O,OH,H_2O)_3(F,Cl)_2$
Georgbarsanovite	$Na_{12}(Mn,Sr,REE)_3Ca_6Fe^{2+}{}_3Zr_3NbSi_{25}O_{76}Cl_2 \cdot H_2O$
Ikranite	$(Na,H_3O)_{15}(Ca,Mn,REE)_6Fe^{3+}{}_2Zr_3(\square,Zr)(\square,Si)Si_{24}O_{66}(O,OH)_6Cl \cdot 2–3H_2O$
Kentbrooksite	$Na_{15}Ca_6Mn_3Zr_3Nb(Si_{25}O_{73})(O,OH,H_2O)_3(F,Cl)_2$
Khomyakovite	$Na_{12}Sr_3Ca_6Fe_3Zr_3W(Si_{25}O_{73})(O,OH,H_2O)_3(Cl,OH)_2$
Labyrinthite	$(Na,K,Sr)_{35}Ca_{12}Fe_3Zr_6TiSi_{51}O_{144}(O,OH,H_2O)_9Cl_3$
Manganokhomyakovite	$Na_{12}Sr_3Ca_6Mn_3Zr_3W(Si_{25}O_{73})(O,OH,H_2O)_3(Cl,OH)_2$
Oneillite	$Na_{15}Ca_3Mn_3Fe_3Zr_3Nb(Si_{25}O_{73})(O,OH,H_2O)_3(OH,Cl)_2$
Raslakite	$Na_{15}(Ca_3Fe_3)(Na,Zr)_3(Si,Nb)(Si_{25}O_{73})(OH,H_2O)_3(Cl,OH)_2$
Rastsvetaevite	$Na_{27}K_8Ca_{12}Fe_3Zr_6Si_{52}O_{144}(O,OH,H_2O)_6Cl_2$
Taseqite	$Na_{12}Sr_3Ca_6Fe_3Zr_3NbSi_{25}O_{73}(O,OH,H_2O)_3Cl_2$
Zirsilite-(Ce)	$(Na,\square)_{12}(REE,Na)_3Ca_6Mn_3Zr_3Nb(Si_{25}O_{73})(OH)_3(CO_3) \cdot H_2O$

Other minerals. Batisite (Kravchenko et al. 1960) and shcherbakovite (Es'kova and Kazakova 1957) are isostructural MHM. In their crystal structures (Nikitin and Belov 1962; Schmahl and Tillmanns 1987; Rastsvetaeva et al. 1997; Uvarova et al. 2003), the SiO_4 tetrahedra link together to form a Si_4O_{12} chain. The TiO_6 octahedra link via common vertices to form straight TiO_5 chains. These two types of [100] chains link together to form a mixed tetrahedral-octahedral framework (Figs. 12 and 13). Each TiO_6 octahedron is connected through common vertices to four SiO_4 tetrahedra, and each SiO_4 tetrahedron is connected to two octahedra and two tetrahedra. The octahedral site is split into two subsites separated by 0.48 Å; in the octahedral subsites Ti can be partly substituted by Nb and Fe^{3+}. In these structures there are three types of interstitial cages significantly different in size. The largest cage is populated by Ba and K; the intermediate cage by K, Na and minor Ba; the smallest cage contains Na only. For the so-called "noonkanbahite" (Prider 1965) ordering of all main extra-framework is suggested with prevailing Ba, K and Na in the three cages, in the order (Uvarova et al. 2003).

Besides the above described mixed-framework zeolite-like minerals, there are species with unknown crystal structures that presumably belong to the MHM family. Candidates

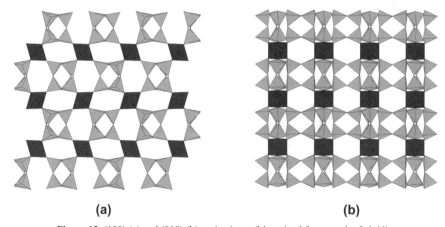

Figure 12. (100) (a) and (010) (b) projections of the mixed framework of elpidite.

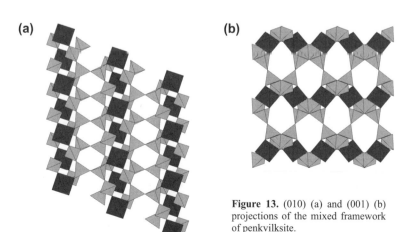

Figure 13. (010) (a) and (001) (b) projections of the mixed framework of penkvilksite.

are: laplandite-(Ce), $Na_4CeTiPSi_7O_{22} \cdot 5H_2O$; ilmajokite, $(Na,Ce,Ba)_2TiSi_3O_5(OH)_{10} \cdot nH_2O$ (both occur in the Yubileynaya pegmatite); paraumbite, $K_3HZr_2Si_6O_{18} \cdot 7H_2O$; umbozerite, $Na_3Sr_4(Mn,Zn,Fe)ThSi_8O_{24}(OH)$; karnasurtite-Ce, $(Ce,La,Th)(Ti,Nb)(Al,Fe^{3+})(Si,P)_2O_7(OH)_4 \cdot 3H_2O$; mongolite, $Ca_4Nb_6Si_5O_{24}(OH)_{10} \cdot 5-6H_2O$. Most of these minerals were formed in alkaline (usually peralkaline) rocks.

Factors influencing the crystal-chemical diversity of MHM

The preceding discussion of the crystal structures of natural MHM, with special attention to large groups of such minerals (e.g., labuntsovite, hilairite, eudialyte and lovozerite groups), demonstrates their unusual structural and compositional variability. Low values of the force parameters related to the angles O-M-O and O-D-O lead to the observed "flexibility" of mixed frameworks. Indeed, frequencies of normal OMO- and OTO-bending vibrations in silicates and oxides usually do not exceed 300 cm^{-1}, whereas OSiO-bending vibrations occur in the IR range 450-650 cm^{-1} (Plyusnina 1967). As a result, the ratio of corresponding bending force parameters is: $k_{OSiO} : k_{OMO} \approx v_{OSiO}^2 : v_{OMO}^2 \geq 4$; in other words, the deformation of the O-M-O fragment requires approximately 4 times less energy than for O-Si-O.

On the other hand, the presence of cations that are very different by size and force characteristics results in a wide variability of the framework configuration and different mechanisms of isomorphous substitutions that sometimes are exotic or impossible for the minerals with rigid structures. Some examples of isomorphous substitutions which are typical for MHM, although unusual, are listed below.

1. Substitution of a pair of extra-framework C cations by the complex $[D(H_2O)_2]^{2+}$ in the labuntsovite group (see above).

2. Substitution accompanied by a change of coordination in the eudialyte group: $Fe^{2+}O_4$ (square) → $Fe^{3+}O_5$ (tetragonal pyramid) → $Fe^{3+}O_6$ (octahedron).

3. Possibility of isomorphous substitutions between cations very different by charge in eudialytes: $Na^+ \to Al^{3+} \to Si^{4+} \to Nb^{5+} \to W^{6+}$ (at the center of the nine-membered rings); $Na^+ \to Fe^{2+} \to Zr^{4+} \to Ta^{5+}$ (in the position with 4-fold planar coordination linking the rings of the CaO_6 octahedra).

4. Partial occupancy of the A and D sites; possibility of high concentrations of vacancies in the M sites (kapustinite) and even in the T sites [ilímaussite-(Ce) – diversilite-(Ce)].

5. Competition between neighboring extra-framework sites that can not be simultaneously occupied by cations (labuntsovites).

6. Possibility for some cations (Ca, REE etc.) to play either a framework or extra-framework role (labuntsovites, eudialytes, steenstrupine). Rarely Na with 6-fold coordination can be involved in a framework; in sazhinite-(Ce) both framework and extra-framework Na is present.

7. Possibility of selective ordering of cations in different extra-framework sites along with existence of disordered analogues (e.g., cation-ordered labuntsovite-group minerals and their disordered analogues). Such site selectivity is a very important property in view of technological applications. In particular, it occurs in some labuntsovite-group minerals (e.g., lemmleinite-Ba, labuntsovites) and almost all eudialyte-group minerals.

8. Existence of cation-deficient varieties for most MHM (partly because of natural leaching). Strongly decationated members (in which the amounts of the A cations are much lower than ideal contents) are known for eudialytes, labuntsovites and some other MHM. In these cases the charge balance is achieved through enrichment of MHM with hydrogen in the forms of H^+, H_3O^+, and Ti-OH, Nb-OH, Si-OH groups.

9. Unlike zeolites, MHM can incorporate extra-framework cations with charges > 2 (such as REE^{3+}, Th^{4+}, U^{4+}).

10. The substitution of Ce by Fe reported by Krivovichev et al. (2003) for diversilite-(Ce) and by Ferraris et al. (2004) for ilímaussite-(Ce) is not common in minerals.

PROPERTIES AND SOME GENETIC IMPLICATIONS

Ion exchange, cation leaching and gas sorption

Specific "zeolitic" properties (e.g., sorption capability, ion-exchange and catalytic properties) of microporous framework heterosilicates depend on the configuration of the channels and local properties of active centers such as polarity, Lewis acidity, force characteristics (Chukanov et al. 2004). Ion-exchange properties of microporous titano-, niobo- and zirconosilicates are of great interest taking into account their capability to accumulate some radioactive species (cf. Rocha and Lin 2005).

Synthetic compounds. The literature about ion-exchange properties of natural MHM and their synthetic analogues is limited. For synthetic MHM the following examples are of interest. Microporous potassium and sodium titanosilicates obtained in hydrothermal synthesis are shown to accumulate ions of different metals from water solutions (Bortun et al. 1999). The ion exchange behavior of the compounds $K_2MSi_3O_9 \cdot H_2O$ with umbite structure and their protonated endmembers, $K_{0.3}H_{1.7}TiSi_3O_9 \cdot 2H_2O$ and $K_{0.5}H_{1.5}ZrSi_3O_9 \cdot 2H_2O$, towards alkali metal ions has been studied: a clear correlation between the exchanger composition (channels size) and selectivity towards certain ions has been found (Poojary et al. 1997; Clearfield et al. 1998). Zirconium-rich silicates with large channels exhibit affinity for rubidium and cesium ions, whereas titanium-rich compounds with smaller channels show a preference for potassium. Ion exchange experiments were carried out also for the removal of trace cesium (^{137}Cs) and trace strontium (^{89}Sr) isotopes by different cationic forms of a titanosilicate analogue of pharmacosiderite (Dyer et al. 1999). Selectivity factors were estimated by determining batch distribution coefficients (K_d) as a function of Na, K, Mg and Ca concentration. Other effects studied were pH, increasing cesium and strontium ion concentration, hydrolysis, presence of complexing agents, and the resistance of the material to radiation damage. The microporous titanosilicate $K_2TiSi_3O_9 \cdot H_2O$ was subjected to ion exchange of other alkaline cations and NH_4^+ (Valtsev et al. 1999). The material can be totally exchanged by NH_4^+ and to different extents by the alkaline cations. The substitution of K leads to changes in the thermal stability and water adsorption capacity as well as of the cell parameters. The framework of the material is very flexible and is stabilized by the presence of cation-water complexes. Rietveld refinement showed that the two cationic positions are not equally exchangeable. Small and highly hydrated cations occupy preferentially the 8-membered ring channel; large cations occupy the 7-sided window. A microporous synthetic analogue of the mineral penkvilksite-1M has been characterized by ion-exchange and absorption measurements by Liu et al. (1999).

These and some other examples of application of synthetic materials with 6-fold coordinated framework atoms have been reviewed by Chukanov et al. (2004). Along with heterosilicates, microporous oxide materials containing transition elements can show pronounced ion-exchange ability.

Labuntsovite-group minerals. These are typical microporous silicates with various M (Ti, Nb), D (Fe, Mg, Mn, Zn), A (Na, K, Ca, Sr, Ba), symmetry, and A cations arrangement. For single-crystals (0.3–0.7 mm in size) of seven representatives of this group cation exchange has been investigated at room conditions using aqueous solutions of NaCl, Na_2CO_3, K_2CO_3, $CaCl_2$, $SrCl_2$, $BaCl_2$ and CsCl during 5 months (Pekov et al. 2002). Orthorhombic members and most cation-ordered monoclinic members (labuntsovite and lemmleinite) are inactive towards cation-exchange processes. Members of the vuoriyarvite subgroup are instead most effective ion-exchangers owing to the absence of the D cations and low contents and disordered arrangement of the A cations. In each grain ion exchange proceeds in the whole volume and practically does not depend on the distance from the grain surface. Thus, the energetic barrier at the interphase surface more than diffusion within the crystal limits the process (Pekov et al. 2002). An observed lowering of the cation-exchange activity after heating is connected with formation of additional bonds between octahedral chains (Turchkova et al. 2003).

The ion-exchange properties of 0.3–0.8 mm particles of natural penkvilksite, zorite, sazykinaite and elpidite have been investigated in neutral 1 M aqueous solutions of KCl, RbCl, CsCl, $SrCl_2$ and $Pb(NO_3)_2$ at room temperature (authors' unpublished data). Penkvilksite from Yubileinaya pegmatite (Lovozero) shows moderate ion-exchange activity towards Pb and low activity towards alkaline and alkaline-earth elements. The maximal PbO contents in the cores of penkvilksite grains reaches 5.6 wt% after 4 months in contact with a $Pb(NO_3)_2$ solution. The amount of other exchanged cations does not exceed 1.3 wt% under similar conditions. Under the same conditions zorite from Yubileinaya pegmatite accumulates the following quantities of oxides (wt% in core and other parts of the particles respectively): K_2O 17.2, 14.8; Rb_2O 17.8, 17.3; Cs_2O 21.3, 21.7; CaO 4.5, 6.0; SrO 7.7, 12.2; PbO 42.9, 42.9; a corresponding lowering of the Na content was observed (up to 0 in the case of exchange for Pb). For sazykinaite-(Y) from Mt. Koashva (Khibiny) an active ion exchange takes place only with a potassium salt and only in the rim (content of K_2O changes there from 3.2 up to 8.4 wt%). In elpidite from Khan Bogdo massif (Mongolia), the whole volume of the particles shows moderate ion-exchange activity towards potassium and about 3 wt% of K_2O is accumulated (Na_2O content lowers approximately from 9 to 6 wt%). The distribution of exchanged cations can be different (Fig. 14): homogeneous (alkali cations in zorite); mainly in the rim of the crystal (bivalent cations in zorite); near the interfaces of concentric compositional zoning of the crystal (sazykinaite); near the interface between individuals in twins (tsepinite); near micro-fractures (elpidite).

Other MHM. In seidite-(Ce) the Na cations are easily exchanged for larger cations (Tl, K, Rb, Cs and Ba) indicating the microporous behavior of its structure (Ferraris et al. 2003; Ferraris and Gula 2005). Taking into account the observed spatial distribution of isomorphous substitutions, the following scheme of reaction of natural ion-exchange process has been suggested for hilairite (Pekov et al. 2003a): $2Na^+ + H_2O \rightarrow 0.5Ca^{2+} + 1.5\square + H_3O^+$.

According to Konnerup-Madsen et al. (1981), in the course of leaching of uranium from steenstrupine in 1 M solution of Na_2CO_3 (at 500°C, 1000 bar), the content of Na in the solid phase can increase from 2.2 up to 13 wt% in the period of 6 days. Leaching of Na from steenstrupine in water solutions, as well as the reverse process of saturation of steenstrupine by Na cations proceeds very effectively and is not accompanied by substantial changes in the content of other metal cations. By analogy with lovozerite-group minerals, Moore and Shen (1983) suggested that in this process charge balance is achieved by hydration of the six-membered ring: $Si_6O_{18} \rightarrow Si_6O_{12}(OH)_6$.

Boiling kapustinite in water at pH $\approx$ 7 for 3 hours results in leaching of Na accompanied by protonation and formation of $Si-OH^{+\delta}$ and H_3O^+ acidic groups that in the IR spectrum give bands in the ranges of 1400–1900 and 2800–3200 cm^{-1} (Pekov et al. 2003b). The final product of interaction of kapustinite with water is litvinskite. In the course of this process the B cations undergo full leaching, whereas the A cations are involved in leaching only to a little extent. Hydration of kapustinite is accompanied by displacement of C cations. Similar mechanisms of hydration are typical for other lovozerite-group minerals.

Behavior in nature. The processes of ion-exchange and leaching of MHM are probably very common in nature under hypergeneous as well as hydrothermal conditions. In the IR

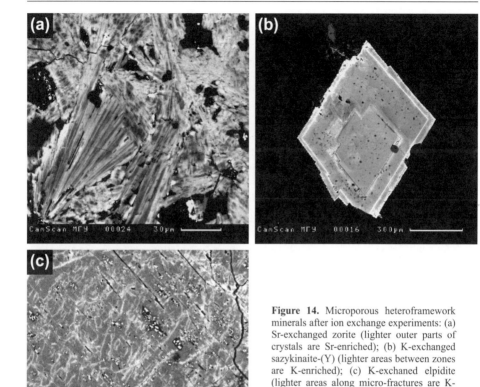

Figure 14. Microporous heteroframework minerals after ion exchange experiments: (a) Sr-exchanged zorite (lighter outer parts of crystals are Sr-enriched); (b) K-exchanged sazykinaite-(Y) (lighter areas between zones are K-enriched); (c) K-exchaned elpidite (lighter areas along micro-fractures are K-enriched).

spectra of "decationated" varieties of different MHM (steenstrupine, lovozerite-, eudialyte- and hilairite-group minerals and some other MHM) broad bands of weakly acidic OH groups (such as H_3O^+ or Si-OH) are often observed in the range 2800–3200 cm^{-1} (Pekov et al. 2003b).

Leaching of A cations from zeolites is hindered owing to charge balance requirements; in fact, charge compensation by the mechanism of protonation of the T-O-T links is energetically unfavorable. For this reason, cation-deficient aluminosilicate zeolites are not common in nature. On the contrary, cation-deficient varieties of MHM with $FD < 23$ are widely present in alkaline rocks. In particular, a deficit of Na (and to a lesser extent of other A cations) is typical for minerals of labuntsovite, hilairite, lovozerite, eudialyte groups, komarovite, vinogradovite, lintisite, sitinakite, steenstrupine and even for fersmanite (with $FD = 22.9$). Leaching of Na from MHM is often accompanied by protonation of the M-O-M links and apical M-O bonds of the framework. In some cases, protonation of the M-O-Si bridges is also possible (Yamnova et al. 2003). The most striking example is lintisite that can lose all A cations without collapsing the crystal structure, to form a phase with the composition $LiTi_2Si_4(O,OH)_{14} \cdot nH_2O$ (Kolitsch et al. 2000).

Among monovalent cations with low force characteristics, Na⁺ has the smallest ionic radius and does not have a stable and bulk solvate shell in aqueous solutions (Breck 1974). As a result, sodium is the most mobile A cation and by leaching and ion-exchange processes is removed from MHM most easily. Ionic conductivity of natural microporous sodium zirconosilicates (catapleiite, lovozerite group minerals: Ilyushin and Demyanets 1986, 1988), effects of migration of sodium in MHM stimulated by electron beam or X-ray irradiation (e.g., vlasovite; Gobetchiya et al. 2003) are also favored by the high mobility of Na⁺.

Sorption of gas. Recently, synthetic microporous Ti- and Nb-silicates have been shown to be effective and selective gas sorbents (Kuznicki et al. 2000; Chukanov et al. 2004). Measurements of gas sorption and desorption by the static volumetric method (using AUTOSORB-1 device, N_2 as adsorbate and He as carrier; authors' unpublished data) show that raw aqueous forms of natural MHM are moderate gas sorbents at room conditions. Their activity towards gas sorption depends on the mean particle size.

For example, specific surface area of hydrothermal elpidite from Lovozero massif is 0.4 m²/g for a massive sample, 0.9 m²/g for long-prismatic crystals (0.1–0.2 mm thick), and 14.6 m²/g for a powder sample. Though mean pore radii for these samples are 424, 173 and 118 Å respectively, the maximum of their pore size distribution densities lies below 10 Å, and the role of internal surface in gas sorption increases by grinding the sample. That means that the free volume is determined mainly by non-structural pores, whereas the specific surface area depends mainly on structural pores. Similarly, it has been observed that the specific surface area of kuzmenkoite-Mn (pseudomorphs after lomonosovite from Lovozero massif) increases after grinding and for a powder sample is 29.1 m²/g, the mean pore radius being 50.9 Å. For the same material, the maximum of the pore size distribution density determined from the sorption isotherm also lies below 10 Å, but the corresponding value determined from the data on desorption is 14.5 Å. The latter value is close to the a parameter of kuzmenkoite-Mn ($a \approx$ 14.4 Å). The low of the sorption-desorption hysteresis for kuzmenkoite-Mn is intermediate between the theoretical lows for cylindrical and conical pore shapes.

Catapleiite is inactive towards gas sorption; its specific surface area even after grinding does not exceed 1.1 m²/g and is determined by the outer surface of the particles.

Other properties

Infrared spectroscopy. In the IR spectra of MHM, the ranges of stretching frequencies involving the H-O, T-O, M-O, D-O and A-O bonds do not overlap; in the order they are: 2000–3700, 850–1100, 550–750, 400–500 and < 400 cm⁻¹. Specific values of the frequencies within each of these ranges are due to several factors, the most important being hydrogen bonding (for the range 2000–3700 cm⁻¹), linkage of framework-forming polyhedra and the kind of dominant species among the T, M and D types. Let us consider some examples.

For silicates, according to the cluster approximation (Chukanov and Kumpanenko 1988), high-frequency Si-O stretching vibrations are independent from other modes. A simple correlation exists between the weighted average frequency $\langle \nu_{\text{Si-O}} \rangle$ of the Si-O-stretching vibrations and the mean number of vertices that a SiO_4 tetrahedron shares with other SiO_4 tetrahedra (Chukanov 1995). For aluminosilicates with the anion stoichiometry $Si_xAl_yO_z$, the correlation is the following (Chukanov 1995):

$$\langle \nu_{\text{Si-O}} \rangle \text{ (cm}^{-1}) = (337.8t + 1827)(0.6428t + 1)^{-1} \text{ where } t = z(x + y/2)^{-1}.$$

Similarly, condensation of MO_6 and DO_6 octahedra leads to changes of the $\langle \nu_{(M,D)-O} \rangle$ values. For manganese oxides with purely octahedral frameworks, a linear correlation exists between wavenumbers of the strongest bands of Mn-O-stretching vibrations and the number n of edges shared per MnO_6 octahedron (Potter and Rossman 1979).

In MHM, linking of framework polyhedra by atoms with lower force characteristics results in smaller relative changes of the $\langle v_{Si-O}\rangle$ or $\langle v_{M-O}\rangle$ values. For example, in terms of cluster approximation and first-order perturbation theory it was shown that the linking of chains of (Ti,Nb)O_6 octahedra (M-octahedra) by additional D-octahedra (D = Fe, Mg, Mn or Zn) in the labuntsovite group results in a linear correlation between $\langle v_{M-O}\rangle$ (in the range 660–700 cm^{-1}) and the number of D atoms per formula unit (Rastsvetaeva and Chukanov 2001, 2002). This band is single and does not depend on the Ti:Nb ratio. Indeed, the position of the Ti(Nb)-O stretching band does linearly correlate with the occupancy of the D site (Fig. 15; when the structure is not known, the occupancy of the D site has been obtained from the chemical data alone as n-8, n being the total number of octahedrally coordinated atoms Ti, Nb, Fe, Mg, Mn, Zn).

This example demonstrates that the D cations give rise to substantial local perturbations of the force field: in the case of complete occupancy of the D site, the observed shifts of the (Ti,Nb)-O stretching bands under the influence of the D cations reach 30 cm^{-1}. These perturbations only slightly depend on the nature of the D cations. Thus, IR spectroscopy enables to obtain information on the distribution of the cations (e.g., Ca, Sr, or Na) that play a dual role in the structures of zeolite-like minerals, without resorting to X-ray diffraction analysis. This is of particular importance when X-ray diffraction studies cannot be performed because of poor crystals.

Along with these cations, some transition-metal cations can also play a dual role in the labuntsovite-like minerals. For example, the total amount of manganese and zinc in some specimens is substantially higher than the theoretical limit for the D cations (i.e., two atoms per 16 Si atoms). However, an excess of Mn + Zn ions does not lead to a substantial

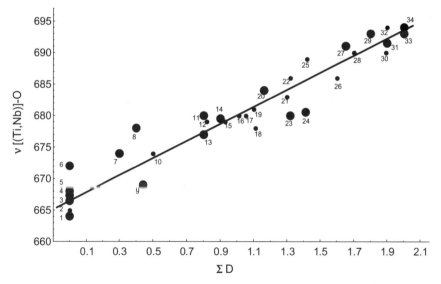

Figure 15. Correlation between the stratching frequency of (Ti,Nb)-O and the content of D cations in labuntsovite-group minerals calculated from structural data (large circles) and as $\Sigma^{VI}M^{2+}$ = Ti + Nb + Mg + Mn + Fe + Zn − 8 for samples with unknown structures (small circles). The numbers correspond to korobitsynite (1-3), nenadkevichite (4, 5), lemmleinite-K (6-8, 10), tsepinite-Na (9), "paralabuntsovite-☐" (11), "labuntsovite-☐" (12, 14, 15), lemmleinite-Ba (13), labuntsovite-Fe (16), labuntsovite-Mg (17, 21, 22, 26), paralabuntsovite-Mg (18), labuntsovite-Mn (19, 21, 23), vuoriyarvite-K (20), karupmøllerite-Ca (24), kuzmenkoite-Mn (25, 27, 30), kuzmenkoite-Zn (28), organovaite-Mn (29), gutkovaite-Mn (31), organovaite-Zn (32, 33) and parakuzmenkoite-Fe (34).

increase in the $v_{(Ti,Nb-O)}$ frequency; hence, the excess ions act as typical extraframework ions. Apparently, in this case, the potassium or barium ions in the C site are replaced by complexes like $[Zn(H_2O)_{1-2}]^{2+}$ (Chukanov et al. 2003b; Raade et al. 2004).

Heating of labuntsovite-group minerals results in formation of additional links between neighboring octahedral chains which can be detected by a shift of the band of (Ti,Nb)-O stretching vibrations towards short-wave region. Additional bonding of octahedral chains leads to a substantial lowering of the cation-exchange activity (Turchkova et al. 2003).

Optical properties. The relationships between chemical composition and optical properties of zeolite-like minerals have been investigated by Zadov et al. (2003). In particular, it was demonstrated that heterovalent isomorphous substitutions of framework cations by cations with lower valences (e.g., Si by Al, Nb by Ti, or Be by Li) lead to an increase of the refractive indices due to the increasing of the total number of A and D cations per unit cell which is required for the charge compensation via the following scheme: $(T,M)^{q+} + x \cdot (A,D) \leftrightarrow (T,M)^{(q+1)+} + (H_2O, \square)$.

Thus, the optical properties of the microporous silicate minerals are not sensitive to the nature of relative light A cations (Na, K, Ca, Sr) as well as D cations. Among A cations, only Ba and Cs have been found to substantially influence the optical properties of zeolites and MHM. In the case of labuntsovite-group minerals, the refractive index γ is more sensitive to the content of barium than α and β (Zadov et al. 2003).

Interaction of MHM with organic substance in nature

In alkaline and especially peralkaline rocks most natural MHM form mainly under hydrothermal conditions. Up to the last decade, modeling of mineral formation processes was the main purpose of the synthesis of alkaline silicates containing transition metals. In this way, several analogues of minerals with microporous zeolite-like structures have been obtained. In particular (see Chukanov et al. 2004), eudialyte was synthesized under moderately alkaline conditions in the system $6SiO_2 + ZrO_2 + 6Na_2CO_3 + CaCO_3 + FeCl_2 + 4H_2O$, with catalytic amounts of Na_2SiF_6 or K_2SiF_6, at 450–550°C and 85 to 700 bars of partial water pressures; khibinskite was obtained from a melt in the system $KOH + K_2CO_3 + SiO_2 + ZrSiO_4$ by gradual cooling from 1250°C to room temperature; parakeldyshite, has can be obtained under hydrothermal conditions in the system $ZrO_2 + SiO_2 + Na_2O + H_2O$. Synthetic analogues of zirconosilicate minerals petarasite, gaidonnayite, umbite, kostylevite and wadeite have been obtained mainly under hydrothermal conditions (Döbelin and Armbruster 2003; Rocha and Lin 2005).

The above examples demonstrate that zeolite-like amphoterosilicates can be formed under a wide range of conditions. Sometimes organic substances can serve as promoters in the synthesis of these materials. For example, ETS-10 has been obtained from titanosilicate gels in the presence of an organic templating agent, both in the presence and absence of ETS-4 (Das et al. 1996). Similar processes can proceed in nature. Bituminous substances in pegmatites of Khibiny massif are effective concentrators of Th, U, rare-earth and alkaline-earth elements. These elements are present in organic phases as aromatic carboxylate salts as well as in microscopic inclusions of mineral phases formed after the transformation of metallorganic phase.

Phase separation in bitumen-like substances containing metalloorganic complexes, leads to the formation of titano- and niobo-silicates of Th, Ca, Ln, Y, Na and K as microphases. Along with known minerals, the crystallization of new amphoterosilicate phases such as (Th,*REE*)(Ti,Nb)(Si,P)$O_6 \cdot nH_2O$, an analog of seidite-(Ce) with Th > *REE*, of a niobosilicate with (Nb+Ti):Si ≈ 3:2 and other MHM has been detected within segregations of bitumens or in close associations with natural organic substances in Lovozero and Khibiny massifs

(Ermolaeva et al. 2005). On the other hand, results of recent investigations (Chukanov et al. 2005) demonstrate a significant geochemical role of these minerals as ion-exchangers and promotors of abiogenic synthesis of organic compounds at the late stages of the formation of alkaline massifs. In their turn, the oxygen-bearing aromatic organic compounds are complexing agents and perform transport and concentration of rare and radioactive elements (e.g., Y, Ln, Th, U).

Solid organic substances. Bitumens and bitumen-like solid organic substances (OS) are common components of peralkaline rocks and related pegmatites and of hydrothermalites in Khibiny and Lovozero massifs, Kola Peninsula. Their most abundant accumulation is observed usually in low-temperature hydrothermal parageneses together with zeolites and especially MHM (e.g., labuntsovite-group minerals, elpidite, hilairite, vinogradovite). In these parageneses, solid OS often form segregations reaching several millimeters in size.

Most investigated samples of OS from Khibiny massif (including so-called "carbocer"; see Chukanov et al. 2003a) contain polycyclic aromatic compounds with oxygen-bearing groups. The prevalent presence of condensed aromatic structures in carbocer was demonstrated recently by IR spectroscopy and X-ray powder diffraction patterns (Chukanov et al. 2005). Along with carboxylate, carboxyl and/or carbonyl groups are present in all investigated samples. Abundant bitumen-like substances similar to carbocer have been found in a number of peralkaline pegmatites on the Mts. Koashva and Kukisvumchorr (Khibiny massif) together with minerals formed at the hydrothermal stage. OS's enriched in oxygen and water are common components of late parageneses of Khibiny massif. In most cases, solid OS occur in close association with MHM. In low-temperature parageneses of the peralkaline Lovozero massifs, the highest concentrations of bitumen-like substrates are clearly confined in large accumulations of microporous Ti-, Nb- and Zr-silicate minerals: elpidite, labuntsovite-group minerals and others. Often bitumens are observed in direct contact with microporous Ti-, Nb- and Zr-silicates. Abundant OS are present in cavities with MHM whereas in neighboring cavities without such minerals bituminous substances are absent (Chukanov et al. 2005). Instead, a connection of OS with alkaline heterophyllosilicates (murmanite, lomonosovite, astrophyllite- and lamprophyllite-groups minerals) is not so obvious. Therefore, the available data likely indicate a promoting role of microporous Ti-, Nb- and Zr-silicates in formation and conversions of organic substances in alkaline pegmatites, hydrothermalites and residual interstitial fluids.

Synthetic micro- and mesoporous materials bearing transition metals, and especially titanosilicates, are known as promising heterogeneous catalysts and catalyst carriers for selective oxidation, reforming, polymerization and other reactions in organic chemistry (Chukanov et al. 2004; Rocha and Lin 2005). In particular, modified microporous titanosilicates ETS-10 and ETAS-10 are efficient catalysts of n-hexane reforming (Philippou et al. 1998). ETS-10 was used as catalyst in alkyne polymerization (Zecchina et al. 2001). Different reactions of selective oxidation of organic substrates on synthetic microporous Ti- and Nb-silicate catalysts have been described recently (Chukanov et al. 2004). Similar catalytic processes could lead to the formation of oxygen-bearing derivatives of naphthalene series hydrocarbons present in alkaline rocks. Light hydrocarbons (methane, ethane, ethylene) as common components of inclusions in minerals from alkaline rocks (Ikorskii 1977; Nivin 1985), can be possible precursors of bitumen-type substances. On the other hand, microporous amphoterosilicate catalysts bearing basic active sites can absorb small molecules from gases and solutions and concentrate them near catalytic sites. For example, ETS-10 and its ion-exchanged derivatives have been found to absorb H_2, CO, N_2, NO, C_2H_4 and NH_3 (Zecchina et al. 1999; Gervasini et al 2000; Otero Arean et al. 2000). As it was demonstrated above, elpidite and kuzmenkoite-Mn (i.e., MHM that are often accompanied by bitumen-type substances in alkaline pegmatites) behave as gas sorbents even under mild conditions. Anyhow, regularly observed neighborhood of OS with microporous titano-, niobo- and zirconosilicates in hydrothermal associations of alkaline pegmatites is undoubtedly not an occasional, but a genetic caused phenomenon.

CONCLUSIONS

1. The diversity of structures, chemical compositions and properties of microporous heterosilicate minerals indicates them as possible prototypes of technological zeolite-like materials.

2. Force constants k_{ii}, $T:M$ ratio and framework density FD (related respectively to the framework flexibility, the framework topology and porosity) give the basis for simple empirical classification of MHM.

3. Heterosilicates with very loose frameworks ($FD < 16$) are important for technological applications.

4. The presence of cations that are very different by size and force characteristics results in wide variability of the framework configuration and different mechanisms of isomorphous substitutions; the latter are sometimes exotic or impossible for minerals with rigid structures.

5. Topochemical mechanisms of ion-exchange in crystals of MHM are very diverse. Ion-exchange and leaching of MHM are very common in nature under hypergene as well as hydrothermal conditions.

6. Additional condensation and heterovalent substitutions of framework cations influence IR and optical properties of MHM.

7. In late derivatives of alkaline rocks, there is an interconnection between MHM and condensed organic substance.

ACKNOWLEDGMENTS

The authors are grateful to G. Ferraris and S. Merlino for constructively reading the manuscript and to A. Gula for the help with bibliography. This work was supported in part by the grant No. 04-05-64085 of the Russian Foundation of Basic Researches (RFBR), the grant No. 1087-2003-5 of the Fundamental Science School Program of the Russian Academy of Sciences and the joint grant No. 03-05-20011 of RFBR and BNTS (Austria).

REFERENCES

Ageeva OV, Borutsky BE, Chukanov NV, Sokolova MN (2002) Alluaivite and genetic aspect of formation of titanium-enriched eudialytes in Khibiny massif. Zap Vseross Mineral Obs 131(1):99-106 (in Russian)

Anderson MW, Terasaki O, Ohsuna T, O'Malley PJ, Philippou A, MacKay SP, Ferreira A, Rocha J, Lidin S (1995) Microporous titanosilicate ETS-10: a structural survey. Philos Mag B71:813-841

Azarova YuV, Pekov IV, Chukanov NV, Zadov AE (2002) Products and processes of vuonnemite alteration at low-temperature transformation of hyperagpaitic pegmatites. Zap Vseross Mineral Obs 131(5):112-121 (in Russian)

Balić-Žunić T, Petersen OV, Bernhardt H-J, Micheelsen HI (2002) The crystal structure and mineralogical description of a Na-dominant komarovite from the Ilímaussaq alkaline complex, South Greenland. N Jb Mineral Mh 2002:497-514

Balzer B, Langbein H (1996) Formation of pyrochlore phases in the $PbO-Nb_2O_5(-Fe_2O_3)$ system. J Alloys Compd 244:1-6

Basso R, Dal Negro A, Della Giusta A, Ungaretti L (1976) The crystal structure of naujakasite. Geol Sur Greenl Rep 116:11-24
Bish DL, Ming DW (eds) (2001) Natural zeolites: occurrence, properties, applications. Rev Mineral Geochem, vol 45, Mineral Soc America, Washington
Blinov VA, Shumyatskaya NG, Voronkov AA, Ilyukhin VV, Belov NV (1977) Refinement of the crystal structure of wadeite $K_2Zr[Si_3O_9]$ and its relationship to kindred structural types. Sov Phys Crystallogr 22:31-35
Boggs RP (1988) Calciohilairite, $CaZrSi_3O_9 \cdot 3H_2O$, the calcium analogue of hilairite from Golden Horn Batolith, Northern Cascades, Washington. Am Mineral 73:1191-1194
Bortun AI, Bortun LN, Khainakov SA, Clearfield A, Trobajo C, Garcia JR (1999) Hydrothermal synthesis and ion exchange properties of the novel framework sodium and potassium niobium silicates. Solvent Extr Ion Exch 17:649-675
Breck DW (1974) Zeolites, molecular sieves: Structure, chemistry and use. John Wiley & Sons Inc, New York
Brunowsky B (1936) Die struktur des katapleits ($Na_2ZrSi_3O_9 \cdot 2H_2O$). Acta Physicochim URSS 5(6):863-892
Bussen IV, Men'shikov YuP, Mer'kov AN, Nedorezova AP, Uspenskaya YeI, Khomyakov AP (1974) Penkvilksite, a new titanium-sodium hydrosilicate. Sci USSR, Earth Sci 217:126-128
Cannillo E, Rossi G, Ungaretti L (1973) The crystal structure of elpidite. Am Mineral 58:106-108
Chao GY (1985) The crystal structure of gaidonnayite $Na_2ZrSi_3O_9 \cdot 2H_2O$. Can Mineral 23:11-16
Chao GY, Watkinson DH, Chen VV (1974) Hilairite, $Na_2ZrSi_3O_9 \cdot 3H_2O$, a new mineral from Mont St. Hilaire, Quebec. Can Mineral 12:237-240
Chaudhari K, Bal R, Srinivas D, Chandwadkar AJ, Sivasanker S (2001) Redox behavior and selective oxidation properties of mesoporous titano- and zirconosilicate MCM-41 molecular sieves. Microporous Mesoporous Mater 50:209-218
Chelishchev NF (1972) Ion-exchange properties of astrophyllites under supercritical conditions. Geokimiya 7:856-860 (in Russian)
Chen G, Takenoshita H, Kasuya M, Satoh H, Kamegashira N (1995) Synthesis and properties of a new complex oxide $Dy_2MnTa_{1+x}O_{7+\delta}$ with a pyrochlore-related structure. J Alloys Compd 228:127-131
Chernitsova NM, Pudovkina ZV, Voronkov AA, Pyatenko YuA (1980) Crystal structure of koashvite $Na_6(Ca, Mn)_{1+0.5x}(Fe^{3+}_xTi_{1-x})[Si_6O_{18}]$. Mineral Zh 2(5):40-44 (in Russian)
Chihara K, Komatsu M, Mizota T (1974) A joaquinite-like mineral from Ohmi, Niigata prefecture, Central Japan. Mineral J 7:395-399
Chukanov NV (1995) On infrared spectra of silicates and aluminosilicates. Zap Vseross Mineral Obs 124(3): 80-85 (in Russian)
Chukanov NV, Kumpanenko IV (1988) Cluster approach in the vibrational spectroscopy of polymers. Chem Phys Lett 146:211-215
Chukanov NV, Pekov IV, Khomyakov AP (2002) Recommended nomenclature for labuntsovite-group minerals. Eur J Mineral 14:165-173
Chukanov NV, Pekov IV, Rastsvetaeva RK (2004) Crystal chemistry, properties and synthesis of microporous silicates containing transition elements. Russ Chem Rev 73(3):205-223
Chukanov NV, Pekov IV, Rastsvetaeva RK, Nekrasov AN (1999) Labuntsovite: solid solutions and features of the crystal structure. Can Mineral 37:901-910
Chukanov NV, Pekov IV, Sokolov SV, Nekrasov AN, Chukanova VN (2003a) On the nature of "carbocer" from Khibiny and state of thorium and rare-earth elements present in it. Proc. of XXI All-Russian Seminar "Alkaline magmatism of the Earth". Apatity:165-166 (in Russian).
Chukanov NV, Pekov IV, Sokolov SV, Nekrasov AN, Chukanova VN, Naumova IS (2005) On the formation and geochemical role of bituminous substances in pegmatites of Khibiny and Lovozero alkaline massifs, Kola peninsula, Russia. Geochem Intern (in Russian) in press
Chukanov NV, Pekov IV, Zadov AE, Rozenberg KA, Rastsvetaeva RK, Krivovichev SV (2003b) New minerals tsepinite-K, $(K,Ba,Na)_2(Ti,Nb)_2(Si_4O_{12})(OH,O)_2 \cdot 3H_2O$, and paratsepinite-Ba, $(Ba,Na,K)_{2-x}(Ti,Nb)_2(Si_4O_{12})(OH,O)_2 \cdot 4H_2O$, and their relationships with other representatives of the labuntsovite group. Zap Vseross Mineral Obshch 132(1):38-51 (in Russian)
Chukanov NV, Pekov IV, Zadov AE, Voloshin AV, Subbotin VV, Sorokhtina NV, Rastsvetaeva RK, Krivovichev SV (2003c) Labuntsovite-Group Minerals. Moscow, Nauka (in Russian)
Clearfield A, Bortun AI, Poojary DM, Khainakov SA, Bortun LN (1998) On the selectivity regulation of $K_2ZrSi_3O_9 \cdot H_2O$-type ion exchangers. J Molec Struct 470:207-213
Das TK, Chandwadkar AJ, Sivasanker S (1996) Studies on the synthesis, characterization and catalytic properties of the large pore titanosilicate, ETS-10. J Molec Catalysis A-Chem 107:199-205
Döbelin N, Armbruster Th (2003) Synthesis and structure analysis of the microporous titanosilicate $K_2TiSi_3O_9 \times H_2O$. XV International Conf. X-Ray Crystallography and Crystal Chemistry of Minerals. Saint-Petersburg, p 80-81

Dowty E (1975) Crystal structure of joaquinite. Am Mineral 60:872-878
Dyer A, Pillinger M, Newton JA, Harjula RO, Moller JT, Tusa EH, Suheel A, Webb M (1999) Ion exchange material. Patent WO 9958243 (GB)
Ermolaeva VN, Chukanov NV, Pekov IV, Nekrasov AN, Sokolov SV, Kogarko LN (2005) On the role of organic substances in transport and concentration of thorium and other rare elements in alkaline pegmatites of Lovozero and Khibiny peralkaline massifs. Geochem Intern (in Russian) in press
Es'kova EM, Dusmatov VD, Rastsvetaeva RK, Chukanov NV, Voronkov AA (2003) Surkhobite (Ca,Na)(Ba,K)(Fe^{2+},Mn)$_4$Ti$_2$(Si$_4$O$_{14}$)O$_2$(F,OH,O), the new mineral (the Alai ridge, Tadjikistan). Zap Vseross Mineral Obs 132(2):60-67 (in Russian)
Es'kova EM, Kazakova ME (1957) Shcherbakovite – a new mineral. Dokl Akad Nauk SSSR 99:837-840 (in Russian)
Es'kova EM, Semenov EI, Khomyakov AP, Kazakova ME, Shumyatskaya NG (1974) Sazhinite, a new mineral. Zap Vses Mineral Obs 103(3):338-341 (in Russian)
Ferraris G, Belluso E, Gula A, Soboleva SV, Khomyakov AP (2003) The crystal structure of seidite-(Ce), Na$_4$(Ce,Sr)$_2${Ti(OH)$_2$(Si$_8$O$_{18}$)}(O,OH,F)$_4$·5H$_2$O, a modular microporous titanosilicate of the rhodesite group. Can Mineral 41:1183-1192
Ferraris G, Gula A (2005) Polysomatic aspects of microporous minerals –heterophyllosilicates, palysepioles and rhodesite-related structures. Rev Mineral Geochem 57:69-104
Ferraris G, Gula A, Zubkova NV, Pushcharovsky DYu, Gobetchiya ER, Pekov IV, Eldjarn K (2004) The crystal structure of ilímaussite-(Ce), (Ba,Na)$_{10}$K$_3$Na$_{4.5}$Ce$_5$(Nb,Ti)$_6$[Si$_{12}$O$_{36}$][Si$_9$O$_{18}$(O,OH)$_{24}$]O$_6$, and the "ilímaussite" problem. Can Mineral 42:787-795
Ferraris G, Ivaldi G, Khomyakov AP (1995) Altisite Na$_3$K$_6$Ti$_2$[Al$_2$Si$_8$O$_{26}$]Cl$_3$, a new hyperalkaline silicate from Kola Peninsula (Russia) related to lemoynite: crystal structure and thermal evolution. Eur J Mineral 7: 537-546
Ferraris G, Ivaldi G, Pushcharovsky DYu, Zubkova NV, Pekov IV (2001) The crystal structure of delindeite, Ba$_2${(Na,K,☐)$_3$(Ti,Fe)[Ti$_2$(O,OH)$_4$Si$_4$O$_{14}$](H$_2$O,OH)$_2$}, a member of the mero-plesiotype bafertisite series. Can Mineral 39:1307-1316
Fisher K (1969) Verfeinerung der Kristallstruktur von Benitoit BaTi[Si$_3$O$_9$]. Z Kristallogr 129:222-243
Fleet SG (1965) The crystal structure of dalyite. Z Kristallogr 121:349-368
Gault RA, Ercit TS, Grice JD, Van Velthuizen J (2004) Manganokukisvumite, a new mineral species from Mont Saint-Hilaire, Quebec. Can Mineral 42:781-785
Gerasimovskiy VI (1962) Keldyshite, a new mineral. Dokl Akad Sci USSR, Earth Sci 142:123-125
Gerasimovskiy VI (1940) Lovozerite, a new mineral from the Lovozero Tundras. Trudy Instituta Geologicheskikh Nauk 31:9-15 (in Russian)
Gervasini A, Picciau C, Auroux A (2000) Characterization of copper-exchanged ZSM-5 and ETS-10 catalysts with low and high degrees of exchange. Microporous Mesoporous Mater 35-36:457-469
Ghose S, Thakur P (1985) The crystal structure of georgechaoite NaKZrSi$_3$O$_9$·2H$_2$O. Can Mineral 23:5-10
Ghose S, Wan Ch, Chao GY (1980) Petarasite, Na$_5$Zr$_2$Si$_6$O$_{18}$(Cl,OH)·2H$_2$O, a zeolite-type zirconosilicate. Can Mineral 18:503-509
Gobechiya ER, Pekov IV, Pushcharovsky DYu, Ferraris G, Gula A, Zubkova NV, Chukanov NV (2003) New data on vlasovite: refinement of the crystal structure and the radiation damage of the crystal during the X-ray diffraction experiment. Crystallogr Rep 48:750-754
Grim RE (1953) Clay mineralogy. McGraw-Hill Book Co, New York
Gross EB, Wainwright JEN, Evans BW (1965) Pabstite, the tin analogue of benitoite. Am Mineral 50:1164-1169
Hawthorne FC (1987) The crystal chemistry of the benitoite group minerals and structural relations in (Si$_3$O$_9$) ring structures. N Jb Mineral Mh 1987:16-30
Henshaw DE (1955) The structure of wadeite. Mineral Mag 30:585-595
Howard TE (1973) The crystal structures of cavansite and pentagonite. Am Mineral 58:412-424
Ikorskii SV (1977) About laws of distribution and time of accumulation of carconic gases in rocks of the Khibiny alkaline massif. Geokhimiya 11:1625-1634 (in Russian)
Ilyukhin VV, Belov NV (1960) The crystal structure of lovozerite. Dokl Akad Sci USSR, Earth Sci 131:379-381
Ilyushin G (1993) New data on crystal structure of umbite K$_2$ZrSi$_3$O$_9$·H$_2$O. Inorg Mater 27:1128-1133
Ilyushin GD, Demyanets LN (1986) Ion conductors in the class of Na,Zr-silicates. New family of 3-dimensional conductors-crystals of the lovozerite type Na$_{8-x}$H$_x$ZrSi$_6$O$_{18}$. Sov Phys Crystallogr 31:41-44
Ilyushin GD, Demyanets LN (1988) Crystal-structural features of ion transport in new OD structures: catapleiite, Na$_2$ZrSi$_3$O$_9$·H$_2$O, and hilairite Na$_2$ZrSi$_3$O$_9$·3H$_2$O. Sov Phys Crystallogr 33:383-387
Ilyushin GD, Khomyakov AP, Shumyatskaya NG, Voronkov AA, Nevsky NN, Ilyukhin VV, Belov NV (1981a) Crystal structure of a new natural zirconium silicate K$_4$Zr$_2$Si$_6$O$_{18}$·2H$_2$O. Sov Phys Dokl 26:118-120

Ilyushin GD, Voronkov AA, Ilyukhin VV, Nevsky NN, Belov NV (1981b) Crystal structure of natural monoclinic catapleiite Na$_2$ZrSi$_3$O$_9$·2H$_2$O. Sov Phys Dokl 26(9):808-810

Ilyushin GD, Voronkov AA, Nevsky NN, Ilyukhin VV, Belov NV (1981c) Crystal structure of hilairite Na$_2$ZrSi$_3$O$_9$·3H$_2$O. Sov Phys Dokl 26:916-917

Johnsen O, Ferraris G, Gault RA, Grice JD, Kampf AR, Pekov IV (2003) The nomenclature of eudialyte-group minerals. Can Mineral 41:785-794

Johnsen O, Gault RA (1997) Chemical variation of eudialyte. N Jb Mineral Abh 171:215-237

Johnsen O, Grice JD (1999) The crystal chemistry of the eudialyte group. Can Mineral 37:865-891

Johnsen O, Grice JD, Gault RA, Ercit TS (1999) Khomyakovite and manganokhomyakovite, two new members of the eudialyte group from Mont Saint-Hilaire, Quebec, Canada. Can Mineral 37:893-899

Kabalov YuK, Zubkova NV, Pushchrovsky DYu, Schneider J, Sapozhnikov AN (2000) Powder Rietveld refinement of armstrongite, Ca[ZrSi$_3$O$_9$]·3H$_2$O. Z Kristallogr 12:757-761

Kalinin VV, Pushcharovsky DYu,Yamnova NA, Dikov YuP, Borisovsky SE (1994) Strakhovite, NaBa$_3$(Mn^{2+}, Mn^{3+})$_4$Si$_6$O$_{19}$(OH)$_3$, a new sodium-bearing silicate of barium and manganese. Zap Vseross Mineral Obs 123(4):94-97 (in Russian)

Kalsbeek N, Rønsbo JG (1992) Refinement of the vinogradovite structure, positioning of Be and excess Na. Z Kristallogr 200:237-245

Kampf AR, Khan AA, Bauer WH (1973) Barium chloride silicate with an open framework: verplanckite. Acta Crystallogr B29:2019-2021

Kapustin YuL, Pudovkina ZV, Bykova AV (1974a) Zirsinalite, a new mineral. Zap Vses Mineral Obs 103(5): 551-558 (in Russian)

Kapustin YuL, Pudovkina ZV, Bykova AV (1980) Tisinalite, Na$_3$H$_3$(Mn,Ca,Fe)TiSi$_6$(O,OH)$_{18}$·2H$_2$O, a new mineral of the lovozerite group. Zap Vses Mineral Obs 109(2):223-229 (in Russian)

Kapustin YuL, Pudovkina ZV, Bykova AV, Lyubomilova GV (1974b) Koashvite, a new mineral. Zap Vses Mineral Obs 103(5):559-566 (in Russian)

Khalilov AD, Khomyakov AP, Makhmudov SA (1978) Crystal structure of keldyshite, NaZr[Si$_2$O$_6$OH]. Sov Phys Dokl 23:8-10

Khomyakov AP (1977) Parakeldyshite, a new mineral. Dokl Akad Nauk SSSR 237:703-705 (in Russian)

Khomyakov AP, Chernitsova NM, Sandomirskaya SM, Vasil'eva GL (1979) Imandrite, a new mineral of the lovozerite family. Mineral Zh 1:89-93 (in Russian)

Khomyakov AP, Ferraris G, Belluso E, Britvin SN, Nechelyustov GN, Soboleva SV (1998) Seidite-(Ce), Na$_4$SrCeTiSi$_8$O$_{22}$F·5H$_2$O, a new mineral with zeolitic properties. Zap Vseross Mineral Obs 127(4):94-100 (in Russian)

Khomyakov AP, Kulikova IE, Sokolova E, Hawthorne FC, Kartashov PM (2003a) Paravinogradovite, (Na,☐)$_2$[(Ti^{4+},Fe^{3+})$_4${Si$_2$O$_6$}$_2${Si$_3$AlO$_{10}$}(OH)$_4$]·H$_2$O, a new mineral species from the Khibina alkaline massif, Kola peninsula, Russia: Description and crystal structure. Can Mineral 41:989-1002

Khomyakov AP, Nechelyustov GN, Ferraris G, Ivaldi G (2000) Manganonaujakasite, Na$_6$(Mn,Fe)Al$_4$Si$_8$O$_{26}$, a new mineral from the Lovozero alkaline massif, Kola peninsula. Zap Vseross Mineral Obs 129(4):48-53 (in Russian)

Khomyakov AP, Nechelyustov GN, Rastsvetaeva RK (1996) Pyatenkoite-(Y), Na$_5$(Y,Dy,Gd)TiSi$_6$O$_{18}$·6H$_2$O – a new mineral. Zap Vseross Mineral Obs 125(4):72-79 (in Russian)

Khomyakov AP, Nechelyustov GN, Rastsvetaeva RK, Ma Zhe Xiong (2003b) Diversilite-(Ce), Na$_2$(Ba,K)$_6$Ce$_2$Fe^{2+}Ti$_3$(Si$_3$O$_9$)$_3$[SiO$_3$OH]$_3$(OH,H$_2$O)$_9$, a new silicate with heterogeneous tetrahedral complexes from the Khibiny alkaline massif, Kola peninsula, Russia. Zap Vseross Mineral Obs 132(5):34-39 (in Russian)

Khomyakov AP, Nechelyustov GN, Rastsvetaeva RK (1993) Sazykinaite-(Y), Na$_5$YZrSi$_6$O$_{18}$·6H$_2$O – a new mineral. Zap Vseross Mineral Obs 122(5):76-82 (in Russian)

Khomyakov AP, Polezhaeva LI, Merlino S, Pasero M (1990) Lintisite, Na$_3$LiTi$_2$Si$_4$O$_{14}$·2H$_2$O, a new mineral. Zap Vses Mineral Obs 119(3):76-80 (in Russian)

Khomyakov AP, Semenov EI, Es'kova EM, Voronkov AA (1974a) Kazakovite, a new mineral of the lovozerite group. Zap Vses Mineral Obs 103(3):342-345 (in Russian)

Khomyakov AP, Semenov EI, Voronkov AA, Nechelyustov GN (1983a) Terskite, Na$_4$ZrSi$_6$O$_{16}$·2H$_2$O, a new mineral. Zap Vses Mineral Obs 112(2):226-232 (in Russian)

Khomyakov AP, Voronkov AA Polezhaeva LI, Smol'yaninova NN (1983b) Kostylevite, K$_4$Zr$_2$[Si$_6$O$_{18}$]·2H$_2$O, a new mineral. Zap Vses Mineral Obs 112(4):469-474 (in Russian)

Khomyakov AP, Voronkov AA, Kobyashev YuS, Polezhaeva LI (1983c) Umbite and paraumbite, new potassium zirconosilicates from the Khibiny alkaline massif. Zap Vses Mineral Obs 112(4):462-469 (in Russian)

Khomyakov AP, Voronkov AA, Lebedeva SI, Bykov VN, Yurkina KV (1974b) Khibinskite, a new mineral. Zap Vses Mineral Obs 103(1):110-116 (in Russian)

Kolitsch U, Pushcharovsky DYu, Pekov IV, Tillmanns E (2000) A new lintisite-related titanosilicate mineral from Russia: crystal structure, occurrence and properties. ECM-19 Nancy, p 363

Konnerup-Madsen J, Holm PM, Rose-Hansen J (1981) An experimental study of the decomposition of steenstrupine in Na_2CO_3 solution. Rapp Gronl Geol Unders 103:113-118

Kossiakoff AK, Leavens PB (1976) The crystal structure of eakerite, a calcium-tin silicate. Am Mineral 61: 956-962

Kravchenko SM, Vlasova YeV, Pinevich NG (1960) Batisite, a new mineral. Dokl Akad Sci USSR, Earth Sci 133:805-808

Krivokoneva GK, Portnov AM, Semenov YeI, Dubakina LS (1979) Komarovite - silificied pyrochlore. Dokl Akad Sci USSR, Earth Sci 248:127-130

Krivovichev S (2005) Topology of microporous structures. Rev Mineral Geochem 57:17-68

Krivovichev SV, Armbruster Th (2003) The crystal structure of jonesite, $Ba_2(K,Na)[Ti_2(Si_5Al)O_{18}(H_2O)](H_2O)_n$: a first example of titanosilicate with porous double layers. Am Mineral 89:314-318

Krivovichev SV, Yakovenchuk VN, Armbruster Th, Pakhomovsky YaA, Depmeier W (2003) Crystal structure of the K, Ti analogue of ilímaussite-(Ce), $(Ba,K,Na,Ca)_{11-12}(REE,Fe,Th)_4(Ti,Nb)_6(Si_6O_{18})_4(OH)_{12} \cdot 4.5H_2O$: revision of structure model and structural formula. Z Kristallogr 218:392-396

Kupriyanova II, Stolyarova TI, Sidorenko GA (1962) Thorosteenstrupine, a new thorium silicate. Zap Vses Mineral Obs 91(3):325-330 (in Russian)

Kuznicki SM, Bell VA, Petrovic I, Desai BT (2000) Small-pored crystalline titanium molecular sieve zeolites and their use in gas separation processes. U.S. Patent 6068682

Lazebnik KA, Lazebnik YuD, Makhotko VF (1984) Davanite, $K_2TiSi_6O_{15}$, a new alkaline titanosilicate. Zap Vses Mineral Obs 113(1):95-97 (in Russian)

LePage Y, Perrault G (1976) Structure cristalline de la lemoynite, $(Na,K)_2CaZr_2Si_{10}O_{26} \cdot 5$-$6H_2O$. Can Mineral 14:132-138

Liu YL, Du HB, Xu Y, Ding H, Pang WQ, Yue Y (1999) Synthesis and characterization of a novel microporous titanosilicate with a structure of penkvilksite-1M. Microporous Mesoporous Mater 28:511-517

Loghinov AP, Kozyrenko VN, Mikhailov ID, Chukanov NV, Kumpanenko IV (1979) Generalized coupled oscillator model for defect polymers. I. Calculation of frequency branches of n-paraffins, fatty acids and glymes. Chem Phys 36:187-196

Makovicky E, Karup-Møller S (1981) Crystalline steenstrupine from Tunugdliarfik in the Ilímaussaq alkaline intrusion, South Greenland. N Jb Mineral Mh 1981:300-330

Malinovskii YuA, Pobedimskaya EA, Belov NV (1976) Crystal structure of muirite $Ba_9(Ca,Ba)(Ca,Ti)_4(OH)_8[Si_8O_{24}](Cl,OH)_8$. Sov Phys Dokl 20:163-164

Mao Y, Li G, Wei Xu, Feng S (2000) Hydrothermal synthesis and characterization of nanocrystalline pyrochlore oxides $M_2Sn_2O_7$ (M = La, Bi, Gd or Y). J Mater Chem 10:479-482

Matsubara F, Mandarino JA, Semenov EI (2001) Redefinition of a mineral of the joaquinite group: Orthojoaquinite-(La). Can Mineral 39:757-760

McCusker LB (2005) IUPAC nomenclature for ordered microporous and mesoporous materials and its application to non-zeolite microporous mineral phases. Rev Mineral Geochem 57:1-16

McDonald AM, Chao GY (2001) Natrolemoynite, a new hydrated sodium zirconosilicate from Mont Saint-Hilaire, Quebec: description and structure determination. Can Mineral 39:1295-1306

Mellini M, Merlino S (1981) The crystal structure of jagoite. Am Mineral 66:852-858

Menshikov YuP, Sokolova EV, Egorov-Tismenko Yu, Khomyakov AP, Polezhaeva LI (1992) Sitinakite, $Na_2KTi_4Si_2O_{13}(OH) \cdot 4H_2O$ - a new mineral. Zap Vseross Mineral Obs 121(1):94-99 (in Russian)

Mer'kov AN, Bussen IV, Goiko EA, Kul'chitskaya EA, Men'shikov YuP, Nedorezova AP (1973) Raite and zorite, new minerals from the Lovozero Tundras. Zap Vses Mineral Obs 102(1):54-62 (in Russian)

Merlino S, Pasero M, Artioli G, Khomyakov AP (1994) Penkvilksite, a new kind of silicate structure: OD character, X-ray single crystal (1M) and powder Rietveld (2O) refinements of two MDO polytypes. Am Mineral 79:1185-1193

Merlino S, Pasero M, Bellezza M, Pushcharovsky DYu, Gobetchiya ER, Zubkova NV, Pekov IV (2004) Crystal structure of calcium catapleiite. Can Mineral 42:1037-1045

Merlino S, Pasero M, Ferro O (2000) The crystal structure of kukisvumite, $Na_6ZnTi_4(Si_2O_6)_4O_4 \cdot 4H_2O$. Z Kristallogr 193:137-148

Merlino S, Pasero M, Khomyakov AP (1990) The crystal structure of lintisite, $Na_3LiTi_2(Si_2O_6)_2O_2 \cdot 18H_2O$, a new titanosilicate from Lovozero (USSR). Z Kristallogr 193:137-148

Möller T, Clearfield A, Harjula R (2002) Preparation of hydrous mixed metal oxides of Sb, Nb, Si, Ti and W with a pyrochlore structure and exchange of radioactive cesium and strontium ions into the materials. Microporous Mesoporous Mater 54:187-199

Moore PB, Shen I (1983) Crystal structure of steenstrupine: a rod structure of unusual complexity. Tsch Mineral Petrogr Mitt 31:47-67

Neronova NN, Belov NV (1963) Crystal structure of elpidite, $Na_2Zr[Si_6O_{15}]\cdot 3H_2O$, and dimorphism of $[Si_6O_{15}]$ dimetasilicate radicals. Dokl Akad Sci USSR, Earth Sci 150:115-118

Nikitin AN, Belov NV (1962) Crystal structure of batisite, $Na_2BaTi_2Si_4O_4 = Na_2BaTi_2O_2[Si_4O_{12}]$. Dokl Akad Sci USSR, Earth Sci 146:142-143

Nivin VA (1985) Structure and distribution of a gas phase in rocks of the Lovozero deposit. Geologiya Rudnykh Mestorozhdenii 27(3):79-83 (in Russian)

Otero Arean C, Turnes Palomino G, Zecchina A, Bordiga S, Llabres i Xamena FX, Paze C (2000) Vibrational spectroscopy of carbon monoxide and dinitrogen adsorbed on magnesium-exchanged ETS-10 molecular sieve. Catal Lett 66:231-235

Pasero M (2005) A short outline of the tunnel oxides. Rev Mineral Geochem 57:291-306

Pattison P, Nastro A, De Luca P, Cruciani G (1998) Rietveld refinement of the zorite structure of ETS-4 molecular sieves. Microporous Mesoporous Mater 21:143-153

Pekov IV, Azarova YuV, Chukanov NV (2004) New data on komarovite series minerals. New Data on Minerals 39:5-13

Pekov IV, Belovitskaya YuV, Kartashov PM, Chukanov NV, Yamnova NA, Egorov-Tismenko YuK (1999) New data on perraultite. Zap Vseross Mineral Obs 128(3):112-120 (in Russian)

Pekov IV, Chukanov NV (2005) Microporous framework silicate minerals with rare and transition elements: minerogenetic aspects. Rev Mineral Geochem 57:145-172

Pekov IV, Chukanov NV, Kononkova NN, Pushcharovsky DYu (2003a) Rare-metal "zeolites" of the hilairite group. New Data on Minerals 38:20-33

Pekov IV, Chukanov NV, Yamnova NA, Egorov-Tismenko YuK, Zadov AE (2003b) Kapustinite $Na_{5.5}Mn_{0.25}ZrSi_6O_{16}(OH)_2$, a new mineral from Lovozero massif, Kola peninsula, and new data on genetic crystallochemistry of the lovozerite group. Zap Vseross Mineral Obs 132(6):1-14 (in Russian)

Pekov IV, Ekimenkova IA, Chukanov NV, Zadov AE, Yamnova NA, Egorov-Tismenko YuK (2000) Litvinskite, $Na_2(\square,Na,Mn)Zr[Si_6O_{12}(OH,O)_6]$, a new mineral of the lovozerite group. Zap Vseross Mineral Obs 129(1):45-53 (in Russian)

Pekov IV, Ekimenkova IA, Kononkova NN (1997) Thorosteenstrupine from Lovozero massif and steenstrupine-(Ce) - thorosteenstrupine an isomorphous series. Zap Vseross Mineral Obs 126(6):35-44 (in Russian)

Pekov IV, Turchkova AG, Kononkova NN, Chukanov NV (2002) Investigation of cation-exchange properties of labuntsovite-group minerals. I. Experiments in aqueous solutions at normal conditions. Proc. of XX All-Russian Seminar "Alkaline magmatism of the Earth". Moscow, GEOKHI RAS, p 76 (in Russian)

Philippou A, Naderi M, Pervaiz N, Rocha J, Anderson AW (1998) n-hexane reforming reactions over basic Pt-ETS-10 and Pt-ETAS-10. J Catal 178:174-185

Piilonen PC, Lalonde AE, McDonald AM, Gault R, Larsen AO (2003a) Insights into astrophyllite-group minerals. I. Nomenclature, composition and development of a standartized general formula. Can Mineral 41:1-26

Piilonen PC, McDonald AM, Lalonde AE (2003b) Insights into astrophyllite-group minerals. II. Crystal chemistry. Can Mineral 41:27-54

Plyusnina II (1967) Infrared spectra of silicates. Moscow University (ed), Moscow (in Russian)

Poojary D, Bortun A, Bortun L, Clearfield A (1997) Synthesis and X-ray powder structures of $K_2(ZrSi_3O_9)\cdot H_2O$ and its ion-exchanged phases with Na and Cs. Inorg Chem 36:207-213

Portnov AM (1964) Calcium catapleiite, a new variety of catapleiite. Dokl Akad Sci USSR, Earth Sci 154:98-100

Portnov AM, Krivokoneva GK, Stolyarova TI (1971) Komarovite, a new niobosilicate of calcium and manganese. International Geological Review, 14:488-490. Translated from Zap Vses Mineral Obs 100(5):599-602

Potter RM, Rossman GR (1979) The tetravalent manganese oxides: identification, hydration, and structural relationships by infrared spectroscopy. Am Mineral 64:1199-1218

Prider RT (1965) Noonkanbahite, a potassic batisite from the lamproites of western Australia. Mineral Mag 34:403-405

Pudovkina ZV, Chernitsova NM (1991) Crystal structure of terskite, $Na_4Zr[H_4Si_6O_{18}]$. Sov Phys Dokl 36:201-203

Pudovkina ZV, Chernitsova, NM, Voronkov AA, Pyatenko YuA (1980) Crystal structure of zirsinalite $Na_6Ca\{Zr[Si_6O_{18}]\}$. Sov Phys Dokl 25:69-70

Pushcharovsky DYu, Pekov IV, Pasero M, Gobechiya ER, Merlino S, Zubkova NV (2002) Crystal structure of cation-deficient calciohilairite and possible mechanisms of decationization in mixed-framework minerals. Crystallogr Rep 47:748-752

Pyatenko JA, Kurova TA, Chernitsova NM, Pudovkina ZV, Blinov VA, Maksimova NV (1999) Niobium, Tantal and Zirconium in Minerals. Nauka, Moscow (in Russian)

Pyatenko JA, Voronkov AA (1977) Comparative characteristics of titanium and zirconium crystallochemical functions in structures of minerals. Izvestiya AN SSSR, Geological series 9:77-88 (in Russian)

Raade G, Chukanov NV, Kolitsch U, Möckel S, Zadov AE, Pekov IV (2004) Gjerdingenite-Mn from Norway – a new mineral species in the labuntsovite group: descriptive data and crystal structure. Eur J Mineral 16:979-987

Rastsvetaeva RK, Andrianov VI (1984) Refined crystal structure of vinogradovite. Sov Phys Crystallogr 29:403-406

Rastsvetaeva RK, Chukanov NV (2001) Effect of force characteristics of cations on their ordering in the structures of labuntsovite-group minerals. Proc Int Symp OMA-2001, Sochi-Lazarevskoe, p 244-254 (in Russian)

Rastsvetaeva RK, Chukanov NV (2002) X-ray diffraction and IR spectroscopy study of labuntsovite-group minerals. Crystallogr Rep 47:939-945

Rastsvetaeva RK, Khomyakov AP (1992) Crystal structure of rare-earth analogue of hilairite. Sov Phys Crystallogr 37:845-847

Rastsvetaeva RK, Khomyakov AP (1996) Pyatenkoite-(Y) $Na_5YTiSi_6O_{18}·6H_2O$, a new mineral of the hilairite group: crystal structure. Dokl Chem 351:283-286

Rastsvetaeva RK, Khomyakov AP, Rozenberg KA, Ma ZX, Shi N (2003) Crystal structure of the K,Ti analogue of ilímaussite, the first representative of silicates with isolated $[Si_3O_9]$ rings and $[SiO_3OH]$ tetrahedra. Dokl Chem 388:9-13

Rastsvetaeva RK, Pushcharovsky DYu, Konev AA, Evsyunin VG (1997) Crystal structure of K-containing batisite. Crystallogr Rep 42:770-773

Rocha J, Lin Z (2005) Microporous mixed octahedral-pentahedral-tetrahedral framework silicates. Rev Mineral Geochem 57:173-202

Rozenberg KA, Rastsvetaeva RK, Verin IA (2003) Crystal structure of surkhobite: new mineral from the family of titanosilicate micas. Crystallogr Rep 48:384-389

Rudenko VN, Rozhdestvenskaya IV, Nekrasov IYa, Dadze TP (1983) Crystal structure of Sn-wadeite. Mineral Zh 5(6):70-72 (in Russian)

Sandomirsky PA, Belov NV (1979) The OD structure of zorite. Sov Phys Crystallogr 24:686-693

Schmahl WW, Tillmanns E (1987) Isomorphic substitutions, straight Si-O-Si geometry, and disorder of tetrahedral tilting in batisite, $(Ba,K)(K,Na)Na(Ti,Fe,Nb,Zr)_2Si_4O_{12}$. N Jb Mineral Mh 1987:107-118

Semenov EI, Bonshtedt-Kupletskaya EM, Moleva VA, Sludskaya NN (1956) Vinogradovite, a new mineral. Dokl Akad Nauk SSSR 109:617-620 (in Russian)

Semenov EI, Kazakova ME, Bukin VI (1968) Ilímaussite, a new rare earth – niobium – barium silicate from Ilímaussaq, South Greenland. Medd Grønland 181(7):3-7

Shariatmadari H, Mermut AR, Benke MB (1999) Sorption of selected cations and neutral organic molecules on palygorskite and sepiolite. Clays and Clay Minerals 47:44-53

Shpanov EP, Nechelustov GN, Baturin SV, Solntseva LS (1989) Byelorussite-(Ce), $NaMnBa_2Ce_2Ti_2Si_8O_{26}(F,OH)·H_2O$, a new joaquinite-group mineral. Zap Vses Mineral Obs 118(5):100-107 (in Russian)

Shumyatskaya NG, Voronkov AA, Pyatenko YaA (1980) Sazhinite, $Na_2Ce[Si_6O_{14}(OH)]·nH_2O$, a new member of dalyite family in crystal chemistry. Sov Phys Crystallogr 25:419-423

Sizova RG, Voronkov AA, Khomyakov AP (1974) Refinement of crystal structure of triclinic modification of $Na_2ZrSi_2O_7$. Struktura i Svoistva Kristallov. Vladimir, 2:30-42 (in Russian)

Sokolova E, Hawtorne FC, Khomyakov AP (2002) The crystal chemistry of fersmanite, $Ca_4(Na,Ca)_4(Ti,Nb)_4(Si_2O_7)_2O_8F_3$. Can Mineral 40:1421-1428

Sokolova EV, Arakcheeva AV, Voloshin AV (1991) Crystal structure of komkovite. Sov Phys Dokl 36:666-668

Sokolova EV, Rastsvetaeva RK, Andrianov VI, Egorov-Tismenko YuK, Men'shikov YuP (1989) The crystal structure of a new natural sodium titanosilicate. Sov Phys Dokl 34:583-585

Subbotin VV, Merlino S, Pushcharovsky DYu, Pakhomovsky YA, Ferro O, Bogdanova AN, Voloshin AV, Sorokhtina NV, Zubkova NV (2000) Tumchaite $Na_2(Zr,Sn)Si_4O_{11}·2H_2O$ – a new mineral from carbonatites of the Vuoriyarvi alkali-ultrabasic massif, Murmansk region, Russia. Am Mineral 85:1516-1520

Tamazyan RA, Malinovsky YuA (1990) Crystal chemistry of silicates of the lovozerite family. Sov Phys Crystallogr 35:227-232

Teraoka Y, Torigoshi K-I, Yamaguchi H, Ikeda T, Kagawa S (2000) Direct decomposition of nitric oxide over stannate pyrochlore oxides: relationship between solid-state chemistry and catalytic activity. J Molec Catal A: Chemical 155:73-80

Tikhonenkova RP, Kazakova ME (1961) Vlasovite, a new zirconium silicate from the Lovozero massif. Dokl Akad Sci USSR, Earth Sci 137:451-452

Turchkova AG, Pekov IV, Chukanov NV, Kononkova NN (2003) Investigation of cation-exchange properties of labuntsovite-group minerals. II. Lowering of exchange capacity after thermal modification and its mechanism. Proc. of XXI All-Russian Seminar "Alkaline magmatism of the Earth". Apatity, p 155-156 (in Russian)

Uvarova YuA, Sokolova E, Hawthorne FC (2003) The crystal chemistry of shcherbakovite from the Khibina massif, Kola peninsula, Russia. Can Mineral 41:1193-1201

Valtchev V, Paillaud JL, Mintova S, Kessler H (1999) Investigation of the ion-exchanged forms of the microporous titanosilicate $K_2TiSi_3O_9 \cdot H_2O$. Microporous Mesoporous Mat 32:287-296

Voloshin AV, Pakhomovskii YaA, Men'shikov YuP, Sokolova EV, Egorov-Tismenko YuK (1990) Komkovite, a new hydrous barium zirconosilicate from Vuoriyarvi carbonatites, Kola peninsula. Mineral Zh 12(3): 69-73 (in Russian)

Voloshin AV, Subbotin VV, Pakhomovskii YaA, Bakhchisaraytsev AYu, Yamnova NA (1991) Belkovite - a new barium-niobium silicate from carbonatites of the Vuoriyarvi massif (Kola Peninsula, USSR). N Jb Mineral Mh 1991:23-31

Vorma A (1963) Crystal structure of stokesite, $CaSnSi_3O_9 \cdot 4H_2O$. Mineral Mag 33:615-617

Voronkov AA, Ilyukhin VV, Belov NV (1975) Chemical crystallography of mixed frameworks: formation principles. Sov Phys Crystallogr 20:340-345

Voronkov AA, Pudovkina ZV, Blinov VA, Ilyukhin VV, Pyatenko YuA (1979) Crystal structure of kazakovite $Na_6Mn\{Ti[Si_6O_{18}]\}$. Sov Phys Dokl 24:132-134

Voronkov AA, Pyatenko YuA (1961) Crystal structure of vlasovite. Sov Phys Crystallogr 6:755-760

Voronkov AA, Shumyatskaya NG, Pyatenko YuA (1970) On crystal structure of new natural modification of $Na_2Zr[Si_2O_7]$. Zh Strukt Khim 11:932-933 (in Russian)

Wagner C, Parodi GC, Semet M, Robert J-L, Berrahma M, Velde D (1991) Crystal chemistry of narsarsukite. Eur J Mineral 3:575-585

Wang X, Jacobson AJ (1999) Crystal structure of the microporous titanosilicate ETS-10 refined from single crystal X-ray diffraction data. Chem Commun p 973-974

Wise WS (1982) Strontiojoaquinite and bario-orthojoaquinite: two new members of the joaquinite group. Am Mineral 67:809-816

Yakovenchuk VN, Pakhomovskii YaA, Bogdanova AN (1991) Kukisvumite, a new mineral from alkaline pegmatites of the Khibiny massif, Kola Peninsula. Mineral Zh 13(2):63-67 (in Russian)

Yamnova NA, Egorov-Tismenko YuK, Pekov IV (1998) The crystal structure of perraultite from the coastal region of the Sea of Azov. Crystallogr Rep 43:401-410

Yamnova NA, Egorov-Tismenko YuK, Pekov IV (2001a) Refined crystal structure of lovozerite, $Na_2CaZr[Si_6O_{12}(OH,O)_6]\cdot H_2O$. Crystallogr Rep 46:937-941

Yamnova NA, Egorov-Tismenko YuK, Pekov IV, Ekimenkova IA (2001b) Crystal structure of litvinskite: a new natural representative of the lovozerite group. Crystallogr Rep 46:190-193

Yamnova NA, Egorov-Tismenko YuK, Pekov IV, Shchegol'kova LV (2003) Crystal structure of tisinalite $Na_2(Mn,Ca)_{1-x}(Ti,Zr,Nb,Fe^{3+})[Si_6O_8(O,OH)_{10}]$. Crystallogr Rep 48:551-556

Young BR, Hawkes JR, Merriman RJ, Styles MT (1978) Bazirite, $BaZrSi_3O_9$, a new mineral from Rockall Island, Inverness-shire, Scotland. Mineral Mag 42:35-40

Zadov AE, Chukanov NV, Pekov IV, Lovskaya EV, Organova NI (2003) On relationships between refraction indices and chemical compositions of zeolite-type minerals. Zap Vseross Mineral Obs 132(4):70-77 (in Russian)

Zecchina A, Llabrés i Xamena FX, Pazè C, Turnes Palomino G, Bordiga S, Otero Areán C (2001) Alkyne polymerization on the titanosilicate molecular sieve ETS-10. Phys Chem Chem Phys 3:1228-1231

Zecchina A, Otero Areán C, Turnes Palomino G, Geobaldo F, Lamberti C, Spoto G, Bordiga S (1999) The vibrational spectroscopy of H_2, N_2, CO and NO adsorbed on the titanosilicate molecular sieve ETS-10. Phys Chem Chem Phys 7:1649-1657

Zhao H, Feng S, Xu W, Shi Y, Mao Y, Zhu X (2000) A rapid chemical route to niobates: hydrothermal synthesis and transport properties of ultrafine $Ba_5Nb_4O_{15}$. J Mater Chem 10:965-968

Zubkova NV, Pushcharovsky DYu, Giester G, Tillmans E, Pekov IV, Krotova OD (2004) Crystal structure of byelorussite-(Ce), $NaMnBa_2Ce_2(TiO)_2[Si_4O_{12}]_2(F,OH)\cdot H_2O$. Crystallogr Rep 49:964-968

Microporous Framework Silicate Minerals with Rare and Transition Elements: Minerogenetic Aspects

Igor V. Pekov[1] and Nikita V. Chukanov[2]

[1]*Faculty of Geology, Moscow State University*
119899 Moscow, Russia
igorpekov@mtu-net.ru

[2]*Institute of Problems of Chemical Physics,*
142432 Chernogolovka, Moscow Oblast, Russia

INTRODUCTION

The chapter deals with relations between genesis and crystal-chemical aspects of microporous heterosilicate minerals (MHM) with mixed octahedral-tetrahedral frameworks and containing 6 or 5 coordinated transition elements (mainly Ti, Nb, Zr, Fe, Mn, Zn) which have been reviewed and discussed by Chukanov and Pekov (2005).

Natural occurrences of microporous silicates with transition elements are very localized: 113 out 122 known MHM (Chukanov and Pekov 2005, Tables 2–4) occur in postmagmatic derivatives of peralkaline rocks. Most of them are known only in this geological setting together with zeolites and zeolite-like beryllo- and borosilicates. In alkaline pegmatites and hydrothermalites, zeolites and MHM may represent up to 90–95% of a rock. Similar diversity and concentrations of microporous silicates are unknown in other geological situations.

Almost all chemical elements present in high-alkaline systems can be incorporated in MHM as either species-forming or important components of isomorphous substitutions; these elements enter into the structure either as framework or extra-framework constituents. The following elements are known as species-forming constituents: O, H, Si, Al, Be, B, P, Zr, Ti, Nb, Sn, Fe, Mn, Zn, Mg, Li, Na, K, Cs, Ca, Sr, Ba, Y, Ce, La, Th, W, F, Cl, C; Ta, Hf, Rb; Nd, Sm, Gd, Dy, Er, Yb, U, Pb, S can be present in MHM with concentrations higher than 1 wt%. In alkaline rocks, relatively high concentrations of some rare elements can be achieved only in microporous minerals thanks to the topological and compositional variety of their structural frameworks and cavities. However, characteristics such as chemical bonds polarization and Lewis acidity of active centers play a role too. As a result, sites having high selectivity towards certain elements occur in the crystal structures of MHM and determine the important role of these minerals for the geochemistry of rare and transition elements in the high-alkaline massifs.

After the completion of crystallization, the geochemical importance of MHM maintains due to properties like cation leaching, ion exchange and catalytic activity towards different reactions. In fact, these minerals are very sensitive to the activity of alkalis and water; thus, through various ion-exchange mechanisms their composition varies with the concentration of large cations in the late low-temperature solutions. The so-called "transformation mineral species" (Khomyakov 1995) form only in the course of solid-state transformations of the corresponding primary minerals. Besides, MHM are not only sensitive to the minerogenetic conditions, but are also active catalysts of natural reactions involving carbon-bearing substances. The latter process leads to the abiogenic formation of complex organic compounds that, on their turn, can act as complexing agents, carriers and concentrators of different rare elements (see below).

The understanding of the mineralogical and crystal-chemical principles ruling the genesis of MHM could usefully suggest paths to the synthesis of mixed-framework microporous materials.

DISTRIBUTION AND GENETIC FEATURES

Occurrences and genetic types

The typical genetic ambience and occurrences of MHM are summarized in Table 1. The data show that most microporous heterosilicates and all significant concentrations of MHM are very rare and confined to high-alkaline rocks and their derivatives: pegmatites, hydrothermalites and metasomatites. In the agpaitic complexes, MHM are formed in a wide range of temperatures and silica concentrations, in agreement with available data on the synthesis of related microporous silicate materials (Chukanov et al. 2004).

Agpaitic magmatic rocks. The variety of MHM in magmatic rocks is small, but the magmatogenic zirconosilicates (eudialyte, lovozerite, elpidite, parakeldyshite), steenstrupine-(Ce) and naujakasite can occur as important accessory and even rock-forming minerals.

Among MHM, eudialyte forms the largest deposits; well known are those in Ilímaussaq (red kakortokite) and in Lovozero (lodes of eudialyte lujavrites and eudialytite up to several hundreds meters, with eudialyte contents up to 90 vol%). Large bodies of eudialyte-rich rocks are known in Pajarito Mountain (New Mexico), Khibiny, Pilanesberg (South Africa) and other agpaitic complexes (Vlasov et al. 1966; Bussen and Sakharov 1972; Sørensen 1992, 1997; Pekov 2000).

Rock-forming lovozerite-group minerals (usually lovozerite) are known in porphyraceous lujavrites of Lovozero. Likely, lovozerite was formed as pseudomorph after primary magmatic zirsinalite under action of relatively low-alkaline solutions. Drill cores from the western part of Lovozero massif contain as main component a Ca- and Mn-deficient zirsinalite-like mineral (M40 in Khomyakov 1995) that is trigonal and chemically close to monoclinic kapustinite (Pekov et al. 2003e).

In underground mine works at deep levels of the Lovozero massif we have discovered large malignite lodes in which parakeldyshite is practically the only zirconium mineral and represents at least 3–5 vol% of the rock. The synthesis of parakeldyshite (Ilyushin et al. 1983) has been obtained only at temperatures above 450°C.

Elpidite is the Si-richest hydrous alkaline zirconosilicate and its most significant concentrations occur in agpaitic granites and related pegmatites. Large deposits of elpidite-rich (up to 20 vol% and more) rocks have been discovered in two alkaline-granitic massifs of Western Mongolia: Khan Bogdo (Vladykin et al. 1972) and Khaldzan Buregteg (P.M. Kartashov, personal communication). Pegmatites of alkaline granites are usually richer than their mother rocks in elpidite and, at Khan Bogdo, armstrongite, a related calcium zirconosilicate. Elpidite is also a common mineral in high-alkaline granites of the massifs at Strange Lake (Quebec-Labrador; Birkett et al. 1992), Gjerdingen (South Norway; Raade and Haug 1982), and some other localities. It is a highly hydrated mineral, and its presence is an indicator of high water activity in the course of the granites formation. Heated in dry air, elpidite transforms into $Na_2ZrSi_6O_{15}$ a compound with dalyite-type structure unknown in nature (our unpublished data).Wadeite and dalyite are common minor minerals in high-potassic intrusive rocks of Murun massif (Konev et al. 1996). These anhydrous potassium

Table 1. Occurrence of microporous heterosilicate minerals with mixed frameworks (MHM).

Genetic types	Minerals	Locality (examples)	Additional comments
I. ALKALINE FORMATIONS			
I.1. Magmatic rocks			
I.1.1. Agpaitic feldspathoid syenites and closely related peralkaline rocks	eudialyte gr.m.*, lovozerite gr.m.*, steenstrupine-(Ce)*, naujakasite*, baotite*, wadeite*, parakeldyshite, dalyite, narsarsukite	Lovozero and Khibiny (Kola Peninsula, Russia); Ilímaussaq (South Greenland); Pilanesberg (South Africa); Los Archipelago (Guinea); Tamazeght (Morocco); Murun (Siberia, Russia)	The largest concentrations of MHM: huge deposits of eudialyte and steenstrupine; large amounts of lovozerite and naujakasite
I.1.2. Peralkaline ultrabasic rocks	eudialyte gr.m.*, parakeldyshite*, wadeite	Lovozero and Khibiny (Kola Peninsula, Russia)	Large deposits of parakeldyshite
I.1.3. Peralkaline granites and quartz syenites	elpidite*, eudialyte gr.m.*, armstrongite, dalyite, bazirite	Khan Bogdo and Khaldzan Buregteg (Western Mongolia); Pajarito Mountain (New Mexico, USA); Strange Lake (Quebec and Labrador, Canada); Gjerdingen (Norway); Rockall Island (North Atlantic)	Large deposits: elpidite in peralkaline granites and eudialyte in quartz syenites
I.1.4. Lamproites	wadeite, shcherbakovite ("noonkanbahite")	West Kimberley (Western Australia)	Rare accessory minerals
I.2. Pegmatites, metasomatites and hydrothermalites			
I.2.1. Fenites and other metasomatites related to agpaitic feldspathoid syenites and peralkaline ultrabasic rocks	eudialyte gr.m.*, vlasovite, catapleiite, wadeite*, dalyite, narsarsukite*, thorosteenstrupine	Lovozero, Khibiny and Turii Cape (Kola Peninsula, Russia); Murun, Burpala and Chergilen (Siberia, Russia)	Sometimes eudialyte-group minerals and wadeite give large deposits
I.2.2. Hydrothermalites related to miaskites	labuntsovite gr.m., catapleiite	Vishnevye Gory (South Urals, Russia)	Extremely rare occurrences
I.2.3. Hydrothermalites related to carbonatites	labuntsovite gr.m., catapleiite, belkovite, gaidonnayite, georgechaoite, hilairite, komkovite, tumchaite	Vuoriyarvi and Sallanlatva (North Karelia, Russia); Kovdor and Sebl'yavr (Kola Peninsula, Russia)	Small amounts

Continued on following page

Table 1 continued. Occurrence of microporous heterosilicate minerals with mixed frameworks (MHM).

Genetic types	Minerals	Locality (examples)	Additional comments
I.2.4. Pegmatites, metasomatites and hydrothermalites related to peralkaline granites and granosyenites	elpidite*, eudialyte gr.m., labuntsovite gr.m., narsarsukite, armstrongite, vlasovite, baotite, calcium catapleiite, calciohilairite, bazirite, pabstite, byelorussite-(Ce)	Khan Bogdo and Khaldzan Buregteg (Western Mongolia); Strange Lake (Quebec and Labrador, Canada); Golden Horn (Washington, USA); Gjerdingen (Norway); Verkhnee Espe (Kazakhstan); Dara-i-Pioz (Tadjikistan); Diabazovoe (Belarus)	Sometimes elpidite gives large concentrations
I.2.5. Pegmatites and hydrothermalites related to agpaitic feldspathoid syenites and peralkaline ultrabasic rocks	*almost all known Na- and K-bearing MHM and some Ca- and Ba-dominant species*: eudialyte gr.m.*, lovozerite gr.m.*, labuntsovite gr.m.*, hilairite gr.m., steenstrupine-(Ce)*, thorosteenstrupine, parakeldyshite*, keldyshite*, khibinskite, catapleiite*, calcium catapleiite, wadeite, gaidonnayite, georgechaoite, elpidite, terskite, vlasovite, narsarsukite, shcherbakovite, batisite, zorite, Na-komarovite, komarovite, penkvilksite, sazhinite-(Ce), vinogradovite, paravinogradovite, dalyite, davanite, seidite-(Ce), ilímaussite-(Ce), diversilite-(Ce), naujakasite, manganonaujakasite, fersmanite, sitinakite, umbite, kostyleite, lemoynite, natrolemoynite, altisite, lintisite, kukisvumite, manganokukisvumite, tumchaite, baotite, joaquinite-(Ce), byelorussite-(Ce), orthojoaquinite-(La)	Lovozero, Khibiny and Kovdor (Kola Peninsula, Russia); Ilímaussaq and Narssârssuk (South Greenland); Mont Saint-Hilaire and Saint-Amable (Quebec, Canada); Langesundsfjord and Tvedalen (Norway); Burpala, Inagli and Korgeredaba (Siberia, Russia); Magnet Cove (Arkansas, USA); Tamazeght (Morocco)	Largest diversity of MHM; some Zr-, Ti- and Nb-silicates give extremely rich concentrations in large agpaitic pegmatites
I.2.6. Late mineralization in alkaline volcanic rocks	labuntsovite gr.m., sazhinite-(Ce), batisite, fersmanite	Aris (Namibia); Eifel volcanic district (Germany)	Rare occurrences
I.2.7. Metamorphogene (?) alkaline hydrothermalites situated within serpentinites	*only Ba-bearing MHM*: benitoite, joaquinite gr.m., baotite	several localities in San Benito County (California, USA); Ohmi (Niigata Prefecture, Japan)	Small amounts
I.2.8. Hydrothermally altered soda-bearing sedimentary rocks	labuntsovite gr.m., elpidite, catapleiite, vinogradovite	Green River Formation (Wyoming and Utah, USA)	Small amounts

Table 1 continued. Occurrence of microporous heterosilicate minerals with mixed frameworks (MHM).

Genetic types	Minerals	Locality (examples)	Additional comments
II. NON-ALKALINE FORMATIONS			
II.1. Barium-enriched metamorphic rocks (sanbornite-bearing rocks; Ba, Mn oxide-silicate rocks)	*only Ba-bearing MHM*: traskite, muirite, verplanckite, benitoite, bazirite, pabstite, strakhovite	several localities within Ba-rich metamorphic formation in Fresno, Mariposa and Santa Cruz Counties (California, USA); Itsy Mts. (Yukon Territory, Canada); Ir-Nimi (Khabarovsk Territory, Russia); Broken Hill (New South Wales, Australia)	Small amounts
II.2. Granitic pegmatites	*only Sn-bearing MHM*: stokesite, eakerite	Urucum (Minas Gerais, Brazil); Himalaya mine (California, USA); Věžná and Ctidružice (Czech Republic); Voron'i Tundry (Kola Peninsula, Russia); Kings Mountain (North Carolina, USA)	Rare occurrences
II.3. Skarns and skarnoids	jagoite, stokesite, eakerite	Långban (Sweden); Pitkäranta and Kitel'skoe (South Karelia, Russia); Roscommon Cliff and Halvosso (Cornwall, England)	Extremely rare occurrences
II.4. Hydrothermalites related to basic effusive rocks	*only V-bearing MHM*: cavansite, pentagonite	Owyhee Dam and Chapman quarry (Oregon, USA); Poona district (Maharashtra, India)	Small amounts

* Minerals abundant in this type of formation (rock-forming species are underlined); gr.m. means "group minerals".

zirconosilicates are known also as accessory minerals in magmatic rocks of other alkaline massifs (Kostyleva-Labuntsova et al. 1978; Linthout et al. 1988).

Early magmatic steenstrupine-(Ce) is a typical accessory mineral of different peralkaline rocks at Ilímaussaq where has been found as an ore mineral in a complex deposit of rare metals and uranium (Sørensen 1962; Sørensen et al. 1974). Late magmatic accessory steenstrupine is typical in foyaites and eudialyte lujavrites of Lovozero (Chukanova et al. 2004).

Pegmatites and hydrothermalites related to peralkaline rocks. The maximal variety of MHM is realized in pegmatites and hydrothermalites of peralkaline rocks where practically all Na or K MHM (except strakhovite and several joaquinite-group minerals) are found. The majority of MHM is known only from late derivatives of peralkaline rocks. The most important pegmatite deposits of MHM are situated in Khibiny and Lovozero, which are the two largest complexes of agpaitic nepheline syenites. In Lovozero, pegmatites containing about 50 vol% eudialyte, often replaced by catapleiite and/or elpidite, are common. Masses of monomineral aggregates of these two minerals can reach several tens or even several hundreds kilograms (Pekov 2000). In other pegmatites, abundant eudialyte is replaced by zirsinalite and lovozerite.In khibinite-related pegmatites at Mt. Takhtarvumchorr, Khibiny, parakeldyshite and products of its hydrolytic alteration (keldyshite and "hydrokeldyshite" or M34 in Khomyakov 1995) are abundant.

The largest segregations of titanium and niobium labuntsovite-group minerals are confined respectively to pegmatites of Lovozero (Mt. Karnasurt) and Khibiny (Mt. Suolaiv and Mt. Koashva). The content of labuntsovite-group minerals in some parts of the large pegmatites can reach at least 10 vol% (Chukanov et al. 2003c). In Lovozero, ussingite and natrolite pegmatites containing up to 20 vol% of steenstrupine-thorosteenstrupine minerals are known. In naujaitic pegmatites of Ilímaussaq, high concentrations of steenstrupine have been observed.

In late hydrothermal parageneses of alkaline pegmatites, a variety of highly hydrated MHM with wide structural pores abundantly occur. Partly these minerals formed by replacement of earlier Ti-, Nb- and Zr-silicates; often, different late microporous silicates occur in close association. For example, in the hydrothermally altered zone of the Yubileinaya pegmatite lode at Mt. Karnasurt, Lovozero, five MHM have been first described: zorite, penkvilksite, terskite, sazhinite-(Ce) and seidite-(Ce). These minerals are associated with narsarsukite and nenadkevichite; earlier MHM are represented by eudialyte and steenstrupine-(Ce) (Pekov 2000). In the hydrothermally altered Hilairitovoe pegmatite at Mt. Kukisvumchorr, Khibiny, fourteen MHM have been recently reported: eudialyte, catapleiite, gaidonnayite, hilairite, elpidite, fersmanite, narsarsukite, vinogradovite, kukisvumite, labuntsovite-Mg, labuntsovite-Mn, nenadkevichite, tsepinite-K and vuoriyarvite-K (Pekov and Podlesnyi 2004). Along with alkaline MHM, the following calcium minerals, which probably are formed by ion-exchange processes, sometimes occur in agpaitic pegmatites: tsepinite-Ca (Pekov et al. 2003b), calciohilairite (Pekov et al. 2003c), calcium catapleiite and armstrongite. Some barium MHM (baotite, bazirite, pabstite) are also known.

Metasomatites of agpaitic massifs. High contents of microporous silicates bearing transition elements, especially low-hydrated and anhydrous MHM, are associated with metasomatites of agpaitic massifs. Unique hyperpotassic metasomatites containing rock-forming wadeite (up to 50 vol%) and accessory dalyite are reported in Murun massif (Konev et al. 1996). The world largest segregation of narsarsukite (up to 80 vol%) occurs in quartz-bearing fenites of Lovozero (Pekov 2000).

Carbonatites. A specific late alkaline mineralization is connected with carbonatites of the alkaline-ultrabasic massifs of Northern Karelia and Kola Peninsula, Russia. Typically, several associations of MHM are known in Vuoriyarvi (North Karelia) where the following minerals have been discovered: the barium Nb-oxysilicate belkovite (Voloshin et al. 1991), the barium Zr-silicate komkovite (Voloshin et al. 1990) and the Sn-bearing sodium Zr-silicate tumchaite (Subbotin et al. 2000). Alkaline hydrous zirconosilicates and labuntsovite-group minerals are widespread in carbonatites of Kovdor and Vuoriyarvi where they form small but rich segregations (Voloshin et al. 1989; Chukanov et al. 2003c).

Non-alkaline rocks. Relatively large deposits of MHM are connected with alkaline rocks and high-barium metamorphites only. Here all MHM contain barium as main large cation. However, the total amount of MHM in the metamorphites is insignificant and the number of species is relatively small even if several minerals with unique structure types have been discovered there (in particular, traskite, a MHM with extremely wide pores; Malinovskii et al. 1976). Traskite, muirite and verplanckite are typical accessory minerals of a specific sanbornite quarzite occurring in a metamorphic formation in California (Alfors et al. 1965).

Crystal chemical aspects of minerogenesis

As shown by the survey given above, MHM are connected with environments that are strongly enriched in Na, K or Ba. This enrichment is the main factor controlling the formation of microporous heterosilicates in natural processes. It is observed, instead, that in absence of Na, K and Ba, only Ca and Sr do not act as promoters of MHM formation. Likely, the capability of cations to stimulate the formation of mixed microporous frameworks is governed by the ratio of the ion charge (z) to the ion radius (r), i.e., the Cartledge ion potential Φ (see, e.g., Urusov 1975). For coordination number 8 (ionic radii from Shannon and Prewitt 1969), the following values of $\Phi = z{:}r$ are obtained: 0.66 for K^+; 0.86 for Na^+; 1.41 for Ba^{2+}; 1.59 for Sr^{2+}; 1.79 for Ca^{2+}. Unlike MHM, zeolites can crystallize in absence of Na, K and Ba; in this case, Ca and/or Sr are able to influence strongly the type of structure. For comparison, oxides with low framework densities (FD = number of framework knots per 1000 Å^3), such as pyrochlore (FD = 14), require high activity of Na.

The agpaitic alkaline rocks are the most abundant silicate rocks that are highly rich of Na and/or K. The same rocks and even more their late derivatives, i.e., pegmatites and hydrothermalites, are also rich of rare and transition elements, in particular the high-charge cations Ti^{4+}, Zr^{4+} and Nb^{5+}. Consequently, the largest variety and abundance of microporous heteroframework silicates occur in peralkaline rocks.

Index of alkalinity. Ussing (1912) defined the agpaitic rocks of the Ilímaussaq complex as follows: "..if *na, k* and *al* are the relative amounts of Na-, K- and Al-atoms in the rock, the agpaites may be characterized by the equation $(na + k)/al \geq 1.2$, whereas in ordinary nepheline syenites this ratio does not exceed 1.1." The value of agpaicity index is the basis of various classification schemes that are used to describe feldspathoid-bearing magmatic rocks and related pegmatites, hydrothermalites and metasomatites. Detailed surveys of such classifications are given by Khomyakov (1995) and Sørensen (1997).

Along with the term "agpaitic," the related term "peralkaline" is used in petrological and mineralogical literature. In particular, the presence of eudialyte and alkaline Ti-silicates (e.g., rinkite), instead of zircon and ilmenite as concentrators of Zr and Ti respectively, has been considered as a mineralogical criterion to determine peralkaline rocks (LeMaitre 1989). In the detailed classification suggested by Khomyakov (1995), the value of the alkalinity modulus $K_{alk} = 100x/(x + y + p)$ has been used as a characteristic of alkalinity. x, y and p are the coefficients in the general formula $A_xM_ySi_pO_q$ for an alkaline "amphoterosilicate"; A = Na, K; M = Ti, Nb, Zr, Be... Four types of alkaline rocks are recognized according to the value of K_{alk}: hyperagpaitic ($K_{alk} > 40$), highly agpaitic ($K_{alk} \approx 35$–40), medium-agpaitic ($K_{alk} \approx 25$–35), low agpaitic ($K_{alk} \approx 15$–25), and miaskitic ($K_{alk} < 15$).

We consider the definition of agpaicity given by Ussing (1912) as the most adequate from a crystal-chemical viewpoint. In fact, Ussing agpaicity index is directly related to parameters governing topological characteristics of the minerals that contain large cations and form in a wide temperature range. In particular, framework and layer silicates are very sensitive to changes of the (Na+K):Al ratio and a crucial (especially in structural sense) change of mineral assemblages occurs at values of the Ussing index close to 1.2 but not at ≈ 1 as suggested by Marakushev et al. (1981).

The role of Ussing agpaicity index. Al is the only geochemically important element that can substitute Si in tetrahedral sites and cannot be ignored in genetic schemes. (Na+K):Al = 1 occurs in the main leucocratic minerals of feldspathoid rocks and their derivatives (i.e., K,Na-feldspars, nepheline, kalsilite, zeolites) as well as in micas of the biotite series and some amphiboles like katophorite and edenite. An accessory Ti, Zr, Nb, *REE* mineralization of such rocks represented by zircon, ilmenite, titanite, pyrochlore, allanite, monazite, bastnäsite, chevkinite, aeschynite, britholite etc., is usually considered as an indicator of miaskitic-type rocks. However it is important to note that this rare-metal mineralization occurs also at somewhat higher (Na + K)/Al values; for example, in miaskitic pegmatites containing sodalite, Na-rich members of the cancrinite group, aegirine-salite and high-alkali amphiboles. The total amount of the latter minerals can reach 30% and increase the agpaicity index up to ~1.1. In parageneses of this type, framework minerals other than aluminosilicates and pyrochlore are absent; layer silicates are represented by micas.

As mentioned above, at values of the agpaicity index near to 1.2, qualitative changes of crystal-chemical (in particular, topological) characteristics occur in most of minerals of Ti and rare elements. At this value, the main phases containing Ti, Zr, Nb, Ta, Be, Ba, Th, U, as well as the phases that are important concentrators of *REE*, Li and Zn, are represented by Na- and/or K-rich minerals having low-density framework or various layer structures. Over this border value of the agpaicity index, a large number of endemic MHM and layer (Ti,Nb)-silicates are formed; thus, the Ussing index can be considered as a crystal-chemical criterion for the determination of the agpaitic alkaline rocks.

In an agpaitic rock, the framework aluminosilicates preserve their role as main rock-forming minerals, but, owing to lack of alumina, the high amount of alkali cations still present in the fluid at late stages of crystallization cannot be fully incorporated by these phases. In accordance with crystal-chemical criteria (in the first place, with the first and second Pauling rules), the excess Na and K is incorporated in phases with low densities, i.e., framework and layer heterosilicates in our case. Lack of Al and increasing role of rare elements result in the formation of specific silicates, in which the framework can contain tetrahedrally coordinated P, B, Be, Zn, Li, on one side, and/or Ti, Nb, Zr, Fe, Mn, Zn, Mg and even Ca and *REE* with coordination numbers 6 and rarely 5. In phyllosilicates and heterophyllosilicates (cf. Ferraris and Gula 2005) of the agpaitic rocks, both small (Ti and Nb) and large (Na) cations, can be present with coordination numbers 5 or 6 in the layers, along with more common elements like Fe, Mn, Mg, Al, Li (lamprophyllite and lomonosovite groups, delindeite and other heterophyllosilicates, and even the mica-group mineral shirokshinite—see Ferraris et al. 2001; Pekov et al. 2003e). Rarely 5- or 6-coordinated Na can be included into a mixed framework; other examples are joaquinite-group minerals and sazhinite (Dowty 1975; Shumyatskaya et al. 1980; Zubkova et al. 2004).

The role of Si. In wide ranges of temperature and activity of Si, heteroframework silicates and heterophyllosilicates are very important concentrators of rare elements in the agpaitic rocks. Unlike zeolites, MHM can be formed in derivatives of alkaline granitoids. At lowest concentrations of Si, i.e., in alkaline-ultrabasic rocks and their derivatives, oxide minerals (loparite, Nb-bearing perovskite, pyrochlore) become the main concentrators of Ti and Nb. In alkaline granites, the role of the microporous framework Ti-silicates (except narsarsukite) is insignificant. On the contrary, microporous Zr-silicates are the main concentrators of Zr in all agpaitic formations. The content of Si determines the type of mixed framework in Zr-silicates: agpaitic granites and their derivatives contain elpidite with infinite bands of Si tetrahedra; in porphyraceous lujavrites of Lovozero characterized by intermediate Si contents, the main zirconosilicates are eudialyte and lovozerite-group minerals, where rings of Si tetrahedra occur; Si-poor malignites of Lovozero bear parakeldyshite, whose structure contains isolated Si_2O_7 groups.

Behavior of Ca and Sr. The crystal-chemical behavior of Ca and Sr in high-alkaline rocks substantially differ from that of alkaline cations. Major amounts of Ca and Sr are concentrated in apatite, titanite, rinkite, amphiboles and pyroxenes, but not in framework and layer heterosilicates (eudialyte and lamprophyllite seem exceptions). Ba, unlike other alkaline-earth elements, behaves like alkali cations (especially K) and concentrates mainly in framework minerals and in various phyllosilicates. In particular, evident differences in the crystal-chemical role of various alkali-earth elements are proved for high-alkaline late-carbonatite formations (Vuoriyarvi, Kovdor etc.): there, Ca and Sr concentrate preferably in carbonates and phosphates, whereas Ba enters microporous silicates together with K and Na.

Zr vs Ti. The existence of huge deposits of microporous Zr-silicates compared with the absence of similar deposits for Ti-silicates may seem paradoxical. The main reason is of crystal-chemical nature. Unlike TiO_6 octahedra, the ZrO_6 octahedra do not show tendency to condensation (see, e.g., Pyatenko et al. 1999). As a result, in peralkaline rocks, the incorporation of Zr in important early Zr-silicates is minor than that of Ti in early Ti minerals (silicates and oxides). Mean contents of ZrO_2 (wt%) in early Zr-silicates are: eudialyte – 13, zirsinalite – 12, kapustinite – 16, elpidite – 21 and parakeldyshite – 38. Mean contents of TiO_2 (wt%) in early Ti-silicates are: titanite – 36, lamprophyllite – 29, lorenzenite – 45, lomonosovite – 25, natisite – 39, astrophyllite – 10 and rinkite – 8. For this reason, the formation of early loparite, lamprophyllite, lomonosovite etc. in the course of magma crystallization leads to additional effective consumption of titanium, whereas consumption of Zr is less significant. An additional reason of consumption of Ti during all stages of minerogenesis is its incorporation, unlike Zr, as minor constituent in aegirine and other important Fe-bearing rock-forming minerals (amphiboles, micas).

The increase of Si content in a mineral-forming medium leads to a higher Si:Zr ratio in Zr-silicates and, consequently, to the increase of the relative contents of these silicates in the rocks. In malignites of Lovozero, the earliest Zr-silicate parakeldyshite (~ 40 wt% ZrO_2) constitutes 5–7% of the rock. On later stages, the concentration of Si increases, and eudialyte and lovozerite-group minerals (10–17% ZrO_2) become important rock-forming minerals and constitute up to several tens percent of the rock. Zirconosilicates (mainly eudialyte and members of the lovozerite group) remain the main MHM on the early stages of formation of hyperagpaitic pegmatites at Khibiny and Lovozero, whereas maximal concentrations of Ti(Nb)-silicates are realized in late-pegmatite parageneses.

For most MHM, the late alkaline pegmatite-hydrothermal fluid is the only medium from where these phases can arise. Usually, postmagmatic assemblages are enriched by Na and K as compared with their parent rocks. Therefore, MHM can occur even in pegmatites and hydrothermalites related to non-agpaitic alkaline magmatic rocks, as happens, e.g., in Kovdor, Vishnevye Gory, etc.

The role of H_2O. Most of the known MHM are hydrous minerals, and the stabilizing role of the water molecules is important to form framework structures. For the synthesis of zeolites, no adequate alternatives to aqueous solutions are known (Barrer 1982). The same is valid for the synthesis of hydrous MHM (see Ilyushin et al. 1983; Rocha et al. 1996; Ilyushin and Demyanets 1997; Chukanov et al. 2004).

Aluminosilicate zeolites can form in the whole temperature range corresponding to hydrothermal and supergene processes, i.e., from the critical point for water down to 0°C, typically under low pressure (Breck 1974; Barrer 1982). Instead, the lowest temperature limit at which MHM can crystallize likely lies above 100°C. This conclusion follows from the experimental data and from the fact that MHM are unknown in supergene parageneses. This difference between MHM and zeolites is likely connected with the low mobility of Ti, Nb and Zr at low temperature. Another possible reason is that low-temperature zeolites can crystallize as Al,Si-disordered phases (Breck 1974; Rabo 1976; Pekov et al. 2004a), so that the process of their formation is entropy-favored. Instead, the microporous heterosilicate minerals are always M,Si-ordered.

EVOLUTION OF PARAGENESES: STRUCTURAL ASPECTS

Evolution series

Pseudomorphs and epitactic intergrowths. Generally, in agpaitic complexes the variety of species and structures of microporous minerals, especially MHM, becomes significantly wider moving from magmatic to late-pegmatitic and hydrothermal stages. Early pegmatitic Ti- and Nb-silicates show mainly either chain and layer or cuspidine-like structures (e.g., rinkite and götzenite). At the hydrothermal stage, these minerals are replaced by Ti- and Nb-silicates with mixed frameworks. The following pseudomorphs are well documented at Lovozero, Khibiny, Ilímaussaq and Mont Saint-Hilaire: labuntsovite group members after lomonoso-vite, vuonnemite, astrophyllite, lamprophyllite and rinkite; zorite and Na-komarovite after vuonnemite; sitinakite and narsarsukite after lomonosovite; vinogradovite and kukisvumite after lamprophyllite; vinogradovite and lintisite after lorenzenite. In the pegmatites of Djelisu (Kyrgyzstan) and Darai-Pioz (Tajikistan), quasi-layer framework Ti-silicates (cf. heterophyl-losilicates in Ferraris and Gula (2005)) of the jinshajiangite – perraultite series epitactically overgrow on close related true-layer minerals of the bafertisite – hejtmanite series.

Zeolites and other tecto-aluminosilicates. It seems interesting to give a brief survey of the structural evolution of both aluminosilicate and non-aluminosilicate framework minerals that occur in derivatives of the alkaline complexes. In these derivatives, the mineralization of zeolites develops according to a general path of structure types and evolution of the extra-framework composition. Late major zeolites have lower framework density as compared with early ones according to the following evolution series: analcime → natrolite → (thomsonite / edingtonite / paranatrolite) → (phillipsite - harmotome / gmelinite) → chabazite (Pekov et al. 2004a).

The tendency to a lower framework density with decreasing temperature is typical also for other tecto-aluminosilicates, first the series of nepheline alteration. From a crystal-chemical viewpoint, the process of transforming nepheline into other tecto-aluminsilicates requires a re-building of the Al,Si-framework which becomes more open; sometimes the increase of the Si:Al ratio is accompanied by changes in the extra-framework composition. A distinctive characteristic among minerals formed under different conditions is the content of their zeolite-like cages and channels. At low temperatures (< 200–300°C), nepheline alters to zeolites, usually natrolite. At higher temperatures and/or H_2O deficiency conditions, nepheline is replaced by sodalite and cancrinite group members, anhydrous or water-poor zeolite-like minerals with Na, Ca, K, Cl, CO_3 and SO_4 in their pores. That is in good agreement with the experimental data showing that occlusion of salts by dehydrated zeolites is a main process in water-deficient and/or high-temperature systems but it is depressed by ion exchange and H_2O absorption in low-temperature hydrothermal systems (Rabo 1976). In alkaline rocks and pegmatites, cancrinite group members usually replace the higher-FD sodalite-like minerals (typically in Kola complexes: Lovozero, Khibiny and Turiy Cape). The tendency to a lower FD with decreasing temperature is general for tecto-aluminosilicates; the evolution tendency for other minerals, including MHM, is more complicated and strongly depends on the alkalinity (term "alkalinity" here and below means joint activity of Na and K but not pH) during each stage.

Beryllosilicates. In the evolution series of beryllosilicates, a low FD occurs only under conditions of high alkalinity. The high-temperature alteration of chkalovite, $Na_2[BeSi_2O_6]$ (FD = 23.4), in agpaitic pegmatites gives tugtupite also known as "beryllosodalite", $Na_4[BeAlSi_4O_{12}]Cl$ (FD = 18.5); at lower temperature, the alkaline Be,Si-zeolites lovdarite, $K_4Na_{12}[Be_8Si_{28}O_{72}]\cdot 18H_2O$, and nabesite, $Na_2[BeSi_4O_{10}]\cdot 4H_2O$, ($FD$ = 18.4 and 16.9, in the order) are formed instead. If the alkalinity decreases, the evolution of a beryllium mineralization differentiates not only in its mineralogical but also in crystal-chemical sense; in fact, a polymerization of Be tetrahedra accompanied by condensation of the structure takes place. At first, the replacement of the tectosilicate chkalovite, which contains isolated Be tetrahedra, by epididymite and/or eudidymite (dimorphous $NaBeSi_3O_7(OH)$ phases), which contain Be_2O_6 groups of edge-connected tetrahedra, takes place. The next step is the formation of sphaerobertrandite, $Be_3SiO_4(OH)_2$ that replaces epidydimite and eudidymite. Sphaerobertrandite is the Be-richest natural silicate; it contains tetrahedral Be_2O_6 dimers connected by BeO_4 tetrahedra to form infinite chains (Pekov et al. 2003d).

Niobates. The evolution series of niobates show the same dual behavior as that of beryllosilicates: in the high-alkaline late hydrothermalites, franconite group minerals ($ANb_4O_{11}\cdot 8$–$10H_2O$; A = Na_2, Ca, Mg) with few Nb atoms per 1000 $Å^3$ (8.8–9.6) can be formed after pyrochlore; instead, if the alkalinity decreases, pyrochlore alters to fersmite and further to columbite, which are minerals with dense octahedral packing.

Dual character of MHM evolution series

Strong tendency to "loosening" of the structures is observed for the evolution series of MHM that develop in peralkaline conditions. This is particularly evident for the zirconosilicate series where initial minerals are usually eudialyte group members (FD = 20.5–21). Later minerals replacing eudialytes, like hilairite group members, catapleiite, zirsinalite, wadeite, gaidonnayite, georgechaoite, terskite, elpidite, umbite and kostylevite are characterized by lower FD, i.e., higher microporous character (see Table 2 in Chukanov and Pekov 2005). In the labuntsovite group, later minerals practically always have lower FD values as compared with earlier ones, including the cases of epitactic intergrowths (Chukanov et al. 2003c). Zeolite-like titanosilicates, e.g., members of the labuntsovite and lintisite groups and vinogradovite, crystallize later than Ti minerals with higher FD such as lorenzenite (Fig. 1) and titanite.

If alkalinity decreases during hydrothermal stages, late MHM are not formed but are substituted by minerals with dense packed cation polyhedra. This is especially true for microporous Ti- and Nb-silicates, which can be replaced by titanite and oxides like anatase, brookite, ilmenite, pyrophanite, aeschynite. For example, in Zr mineralization at Kovdor we observed baddeleyite pseudomorphs after catapleiite in cavities of carbonatites. Significant "compression" of the framework can also be a consequence of the decationization of MHM when alkalinity decreases, at the hydrothermal stages. The framework collapse is a result of the "thrust extraction", i.e., Na (rarely K) leaching. It is evident in the following examples of solid-state transformations of MHM (for the mechanisms see below): zirsinalite → lovozerite (FD 19.9 → 20.4); kazakovite → tisinalite (20.4 → 21.4); kapustinite → litvinskite (17.9 → 18.5); parakeldyshite → keldyshite (19.8 → 20.0); Na-komarovite → komarovite (15.0 → 15.6); lintisite → Na-free Li,Ti-silicate $LiTi_2Si_4(O,OH)14\cdot nH_2O$ (21.8 → 22.9).

Figure 1. Epitactic growth of vinogradovite (grey) at the top of lorenzenite (dark) within a matrix of natrolite (white). Mt. Kukisvumchorr, Khibiny alkaline complex, Kola Peninsula. Specimen 3.5 × 2.5 cm. A.S. Podlesnyi collection. Photo: N. Pekova.

Thus, in pegmatites and hydrothermalites of alkaline complexes the evolution proceeds in the direction of increasing the molar volume if the process (with temperature lowering) proceeds under high-alkaline conditions and, in the direction of a condensation of structures if alkalinity decreases. In series of structurally related framework minerals, the increasing of molar volume can be realized via the straightening of bridges in the framework, i.e., increasing of the valence angles M-O-M, Si-O-Si and others (e.g., the labuntsovite and lovozerite groups).

Role of epitaxy

Some earlier MHM favor nucleation of others, in particular by epitaxy; some examples follow.

Eudialyte and lovozerite groups. Epitactic relations between different eudialyte group minerals are typical in the alkaline massifs of Kola Peninsula: parallel intergrowths of eudialyte with alluaivite (Khomyakov et al. 1990a; Ageeva et al. 2002) and overgrowths of eudialyte by rims of taseqite (Azarova 2003), ferrokentbrooksite or feklichevite have been observed. The case of alluaivite, the Ti analogue of eudialyte, is especially interesting from a crystal-chemical viewpoint. The rarity of alluaivite seems to depend from the presence of isolated Ti octahedra in its structure, an uncommon feature for Ti-silicates (Pyatenko et al. 1999). In the case of epitactic growth of alluaivite on eudialyte, the energy advantage allows to overcome the tendency of Ti octahedra to condensation. A similar case was observed for the minerals of the lovozerite group: crystals of kapustinite, a Zr member of the group, are surrounded by an epitactic rim of kazakovite, a Ti member of the group (Pekov et al. 2003e).

Labuntsovite group. Labuntsovite group minerals often form epitactic intergrowths with elpidite which is always an earlier phase that favors the nucleation of labuntsovite-type minerals. In pegmatites of Lovozero, korobitsynite almost always overgrows elpidite (Pekov et al. 1999b). Epitaxy of nenadkevichite, vuoriyarvite-K, kuzmenkoite-Mn, organovaite-Mn, organovaite-Zn, gjerdingenite-Fe, gjerdingenite-Mn and tsepinite-K (Fig. 2) on elpidite is typical in numerous localities at Lovozero, Khibiny, Vuoriyarvi, Mont Saint-Hilaire, Narssarssuk and Gjerdingen. Epitaxy of these minerals is due to very close (or multiple) values of their cell dimensions. In all intergrowths of monoclinic labuntsovite-type minerals with elpidite, the b axis of formers is parallel to the c axis of the latter (Chukanov et al. 2003c). Probably the energy advantage provided by the epitaxy on elpidite favors the wide presence

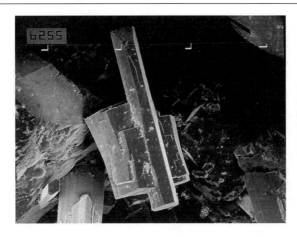

Figure 2. Epitactic growth of tabular tsepinite-K around a prismatic crystal of elpidite 1 mm long. Mt. Kukisvumchorr, Khibiny alkaline complex, Kola Peninsula. SEM photo.

of labuntsovite group minerals as compared with other Ti- and Nb-silicates in Si-enriched peralkaline hydrothermalites. Epitactic (parallel) intergrowths of different labuntsovite-group minerals (Figs. 3-4) including micro-intergrowths with sharp phase boundaries (some of them can be also considered as complex crystals with sharp zoning), are very common.

Inheritance of structural fragments. Topotactic replacements with inheritance of structural fragments of the earlier minerals are typical for vinogradovite and lintisite formed after lorenzenite. These minerals have two close periodicities corresponding to the directions of pyroxene-like and Ti octahedral chains.

The evolution series involving oxysilicates of the komarovite series is of particular interest (Azarova et al. 2002; Pekov et al. 2004b). Na-komarovite and komarovite are intermediate members of two oppositely directed evolution series: from silicates of the labuntsovite group to oxides of the pyrochlore group (Lovozero), and vice versa (Khibiny). Komarovites contain "ready building blocks": labuntsovite-type silicate rings and pyrochlore-type (Nb,Ti),O modules (Balič-Žunič et al. 2002); i.e., they are an "intermediate link" between the labuntsovite and pyrochlore groups in both chemical and structural aspects (Fig. 5). Thus, transformation of komarovite to labuntsovite-type minerals or to pyrochlore seems favored by the inheritance of stable structural fragments.

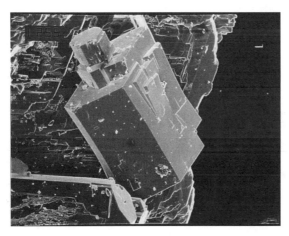

Figure 3. Epitactic growth of tsepinite-K around prismatic nenadkevichite. The size of tsepinite-K crystal is 0.8 × 0.7 mm. Mt. Kukisvumchorr, Khibiny alkaline complex, Kola Peninsula. SEM photo.

Figure 4. Auto-epitactic growth of two generations of kuzmenkoite-Mn crystals: two small crystals of a Nb-poor variety overgrow a larger crystal of a Nb-rich variety. The large Nb-rich crystal is 1.5 mm long. Mt. Flora, Lovozero alkaline complex, Kola Peninsula. SEM photo.

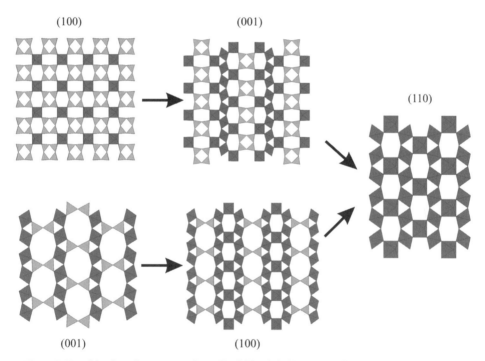

Figure 5. Transition from the structure of nenadkevichite (left side) to that of Na-komarovite (center) and further into that of pyrochlore (right side). Only framework polyhedra are shown.

The most abundant products of the eudialyte alteration in potassium-rich pegmatites of Khibiny is wadeite and the most typical Na,Zr silicate formed after eudialyte in various agpaitic complexes is catapleiite. Probably that is a result of the inheritance of an important fragment of the eudialyte framework: the three-membered ring of Si tetrahedra connected with three isolated Zr octahedra. This group occurs also in the wadeite – catapleiite framework and can offer energy advantage to the formation of catapleiite and wadeite as compared with other Na- and K-zirconosilicates that may replace eudialyte under the same conditions.

SOLID-STATE TRANSFORMATIONS WITH THE PRESERVATION OF THE FRAMEWORK

Natural solid-state transformations with the preservation of the mixed framework as the most stable fragment of the structure are very typical for MHM. During these processes, changes in the framework are insignificant, whereas significant variations concern the extra-framework cations. Decationization (cation leaching), hydration and ion exchange are the most important processes occurring in this type of transformation.

Decationization and hydration

Fresh and hydrothermally altered MHM in peralkaline rocks offer very good examples to study natural solid-state transformations of minerals. Basing on the lomonosovite group (Khomyakov 1976b), the lovozerite group (Chernitsova et al. 1975; Khomyakov et al. 1978), the keldyshite group (Khomyakov 1976a; Khalilov et al. 1978) and in general on the inheritance principle in minerogenesis (Yushkin 1977; Yushkin et al. 1984), Khomyakov (1995) defined the "transformation minerals" as a specific genetic group of secondary that contain H instead of alkali cations. Unlike "ordinary" minerals, these secondary minerals cannot crystallize from melts or solutions; they are formed only as pseudomorphs after the corresponding precursor minerals (Khomyakov 1995).

We have found that the number of minerals formed by such process is significantly larger than suspected before. Typical transformation minerals belong to the pyrochlore, eudialyte and astrophyllite groups and the komarovite series. In other cases, partially decationized and/or ion-exchanged forms are observed. Among MHM, typical transformation species are keldyshite, lovozerite, tisinalite (Khomyakov 1995), litvinskite, komarovite, ikranite and aqualite. FD mainly determines the capability of ion leaching and exchanging. MHM with wide pores are sensitive to the content of large cation modifiers in the mineral-forming medium (Na and K in agpaitic systems) not only at the nucleation stage but also during further solid-state transformations.

Endemic MHM of agpaitic and, especially, hyperagpaitic systems can crystallize from solution or melt only as phases with the Na and/or K sites fully occupied. Alkali-depleted MHM and their varieties enriched by higher-charge cations (Ca, Sr, Pb, U, etc.) are usually a result of late solid-state transformations under hydrothermal or supergene conditions. These transformations are caused by the activity of low-alkali solutions and proceed according to following ideal schemes:

$$nA^+ \to A'^{n+} + (n-1)\square$$

$$nA^+ \to A'^{n+} + (n-1)H_2O$$

$$A^+ \to (H_3O)^+$$

When the framework density decreases, the properties of MHM change gradually. At first, the possibility of decationization appears (at $FD = 19-23$); with further decreasing of FD, the possibility of ion exchange appears at room conditions ($FD < 17-19$), as confirmed by experimental data (see Chukanov and Pekov 2005) and the existence of cation-deficient varieties of corresponding minerals; the same properties do not appear in minerals with higher FD. The same tendency is typical for framework oxides (perovskites, pyrochlores, etc.).

Na is practically the only cation leached by MHM in nature; leaching of other cations is very rare. We have observed significant leaching of K together with Na in sitinakite and of Ca probably in fersmanite. Na has not a permanent solvate shell in water solutions (Breck 1974). It causes ion conductivity in zeolite-like zirconosilicates (catapleiite, hilairite, lovozerite; Ilyushin and Demyanets 1986, 1988) and actively migrates in microporous crystals under electron beam and X-ray irradiation (Gobechiya et al. 2003), a property likely related to ion conductivity. The degree of alteration of the agpaitic pegmatites by low-alkaline solutions is related to the degree of Na deficiency in MHM. In labuntsovite-group minerals, the total content of large cations correlates with FD: the members of the nenadkevichite and vuoriyarvite subgroups with the lowest FD values are characterized by the lowest amount of extra-framework cations (Chukanov et al. 2003c). The appearance of decationized phases is probable for MHM containing M-O-M bridges between octahedra, an aspect that supports the substitution of O^{2-} by OH^- in parallel with the leaching of large cation according to the following reaction:

$$Na^+ + (M\text{-O-}M) \rightarrow (H_2O,\square) + (M\text{-OH-}M)$$

Anhydrous crystals with interrupted frameworks are the best candidates for decationization because of the easy substitution of O^{2-} by OH^- on "hanging" apices of framework polyhedra according to the following reaction: $Na^+ + (Si\text{-O}) \rightarrow (H_2O,\square) + (Si\text{-OH})$. The lovozerite group minerals represent the best example (Chernitsova et al. 1975; Tamazyan and Malinovskii 1990; Pyatenko et al. 1999; Pekov et al. 2003e).

The mechanisms of decationization and hydration of the lovozerite group minerals have been studied for three evolution series: kapustinite → litvinskite, zirsinalite → lovozerite, and kazakovite → tisinalite (Yamnova et al. 2001a,b, 2003; Pekov et al. 2003e). The minerals of the group have general formula $A_3B_3C_2\{M[Si_6O_9(O,OH)_9]\}$ where M = Zr, Ti, Fe^{3+}, C = Ca, Mn^{2+}, A and B = Na, H_2O (for framework description see Chukanov and Pekov (2005); the extra-framework A and B sites alternate in zeolite-like channels). Table 2 and Figure 6 show that B cations are leached completely, whereas A cations slightly. Zirconian members of the group show also that ordering of cations in the C octahedra proceeds together with hydration: cations move into C' site whereas C'' octahedron becomes empty.

A different mechanism of decationization and hydration has been proposed for lintisite. In a hydrothermally altered pegmatite (Lovozero) we have found a potentially new Li,Ti-silicate, $LiTi_2Si_4(O,OH)14 \cdot nH_2O$, that can be considered a completely decationized lintisite with OH groups substituting O atoms at bridges in the framework. The peculiarity of this phase consists in a practically complete absence of large cations, as confirmed by the structure refinement (Kolitsch et al. 2000).

Cation-deficient calciohilairite from Lovozero shows a mechanism of partial decationization and hydration involving hydronium. This trigonal mineral differs from all other hilairite-roup members not only in its halved c parameter (see Chukanov and Pekov 2005: Table 2) but also in the occurrence of a site partially occupied by $(H_3O)^+$ and water molecules. As confirmed by local bond-valence balance, hydronium form hydrogen bonds with O atoms at 2.99 Å; (Pushcharovsky et al. 2002). The scheme of isomorphous substitutions which can be also considered a scheme of the transformation (decationization + ion exchange) of hilairite to the cation-deficient calciohilairite is the following: $2Na^+ + H_2O \rightarrow 0.5Ca^{2+} + 1.5\square + (H_3O)^+$ (Pekov et al. 2003c). The presence of hydronium was inferred from X-ray and IR investigations in three other cation-deficient MHM: tsepinite-Na (Rastsvetaeva et al. 2000), ikranite (Ekimenkova et al. 2000), and aqualite (Rastsvetaeva and Khomyakov 2002). Cation deficiency is also probable in komarovite, vinogradovite, sitinakite and fersmanite.

The zeolite-like zirconosilicates contain isolated Zr octahedra. The substitution $O^{2-} \rightarrow OH^-$ on M-O-Si bridges is not common but has been reported in few cases (Khalilov et al. 1978; Yamnova et al. 2003) together with presence of hydronium. Such combination apparently allows compensating a deficiency of Na in terskite, elpidite, gaidonnayite and Ca-poor catapleiite.

Minerogenetic Aspects of Microporous Framework Silicate Minerals   161

Table 2. Composition of structural sites (A, B, C, M) and fragment $[T_6(O,OH)_{18}]$ in the following three cases of mineral formation via decationization and hydration in lovozerite-group minerals: kapustinite → litvinskite, zirsinalite → lovozerite, and kazakovite → tisinalite.

Site	$A(1)$	$A(2)$	$B(1)$	$B(2)$	C'	C''	M	$T_6(O,OH)_{18}$
Kapustinite	Na	Na$_2$	Na$_{0.8}\square_{0.2}$	Na$_{1.6}\square_{0.4}$	$\square_{0.7}$Zr$_{0.15}$Mn$_{0.15}$	$\square_{0.7}$Zr$_{0.15}$Mn$_{0.15}$	Zr$_{0.7}\square_{0.3}$	Si$_6$O$_{16}$(OH)$_2$
Litvinskite	Na$_{0.8}\square_{0.2}$	Na$_{1.6}$(H$_2$O)$_{0.4}$	$\square$	$\square_2$	$\square_{0.6}$Na$_{0.2}$Mn$_{0.2}$	$\square$	Zr	Si$_6$O$_{13}$(OH)$_5$
Zirsinalite	Na	Na$_2$	Na	Na$_2$	Ca$_{0.5}\square_{0.5}$	Ca$_{0.5}\square_{0.5}$	Zr	Si$_6$O$_{18}$
Lovozerite	Na$_{0.8}\square_{0.2}$	Na$_{1.6}\square_{0.4}$	$\square_{0.7}$(H$_2$O)$_{0.3}$	$\square_{1.4}$(H$_2$O)$_{0.6}$	Ca$_{0.4}$Na$_{0.2}$Mn$_{0.1}\square_{0.3}$	$\square$	Zr$_{0.8}$Fe$_{0.1}\square_{0.1}$	Si$_6$O$_{13}$(OH)$_5$
Kazakovite	Na	Na$_2$	Na	Na$_2$	Mn$_{0.5}\square_{0.5}$	Mn$_{0.5}\square_{0.5}$	Ti	Si$_6$O$_{18}$
Tisinalite	Na$_{0.7}\square_{0.3}$	Na$_{1.4}\square_{0.6}$	$\square$	$\square_2$	$\square_{0.7}$Mn$_{0.2}$Ca$_{0.1}$	$\square_{0.75}$Mn$_{0.25}$	$M1$: (Ti$_{0.65}$Fe$_{0.35}$)$_{0.5}$ $M2$: (Ti$_{0.4}$Zr$_{0.2}$Nb$_{0.2}$)$_{0.5}$	Si$_6$O$_{13}$(OH)$_5$

NOTE: the data on the site occupancy in zirsinalite (Pudovkina et al. 1980) and kazakovite (Voronkov et al. 1979) are given in simplified form (main cations only); the data for other minerals are given in accordance with real chemical compositions (Yamnova et al. 2001a,b, 2003; Pekov et al. 2000, 2003e).

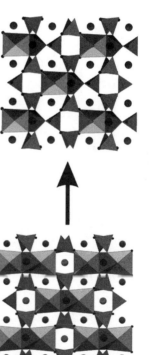

Figure 6. Transition from the structure of kapustinite (left side) to that of litvinskite (right side). Both structures are projected into the (001) plane.

Natural ion exchange

Ion exchange properties of MHM are discussed in Chukanov and Pekov (2005); it is unquestionable that they occur in nature. Specific kinetic parameters of the ion exchange reactions are low values of activation energy and high reaction velocities (see, e.g., Chelishchev 1973). Post-crystallization interactions with solutions via ion exchange seem the easiest way to increase the stability of MHM under changing conditions. Alkaline complexes, especially large ones, are long-living geological systems. Some of them are Paleozoic (e.g., Kola, South Urals and Aldan Provinces, Russia) or even Pre-Cambrian (e.g., Gardar Province, South Greenland). The postmagmatic evolution of these complexes can be continuous in time; that significantly increases the chance of ion exchange for microporous crystals, especially if they occur in fractures and cavities of the rocks, where they are better exposed to the circulating fluids.

In many cases, crystals of minerals of the labuntsovite and the hilairite groups and the steenstrupine-(Ce) – thorosteenstrupine series are heterogeneous chemically: they show "spotty" zones characterized by strong differences in extra-framework cation composition (Pekov et al. 1997, 2003c; Chukanov et al. 2003c). The distribution of zones is usually irregular; both sharp interfaces and gradual transitions have been observed. Boundary zones, including those between twins, are typically enriched by large cations (Fig. 7). Such heterogeneity is probably the result of an interruption of the post-crystallization cation exchange before reaching equilibrium. This hypothesis is experimentally confirmed by Chukanov and Pekov (2005). Concentric zoning with gradual transition in composition is observed in crystals of cation-deficient hilairite group minerals from Lovozero (Fig. 8). Border parts of these crystals are depleted by Na and enriched by Ca, K, Sr and Ba. Apparently the latter cations substituted Na by cation exchange (Pekov et al. 2003c). Likely, the presence of hydronium in MHM is an indicator of ion exchange with water solutions: it is known (Breck 1974) that the substitution of metal cations by H^+ is an intermediate step in the exchange reactions occurring in zeolites.

The following MHM that occur in hydrothermally altered agpaitic pegmatites are probable products of cation exchange: calciohilairite and its Ba-rich variety from Lovozero (Pekov et al. 2003c); K-rich sazykinaite-(Y) from Khibiny (Pekov 1998); Ba-, Sr- and Ca-

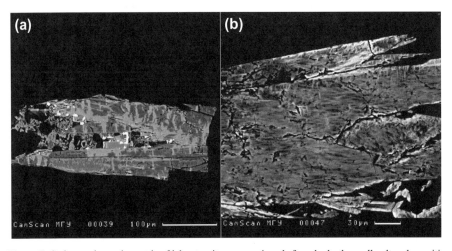

Figure 7. Cation-exchanged crystals of labuntsovite-group minerals from hydrothermally altered agpaitic pegmatites. Ba-rich light areas are shown in (a) tsepinite-Na (Mt. Lepkhe-Nelm, Lovozero alkaline complex, Kola Peninsula) and (b) in the rim of tsepinite-Sr (Mt. Eveslogchorr, Khibiny alkaline complex, Kola Peninsula).

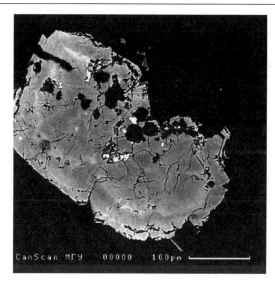

Figure 8. Zoning in a crystal a hilairite group from hydrothermally altered agpaitic pegmatite - the result of natural cation exchange. The core (hilairite) is enriched by residual Na, whereas the rim (cation-deficient calciohilairite) is Na-depleted and enriched by Ca, K, Sr and Ba (Mt. Flora, Lovozero alkaline complex, Kola Peninsula).

rich minerals of the vuoriyarvite and tsepinite subgroups (crystals with "spotty" zoning) from Lovozero and Khibiny (Chukanov et al. 2003c); K- and Th-rich steenstrupine from Khibiny (Chukanova et al. 2004); Pb- and Ba-rich komarovite from Khibiny (Pekov et al. 2004b); K- and Ca-rich gaidonnayite and Na-deficient georgechaoite from Khibiny; Ca-enriched varieties of catapleiite and paraumbite from Lovozero and Khibiny; Ba-, Th- and U-rich Ca-deficient hydrated fersmanite from Khibiny; Ba-, K- and Sr-rich aqualite from Kovdor. It is not excluded that unusual SO_4-bearing Cl- and F-deficient hydrated eudialyte-group minerals recently found in hydrothermally altered pegmatites of Lovozero and Kovdor are a product of natural anion exchange.

MHM AS GEOCHEMICAL AND MINEROGENETIC INDEX MINERALS

The specific structure features and related properties assign to MHM an important role as index minerals in the minerogenesis and geochemistry of the rare and transition elements in the agpaitic complexes. At the same time, MHM can be used in genetic mineralogy as indicators of specific geological situations.

MHM as selectors and concentrators of litophile rare elements

The pegmatites and hydrothermalites related to agpaitic rocks, at first nepheline syenites, are characterized by a unique mineral diversity: up to 40–50 and in several cases up to 75–80 mineral species have been found within single pegmatite bodies that usually do not exceed 20–30 m in length and 1–2 m in thickness. Pegmatites of this type are: Hilairitovoye in Khibiny (78 mineral species!); Yubileinaya, Palitra, Shkatulka and Shomiokitovoye in Lovozero; large hydrothermally altered pegmatites at Mont Saint-Hilaire and Tvedalen (Horvath and Gault 1990; Andersen et al. 1996; Pekov 2000; Pekov and Podlesnyi 2004; Pekov 2005). The occurrence of chemically complex minerals with highly ordered structures containing up to 4–5 or even more independent sites that may be occupied by different cations is typical in these pegmatites. Such minerals are major selective concentrators of rare elements. Among them, MHM play a major role in the concentration of lithophile rare metals: Zr, Hf, Nb, W, Ba, Y, *HREE* and sometimes also *LREE*, Th, U, Sr, Zn, Li, Sn, Ta. The complex structures of several MHM, typically characterized by a strict cation ordering, are highly selective relatively to these elements. The contents of different sites are separated by hyphens in the following examples of MHM showing ordered cations.

- Labuntsovite group: alsakharovite-Zn, (Na,Ca)-Sr-K-Ba-Zn-(Ti,Nb)-Si (Pekov et al. 2003f); lemmleinite-Ba, Na-K-Ba-Mn-Ti-Si (Chukanov et al. 2001); gutkovaite-Mn, Ca-K-Mn-Ti-Si (Pekov et al. 2002a).
- Eudialyte group: manganokhomyakovite, Na-Sr-Ca-Mn-Zr-W-Si (Johnsen et al. 1999); feklichevite, Na-Ca-Fe-Zr-Nb-Si (Pekov et al. 2001).
- Hilairite group: sazykinaite-(Y), Na-(Y,*HREE*)-Zr-Si (Khomyakov et al. 1993).
- Joaquinite group: byelorussite-(Ce), Ba-*LREE*-Na-Mn-Ti-Si (Shpanov et al. 1989; Zubkova et al. 2004).
- Steenstrupine-(Ce), Na-(*LREE*,Th,U)-Mn-Fe-(Zr,Ti)-Si-P (Moore and Shen 1983).
- Ilímaussite-(Ce), Ba-K-Na-*LREE*-Nb-Si (Semenov et al. 1968; Ferraris et al. 2004).
- Lintisite, Na-Li-Ti-Si (Khomyakov et al. 1990b).

Among the listed examples, alsakharovite-Zn is of particular interest. Its crystal-chemical formula is $[(Na,Ca)_{2-x}(Sr,Na)_{2-x}][(H_2O)_2(K,H_2O)_2][Ba_{0.2}(H_2O)_{3.6}]\{Zn(Ti_{5.7}Nb_{2.3}(OH_{4.8}O_{3.2}))(Si_4O_{12})_4\}\cdot[7H_2O]$; the contents of extra-framework sites are given in square brackets and the composition of the framework is within braces. Alsakharovite-Zn represents a unique case of selectivity by ordering alkaline and earth alkaline elements within the same crystal: (Na + Ca), Sr, K and Ba (all in significant concentrations) are ordered in four independent sites within channels. Taking into account the mixed framework formed by the Si tetrahedra, (Ti,Nb) octahedral chains and isolated Zn octahedra, seven independent sites occupied by different cations occur.

Minerals of the labuntsovite group show clear dependence of their cation composition on the structural type. The members of the kuzmenkoite and organovaite subgroups mainly accumulate Zn; those of the vuoriyarvite subgroup accumulate instead Ca and Sr; finally, the members of the labuntsovite subgroup are poor of these three elements and Nb but are rich of Mg and Fe. In particular, that is observed in syntactic and epitactic intergrowths of representatives of different subgroups of the labuntsovite group. In the most common intergrowths of minerals of the kuzmenkoite and vuoriyarvite subgroups, the former are selective for Ti, Mn, Zn, and K and the latter for Nb, Sr, Ca and Na (Chukanov et al. 2003c).

Eudialyte group members not only represent main phases for Zr and Hf but also are effective concentrators of elements that are exotic for minerals of the nepheline syenite complexes such as Y, *HREE*, W and Ta. W, like Nb, is selectively concentrated in a specific octahedral site that center nine-membered rings of Si tetrahedra (Johnsen et al. 1999, 2003); Ta, instead, (unlike Nb) occupies a quadratic site together with Fe (Rastsvetaeva et al. 2003a). In the case of large deposits, eudialytes play also a significant role in the geochemistry of Sr, *LREE* and Nb. Some of the mentioned elements (Hf, Nb, Y, *HREE*) can be found also in products of the eudialyte alteration, i.e., late microporous zirconosilicates. The most interesting example is sazykinaite-(Y), $Na_5(Y,HREE)ZrSi_6O_{18}\cdot 6H_2O$, that crystallized in cavities following the eudialyte dissolution; this mineral is a main concentrator of Y and *HREE* in the hydrothermal parageneses of pegmatites in the Koashva area at Khibiny (Pekov 1998).

Some MHM are selective concentrators of Sn, an element very uncommon in the agpaitic complexes. That is due to the close crystal-chemical relationship of Sn^{4+} with Ti^{4+} and Zr^{4+}.

Significant Sn substituting Zr was found in tumchaite from Vuoriyarvi (Subbotin et al. 2000). An almost Ti-free pabstite from Darai-Pioz contains 30 wt% SnO_2: it is the only tin mineral of this alkaline massif (Pautov 2003).

The formation of some MHM in alkaline pegmatites and hydrothermalites is related to the presence of species-forming Ba, the largest of the geochemically significant bivalent cations. The relevant species are baotite, belkovite, minerals of the joaquinite group and of the benitoite and the ilímaussite series. Sr, unlike Ba, is not a typical constituent of MHM. In aluminosilicate zeolites, Sr is common as a species-forming cation (Pekov et al. 2004a) but MHM usually contain it only as minor element. Minerals with Sr as species-forming element are known in the labuntsovite, eudialyte and joaquinite groups but are extremely rare.

Members of the steenstrupine-(Ce) – thorosteenstrupine series are very important concentrators of *LREE*, Th and U in complexes of hypersodic agpaitic rocks, namely Ilímaussaq and Lovozero (Bussen and Sakharov 1972; Sørensen et al. 1974; Pekov et al. 1997; Chukanova et al. 2004). These high-charge and large cations occupy extra-framework sites in minerals of the steenstrupine series. Note that, because of their metamictic state, the crystal chemistry of thorium minerals occurring in agpaitic complexes is insufficiently studied. Likely Th plays a role similar to that of *LREE* and in particular can be species-forming element in some amorphous phases that are similar to known MHM in their stoichiometry and IR spectra. Umbozerite, $Na_3Sr_4Th(Mn,Zn,Fe)Si_8O_{24}(OH)$, and a recently found Na,Th,Ti-silicate (Ermolaeva et al. 2005) are examples; both of them have features in common with seidite-(Ce).

The index role of decationized MHM

The importance of "transformation" (i.e., Na-depleted) mineral species for paleo-mineralogical reconstructions has been proved for lovozerite, keldyshite and lomonosovite group members (Khomyakov 1995). The presence of hydrated low-alkaline members of these and other evolution series (like decationized MHM of the eudialyte and the hilairite groups, the komarovite and the lintisite series) is a clear sign that the rock passed through the hyperagpaitic stage. On the other hand, these minerals show that the rock was altered by low-alkaline and, as a rule, low-temperature solutions.

The minerogenetic role of ion exchange processes

The role of ion exchange processes in the minerogenesis seems very important; however, it is little discussed in the literature. An ion-exchanged MHM, as a decationized one, can be considered a pseudomorph that inherits from its precursor the most stable part of the structure, i.e., the framework. The importance of ion-exchanged phases for geochemistry and genetic mineralogy resides mainly in the two following unique features.

- Ion exchange allows microporous minerals to enrich in some elements under conditions not favorable to the direct crystallization from solutions or melts (that happens under low temperature and/or deficiency of framework-forming elements in the medium).

- Many microporous phases have ultra-high selectivity with respect to some elements, including very rare ones. That allows catching and concentrating these elements from solutions, even in traces. It is important that a crystal can act as ion exchanger for an unlimited time, whereas its action as element incorporator during growing is limited in time; thus, such crystals can extract selected elements from solutions for a long time, playing a role of filtering.

From these two features confirmed by experiments on natural and synthetic microporous materials (see Chukanov et al. 2004), several consequences result. At first, the ion-exchange properties of minerals give a unique possibility of knowing the chemistry of late residual solutions that did not produce their own minerals. For example, the cation composition of late wide-pore zeolites show that residual solutions in Kola alkaline complexes (including hypersodic Lovozero massif) were rich of K and Ca, unlike at Mont Saint-Hilaire and Ilímaussaq where they were Na-rich (Pekov et al. 2004a). The ability of microporous minerals to capture selectively trace elements is extremely important for the geochemistry of rare metals. In alkaline complexes, the evolution towards a large-cation composition in pyrochlore group minerals at the late hydrothermal stages is the best example. Owing to ion exchange, ordinary pyrochlore transforms to members of the same group with an uncommon composition for a nepheline syenite formation: U, Y, Pb, Ba and Sr become prevailing extra-framework cations instead of Ca and Na. This process is not so local and sometimes leads to the formation of industrially important deposits of uranpyrochlore (Lovozero; Pekov 2000) and even of huge deposits of bariopyrochlore as the well known Araxá and Catalão II in Brazil. MHM containing the pyrochlore modulus, like komarovites and fersmanite (Balič-Žunič et al. 2002; Sokolova et al. 2002) present selectivity with respect to the same large cations. Clearly, the study of ion exchange properties of MHM can give suggestions for the synthesis of novel microporous materials with high cations selectivity.

The catalytic role of MHM in nature

The catalytic activity of MHM in the pegmatitic-hydrothermal rare-element minerogenesis in agpaitic complexes is discussed by Chukanov and Pekov (2005). In our opinion, MHM as catalysts can promote and accelerate reactions leading to the formation and transformation of many organic substances including complex ones: alkyl-aromatic, polycyclic, O-, N- and S-bearing compounds. The catalytic properties of microporous phases are mainly ascribed to OH-groups of the framework that operate as Broensted acid centers and exchange protons with adsorbed molecules, oxygen vacancies in the framework that operate as Lewis acids owing to their positive charge, and electrostatic field gradients that polarize (also as Lewis acid) bonds in adsorbed molecule. Synthetic zeolite-like Ti-, Nb- and Zr-silicates structurally identical or close to MHM have been successfully used in catalysis of selective oxidation, polymerization and reforming of organic substances (cf. Chukanov et al. 2004).

In their turn, organic compounds under hydrothermal conditions can act as complexing agents carrying out extraction, transport and selective concentration of some rare elements at relatively low temperatures. That is especially important for the geochemistry of Th, U^{4+} and *REE* characterized by a low mobility of their ions in water solutions. For example, oxidized derivatives of polynuclear aromatic hydrocarbons found in bitumen of Khibiny pegmatites (Chukanov et al. 2003a) are known as complexing agents for Th, *REE*, Ba, Sr and Ca. These are analogues of high-molecular heterocyclic constituents of petroleum and, in particular, of polynaphthenic acids. The latter are very effective complexing agents and sorbents for Th and U, thus easily extract these elements from water solutions, even in traces (Efendiev et al. 1964). In conclusion, the following genetic chain can be considered for the agpaitic pegmatites and hydrothermalites: MHM (*catalysis in reactions of organic substances*) → complex organic compounds (*extraction, transport, fractionation and concentration of rare elements*) → silicates, phosphates, carbonates and oxides of *REE*, Th, U etc. (*formation of rich late rare-metal mineralization*).

CONCLUSIONS

Most microporous heterosilicates (113 out of 122 known MHM) and all their significant concentrations are confined to high-alkaline rocks and especially their derivatives. The magmatogenic zirconosilicates (eudialyte, elpidite, lovozerite, parakeldyshite), steenstrupine and naujakasite can occur as accessory and even rock-forming minerals. Sometimes they give huge industrially important deposits (Ilímaussaq and Lovozero complexes; agpaitic granite massifs of Western Mongolia).

The enrichment of the mineral-forming medium in Na, K or Ba is the main factor controlling the formation of MHM in nature. Ca and Sr do not act as promoters of MHM formation in absence of Na, K and Ba. The Ussing agpaicity index can be considered as a crystal chemical criterion to determine the agpaitic formation. At values of this index near to 1.2, qualitative changes of crystal chemical (in particular, topological) characteristics occur in most of minerals of Ti and rare elements: the main phases containing Ti, Zr, Nb, Ta, Be, Ba, Th, U are represented by Na- and/or K-rich minerals having low-density framework or various layer structures.

The universal tendency to a lower *FD* with decreasing temperature is typical for the evolution series of tecto-aluminosilicates. Unlike, the evolution series of MHM, beryllosilicates and niobates show dual behavior: strong tendency to "loosening" of the structures along the evolution series that develop in peralkaline conditions; but if the alkalinity decreases during hydrothermal stages, minerals with open structures are substituted by minerals with dense packed cation polyhedra.

Natural solid-state transformations with the preservation of the mixed framework are very typical for MHM: changes in the framework are insignificant, whereas significant variations concern the extra-framework cations. Decationization (cation leaching), hydration and ion exchange are the most important processes. Alkali-depleted MHM and their varieties enriched by higher-charge cations (Ca, Sr, Pb, U, etc.) are usually a result of late solid-state transformations under hydrothermal or supergene conditions. When *FD* decreases, the properties of MHM change: at first, the possibility of decationization appears; with its further decreasing, the possibility of ion exchange appears.

The specific structure features and related properties assign to MHM an important role as index minerals in the minerogenesis and geochemistry of the rare and transition elements in the agpaitic complexes. MHM play a major role in the concentration of Zr, Hf, Nb, W, Ba, Y, *HREE* and sometimes also *LREE*, Th, U, Sr, Zn, Li, Sn, Ta. The ion exchange allows microporous minerals to enrich in some elements under conditions not favorable to the direct crystallization from solutions or melts. Many microporous phases have ultra-high selectivity with respect to some elements that allows to catch and concentrate these elements from solutions, even in traces; a crystal can act as ion exchanger for an unlimited time. The ion-exchange properties of minerals give a unique possibility of knowing the chemistry of late residual solutions that did not produce their own minerals.

MHM as catalysts can promote and accelerate reactions leading to the formation and transformation of many organic substances.

ACKNOWLEDGMENTS

We are grateful to G. Ferraris for valuable discussion, comments and significant editor's work on the manuscript and to A. Gula for the help with the bibliography. This work was supported in part by the joint grant No. 03-05-20011 of the Russian Foundation of Basic Research (RFBR) and BNTS (Austria), the grant No. 04-05-64085 of the Russian Scientific Foundation and the grant No. 1087-2003-5 of the Fundamental Science School Program of the Russian Academy of Sciences.

REFERENCES

Ageeva OV, Borutsky BE, Chukanov NV, Sokolova MN (2002) Alluaivite and genetic aspect of formation of titanium-enriched eudialytes in Khibiny massif. Zap Vseross Mineral Obs 131(1):99-106 (in Russian)

Alfors JT, Stinson MC, Matthews RA, Pabst A (1965) Seven new barium minerals from eastern Fresno County, California. Am Mineral 50:314-340

Andersen F, Berge SA, Burvald I (1996) Die Mineralien des Langesundsfjords und des umgebenden Larvikit-Gebietes, Oslo-Region, Norwegen. Mineralien-Welt 4:21-100

Azarova YuV (2003) On the bahaviour of Ba, Sr and Ca in rocks of the lujavrite-malignite complex of Khibiny massif at late stages of mineral formation. Proc. of XXI All-Russian Seminar "Alkaline magmatism of the Earth". Apatity, p 12-13 (in Russian)

Azarova YuV, Pekov IV, Chukanov NV, Zadov AE (2002) Products and processes of vuonnemite alteration at low-temperature transformation of hyperagpaitic pegmatites. Zap Vseross Mineral Obs 131(5):112-121 (in Russian)

Balić-Žunić T, Petersen OV, Bernhardt H-J, Micheelsen HI (2002) The crystal structure and mineralogical description of a Na-dominant komarovite from the Ilímaussaq alkaline complex, South Greenland. N Jb Mineral Mh 2002:497-514

Barrer RM (1982) Hydrothermal Chemistry of Zeolites. Academic, London.

Birkett TC, Miller RR, Roberts AC, Mariano AN (1992) Zirconium-bearing minerals from the Strange Lake intrusive complex, Quebec-Labrador. Can Mineral 30:191-205

Breck DW (1974) Zeolites, molecular sieves: Structure, chemistry and use. New York

Bussen IV, Sakharov AS (1972) Petrology of Lovozero alkaline massif. Nedra, Leningrad

Chelishchev NF (1973) Ion-exchange properties of minerals. Nauka, Moscow (in Russian)

Chernitsova NM, Pudovkina ZV, Voronkov AA, Kapustin YuL, Pyatenko YuA (1975) On a new lovozerite crystal-chemical femily. Zap Vses Mineral Obs 104(1):18-27 (in Russian)

Chukanov NV, Pekov IV (2005) Heterosilicates with tetrahedral-octahedral frameworks: mineralogical and crystal-chemical aspects. Rev Mineral Geochem 57:105-144

Chukanov NV, Pekov IV, Rastsvetaeva RK (2004) Crystal chemistry, properties and synthesis of microporous silicates containing transition elements. Russ Chem Rev 73(3):205-223

Chukanov NV, Pekov IV, Rastsvetaeva RK, Zadov AE, Nedelko VV (2001) Lemmleinite-Ba, $Na_2K_2Ba_{1+x}Ti_4[Si_4O_{12}]_2(O,OH)_4 \cdot 5H_2O$, a new labuntsovite-group mineral. Zap Vseross Mineral Obs 130(3):36-43 (in Russian)

Chukanov NV, Pekov IV, Sokolov SV, Nekrasov AN, Chukanova VN (2003a) On the nature of "carbocer" from Khibiny and state of thorium and rare-earth elements present in it. Proc. of XXI All-Russian Seminar "Alkaline magmatism of the Earth." Apatity:165-166 (in Russian).

Chukanov NV, Pekov IV, Zadov AE, Voloshin AV, Subbotin VV, Sorokhtina NV, Rastsvetaeva RK, Krivovichev SV (2003b) Labuntsovite-Group Minerals. Moscow, Nauka (in Russian)

Chukanova VN, Kogarko LN, Williams CT, Pekov IV, Chukanov NV (2004) Compositional features and genesis of steenstrupine from igneous rocks of the Lovozero alkaline massifs, Kola peninsula. Geochem Intern 42:295-308

Dowty E (1975) Crystal structure of joaquinite. Am Mineral 60:872-878

Efendiev GKh, Alekperov RA, Nuriev AN (1964) Problems of geochemistry of radioactive elements of oil deposits. Edition of Academy of Sciences of Azerbaidjan, Baku

Ekimenkova IA, Rastsvetaeva RK, Chukanov NV (2000) Crystal structure of hydronium-bearing analog of eudialyte. Dokl RAS 371:625-628 (in Russian)

Ermolaeva VN, Chukanov NV, Pekov IV, Nekrasov AN, Sokolov SV, Kogarko LN (2005) On the role of organic substances in transport and concentration of thorium and other rare elements in alkaline pegmatites of Lovozero and Khibiny peralkaline massifs. Geochem Intern (in Russian) in press

Ferraris G, Gula A (2005) Polysomatic aspects of microporous minerals –heterophyllosilicates, palysepioles and rhodesite-related structures. Rev Mineral Geochem 57:69-104

Ferraris G, Gula A, Zubkova NV, Pushcharovsky DYu, Gobetchiya ER, Pekov IV, Eldjarn K (2004) The crystal structure of ilímaussite-(Ce), $(Ba,Na)_{10}K_3Na_{4.5}Ce_5(Nb,Ti)_6[Si_{12}O_{36}][Si_9O_{18}(O,OH)_{24}]O_6$, and the "ilímaussite" problem. Can Mineral 42:787-795

Ferraris G, Ivaldi G, Pushcharovsky DYu, Zubkova NV, Pekov IV (2001) The crystal structure of delindeite, $Ba_2\{(Na,K,\square)_3(Ti,Fe)[Ti_2(O,OH)_4Si_4O_{14}](H_2O,OH)_2\}$, a member of the mero-plesiotype bafertisite series. Can Mineral 39:1307-1316

Gobechiya ER, Pekov IV, Pushcharovsky DYu, Ferraris G, Gula A, Zubkova NV, Chukanov NV (2003) New data on vlasovite: refinement of the crystal structure and the radiation damage of the crystal during the X-ray diffraction experiment. Crystallogr Rep 48:750-754

Horvath L, Gault RA (1990) The Mineralogy of Mont Saint-Hilaire, Quebec. Mineral Record 4:284-359

Ilyushin GD, Demyanets LN (1986) Ion conductors in the class of Na,Zr-silicates. New family of 3-dimensional conductors-crystals of the lovozerite type $Na_{8-x}H_xZrSi_6O_{18}$. Sov Phys Crystallogr 31:41-44

Ilyushin GD, Demyanets LN (1988) Crystal-structural features of ion transport in new OD structures: catapleiite, $Na_2ZrSi_3O_9 \cdot H_2O$, and hilairite $Na_2ZrSi_3O_9 \cdot 3H_2O$. Sov Phys Crystallogr 33:383-387

Ilyushin GD, Demyanets LN (1997) Hydrothermal system $KOH - ZrO_2 - SiO_2 - H_2O$: Synthesis of potassium zirconosilicates and crystal chemical correlations. Crystallogr Rep 42:1047-1052

Ilyushin GD, Demyanets LN, Ilyukhin VV, Belov NV (1983) Formation of the structure of analogs of natural minerals and synthetic phases in the hydrothermal system $NaOH - ZrO_2 - SiO_2 - H_2O$. Sov Phys Dokl 28:603-604

Johnsen O, Ferraris G, Gault RA, Grice JD, Kampf AR, Pekov IV (2003) The nomenclature of eudialyte-group minerals. Can Mineral 41:785-794

Johnsen O, Grice JD, Gault RA, Ercit TS (1999) Khomyakovite and manganokhomyakovite, two new members of the eudialyte group from Mont Saint-Hilaire, Quebec, Canada. Can Mineral 37:893-899

Khalilov AD, Khomyakov AP, Makhmudov SA (1978) Crystal structure of keldyshite, $NaZr[Si_2O_6OH]$. Sov Phys Dokl 23:8-10

Khomyakov AP (1976a) Homoaxial pseudomorphs of titano- and zirconosilicates as indicators of physicochemical conditions of mineral formation in alkaline massifs. *In*: Problems of genetic information in mineralogy. Komi filial of the Academy of Sciences of USSR, Syktyvkar, p 68

Khomyakov AP (1976b) Constitution and typochemical features of minerals of the lomonosovite group. *In*: Constitution and Properties of Minerals. Vol 10. Dumka N (ed) Kiev, p 96-104

Khomyakov AP (1995) Mineralogy of Hyperagpaitic Alkaline Rocks. Clarendon Press, Oxford, UK

Khomyakov AP, Kaptsov VV, Shchepochkina NI, Rudnitskaya ES, Krutetskaya LM (1978) High-rate hydrolysis of ultra-alkaline titanium- and zirconium-silicates. Experimental verification. Dokl Akad Sci USSR, Earth Sci 243:152-155

Khomyakov AP, Nechelyustov GN, Rastsvetaeva RK (1990a) Alluaivite, $Na_{19}(Ca,Mn)_6(Ti,Nb)_3Si_{26}O_{74}$ $Cl \cdot 2H_2O$, a new titanosilicate with eudialyte-type structure. Zap Vses Mineral Obs 119(1):117-120 (in Russian)

Khomyakov AP, Nechelyustov GN, Rastsvetaeva RK (1993) Sazykinaite-(Y) $Na_5YZrSi_6O_{18} \cdot 6H_2O$ – a new mineral. Zap Vseross Mineral Obs 122(5):76-82 (in Russian)

Khomyakov AP, Polezhaeva LI, Merlino S, Pasero M (1990b) Lintisite, $Na_3LiTi_2Si_4O_{14} \cdot 2H_2O$, a new mineral. Zap Vses Mineral Obs 119(3):76-80 (in Russian)

Kolitsch U, Pushcharovsky DYu, Pekov IV, Tillmanns E (2000) A new lintisite-related titanosilicate mineral from Russia: crystal structure, occurrence and properties. ECM-19 Nancy, p 363

Konev AA, Vorobyev EI, Lazebnik KA (1996) Mineralogy of Murun alkaline massif. Edition of the Siberian Branch of RAS, Novosibirsk

Kostyleva-Labuntsova EE, Borutsky BE, Sokolova MN, Shlyukova ZV, Dorfman MD, Dudkin MD, Kozyreva LV (1978) Mineralogy of Khibiny massif. Vol 2. Nauka, Moscow

LeMaitre RW (ed) (1989) A Classification of Igneous Rocks and Glossary of Terms. Blackwell, Oxford

Linthout K, Nobel FA, Lustenhouwer WJ (1988) First occurrence of dalyite in extrusive rock. Mineral Mag 52:705-708

Malinovskii YuA, Pobedimskaya EA, Belov NV (1976) Crystal structure of muirite $Ba_9(Ca,Ba)(Ca,Ti)_4(OH)_8$ $[Si_8O_{24}](Cl,OH)_8$. Sov Phys Dokl 20:163-164

Marakushev AA, Emelyanenko PF, Kuznetsov IE, Rakcheev AD, Sobolev RN, Frolova TI, Feldman VI, Fenogenov AN, Shteinberg DS, Yakovleva EB, Gushchin AV (1981) Petrography. Part II. Edition of MSU, Moscow

Moore PB, Shen I (1983) Crystal structure of steenstrupine: a rod structure of unusual complexity. Tsch Mineral Petrogr Mitt 31:47-67

Pautov LA (2003) Pabstite from the Dara-i-Pioz moraine (Tadjikistan). New Data on Minerals 38:15–19

Pekov IV (1998) Yttrium mineralization in Khibiny-Lovozero alkaline complex, Kola peninsula. Zap Vseross Mineral Obs 127(5):66-85 (in Russian)

Pekov IV (2000) Lovozero Massif: History, Pegmatites, Minerals. Ocean Pictures, Moscow

Pekov IV (2005) The Palitra, a new hyperalkaline pegmatite in the Lovozero massif, Kola peninsula, Russia. Mineral Rec (in press)

Pekov IV, Azarova YuV, Chukanov NV (2004a) New data on komarovite series minerals. New Data on Minerals 39:5-13

Pekov IV, Chukanov NV, Ferraris G, Gula A, Puscharovsky DYu, Zadov AE (2003a) Tsepinite-Ca, $(Ca,K,Na,\square)_2(Ti,Nb)_2(Si_4O_{12})(OH,O)_2 \cdot 4H_2O$, a new mineral of the labuntsovite group from the Khibiny alkaline massif, Kola Peninsula - Novel disordered sites in the vuoriyarvite-type structure. N Jb Mineral Mh 2003:461-480

Pekov IV, Chukanov NV, Ferraris G, Ivaldi G, Pushcharovsky DYu, Zadov AE (2003b) Shirokshinite, $K(NaMg_2)Si_4O_{10}F_2$, a new mica with octahedral Na from Khibiny massif, Kola Peninsula: descriptive data and structural disorder. Eur J Mineral 15:447-454

Pekov IV, Chukanov NV, Khomyakov AP, Rastsvetaeva RK, Kucherinenko YaV, Nedel'ko VV (1999) Korobitsynite, $Na_{3-x}(Ti,Nb)_2[Si_4O_{12}](OH,O)_2 \cdot 3\text{-}4H_2O$, a new mineral from Lovozero massif, Kola Peninsula. Zap Vseross Mineral Obs 128(3):72-79 (in Russian)

Pekov IV, Chukanov NV, Kononkova NN, Pushcharovsky DYu (2003c) Rare-metal "zeolites" of the hilairite group. New Data on Minerals 38:20-33

Pekov IV, Chukanov NV, Larsen AO, Merlino S, Pasero M, Pushcharovsky DYu, Ivaldi G, Zadov AE, Grishin VG, Asheim A, Taftoe J, Chistyakova NI (2003d) Sphaerobertrandite, $Be_3SiO_4(OH)_2$: new data, crystal structure and genesis. Eur J Mineral 15:157-166

Pekov IV, Chukanov NV, Rastsvetaeva RK Zadov AE, Kononkova NN (2002) Gutkovaite-Mn, $CaK_2Mn(Ti,Nb)_4(Si_4O_{12})_2(O,OH)_4 \cdot 5H_2O$, a new labuntsovite-group mineral from Khibiny massif, Kola peninsula. Zap Vseross Mineral Obs 131(2):51-57 (in Russian)

Pekov IV, Chukanov NV, Yamnova NA, Egorov-Tismenko YuK, Zadov AE (2003e) Kapustinite $Na_{5.5}Mn_{0.25}ZrSi_6O_{16}(OH)_2$, a new mineral from Lovozero massif, Kola peninsula, and new data on genetic crystallochemistry of the lovozerite group. Zap Vseross Mineral Obs 132(6):1-14 (in Russian)

Pekov IV, Chukanov NV, Zadov AE, Rozenberg KA, Rastsvetaeva RK (2003f) Alsakharovite-Zn, $NaSrKZn(Ti,Nb)_4[Si_4O_{12}]_2(O,OH)_4 \cdot 7H_2O$, a new labuntsovite-group mineral from Lovozero massif, Kola peninsula. Zap Vseross Mineral Obs 132(1):52-58 (in Russian)

Pekov IV, Ekimenkova IA, Chukanov NV, Rastsvetaeva RK, Kononkova NN, Pekova NA, Zadov AE (2001) Feklichevite, $Na_{11}Ca_9(Fe^{3+},Fe^{2+})_2Zr_3Nb[Si_{25}O_{73}](OH,H_2O,Cl,O)_5$, a new eudialyte-group mineral from Kovdor massif, Kola peninsula. Zap Vseross Mineral Obs 130(3):55-65 (in Russian)

Pekov IV, Ekimenkova IA, Chukanov NV, Zadov AE, Yamnova NA, Egorov-Tismenko YuK (2000) Litvinskite, $Na_2(\square,Na,Mn)Zr[Si_6O_{12}(OH,O)_6]$, a new mineral of the lovozerite group. Zap Vseross Mineral Obs 129(1):45-53 (in Russian)

Pekov IV, Ekimenkova IA, Kononkova NN (1997) Thorosteenstrupine from Lovozero massif and steenstrupine-(Ce) - thorosteenstrupine an isomorphous series. Zap Vseross Mineral Obs 126(6):35-44 (in Russian)

Pekov IV, Podlesnyi AS (2004) Kukisvumchorr Deposit: Mineralogy of Alkaline Pegmatites and Hydrothermalites. Mineral Almanac 7:1-168

Pekov IV, Turchkova AG, Lovskaya EV, Chukanov NV (2004b) Zeolites of alkaline massifs. Ekost, Moscow (in Russian)

Pudovkina ZV, Chernitsova, NM, Voronkov AA, Pyatenko YuA (1980) Crystal structure of zirsinalite $Na_6Ca\{Zr[Si_6O_{18}]\}$. Sov Phys Dokl 25:69-70

Pushcharovsky DYu, Pekov IV, Pasero M, Gobechiya ER, Merlino S, Zubkova NV (2002) Crystal structure of cation-deficient calciohilairite and possible mechanisms of decationization in mixed-framework minerals. Crystallogr Rep 47:748-752

Pyatenko JA, Kurova TA, Chernitsova NM, Pudovkina ZV, Blinov VA, Maksimova NV (1999) Niobium, tantal and zirconium in minerals. Nauka, Moscow (in Russian)

Raade G, Haug J (1982) Gjerdingen, Fundstelle seltener Mineralien in Norwegen. Lapis 7(6):9-15

Rabo JA (ed) (1976) Zeolite Chemistry and Catalysis. Am Chem Soc, Washington

Rastsvetaeva RK, Chukanov NV, Möckel S (2003) Characteristic structural features of a tantalum-rich eudialyte variety from Brazil. Crystallogr Rep 48:216-221

Rastsvetaeva RK, Khomyakov AP (2002) Structural features of the Na,Fe-decationated eudialyte with the symmetry R3. Crystallogr Rep 47:267-271

Rastsvetaeva RK, Organova NI, Rozhdestvenskaya IV, Shlyukova ZV, Chukanov NV (2000) Crystal structure of hydronium mineral of the nenadkevichite-labuntsovite group from Khibiny massif. Dokl Chem 371: 52-56

Rocha J, Brandao P, Lin Z, Kharlamov A, Anderson MW (1996) Novel microporous titanium-niobium-silicates with the structure of nenadkevichite. Chem Commun 5:669-670

Semenov EI, Kazakova ME, Bukin VI (1968) Ilímaussite, a new rare earth – niobium – barium silicate from Ilímaussaq, South Greenland. Medd Grønland 181(7):3-7

Shannon RD, Prewitt CT (1969) Effective ionic radii in oxides and fluorides. Acta Crystallogr 25:925-945

Shpanov EP, Nechelustov GN, Baturin SV, Solntseva LS (1989) Byelorussite-(Ce), $NaMnBa_2Ce_2Ti_2Si_8O_{26}(F,OH) \cdot H_2O$, a new joaquinite-group mineral. Zap Vses Mineral Obs 118(5):100-107 (in Russian)

Shumyatskaya NG, Voronkov AA, Pyatenko YaA (1980) Sazhinite, $Na_2Ce[Si_6O_{14}(OH)] \cdot nH_2O$, a new member of dalyite family in crystal chemistry. Sov Phys Crystallogr 25:419-423

Sokolova E, Hawtorne FC, Khomyakov AP (2002) The crystal chemistry of fersmanite, $Ca_4(Na,Ca)_4(Ti,Nb)_4(Si_2O_7)_2O_8F_3$. Can Mineral 40:1421-1428

Sørensen H (1962) On the occurrence of steenstrupine in the Ilimaussaq massif, Southwest Greenland. Groenlands Geologiske Undersogelse Bulletin No 32, 251 pp

Sørensen H (1992) Agpaitic nepheline syenites: a potential source of rare elements. Appl Geochem 7:417-427

Sørensen H (1997) The agpaitic rocks – an overview. Mineral Mag 61:485-498
Sørensen H, Rose-Hansen J, Nielsen BL Loevborg L, Soerensen E, Lundgaard T (1974) The uranium deposit at Kvanefjeld, the Ilimaussaq intrusion, South Greenland: geology, reserves and beneficiation Rapp Groenl Geol Unders 60, 54 pp
Subbotin VV, Merlino S, Pushcharovsky DYu, Pakhomovsky YA, Ferro O, Bogdanova AN, Voloshin AV, Sorokhtina NV, Zubkova NV (2000) Tumchaite $Na_2(Zr,Sn)Si_4O_{11} \cdot 2H_2O$ – a new mineral from carbonatites of the Vuoriyarvi alkali-ultrabasic massif, Murmansk region, Russia. Am Mineral 85:1516-1520
Tamazyan RA, Malinovsky YuA (1990) Crystal chemistry of silicates of the lovozerite family. Sov Phys Crystallogr 35:227-232
Urusov VS (1975) Energetic Crystal Chemistry. Nauka, Moscow (in Russian)
Ussing NV (1912) Geology of the country around Juleanehaab, Greenland. Medd Grønl 38:1-426
Vladykin NV, Kovalenko VI, Lapides IL, Sapozhnikov AN, Pisarskaya VA (1972) First find of elpidite in Mongolia. *In*: Problems of mineralogy of rocks and ores of Western Siberia. Siberian branch of Academy of Sciences of the USSR, Irkutsk, p 6-14
Vlasov KA, Kuz'menko MV, Es'kova EM (1966) The Lovozero Alkali Massif. Hafner Publishing, New York
Voloshin AV, Pakhomovskii YaA, Men'shikov YuP, Sokolova EV, Egorov-Tismenko YuK (1990) Komkovite, a new hydrous barium zirconosilicate from Vuoriyarvi carbonatites, Kola peninsula. Mineral Zh 12(3):69-73 (in Russian)
Voloshin AV, Subbotin VV, Pakhomovskii YaA, Bakhchisaraytsev AYu, Yamnova NA (1991) Belkovite - a new barium-niobium silicate from carbonatites of the Vuoriyarvi massif (Kola Peninsula, USSR). N Jb Mineral Mh 1991:23-31
Voloshin AV, Subbotin VV, Pakhomovskii YaA, Menshikov YuP (1989) Sodium zirconosilicates from carbonatites of Vuoriyarvi, Kola peninsula. New Data on Minerals 36:3-12
Voronkov AA, Pudovkina ZV, Blinov VA, Ilyukhin VV, Pyatenko YuA (1979) Crystal structure of kazakovite $Na_6Mn\{Ti[Si_6O_{18}]\}$. Sov Phys Dokl 24:132-134
Yamnova NA, Egorov-Tismenko YuK, Pekov IV (2001a) Refined crystal structure of lovozerite, $Na_2CaZr[Si_6O_{12}(OH,O)_6] \cdot H_2O$. Crystallogr Rep 46:937-941
Yamnova NA, Egorov-Tismenko YuK, Pekov IV, Ekimenkova IA (2001b) Crystal structure of litvinskite: a new natural representative of the lovozerite group. Crystallogr Rep 46:190-193
Yamnova NA, Egorov-Tismenko YuK, Pekov IV, Shchegol'kova LV (2003) Crystal structure of tisinalite $Na_2(Mn,Ca)_{1-x}(Ti,Zr,Nb,Fe^{3+})[Si_6O_8(O,OH)_{10}]$. Crystallogr Rep 48:551-556
Yushkin NP (1977) Theory and Methods of Mineralogy. Nauka, Leningrad (in Russian)
Yushkin NP, Khomyakov AP, Evzikova NZ (1984) Principle of inheritance in minerogenesis. Preprint #93. Komi filial of the Academy of Sciences of USSR, Syktyvkar (in Russian)
Zubkova NV, Pushcharovsky DYu, Giester G, Tillmans E, Pekov IV, Krotova OD (2004) Crystal structure of byelorussite-(Ce), $NaMnBa_2Ce_2(TiO)_2[Si_4O_{12}]_2(F,OH) \cdot H_2O$. Crystallogr Rep 49:964-968

Microporous Mixed Octahedral-Pentahedral-Tetrahedral Framework Silicates

João Rocha and Zhi Lin

Department of Chemistry, CICECO
University of Aveiro
3810-193 Aveiro, Portugal
rocha@dq.ua.pt

INTRODUCTION

The frameworks of zeolites and related crystalline microporous oxide materials, such as aluminophosphates, silicoaluminophosphates and metaloaluminophosphates, are built up of tetrahedrally coordinated (e.g., Si, Al, P) atoms. Microporous structures comprised entirely of octahedral sites are also known, encompassing manganese oxides (OMS materials, Suib 1998) and phases with composition $Na_2Nb_{2-x}M_xO_{6-x}(OH)x \cdot H_2O$, where M = Ti, Zr, and $0 < x \leq 0.4$ (SOMS solids; Nyman et al. 2001, 2002). Mixed octahedral-pentahedral-tetrahedral microporous (OPT) siliceous frameworks have been much studied since the early 1990s (Rocha and Anderson 2000; Anderson and Rocha 2002). Comprehensive research into synthetic OPT materials was initiated by the seminal contributions of Kuznicki (1989, 1990) (titanosilicates ETS-4 and ETS-10), Chapman (1990); Chapman and Roe (1990) (ETS-4, and titanosilicate analogues of minerals vinogradovite and pharmacosiderite) and Anderson et al. (1994, 1995b) (ETS-10), and the early work of the groups of Raveau (Choisnet et al. 1976, 1977) (tantalo and niobosilicates), Corcoran and Vaughan (1989) and Corcoran et al. (1989, 1992) (stannosilicates).

Much research on the synthesis of OPT materials has been inspired and motivated by the many examples of such solids provided by Nature and brought to light by a plethora of mineralogists. As an example, consider the titanosilicate mineral zorite, discovered in 1973 on the Lovozero Tundra (Mer'kov et al. 1973). The structure solution of this mineral was published six years later by Sandomirskii and Belov (1979). Zorite is important because it was one of the first examples of microporous titanosilicates to be prepared in the laboratory (Kuznicki 1989; Chapman 1990; Chapman and Roe 1990). The details of the structure of the synthetic analogue of zorite, known as ETS-4, remained somewhat elusive until a single-crystal x-ray diffraction study published in 2001 settled the matter and clarified the differences between natural and synthetic materials (Nair et al. 2001). In the mean time, the desire to better understand the structure and properties of ETS-4 stimulated a considerable amount of work. The search for OPT silicates of metals other than titanium has also been guided by examples drawn from Nature. For instance, Bortun et al. (1997); Lin et al. (1999a,b); and Jale et al. (1999), extended the OPT field to zirconosilicates, reporting the synthesis of a synthetic analogue of the mineral gaidonnayite ($Na_2ZrSi_3O_9 \cdot 2H_2O$). Until the end of the last century, many microporous OPT minerals containing other metals, such as Nb, V, Sn and Ca, have been reported (Rocha and Anderson 2000; Anderson and Rocha 2002). Following the synthesis and structural characterization work, the potential applications of these novel materials have been evaluated, particularly in the areas of catalysis, separation and ion-exchange. It is appropriate to note here that, so far, most of this applied work concentrates on ETS-10 (more than 200 studies published) and, to a lesser extent, on ETS-4.

With the advent of the nanotechnology era, and the increasing interest in the use of molecular sieves for device applications (Tsapatsis 2002), the constituent elements of OPT materials have been further extended to the realm of lanthanide metals, exploring properties like photoluminescence (Rocha and Carlos 2003). In addition, at present, materials with interesting magnetic properties are also being sought after and this motivated, for example, the synthesis of copper silicates (Wang et al. 2003).

The field of zeolite membrane reactors has witnessed an intense research activity in the last ten years or so (Coronas and Santamaria 2004). However, as far as OPT materials are concerned, only a few studies are available on ETS-4 (Braunbarth et al. 2000a) and ETS-10 (Lin et al. 2004) membranes. Since the applications of molecular sieves in devices often require membranes, not powders, much work is now in progress in this area.

Advances in the field of OPT microporous materials have been reviewed by Rocha and Anderson (2000), Anderson and Rocha (2002) and Chukanov et al. (2004). Because significant progress has been made in the last three years, we believe it is now timely to appraise this recent work. Although clays and related materials, in some way, fall into the category of OPT materials, they only possess two-dimensional covalent connectivity and are out of the scope of this review. We concentrate on siliceous materials and, thus, overlook, for example, the important porous phosphates.

SYNTHESIS

Zeolites and OPT siliceous materials, particularly titanosilicates (Kuznicki 1989, 1990; Chapman 1990; Chapman and Roe 1990), are in general prepared under hydrothermal conditions, in Teflon-lined autoclaves, at temperatures ranging from *ca.* 120 to 230°C and times varying between a few hours and *ca.* 30 days. In a few cases, the hydrothermal synthesis was carried out at relatively high (e.g., 750°C) temperatures (Harrison et al 1995). For OPT materials, the pH of the synthesis gel is normally high, in the range 10–13 (after 1: 100 dilution). Seeding the gel with a little amount of the desired phase (or even with a related solid) is a common practice. In general, stirring of the gel is not used, although a recent study of the synthesis of ETS-4 found that, under these conditions, the crystal morphology is quite different from that of materials prepared with no agitation. The syntheses are, in general, kinetically controlled, with different metastable phases being formed with time. Although most research has concentrated on the synthesis of microporous titanosilicates considerable experience has been gained into other systems, particularly, Zr, Nb, V and Sn silicates.

ETS-10 was first prepared by Kuznicki (1989), using $TiCl_3$ as the titanium source. A slight modification of this method afforded highly crystalline and pure ETS-10 suitable for structural investigation (Anderson et al. 1994, 1995b). $TiCl_4$ is another much used (and cheaper) titanium source (Das et al. 1996b) as is TiF_4 (Yang et al. 2001) a compound which has the advantage of not being sensitive to ambient air moisture. Anatase and rutile (TiO_2 phases, Liu and Thomas 1996) and $Ti(SO_4)_2$ (Kim et al. 2002) may also be used as suitable titanium sources for the synthesis of porous titanosilicates. The synthesis of ETS-10 has been carried out at 200°C using a 2^3 factorial method to optimize the overall composition of the reaction mixture to produce a pure material within a short (18 h) time. Kinetic studies on ETS-10 and ETS-4 have been performed using the optimum compositions and the apparent activation energies were calculated as 66.78 and 14.47 $kJmol^{-1}$, respectively (Kim et al. 2000). Thermodynamically, ETS-4 is more stable than ETS-10 with respect to the oxides (or the elements) at 298 K and 1 atm, probably because of the relatively higher degree of hydration of the former (Xu et al. 2001). This result has been obtained by high-temperature drop solution calorimetry, which allowed the measurement of the pertinent standard enthalpies of formation from the oxides and from the elements.

A few organotitanium compounds, such as $Ti(OC_2H_5)_4$ (Chapman and Roe 1990) and titanium isopropoxide (Poojary et al. 1994) have also been used to prepare microporous titanosilicates. Because they are very sensitive to moisture, these compounds are difficult to handle and require special facilities for their transfer during the gel preparation. Sodium silicate solutions, fumed and colloidal silica are adequate silicon sources. Organosilicon compounds, such as tetraethylorthosilicate have also been used (Poojary et al. 1994). Although ETS-10 usually (although not always) crystallizes in the presence of Na^+ and K^+ ions, other phases, such as some of the AM-n (Aveiro-Manchester) microporous solids (Rocha et al. 1996a), are produced pure only when a single type of alkaline cation is present. Potassium fluoride is often used for preparing ETS-10 but its presence is not crucial. Comprehensive studies of the hydrothermal synthesis conditions, which yield pure and highly crystalline ETS-10 are now available (Rocha et al. 1998d; Das et al. 1995).

Although most OPT materials can be prepared without the addition of any organic templates, several groups have prepared ETS-10 with a range of templates: tetramethylammonium chloride (Valtchev and Mintova 1994; Valtchev 1994), pyrrolidine, tetraethylammonium chloride, tetrapropylammonium bromide,1,2-diaminoethane (Valtchev 1994), choline chloride and the bromide salt of hexaethyl diquat-5 (Das et al. 1996a), ethanolamine (Kim et al. 2002). The latter work also reports kinetic studies, performed at different temperatures, where the interpretation of the data was carried out with a modified Avrami-Erofeev equation and the activation energies for nucleation, transition and crystal growth stages were calculated.

The framework substitution (doping) of titanium and silicon by other elements requires a judicious choice of the respective source which is usually introduced in the parent synthesis gel. The following sources of Al, Ga, Nb and B for element insertion in the ETS-10 framework have been reported: $NaAlO_2$ (Anderson et al. 1995a) $GaCl_3$ (Rocha et al. 1995) $Nb(HC_2O_4)_5$ (Rocha et al. 1999) and $Na_2B_2O_4$ (Rocha et al. 1998a; see also Kuznicki et al. 1991; Kuznicki and Trush 1991a, 1993). In the synthesis of microporous zirconosilicates $ZrCl_4$ (Jale et al. 1999; Rocha et al. 1998c; Lin et al. 1999a,b) and $Zr(OC_3H_7)_4$ (Bortun et al. 1997) have been used as the Zr sources, while the synthesis of lanthanide (Ln) silicates normally requires $LnCl_3$ or $Ln(NO_3)_3$ (Ananias et al. 2001).

A few authors have been devoting considerable attention to the study of the crystal growth mechanisms of ETS-4 and ETS-10. In view of the defect structures of these materials, this is a highly challenging task, which is increasing our understanding of the synthesis mechanisms of OPT materials and zeolites. Here we can only give a succinct outline of the subject. Miraglia et al. (2004) studied the hydrothermal synthesis of ETS-4 using organic precursors. The large, rectangular crystals and small, intergrown plates (cuboids), present were examined by AFM and FE-SEM, revealing a layered growth mechanism on the largest and slowest growing faces. These faces contain steps emanating from spiral hillocks with a characteristic height of one half of the orthorhombic ETS-4 unit cell parameter a. The disparity between the cuboid and rectangular crystal morphology indicates localized variations in Ti incorporation, nucleation and growth rate of ETS-4. Thus, to obtain uniform morphologies and minimize defects, solution chemistries need to be developed to improve miscibility and homogeneity. The group of Anderson has also much contributed for the elucidation of crystal growth mechanisms of zeolites (Anderson et al. 2001a,b). For example, they have shown that in ETS-10 the intergrowth structure and defects are also a consequence of a layer-by-layer growth mechanism which is described. With this knowledge of the crystal growth of ETS-10 strategies may now be developed to control the intergrowth structure and the defect density.

STRUCTURE

Titanosilicates

ETS-10 and ETS-4. ETS-10 and ETS-4 were prepared first at Engelhard Corporation (Kuznicki 1989, 1990). At about the same time Chapman (1990) and Chapman and Roe (1990), published the synthesis of a microporous titanosilicate material, whose structure apparently resembled the structure of the natural mineral zorite by comparison of the diffraction patterns. Owing to its wide-pore nature and thermal stability ETS-10 is the most important OPT microporous titanosilicate prepared so far. Because much has been written about the structure of ETS-10 and ETS-4, here we shall only summarize the main structural features of these materials.

The structure of ETS-10 [$(Na,K)_2Si_5TiO_{13} \cdot xH_2O$] was solved and reported briefly by Anderson et al. (1994) and later described fully by Anderson et al. (1995b) (Fig. 1). In 1999, a single crystal study of ETS-10 has refined the superposition structure, corresponding to an averaged combination of all the ordered variants, confirming substantially all previous structural work (Wang and Jacobson 1999). The structure comprises corner-sharing SiO_4 tetrahedra and TiO_6 octahedra linked through bridging oxygen atoms. The pore structure of ETS-10 contains 12-rings in all three dimensions; these are straight along [100] and [010] and crooked along the direction of disorder. Only a handful of microporous zeolitic materials with a three-dimensional 12-ring pore system are known and in this aspect ETS-10 has excellent diffusion characteristics. It is important to stress that the structure is inherently an intergrowth structure, consisting of randomly stacked layers. The basic unit is $Si_{40}Ti_8O_{104}^{16-}$ which is counterbalanced by 16 monovalent cations. Many ordered variants of ETS-10 exist, some of which are chiral (Fig. 1b). An important aspect of the structure of ETS-10 is that it contains –O–Ti–O–Ti–O– chains (with alternating long-short bonds, according to Sankar et al. (1996) which are surrounded by a silicate ring structure. These combine to make up a rod and it is this rod nature which imparts ETS-10 some interesting physical properties. Adjacent layers of rods are stacked orthogonal to each other.

ETS-4 is essentially the synthetic analogue of the mineral zorite, possessing a mixed tetrahedral (Si)-pentahedral (Ti2)-octahedral (Ti1) framework (Sandomirskii and Belov 1979;

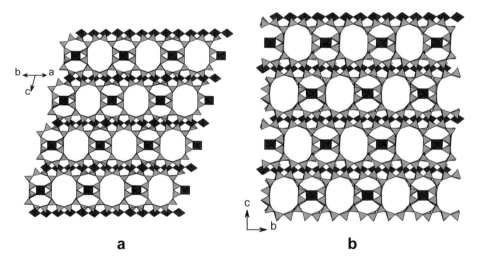

Figure 1. Titanosilicate ETS-10 projections: (a) down [110] direction for polymorph B (space group $C2/c$); (b) down [100] direction for polymorph A (space group $P4_1$). Dark TiO_6 octahedra, light grey SiO_4 tetrahedra.

Philippou and Anderson 1996; Cruciani et al. 1998, Braunbarth et al. 2000b; Nair et al. 2001). It is faulted in the [100] and [001] directions and can be described as a random intergrowth of four pure polytypes, which differ in the arrangement of the titanosilicate bridging units. Although larger openings are present in its structure, faulting ensures that access to the crystal interior of ETS-4 occurs through the relatively narrow 8-membered rings (Nair et al. 2001). A distinct feature of ETS-4 is the presence of semioctahedra (five-coordinated Ti with a titanyl, Ti=O, linkage to the apical oxygen) that are connected to framework silicon atoms through only four oxygen bridges, which gives rise to a planar connectivity. An intriguing difference between the structures of ETS-4 and zorite is the rotation of the TiO_5 square pyramid in zorite, an effect that is absent in ETS-4 for reasons which have been discussed (Nair et al. 2001). Kuznicki et al. (2001) have shown that the ETS-4 framework may be systematically contracted through dehydration at elevated temperatures to tune the effective size of the pores giving access to the crystal interior. This has been named 'molecular gate' effect and may be used to tailor the adsorption properties of the material to give size-selective adsorbents suitable for commercially important separations of gas mixtures of molecules with similar size in the 3.0–4.0 Å range, such as N_2/CH_4, Ar/O_2 and N_2/O_2. A pilot plant has been in use and successfully purified nitrogen contaminated natural gas by pressure swing adsorption at well-head pressures.

Recently, McDonald and Chao (2004) reported the structure of haineaultite, a new hydrated sodium calcium titanosilicate from Mont Saint-Hilaire (Quebec) with ideal formula $(Na,Ca)_5Ca(Ti,Nb)_5(Si,S)_{12}O_{34}(OH,F)_8 \cdot 5H_2O$. The structure of this mineral consists of 8-rings of SiO_4 tetrahedra, linked to adjacent rings to form vierer double chains along [001], which are cross-linked by TiO_6 octahedra. Channels running parallel to [100] are occupied by Ca and H_2O, with Na residing in channels parallel to [001]. Haineaultite possesses an OD structure, exemplified by disordering of framework and interframework ions. The structures of haineaultite and zorite are similar. The main differences are: the species occupying the channels are not the same; the Ti polyhedra that cross-link the silicate chains are disordered TiO_5 polyhedra, in zorite, and ordered TiO_6 octahedra, in haineaultite. There are similarities between the crystal structure of haineaultite and mineral belonging to the rhodesite group (see the section on Rare-earth silicates), such as seidite-(Ce).

Like with conventional zeolites, the insertion of heteroatoms in the framework of OPT materials is an important process because it allows the fine-tuning of their properties (for example, substitution of Si by Al may generate sites for zeolite-type acidity). The studies on framework insertion of aluminum, gallium, boron, niobium, vanadium and chromium on ETS-10 and a few other materials have been reviewed (Rocha and Anderson 2000; Anderson and Rocha 2002).

Other titanosilicates. Although most research on OPT materials has been focused on ETS-4 and ETS-10, many other titanosilicate microporous materials are now known, most of which are synthetic counterparts of rare minerals. Importantly, these solids are in general small-pore silicates or tunnel structures, their porosity being only accessible to small molecules. As a result, the studies exploring their potential applications concentrate on ion-exchange properties.

Much has been learned from trying to mimic mineral pharmacosiderite, a non-aluminosilicate OPT molecular sieve with framework composition $KFe_4(OH)_4(AsO_4)_3 \cdot 6H_2O$. Chapman and Roe (1990) prepared titanosilicate analogues of pharmacosiderite while, later, Behrens et al. (1996) studied in detail the structure of analogues with composition $HM_3Ti_4O_4(SiO_4)_3 \cdot 4H_2O$, where $M = H^+$, K^+, Cs^+. Harrison et al. (1995) grew single crystals of the same cesium phase and solved its structure by single-crystal methods. Pharmacosiderite materials possess a most interesting structure built up from TiO_6 octahedra sharing faces to form Ti_4O_4 cubes placed at the corners of the cubic unit-cell; silicate tetrahedra join the titanium octahedra to form a three-dimensional framework. Extra-framework Cs^+ species occupy sites slightly displaced from the centers of the intercage eight-ring windows, and also make $Cs-OH_2$ bonds to the water molecules which reside in the spherical cages.

Poojary et al. (1994) reported the synthesis, crystal structure and ion-exchange properties of a novel porous titanosilicate of ideal composition $Na_2Ti_2O_3SiO_4 \cdot 2H_2O$ and whose structure is reminiscent of the structure of pharmacosiderite. The titanium atoms are octahedrally coordinated by oxygen atoms and occur in clusters of four, grouped about the 4_2 axis. The silicate groups link the titanium clusters into groups of four arranged in a square of about 7.8 Å in length. These squares are linked to similar squares in the c direction by sharing corners to form a framework enclosing a tunnel. Half the Na^+ ions are located in the framework, coordinated by silicate oxygen atoms and water molecules, while the remaining Na^+ ions reside in the cavity. The Na^+ ions within the tunnels are exchangeable, particularly by Cs^+ ions.

Another example of how the chemist draws lessons from Nature to engineer OPT materials is illustrated by the research carried out on the synthesis of analogues of nenadkevichite, a rare mineral first found in the Lovozero massif (Kola peninsula, Russia) and later in Mont Saint-Hilaire, Quebec (Canada), with the composition $(Na,Ca)(Nb,Ti)Si_2O_6(O,OH) \cdot 2H_2O$. Rocha et al. (1996a,b) prepared a series of synthetic analogues of nenadkevichite with Ti/Nb molar ratios ranging from 0.8 to 17.1 and a purely titaneous sample. The structure of nenadkevichite consists of square rings of silicon tetrahedra Si_4O_{12} in the (100) plane joined together by chains of $(Nb,Ti)O_6$ octahedra in the [100] direction (Perrault et al. 1973). The pores accommodate Na^+ and water molecules.

Many porous framework titanosilicates contain Ti-O-Ti linkages which often form infinite chains. Interestingly, the materials known as AM-2 (Dadachov and Le Bail 1997; Lin et al. 1997, 1999a; Poojary et al. 1997; Jale et al. 1999), AM-3 and UND-1 (Liu et al. 1997a) do not contain any such linkages. AM-2 is a synthetic potassium titanosilicate analogue of the mineral umbite, a rare zirconosilicate found in the Khibiny alkaline massif, Kola Peninsula, Russia (Ilyushin 1993). Although the ideal formula of umbite is $K_2ZrSi_3O_9 \cdot H_2O$, a pronounced substitution of titanium for zirconium occurs. However, the natural occurrence of purely titaneous umbite is unknown. The successful synthesis of umbite materials with different levels of titanium for zirconium substitution has been reported (Lin et al. 1999b). AM-3 is the synthetic counterpart of the mineral penkvilksite, found at Mont Saint-Hilaire and Kola Peninsula, with formula $Na_4Ti_2Si_8O_{22} \cdot 5H_2O$. Penkvilksite occurs in two polytypic modifications, orthorhombic (penkvilksite-2*O*) and monoclinic (penkvilksite-1*M*), which have been described according to the OD theory as two of the four possible polytypes with the maximum degree of order within a family of OD structures formed by two OD layers (Merlino et al. 1994). Despite the different space group symmetries the 2*O* (*Pnca*) and 1*M* ($P2_1/c$) polytypes have the same atoms, labeled in the same way, in the asymmetric unit. They differ only in the stacking of the same building blocks. AM-3 is the synthetic analogue of penkvilksite-2*O* (Lin et al. 1997). The synthesis of the 1*M* polytype has been reported by Liu et al. (1997b, 1999). The material known as UND-1 ($Na_{2.7}K_{5.3}Ti_4Si_{12}O_{36} \cdot 4H_2O$) is the third example known of a porous framework titanosilicate containing *no* Ti-O-Ti linkages (Liu et al. 1997a).

A few other OPT titanosilicates, with a known structure, are available, for example the synthetic analogue of the mineral vinogradovite with composition $Na_5Ti_4Si_7AlO_{26} \cdot 3H_2O$ (Chapman and Roe 1990; Chapman 1990; Rastsvetaeva and Andrianov 1984). The structure of the mineral is composed of pyroxene chains joined to edge-sharing TiO_6 octahedra which form brookite columns. Vinogradovite chains Si_3AlO_{10} consisting of four-membered rings connected through the common vertices are also involved in the framework of vinogradovite (the crystal structures of vinogradovite and related minerals are described in this volume by

Chukanov and Pekov 2005). These polyhedra define one-dimensional (4 Å free) channels containing zeolitic water. Examples of medium-size or large-pore titanosilicates whose structures are, as yet, unknown, have been reported (e.g., JLU-1 by Liu et al. 2000 and AM-18 by Brandão et al. 2002c).

A number of minerals not yet prepared in the laboratory exhibit very interesting and complexly connected frameworks (for a useful review see Smith 1988). A case in point is mineral verplanckite $[(Mn,Ti,Fe)_6(OH,O)_2(Si_4O_{12})_3]Ba_{12}Cl_9\{(OH,H_2O)_7\}$, which has a framework with triple units of (Mn, Ti, Fe) in square-pyramidal coordination and four-rings of silicon tetrahedra (Kampf et al. 1973). The voids have a free diameter of 7 Å.

When attempting to prepare novel OPT titanosilicates, layered materials are sometimes obtained (e.g., AM-1, Anderson et al 1995b; Lin et al. 1997, or JDF-L1, Du et al. 1996; Roberts et al. 1996, and AM-4 Lin et al. 1997; Dadachov et al. 1997), displaying interesting and unusual structures and presenting potential for being used in a number of applications such as ion exchange. An interesting future development in the OPT field is the synthesis of layered materials possessing microporous layers. These solids are of fundamental interest in exploring the connection between layered and framework silicates. An intriguing example is provided by Jeong et al. (2003) with a layered silicate know as AMH-3 ($Na_8Sr_8Si_{32}O_{76} \cdot 16H_2O$).

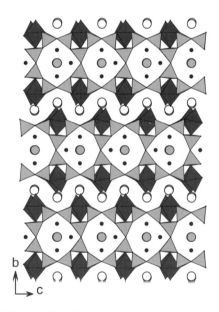

This material is composed of silicate layers containing 8-rings in all three principal crystal directions, and spaced by Sr^+, Na^+ and water molecules. Nature is an endless source of inspiration: Krivovichev and Armbruster (2004) have recently reported the first example of a titanosilicate mineral (jonesite) with porous double layers (Fig. 2). The structure of jonesite, $Ba_2(K,Na)[Ti_2(Si_5Al)O_{18}(H_2O)](H_2O)_n$, from the Benitoite Gem Mine, San Benito County, California, is based upon porous double layers of distorted $Ti\Phi_6$ octahedra ($\Phi = O$, H_2O) and TO_4 tetrahedra (T = Si,Al) parallel to (010). The layers consist of two sheets of corner-sharing $Ti\Phi_6$ octahedra and Si_2O_7 groups each. The two adjacent sheets are linked along b by T_4O_{12} tetrahedral rings that are disordered over two positions. The double layer has an open structure characterized by 8-membered tetrahedral rings with free diameters of 3.37×3.37 Å^2 and 3.33×3.33 Å^2, for two symmetrically non-equivalent rings. K^+ and water molecules reside in the pores of the double layers. Ba^{2+} cations are located between the double layers and provide their linkage into a three-dimensional structure.

Figure 2. Crystal structure of jonesite: projection along the a axis showing the titanosilicate layers parallel to (010) with pores filled with (K,Na) and water molecules, linked by Ba. White circles Ba^{2+} ions, grey circles K^+ cations, dark circles H_2O.

Zirconosilicates

Zirconium silicates occur extensively in Nature. Their formation under hydrothermal conditions (from *ca.* 300 to 550°C) has been much studied, though mainly for the solution of general geophysical and mineralogical problems (Bortun et al. 1997). Maurice (1949) performed some of the first hydrothermal syntheses of zirconosilicates. Baussy et al. (1974)

summarized the early work in this field and report the hydrothermal synthesis of analogues of minerals such as catapleiite ($Na_2ZrSi_3O_9 \cdot 2H_2O$) and elpidite ($Na_2ZrSi_6O_{15} \cdot 3H_2O$). Some interesting aspects of the chemistry of sodium zirconosilicates have been highlighted by the work of Bortun et al. (1997) reporting the synthesis, characterization and properties of three novel layered materials and five other zirconosilicates, for example a synthetic analogue of the mineral gaidonnayite ($Na_2ZrSi_3O_9 \cdot 2H_2O$). This material has also been synthesized by Lin et al. (1999a) (AV-4) and by Jale et al. (1999). The framework of gaidonnayite is composed of sinusoidal single chains of SiO_4 tetrahedra, repeating every six tetrahedra (Chao 1985). The chains extend alternately along [011] and [01$\bar{1}$] and are cross-linked by a ZrO_6 octahedron and two distorted NaO_6 octahedra.

Petarasite [$Na_5Zr_2Si_6O_{18}(Cl,OH) \cdot 2H_2O$] is another OPT zirconosilicate and its synthetic analogue is known as AV-3 (Lin et al. 1999a,b; Rocha et al. 1998c). The structure of this rare mineral is unusual, consisting of an open three-dimensional framework built of corner-sharing six-membered rings and ZrO_6 octahedra (Ghose et al. 1980). Elliptical channels (3.5 × 5.5 Å^2) defined by mixed six-membered rings, consisting of pairs of SiO_4 tetrahedra linked by zirconium octahedra, run parallel to the b and c axes. Other channels limited by six-membered silicate rings run parallel to the c axis. The Na^+, Cl^-, OH^- ions and the water molecules reside within the channels. The framework collapses only after the release of Cl at ca. 800°C.

Synthetic umbite-type materials have already been alluded to in this review. The synthesis and characterization of AV-8 ($Na_{0.2}K_{1.8}ZrSi_3O_9 \cdot H_2O$), an analogue of the mineral kostylevite, have been reported (Ferreira et al. 2001b). Kostylevite and umbite are the monoclinic and orthorhombic polymorphs of $K_2ZrSi_3O_9 \cdot H_2O$, respectively. The two minerals exhibit the same octagonal, heptagonal and hexagonal distorted tunnels and windows, delimited by edges from tetrahedra and octahedra alternating in exactly the same way. However, kostylevite is a cyclohexasilicate, while umbite is a long-chain polysilicate (Dadachov and Le Bail 1997). Titanosilicate UND-1 (Liu et al. 1997a) is a titaneous analogue of kostylevite and, thus, also an analogue of AV-8. The thermal transformations of zirconosilicates AV-3 (synthetic petarasite) and AM-2 (analogue of umbite) yield analogues of the dense minerals parakeldyshite and wadeite, respectively.

Recently, Ferreira et al. (2003b) reported a small-pore zirconosilicate known as Zr-AV-13 ($Na_{2.29}ZrSi_3O_9Cl_{0.27} \cdot 2.5H_2O$) and tin and hafnium analogues of this solid. The structure of these materials is unprecedented, consisting of SiO_4 tetrahedra forming six-membered [Si_6O_{18}]$^{12-}$ rings which are interconnected by corner-sharing MO_6 (M = Zr, Hf, Sn) octahedra. The structure is better understood by considering a three-dimensional "knots-and-crosses" lattice. In a given layer, successive distorted-cube M_8 cages contain [$Na_{6-x}(H_2O)_x$](H_2O,Cl^-) octahedral (knots) and cyclohexasilicate (crosses) units. While the former are extra-framework species, the 6-rings are part of the framework. The cages are accessed via seven-membered [$M_3Si_4O_{27}$]$^{26-}$ windows (free aperture 2.3 × 3.3 Å^2), one per each pseudo-cube face. Pilling up layers generates the structure, with knots-and-crosses alternating. This unusual structure shows that it is worth to further explore the chemistry of OPT zirconosilicates. Nature provides some interesting starting points for this endeavour. For example, mineral lemoynite, $(Na,K)_2CaZr_2Si_{10}O_{26} \cdot 5-6H_2O$, possesses a $ZrSi_5O_{13}$ framework with open channels where Na^+, K^+, Ca^{2+} and water molecules reside (Le Page and Perrault 1976). This framework comprises thick (7 Å) layers of hexagons of silicate groups. The sheets are bound together by six-coordinated zirconium atoms. The hexagons are tilted with respect to the (001) layer and the architecture of these layers is new. Another case in point is altisite, $Na_3K_6Ti_2[Al_2Si_8O_{26}]Cl_3$, an hyperalkaline aluminosilicate from the Kola Peninsula (Russia) related with lemoynite (Ferraris et al. 1995).

Vanadosilicates and niobosilicates

Some examples of OPT vanadosilicates have been reported (Evans 1973; Rinaldi et al. 1975; Rocha et al. 1997a; Wang et al. 2001, 2002c; Huang et al. 2002; Li et al. 2002). Among minerals, two solids stand out, cavansite and pentagonite, the dimorphs of $Ca(VO)(Si_4O_{10}) \cdot 4H_2O$. The tetrahedral layers consist of 6-rings in pentagonite and alternating 4- and 8-rings in cavansite, connected vertically by V(IV) cations in a square-pyramidal coordination (Evans 1973; Rinaldi et al. 1975). Replacement of the VO_5 groups by two bridging oxygens would produce a tetrahedral framework topologically identical to that of zeolite gismondine. The Ca^{2+} cations and the water molecules reside in the channels formed by the 8-rings and between the silicate layers. Cavansite has channels running parallel to the c direction with a free diameter of only 3.3 Å in the hydrated state. The first synthetic large-pore OPT vanadosilicate reported (AM-6) contains octahedral vanadium and possesses the structure of titanosilicate ETS-10 (Rocha et al. 1997a). The presence of stoichiometric amounts of vanadium in the framework of AM-6 affords this material potential for applications as a catalyst, sorbate or functional material.

The compelling work of Jacobson and co-workers (Wang et al. 2001, 2002c) shows that much remains to be learned in the field of OPT vanadosilicates: ten distinct framework types have been identified, all having structures based on cross-linking single silicate sheets with $V^{IV}O_5$ tetragonal pyramids or (H_2O)-$V^{IV}O_5$ distorted octahedra, giving compounds with formula $A_r[(VO)_s(Si_2O_5)_p(SiO_2)_q]_t \cdot H_2O$ (A = Na, K, Rb, Cs or a combination). These, so-called, VSH-n structures have free channel diameters up to 6.5 Å and exhibit good thermal stability, absorption and ion-exchange properties. The structures of VSH-1K and VSH-2Cs are closely related to those of cavansite and pentagonite. Unlike other VSH-n materials which contain V(IV), VSH-16Na (Huang et al. 2002) is built up of $V^{III}O_6$ octahedra which connect unbranched $[Si_8O_{22}]^{12-}$ chains (Fig. 3). Brandão et al. (2002a, 2003) reported several new large-pore vanadosilicates with unknown structure.

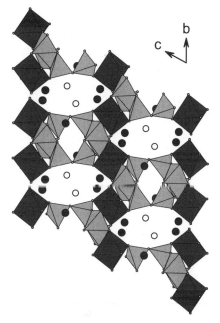

The structural diversity may be further expanded by substitution of the vanadyl groups with other components, such as $[ZrF_2]^{2+}$, $[NbOF]^{2+}$ and UO_2^{2+} (Wang et al. 2002c). For example, a series of uranium silicates have also been prepared by replacing the VO^{2+} cation in the vanadium(IV) silicates with UO_2^{2+} as the bridging metal species (Wang et al. 2002 a,b; Huang et al. 2003). The structures of USH-n, with n = 2, 4, 5, materials contain UO_6 tetragonal bipyramidal units connected by double chains, 4-rings and single chains of SiO_4 tetrahedra, respectively.

Comparatively few studies are available on OPT niobosilicates. The work carried out on nenadkevichite analogues containing titanium and niobium has already been mentioned in this review (Rocha et al. 1996a,b). The hydrothermal synthesis and evaluation of the catalytic properties of a microporous sodium niobosilicate whose structure is, at present, unknown have

Figure 3. View of the structure of vanadosilicate VSH-16Na along [100]. Dark VO_6 octahedra, light grey SiO_4 tetrahedra, dark circles Na^+ cations, white circles H_2O.

been described (Rocha et al. 1998b; Philippou et al. 2001; Brandão et al. 2001a, 2002b). Francis and Jacobson (2001) reported the first organically templated OPT niobosilicate, [(C$_4$N$_2$H$_{11}$)Nb$_3$SiO$_{10}$] (NSH-1). The structure of this material is built up of two edge-shared NbO$_6$ octahedra that are further connected by corner-shared O atoms to form infinite double columns running along the [010] direction (Fig. 4). A second chain of [NbO$_6$-Si(Ge)O$_4$-NbO$_6$]$_n$ polyhedra is formed from corner-sharing NbO$_6$ octahedra and runs along the [100] direction. The two chains are cross-linked through O atoms to form the open-framework structure. The framework of NSH-1 consists of a network of three interconnecting one-dimensional channels, two consisting of 6-rings of polyhedra and the other an eight-membered ring. The two 6-ring channels are formed from six NbO$_6$ corner-linked units. The first runs along the [100] direction; the second along the [111] direction. The 8-ring channels are constructed from six NbO$_6$ and two SiO$_4$ units and run along the [010] direction perpendicular to one of the 6-ring channels and at an angle of 55° to the other. The three channels intersect forming an irregular, roughly triangular shaped cavity, where the piperazinium cations reside. The 8-ring channels are elliptical in shape with an approximate free-pore diameter of 3.4 × 5 Å^2. Another interesting example of a niobosilicate, which is probably a tunnel structure, was reported recently (Kao and Lii 2002).

Stannosilicates and indosilicates

Although Dyer and Jáfar were probably the first authors to report the hydrothermal synthesis of an OPT stannosilicate (as a poster presented in 1987 at the Innovations in Zeolite Materials Science International Conference in Nieuwpoort, Belgium; pers. comm.), the early work in this field is due to Corcoran et al. (1989) and Corcoran and Vaughan (1989) carried out at Exxon Research. Some of these materials are useful sorbents, for example for the separation of hydrogen sulfides from gas streams containing hydrogen contaminated with hydrogen sulfides or oxysulfides (Corcoran et al. 1992). Later, Dyer and Jáfar (1990, 1991) also reported a microporous sodium stannosilicate and studied its ion-exchange properties for the replacement of Na$^+$ by a range of monovalent and divalent ions. However, the structures of all these materials have not been determined. Recently, we reported the synthesis and structures of microporous stannosilicates AV-6 and AV-7 analogues of, respectively, zirconosilicate minerals umbite (Lin et al. 1999b) and kostylevite (Lin et al. 2000). Two other materials were also

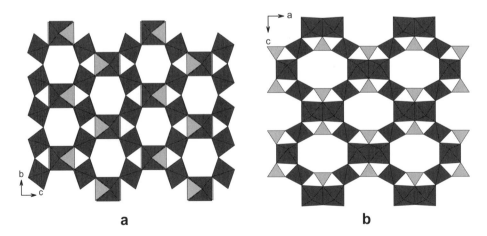

Figure 4. Structure of niobosilicate NSH-1 viewed (a) along the *a* axis, showing the 6-ring channels, and (b) along the *b* axis, showing the larger 8-ring channels. Dark NbO$_6$ octahedra, light grey SiO$_4$ tetrahedra. The template molecules are omitted for clarity.

reported, AV-10 (Ferreira et al. 2001a) and AV-13 (a tin analogue of the zircosilicate AV-13 described above, Ferreira et al. 2003b). The powder XRD patterns of AV-6, AV-10 and AV-13 are similar to the patterns of, respectively, phases G, A and B reported by Corcoran et al. (1992). The very unusual structure of AV-10 (Ferreira et al. 2001a), $Na_2SnSi_3O_9 \cdot 2H_2O$ (chiral space group $C222_1$), is composed of corner sharing SnO_6 octahedra and SiO_4 tetrahedra, forming a three-dimensional framework structure (Fig. 5). The SiO_4 tetrahedra form helix chains along [001] interconnected by SnO_6 octahedra. The SnO_6 octahedra are isolated by SiO_4 tetrahedra and, thus, there are no Sn-O-Sn linkages. The zeolitic water of AV-10 is reversibly lost.

Only a few examples of OPT indosilicates were reported and they were synthesized hydrothermally at high temperature (600°C) and pressure (170 MPa). For example, the structure of $K_2In(OH)Si_4O_{10}$ consists of unbranched vierer 4-fold chains of corner-sharing SiO_4 tetrahedra running along the b axis linked together via corner sharing by chains of *trans*-corner-sharing $InO_4(OH)_2$ octahedra to form a three-dimensional framework which delimits 8-ring and 6-ring channels to accommodate K^+ cations (Hung et al. 2003). Another case in point is $Na_5InSi_4O_{12}$ (Hung et al. 2004) isotypic with $Na_5ScSi_4O_{12}$ (Merinov et al. 1980). The structure consists of 12-membered single rings of corner-sharing SiO_4 tetrahedra linked together via corner sharing by single InO_6 octahedra to form a three-dimensional framework delimiting two types of channels along the c-axis. Two types of Na^+ ions are present: those within the 12-ring channels are immobile while the Na^+ ions within 7-ring channels are highly mobile.

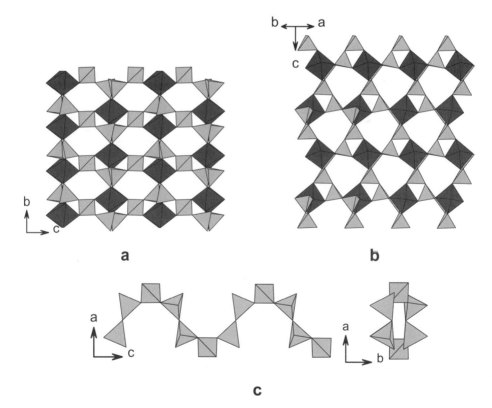

Figure 5. (a,b) Polyhedral representation of the structure of stannosilicate AV-10; (c) the helix chains of SiO_4 tetrahedra in AV-10. Dark SnO_6 octahedra, light grey SiO_4 tetrahedra. Cations and water molecules are omitted for clarity.

Rare-earth silicates

Research into OPT rare-earth silicates is, at present, a very active field of research because the presence of lanthanide ions affords the materials interesting photoluminescence properties. As an example, consider AV-1 ($Na_4K_2Y_2Si_{16}O_{38} \cdot 10H_2O$), the synthetic analogue of the rare mineral monteregianite-(Y) (Rocha et al. 1997b; Rocha et al. 1998e). This yttrium silicate possesses an unusual structure (Fig. 6) consisting of two different types of layers alternating along the [010] direction (Ghose et al. 1987): (a) a double silicate sheet, where the single silicate sheet is of the apophyllite type with 4- and 8- rings, and (b) an open octahedral sheet composed of YO_6 and three distinct $[NaO_4(H_2O)_2]$ octahedra. The layers are parallel to the (010) plane. The K^+ ions are ten-coordinate and the six water molecules are located within large channels formed by the planar eight-membered silicate rings. The structure of the alkali calcium silicate mineral rhodesite, $HKCa_2Si_8O_{19} \cdot 6H_2O$ (Hesse et al. 1992) and its synthetic analogue AV-2 (Rocha et al. 2000b) is closely related to that of monteregianite. Rhodesite and monteregianite-(Y) have double silicate layers of the same topology. In fact, other minerals such as delhayelite, hydrodelhayelite and macdonaldite, all have similar double silicate sheets. In rhodesite, these layers possess the maximum topological symmetry (*P2mm*) while in monteregianite-(Y) symmetry has been lost.

Rocha et al. (2000b) reported the first microporous sodium cerium silicate (AV-5) possessing the structure of mineral monteregianite-(Y). Recently, the same group reported the synthesis and *ab initio* structure determination of the first sodium potassium microporous

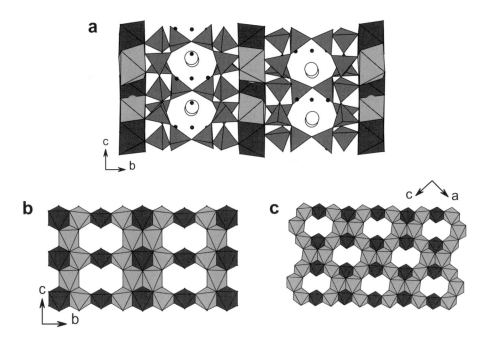

Figure 6. (a) Crystal structure of monteregianite-type materials showing two types of alternating layers: a double silicate sheet, and (b,c) an open octahedral sheet composed of LnO_6 and $[NaO_4(H_2O)_2]$ octahedra. The tetrahedral layers of all these solids are similar but this is not the case with the octahedral layers. For example, monteregianite and its Y^{3+} and Ce^{3+} analogues contain a single type of Ln^{3+} facing the pores (c), while the Eu^{3+} and Tb^{3+} analogues contain two types of Ln^{3+}, one is isolated by $[NaO_4(H_2O)_2]$ octahedra and the other faces the pores (b). Dark LnO_6 octahedra, light grey $[NaO_4(H_2O)_2]$ octahedra, medium grey SiO_4 tetrahedra, white circles K^+ ions, dark circle H_2O.

europium and terbium silicates, named AV-9 (Ananias et al. 2001). The structures of AV-9 materials (and related materials containing Er, Tb, Ananias et al. 2004, and Nd, Rocha et al. 2004a) and mineral monteregianite-(Y) are related. However, although the tetrahedral layers of AV-9 solids and monteregianite-(Y) are similar this is not the case with the octahedral layers. Perhaps the most important difference between the octahedral layers of AV-9 and monteregianite lies in the fact that the latter contains a single kind of Y(III), facing the pores, while AV-9 contains two kinds of Eu(III), Tb(III): one is isolated by [NaO$_4$(H$_2$O)$_2$] octahedra while the other is facing the pores. This is expected to influence, for example, the luminescence behavior of these two Eu(III), Tb(III) centers. Rhodesite-type minerals are also discussed by Ferraris and Gula (2005) in this volume.

Another interesting system, is based on mineral sazhinite-(Ce) Na$_2$(CeSi$_6$O$_{14}$)(OH)·1.5H$_2$O (Shumyatskaya et al. 1980). For example, AV-21 [Na$_3$(EuSi$_6$O$_{15}$)·2H$_2$O] contains two crystallographically distinct Eu^{3+} centers coordinated to six different SiO$_4$ tetrahedra, in a distorted octahedral coordination (Rocha et al. 2004b). The crystal structures of AV-21, cerium silicate Na$_3$(CeSi$_6$O$_{15}$)·2H$_2$O recently reported by Jeong et al. (2002a), mineral sazhinite-(Ce), and the α and β forms of K$_3$(NdSi$_6$O$_{15}$)·xH$_2$O (Haile and Wuensch 2000a,b) show several remarkable similarities. All structures are formed by undulated [Si$_2$O$_5^{2-}$]$_\infty$ layers (in the bc plane, for AV-21, Fig. 7b) in which all the SiO$_4$ tetrahedra are formed by three bridging O-atoms (to neighboring intra-layer Si-atoms) and one terminal O-atom coordinated to one Eu^{3+} ion (tetrahedral vertices pointing "upwards" and "downwards" in Fig. 7). Although in all the structures the layers are formed by the condensation of adjacent xonotlite-type ribbons the way they coalesce in AV-21 and Na$_3$(CeSi$_6$O$_{15}$)·2H$_2$O is different from what is observed for

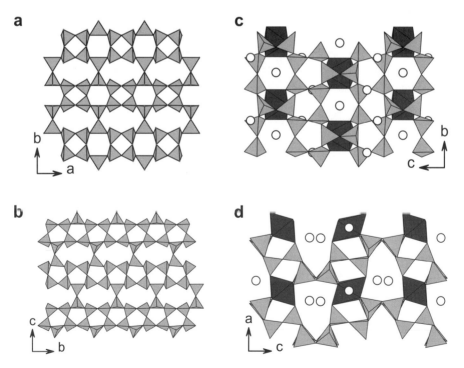

Figure 7. [Si$_2$O$_5^{2-}$]$_\infty$ layers present in (a) mineral sazhinite-(Ce) [and in the α and β forms of K$_3$(NdSi$_6$O$_{15}$)·xH$_2$O] and (b) sodium europium silicate AV-21 [similar to layers in Na$_3$(CeSi$_6$O$_{15}$)·2H$_2$O]. (c,d) Polyhedral representation of the structure of AV-21.

sazhinite and the α and β forms of $K_3(NdSi_6O_{15})·xH_2O$: while in the mineral (and related structures) the coalescence of xonotlite-type ribbons leads to the formation of alternating 4- and 6-rings in the interface between ribbons (Fig. 7a), for AV-21 and $Na_3(CeSi_6O_{15})·2H_2O$ only 5-rings are observed in the interface due to an alternate way of connection, with relative shift of adjacent xonotlite ribbons (Fig. 7b). Adjacent $[Si_2O_5^{2-}]_\infty$ layers are pillared by the Eu^{3+} ions, connected through coordinative interactions to the terminal SiO_4 oxygen atoms, forming a porous three-dimensional network (Fig. 7c,d). Sazhinite-(Ce) contains channels running along each crystallographic direction of the unit cell. The largest channels are formed by the hills and valleys of the corrugated $[Si_2O_5^{2-}]_\infty$ sheets, which leads to the formation of two different types of irregular 8-membered tunnels (cross-sections of *ca.* 4.2×3.2 Å² and *ca.* 3.0×2.0 Å² for the channels which contain water molecules and Na^+, respectively) (Shumyatskaya et al. 1980). The same structural type of 8-membered channel is also observed for AV-21 (running parallel to the *b* axis with cross-section of *ca.* 3.0×1.8 Å²; Fig. 7d), $Na_3(CeSi_6O_{15})·2H_2O$, and the α and β forms of $K_3(NdSi_6O_{15})·xH_2O$. It is interesting to note that in these compounds only one type of channel is observed, with shapes best described as intermediate between the shapes of the two different channels present in sazhinite-(Ce). In AV-21 and $Na_3(CeSi_6O_{15})·2H_2O$, the alternate way for the xonotlite-type ribbons in the $[Si_2O_5^{2-}]_\infty$ layers also has the peculiar feature of forming channels only in two crystallographic directions of the unit cell (in AV-21, channels are present along the *a* and *b* axes, Fig. 7c,d). Na^+ ions and water molecules are located within the channels of the open anionic frameworks.

A common feature of the tobermorite (hydrated calcium silicates) mineral family is the layer built up by seven-coordinated (mono-capped trigonal prisms) calcium cations, as described by Bonaccorsi and Merlino (2005) in this volume. Tobermorites are, thus, not OPT materials but they deserve a reference here because the interesting photoluminescence properties of AV-20 materials (Ferreira et al. 2003a) and the rare-earth OPT silicates described above are somewhat similar. The structure of Eu-AV-20 is closely related with the structure of tobermorite 11 Å minerals. In AV-20, the substitution $2\ Ca^{2+} \leftrightarrow Eu^{3+} + Na^+$ occurs, with Eu^{3+} and Na^+ being seven-coordinated. In addition, some (*ca.* 10%) Eu^{3+} ions are disordered over the Na^+ sites. Other lanthanide ions may be introduced into the structure.

PROPERTIES AND APPLICATIONS

Catalysis and sorption

So much work is now available on the evaluation of the catalytic properties of ETS-10 and related materials (and to some extent on ETS-4) that here only some key features will be highlighted. The catalytic applications have focused on the following attributes of ETS-10: high cation exchange capacity leading to many possibilities particularly in base catalysis; facile metal loading for bi-functionality; very low acidity; possible chiral activity; photocatalytic properties. On the other hand, the titanosilicates with octahedrally-coordinated titanium do not show good properties for oxidation catalysis in a similar manner to four-coordinate titanium. This is due to the ligand saturation which prevents further attachment by oxidation agents such as hydrogen peroxide.

At present, the study of the photo-reactivity of ETS-10 (Howe and Krisnandi 2001) and the evaluation of this material in photo-catalysis (Calza et al. 2001) is an area of intense research. For example, the activity of ETS-10 for the photo-catalytic degradation of cyclohexanol, cyclododecanol, 2-hexanol, and benzyl alcohol was compared with TiO_2 particles included within small and large pore zeolitic supports suspended in acetonitrile by Fox et al. (1994). Southon and Howe (2002) reported that the disordered stacking of repeat units in the structure of ETS-10 gives rise to defects that strongly influence the reactivity. Changes in the ultraviolet absorbance spectrum indicated disruptions of the delocalized charge-transfer transitions along the –Ti–O–Ti– chains. The stacking defects interrupt these chains and hydroxyl groups are associated with these defects. The same group later reported that the photo-reactivity and photo-catalytic activity of ETS-10 are strongly influenced by the defects in the structure (Krisnandi et al. 2003). A relatively defect free material catalyses the photo-polymerization of ethane and in the presence of oxygen catalyses the partial oxidation of ethene to acetic acid and acetaldehyde, which remain strongly adsorbed in the pores. A more defective material is photo-reduced when irradiated in the presence of adsorbed ethene, and catalyses the complete oxidation of ethene to carbon dioxide and water in the presence of oxygen. Although Calza et al. (2001) have shown that ETS-10 can be used as a shape-selective photocatalyst for the decomposition of aromatic molecules, the actual use of this material is discouraged by its low activity, when compared with that of TiO_2. Later, the same group showed that a mild treatment with HF enhances the activity of ETS-10 in the photo-degradation of large aromatic molecules that are unable to penetrate inside the pores, such as 2,5-dichlorophenol, 2,4,5-trichlorophenol, 1,3,5-trihydroxybenzene, and 2,3-dihydroxynaphthalene (Xamena et al. 2003). The photo-activity of the acid-treated materials is comparable to, or greater than, that of the nonselective TiO_2 catalyst. Moreover, the enhancement of the photoactivity is accompanied by a parallel increase of the shape selectivity, particularly toward dihydroxynaphthalene. Uma et al. (2004) explored the photo-catalytic decomposition of acetaldehyde over ETS-10 and transition metal ion exchanged ETS-10 decomposed acetaldehyde under UV radiation and the rate constant for the decrease of acetaldehyde is comparable to that of pure TiO_2. Interesting visible ($\lambda > 420$ nm) and UV light activity was found for ETS-10 samples incorporated with Cr and Co (Cr, Co/Ti = 0.05) and for materials synthesized by ion-exchange reactions such as Co-ETS-10 and Ag-ETS-10. All other ion-exchanged $(Na,K)_{(2-x)}M'_xTiSi_5O_{13}$ [M' = Cr(III), Mn(II), Fe(III), Co(II), Ni(II), and Cu(II)] samples were found to be active photo-catalysts that can decompose acetaldehyde under UV radiation. So far, the photo-catalytic activity of ETS-4 deserved much less attention. A recent study by Guan et al. (2004) explored a CdS/ETS-4 composite for hydrogen production from water under visible light irradiation ($\lambda > 420$ nm). It was found that nano-sized CdS particles embedded in the ETS-4 nano-pores show stable photo-catalytic activity in an aqueous solution containing Na_2S and Na_2SO_3 electron donors and the energy conversion efficiency was improved by combining CdS with ETS-4.

Bianchi and Ragaini 1997; Bianchi et al. (1996, 1998a,b) report on Fischer-Tropsch chemistry over Co and Ru exchanged ETS-10. This material was found to be a suitable support for metal catalysts, having high surface area, high ion-exchange capability and no acidic function. The importance of alpha-olefin readsorption within the catalyst is discussed and the nature of this readsorption is tailored by effective control of the metal distribution inside the pores of ETS-10. The CO conversion and selectivity obtained also varies depending upon whether the active metal is introduced in the ETS-10 cages by ion-exchange or simply by impregnation.

Philippou et al. (1998a), Waghmode et al. (1999) and Das et al. (1997, 1998) have studied the bifunctional reforming reaction of hexane to benzene over Pt-supported basic ETS-10. The basicity of the titanosilicate can be controlled through samples exchanged with different alkali metals (M = Li, Na, K, Rb, or Cs). A distinct relationship between the intermediate electronegativity (S-int) of the different metal-exchanged ETS-10 samples and benzene yield is reported, suggesting the activation of Pt by the basicity of the exchanged metal. ETS-10 samples exhibit greater aromatization activities than related Pt-Al_2O_3 catalysts. The very high basicity of ETS-10 in comparison with, for example zeolite X, is illustrated by Philippou et al. (1999) in a paper which monitors the relative conversion of isopropanol to acetone. The same group has also demonstrated how these same properties are effective in aldol chemistry (Philippou and Anderson 2000) and dehydration of *t*-butanol (Philippou et al. 1998b;

Das et al. 1996c). In this latter reaction conversions and selectivities close to 100% are observed at relatively modest temperatures.

Acylation of alcohols with acetic acid can be carried out efficiently in the liquid phase over ETS-10 exchanged with several ions (Waghmode et al. 2001). The best activity for acylation of primary alcohols is found for H, Rb and Cs-ETS-10. ETS-10 may also be used for the acylation of secondary alcohols and esterification with long chain carboxylic acids.

Pd-loaded ETS-10 was used as a catalyst in Heck reaction (Waghmode et al. 2003). The catalyst exhibits high activity and selectivity towards the carbon-carbon coupling of aryl halides with olefins. In the case of the coupling of ethyl acrylate with iodobenzene, 96% conversion of iodobenzene with greater than 98% selectivity was obtained within one hour over a 0.2 wt% Pd-loaded catalyst. The catalyst activates aryl bromide and chloride substrates, and appears to be heterogeneous.

ETS-10 was shown to be a good catalyst for the transesterification of soybean oil with methanol, providing higher conversions than zeolite X (Suppes et al. 2004). The increased conversions were attributed to the higher basicity of ETS-10 and larger pore structure that improved intra-particle diffusion.

The catalytic activity of a series of Cr-containing ETS-10 samples with different Cr/Ti was characterized by standard probe reactions (Brandão et al. 2001b): isopropanol conversion (to propene and acetone), t-butanol conversion (to iso-butene) and ethanol oxidation (to acetaldehyde and ethyl acetate).

The gas-phase oxidative dehydrogenation of cyclohexanol with air using ETS-10 materials has been studied (Valente et al. 2001). At reaction temperatures below 200°C ETS-10 is 100% selective to cyclohexanone and 75% cyclohexanol conversion is achieved. The introduction of Cr, Fe and K, Cs in ETS-10 changes the stability and decreases the conversion and selectivity towards cyclohexanone.

The de-NOx activity of Cu-based catalysts prepared by dispersing CuO on ETS-10 was studied by Gervasini and Carniti (2002). The activity of NO reduction with ethylene was related to the morphological and chemical properties of the catalysts. The results are consistent with a value of about 2.5–3 atom(Cu) nm^{-2} for the maximum dispersion capacity of CuO on the ETS-10 matrix. The amount of Cu deposited on ETS-10 affects the activity of catalysts towards NO reduction. The turnover frequencies per Cu site calculated as a function of Cu concentration showed a clear decreasing trend starting from 0.7 to 3.5 atom(Cu) nm^{-2}.

Bal et al. (2000) carried out an EPR study of Ti(III) in ETS-4, ETS-10, TS-1 and TiMCM-41. Ti(III) was obtained by reduction of Ti(IV) by dry hydrogen at temperatures above 673 K. Interaction of tetrahedrally-coordinated Ti(III) (in TS-1 and TiMCM-41) with O_2 and H_2O_2 results in a diamagnetic Ti(IV) hydroperoxo species. Under the same conditions, octahedrally-coordinated Ti(III) (in ETS-4 and ETS-10) forms a paramagnetic Ti superoxo species. The poor activity in selective oxidation reactions of ETS materials has been attributed to the absence of formation of titanium hydroperoxo species.

Bodoardo et al. (2000) studied the status of Ti(IV) in ETS-10 and TS-1 by voltammetry. Both, tetrahedral and octahedral, TS-1 Ti(IV) species show electrochemical response. It was shown that the use of acid solutions allows discrimination between Ti(IV) ions in TS-1 and in ETS-10, since only the former is able to coordinate water molecules.

Zecchina et al. (2001) investigated the interaction of methyl and ethyl acetylene with the acidic form of ETS-10. In the first adsorption step, at room temperature, π hydrogen-bonded adducts are formed between the alkyl acetylene and ETS-10 hydroxyl groups. These hydrogen-bonded species act as precursors in the second step where oligomerization takes place, leading to formation of carbocationic double-bond conjugated systems of increasing length. The same group also reported on the adsorption of carbon dioxide on Na- and Mg-exchanged ETS-10 (Xamena and Zecchina 2002). CO_2 was found to adsorb reversibly onto the (M) charge-balancing cations with formation of linear end-on adducts of the type M...OCO. Carbon dioxide also forms different carbonate-like species upon adsorption on the Na(Mg)-ETS-10, which are the origin of the activity of ETS-10 in heterogeneous base-catalyzed processes.

Gervasini et al. (2000) studied the adsorption and surface properties of Cu-exchanged ETS-10 by calorimetry of adsorbed NO, CO, C_2H_4 and NH_3 probe molecules. Cu(I) is the prominent species in Cu-ETS-10 and the number of Cu(I) species increased as the level of copper loading increases. Unfortunately, the structure of ETS-10 collapses at high copper loadings. Adsorption of several alkylamines by ETS-10 was also studied by Ruiz et al. (2004a,b), while Bagnasco et al. (2003) reported on the adsorption of water and ammonia on this material.

Serralha et al. (2000) showed that ETS-10 materials are good supports for an enzymatic alcoholysis reaction. The recombinant cutinase *Fusarium solani pisi* was immobilized by adsorption on ETS-10, ETAS-10 and vanadosilicate analogue of ETS-10, AM-6. The enzymatic activity in the alcoholysis of butyl acetate with hexanol, in isooctane, was measured as a function of the water content and water activity.

The synthesis of the pure chiral polymorph of ETS-10 would offer the potential for chiral catalysis or asymmetric synthesis. Along with zeolite beta, ETS-10 is currently the only known wide-pore microporous material which possesses a chiral polymorph with a spiral channel. The as-prepared materials of both these structures contain an equal proportion of intimately intergrown enantiomorphs and are consequently achiral. The task of preparing a pure chiral polymorph remains a formidable challenge.

Much research is now being carried out aimed at preparing zeolites membranes for applications in membrane reactors (Coronas and Santamaria 2004). Although most of the work has concentrated on "traditional" zeolites a few studies are available on microporous titanosilicate membranes, particularly on ETS-4 (Braunbarth et al. 2000a; Guan et al. 2001, 2002). The heteroepitaxial growth of ETS-10 on ETS-4 was described by Jeong et al. (2002b). Lin et al. (2004) have recently prepared a pure, 5 μm thick ETS-10 membrane exhibiting a good degree of crystal intergrowth. The development of titanosilicate-based membranes is likely to continue on account of some key advantages related to their preparation and properties, namely: (a) in general, a pure phase can be obtained in the absence of costly organic templates, thus avoiding calcination, which often leads to defects and loss of active surface groups (Dong et al. 2000); (b) titanosilicates are prepared under relatively mild pH conditions, reducing the chemical attack on the support and synthesis equipment used; (c) OPT materials present novel possibilities for isomorphous framework substitution, which allow the fine tuning of the catalytic and adsorption properties of a given membrane, while preserving its microporous structure; (d) these materials often exhibit strong basicity that provides an alternative to the acid properties of classic zeolites and opens up new application possibilities. Although ETS-10 has no natural counterpart, its catalytic properties are much studied and, thus, it is a convenient case in point. ETS-10 catalytic membrane could perform the gas-phase oxidative dehydrogenation of cyclohexanol (Valente et al. 2001) while simultaneously removing the desired product from the reaction environment. Another example would be the use of an ETS-10 membrane to catalyse the liquid-phase acylation of alcohol (Waghmode et al. 2001) and the aldol condensation of acetone (Philippou and Anderson 2000) and to remove *in situ* the product water.

Cation exchange

OPT materials present new opportunities for ion-exchange applications. Clearfield and co-workers have shown that the unique selectivity of titano-and zirconosilicates is due to the correspondence of the geometrical parameters of their ion exchange sites (channels, cavities) to the size of the selectively sorbed ions (Harrison 1995; Poojariy et al. 1994; Behrens et al. 1996; Poojary et al. 1996). Because of the small size of the channels or cavities, the framework acts as a coordinating ligand to the cations.

Much attention has been focused on two structurally-related titanosilicates, $Na_2Ti_2O_3SiO_4·2H_2O$ and synthetic pharmocosiderite, $HM_3Ti_4O_4(SiO_4)_3·4H_2O$. The former displays great affinity for Cs^+ (Poojary et al. 1994; Bortun et al. 1996) while the latter is selective for ppm concentrations of Cs^+ and Sr^{2+} in the presence of ppm levels of Na^+, K^+, Mg^{2+} and Ca^{2+}cations at pH 7 (Behrens and Clearfield 1997; Behrens et al. 1998; Puziy 1998). Dyer et al. (1999) have also reported on the removal of trace ^{137}Cs and ^{89}Sr by different cationic forms of synthetic pharmacosiderite. Other recent reports on $M_2Ti_2O_3SiO_4·2H_2O$ further show the importance of this material as a Cs^+ and Sr^{2+} exchanger (Sylvester et al. 1999; Clearfield et al. 2000; Clearfield 2001; Solbra et al. 2001) and its relevance to treat radioactive wastes containing ^{241}Am and ^{236}Pt (Al-Attar et al. 2003) and ^{134}Cs and ^{89}Sr and ^{57}Co (Moller et al. 2002).

The synthetic analogues of mineral umbite, AM-2 materials ($K_2MSi_3O_9·2H_2O$, M = Ti, Zr) (Lin et al. 1999a) were shown to be good exchangers for Rb^+, Cs^+ and K^+ cations (Bortun et al. 2000; Valtchev et al. 1999). The ion-exchange behavior of a series of mixed Zr, Ti AM-2 materials containing different amounts of Ti and Zr was studied. It was found that Zr-rich silicates with large channels exhibit affinity for Rb^+ and Cs^+ cations, whereas Ti-rich compounds with much smaller channels show a preference for the K^+ ion (Clearfield et al. 1998). These data suggest that the chemical alteration of the structure of exchangers may be a promising way for tailoring their selectivity.

The ion-exchange for uranium in microporous titanosilicates ETS-10, ETS-4, and $Na_2Ti_2O_3SiO_4·2H_2O$ has been studied by Al-Attar et al. (2000) and Al-Attar and Dyer (2001). The difference in their ability to take up uranium has been discussed in terms of their crystal structure and the determination of their cation exchange capacity. Pavel et al. 2002 studied the sorption of ^{137}Cs and ^{60}Co by ETS-10 and ETS-4, while Al-Attar et al. 2003 examined the potential of these materials for treating radioactive wastes containing ^{241}Am and ^{236}Pu. ETS-10 was shown to be particularly selective for Pb^{2+} ions (Kuznicki and Trush 1991b; Zhao et al. 2003).

Koudsy and Dyer (2001) studied the sorption of radioactive ^{60}Co by the synthetic titanosilicate analogue of mineral penkvilsite-2O (AM-3). Synthetic penkvilksite was also shown to be particularly selective for Li^+ cations, suggesting possible applications in battery technology (Kuznicki et al. 1999).

Optical and magnetic properties

The presence of stoichiometric quantities of transition metals in the OPT frameworks may confer these materials interesting optical properties (De Man and Sauer 1996; Mihailova et al. 1996; Xu et al. 1997; Borello et al. 1997; Lamberti 1999). The structure of ETS-10 contains –O–Ti–O–Ti–O– ("TiO") chains, with alternating long-short bonds, which are effectively isolated from each other by a silicate sheet. In other words, ETS-10 contains "TiO"-wires embedded in an insulating SiO_2 matrix, which leads to a one-dimensional quantum confinement of electrons or holes within this wire and a band-gap blue shift. The effective reduced mass, μ, of electrons and holes within such a wire is calculated to be

1.66 M(e) < μ < 1.97 M(e), which is consistent with a band gap of 4.03 eV, much different from that of bulk TiO_2. *Ab initio* calculations on a "TiO"-chain embedded in an envelope of SiO_4 tetrahedra, mimicking the structure of ETS-10, confirm that the peculiar optical properties of solid are associated with the presence of such linear chains (Bordiga et al. 2000; Damin et al. 2004). The UV-Vis spectra and the electron spin resonance properties of the chains may be modified by adsorbing Na vapors: Ti(IV) + Na → Ti(III) + Na^+, thus generating both unpaired electrons within the titanate chain, and extra cation sites. After this reduction the material is not air stable and is easily reoxidized, generating additional Na_2O within the channels. Such redox couples give strong indications of possible applications in battery technology. These studies also highlight the possible future role of OPT materials in optoelectronic and non-linear optical applications.

Lanthanide (Ln) containing materials emit over the entire spectral range: near infrared (Nd^{3+}, Er^{3+}), red (Eu^{3+}, Pr^{3+}, Sm^{3+}), green (Er^{3+}, Tb^{3+}), and blue (Tm^{3+}, Ce^{3+}). Their optical transitions take place only between 4f orbitals, which are well shielded from their chemical environment by $5s^2$ and $5p^6$ electrons (Rocha and Carlos 2003). As a consequence, atomic-like emission spectra displaying characteristic sharp lines may be observed. These materials find applications in many important devices, for example in cathode ray tubes, projection televisions, fluorescent tubes and x-ray detectors, and in photonics and optical communications.

In the last few years, much research has been carried out on materials comprising lanthanide guests in microporous zeolite-type hosts, systems which often exhibit photoluminescent properties; this field has been recently reviewed by Rocha and Carlos (2003). Microporous materials hosting lanthanide luminescent centers encompass: (a) zeolites doped via ion exchanging the extra-framework cations by Ln^{3+}; (b) lanthanide complexes with organic ligands, enclosed in the zeolitic pores and channels; (c) open-framework coordination polymers (organic-inorganic hybrids); and (d) novel lanthanide silicates, with stoichiometric amounts of framework Ln^{3+} ions (see section on Rare-earth silicates).

ETS-10 has been doped with Eu^{3+} (Rainho et al. 2000) and Er^{3+} (Rocha et al. 2000a) by ion-exchanging the Na^+ and K^+ cations, and studied the luminescence properties of these materials. Only Eu^{3+}-doped ETS-10 is optically active, even at 300 K. However, the presence of water molecules within the channels of ETS-10, coordinating Eu^{3+}, partially quenches the radiative process, reducing photoluminescence. The emission spectrum indicates the presence of, at least, two local environments for the Eu^{3+} ions in ETS-10. Similar studies were carried out on titanosilicates AM-2 and AM-3 (Rocha et al. 2004a) and zirconosilicate analogue of AM-2 (Rainho et al. 2004). One possible way to circumvent the quenching of photoluminescence by water molecules and hydroxyl ions in these materials consists in collapsing the microporous framework into more dense structures. For example, Rocha and co-workers prepared luminescent dense materials from microporous precursors (Rainho et al. 2000, 2002; Rocha et al. 2000a). Upon calcinations at temperatures in excess of 700°C, Eu^{3+}- and Er^{3+}-doped ETS-10 transform into analogues of the mineral narsarsukite, which display very interesting luminescence properties. Er^{3+}-doped narsarsukite, in particular, exhibits a high and stable room-temperature emission in the visible and infrared spectral regions (Rocha et al. 2000a). An efficient energy transfer between the narsarsukite skeleton and the Er^{3+} centers shows that the ion-lattice interactions can play an important role in the luminescence properties of these titanosilicates.

Embedding lanthanide ions in the framework is a novel approach proposed by Rocha and co-workers for preparing photoluminescent microporous material, which contain stoichiometric amounts of Ln (Rocha and Carlos 2003). These materials are important because they are multifunctional, combining in a single, stable solid microporosity and photoluminescence. Rocha et al. (2000b) reported Ce^{3+}-monteregianite-type materials (AV-5). The Ce(III)/Ce(IV) ratio in AV-5 may be controlled by oxidation/reduction in an appropriate atmosphere, allowing the fine tuning of the luminescence, adsorption and ion exchange properties of the material. Eu- and Tb-AV-9 monteregianite-type materials are photoluminescent in the visible region (Ananias et al. 2001) while dehydrated Er-AV-9 (Ananias et al. 2004) and Nd-AV-9 (Rocha et al. 2004a) are room-temperature infrared emitters. Tb-AV-9 is also the first example of a microporous siliceous x-ray scintillator (Ananias et al 2004), i.e., it emits in the visible region when excited with X-ray (CuKα) radiation (integrated intensity 60% of that of standard material Gd_2O_2 S:Tb). The sazhinite-type material AV-21 [$Na_3(EuSi_6O_{15})\cdot 2H_2O$] is photoluminescent in the visible region. When excited at 394 nm, the room-temperature emission spectra of hydrated and dehydrated AV-21 exhibit subtle differences, particularly a shift in the $^5D_0 \rightarrow {}^7F_0$ transition lines. This observation raises the possibility of using the photoluminescence of this material to sense the presence of other molecules (Rocha et al. 2004b).

Tobermorite-like, AV-20 materials ($Na_{1.08}K_{0.5}Ln_{1.14}Si_3O_{8.5}\cdot 1.78H_2O$, Ln = Eu, Tb, Sm, Ce, Gd) have very interesting photoluminescent properties (Ferreira et al. 2003a). In particular, introducing a second type of Ln^{3+} ion in the framework allows the tuning of photoluminescence: the gathered evidence clearly shows the occurrence of energy transfer from Tb^{3+} to Eu^{3+} (in Eu/Tb-AV-20) and Gd^{3+} to Tb^{3+} (in Tb/Gd-AV-20).

Engineering the magnetic properties of OPT materials is an almost completely unexplored field with tremendous potential for applications. Some reference to the magnetic behavior (magnetic susceptibility as a function of temperature) of vanadosilicates is available from the work of Jacobson and co-workers (Wang et al. 2001, 2002c; Huang et al. 2002) and Li et al. (2002). As the valence state of vanadium can be easily manipulated novel magnetic properties may be anticipated. In practice, so far, the materials are paramagnetic down to very low temperatures (a few degrees K) when they may exhibit weak pairwise antiferromagnetic interactions. Brandão et al. (2004) reported a large-pore paramagnetic cromosilicate material (AV-15, $Na_3CrSi_6O_{15}\cdot 4H_2O$) with a pseudo-Curie temperature of 0.71 K and a magnetic moment of 3.69 μ_B, indicating the presence of Cr(III).

Perhaps the most interesting attempt to engineer magnetic centers in OPT materials is the synthesis by Wang et al. (2003) of the copper silicate system CuSH-1A (A = Na, K, Rb, Cs, $x \approx 1$, $y \approx 1$, $z \approx 6$): $Na_4[Cu_2Si_{12}O_{27}(OH)_2][(AOH)_x(NaOH)_y(H_2O)_z]$. These structures are unusual in several respects: they are the first examples to contain double layers of silicate tetrahedra that define straight 12-ring channels, analogous to those found in LTL-type zeolites. Unlike LTL, however, the framework composition $Na_4[Cu_2Si_{12}O_{27}(OH)_2]$ that defines the 12-ring channels is neutral, and as a consequence only neutral MOH and H_2O species fill these channels. The porous framework is built from silicate double layers that are arranged perpendicular to the [100] direction, and are cross-linked by interlayer Cu-O and Na-O bonds (Fig. 8). The silicate double layer contains channels defined by 12-rings of tetrahedra that run along the [001] direction and have an aperture of 7.3 × 4.4 Å^2, and 8-ring channels along the [011] and [01$\bar{1}$] directions, with an aperture of 3.8 × 2.5 Å^2. Neighboring silicate double layers are intercon-

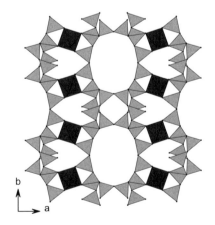

Figure 8. Structure of copper silicate CuSH-1Cs viewed along the [001] direction showing the 12-ring channels. Dark CuO_6 octahedra, light grey SiO_4 tetrahedra. Na^+, Cs^+ and water molecules are omitted for clarity.

nected by CuO_4 squares that share corners with SiO_4 tetrahedra, with Cu-O bond lengths of 1.95 Å. The copper-ion coordination is completed by interlayer water molecules to form single chains of $CuO_4(H_2O)_2$ elongated octahedra with $Cu-OH_2$ bond lengths of 2.70 Å.

CONCLUSIONS AND OUTLOOK

The research into mixed octahedral-pentahedral-tetrahedral microporous silicates is now a well established field. The initial impetus on titanosilicates has been followed by a comprehensive synthesis effort leading to many new transition metal silicates. A few examples of calcium, tin, indium and uranium silicates have also been reported. At present, the synthesis effort is also being directed towards lanthanide silicates.

The study of OPT materials brought to light new molecular architectures and has broadened considerably our understanding of the chemistry (and, to some extent, the physics) of microporous zeolite-type materials. So far, the search for potential applications of OPT materials has been concentrated on titanosilicate ETS-10 and heterogeneous catalysis, exploring the intrinsic basic properties and photocatalytic activity of this material. An important body of work is also available on the use of small-pore OPT titanosilicates and zirconosilicates as ion-exchangers, particularly for the treatment of radioactive wastes, and in the field of adsorption and separation of small molecules.

In the near future it is anticipated that the catalysis applications will become even more important, clearly showing that the research into OPT materials is not simply an academic exercise (even if a most rewarding one). We feel, however, that perhaps the real potential of these new materials lies not in the "conventional" areas of use of zeolites but rather in totally new fields, such as optoelectronics and non-linear optics, batteries, magnetic materials, sensors and other nanotechnology device applications. The new research into photoluminescent lanthanide silicates and the very recent development of (potentially) magnetic copper silicates provides the first hints that some of the most interesting properties of OPT materials remain largely uncovered.

The expedition into the exotic world of OPT materials has been motivated, in no small measure, by the sense of wonder conveyed by the amazing mineral structures provided by Nature. Clearly, this intellectual endeavor overrides any artificial borders scientists may have built up in the past between Chemistry, Physics and Mineralogy. In this sense, the field of OPT materials also presents excellent scientific case studies for undergraduate and graduate students, no longer being in the strict possession of a number of (excited) scientists.

ACKNOWLEDGMENTS

We thank financial support from FCT, POCTI and FEDER.

REFERENCES

Al-Attar L, Dyer A (2001) Sorption of uranium onto titanosilicate materials. J Radioanal Nucl Chem 247:121-128
Al-Attar L, Dyer A, Blackburn R (2000) Uptake of uranium on ETS-10 microporous titanosilicate. J Radioanal Nucl Chem 246:451-455
Al-Attar L, Dyer A, Harjula R (2003) Uptake of radionuclides on microporous and layered ion exchange materials. J Mater Chem 13:2963-2968
Ananias D, Ferreira A, Rocha J, Ferreira P, Rainho JP, Morais C, Carlos LD (2001) Novel microporous framework europium and terbium silicates. J Am Chem Soc 123:5735-5742

Ananias D, Rainho JP, Ferreira A, Rocha J, Carlos LD (2004) The first examples of X-ray phosphors, and C-band infrared emitters based on microporous lanthanide silicates. J Alloys Compd 374:219-222

Anderson MW, Agger JR, Hanif N, Terasaki O (2001a) Growth models in microporous materials. Microporous Mesoporous Mater 48:1-9

Anderson MW, Agger JR, Hanif N, Terasaki O, Ohsuna T (2001b) Crystal growth in framework materials. Solid State Sci 3:809-819

Anderson MW, Philippou A, Lin Z, Ferreira A, Rocha J (1995a) Al, Ti, avoidance in the microporous titano-aluminosilicate ETAS-10. Angew Chem, Int Ed Engl 34:1003-1005

Anderson MW, Rocha J (2002) Synthesis of titanosilicates and related materials. *In:* Handbook of porous solids. Vol 2. Schüth F, Sing KSW, Weitkamp J (eds) Wiley-VCH, Weinheim, p 876-903

Anderson MW, Terasaki O, Oshuna T, O'Malley PJ, Philippou A, Mackay SP, Ferreira A, Rocha J, Lidin S (1995b) Microporous titanosilicate ETS-10: a structural survey. Philos Mag B 71:813-841

Anderson MW, Terasaki O, Oshuna T, Philippou A, Mackay SP, Ferreira A, Rocha J, Lidin S (1994) Structure of the microporous titanosilicate ETS-10. Nature 367:347-351

Bagnasco G, Turco M, Busca G, Armaroli T, Nastro A, De Luca P (2003) Characterization of the structural and gas adsorption properties of ETS-10 molecular sieve. Adsorpt Sci Technol s21:683-696

Bal R, Chaudhari K, Srinivas D, Sivasanker S, Ratnasamy P (2000) Redox and catalytic chemistry of Ti in titanosilicate molecular sieves: an EPR investigation. J Mol Catal A: Chem 162:199-207

Baussy G, Caruba R, Baumer A, Turco G (1974) Experimental mineralogy in ZrO_2-$SiO_2$$Na_2O$-$H_2O$ system – petrogenetic correlations. Bull Soc Fr Minéral Crystallogr 97:433-444

Behrens EA, Clearfield A (1997) Titanium silicates, $M_3HTi_4O_4(SiO_4)_3 \cdot 4H_2O$ (M = Na^+, K^+), with three-dimensional tunnel structures for the selective removal of strontium and cesium from wastewater solutions. Microporous Mater 11:65-75

Behrens EA, Poojary DM, Clearfield A (1996) Syntheses, crystal structures, and ion-exchange properties of porous titanosilicates, $HM_3Ti_4O_4(SiO_4)_3 \cdot 4H_2O$ (M=H^+, K^+, Cs^+), structural analogues of the mineral pharmacosiderite. Chem Mater 8:1236-1244

Behrens EA, Sylvester P, Clearfield A (1998) Assessment of a sodium nonatitanate and pharmacosiderite-type ion exchangers for strontium and cesium removal from DOE waste simulants. Environ Sci Technol 32: 101-107

Bianchi CL, Ardizzone S, Ragaini V (1998a) Synthesis gas to branched hydrocarbons: a comparison between Ru-based catalysts supported on ETS-10 and on Al_2O_3 (doped with sulfated zirconia). *In*: Natural gas conversion V. Stud Surface Sci Catal Vol 119. Parmaliana A, Sanfilippo D, Frusteri F, Vaccari A, Arena F (eds) Elsevier, Amsterdam, p 173-178

Bianchi CL, Carli R, Merlotti S, Ragaini V (1996) Fischer-Tropsch synthesis on Co and Co(Ru-doped) ETS-10 titanium silicate catalysts. Catal Lett 41:79-82

Bianchi CL, Ragaini V (1997) Experimental evidence of alpha-olefin readsorption in Fischer-Tropsch synthesis on ruthenium-supported ETS-10 titanium silicate catalysts. J Catal 168:70-74

Bianchi CL, Vitali S, Ragaini V (1998b) Comparison between Co and Co(Ru-promoted)-ETS-10 catalysts prepared in different ways for Fischer-Tropsch synthesis. *In*: Natural gas conversion V. Stud Surface Sci Catal Vol 119. Parmaliana A, Sanfilippo D, Frusteri F, Vaccari A, Arena F (eds) Elsevier, Amsterdam, p 167-172

Bodoardo S, Geobaldo F, Penazzi N, Arrabito M, Rivetti F, Spano G, Lamberti C, Zecchina A (2000) Voltammetric characterization of structural titanium species in zeotypes. Electrochem Commun 2:349-352

Bonaccorsi E, Merlino S (2005) Modular microporous minerals: cancrinite-davyne group and C-S-H phases. Rev Mineral Geochem 57:241-290

Bordiga S, Palomino GT, Zecchina A, Ranghino G, Giamello E, Lamberti C (2000) Stoichiometric and sodium-doped titanium silicate molecular sieve containing atomically defined -OTiOTiO- chains: quantum *ab initio* calculations, spectroscopic properties, and reactivity. J Chem Phys 112:3859-3867

Borello E, Lamberti C, Bordiga S, Zecchina A, Arean CO (1997) Quantum-size effects in the titanosilicate molecular sieve. Appl Phys Lett 71:2319-2321

Bortun AI, Bortun LN, Clearfield A (1996) Ion exchange properties of a cesium ion selective titanosilicate. Solvent Extr Ion Exch 14:341-354

Bortun AI, Bortun LN, Clearfield A (1997) Hydrothermal synthesis of sodium zirconium silicates and characterization of their properties. Chem Mater 9:1854-1864

Bortun AI, Bortun LN, Poojary DM, Xiang O, Clearfield A (2000) Synthesis, characterization, and ion exchange behavior of a framework potassium titanium trisilicate $K_2TiSi_3O_9 \cdot H_2O$ and its protonated phases. Chem Mater 12:294-305

Brandão P, Philippou A, Hanif N, Claro PC, Rocha J, Anderson MW (2002a) Synthesis and characterization of two novel large-pore crystalline vanadosilicates. Chem Mater 14:1053-1057

Brandão P, Philippou A, Rocha J, Anderson MW (2001a) Gas-phase synthesis of MTBE from methanol and t-butanol over the microporous niobium silicate AM-11. Catal Lett 73:59-62

Brandão P, Philippou A, Rocha J, Anderson MW (2001b) Synthesis and characterization of chromium-substituted ETS-10. Phys Chem Chem Phys 1773-1777

Brandão P, Philippou A, Rocha J, Anderson MW (2002b) Dehydration of alcohols by microporous niobium silicate AM-11. Catal Lett 80:99-102

Brandão P, Valente A, Ferreira A, Amaral VS, Rocha J (2004) First stoichiometric large-pore chromium(III) silicate catalyst. Microporous Mesoporous Mater 69:209-215

Brandão P, Valente A, Philippou A, Ferreira A, Anderson MW, Rocha J (2002c) Novel large-pore framework titanium silicate catalyst. J Mater Chem 12:3819-3822

Brandão P, Valente A, Philippou A, Ferreira A, Anderson MW, Rocha J (2003) Hydrothermal synthesis and characterization of two novel large-pore framework vanadium silicates. Eur J Inorg Chem 1175-1180

Braunbarth CM, Boudreau LC, Tsapatsis M (2000a) Synthesis of ETS-4/TiO_2 composite membranes and their pervaporation performance. J Membr Sci 174:31-42

Braunbarth CM, Hillhouse HW, Nair S, Tsapatsis M, Burton A, Lobo RF, Jacubinas RM, Kuznicki SM (2000b) Structure of strontium ion-exchanged ETS-4 microporous molecular sieves. Chem Mater 12:1857-1865

Calza P, Paze C, Pelizzetti E, Zecchina A (2001) Shape-selective photocatalytic transformation of phenols in an aqueous medium. Chem Commun 2130-2131

Chao GY (1985) The crystal-structure of gaidonnayite $Na_2ZrSi_3O_9.2H_2O$. Can Mineral 23:11-15

Chapman DM (1990) Crystalline group IVA metal-containing molecular sieve compositions. US Patent 5 015 453

Chapman DM, Roe AL (1990) Synthesis, characterization and crystal-chemistry of microporous titanium-silicate materials. Zeolites 10:730-737

Choisnet J, Nguyen N, Groult D, Raveau B (1976) New oxides in lattice of octahedra NbO_6 (TaO_6) and Si_2O_7 groups – phases $A_3Ta_6Si_4O_{26}$ (A=Ba, Sr) and $K_6M_6Si_4O_{26}$ (M=Nb,Ta). Mater Res Bull 11:887-894

Choisnet J, Nguyen N, Groult D, Raveau B (1977) Nonstoichiometric silicotantalates and siliconiobates. Mater Res Bull 12:91-96

Chukanov NV, Pekov IV (2005) Heterosilicates with tetrahedral-octahedral frameworks: mineralogical and crystal-chemical aspects. Rev Mineral Geochem 57:105-144

Chukanov NV, Pekov IV, Rastsvetaeva RK (2004) Crystal chemistry, properties and synthesis of microporous silicates containing transition elements. Russ Chem Rev 73:227-246

Clearfield A (2001) Structure and ion exchange properties of tunnel type titanium silicates. Sol State Sci 3: 103-112

Clearfield A, Bortun LN, Bortun AI (2000) Alkali metal ion exchange by the framework titanium silicate $M_2Ti_2O_3SiO_4·nH_2O$ (M = H, Na). React Funct Polym 43:85-95

Clearfield A, Bortun LN, Bortun AI, Poojary DM, Khainakov SA (1998) On the selectivity regulation of $K_2ZrSi_3O_9.H_2O$-type ion exchangers. J Mol Struct 470:207-213

Corcoran Jr EW, Newsam JM, King Jr HE, Vaughan DEW (1989) Phase identification of hydrothermal crystallization products from $M_2O-SiO_2-SnO_2-H_2O$ gels or solutions. In: Zeolite Synthesis. ACS Symposium Series #398. Occelli ML Robson HE (eds) American Chemical Society, Washington D.C., 603

Corcoran Jr EW, Vaughan DEW (1989) Hydrothermal synthesis of mixed octahedral-tetrahedral oxides – synthesis and characterization of sodium stannosilicates. Solid State Ionics 32/33:423-249

Corcoran Jr EW, Vaughan DEW, Eberly Jr PE (1992) Stannosilicates and preparation thereof. US patent 5 110 568

Coronas J, Santamaria J (2004) State-of-the-art in zeolite membrane reactors. Topics Catal 29:29-44

Cruciani G, De Luca P, Nastro A, Pattison P (1998) Rietveld refinement of the zorite structure of ETS-4 molecular sieve. Microporous Mesoporous Mater 21:143-153

Dadachov MS, Le Bail A (1997) Structure of zeolitic $K_2TiSi_3O_9.H_2O$ determined *ab initio* from powder diffraction data. Eur J Solid State Inorg Chem 34:381-390

Dadachov MS, Rocha J, Ferreira A, Anderson MW (1997) *Ab initio* structure determination of layered sodium titanium silicate containing edge-sharing titanate chains (AM-4) $Na_3(Na,H)Ti_2O_2[Si_2O_6]·2.2H_2O$. Chem Commun 2371-2372

Damin A, Xamena FXL, Lamberti C, Civalleri B, Zicovich-Wilson CM, Zecchina A (2004) Structural, electronic, and vibrational properties of the Ti-O-Ti quantum wires in the titanosilicate ETS-10. J Phys Chem B 108:1328-1336

Das TK, Chandwadkar AJ, Belhekar AA, Sivasanker S (1996a) Studies on the synthesis of ETS-10. II. Use of organic templates. Microporous Mater 5:401-410

Das TK, Chandwadkar AJ, Budhkar AP, Belhekar AA, Sivasanker S (1995) Studies on the synthesis of ETS-10. I. Influence of synthesis parameters and seed content. Microporous Mater 4:195-203

Das TK, Chandwadkar AJ, Sivasanker S (1996b) A rapid method of synthesizing the titanium silicate ETS-10. Chem Commun 1105-1106

Das TK, Chandwadkar AJ, Sivasanker S (1998) Aromatization of n-hexane over platinum alkaline ETS-10. In: Recent advances in basic and applied aspects of industrial catalysis. Stud Surface Sci Catal 113. Rao TSRP, Dhar GM (ed) Elsevier, Amsterdam, p 455-462

Das TK, Chandwadkar AJ, Soni HS, Sivasanker S (1996c) Studies on the synthesis, characterization and catalytic properties of the large pore titanosilicate, ETS-10. J Mol Catal A: Chem 107:199-205

Das TK, Chandwadkar AJ, Soni HS, Sivasanker S (1997) Hydroisomerization of n-hexane over Pt-ETS-10. Catal Lett 44:113-117

De Man AJM, Sauer J (1996) Coordination, structure, and vibrational spectra of titanium in silicates and zeolites in comparison with related molecules. An ab $initio$ study. J Phys Chem 100:5025-5034

Dong J, Lin YS, Hu MZC, Peascoe RA, Payzant EA (2000) Template-removal-associated microstructural development of porous-ceramic-supported MFI zeolite membranes. Microporous Mesoporous Mater 34: 241-253

Du H, Chen J, Pang W (1996) Synthesis and characterization of a novel layered titanium silicate JDF-L1. J Mater Chem 6:1827-1830

Dyer A, Jáfar JJ (1990) Novel stannosilicates. I. Synthesis and characterization. J Chem Soc Dalton Trans 3239-3242

Dyer A, Jáfar JJ (1991) Novel stannosilicates. II. Cation-exchange studies. J Chem Soc Dalton Trans 2639-2642

Dyer A, Pillinger M, Amin S (1999) Ion exchange of cesium and strontium on a titanosilicate analogue of the mineral pharmacosiderite. J Mater Chem 9:2481-2487

Evans Jr HT (1973) Crystal-structures of cavansite and pentagonite. Am Mineral Sect B 58:412-424

Ferraris G, Gula A (2005) Polysomatic aspects of microporous minerals –heterophyllosilicates, palysepioles and rhodesite-related structures. Rev Mineral Geochem 57:69-104

Ferraris G, Ivaldi G, Khomyakov AP (1995) Altisite $Na_3K_6Ti_2[Al_2Si_8O_{26}]Cl_3$ a new hyperalkaline aluminosilicate from Kola Peninsula (Russia) related to lemoynite: crystal structure and thermal evolution. Eur J Mineral 7:537-546

Ferreira A, Ananias D, Carlos LD, Morais CM, Rocha J (2003a) Novel microporous lanthanide silicates with tobermorite-like structure. J Am Chem Soc 125:14573-14579

Ferreira A, Lin Z, Rocha J, Morais C, Lopes M, Fernandez C (2001a) Ab $initio$ structure determination of a small-pore framework sodium stannosilicate. Inorg Chem 40:3330-3335

Ferreira A, Lin Z, Soares MR, Rocha J (2003b) Ab $initio$ structure determination of novel small-pore metal-silicates: knots-and-crosses structures. Inorg Chim Acta 356:19-26

Ferreira P, Ferreira A, Rocha J, Soares MA (2001b) Synthesis and structural characterization of zirconium silicates. Chem Mater 13:355-363

Fox MA, Doan KE, Dulay MT (1994) The effect of the inert support on relative photocatalytic activity in the oxidative decomposition of alcohols on irradiated titanium-dioxide composites. Research Chem Intermediates 20:711-722

Francis RJ, Jacobson AJ (2001) The first organically templated open-framework niobium silicate and germanate phase: low-temperature hydrothermal syntheses of $[(C_4N_2H_{11})Nb_3SiO_{10}]$ (NSH-1) and $[(C_4N_2H_{11})Nb_3GeO_{10}]$ (NGH-1). Angew Chem Int Ed 40:2879-2881

Gervasini A, Carniti P (2002) CuO_x sitting on titanium silicate (ETS-10): influence of copper loading on dispersion and redox properties in relation to de-NO_x activity. Catal Lett 84:235-244

Gervasini A, Picciau C, Auroux A (2000) Characterization of copper-exchanged ZSM-5 and ETS-10 catalysts with low and high degrees of exchange. Microporous Mesoporous Mater 35-36:457-469

Ghose S, Gupta PKS, Campana CF (1987) Symmetry and crystal-structure of montereginite, $Na_4K_2Y_2Si_{16}O_{38}\cdot10H_2O$, a double-sheet silicate with zeolitic properties. Am Mineral 72:365-374

Ghose S, Wan C, Chao GY (1980) Petarasite, $Na_5Zr_2Si_6O_{18}(Cl,OH)\cdot2H_2O$, a zeolite-type zirconosilicate. Can Mineral 18:503-509

Guan G, Kusakabe K, Morooka S (2001) Synthesis and permeation properties of ion-exchanged ETS-4 tubular membranes. Microporous Mesoporous Mater 50:109-120

Guan G, Kusakabe K, Morooka S (2002) Separation of nitrogen from oxygen using a titanosilicate membrane prepared on a porous alpha-alumina support tube. Separ Sci Techol 37:1031-1039

Guan GK, Kida T, Kusakabe K, Kimura K, Fang XM, Ma TL, Abe E, Yoshida A (2004) Photocatalytic H_2 evolution under visible light irradiation on CdS/ETS-4 composite. Chem Phys Lett 385:319-322

Haile SM, Wuensch BJ (2000a) Structure, phase transitions and ionic conductivity of $K_3NdSi_6O_{15}\cdot xH_2O$. I. α-$K_3NdSi_6O_{15}\cdot2H_2O$ and its polymorphs. Acta Crystallogr Sect B: Struct Sci 56:335-348

Haile SM, Wuensch BJ (2000b) Structure, phase transitions and ionic conductivity of $K_3NdSi_6O_{15}\cdot xH_2O\cdot$II. Structure of β-$K_3NdSi_6O_{15}$. Acta Crystallogr Sect B: Struct Sci 56:349-362

Harrison WTA, Gier TE, Stucky GD (1995) Single-crystal structure of $Cs_3HTi_4O_4(SiO_4)_3 \cdot 4H_2O$, A titanosilicate pharmacosiderite analog. Zeolites 15:408-412

Hesse K-F, Liebau FZ, Merlino S (1992) Crystal structure of rhodesite, $HK_{1-X}Na_{X+2Y}Ca_{2-Y}[Si_8O_{19}] \cdot (6-Z)H_2O$, from three localities and its relation to other silicates with dreier double layers. Kristallogr 199:25-48

Howe RF, Krisnandi YK (2001) Photoreactivity of ETS-10. Chem Commun 1588-1589

Huang J, Wang X, Jacobson AJ (2003) Hydrothermal synthesis and structures of the new open-framework uranyl silicates $Rb_4(UO_2)_2(Si_8O_{20})$ (USH-2Rb), $Rb_2(UO_2)(Si_2O_6) \cdot H_2O$ (USH-4Rb) and $A_2(UO_2)(Si_2O_6) \cdot 0.5H_2O$ (USH-5A; A = Rb, Cs). J Mater Chem 13:191-196

Huang J, Wang XQ, Liu LM, Jacobson AJ (2002) Synthesis and characterization of an open framework vanadium silicate (VSH-16Na). Solid State Sci 4:1193-1198

Hung L-I, Wang S-L, Kao H-M, Lii K-H (2003) Hydrothermal synthesis, crystal structure, and solid-state NMR spectroscopy of a new indium silicate: $K_2In(OH)(Si_4O_{10})$. Inorg Chem 42:4057-4061

Hung L-I, Wang S-L, Szu S-P, Hsieh C-Y, Kao H-M, Lii K-H (2004) Hydrothermal synthesis, crystal structure, solid-state NMR spectroscopy, and ionic conductivity of $Na_5InSi_4O_{12}$, a silicate containing a single 12-membered ring. Chem Mater 16:1660-1666

Ilyushin GD (1993) New data on the crystal structure of umbite $K_2ZrSi_3O_9 \cdot H_2O$. Inorg Mater 29:853-857

Jale SR, Ojo A, Fitch FR (1999) Synthesis of microporous zirconosilicates containing ZrO_6 octahedra and SiO_4 tetrahedra. Chem Commun 411-412

Jeong H-K, Chandrasekaran A, Tsapatsis M (2002a) Synthesis of a new open framework cerium silicate and its structure determination by single crystal X-ray diffraction. Chem Commun 2398-2399

Jeong H-K, Krohn J, Sujaoti K, Tsapatsis M (2002b) Oriented molecular sieve membranes by heteroepitaxial growth. J Am Chem Soc 124:12966-12968

Jeong H-K, Nair S, Vogt T, Dickinson LC, Tsapatsis M (2003) A highly crystalline layered silicate with three-dimensionally microporous layers. Nature Mater 2:53-58

Kampf AR, Khan AA, Baur WH (1973) Barium chloride silicate with an open framework – verplanckite. Acta Crystallogr Sect B: Struct Sci B29:2019-2021

Kao H-M, Lii K-H (2002) The first observation of heteronuclear two-bond J-coupling in the solid state: Crystal structure and solid-state NMR spectroscopy of $Rb_4(NbO)_2(Si_8O_{21})$. Inorg Chem 41:5644-5646

Kim WJ, Kim SD, Jung HS, Hayhurst DT (2002) Compositional and kinetic studies on the crystallization of ETS-10 in the presence of various organics. Microporous Mesoporous Mater 56:89-100

Kim WJ, Lee MC, Yoo JC, Hayhurst DT (2000) Study on the rapid crystallization of ETS-4 and ETS-10. Microporous Mesoporous Mater 41:79-88

Koudsi UY, Dyer A (2001) Sorption of ^{60}Co on a synthetic titanosilicate analogue of the mineral penkvilksite-2O and antimonysilicate. J Radioanal Nucl Chem 247:209-218

Krisnandi YK, Southon PD, Adesina AA, Howe RF (2003) ETS-10 as a photocatalyst. Intl J Photoenergy 5: 131-140

Krivovichev SV, Armbruster T (2004) The crystal structure of jonesite, $Ba_2(K,Na)[Ti_2(Si_5Al)O_{18}(H_2O)](H_2O)_n$: a first example of titanosilicate with porous double layers. Am Mineral 89:314-318

Kuznicki SM (1989) Large-pored crystalline titanium molecular sieve zeolites. US Patent 4 853 202

Kuznicki SM (1990) Preparation of small-pored crystalline titanium molecular sieve zeolites. US Patent 4 938 939

Kuznicki SM, Bell VA, Nair S, Hillhouse HW, Jacubinas RM, Braunbarth CM, Toby BH, Tsapatsis M (2001) A titanosilicate molecular sieve with adjustable pores for size-selective adsorption of molecules. Nature 412:720-724

Kuznicki SM, Curran JS, Yang X (1999) ETS-14 crystalline titanium silicate molecular sieves, manufacture and use thereof. US Patent 5 882 624

Kuznicki SM, Madon RJ, Koermer GS, Thrush KA (1991) Large-pored molecular sieves and uses thereof. EPA 0405978A1

Kuznicki SM, Thrush AK (1991a) Large-pored molecular sieves containing at least one octahedral site and tetrahedral sites of at least one type. WO91/18833

Kuznicki SM, Thrush AK (1991b) Removal of heavy metals, especially lead, from aqueous systems containing competing ions utilizing wide-pored molecular sieves of the ETS-10 type. US Patent 4 994 191

Kuznicki SM, Thrush AK (1993) Large-pored molecular sieves with charged octahedral titanium and charged tetrahedral aluminum sites. US Patent 5 244 650

Lamberti C (1999) Electron-hole reduced effective mass in monoatomic ...-O-Ti-O-TiO- ... quantum wires embedded in the siliceous crystalline matrix of ETS-10. Microporous Mesoporous Mater 30:155-163

Le Page Y, Perrault G (1976) Structure Cristalline de la lemoynite $(Na,K)_2CaZr_2Si_{10}O_{26} \cdot (5-6)H_2O$. Can Mineral 14:132-138

Li C-Y, Hsieh CY, Lin H-M, Kao H-M, Lii H-K (2002) High-temperature, high pressure hydrothermal synthesis, crystal structure, and solid state NMR spectroscopy of a new vanadium(IV) silicate: $Rb_2(VO)(Si_4O_{10}) \cdot xH_2O$. Inorg Chem 41:4206-4210

Lin Z, Rocha J, Brandão P, Ferreira A, Esculcas AP, Pedrosa de Jesus JD, Philippou A, Anderson MW (1997) Synthesis and structural characterization of microporous umbite, penkvilksite and other titanosilicates. J Phys Chem 101:7114-7120

Lin Z, Rocha J, Ferreira P, Thursfield A, Agger JR, Anderson MW (1999a) Synthesis and characterization of microporous zirconium silicates. J Phys Chem B 103:957-963

Lin Z, Rocha J, Navajas A, Téllez C, Coronas J, Santamaria J (2004) Synthesis and characterization of ETS-10 membranes. Microporous Mesoporous Mater 67:79-86

Lin Z, Rocha J, Pedrosa de Jesus JD, Ferreira A (2000) Synthesis and structure of novel microporous framework stannosilicate. J Mater Chem 10:1353-1356

Lin Z, Rocha J, Valente A (1999b) Synthesis and characterization of a framework microporous stannosilicate. Chem Commun 2489-2490

Liu X, Shang M, Thomas JK (1997a) Synthesis and structure of a novel microporous titanosilicate (UND-1) with a chemical composition of $Na_{2.7}K_{5.3}Ti_4Si_{12}O_{36} \cdot 4H_2O$. Micropor Mater 10:273-281

Liu X, Thomas JK (1996) Synthesis of microporous titanosilicates ETS-10 and ETS-4 using solid TiO_2 as the source of titanium. Chem Commun 1435-1436

Liu Y, Du H, Xiao F, Zhu G, Pang W (2000) Synthesis and characterization of a novel microporous titanosilicate JLU-1. Chem Mater 12:665-670

Liu Y, Du H, Xu Y, Ding H, Pang W, Yue Y (1999) Synthesis and characterization of a novel microporous titanosilicate with a structure of penkvilksite-*1M*. Microporous Mesoporous Mater 28:511-517

Liu Y, Du H, Zhou F, Pang W (1997b) Synthesis of a new titanosilicate: an analogue of the mineral penkvilksite. Chem Commun 1467-1468

MacDonald AM, Chao GY (2004) Haineaultite, a new hydrated sodium calcium titanosilicate from Mont Saint-Hilaire, Quebec: description, structure determination and genetic implications. Can Mineral 42:769-780

Maurice OD (1949) Transport and deposition of the non-sulfide vein minerals. V. Zirconium minerals. Econ Geol 44:721-731

Merinov BV, Maksimov BA, Belov NV (1980) The crystal structure of sodium scandium silicate $Na_5ScSi_4O_{12}$. Doklady Akademii Nauk SSSR 255:577-582 (in Russian)

Mer'kov AN, Bussen IV, Goiko EA, Kul'chitskaya EA, Men'shikov YuP, Nedorezova AP (1973) Raite and zorite, new minerals from the Lovozero tundra. Zap Vses Mineral Obs 102:54-62 (in Russian)

Merlino S, Bonaccorsi E, Armbruster T (2001) The real structure of tobermorite 11 Å: normal and anomalous forms, OD character and polytypic modifications. Eur J Mineral 13:577-590

Merlino S, Pasero M, Artioli G, Khomyakov AP (1994) Penkvilksite, a new kind of silicate structure – OD character, X-ray single-crystal (1M), and powder rietveld (2*O*) refinements of 2 MDO polytypes. Am Mineral 79:1185-1193

Mihailova B, Valtchev V, Minatova S, Konstantinov L (1996) Vibrational spectra of ETS-4 and ETS-10. Zeolites 16:22-24

Miraglia PQ, Yilmaz B, Warzywoda J, Bazzana S, Sacco Jr A (2004) Morphological and surface analysis of titanosilicate ETS-4 synthesized hydrothermally with organic precursors. Microporous Mesoporous Mater 69:71-76

Moller T, Harjula R, Lehto J (2002) Ion exchange of ^{85}Sr, ^{134}Cs and ^{57}Co in sodium titanosilicate and the effect of crystallinity on selectivity. Separation Purification Technol 28:13-23

Nair S, Jeong HK, Chandrasekaran A, Braunbarth CM, Tsapatsis M, Kuznicki SM (2001) Synthesis and structure determination of ETS-4 single crystals. Chem Mater 13:4247-4254

Nyman M, Tripathi A, Parise JB, Maxwell RS, Harrison WTA, Nenoff TM (2001) A new family of octahedral molecular sieves: Sodium Ti/Zr-IV niobates. J Am Chem Soc 123:1529-1530

Nyman M, Tripathi A, Parise JB, Maxwell, RS, Nenoff TM (2002) Sandia octahedral molecular sieves (SOMS): Structural and property effects of charge-balancing the M-IV-substituted (M = Ti, Zr) niobate framework. J Am Chem Soc 124:1704-1713

Pavel CC, Vuono D, Nastro A, Nagy JB, Bilba N (2002) Synthesis and ion exchange properties of the ETS-4 and ETS-10 microporous crystalline titanosilicates. *In*: Impact of Zeolites and Other Porous Materials on the New Technologies at the Beginning of then Millennium. Studies Surface Sci Catal Vol 142 (A,B). Aiello R, Giordano G, Testa F (eds) Elsevier, Amsterdam, p 295-302

Perrault PG, Boucher C, Vicat J, Cannillo E, Rossi G (1973) Crystal-structure of nenadkevichite - $(Na,K)_{2-x}(Nb,Ti)(O,OH)Si_2O_6 \cdot 2H_2O$. Acta Crystallogr Sect B: Struct Sci 29:1432-1438

Philippou A, Anderson MW (1996) Structural investigation of ETS-4. Zeolites 16:98-107

Philippou A, Anderson MW (2000) Aldol-type reactions over basic microporous titanosilicate ETS-10 type catalysts. J Catal 189:395-400

Philippou A, Brandão P, Ghanbari-Siakhali A, Dwyer J, Rocha J, Anderson MW (2001) Catalytic studies of the novel microporous niobium silicate AM-11. Appl Catal A 207:229-238

Philippou A, Naderi M, Pervaiz N, Rocha J, Anderson MW (1998a) n-hexane reforming reactions over basic Pt-ETS-10 and Pt-ETAS-10. J Catal 178:174-181

Philippou A, Naderi M, Rocha J, Anderson MW (1998b) Dehydration of t-butanol over basic ETS-10, ETAS-10 and AM-6 catalysts. Catal Lett 53:221-224

Philippou A, Rocha J, Anderson MW (1999) The strong basicity of the microporous titanosilicate ETS-10. Catal Lett 57:151-153

Poojary DM, Bortun AI, Bortun LN, Clearfield A (1996) Structural studies on the ion-exchanged phases of a porous titanosilicate, $Na_2Ti_2O_3SiO_4 \cdot 2H_2O$. Inorg Chem 35:6131-6139

Poojary DM, Bortun AI, Bortun LN, Clearfield A (1997) Syntheses and X-ray powder structures of $K_2ZrSi_3O_9 \cdot H_2O$ and its ion-exchanged phases with Na and Cs. Inorg Chem 36:3072-3079

Poojary DM, Cahill RA, Clearfield A (1994) Synthesis, crystal-structures, and ion-exchange properties of a novel porous titanosilicate. Chem Mater 6:2364-2368

Puziy AM (1998) Cesium and strontium exchange by the framework potassium titanium silicate $K_3HTi_4O_4(SiO_4)_3 \cdot 4H_2O$. J Radioanal Nucl Chem 237:73-79

Rainho JP, Ananias D, Lin Z, Ferreira A, Carlos LD, Rocha J (2004) Photoluminescence and local structure of Eu(III)-doped zirconium silicates. J Alloys Compd 374:185-189

Rainho JP, Carlos LD, Rocha J (2000) New phosphors based on Eu^{3+}-doped microporous titanosilicates. J Lumin 87-89:1083-1086

Rainho JP, Pillinger M, Carlos LD, Ribeiro SJL, Almeida RM, Rocha J (2002) Local Er(III) environment in luminescent titanosilicates prepared from microporous precursors. J Mater Chem 12:1162-1168

Rastsvetaeva RK, Andrianov VI (1984) Refined crystal-structure of vinogradovite. Sov Phys Crystallogr 29:403-406

Rinaldi R, Pluth JJ, Smith JV (1975) Crystal-structure of cavansite dehydrated at 220°C. Acta Crystallogr Sect B: Struct Sci 31:1598-1602

Roberts MA, Sankar G, Thomas JM, Jones RH, Du H, Fang M, Chen J, Pang W, Xu R (1996) Synthesis and structure of a layered titanosilicate catalyst with five-coordinate titanium. Nature 381:401-404

Rocha J, Anderson MW (2000) Microporous titanosilicates and other novel mixed octahedral-tetrahedral framework oxides. Eur J Inorg Chem 801-818

Rocha J, Brandão P, Anderson MW, Ohsuna T, Terasaki O (1998a) Synthesis and characterization of microporous titano-borosilicate ETBS-10. Chem Commun 667-668

Rocha J, Brandão P, Lin Z, Anderson MW, Alfredsson V, Terasaki O (1997a) The First example of a microporous framework vanadosilicate containing hexacoordinated vanadium. Angew Chem, Int Ed Engl 36:100-102

Rocha J, Brandão P, Lin Z, Esculcas AP, Ferreira A, Anderson MW (1996a) Synthesis and structural studies of microporous titanium-niobium-silicates with the structure of nenadkevichite. J Phys Chem 100:14978-14983

Rocha J, Brandão P, Lin Z, Kharlamov A, Anderson MW (1996b) Novel microporous titanium-niobium-silicates with the structure of nenadkevichite. Chem Commun 669-670

Rocha J, Brandão P, Pedrosa de Jesus JD, Philippou A, Anderson MW (1999) Synthesis and characterization of microporous titanoniobosilicate ETNbS-10. Chem Commun 471-472

Rocha J, Brandão P, Philippou A, Anderson MW (1998b) Synthesis and characterization of a novel microporous niobium silicate catalyst. Chem Commun 2687-2688

Rocha J, Carlos LD (2003) Microporous materials containing lanthanide metals. Curr Opinion Solid State Mater Sci 7:199-205

Rocha J, Carlos LD, Ferreira A, Rainho J, Ananias D, Lin Z (2004a) Novel microporous and layered luminescent lanthanide silicates. Mater Sci Forum 455 1561527 531

Rocha J, Carlos LD, Paz FAA, Ananias D, Klinowski J (2004b) Novel microporous luminescent europium(III) silicate. In: Recent advances in the science and technology of zeolites and related materials. Studies Surface Sci Catal Vol 154 (A,B,C) Van Steen E, Callanan LH, Claeys M (eds) Elsevier, Amsterdam, p 3028-3034

Rocha J, Carlos LD, Rainho JP, Lin Z, Ferreira P, Almeida RM (2000a) Photoluminescence of new Er^{3+}-doped titanosilicate materials. J Mater Chem 10:1371-1375

Rocha J, Ferreira A, Lin Z, Agger JR, Anderson MW (1998c) Synthesis and characterization of a microporous zirconium silicate with the structure of petarasite. Chem Commun 1269-1270

Rocha J, Ferreira A, Lin Z, Anderson MW (1998d) Synthesis of microporous titanosilicate ETS-10 from $TiCl_3$ and TiO_2: A comprehensive study. Microporous Mesoporous Mater 23:253-263

Rocha J, Ferreira P, Carlos LD, Ferreira A (2000b) The first microporous framework cerium silicate. Angew Chem Int Ed 39:3276-3279

Rocha J, Ferreira P, Lin Z, Brandão P, Ferreira A, Pedrosa de Jesus JD (1997b) The first synthetic microporous yttrium silicate containing framework sodium ions. Chem Commun 2103-2104

Rocha J, Ferreira P, Lin Z, Brandão P, Ferreira A, Pedrosa de Jesus JD (1998e) Synthesis and characterization of microporous yttrium and calcium silicates. J Phys Chem 102:4739-4744

Rocha J, Lin Z, Ferreira A, Anderson MW (1995) Ga, Ti avoidance in the microporous titanogallosilicate ETGS-10. J Chem Soc Chem Commun 867-868

Ruiz JAC, Ruiz VSO, Airoldi C, Pastore HO (2004a) Thermochemistry of n-alkylamines interaction with ETS-10 titaniunsilicate. Thermochim Acta 411:133-138

Ruiz JAC, Ruiz VSO, Airoldi C, Pastore HO (2004b) Total acidity calculation for ETS-10 by calorimetry, thermogravimetry and elemental analysis. Appl Catal A-General 261:87-90

Sandomirskii PA, Belov NV (1979) The OD structure of zorite. Sov Phys Crystallogr 24:686-693

Sankar G, Bell RG, Thomas JM, Anderson MW, Wright PA, Rocha J, Ferreira A (1996) Determination of the structure of distorted TiO_6 units in the titanosilicate ETS-10 by a combination of X-ray absorption spectroscopy and computer modeling. J Phys Chem 100:449-452

Serralha FN, Lopes JM, Lemos F, Prazeres DMF, Aires-Barros MR, Rocha J, Cabral JMS, Ramôa Ribeiro F (2000) Titanosilicates as supports for an enzymatic alcoholysis reaction. React Kinet Catal Lett 69: 217-222

Shumyatskaya NG, Voronkov AA, Pyatenko YA (1980) Sazhinite $Na_2Ge[Si_6O_{14}(OH)]\cdot nH_2O$: a new representative of the dalyite family in crystal chemistry. Sov Phys Crystallogr 25:419-423

Smith JV (1988) Topochemistry of zeolites and related materials. I. Topology and geometry. Chem Rev 88: 149-182

Solbra S, Alison N, Waite S, Mikhalovsky SV, Bortun AI, Bortun LN, Clearfield A (2001) Cesium and strontium ion exchange on the framework titanium silicate $M_2Ti_2O_3SiO_4\cdot nH_2O$ (M = H, Na). Environ Sci Technol 35:626-629

Southon PD, Howe RF (2002) Spectroscopic studies of disorder in the microporous titanosilicate ETS-10. Chem Mater 14:4209-4218

Suib SL (1998) Microporous manganese oxides. Curr Opinion Solid State Mater Sci 3:63-70

Suppes GJ, Dasari MA, Doskocil EJ, Mankidy PJ, Goff MJ (2004) Transesterification of soybean oil with zeolite and metal catalysts. Appl Catal A-General 257:213-223

Sylvester P, Behrens EA, Graziano GM, Clearfield A (1999) An assessment of inorganic ion-exchange materials for the removal of strontium from simulated Hanford tank wastes. Sep Sci Technol 34:1981-1992

Tsapatsis M (2002) Molecular sieves in the nanotechnology era. AIChE Journal 48:654-660

Uma S, Rodrigues S, Martyanov IN, Klabunde KJ (2004) Exploration of photocatalytic activities of titanosilicate ETS-10 and transition metal incorporated ETS-10. Microporous Mesoporous Mater 67: 181-187

Valente A, Lin Z, Brandão P, Portugal I, Anderson MW, Rocha J (2001) Gas phase oxidative dehydrogenation of cyclohexanol over ETS-10 and related materials. J Catal 200:99-105

Valtchev V, Mintova S (1994) Synthesis of titanium silicate ETS-10 – the effect of tetramethylammonium on the crystallization kinetics. Zeolites 14:697-700

Valtchev V, Paillaud J-L, Mintova S, Kessler H (1999) Investigation of the ion-exchanged forms of the microporous titanosilicate $K_2TiSi_3O_9\cdot H_2O$. Microporous Mesoporous Mater 32:287-296

Valtchev VP (1994) Influence of different organic-bases on the crystallization of titanium silicates ETS-10. J Chem Soc Chem Commun 261-262

Waghmode SB, Das TK, Vetrivel R, Sivasanker S (1999) Influence of the nature of the exchanged ion on n-hexane aromatization activity of Pt-M-ETS-10: *Ab initio* calculations on the location of Pt. J Catal 185: 265-271

Waghmode SB, Thakur VV, Sudalai A, Sivasanker S (2001) Efficient liquid phase acylation of alcohols over basic ETS-10 molecular sieves. Tetra Lett 42:3145-3147

Waghmode SB, Wagholikar SG, Sivasanker S (2003) Heck reaction over Pd-loaded ETS-10 molecular sieve. Bull Chem Soc Japan 76:1989-1992

Wang X, Huang J, Liu L, Jacobson AJ (2002a) The novel open-framework uranium silicates $Na_2(UO_2)(Si_4O_{10})\cdot 2.1H_2O$ (USH-1) and $RbNa(UO_2)(Si_2O_6)\cdot H_2O$ (USH-3). J Mater Chem 12:406-410

Wang X, Huang J, Liu L, Jacobson AJ (2002b) $[(CH_3)_4N][(C_5H_5NH)_{0.8}(CH_3)_3(NH)_{0.2}]U_2Si_9O_{23}F_4$ (USH-8): An organically templated open-framework uranium silicate. J Am Chem Soc 124:15190-15191

Wang XQ, Jacobson AJ (1999) Crystal structure of the microporous titanosilicate ETS-10 refined from single crystal X-ray diffraction data. Chem Commun 973-974

Wang XQ, Liu LM, Jacobson AJ (2001) The novel open-framework vanadium silicates $K_2(VO)(Si_4O_{10})\cdot H_2O$ (VSH-1) and $Cs_2(VO)(Si_4O_{10})\cdot H_2O$ (VSH-2). Angew Chem Int Ed 40:2174-2176

Wang XQ, Liu LM, Jacobson AJ (2002c) Open-framework and microporous vanadium silicates. J Am Chem Soc 124:7812-7820

Wang XQ, Liu LM, Jacobson AJ (2003) Nanoporous copper silicates with one-dimensional 12-ring channel system. Angew Chem Int Ed 42:2044-2047

Xamena FXLI, Calza P, Lamberti C, Prestipino C, Damin A, Bordiga S, Pellizzetti E, Zecchina A (2003) Enhancement of the ETS-10 titanosilicate activity in the shape-selective photocatalytic degradation of large aromatic molecules by controlled defect production. J Am Chem Soc 125:2264-2271

Xamena FXLI, Zecchina A (2002) FTIR spectroscopy of carbon dioxide adsorbed on sodium- and magnesium-exchanged ETS-10 molecular sieves. Phys Chem Chem Phys 4:1978-1982

Xu H, Zhang YP, Navrotsky A (2001) Enthalpies of formation of microporous titanosilicates ETS-4 and ETS-10. Microporous Mesoporous Mater 47:285-291

Xu Y, Ching WY, Gu ZQ (1997) Electronic structure of microporous titanosilicate ETS-10. Ferroelectrics 194:219-226

Yang X, Paillaud JL, Breukelen HFWJ, Kessler H, Duprey E (2001) Synthesis of microporous titanosilicate ETS-10 with TiF_4 or TiO_2. Microporous Mesoporous Mater 46:1-11

Zecchina A, Xamena FXLI, Paze C, Palomino GT, Bordiga S, Arean CO (2001) Alkyne polymerization on the titanosilicate molecular sieve ETS-10. Phys Chem Chem Phys 3:1228-1231

Zhao GXS, Lee JL, Chia PA (2003) Unusual adsorption properties of microporous titanosilicate ETS-10 toward heavy metal lead. Langmuir 19:1977-1979

The Sodalite Family – A Simple but Versatile Framework Structure

Wulf Depmeier

Institut für Geowissenschaften
Universität Kiel
Olshausenstr. 40, D-24098 Kiel, Germany
wd@min.uni-kiel.de

ABSTRACT

A particular family of crystalline microporous solids, the so-called sodalite family, is presented in various chemical, topological, geometrical, and crystallographic aspects. It is shown that the sodalite structure can be decomposed into three partial structures, namely i) the sodalite framework of all-corner connected TO_4 tetrahedra, ii) a virtual lattice of cage cations, and iii) isolated cage anions. Important properties and features can be explained by interactions of the partial structures. In some cases the interactions are frustrated, resulting in the formation of modulated structures. Substitution effects occur on all three partial structures and occasionally have dramatic effects on phase transitions and phase diagrams.

INTRODUCTORY REMARKS

A functional chemical formula

In classical textbooks on mineralogy or chemistry the chemical formula for the mineral sodalite can be found written in different ways. Usually, it is given in an idealized way, i.e., substitution effects are neglected and the stoichiometric coefficients refer to a cubic unit cell having a lattice parameter of about 9 Å. The formula thus given may then read such as $Na_8(AlO_2)_6(SiO_2)_6Cl_2$, or $Na_8Cl_2Al_6Si_6O_{24}$, or, especially in the more recent literature, as $Na_8[Al_6Si_6O_{24}]Cl_2$. Whereas the former formulae represent merely a kind of inventory of the chemical species present in the unit cell, the latter notation provides considerably more information as it assigns a functionality to the different chemical species. The square brackets tell there is a framework present with a composition given within the brackets. The chemical symbols outside of the brackets pertain to species which are not part of the framework. They are often referred to as guests, because they reside in the pores of the framework which in turn is considered the host. In the particular case considered here it is only from the context that one knows the framework present is that of sodalites.

The sodalite framework is a tetrahedral framework—it is built from all-corner-connected tetrahedra. Tetrahedra frameworks exist with many different linkage patterns. The Atlas of Zeolite Framework Types (website maintained by Baerlocher and McCusker; http://www.iza-structure.org/databases/), hereafter the "ATLAS" for short, lists not less than 152 different topologies ("zeolite framework types"), at the time of writing. Of these a certain number may exist with the same host composition $[Al_6Si_6O_{24}]^{6-}$. Besides sodalite, these are e.g., zeolite A, cancrinite, zeolite P or zeolite X. There is thus an ambiguity with respect to the framework type for a given composition. This can be resolved by suitably specifying the framework topology. A convenient way of doing so is by labeling the formula with the three-letter IZA code which is also used as entry in the ATLAS. This code is attributed to each unique zeolite framework type upon acceptance by the IZA, the International Zeolite Association. Thus, Na$_8$[Al$_6$Si$_6$O$_{24}$]Cl$_2$-**SOD** indicates the mineral sodalite sufficiently, at least up to a certain level of approximation. In an effort to find a remedy also for other kinds of deficiencies when describing framework types, a commission was charged to define a stringent nomenclature for ordered microporous and mesoporous materials, because expressions such as cages, pores, channels etc. had previously been used in a rather loose manner (McCusker et al. 2001). The commission was also charged to improve the rudimentary functional crystal chemical formula as given above. The results are discussed in detail in this volume (McCusker 2005), which allows us to restrict ourselves to a minimum of explanation.

In these pseudo-chemical formulae bold letters, such as **T**, are used as so-called structure-site symbols. These symbols indicate the occupation of specific sites in a structure type without explicitly specifying the exact chemical nature of the respective species. Such a structure-site can be occupied by various chemical species, occasionally, but not necessarily, in form of a substitution. By way of contrast, normal letters indicate specific chemical elements. Different typographical symbols are used to separate the guest species from the host. The possible guest species are cations **A**, anions **X** and neutral molecules **M**; they are listed inside vertical bars together with their stoichiometric coefficients: $|A_aX_xM_m|_n$. The part of the formula pertaining to the host species is enclosed in square brackets. Central and peripheral host atoms, $_{ce}$**H** and $_{pe}$**H**, are distinguished. In the sodalite structure type these are normally the **T** and O atoms, respectively. **T** atoms are tetrahedrally coordinated, small, highly charged atoms such as Si^{4+} or Al^{3+}.

In some frameworks the host may also contain interstitial species ($_i$**A**, $_i$**X**, $_i$**M**), where **A**, **X**, **M** have the same meaning as before, and the subindex i indicates interstitial. Such species are by definition located in voids of the framework having a diameter of less than 2.5 Å. Interstitial species are not essential for the host function of the framework, and very often they are simply absent. This seems to be the case for most, if not all, members of the sodalite family, because the sodalite framework does not provide the required voids *within* its thin wall, the thickness of which corresponds to the diameter of only one **T**O$_4$ tetrahedron. Of course, the wall encloses the much larger sodalite cages, to be discussed further on.

If desired or deemed useful, additional information can be added to the functional formula, e.g., on oxidation states, coordination numbers, or other meaningful crystallographic or crystal chemical descriptors. However, in order to avoid information overflow, the formula is preferably restricted to a minimum of relevant information for the respective purpose. According to these rules the mineral sodalite is denoted as |Na$_8$Cl$_2$|[Al$_6$Si$_6$O$_{24}$]-**SOD**, and the formula of template-free silica sodalite is [SiO$_2$]-**SOD**.

Poroates, zeoates, clathrates

Some people consider the sodalite framework as an archetypical zeolite framework. However, there are scientific communities which feel impelled to exclude sodalites from their classification of zeolite-type structures for reasons of definitions given by themselves, and much to their regret as it seems to have happened in some cases. The problem arises when definitions are used which are based on the traditional classification of natural zeolites in which the release or uptake of water, or other species, or associated topological features (apertures), are decisive. In some cases the rather curious situation arose that the sodalite structure type in effect should have been excluded on grounds of defining conditions, but was accepted anyhow; perhaps as a kind of honorary zeolite framework type? The various definitions, and differences between them, have been discussed elsewhere (Depmeier 2002).

While being basically logic and also versatile, some people feel that the new IUPAC definition and nomenclature suffers from at least one inconsistency. This is due to the fact that it is biased in aid of the traditional zeolites, admittedly in recognition of their long-standing scientific history and technical importance. While agreeing with most aspects of the proposed nomenclature and the construction of the functionalized formula, Liebau (2003) took issue with the bias. In particular, he argued that framework structures which do, in fact, contain guest species in their micropores, but do not allow for their diffusion because the guests are trapped in their respective cages (where they might be highly mobile anyhow without being able to leave the cage), may claim scientific and technical interest in their own right. Such compounds are sometimes called inclusion compounds, especially in organic chemistry. Liebau (2003) proposed the generic name poroates for all structures holding pores with diameters exceeding 2.5 Å. He further subdivided poroates into zeoates and clathrates with the distinction that the pores of the former should be accessible for migrating ions or molecules, whereas those of the latter should not. Liebau (2003) argued that the importance and numerous applications of zeoates—for catalysis, crude oil cracking, shape selective reactions, as molecular sieves, drying agents, etc., and mostly under the popular name zeolites—call for the accessibility of their pore systems, whereas clathrates bear potential for applications exploiting exactly the inaccessibility of their pores. This feature makes it possible to encapsulate, and thus rendering harmless, dangerous species, such as radioactive isotopes, or to use organic clathrates for the controlled sustained release of drugs. Furthermore, Liebau (2003) proposed a nomenclature which allows the easy identification of the framework composition from the name. Thus, porosils and poroals denote poroates which contain exclusively Si or Al, respectively, as **T** cations in their oxidic framework, whereas porolites contain both, Si and Al. According to this nomenclature, the familiar name zeolite should be reserved for microporous oxidic framework structures having accessible pores and Al and Si as **T** cations. Because of its, in general, inaccessible cages $|Na_8Cl_2|[Al_6Si_6O_{24}]$-**SOD**, would be called a clathralite, and aluminate sodalites, e.g., $|Ca_8(WO_4)_2|[Al_{12}O_{24}]$-**SOD**, as clathrals.

Very recently, Liebau (2004) recognized that the comprehension of certain host-guest structures may benefit from their description as a hierarchy of hosts and guests, and denoted such materials as n^{th}-order poroates. This expression should emphasize that a given host structure accommodates guests which in turn serve as host for a second generation of guests, and so forth. In an nth-order poroate $(n-1)$ generations are porous, whereas the nth generation is non-porous. The functionalized IUPAC-formula can be easily extended to cover n^{th}-order poroates. Some sodalite framework structures can be considered as higher order clathrates (see below). Liebau (2004) also suggested a notation corresponding to the one which is used for endofullerenes. Thus, the "@" symbol used emphasizes both a structurally based functionality and possible applications.

Basic structural features and chemistry of sodalite-type structures

Sodalites belong to the class of crystalline microporous tetrahedral framework structures, which is another way of saying that $T_{pe}H_4$-tetrahedra are connected via common $_{pe}H$ atoms at the corners of tetrahedra centered by **T** which span a 3D framework structure enclosing voids of at least 2.5 Å diameter Ø. In "normal" sodalites, such as oxidic sodalites, **T** represents small, usually highly charged cations, mostly Si^{4+} or Al^{3+}. $_{pe}H$ is, of course, O^{2-}.

The lower limit of Ø has been chosen because only voids having at least this diameter allow the accommodation of small molecules like H_2O, whereas those with a smaller diameter do not. According to a IUPAC proposal, voids with 2.5 Å < Ø < 20 Å are called micropores; mesopores have 20 Å ≤ Ø ≤ 500 Å, and the diameters of macropores exceed 500 Å (Rouquerol et al. 1994). The pores in sodalite-type structures have typical diameters of about 4 Å, they are thus called micropores. In the professional jargon these pores are often denoted as sodalite cages or β-cages. As a first approximation the sodalite framework can be thought of as being composed of a space-filling arrangement of these sodalite-cages joined via common 4-rings

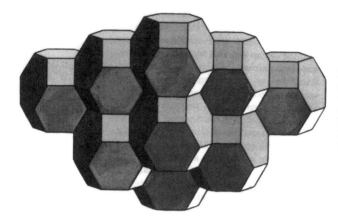

Figure 1. The space-filling packing of regular truncated octahedra is a model of the sodalite framework. [Used by permission of Princeton University Press, from Weyl (1952), Fig. 56, p. 92.]

of **T** atoms along the cubic <100>, and via common 6-rings along the cubic <111> directions. An *n*-ring consists of *n* **T** atoms linked by O atoms, so that it contains 2*n* atoms. This simple, highly symmetric, straightforward spatial arrangement has fascinated many people, last but not least because of its aesthetic appeal, and numerous illustrations can be found in the literature. An example from Weyl's (1952) famous book on symmetry is given in Figure 1.

The sodalite cages completely define the 3D sodalite framework in the sense that the latter is identical with the walls of the cages. It depends, in fact, on the purpose whether one considers the sodalite framework as a solid framework of corner-connected tetrahedra enclosing the sodalite cages, or as an arrangement of cages separated by the thin-walled tetrahedral framework. Both views are, of course, complementary to each other.

As mentioned before, most sodalite frameworks contain guests. For some purposes it is convenient to consider the guests as self-contained entities, and the entire (classical) sodalite structure as being composed of three partial structures—the **SOD** framework, a virtual lattice spanned by the **A** cations, and the guest anions **X**. Figure 2 shows a regular truncated octahedron representing the **SOD** framework interpenetrated by the **A** cation lattice (Depmeier 1983). Of course, the differences between the **SOD** framework and the **A** cation lattice should be obvious, as the **SOD** framework is hold together by strong chemical bonds, whereas the **A** cations do not form bonds between each other. If one accepts the existence of an **A** cation lattice and tries to rank the partial structures in the sense of Liebau's (2004) n^{th}-order poroates, one encounters an interesting situation, because the lattice of **A** cations and the **SOD** framework interpenetrate each other without revealing any priority, and the anions **X** have to be considered as common guests in two host lattices.

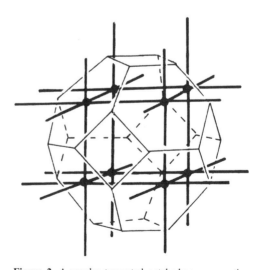

Figure 2. A regular truncated octahedron representing the topological sodalite framework, interpenetrated by a lattice of cage cations.

In most cases of oxidic sodalites the framework is negatively charged, only in the case of silica sodalite, [Si$_{12}$O$_{24}$]-**SOD**, or [Al$_6$P$_6$O$_{24}$]-**SOD**, the framework is electrically neutral. Usually, the sodalite cages are occupied by various guests. Guest species can be simple or complex ions, neutral or charged molecules, or even small atomic clusters. One of the most remarkable features of the sodalite family is that it represents one of the few examples of microporous structures where negatively charged species occur as possible guests, other examples being cancrinite, tschoertnerite and scapolite. In this respect it is also interesting that not only ordinary anions, such as Cl$^-$ or SO$_4^{2-}$, occur as guests, but that one may also encounter rather exotic species, such as otherwise unstable radical anions, or even unpaired electrons.

Sodalites with empty cages have been prepared by the calcination of silica sodalites containing organic templates (Braunbarth et al. 1997). The preparation is not straightforward because the products of oxidation or evaporation of the organic templates do not easily pass through the narrow apertures of the sodalite cages, in accordance with the clathrate character of sodalite.

The compactness of the sodalite structure is, in fact, a matter of particular importance for its properties, and sets it apart from more open microporous, let alone mesoporous, structures. The narrow width of the pores results in short interatomic distances between the framework and its guests and, hence, in relatively strong interactions. Furthermore, it implies a strongly corrugated potential hyperplane of the framework which means that the guests have to match more or less exactly the rigid constraints set by the host's geometry and charge. Thus, rather strong boundary conditions determine possible combinations of host and guest species. The interactions require sometimes the adaptation of the framework geometry to that of the guest species, which may, or may not, break the latent cubic symmetry of the sodalite concerned.

The forces which bind the guests **A**, **M** and **X** to their sodalite host are usually weaker than the strong chemical bonds within the framework which are normally of ionic-covalent character. As a consequence, the cohesion between framework and guests is rather easily broken, notably upon heating. When this happens the guests gain kinetic energy and become mobile. However, because the small aperture of the sodalite cages prevents the guests largely from translational movements, such as migration, they remain trapped in their respective micropores. In these sodalite cages, the guests may use their kinetic energy to vibrate, jump or pivot about their center according to the temperature and depending on the available free space. Despite the small aperture of the sodalite cages, and somehow in contradiction with the classification of sodalites as clathrates, some ion exchange has been observed indeed, but only to a relatively small extent at elevated temperatures and virtually negligible at room temperature. Table 1 provides an exemplary, rather than exhaustive, listing of chemical species found as hosts and guests in sodalite-type structures.

Substitution has been observed in all three partial structures. The particular effect of Al-Si order-disorder, and the so-called Loewenstein's rule will be discussed further below. Substitution on the **A** or **X** sites may change substantially the delicate equilibrium of interactions between the partial structures, and thus result in complex symmetry relations and phase transitions. It happens quite often that the symmetry of the guest species and the site symmetry in the overall space group do not match, resulting in static or dynamic disorder of the guests.

"Classical" sodalites are built from oxidic frameworks, where the peripheral framework atoms, $_{pe}$**H**, are O atoms. As shown before, there exist also sodalites with other kinds of $_{pe}$**H** atoms. Some new compositions have been prepared on purpose. One of the reasons is probably this: In the past few decades, oxides have found much more attention, and hence were much more intensively studied than the heavier homologues, such as sulfides. As a consequence, our present knowledge of the higher chalcogenides lags behind that of oxides. Two major reasons

Table 1. An exemplary, rather than exhaustive, listing of chemical species found as hosts and guests in sodalite-type structures.

Oxidic frameworks (partial list only):

$[Al_6Si_6O_{24}]^{6-}$ $[Al_6Ge_6O_{24}]^{6-}$
$[Al_2Si_{10}O_{24}]^{2-}$ $[Al_3Ga_3Si_6O_{24}]^{6-}$
$[Al_8Si_4O_{24}]^{8-}$ $[Ga_6Ge_6O_{24}]^{6-}$
$[Al_{8+x}Si_{4-x}O_{24}]^{(8+x)-}$ $[Al_2Be_2Si_8O_{24}]^{6-}$
$[Si_{12}O_{24}]$ $[Be_6Si_6O_{24}]^{12-}$
$[Al_{12}O_{24}]^{12-}$ $[Ga_6Si_6O_{24}]^{6-}$
$[B_{12}O_{24}]^{12-}$ $[Al_6P_6O_{24}]$ and variants

Non-oxidic frameworks:

$[P_{12}N_{24}]^{12-}$ $[P_{12}N_{18}O_6]^{6-}$ $[Cu_2Zn_{10}Cl_{24}]^{2-}$

Hydrogen-bonded framework:

Tetramethylammonium hydroxide pentahydrate, $(CH_3)_4NOH \cdot 5H_2O$, (McMullan et al. 1966)

Guest cations:

Na^+, Ca^{2+}, Li^+, K^+, Zn^{2+}, Fe^{2+}, Mn^{2+}, Cd^{2+},
Ag^+, REE (rare earth elements), $(CH_3)_4N^+$, $(CH_3)_3NH^+$

Guest anions:

Cl^-, Br^-, I^-, SO_4^{2-}, WO_4^{2-}, MoO_4^{2-}, CrO_4^{2-}, S_2^-, S_3^-,
CO_3^{2-}, OH^-, ClO_3^-, MnO_4^-, SCN^-, NO_2^-, HCO_2^-, BO_2^-, $B(OH)_4^-$

Unpaired electrons as guests:

e^-

Neutral guest species:

H_2O, Pb_4O_4 (Scheikowski and Müller-Buschbaum 1993), ethylene glycol (Bibby and Dale 1985) 1,3,5-trioxane (Keijsper et al. 1989), 1,3-dioxolane (van de Goor et al. 1994) ethanol amine (Braunbarth et al. 1996), ethylene diamine (Braunbarth et al. 1996), etc.

---

for the different intensity with which the different classes were studied may be i) the much more widespread use of oxides in traditional techniques such as catalysis, ceramics or glasses, and ii) probably the proven potential of many oxides for uses in modern high technology applications, exploiting physical phenomena such as high temperature supraconductivity, ionic conductivity, ferroelectricity, ferroelasticity, or giant magnetic resistance, to mention just a few. The great demand has required the employment of the full range of synthetic and analytical tools and methods of solid state sciences; some methods were even specifically developed for that purpose.

At present, there is growing evidence for increasingly higher barriers, or even possible dead ends, on the oxide route, despite the still many successful applications, and continuing high level of research activity. In order to cope with future needs of modern civilization, it is appropriate to study other classes of materials with an adequate effort.

Sulfur is the higher homologue of the lightest chalcogen, viz. oxygen. Despite their affiliation to the same group of the periodic table a barrier divides the oxides from the sulfides (and selenides). This is demonstrated by the fact that the structure of most sulfides, while being isotypic with that of the corresponding selenides, normally differs fundamentally from that of the corresponding oxide. This is attributable to the more covalent character of the chemical bonds in sulfides/selenides, compared to the more ionic nature of the corresponding oxides. The two classes differ also with respect to their electronic band structures. The higher chalcogenides have the tendency to be in, or to transform more easily into, the electronically semi-conducting or even conducting state, whereas the oxides are chiefly insulators, or ionic conductors. Of course, many other physical properties differ as well. Because of the similarities and despite of the differences, sulfides represent a natural choice for the search of possible alternatives to the oxide route. Some activities in the field of metal sulfide-based microporous solids have already been performed (Bedard et al. 1989). New preparation strategies have been developed over the past few years, for example solvothermal synthesis, possibly clearing the way to the synthesis of potential new microporous structures on the basis of sulfides (Stoll et al. 1998).

A natural example of a sulfidic **SOD** framework is the mineral tetrahedrite, $Cu_{12}Sb_4S_{13}$, (Wuensch 1964; Koch and Hellner 1981; Nyman and Hyde 1981). The stoichiometric relationship with sodalite becomes obvious by rearranging the chemical formula to yield $|Cu_{12}Sb_8S_2|[Cu_{12}S_{24}]$. In fact, tetrahedrite is made up of a sodalite-like framework of corner-connected CuS_4-tetrahedra with cages containing S-centered Cu_6-octahedra, encircled by a Sb_4-tetrahedron. According to Liebau (2004) tetrahedrite could be considered as a 4th-order poroate, if one again ignores the lack of chemical bonds within the 2nd- and 3rd-order hosts.

Another possible way of obtaining new microporous structures is by replacing O by N. Since both elements and their chemistry differ much more than O and S, new techniques for synthesis and preparation are required, and it can be anticipated that the stabilities of the compounds will be rather different from oxidic sodalites, and the structures may show new particularities. Schnick and co-workers successfully prepared the compound $Zn_7[P_{12}N_{24}]Cl_2$, which indeed turned out to be of the sodalite-type (Schnick and Lücke 1992). This result proved the analogies between nitridophosphates and oxosilicates. Applying the IUPAC formula to this compound, it reads $|Zn_7Cl_2|[P_{12}N_{24}]$-**SOD**. Considerable variability was found for the extra-framework cations and anions. The host-guest character of these nitridophosphates was demonstrated when the same authors were able to prepare a halogen-free sodalite of composition $|Zn_6|[P_{12}N_{24}]$-**SOD**, which exhibits typical zeoate reactions with reversible hydrogen intake and release. Besides, this observation demonstrates the inadequacy of a definition based on transport properties. With respect to hydrogen the said compound is a zeoate, but for larger species it would rather be called a clathrate.

Replacing one quarter of all bridging N atoms in PN sodalites by oxygen, oxonitridophosphate sodalites $|A_{8-m}H_mX_2|[P_{12}N_{18}O_6]$ with $X = Cl^-$, Br^- or I^- could also be prepared. Monovalent guest cations $A = Li^+$ or Cu^+ make these materials possible candidates for ionic conductivity (Stock et al. 1998). MAS-NMR-investigations and neutron powder diffraction indicate that the sodalite-type framework consists of PON_3 tetrahedra with disordered N and O atoms.

In still another approach to enlarge the accessible range of chemical elements for the preparation of framework structures, metal halide analogs of zeolites could be prepared. In particular, a sodalite-type compound with a framework of composition $[Cu_2Zn_{10}Cl_{24}]^{2-}$ and $(CH_3)_3NH^+$ guests could be obtained (Martin and Greenwood 1997).

SODALITES AS FUNCTIONAL MATERIALS

Sodalites as a border-line case?

In a recent publication (Weller 2000), interesting ideas were developed which seem to be somehow along the lines of Liebau's (2003) thinking in that they combine aspects of structure, chemistry and function as criteria for a classification of inorganic, non-metallic, mostly oxidic, "useful" materials. Weller distinguished two classes of materials for which it is obvious that they may claim interest in this respect. The first class is what he calls complex metal oxides. These are dense structures, often based on closest sphere packings, or derivatives thereof. The cations, often transition or post-transition metals, are located in coordination polyhedra of various geometries, often tetrahedra, octahedra, trigonal prisms, pyramids or square planes, which are composed and connected in different ways, in accordance with Pauling's classical rules. Examples of such materials are perovskites, layered perovskites of the K_2NiF_4 structure type, garnets, or quartz. Supraconductivity, semiconductivity, ionic conductivity, primary and secondary ferroic properties, such as ferro- and antiferroelectricity, pyroelectricity, piezoelectricity, ferro-, ferri- or antiferromagnetism, ferroelasticity, giant magnetic resistance, or nonlinear optical behavior, are among the physical properties which are typically found in dense complex oxides. Many technical applications are based on these materials and their properties. Their importance for the modern civilization cannot be overestimated. Energy transfer, communication techniques, signal transformation, sensors and actuators are just a few examples to illustrate the fields of possible applications, where such materials are in use. Many of the listed properties are associated with structural phase transitions, often describable in terms of phenomenological, e.g., Landau-type, theories. The latter argument implies that a kind of "mean field behavior" is involved, meaning that correlation lengths in the vicinity of the phase transitions tend to extend to infinity. Note that quartz is considered a dense structure despite the fact that it consists of a framework of corner-connected tetrahedra, just like sodalite.

On the other side of his spectrum Weller puts open, porous structures, where typically (not exclusively, however) pre-transition metals are found, especially for those elements which together with O form 3D tetrahedral frameworks. The frameworks enclose channels and cages which can be filled by a large variety of other species. Typical representatives are, of course, zeolites (in the classical notation). The filling of the zeolite pore systems by guest species is often rather unspecific, as regards both, the composition and the positions of the guest atoms. The guest species may relatively easily enter or leave the channel or cage systems and give the structures a touch of indeterminacy. This prevents or at least hinders many of the cooperative phenomena described above for dense oxides from occurring in open structures. On the other hand, the possible applications of open structures rely exactly on the open inner space. The contents of the inner space can easily be exchanged provided the mobile species can pass through the apertures of the cages and channels which connect the inner with the outer space, thus giving rise to the typical applications of zeoates as mentioned before.

The high scientific interest and technical potential of both classes of materials, dense and open, has led to intensive and successful research activities on all levels in both fields. It is rather interesting that scientists working in one or the other field belong to almost disjoint communities with their own conferences, journals and even terminology. As stated before, oxides are still and by far the most important materials in both fields, and will probably continue to be so in the foreseeable future, but it is certainly appropriate to check also other than the well-established tracks. Apart from varying the chemical composition there is another way of opening new fields of research and enabling possible new applications. This can be realized by treating the two classes of materials in a more holistic way, viz. by realizing that both classes are not really disjoint, but overlap to a certain extent. Weller remarks that the class of silicate framework structures stretches from very open to dense structures, say from zeolites to quartz. Quartz, while being built essentially from the same chemical and structural elements as open zeolites, is classified with the dense oxides on grounds of its framework density and its properties. If one was able to change the density of the tetrahedra frameworks continuously from that of quartz to that of the classical zeolites, one would anticipate a transition zone where the character transforms smoothly from that of a dense oxide to that of a porous structure. In the transition zone one would thus expect materials which exhibit to a certain extent a combination of the extreme characteristics. In fact, Weller identifies this transition zone with the members of the **SOD** family, in addition to the frameworks labeled by their IZA codes **CAS**, **ANA**, **BIK** and **ABW**, and considers these materials to belong to the class of "semi-condensed" tetrahedra framework structures. One of the arguments used to consider these frameworks as a class of their own follows these lines. A suitable quantity to measure the openness of a tetrahedral framework structure is its so-called framework density given as the number of **T** atoms per 1000 Å^3. The border between dense and open structures is normally taken to be at about 20–21 **T** atoms per 1000 Å^3. **SOD** with a value of around 16.7 T/1000 Å^3, would then be considered open. On the other hand, for practical reasons, the zeolite community has chosen to use structural parameters in order to decide whether a structure should be considered a zeolite or not. Thus, one has decided that only open framework structures having n-rings (apertures) with $n > 6$ should be accepted. Thus, **SOD** which contains only 4- and 6-rings, is excluded from the zeolites whereas the technically very important zeolite ZSM-5 is accepted although its framework density with about 18 **T** atoms per 1000 Å^3 is even higher than that of sodalite. Its zeolitic properties rely on the existence of a wide 3D-channel network with 10-rings as apertures. This conceptual difficulty was also the starting point of Liebau's criticism leading to his differentiation between zeoates and clathrates, as cited above.

In order to illustrate the difference we can resort to images borrowed from the macroscopic world. *Zeoates* with their accessible pores can be compared with a mall, where it is essential for the operating company that many people have free access in order to do their shopping. *Clathrates*, on the other hand, can be compared with the requirement to provide as much space as possible, for instance for the installation of a big machine, such as a generator or a turbine, in a factory. Usually, such big items pass through the gates only once, probably taken to pieces, and will stay in the building normally during all their service life. Therefore the aperture of the gates is of minor importance. Besides, shopping malls have to be controlled sometimes, or even closed for some reason. On the length scale of microporous structures this can be realized by cations which partly or totally block the access to the internal space.

According to Weller the **SOD** family should be considered as a semi-condensed tetrahedral framework structure, exhibiting the properties of both, dense and open structures. This is, indeed, true. For instance, some sodalites definitely exhibit zeolitic properties like ionic exchange (Homeyer et al. 2001), and they may exist as empty silica frameworks (Braunbarth et al. 1997). On the other hand, some sodalites have properties typical of dense structures. They undergo ferroic phase transitions (Depmeier 1992a), exhibit corresponding ferroic properties (Depmeier 1988a) and even form commensurately and incommensurately modulated phases (Depmeier 1999). Weller's classification of sodalites as transitional structure seems therefore to be justified. The clathrate and the zeoate character of sodalites will be exemplified in the following two paragraphs.

Sodalites as clathrates; matrix isolation

The cages in sodalite are about 4–4.5 Å wide. The guest species in adjacent cages are isolated from each other and interact only weakly and indirectly, if at all. This affects the chemical, physicochemical and physical properties of the guest species. The effect is called matrix isolation, and is frequently exploited to study unstable species, mostly by spectroscopic methods. The matrix prevents the enclosed species from undergoing intermolecular reactions, the possibility of applying low temperatures inhibits intramolecular changes. One of the obvious advantages of matrix isolation is that it enables to study the behavior of single molecules, such as H_2O, over much larger temperature or pressure ranges than would be possible for free molecules, or their ensembles. For example, at low temperatures or high pressures a liquid or a gas condenses and becomes a solid. A solid is characterized by additional important intermolecular interactions, and thus radically different behavior than single molecules. Another important advantage of the method is that rare species can be obtained with high concentration, but low interaction of the species. The effect of matrix isolation in sodalites can be illustrated by the following few examples.

$|Ca_8(SO_4)_2|[Al_{12}O_{24}]$-**SOD** forms from CaO, Al_2O_3 and $CaSO_4$ at around 1350°C, a temperature at which $CaSO_4$ decomposes readily. Obviously, the SO_3 is trapped in the forming sodalite cages, without being able to escape.

The bright blue color of the highly appreciated gemstone ultramarine is generally believed to come from the presence of special guests in its sodalite-type framework, viz. radical anions S_2^- or S_3^-. These very reactive species would undergo almost immediate dimerization in any medium which allowed them to approach each other. However, in the cages of the sodalite framework they exist almost infinitely long.

One of the most astounding sodalite species is electride sodalite, $|Na^+{}_8e^-{}_2|[Al_6Si_6O_{24}]$-**SOD**, also known as black sodalite. It can be prepared by absorbing Na vapor in $|Na_6|[Al_6Si_6O_{24}]$-**SOD**, which in turn can be prepared by Soxhlett extraction of $|Na_8(OH)_4 \cdot nH_2O|[Al_6Si_6O_{24}]$-**SOD**, and subsequent dehydration (Felsche et al. 1986). The extra Na atoms in black sodalite loose one electron and become ionized. The free electron e^- is shared between the four Na^+ ions in each cage. This curious structure can be considered as a crystalline, body-centered, high-density array of F color centers, hence the black color. Because of its special features, black sodalite has attracted the attention of many physicists (Windiks and Sauer 1999; Sauer and Windiks 2003). Since the delocalized free electrons are confined to distinct cages, the material is non-metallic.

The clathrate character of sodalites gives also rise to particular applications which are based on exactly these properties. There is a great concern about radioactive isotopes which are the products of many industrial processes such as mining of uranium-bearing minerals, or the radioactive waste management of nuclear fuel, or the degradation from uranium ammunition. A possible way of rendering the dangerous species harmless is by encapsulating them in the cages of a suitable sodalite structure where they are not able to escape. The sodalites can then be molten to form a glass and stored safely (Pentinghaus et al. 1990). By the way, if the isotopes to be enclosed are cations, then it would be advisable to have in the framework of an aluminosilicate sodalite as much Al as possible, as for each Al^{3+} replacing one Si^{4+} in the framework a positive charge in form of a guest cation can be accepted. Of course, if one considers possibilities to apply aluminosilicate sodalites as storage material for such isotopes the stability of the corresponding framework against attack of, e.g., water has to be taken into account. In general, it is believed that the chemical stability of porous aluminosilicates decreases with increasing Al content. This has to be taken duly into account. On the other hand it has been demonstrated that in the series $|(Eu_xCa_{2-x})_4(OH)_8|[(Al_{2+x}Si_{1-x})_4O_{24}]$-**SOD** the stability against thermal degradation increases with increasing Al content (Lars Peters, pers. communication, 2004), thereby lending support to the idea that Al-rich sodalites may indeed be useful for the storage of radioactive waste. Sodalites seem to be especially suited for this purpose because, firstly, they display clathrate character for not too small guest species, second, small cages such as those of sodalites are a prerequisite for a high degree of Al-substitution for Si, and ensuing ability to accommodate larger amounts of guest cations.

A computational study of natural sodalite using quantum-chemical methods and *ab-initio* Density-Functional methods has revealed that the localized optical transmission on Cl⁻ cage anions determines its absorption edge (Sokol and Catlow 1996).

A rather significant screening effect has been observed in synthetic danalite, $|Fe_8X_2|[Be_6Si_6O_{24}]$. Mössbauer spectra and magnetic measurements indicate that despite the divalent iron being in the high spin state and the short separation between the Fe_4X units, no long range magnetic order occurs down to 4.2 K (Armstrong et al. 2003).

From a slightly different point of view the sodalite structure can be regarded as consisting of distinct isolated entities immersed in the infinite sodalite framework, e.g., $Na^+_4e^-$ in black sodalite, Na_4Cl^{3+} in natural sodalite, or Cd_4S^{6+} groups in $|Cd_8S_2|[Be_6Si_6O_{24}]$ (Dann and Weller 1996). The latter groups consist of anion-centered cation tetrahedra, resembling small pieces of the CdS structure type. In materials like bicchulite, $|Ca_8(OH)_8|[Al_8Si_4O_{24}]$-**SOD** (Henmi et al. 1974; Sahl and Chatterjee 1977; Gupta and Chatterjee 1978; Sahl 1980) or $|REE_4(Pb_4O_4)_2|[Al_{12}O_{24}]$-**SOD** (Scheikowski and Müller-Buschbaum 1993), $Ca_4(OH)_4$ or Pb_4O_4 entities, respectively, have the geometry of small chips of a NaCl-type structure.

Sodalites as zeoates; ion exchange

Admittedly, at room temperature and atmospheric pressure the 6-ring windows of the sodalite framework do not allow significant transport of normal guest species. However, if the temperature is raised high enough then cations can migrate quite easily and the sodalite become ion-exchanged. This has been demonstrated, for example, by the observation that a sodalite of normal composition, $|Na_8Cl_2|[Al_6Si_6O_{24}]$-**SOD**, pressed into KBr pellets for high temperature infrared measurements, showed rapid exchange of Na^+ against K^+ already at about 650 K (Homeyer et al. 2001).

True zeolitic behavior is exhibited by sodalites of the series $|Na_6 \cdot nH_2O|[Al_6Si_6O_{24}]$-**SOD**, $n = 0, 4, 8$, and $|Na_8(OH)_2 \cdot nH_2O|[Al_6Si_6O_{24}]$-**SOD**, $n = 0, 4$. They can be prepared by an appropriate hydrothermal treatment starting from kaolinite and NaOH. Upon heating the water molecules are successively driven out, without affecting the integrity of the sodalite framework, in accordance with the classical definition of a zeolite. A rather surprising behavior of the respective cubic lattice parameter was observed during dehydration. Whereas in $|Na_8(OH)_2 \cdot 4H_2O|[Al_6Si_6O_{24}]$-**SOD** the loss of the four water molecules per formula unit results in a contraction of the unit cell volume by 5.1%, $|Na_6 \cdot 8H_2O|[Al_6Si_6O_{24}]$-**SOD** rather *expands* by 9.4% upon dehydration. While the behavior of the former compound is in accordance with the idea that the guest molecules serve as "spacers" for the framework, the behavior of the latter compound is rather counterintuitive. It has been attributed to the detachment of hydrogen bonds acting between the water molecules and the framework oxygens in the cage walls. The hydrogen bonds were considered to exert a kind of negative pressure upon the surrounding framework. Upon release of water the hydrogen bonds and, hence, the negative pressure disappears, accompanied by the framework's relaxation and expansion. The influence of the hydrogen bonds in $|Na_6 \cdot 8H_2O|[Al_6Si_6O_{24}]$-**SOD** becomes also noticeable in its small unit cell parameter 8.848(1) Å, compared with 9.122(1) Å of $|Na_8(OH)_2 \cdot 4H_2O|[Al_6Si_6O_{24}]$-**SOD**, both at room temperature (Felsche and Luger 1986; Felsche et al. 1986).

CRYSTALLOGRAPHY – STRUCTURAL DETAILS

Basic crystallography of sodalite

"Where are the atoms?" For a crystallographer, or for any other solid state scientists, this is often the most basic question from which, if answered, many properties of the solid under study can be predicted or even calculated, at least in principle. The appropriate, and most commonly employed, method to locate atoms in a structure is diffraction of X-rays, neutrons or electrons. The main features of the sodalite structure have been known for almost three quarters of a century from the seminal work of Linus Pauling on the X-ray structure determination of the minerals sodalite, $|Na_8Cl_2|[Al_6Si_6O_{24}]$-**SOD**, and helvite, $|Mn_8S_2|[Be_6Si_6O_{24}]$-**SOD** (Pauling 1930). From rotation and Laue photographs he could determine a cubic unit cell with a lattice parameter of 8.87 Å for sodalite; as space group he assigned $P\bar{4}3n$. He emphasized that only very weak reflections violated the extinction rules for the space group $I\bar{4}3m$, indicating that the deviations of the atoms from a body-centered structure are very small. The main building principles of the structure could be determined from symmetry considerations; more precise information was obtained by calculating the intensities for a small number of reflections and comparing them with the corresponding visually estimated ones. The free parameters were varied stepwise within reasonable ranges until the agreement between observed and calculated intensities was deemed satisfactory. Of course, in those days without computers all the necessary calculations had to be performed by hand, therefore computing-intensive procedures like least-squares refinements, error estimations or the calculation of quantitative criteria for the goodness of the fit, e.g., R-values, could not be performed. These details were left to future generations benefiting from the advancement of instruments and techniques.

Pauling obtained the following atomic coordinates for sodalite, $|Na_8Cl_2|[Al_6Si_6O_{24}]$-**SOD** in $P\bar{4}3n$; for convenience, multiplicities, Wyckoff letters and site symmetries are also listed:

2 Cl in	2a	23.	0,0,0	
8 Na in	8e	.3.	x,x,x	x = 0.175
6 Si in	6d	$\bar{4}$..	0,½,¼	
6 Al in	6c	$\bar{4}$..	¼,½,0	
24 O in	24i	1	x,y,z	x = 0.135, y = 0.440, z = 0.150

Given the rather rough method being at his disposal Pauling's results agree remarkably well with those of subsequent, more elaborate work. For example, a least-squares refinement against film data yielded x(Na) = 0.1777(4), and O in 0.1401(4), 0.4385(3), 0.1487(4) (Löns and Schulz 1967), and a modern refinement using neutron diffractometer data gave x(Na) = 0.1778(2), and O in 0.13925(4), 0.43851(4), 0.14954(4) at 295 K (McMullan et al. 1996). It did not escape Pauling's attention that the observed small deviation from a body-centered structure can be explained by ordering of Al and Si on the *6 c* and *6 d* positions, resulting in small displacements of the O atoms from the mirror planes in x,x,z, because of the different size of Al and Si atoms. Based on the atomic coordinates thus determined Pauling describes the structure as being composed of strictly alternating, all-corner-connected SiO_4- and AlO_4-tetrahedra. Pauling's ability to understand structures and imagine their behavior is best demonstrated by the following passage taken from his work (Pauling 1930):

> "This crystal provides a remarkable example of a framework structure. The forces between the highly charged cations Si^{+4} and Al^{+3} and the oxygen ions are by far the strongest forces in the crystal. They cause the joined tetrahedra to form a strong framework, of composition $Al_6Si_6O_{24}$, extending throughout the crystal and essentially determining its structure. Within the framework are room and passages, spaces which can be occupied by other ions or atoms or molecules, in this case sodium and chlorine ions. The framework, while strong, is not rigid, for there are no strong forces tending to hold it tautly expanded. In sodalite the framework collapses, the tetrahedra rotating about the twofold axes until the oxygen ions come into contact with the sodium ions, which themselves are in contact with the chlorine ions. This partial collapse of the framework reduces the edge of the unit cell from its maximum value, about 9.4 Å, to 8.87 Å."

Pauling's statement is exemplary in both, its lucid style and high information density. Not only does he make a short excursus on the nature of the chemical bonds in sodalite, but he also touches upon two important items, viz. the topology of the framework and its conformational changes.

The topology of the sodalite framework type

The last two sentences of Pauling's statement imply that one can conceive an un-collapsed, or fully-expanded, framework. The transition from the collapsed to the un-collapsed state can be imagined to happen by continuously changing the relative orientation of neighboring, virtually rigid, tetrahedra without destroying the connectivity of the framework. The fully-expanded structure thus obtained can be further idealized until the most regular, highest symmetric, so-called topological structure, is attained, regardless of whether this may exist in reality or not. The idealization can be brought about by making the **T** atoms indistinguishable, and adjusting the TO_4 tetrahedra such that they have the geometry of regular tetrahedra. The idea is similar to that of *aristotypes*, introduced by Megaw (1973). The associated symmetry is called the topological symmetry. In the case of the **SOD** framework its space group is $Im\bar{3}m$. The topological symmetry and the symmetries of really existing structures are usually related by group-subgroup relationships of the corresponding space groups (Wondratschek and Müller 2004). Specifically, $P\bar{4}3n$, as assigned by Pauling (1930), is a subgroup of $Im\bar{3}m$, however, not maximal. This can be easily understood because two different mechanisms of structural deviations from the topological symmetry are involved in the symmetry-reduction, viz. the partial collapse and the order on the **T** sites. A discussion of symmetry relationships in the sodalite family has been published elsewhere (Depmeier 1992b). Even though topological structures are only virtual constructs they are very useful concepts because they facilitate the understanding of the most general features of a structure's architecture, without confusing the reader with too much complexity. Often even simpler models are used to represent the connectivity of frameworks. These models consist of the **T** atoms at the centers of the TO_4 tetrahedra only.

There is no unique way to describe the topological structure of framework types, including that of sodalite. Various descriptions can be found in the literature, the differences usually being due to the respective author's interests. In the following we will briefly present some approaches.

Often structures, and in particular framework structures, are regarded as consisting of rigid building units and the primary interest is directed at a general understanding of how these building units are connected with each other. One of the most important sources of information on the topologies of framework structures, in particular with tetrahedral building units, is beyond doubt the already mentioned ATLAS. As already said, presently it comprises 152 so-called zeolite framework types, each one being identified and labeled by its IZA three-letter code. For each framework type the topological symmetry is given together with idealized cell parameters, coordinates, site multiplicity and the point symmetry of each symmetrically independent **T**-atom. Drawings of the connectivity, i.e., ignoring the O atoms, highlight the linking into a 3D framework. The understanding is greatly enhanced by featuring certain conspicuous building blocks and their assembly into the entire framework. Animated 3D-drawings are a very recent achievement; they allow to zoom in and out or to rotate the framework about various axes, thus imparting a 3D impression of the framework.

The ATLAS characterizes each framework type by various additional items. For **SOD**, in particular, it states that the framework is composed of only 4- and 6-rings. This fact gives the channel system a dimensionality of zero, because in order to be accepted as a channel a passage in a structure by definition must have smallest apertures exceeding that of 6-rings. Three types of secondary building units (SBU's) are distinguished, viz. those which are denoted as 4, 6 and 6-2. The former two SBU's are the already mentioned 4- and 6-rings, whereas the 6-2 unit consists of a pair of a 6- and a 4-ring sharing a common edge. In this context it is worth of notice that SBU's are purely theoretical building units; they are not necessarily involved in the crystallization process. Other items in the ATLAS description are partly based on graph theory. The loop configuration is a graph which shows how the topologically independent **T** atoms are involved in 3- or 4-membered rings. In the **SOD** framework there is only one topologically independent **T** atom which belongs to two adjacent corner-connected 4-rings. These rings make up the graph. The so-called coordination sequence and the vertex symbol together (probably) uniquely characterize a given zeolite framework type.

The **SOD** framework can be embedded into a larger topological class, called the ABC-family. The structures of all members of this family can be thought of to consist of common building blocks. So-called periodic building units consist of planar 6-rings centered on a two-dimensional periodic array. These arrays stack onto each other with adjacent layers possibly being related by discrete shifts with respect to a common origin. In the case of **SOD** successive layers are displaced by (2/3**a** + 1/3**b**), **a** and **b** being the basic vectors of the hexagonal arrays. 6-rings of adjacent layers are finally connected by tilted 4-rings thus completing the 3D dimensionality. The stacking of the hexagonal arrays in **SOD** with the particular shift as given above results in a periodic sequence ...ABC.... This situation resembles closely that of densest sphere packings. In both cases the ...ABC... sequence is unique in its respective stacking series, because it results in cubic symmetry with layers stacked along the symmetrically equivalent cubic <111> directions. Other stacking modes of the hexagonal arrays produce the various members of the ABC-family, viz. **AFG, AFT, AFX, CAN, CHA, EAB, ERI, FRA, GME, LEV, LIO, LOS, OFF** and **SAT** (see ATLAS). This elegant way of integrating the **SOD** framework into a generic system is very appealing, however, it is less appropriate for an immediate and straightforward understanding of the specific features of a given topology. Other approaches are better suited for this purpose.

In a recent tabular work on the topologies and some structural characteristics of tetrahedral frameworks, the sodalite framework is also included (Smith 2000). Besides repeating the well-known fact that the **SOD** framework type consists of 4- and 6-rings, the author decomposes the framework into various building units of different dimensionality, viz. polyhedral, 2D and 1D units. There are various ways of doing so.

The most obvious decomposition of the sodalite framework is that into the so-called sodalite cages, or β–cages (see Fig. 3). In the community interested in microporous framework structures this polyhedron has become quite popular. If one considers only the **T** atoms at the centers of the **T**O$_4$ tetrahedra and idealizes its geometry the polyhedron becomes a regular truncated octahedron. This is the most regular arrangement of the 24 **T** atoms. It consists of two kinds of faces, viz. six squares and eight regular hexagons, it can therefore be described by its face symbol, $4^6 6^8$. Because all vertices are equivalent, $4^6 6^8$ belongs to the class of semi-regular, or Archimedean, solids. Another characterization of the regular truncated octahedron uses its *Schläfli* symbol 4.6.6. This symbol expresses that one square and two hexagons meet at each of the equivalent corners.

We shall now itemize some remarkable features of the $4^6 6^8$ polyhedron. Besides its occurrence as sole cage in the frameworks of the various members of the **SOD** family, the sodalite cage is found combined with other cages in

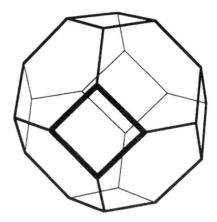

Figure 3. The sodalite cage represented as a regular truncated octahedron, also Smith's polyhedral building unit **toc**. [Used by permission of Springer, Berlin, Heidelberg, from Smith (2000) Fig. 16.3.1, p. 125]

several zeolite frameworks, namely in **EMT**, **FAU**, **LTA**, **LTN**, **TSC**. An atomic arrangement with the shape of 4^66^8 is also found in certain gas-hydrates and in the structure of many alloys, and even the structure of an element, viz. that of α-Mn, exhibits structural motives which resemble closely the sodalite cage (Nyman and Hyde 1981). 4^66^8 belongs to the so-called *Fedorov* polyhedra. These polyhedra cover completely the 3D space without any gaps or overlaps, with just one kind of tile. Of all *Fedorov* polyhedra, 4^66^8 exhibits the smallest value of its surface-to-volume ratio. In a certain sense it can therefore be considered the best Euclidean approximation to a complete covering of 3D space with spheres. It is worth mentioning that the regular truncated octahedron 4^66^8 must not be confused with the cuboctahedron, the latter being characterized by its face symbol 3^88^6 and *Schläfli* symbol 3.8.8.

The packing of the sodalite cages (idealized as 4^66^8 polyhedra) in 3D space is remarkable. Imagine eight cages centered at the corners of a cubic unit cell in such a way that all cages share all their square faces with neighboring cages along <100>. The space omitted will then form another congruent cage centered at ½,½,½. The latter cage is translationally equivalent to the cages at 0,0,0 and the whole arrangement becomes body-centered cubic (bcc), in accordance with the topological symmetry $Im\bar{3}m$ of the **SOD** framework type. Cages at 0,0,0 and ½,½,½ are then connected along <111> via common hexagons. A stereo-view of three sodalite cages packed along a body-diagonal can be enjoyed in Figure 4 (from McMullan et al. 1996). It is well-known that the bcc and the fcc (face-centered cubic) lattices are dual in the sense that the reciprocal lattice of fcc is bcc, and vice versa. The decomposition of the bcc reciprocal lattice into Wigner-Seitz cells leads to a **SOD**-type tiling by 4^66^8 polyhedra. Hence, the shape of the first Brillouin zone of a fcc lattice is described by a 4^66^8 polyhedron. Besides, the shape of the first Brillouin zone of a bcc lattice is that of a rhombic dodecahedron. This is also one of Fedorov's polyhedra; its surface-to-volume ratio comes a close second to that of the regular truncated octahedron.

In many textbooks on mineralogy, solid state chemistry or solid state physics the conspicuous space filling arrangement of sodalite cages is presented as an archetypical example of a zeolitic framework structure, or even of framework silicates in general. The reason for this preference is presumably the high symmetry of the **SOD** framework, and furthermore the fact that only one cage type is sufficient for its description. Both make the architecture of the **SOD** framework very intelligible. Other frameworks often shown are those of **FAU** and **LTA**, both consisting of sodalites cages, but with additional building units, and spatially arranged

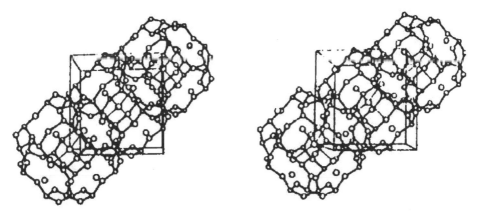

Figure 4. Stereoview illustrating the stacking of sodalite cages along a cubic <111> direction. [Used by permission of The International Union of Crystallography, from McMullan et al. (1996), Acta Crystallogr. B52, Fig. 1, p. 616.]

in different ways. Another possible reason why these three framework types enjoy such a high popularity is their undeniable aesthetic appeal which is almost certainly due to their high symmetry (the topological symmetries of **FAU** and **LTA** are $Fd\bar{3}m$ and $Pm\bar{3}m$, respectively), and to the transparency of the **T** atom skeleton representation of their frameworks.

Coming back to Smith's classification, the regular truncated octahedron is listed as the obvious polyhedral building unit of **SOD** and denoted as **toc**. Less conspicuous than **toc** are the two- and one-dimensional building units. The 2D building units will not be further considered here, but Smith's 1D building units of **SOD** are displayed in Figure 5 (Smith 2000). A quite obvious 1D structural element is a chain of **toc** units connected via common 6-rings along a specific cubic <111> direction. It is denoted as **kda**. If there was any migration of species, say at high temperatures, this would presumably be their pathway. However, as stated earlier, the relatively small aperture of 6-rings represents an unfavorable bottleneck for any migrating guest species. Note, there is a typing error in Smith's Table 16.2.1, as **kdd** has erroneously been given instead of the correct **kda**. The **kfb** unit shows basically a string of **toc** units with common 4-rings along the cubic <100> directions. Because of the small size of the 4-rings this unit is even less likely than **kda** to become a diffusion pathway. The **kgj** units run along <110>. Here, **toc** units are linked into chains by interposed 4-rings.

The **ts** units exhibit their one-dimensional character more clearly than **kda**, **kfb** and **kgj**. They form strings of corner-connected 4-rings which are alternately twisted by 90° with respect to each other in such a way that the planes of the 4-rings are more or less parallel with {100}. The **ts** units run along the cubic <100> directions, and are thus parallel with **kfb**. Each **ts** unit is the intersection of four neighboring **kfb** units.

The framework of close-packed 4^66^8 polyhedra can be considered as the skeleton of the sodalite framework, consisting of only the **T** atoms. It informs well on the connectivity of the 3D framework, but is less suited for an understanding or even description of a framework

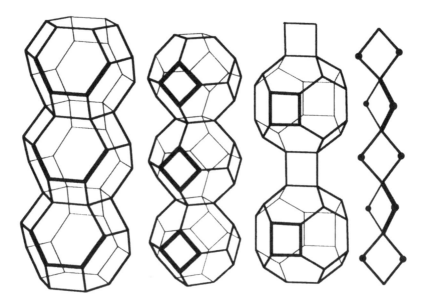

Figure 5. Smith's decomposition of the sodalite framework into one-dimensional building units. From the left: **kda**, **kfb**, **kgj**, **ts**. [Used by permission of Springer, Berlin, Heidelberg, from Smith (2000) Fig. 16.5.1, pp. 217, 221, 225, 246.]

which has undergone conformational changes as originally described by Pauling under the name of partial collapse. The next step in approaching reality consists of putting points representing O atoms midway between the **T** atoms. By doing so, Koch and Hellner (1981) describe the ideal sodalite framework in space group $Im\bar{3}m$ as consisting of corner-connected regular tetrahedra with vertices (O atoms) at Wyckoff position *24h x,x*,0, where $x = 1/4\sqrt{2} = 0.3536$ and centers (**T** atoms) at *12d* ¼,½,0. The non-framework atoms **A** and **X** take also their most regular positions, viz. at *8 c* ¼,¼,¼ and *2a* 0,0,0 , respectively. The *12d, 6b* and *2a* positions correspond to the cubic invariant lattice complexes W*, J* and I, respectively; the lattice complex corresponding to *8c* is denoted ¼¼¼ P$_2$ (Fischer and Koch 2002a). The arrangement of the **A** atoms at *8c* ¼,¼,¼ is such that they form a cubic primitive lattice with half the lattice parameter of the entire sodalite. One quarter of the small cubes are centered by the cage anions **X** at position *2a* 0,0,0, three quarters center at position *6b* 0,½,½. Each small cube at *6b* encloses a 4-ring of the sodalite framework. The before-mentioned **ts** strings run along <100> within tubes formed by the face-sharing small cubes of **A** cations centered at position *6b*. In each small cube two perpendicular **ts** strings intersect in a common 4-ring. The small anion-filled cubes centered at *2a* share common corners and this configuration corresponds to the structure of PtHg$_4$ (see Hyde and Andersson 1989).

The framework of corner-connected rigid tetrahedra as presented by Koch and Hellner (1981) is very useful, e.g., for the discussion of conformational changes of the framework. However, since in the model the O atoms at the corners of the tetrahedra are represented as points this may impart an unrealistic impression, if problems related to space filling have to be considered. Closer to reality comes the replacement of the dimensionless points by spheres of identical size, with the radius suitably chosen such that neighboring spheres touch each other. If for any two spheres in the infinite framework an uninterrupted pathway of touching spheres can be found, this packing is called a sphere packing. It is commonplace in crystallography that the packing density of closest sphere packings is 0.7405 (this is Kepler's conjecture; it has been proven recently; for the story the reader is referred to Szpiro (2003)). The in-depth discussion of closest sphere packings, Kepler's conjecture and its consequences are subjects taught explicitly in introductory courses into solid state sciences. For the present discussion a quite different point of view is probably more appropriate, namely to ask for packings of minimal densities. Sphere packings were extensively studied by W. Fischer and his colleagues (Fischer 1968, 1971, 1973, 1974, 1976, 1991a,b, 1993, 2004; Koch and Fischer 1972, 1978, 1995, 1999a; Koch 1984, 1985; Sowa and Koch 1999, 2001, 2002, 2004a,b; Fischer and Koch 2002b; Sowa et al. 2003; Koch and Sowa 2004). In particular, Fischer studied the minimal densities of cubic sphere packing types and found for the type which he denotes 4/4/c3 a minimum density of 0.30230 (Fischer 2004). 4/4/c3 corresponds to spheres at the positions *24h x,x*,0 of $Im\bar{3}m$ with $x = 1/3$. This comes very close to the coordinates of the sphere packing in the sodalite framework. Even if the O atoms in the real sodalite framework are slightly displaced from this position, the density will not increase dramatically. This demonstrates clearly that there is indeed ample free space in the sodalite framework, and justifies the term "open framework." The question arises whether these roughly 70% can be really empty. In order to avoid the need of any charged guest ions, the framework would have to be electrically neutral. Because of the 1:2 ratio of **T** and O^{2-} atoms, a possible framework composition would then be [SiO$_2$]—the structure in question would be a silica phase. In fact, silica-sodalites have been prepared with various organic molecules as so-called templates (e.g., Braunbarth et al. 1997) and empty frameworks by calcination, as described earlier. If the average formal charge per **T** atom in the framework is less than 4+, charge compensation by guest cations (and guest anions) is required. These must then occupy the empty space. However, in order to do so the non-framework atoms have to comply with the sizes and shapes of the voids within the sphere packing. Koch and Hellner (1981) analyzed the various possibilities by determining the so-called Dirichlet domains for the sodalite structure type. The vertices of the Dirichlet domains correspond to voids in the sphere packing and define possible coordination polyhedra spanned by the framework O atoms. These are shown in Figure 6 (from Koch and Hellner 1981). The tetrahedra (*4t*): *W** accommodate the framework **T** atoms, the big cuboctahedra (*12co*): *I* enclose the cage anions and the trigonal antiprisms (*6a*): P_2' define the coordination polyhedra of the cage cations. Note, while the O atoms of the trigonal antiprisms are arranged in two parallel, but distant planes, the ring of the corresponding **T** atoms is planar. The highly distorted tetragonal antiprisms (*8a*): *I(6o)* are usually unoccupied.

A radically different way of understanding the sodalite framework opens up when one leaves the scope of Euclidean geometry. Periodic minimal surfaces or their derivatives were used by several authors as versatile tools in crystal chemistry (e.g., Hyde et al. 1984; Fischer and Koch 1987, 1989a,b,c, 1990, 1996a,b, 2001; Koch and Fischer 1988, 1989a,b, 1990, 1993a,b, 1999b; Koch 2000a,b, 2001). One of the simplest surface occurring in these studies is the so-called Schwarz primitive surface P* (for an illustration, see http://www.indiana.edu/~minimal/archive/triply/schwarzp/schwarzp.html). The P* surface is a representative of space group $Im\overline{3}m$ and many of its subgroups, and, up to a certain point, can be identified with the ideal sodalite framework, or, more precisely, with the wall separating the sodalite cages from each other. The wall follows smoothly the P* surface. It has been argued that the predominantly covalent chemical bonding within the T-O framework follows closely the P* surface, whereas the lone electron pairs of the O atoms stick out from that surface to form a kind of negatively charged double layer. The latter is believed to make most of the ionic bonding with the cations (von Schnering and Nesper 1987). The P* surface is 3D periodic. It subdivides space into two interpenetrating, but non-intersecting subspaces or labyrinths. Non-intersecting means that if one side is painted black and the other one white, black and white would not mix. The fact that the two labyrinths defined by the P* surface do not intersect can be expressed in a slightly

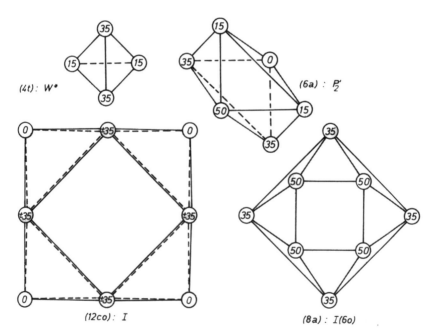

Figure 6. Voids spanned by the O atoms in the fully-expanded sodalite framework. The voids define possible coordination polyhedra. [Used by permission of Akademische Verlagsgesellschaft, from Koch and Hellner (1981), Z. Kristallogr. Vol. 154, Fig. 7, p. 104.]

sentimental way by saying that two creatures existing on either side of a P* surface in their respective labyrinth would never have a chance to meet each other. Together with the small apertures of the sodalite cages the labyrinth-type character of the sodalite framework explains much of the matrix isolation effects as discussed earlier.

An astoundingly simple relationship between the sodalite framework and curved space was found recently (Andersson and Jacob 1997) in an approach to image geometrical features of various crystal structures by mathematical equations. In particular, they discuss the equation $\cos(x) + \cos(y) + \cos(z) = 0$ and call its graphical representation the 3D cosine nodal surface. It is very similar to the P* surface (see Fig. 8 of Andersson and Jacob 1997). It therefore seems that a non-Euclidean representation of the sodalite framework can be represented by the very simple equation as given above.

3-periodic nets have found widespread attention as a versatile tool to analyze existing framework topologies, including that of sodalite, and to predict new ones, e.g., so-called augmented sodalite. Early work by Wells was followed by a great number of studies, too numerous to be cited here. The books by Wells (1977, 1984) give a good introduction into the field. A recent approach is the development of an interactive electronic data base entitled "Reticular Chemistry Structure Resource" (http://okeeffe-ws1.la.asu.edu/RCSR/home.htm). This aborning tool will certainly emerge into a future important source of knowledge.

The classical sodalite tilt system

Pauling (1930) had recognized the sodalite framework's ability to reduce its volume by changing its conformation and called the mechanism "partial collapse." This remarkable property has raised considerable attention in the past and continues to do so. It is generally argued that the partial collapse of the sodalite framework can be understood as its adaptation to the size of the enclosed guests. Further below it will be demonstrated that the sodalite framework is able not only to adapt itself to the size of the guests, but also to their shape.

The mechanism which leads to a conformational change of a framework structure is often called tilting. Tilt systems are material-specific and have to be defined. Usually, a framework of corner-connected, rigid building units is assumed, such as the tetrahedral framework of sodalite. The common corners can be considered to serve as a kind of flexible hinges. In aluminosilicates, the common corners of the tetrahedral building units are O atoms. The most basic way to express the mutual orientation of rigid polyhedra is by quoting the intertetrahedral angles **T-O-T**. From numerous studies, it is well-known that these angles can vary over a large range with little cost of energy. Unfortunately, the **T-O-T** angles are not very practical for descriptive crystal chemistry as they lack clearness and do not have an evident, fixed reference point. For example, the angles **T-O-T** in the tetrahedral framework structures of aluminosilicates scatter around 140–145°, the exact value depending in a complex way on the topology and on various interacting crystal chemical parameters and thermodynamic variables, such as composition, size and charge of the atoms present, temperature and pressure.

In order to overcome the difficulties, tilt systems were devised with the aim to provide better insight into the behavior of the framework structures. A prerequisite is an appropriate definition of a zero-position for the tilt system and the framework type under consideration, whereby often the unit cell edges, or other well-defined directions, serve as reference points. For **SOD** the zero-position is defined in the following way. The site symmetry of the **T** atoms in Wyckoff-position *12d* of the topological symmetry $Im\bar{3}m$ is $\bar{4}2m$ (Hahn 2002). For the zero position of the tilt system those edges of the TO_4 tetrahedra which are perpendicular to the $\bar{4}$-axes of the site symmetry should be parallel with the edges of the cubic unit cell. If the tilt has become active both edges are no longer parallel, and the angle spanned by them is taken as the tilt angle. It is assumed that the shape of the tetrahedra does not change.

One important feature of tilt systems is that the polyhedra keep their connectivity intact and hardly change shape in the course of the tilt. In a linear approach this means that the rotation of a given polyhedron is entirely transferred to its neighbors which have to respond by their own rotations. These may be in-phase, out-of-phase, or in anti-phase with the rotation of the originator. A tilt system is thus a cooperative phenomenon. In 3D framework structures the tilt usually propagates in all three dimensions, however, two- or one-dimensional tilt systems are also possible. Usually, a tilt system tends to extend to infinity, but local tilt systems are also known. The possible tilt systems of a given framework depend on its topology.

Tilt systems have been studied in much detail for the octahedra frameworks of the perovskite family, because of the immense technological importance of this structural family and facilitated by its relatively simple topology which lends itself to a straightforward classification and notation. For tetrahedra frameworks comparably simple mechanisms and classifications are less easily conceived.

When Pauling coined the term "partial collapse" he certainly wanted to emphasize that the volume of the unit cell is affected. Frameworks in which a characteristic tilt system reduces the volume of the unit cell with increasing tilt angle have been termed "collapsible" (Baur 1992). **SOD** belongs to such frameworks. Other structures are called "non-collapsible," because while cooperative rotations of the polyhedra make the corresponding framework flexible, they do not reduce the unit cell volume. This is because the effects of changes of individual polyhedra cancel each other and hence do not result in any significant volume reduction (Baur 1995a,b; Baur et al. 1996). The **LTA** framework of zeolite A is an example for such a flexible, but non-collapsible framework.

How far a collapsible framework can collapse? This point was already addressed by Pauling (1930). Later, Baur (1995b) proposed two processes which in his opinion would be able to stop the collapse of a framework. The first process would become active when framework O atoms came into too close contact with guest species, such as cations, anions, water, or other molecules; in this process the guest species are considered to be a kind of spacer for the framework. The second process would occur for very small cations or for empty frameworks. In this case the framework could stop from collapsing when too small **T-O-T** angles would bring non-bonded O atoms of adjacent tetrahedra into their van-der-Waals distances where the steep increase of the repulsion term would prevent any further approach.

The classical tilt system of the **SOD** framework (e.g., Pauling 1930; Taylor 1972; Depmeier 1984a; Hassan and Grundy 1984) is 3D. It is depicted in Figure 7. Nyman and Hyde (1981) analyzed the geometry of a tilted sodalite framework built from regular tetrahedra. They found the following relationships between the lattice parameter a, the edge length of the regular tetrahedron d and the tilt angle φ. It should be noted that similar relationships were already described earlier (e.g., Taylor 1972):

$$a = 2d(\cos \varphi + 1/\sqrt{2})$$

The coordinates of the framework O atoms in x,x,z, and the tilt angle φ are related as follows:

$$x = \cos \varphi / 4(\cos \varphi + 1/\sqrt{2})$$

$$z = \sin \varphi / 4(\cos \varphi + 1/\sqrt{2})$$

$$\varphi = \arctan(z/x)$$

The z parameter rarely exceeds 0.08 for aluminosilicates, thus restricting the range of the tilt angle φ to less than about 30°. In tetrahedrite the tilt angle is about 50°—this framework is considered as over-collapsed (Nyman and Hyde 1981).

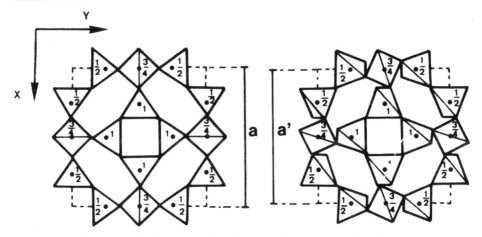

Figure 7. A fully-expanded (left), and a partially-collapsed sodalite framework. [Used by permission of the Mineralogical Society, from Taylor (1972), Mineral. Mag., Vol. 38, Fig. 1, p. 595 (modified).]

If the tilt angle φ = 0°, and only one type of **T** atom is present, the space group of the sodalite framework is $Im\bar{3}m$. When φ deviates from 0°, the inversion center is broken, but the cubic symmetry retained, and the space group becomes $I\bar{4}3m$. In their before-mentioned paper, Koch and Hellner (1981) analyze the changes of the **SOD** framework and its voids brought about by the tilt. As far as the framework is concerned, the **T** atoms occupy invariant positions in $Im\bar{3}m$ (12d) and $I\bar{4}3m$ (12d) and are virtually unaffected by the tilt (apart from the change in the unit cell volume). However, as mentioned earlier, the site symmetry changes and the O atoms become free to move. This has consequences not only for the geometry of the tetrahedra around **T** (see below), but also for the voids in the framework. However, because there are so many geometrical and symmetrical constraints in the sodalite framework the extent of the possible changes is rather limited. For example, it has been shown by Koch and Hellner (1981) that the vertices of the tetrahedra—the O atoms—obey the relationship $8x^2 - 8z^2 - 1 = 0$ on their move from their positions 24h x,x,0 in $Im\bar{3}m$ to 24g x,x,z in $I\bar{4}3m$. Koch and Hellner (1981) demonstrate that this relationship is indeed well fulfilled by the real members of the sodalite family (see Fig. 4 in Koch and Hellner 1981), where small deviations are due to secondary effects such as **T** atom order and deviations from regular geometry.

Since the *Dirichlet* domains are defined as voids in the packing of O atoms, they have to change as well. In fact, the analysis of the *Dirichlet* domains as a function of the parameters x and z revealed significant changes. The voids do not only change their shape, but some of them split up or even disappear. Rather fortunately, however, those voids which are relevant for the host function of the **SOD** framework can be considered to remain virtually unchanged for practical purposes: this concerns the tetrahedral voids around the **T** atoms, and a big 24-vertex polyhedral void around the *I* lattice complex which can be considered to be the result of joining the (12co): *I* polyhedron and the six (distorted) (6a): P_2' trigonal antiprisms as described earlier for z = 0 (see Fig. 6). It accommodates both, the cage cations and the cage anions.

As an interesting side effect, (Koch and Hellner 1981) found out that for the special coordinates x = 3/8 and z = 1/8 the framework can be derived from a cubic closest sphere packing by omitting a certain number of packing spheres. The overcollapsed framework of the mineral tetrahedrite corresponds approximately to this state.

The z coordinates of the framework O atoms, i.e., the degree of tilting, depend on the size of the **A** cations. Because their coordination requirements must be met by the O atoms, feedback effects occur and the cations also move under the influence of the tilt; they move on 8c x,x,x, away from ¼,¼,¼. In the discussion of the topological structure it was described that the **A** cations form a primitive cubic array with half the lattice parameter of the sodalite. Under the influence of the tilt this array distorts as well. Each small cube becomes a *stella quadrangula*. An ideal *stella quadrangula* has $x = 0.187$. It can be described to consist of five regular tetrahedra, with a central tetrahedron sharing each face with one of the remaining four tetrahedra (see Fig. 8 for an ideal *stella quadrangula*). If $x \neq 0.187$, the *stella quadrangula* is distorted. In view of this approach a cube is a highly distorted *stella quadrangula* with $x = 0.25$. It has already been mentioned that for some purposes it is convenient to consider the **A** cations as an independent framework interpenetrating the sodalite framework (Fig. 2). In the case of a tilted sodalite this framework consists of *stellae quadrangulae*. The eight **A** cations within the cubic unit cell at the corners of one of these *stellae quadrangulae* take a special position with respect to the non-Euclidean P* surface. The inner four **A** cations are located in one, say black, labyrinth, the outer four cations in the other, the white one. Von Schnering and Nesper (1987) argued that the ionic part of the **T**-O bonding system resides as a double layer surrounding a preferentially covalently bonded partial structure. If this is really the case, one would expect that it is difficult to switch the orientation of the **A** cations from x,x,x with $x + \delta$, to x,x,x with $x - \delta$, across the P* surface without considerable costs in energy.

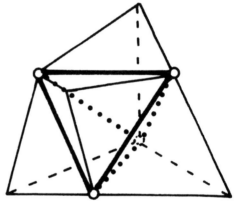

Figure 8. An ideal *stella quadrangula* can be considered to consist of five ideal tetrahedra: a central one shares all its faces with other tetrahedra.

The classical tilt mechanism of the sodalite framework breaks the topological symmetry. With only one kind of **T** atom the space group becomes $I\bar{4}3m$, with two kinds of **T** atoms (e.g., Si^{4+} and Al^{3+}) in a 1:1 ratio and order, the space group becomes $P\bar{4}3n$, which is the space group originally determined by Pauling (1930). Note, that in both cases the cubic symmetry is retained. The site symmetry of the **T** atoms in both, $I\bar{4}3m$ and $P\bar{4}3n$, is $\bar{4}..$, hence the tetrahedra can undergo distortion without breaking the site symmetry.

Pauling (1930) has chosen the coordinates as reported earlier, in particular, he decided to put Al on 6c and Si on 6d. Could he have chosen 6d for Al and 6c for Si just as well? The answer is yes. Several equivalent descriptions exist. The reason is that the Euclidean normalizer of $P\bar{4}3n$, $Im\bar{3}m$ has higher symmetry than the space groups itself (Koch et al. 2002). Fischer and Koch (1983) have discussed the relationships between a given space group and its Euclidean normalizer coordinate sets. In particular, they demonstrate the freedom of decision of a researcher when choosing the coordinates of the first atoms in the course of a structure determination. By way of example they have chosen the description of the sodalite structure using the coordinates as determined by Löns and Schulz (1967). They give the following list of four equivalent coordinate sets:

$P\bar{4}3n$	2a	6c	6d	8e	24i
	0,0,0	¼,½,0	¼,0,½	x,x,x	x,y,z
sodalite	Cl	Al	Si	Na 0.1777	O 0.1487 0.1401 0.4385
sodalite	Cl	Si	Al	Na 0.1777	O 0.1401 0.1487 0.4385
sodalite	Cl	Al	Si	Na −0.1777	O 0.1487 0.1401 −0.4385
sodalite	Cl	Si	Al	Na −0.1777	O 0.1401 0.1487 −0.4385

Note, because $Im\bar{3}m$ is its own Euclidean normalizer, there is a unique coordinate set for the topological structure of sodalite.

As a rule of thumb the relative force constants of **T-O** bonds, **O-T-O** angles and **T-O-T** angles are often assumed to be in a ratio of 1.0:0.1:0.01. From this rule it is clear that much less energy is needed to distort a framework via changes of the **T-O-T** angles than by changing the shape of the tetrahedra. This is the basis for the description of conformational changes by tilt systems. Based on the same ground, but going beyond descriptive crystal chemistry, is the concept of so-called "rigid unit modes" or "RUMs." These are low-, or zero-, energy modes which define possible or probable distortion modes of frameworks. A first model was developed with the aim to explain the phase transition in quartz (Grimm and Dorner 1975). The idea was advanced by Heine, Dove and co-workers and applied to other frameworks, tetrahedral and others (Giddy et al. 1993; Hammonds et al. 1994, 1996; Dove et al. 1995; Hammonds et al. 1997; Dove 1997). They developed a method which allows to calculate the complete RUM spectrum for any framework. The main idea is to split the O atoms which join two rigid TO_4 tetrahedra into halves which are kept together by a strong harmonic force, the force constant being proportional to the rigidity of the TO_4 tetrahedra. The rigid unit modes are precisely those with which the tetrahedra can vibrate without the split O atoms becoming separated. They calculate with zero or very low energy in the phonon spectrum.

The model of rigid unit modes was successfully applied to find answers to a number of puzzling questions related to tetrahedral framework structures. With respect to the sodalite framework, it was possible to explain why sodalites are able to undergo partial collapse at all, whereas other framework structures cannot. A very remarkable result obtained for ideal **SOD** framework in its topological symmetry is that a whole band of rigid unit modes occurs throughout the Brillouin zone. This means that many different superstructures can be formed by condensation of these RUMs, in accordance with observation, notably on aluminate sodalites (e.g., Depmeier 1988a). The model could also answer the question how some frameworks, including **SOD**, fold locally around cations or molecules without distorting the remainder of the structure. This is accomplished by linear combinations of static RUMs. It remains to be studied how far the results can be transferred to sodalites of space groups $I\bar{4}3m$ and $P\bar{4}3n$. However, we know from many studies, especially on aluminate sodalites, that the sodalite host structure is able to accommodate guests the shape of which does not fit that of the surrounding framework. The accommodation is achieved by deformation of the respective sodalite framework, sometimes involving breaking of symmetry. It is also known that sodalites are almost notorious for their ability to accommodate all kind of minor guest species as impurities. In this respect sodalites prove to be closely related to zeolites. RUM studies have shown that such substitutions can easily be brought about by forming localised wave packets of RUMs to form rings or clusters of O atoms around the foreign guests with almost no cost of elastic energy.

This is even true for substitutions within the **SOD** framework as shown in the following. Small amounts of Al atoms replacing Si in the TO_4 tetrahedra in highly silicious zeolites are necessary for their catalytic properties. Because of the different sizes of Al^{3+} and Si^{4+} the substitution produces stress around the AlO_4 tetrahedron. There is an interest to know how the framework relaxes around such a local distortion. Experimentally this kind of information is not easy to obtain. When quantum mechanical computational methods were used to determine the re-orientation potentials of tetramethylammonium cations as guest-molecule in the β-cages of tetramethylammonium sodalite, $|(N(CH_3)_4)_2|[Al_2Si_{10}O_{24}]$-**SOD** (Griewatsch 1998), the desired information was obtained, as a sort of by-product.

The periodic model in space group $P1$ contained three symmetrically independent TO_4 tetrahedra, denoted AlO_4, Si_0O_4 and Si_1O_4. Si_0O_4 has no direct connection with AlO_4, it is considered to represent to a certain extent an undisturbed silica matrix. Si_1O_4, as intermediate, links Si_0O_4 on all four corners with AlO_4 tetrahedra as second-nearest neighbors. Like Si_0O_4, the AlO_4 share corners with Si_1O_4 tetrahedra only. Si_1O_4 has one Si_0O_4, one AlO_4 and two Si_1O_4 as nearest neighbors.

The stress is released not only by conformational changes, but is assisted by significant tetrahedron distortion. The conformational changes do not show the simple features of the classical sodalite tilt system. They concern mainly the AlO_4 tetrahedron which is said to "twist;" the "twist" is zero for Si_0O_4. Both, AlO_4 and Si_0O_4 tetrahedra show only topology-induced tetragonal tetrahedron distortion (see below). The Si_1O_4 tetrahedra as intermediates undergo heavy bond length and angular distortion. As a result one can state that the stress relaxes back to zero over a distance of only three TO_4 tetrahedra, or half the edge length of the cubic unit cell. Furthermore, the example demonstrates clearly that the sodalite framework can relax locally, in accordance with the idea of the formation of bundles of RUMs. On the other hand, the fact that variations of the **T**-O bond lengths and O-**T**-O bond angles are involved in such an important manner questions simplifying approaches such as the model of simple tilt systems of rigid, not to say ideal, tetrahedra. It will be emphasized in a different context (Loewenstein's rule) that structures such as that of sodalite are complex systems. The different partial structures do not behave independently, they rather interact in complicated ways which only in an approximation of lowest order can be considered as independent and linear. It must be kept in mind that Nature does not pay attention to how people decompose a structure into "partial structures." The same is true for items such as atoms, bonds, atomic groups, and so on. All these notions are merely crooks for the restricted human imagination and mind, used to rationalize quantum-mechanical systems. Therefore, one should duly be aware of the limitations when one discusses the behavior of a structure. If this caveat is duly respected, simplifications are, indeed, helpful.

Tetragonal tetrahedron distortion

The geometry of the TO_4 tetrahedra in the sodalite frameworks is not ideal. In light of the reported complexity of the structure this should not come as a big surprise and could not raise much interest, were it not for the fact that some of these distortions occur systematically.

The first one which shall be dealt with is the so-called tetragonal tetrahedron distortion (Depmeier 1984). It was first observed in the structures of aluminate sodalites, in particular |Ca$_8$(WO$_4$)$_2$|[Al$_{12}$O$_{24}$]-**SOD**, where a strong deviation of the geometry of the tetrahedra from regular occurred. The tetrahedra were found to be compressed along the $\bar{4}$-axis of their site symmetry, with two O-**T**-O angles α ~ 120°, and four angles α' ~ 105°. The deviation from regular is much more pronounced than in aluminosilicate sodalites where the corresponding angles amount to only about 112° and 108°. This is probably the reason why this systematic effect had previously been ignored. The tetragonal tetrahedron distortion preserves the cubic symmetry and is even compatible with the body-centering. Formulae were given which express the relations between the cubic lattice parameter, the tilt angle and the tetragonal tetrahedron distortion (Depmeier 1984). The geometrical formulae also allowed to find explicit relationships between the tetrahedron distortion angle α, the tilt angle φ, and the intertetrahedron angle γ. It was shown that the tetragonal tetrahedron distortion having an angle α = 120° reduces the volume of the tetrahedron by 2.5%. In addition, it reduces the volume of the unit cell and also the free volume, i.e., that volume of the unit cell which is not occupied by the framework. It is interesting to note that the lattice parameter depends more strongly on a change in α than on a change in φ.

Structural data on several natural and synthetic sodalites known at that time were plotted into diagrams of e.g., z(O) vs. x(O), or α vs. average **T** atom radius, where **T** is Al and Si, or φ vs. α with the dependent γ as parameter. Some observations in these diagrams called for an explanation. For instance, it was found that with the data at hand the amount of the tetragonal tetrahedron distortion, expressed by α, increases almost linearly with the average **T** atom radius, i.e., the Al-content in the framework composition of the aluminosilicate series [Al$_{12-x}$Si$_x$O$_{24}$]$^{(12-x)-}$. This behavior was rationalized as follows: a sodalite framework consisting of regular tetrahedra and being in its fully expanded state would have a **T**-O-**T** angle of 160.5°. This value is unfavorable because in silicate linkage structures the preferred angle is about 145° (Megaw 1973). Therefore a stress is created and two possibilities exist for the structure to reduce this angle, and thereby, to release the stress, viz. i) by increasing the tilt angle or, ii) by increasing the angle α. The first is the favorite mechanism for the Si-rich side of the series, whereas the second one prevails for Al-rich members. The tilt angle φ is mainly determined by the cage content, whereas α seems to depend on the framework composition, or to be more precise, on the Al content. Since the publication of Depmeier (1984) several new sodalite species have been synthesized and their structure determined. The results with respect to the topological stress, the tetragonal tetrahedron distortion, are less clear than it appeared 20 years ago with a limited data set. While in practically all cases the topology-induced tetrahedron distortion is observed, the linear relationship between the framework composition and the distortion angle found several exceptions, for still unknown reasons.

Nuclear magnetic resonance, NMR, techniques allow the determination of the distortion of the AlO$_4$ tetrahedra on a local length scale. For a series of aluminosilicates the quadrupole coupling constant, QCC, for ^{27}Al, was related to the respective tetrahedron distortion (Ghose and Tsang 1973). Two parameters were introduced, namely a longitudinal strain, and a shear strain. The QCC was found to depend strongly on the latter, but to be almost independent of the former. Shear strain and QCC are linearly related for a whole series of various aluminosilicates. The problem with the standard MAS-NMR technique is that it cannot cope with the second order interactions which produce strong line broadening for the resonances of quadrupolar nuclei, such as ^{27}Al. A special technique, Double-Rotation-NMR, allows to overcome the problem and to register spectra of high resolution from such nuclei (Samoson et al. 1988). One of the first examples where this method could demonstrate its power was |Ca$_8$(WO$_4$)$_2$|[Al$_{12}$O$_{24}$]-**SOD**, which was known to contain seven independent, highly distorted, AlO$_4$ tetrahedra (Depmeier 1984). These features made it a suitable case for testing the new method. The test was successful and the dramatic peak sharpening allowed the easy identification of the expected seven central resonances (Engelhardt et al. 1992). The linear relationship between QCC and the tetrahedron distortion described by Ghose and Tsang (1973) could be confirmed.

Mixed crystal effects could be studied in the system $|(Ca_{1-x}Sr_x)_8(WO_4)_2|[Al_{12}O_{24}]$-**SOD**. Already for $x = 0.1$ heavily broadened ^{27}Al-MAS-NMR spectra were found. The overall shape of the spectra did not change. This is interpreted to be due to superposition of many differently distorted environments of the ^{27}Al-nuclei. Different distortions result from different distributions of non-framework atoms around the AlO$_4$ tetrahedra. NMR spectra in the middle range of the composition are much broader, but otherwise similar to those of the non-cubic end members. The X-ray patterns of these samples appear cubic. This indicates that the crystals are cubic on the characteristic length scale of X-rays (say 1000 Å), but non-cubic on the much shorter 10–20 Å length scale of the NMR technique (Többens 1998).

Twisting and cage cation ordering

A remarkable combination of both, a conformational change of the sodalite framework and a distortion of its constitutents, the **TO**$_4$ tetrahedra, has been observed in a series of rare earth (REE) containing Al-rich sodalites of the formula $|REE_4(MoO_4)_2|[Al_8Si_4O_{24}]$-**SOD** (Bernotat and Pentinghaus 1991; Roth et al. 1989). These compounds have a potential as storage media for radioactive waste materials. While each sodalite cage contains one MoO$_4$ group, only half of the available cage cation positions are actually occupied. For different REE cations in this family a strong dependence of the lattice parameters on the ionic radii was observed. The classical tilt system as described before cannot be active, because it is incompatible with the centrosymmetric space group $Pn\bar{3}m$, observed for all the species. The volume reduction is rather the result of i) a twist-like tetrahedron distortion, and ii) a conformational change of the framework. Both mechanisms together constitute what has become known as "twisting". By i) the **TO**$_4$ tetrahedra are distorted into rhombic disphenoids of point group 222; ii) preserves the inversion and the cubicity, but destroys the body-centering, i.e., **TO**$_4$ tetrahedra related by a +(½,½,½) translation turn in opposite directions. The magnitude of the distortion depends on the size of the REE cations: the smaller the REE, the stronger the twist, and vice versa.

The half-occupancy of the cage cation positions has interesting consequences. In the course of the transformation from $Im\bar{3}m$ to $Pn\bar{3}m$ the position $8c$ ¼,¼,¼ splits into two positions, viz. $4b$ ¼,¼,¼ and $4c$ ¾,¾,¾, only $4b$ being reported to be occupied by REE cations. The arrangement of the REE with respect to each cage is that of a tetrahedron, whereby each REE atom belongs to two adjacent cages, there are thus 4/2 REE cations per cage. The REE tetrahedra in the two cages at 0,0,0 and ½,½,½ are oriented in opposite directions. Each cage is occupied by one MoO$_4$ tetrahedron (disordered) in "cubic" orientation, i.e., local triads point in the cubic <111> directions, but tetrahedra in the cages at 0,0,0 and ½,½,½ have opposite directions, again in accordance with the absence of the body-centering.

The centers of the anions (Mo) at $2a$ and the REE at $4b$ together form an arrangement which is similar to that of the cuprite structure type, i.e., it consists of two interpenetrating cristobalite-type structures with Mo corresponding to O of cuprite and REE to Cu. This raises the question whether a structure with cations at exactly ¼,¼,¼ can be really stable, at least in view of the arguments of (von Schnering and Nesper 1987) as given above. The doubts find support by the observed strongly anisotropic atomic displacement functions of the REE with elongated ellipsoids parallel <111>, probably indicative of disorder of the REE across the 6-ring. It should also be noted that the differently oriented MoO$_4$ tetrahedra render the coordination of the REE at $4b$ asymmetric which lends additional support to the idea of disorder.

The angle of twisting can be defined as the angle between two edges of framework tetrahedra which are perpendicular to each other (in projection) in the topological symmetry and also perpendicular to the -4 axis in the case of tilting. Geometrical relationships between atomic coordinates, lattice parameters and the twisting angle have been worked out (Pentinghaus et al. 1991).

Cage anion ordering and shearing

Cage anion ordering is of paramount importance for aluminate sodalites, e.g., $|Ca_8(WO_4)_2|[Al_{12}O_{24}]$-**SOD** (Depmeier 1984b). Their cubic phases are characterized by dynamical disorder of the tetrahedral cage anions XO_4 about six equivalent "tetragonal orientation states" (Depmeier 1988b). The resulting structure is cubic only on the space/time average. The existence of strong rotational-translational coupling to the framework has been inferred from the anomalous temperature dependence of certain atomic displacement parameters (Depmeier 1988b). At a given temperature this thermally activated process breaks down, the anion disorder condenses, and the cubic symmetry gets lost. This is accompanied by the formation of superstructures, commensurate for end-members, but also incommensurate for mixed crystals (Többens 1998). Structurally the formation of the superstructures are the results of non-homogeneous distortions of the framework, called conformational shearing (for an illustration, see Fig. 3 of Depmeier 1988a). Shearing is a cooperative, but local phenomenon, caused by non-homogeneous distribution of repulsive forces acting between XO_4 groups and O atoms of the framework. Up to now shearing has only been identified in aluminate sodalites (Depmeier 1984b; Depmeier and Bührer 1991), but it can be expected to occur also in the sulphate-bearing aluminosilicate sodalite, hauyne. Shearing is characterized by the following facts: i) It breaks the cubic symmetry, ii) It breaks translational symmetry, such that superstructures are formed, iii) Its occurrence depends on the type, arrangement, orientation and order of the tetrahedral oxyanions in the sodalite cages, iv) The latter depends on the nature of the cage cations, therefore different shearing patterns occur for different compositions. v) It is highly probable that shearing occurs locally and temporally in the cubic phases of aluminate sodalites, as indicated by NMR results (Többens 1998) vi) Substitution on the cage anion sites influence the equilibrium between the three partial structures, but less so than with cage cation substitution.

T site ordering, Loewenstein's rule

T site ordering is important, for example for 1:1 aluminosilicate sodalites with framework composition $[Al_6Si_6O_{24}]^{6-}$, or similar 1:1 sodalites. The ordering often results in alternate occupation of the TO_4 tetrahedra by Al and Si. It destroys the body-centering, but preserves the cubic symmetry and the inversion center, if present. As a matter of fact, however, **T** site ordering seems to have been only observed for tilted aluminosilicate sodalites until now. Correspondingly the highest symmetry reduces from $I\bar{4}3m$ to $P\bar{4}3n$, the space group of the mineral sodalite (Pauling 1930). An untilted, but **T**-ordered sodalite would have space group $Pm\bar{3}n$. Possible sequences of group-subgroup relations are: i) $Im\bar{3}m \to I\bar{4}3m \to P\bar{4}3n$ or ii) $Im\bar{3}m \to Pm\bar{3}n \to P\bar{4}3n$. The corresponding sequences of occupied Wyckoff positions and their site symmetries (in parentheses) for the **T** atoms are: i) $12d\,(\bar{4}m.2) \to 12d\,(\bar{4}..) \to 6c$; $6d$ (both $\bar{4}..$) and ii) $12d\,(\bar{4}m.2) \to 6c$; $6d\,(\bar{4}m.2) \to 6d$; $6c$ (both $\bar{4}..$) (Müller 2004). It should be noticed that the **T** site ordering itself does not affect the symmetry of the TO_4 tetrahedra.

Of course, **T** site ordering is irrelevant for sodalites containing only one type of **T** atom, e.g., in aluminate sodalites, silica sodalites, borate sodalites or nitride sodalites. For aluminosilicate sodalites having an Al:Si ratio different from 1.0, long range order of the **T** atoms does not seem to occur. Examples are $|((CH_3)_4N)_2|[Al_2Si_{10}O_{24}]$-**SOD** (Baerlocher and Meier 1969; Sokolova et al. 1993; Griewatsch et al. 1998), bicchulite $|Ca_8(OH)_8|[Al_8Si_4O_{24}]$-

SOD (Henmi et al. 1974; Sahl and Chatterjee 1977; Gupta and Chatterjee 1978; Sahl 1980), or |REE$_4$(MoO$_4$)$_2$|[Al$_8$Si$_4$O$_{24}$]-**SOD** (Bernotat and Pentinghaus 1991a,b; Roth et al. 1989).

Loewenstein (1954) analyzed the distribution of Al in tetrahedral structures of silicates and aluminates which were known at that time, and formulated the following rules, which were later named after him: *"Whenever two tetrahedra are linked by one oxygen bridge, the center of only one of them can be occupied by aluminium; the other center must be occupied by silicon (...). Likewise, whenever two aluminium ions are neighbors to the same oxygen anion, at least one of them must have a coordination number larger than four (...). These rules explain the maximum substitution of 50% of the silicon in three-dimensional frameworks and plane networks of tetrahedra by aluminium. For 50% substitution, rigorous alternation between silicon and aluminium tetrahedra becomes necessary (...).*"

Loewenstein rationalized these observations on grounds of Pauling's rules and noted that the electrostatic valence rule was not really helpful in order to understand the experimental facts. He rather relied on Pauling's third rule which states that coordination polyhedra of cations with low coordination number tend to avoid direct linkage. If any two of such cations should be obliged to link their coordination polyhedra they would tend to increase their coordination number. This tendency would be supported if the ionic radii of the cations involved would approach the critical value which allows for a change of the coordination number. This should be the case for Al^{3+} which is found equally well in tetrahedral and octahedral oxygen environment. Si^{4+}, on the other side, has practically invariably tetrahedral coordination, at least at ambient and not too high pressure.

Loewenstein's rule, also known as the aluminium avoidance rule, has become both an useful and popular guide in the crystal chemistry of silicates and aluminosilicates. Recently, it has found theoretical support by a number of studies. Lattice energy calculations have shown that Al-O-Al linkages are, indeed, slightly more unfavorable than Si-O-Al, the enthalpy difference being of the order 0.5–1.0 eV. This means that alternating SiO$_4$ and AlO$_4$ tetrahedra are favored in comparison with direct linkage of AlO$_4$ tetrahedra (Dove et al. 1993). Bosenick et al. (2001) determined the exchange energies for the reaction Al-O-Al + Si-O-Si = 2 Al-O-Si for several structures having one-, two- or three-dimensionally linked tetrahedra and found strong variability. This was taken as indication that it is not the difference in formal charge between Al^{3+} and Si^{4+}, but rather the difference in cation size which is responsible for the energy differences (ionic radii are 0.39 Å for Al^{3+} and 0.26 Å for Si^{4+}; Liebau 1985). The different topologies of the tetrahedron linkages in the structures studied allows more or less easy relaxation of the framework in response to the different size of the tetrahedra. From these observations one arrives at the conclusion that Loewenstein's rule is based on the higher energy costs which have to be paid due to the elastic deformation of the structure when larger AlO$_4$ tetrahedra replace SiO$_4$.

The exception proves the rule. One apparent counter-example to his rule was already mentioned in Loewenstein's original paper. In order to explain the existence of the cristobalite-type structure of KAlO$_2$ with exclusively tetrahedrally coordinated Al^{3+}, he argued that this compound could not crystallize in an alternative structure, however, without explaining why the compound should then form at all. Besides KAlO$_2$, several other aluminates with corner-linked AlO$_4$ tetrahedra have been synthesized, and could naively be also considered as exceptions to Loewenstein's rule. This concerns, e.g., earth alkaline aluminates of formula M^{II}Al$_4$O$_7$ and M^{II}Al$_2$O$_4$ (Kahlenberg 2001), tricalcium aluminate and derivatives (Nishi and Takéuchi 1975; Mondal and Jeffrey 1975; Takéuchi et al. 1980), and numerous aluminate sodalites (e.g., Depmeier 1988b). In view of the explanation given above for its energetic grounds, it is clear that Loewenstein's rule is not applicable to structures with just one kind of **T**O$_4$ tetrahedra, and the mentioned aluminates are only apparent exceptions. Besides, it is quite remarkable, and for disciples of the zeolite community perhaps counter-intuitive, that aluminate tetrahedra framework structures often exhibit high thermal stability. Aluminate sodalites, for example, melt at temperatures as high as about 2000°C, or even higher. This stabilization is probably due to the formation of strong ionic bonds between bi- or trivalent cations and the underbonded O atoms in the Al-O-Al linkages.

Loewenstein's rule is normally used to corroborate two assumptions, viz., i) the supposition of strictly alternating AlO_4 and SiO_4 tetrahedra in 1:1 aluminosilicates, and ii) the postulation of an upper limit for the Al:Si ratio of 1:1 in aluminosilicates. This in principle rather useful rule is sometimes misused as a strict law and in unfortunate cases this might lead to wrong conclusions or dubious interpretations. For example, there is ample evidence of deviations from strict alternation of AlO_4 and SiO_4 tetrahedra in 1:1 aluminosilicates. This should not come as a big surprise in view of the relatively small energy differences involved. Simple thermodynamic considerations make it plausible that a 1:1 aluminosilicate which is ordered at room temperature becomes disordered at higher temperatures due to entropic stabilization. The disorder results necessarily in a certain number of direct Al-O-Al linkages and, hence, in local deviations from Loewenstein's rule. On a local length scale we have therefore Al-Al order with temperature-dependent correlation lengths. Upon cooling the (global) Al-Si-disorder can be frozen in and become metastably fixed. Such deviations from Loewenstein's rule at high and low temperatures, their dependence on the topology of the linkages and their dimensionality, as well as their influence on the critical temperature of phase transitions have been studied theoretically and experimentally not only for 1:1 aluminosilicates, but also for aluminosilicates with Al:Si < 1.0 (Bosenick et al. 2001).

While most people seem to agree, at least tacitly, with the temperature-dependent deviations from Loewenstein's rule of aluminosilicates with Al:Si ≤ 1.0, the existence of a strict upper limit for the Al content at a Al:Si ratio of 1:1 does not seem to have been disputed in the community, unless topological particularities can be hold responsible for such a behavior, as for example, in the gehlenite structure (Thayaparam et al. 1994). In some cases graphics were published to illustrate the dependence of certain physical properties as a function of the Al:Si ratio. The diagrams end exactly at the 1:1 ratio, thereby insinuating that beyond that border the area is not just unknown, but inaccessible on principle. This is in particular true for 3D linked tetrahedral frameworks with only one topologically independent **T** atom. Such a framework is of course that of **SOD**. Not surprisingly, entropically stabilized 1:1 aluminosilicate sodalites with Al-Si disorder are known to exist (Tarling et al. 1988). Rather more surprising, and contrary to the expectations of Loewenstein's rule, are reports on the existence of Al-rich aluminosilicate sodalites with a framework composition $[Al_{12-x}Si_xO_{24}]^{(12-x)-}$ with $0 < x < 6$. Such sodalites must break Loewenstein's rule in the sense that they are not in accordance with the idea of inaccessibility of the area with Al:Si > 1.0. The reports concern the rare mineral bicchulite $|Ca_8(OH)_8|[Al_8Si_4O_{24}]$-**SOD**, (Henmi et al. 1974; Sahl and Chatterjee 1977; Gupta and Chatterjee 1978; Sahl 1980), and the even rarer mineral kamaishilite (Uchida and Iiyama 1981). The latter mineral is reported to have the same composition as bicchulite, but to be tetragonal instead of cubic. According to the literature Al and Si in bicchulite are disordered, for kamaishilite the distribution is unknown. Bicchulite can be synthesized from gehlenite, $Ca_2Al[AlSiO_7]$, by hydrothermal methods. Gehlenite crystallizes in space group $P42_1m$ in the melilite structure type. It contains two topologically different tetrahedral positions, one of which is fully occupied by Al^{3+}, whereas the second one contains Al^{3+} and Si^{4+}. Normally, the Al^{3+} molar fraction in the second tetrahedral position does not exceed $x_{Al} = 0.5$, in accordance with Loewenstein's rule. Recently, we have shown that it is possible to substitute much more Al^{3+} for Si^{4+} in the second tetrahedral site than is allowed by Loewenstein's rule. This happens via a coupled substitution of the type $(REE^{3+}+Al^{3+}) \rightarrow (Ca^{2+}+Si^{4+})$. Compounds of the composition $REE_xCa_{2-x}Al[Al_{1+x}Si_{1-x}O_7]$ with $0 < x < 1$ were synthesized at 1773 K and ambient pressure.

Rietveld-refinements of powder diffraction patterns show that the single phased products crystallize in space group $P42_1m$, without changing the Wyckoff-positions of the ions in the melilite structure. Substitutions of a similar type were shown to be possible in the sodalite structure, starting from bicchulite. Note that in contrast to gehlenite the sodalite structure type contains only one topologically independent tetrahedral position. Solid solutions in the compositional range $|Ca_8(OH)_8|[Al_8Si_4O_{24}]$-**SOD** – $|Eu_4Ca_4(OH)_8|[Al_8Si_4O_{24}]$-**SOD** have been prepared hydrothermally at 800–900 K, 0.1 GPa, starting from the melilite-type compounds.

The 2:1 ratio of Al:Si in **SOD**-type structures has also been found in the already mentioned REE-containing structures with half-occupancy of the positions of guest cations, e.g., $|Dy_4(MoO_4)_2|[Al_8Si_4O_{24}]$-**SOD** (Scheikowski and Müller-Buschbaum 1993). Violations of Loewenstein's upper limit assumption in **SOD**-type structures do not seem to be restricted to the conspicuous 2:1 ratio of Al:Si. Löns (1969) reported on the synthesis of Al-rich mixed crystals of hauynite and $CaSO_4$ aluminate sodalites. From structural and optical investigations he concluded that in the synthesized compounds Al-rich framework compositions occur, with Al:Si ratios ranging from 2.0 to 5.0. Independent verifications of the results and characterization of the products employing modern techniques would be desirable.

In conclusion one can state that Loewenstein's rule, also known as the aluminum-avoidance rule, is violated for the **SOD** structure type. Ongoing work is aimed at finding systematically **SOD**-type structures with variable Al:Si > 1.0, and to relate the structural studies with the investigation of properties. An important question is whether other framework types are also capable of breaking Loewenstein's rule. If this is not the case, the question arises what makes **SOD** special with respect to the Al-Si distribution in comparison with other framework types. Possible clues to an answer are the following facts: i) there is only one topologically distinct **T** atom in **SOD**, ii) the high global flexibility of the **SOD** framework, iii) the high local flexibility of the **SOD** framework which allows local deformations by linear combinations of static rigid unit modes, iv) the smallness of the micropores of **SOD** which allows a high local concentration of cationic charge in order to satisfy the bond requirements of the underbonded O in Al-O-Al linkages, v) the fact that the micropores in **SOD** can be filled with guest species excluding water, water being known to weaken the Al-O-Al linkages more than the Al-O-Si, let alone Si-O-Si, v) a lucky combination of circumstances which allows the "right" combination of cage cations and **T** cations to ensure a coupled substitution of the kind as given above and which might not be possible in other structures.

STRUCTURAL DEPENDENCE ON SUBSTITUTION, TEMPERATURE AND PRESSURE

Phase transitions and modulated structures

One of the most appealing properties of the sodalite family is the fact that its members undergo many interesting phase transitions (Depmeier 1988a). Some of these phase transitions have been investigated in detail. Some are of ferroic character and are accompanied by attractive properties, such as ferroelasticity or ferroelectricity (Depmeier 1988a). Furthermore, the phase transitions often result in the formation of modulated structures. This property seems to make the sodalite family unique among microporous structures. The formation, the symmetry and reasons for the formation of modulated structures in sodalites have been discussed in detail elsewhere (Depmeier 1992a). The main arguments are: i) The occurrence of modulated structures seems to be linked to the presence of tetrahedral cage anions in the sodalite cages, therefore modulations are especially frequent in aluminate sodalites; ii) the reason for the occurrence of modulated structures is believed to be structural frustration between the three partial structures, viz., a) the sodalite framework, b) the cage cations, c) the cage anions:

Favorable interactions a) – b) entail unfavorable interactions a) – c), and *vice versa*, hence the frustration; iii) commensurate and incommensurate modulations have been found; iv) (3+1) and (3+3) dimensional superspace groups were assigned; v) substitution on either of the two partial structures pertaining to the guest species, affects the phase transitions and the modulations. The results are complicated phase diagrams with temperature and composition as parameters. The influence of cation substitution has been studied and described in some detail (Többens and Depmeier 1998), and corresponding studies on the effect of anionic substitutions are currently under way.

Thermal expansion

The classical model of anharmonic atomic pair potentials is normally insufficient if the thermal expansion of tetrahedral framework structure is to be described. The distortion mechanisms on the microscopic level must be taken into account, as these may contribute to the expansion. They may also cancel or even result in zero or negative expansion coefficients. The unfolding of a partially collapsed framework will contribute strongly to its expansion, until it comes to a stop, for instance, when the full-expanded state is attained, or if other forces resist further unfolding. At this point a discontinuity will be expected in the thermal expansion curve. If the tetrahedra in a flexible framework become dynamically disordered, librational movements may result in apparent bond shortening and negative expansion for lattice parameters and volume.

One of the most-extensively studied family of tetrahedral frameworks, with regard to thermal expansion behavior, is just that of the sodalites. The thermal expansion of 15 different cubic alkalihalide aluminosilicate sodalites was determined by X-ray powder diffraction experiments (Henderson and Taylor 1978). Some sodalites containing big cage anions exhibited quite unexpected behavior, as they underwent strong thermal expansion (about $22 \cdot 10^{-6}$ K^{-1}) at lower temperatures, but changed discontinuously to a regime of low thermal expansion (about $8 \cdot 10^{-6}$ K^{-1}) at higher temperatures. There was no doubt that unfolding of the framework contributed significantly to the behavior in the high expansion regime, beyond the discontinuity normal anharmonicity was assumed to prevail. The reason for the discontinuity in the slope of the thermal expansion curves has remained doubtful for many years and continues to do so.

Some people believed that the framework attains its fully-expanded state at the discontinuity (Henderson and Taylor 1978). Another hypothesis was that by expansion of the bonds between cage anions and cage cations the latter are pressed against the framework, which expands passively by giving way to this action. The process was believed to come to an end when the cations reach the ¼,¼,¼ position in the cubic sodalite structure. This position was assumed to be a particularly stable one for the cation (Hassan and Grundy 1984; Dempsey and Taylor 1980). This point of view was not in agreement with the earlier mentioned ideas on the sodalite framework as an approximation to the *Schwarz* P* surface with a layered distribution of ionic and covalent bond characteristics (von Schnering and Nesper 1987). On grounds of a geometrical model it was also argued that the ¼,¼,¼ position of the cations could not be particularly stable. It was proposed instead that the topology of the sodalite framework creates a barrier against further unfolding of the framework. The barrier was postulated to be at **T**-**O**-**T** angles of about 160° (Depmeier 1984a).

A recent study employing temperature-dependent neutron powder diffraction and Rietveld refinement revealed the concerted movement of the atoms during the unfolding of the framework of the cubic aluminate sodalite |(Ca$_{0.5}$Sr$_{0.5}$)$_8$(WO$_4$)$_2$|[Al$_{12}$O$_{24}$]-**SOD** (Többens 1998). It could be observed that the cations passed beyond the ¼,¼,¼ position, obviously without stopping, and no discontinuity in the thermal expansion curve was observed. The **T**-**O**-**T** angle remained below 160°. This observation is thus in accordance with (Depmeier 1984a).

The effects of ferroelastic phase transitions have been measured for several aluminate sodalites by the temperature-dependent determination of the lattice parameters (Depmeier 1988a). In $|Sr_8(MoO_4)_2|[Al_{12}O_{24}]$-**SOD**, the c lattice parameter follows a power law behavior below the cubic → tetragonal transition, as does the unit cell volume. Surprisingly, the $a = b$ lattice parameters seem to be unaffected by the phase transition as they follow an almost linear downward slope with decreasing temperature, even across the phase transition. This peculiar behavior could be explained by assuming superposition of two main effects, viz. a spontaneous strain and a volume strain. Both strains cancel in their effects on the a-lattice parameter, but add up for c, thus yielding the observed behavior (Depmeier et al. 1993).

High pressure studies

Several members of the sodalite family, and other microporous structures, have been the subject of high pressure research, both experimental and computational (Hazen and Finger 1982; Kudoh and Takéuchi 1985; Hazen and Sharp 1988; Li et al. 1989; Chaplot and Sikka 1993; Fütterer et al. 1994; Melzer et al. 1995; Többens et al. 1995; Werner and Plech 1995; Werner et al. 1996; Dove 1997; Knorr et al. 2001).

Bulk moduli determined for a number of them are given in Table 2 (modified after Griewatsch 1998). As a first approximation the different values can be rationalized by referring to different lattice energies and cage loadings. For example, simple consideration of the lattice energy would have predicted for helvite, $|Mn_8S_2|[Be_6Si_6O_{24}]$-**SOD**, to be one of the stiffest members of the sodalite family, if not the stiffest. This is indeed the case, as determined by single crystal techniques up to 4.5 GPa (Kudoh and Takéuchi 1985). However, in this and other high pressure studies on members of the sodalite family some remarkable results occur which cast doubt on the reliability of the experimental conditions. In the particular case of helvite, an abrupt shrinkage of the BeO$_4$ tetrahedra and a slight expansion of the SiO$_4$ tetrahedra is reported to occur at about 3 GPa. Similar effects have been observed in other sodalites as well, but have not found a satisfactory explanation up to now. *Ab initio* calculations based on density functional theory lend support to the experimental observations of small discontinuities in the pressure range around 3 GPa (Knorr et al. 2001).

As far as the global mechanism of the compression of the sodalite frameworks is concerned, results of experimental and computational studies seem to agree that it is, to a large extent, the classical tilt system which is responsible for the volume reduction under pressure (Fütterer et al. 1994; Knorr et al. 2001).

Table 2. Bulk moduli for several members of the sodalite family.

Name	Formula	Bulk modulus [GPa]	Reference
TMASOD	$|((CH_3)_4N)_2|[Al_2Si_{10}O_{24}]$	24.6 (7)	Griewatsch 1998
CAM	$|Ca_8(MoO_4)_2|(Al_{12}O_{24}]$	41(2)	Többens 1998
TRSS	$|(C_3H_6O_3)_2|[Si_{12}O_{24}]$	44(2)	Fütterer et al. 1994
Sodalite	$|Na_8Cl_2|[Al_6Si_6O_{24}]$	49(6)	Werner et al. 1996
Sodalite	$|Na_8Cl_2|[Al_6Si_6O_{24}]$	52(8)	Hazen and Sharp 1988
CAW	$|Ca_8(WO_4)_2|[Al_{12}O_{24}]$	59(6)	Többens 1998
Tugtupite	$|Na_8Cl_2|[Al_2Be_2Si_8O_{24}]$	62 no e.s.d	Werner and Plech 1995
SACR	$|Sr_8(CrO_4)_2|[Al_{12}O_{24}]$	78 no e.s.d.	Melzer et al. 1995
Helvite	$|Mn_8S_2|[Be_6Si_6O_{24}]$	111 no e.s.d.	Kudoh and Takéuchi 1985

CONCLUSION

The sodalite family is, indeed, an interesting structural family combining the virtues of microporous materials with those of dense structures. Only few of them could have been covered in this contribution. Future studies will undoubtedly reveal many more attractive features. New synthetic routes will provide us with new members of the family having up to now unheard-of chemical compositions or structural particularities. The sodalite structure lends itself to a decomposition into partial structures. This decomposition allows many basic properties to be explained using relatively simple arguments. However, care must be taken, beyond a certain level of complexity the simple arguments may be misleading.

ACKNOWLEDGMENT

The financial support of many years by the Deutsche Forschungsgemeinschaft is gratefully acknowledged (Contract # De 412/*-*)

REFERENCES

Andersson S, Jacob M (1997) The Mathematics of Structures – The Exponential Scale R. Oldenbourg Verlag, München, Wien

Armstrong JA, Dann SE Neumann K, Marco JF (2003) Synthesis, structure and magnetic behavior of the danalite family of minerals, $Fe_8[BeSiO_4]_6X_2$ (X = S, Se, Te). J Mater Chem 13:1229-1233

Baerlocher C, Meier WM (1969) Synthese und Kristallstruktur von Tetramethylammonium-Sodalith. Helv Chim Acta 52:1853-1860

Baur WH (1992) Self-limiting distortion by antirotating hinges as the principle of flexible but noncollapsible frameworks. J Solid State Chem 97:243-247

Baur WH (1995a) Framework mechanics: limits to the collapse of tetrahedral frameworks. In: Proceedings of the 2nd Polish-German Zeolite Colloquium, Rozwadowski M (ed), Nicholas Copernicus University Press, Toruń, p 52-54

Baur WH (1995b) Why the open framework of zeolite A does not collapse, while the dense framework of natrolite is collapsible. In: Proceedings of the 2nd Polish-German Zeolite Colloquium, Rozwadowski M (ed), Nicholas Copernicus University Press, Toruń, p 171-185

Baur WH, Joswig W, Müller G (1996) Mechanics of the feldspar-type framework and the crystal structure of Li-feldspar. J Solid State Chem 121:12-23

Bedard RL, Wilson ST, Vail LD, Bennett JM, Flanigen EM (1989). The next generation: synthesis, characterization, and structure of metal sulfide-based microporous solids. In: Zeolites: Facts, Figures, Future. Proceedings of the 8th International Zeolite Conference. Jacobs PA, van Santen RA (eds) Elsevier, Amsterdam, p 375-387

Bernotat H, Pentinghaus H (1991) Über neue Glieder der Sodalith-Strukturfamilie: $Se_4Al_8Si_4O_{24}(MoO_4)_2$. Abstracts Gemeinsame Tagung der AGKr und VFK, München

Bibby DM, Dale MP (1985) Synthesis of silica-sodalite from non-aqueous systems. Nature 317:157-158

Bosenick A, Dove MT, Myers ER, Palin EJ, Sainz-Diaz CI, Guiton BS, Warren MC, Craig MS, Redfern SAT (2001) Computational methods for the study of energies of cation distributions: applications to cation-ordering phase transitions and solid solutions. Mineral Mag 65:193-219

Braunbarth CM, Behrens P, Felsche J, van de Goor G (1997) Phase transitions and thermal behavior of silica sodalites. Solid State Ionics 101-103:1273-1277

Braunbarth CM, Behrens P, Felsche J, van de Goor G, Wildermuth G, Engelhardt G (1996) Synthesis and characterization of two new silica sodalites containing ethanol amine or ethylene diamine as guest species: $[C_2H_7NO]_2[Si_6O_{12}]_2$ and $[C_2H_8N_2]_2[Si_6O_{12}]_2$. Zeolites 16:207-217

Chaplot SL, Sikka SK (1993) Molecular-dynamics simulation of pressure-induced crystalline-to-amorphous transition in some corner-linked polyhedral compounds. Phys Rev B 47:5710-5714

Dann SE, Weller MT (1996) Synthesis and structure of cadmium chalcogenide beryllosilicate sodalites. Inorg Chem 35:555-558

Dempsey MJ, Taylor D (1980) Distance least-squares modeling of the cubic sodalite structure and of the thermal expansion of $Na_8(Al_6Si_6O_{24})I_2$. Phys Chem Miner 6:197-208

Depmeier W (1984a) Tetragonal tetrahedra distortions in cubic sodalite frameworks. Acta Cryst B 40:185-191

Depmeier W (1984b) Aluminate sodalite $Ca_8[Al_{12}O_{24}](WO_4)_2$ at room temperature. Acta Cryst C 40, 226-231

Depmeier W (1988a) Aluminate sodalites - a family with strained structures and ferroic phase transitions. Phys Chem Miner 15:419-426

Depmeier W (1988b) The structure of cubic aluminate sodalite $Ca_8[Al_{12}O_{24}](WO_4)_2$ in comparison with its orthorhombic phase and with cubic $Sr_8[Al_{12}O_{24}](CrO_4)_2$. Acta Cryst B 44:201-207

Depmeier W (1992a) Phase transitions and modulated phases in aluminate sodalites. J Alloys Compd 188: 21-26

Depmeier W (1992b) Remarks on symmetries occurring in the sodalite family. Z Kristallogr 199 75-89

Depmeier W (1993) A contribution to the knowledge of aluminate sodalites. Habilitation Thesis, University of Geneva, Switzerland

Depmeier W (1999) Structural distortions and modulations in microporous materials. In: Molecular Sieves - Science and Technology, Vol. 2: Structures and Structure Determination. Karge HG, Weitkamp J (eds.), Springer, New York, Berlin, Heidelberg, p 113-140

Depmeier W (2002) Crystalline microporous solids. Introduction and structure. In: Handbook of Porous Solids. Vol 2. Schüth F, Sing KSW, Weitkamp J (eds) Wiley-VCH, Weinheim, p 699-736

Depmeier W, Bührer W (1991) Aluminate Sodalites: $Sr_8[Al_{12}O_{24}](MoO_4)_2$ (SAM) at 293, 423, 523, 623 and 723 K and $Sr_8[Al_{12}O_{24}](WO_4)_2$ (SAW) at 293 K. Acta Cryst B 47:197–206

Depmeier W, Melzer R, Hu X (1993) The phase transition in $Sr_8[Al_{12}O_{24}](MoO_4)_2$ aluminate sodalite SAM. Acta Crystallogr B 49:483-490

Dove MT (1997) Theory of displacive phase transitions in minerals. Am Mineral 82:213-244

Dove MT, Cool T, Palmer DC, Putnis A, Salje EKH, Winkler B (1993) On the role of Al/Si ordering in the cubic-tetragonal phase transition in leucite. Am Mineral 78:486-492

Dove MT, Heine V, Hammonds KD (1995) Rigid unit modes in framework silicates. Mineral Mag 59:629-639

Engelhardt G, Koller H, Sieger P, Depmeier W, Samoson A (1992) ^{27}Al and ^{23}Na double-rotation NMR of sodalites. Solid State Nucl Magn Res 1:127-135

Felsche J, Luger S (1986) Structural collapse or expansion of the hydro-sodalite series series $Na_8[AlSiO_4]_6(OH)_2 \cdot nH_2O$ and $Na_6[AlSiO_4]_6 \cdot nH_2O$ upon dehydration. Ber Bunsenges Phys Chem 90: 731-736

Felsche J, Luger S, Baerlocher Ch (1986) Crystal structures of the hydro-sodalite $Na_6[AlSiO_4]_6 \cdot 8H_2O$ and of the anhydrous sodalite $Na_6[AlSiO_4]_6$. Zeolites 6:367-372

Fischer W (1968) Kreispackungsbedingungen in der Ebene. Acta Crystallogr A24:67-81

Fischer W (1971) Existenzbedingungen homogener Kugelpackungen in Raumgruppen tetragonaler Symmetrie. Z Kristallogr 133:18-42

Fischer W (1973) Existenzbedingungen homogener Kugelpackungen zu kubischen Gitterkomplexen mit weniger als drei Freiheitsgraden. Z Kristallogr 138:129-146

Fischer W (1974) Existenzbedingungen homogener Kugelpackungen zu kubischen Gitterkomplexen mit drei Freiheitsgraden. Z Kristallogr 140:50-74

Fischer W (1976) Eigenschaften der Heesch-Laves-Packung und ihres Kugelpackungstyps. Z Kristallogr 143: 140-155

Fischer W (1991a) Tetragonal sphere packings. I. Lattice complexes with zero or one degree of freedom. Z Kristallogr 194:67-85

Fischer W (1991b) Tetragonal sphere packings. II. Lattice complexes with two degrees of freedom. Z Kristallogr 194:87-110

Fischer W (1993) Tetragonal sphere packings. III. Lattice complexes with three degrees of freedom. Z Kristallogr 205:9-26

Fischer W (2004) Minimal densities of cubic sphere-packing types. Acta Crystallogr A60:246-249

Fischer W, Koch E (1976) Durchdringungen von Kugelpackungen mit kubischer Symmetrie. Acta Crystallogr A32:225-232

Fischer W, Koch E (1983) On the equivalence of point configurations due to Euclidean normalizers (Cheshire Groups) of space groups. Acta Crystallogr A39:907-915

Fischer W, Koch E (1987) On 3-periodic minimal surfaces. Z Kristallogr 179:31-52

Fischer W, Koch E (1989a) New surface patches for minimal balance surfaces. I. Branched catenoids. Acta Crystallogr A45:166-169

Fischer W, Koch E (1989b) New surface patches for minimal balance surfaces. III. Infinite strips. Acta Crystallogr A45:485-490

Fischer W, Koch E (1989c) Genera of minimal balance surfaces. Acta Crystallogr A45:726-732

Fischer W, Koch E (1990) Crystallographic aspects of minimal surfaces. Coll Phys 51 - C7:131-147

Fischer W, Koch E (1996a) Two 3-periodic self-intersecting minimal surfaces related to the Cr_3Si structure type. Z Kristallogr 211:1-3

Fischer W, Koch E (1996b) Spanning minimal surfaces. Phil Trans R Soc London A354:2105-2142

Fischer W, Koch E (2001) 3-periodic minimal surfaces derivable from Laves nets. Z Anorg Allg Chem 627: 2091-2094

Fischer W, Koch E (2002a) Lattice complexes. *In*: International Tables for Crystallography, Vol. A. Space Group Symmetry, Hahn Th (ed) 5th edition, Published for the International Union of Crystallography by Kluwer Academic Publishers, Dordrecht, Boston, London

Fischer W, Koch E (2002b) Homogeneous sphere packings with triclinic symmetry. Acta Crystallogr A58: 509-513

Fütterer K, Depmeier W, Altorfer F, Behrens P, Felsche J (1994) Compression mechanism in trioxan silica sodalite, [$Si_{12}O_{24}$].$2C_3H_6O_3$. Z Krist 209:517-523

Ghose S, Tsang T (1973) Structural dependence of quadrupole coupling constant e^2qQ/h for ^{27}Al and crystal field parameter D for Fe^{3+} in aluminosilicates. Am Mineral 58:748-755

Giddy AP, Dove, MT, Pawley GS, HeineV (1993) The determination of rigid unit modes as potential soft modes for displacive phase transitions in framework crystal structures. Acta Cryst A49:697-703

Griewatsch C (1998) Über die Dynamik von Tetramethylammonium in sodalith-artigen Strukturen. PhD Dissertation, Institut für Geowissenschaften, Universität Kiel

Griewatsch C, Winkler B, Depmeier W (1998) The high pressure behavior of Tetramethylammonium-Sodalite. Z Kristallogr Supplement Issue 15:198

Grimm H, Dorner B (1975) On the mechanism of the $\alpha\beta$ phase transformation of quartz. Phys Chem Solids 36:407-413

Gupta AK, Chatterjee ND (1978) Synthesis, composition, thermal stability, and thermodynamic properties of bicchulite, $Ca_2[Al_2SiO_6](OH)_2$. Am Mineral 63:58-65

Hahn Th (ed) International Tables for Crystallography, Vol. A. Space Group Symmetry, 5th edition, Published for the International Union of Crystallography by Kluwer Academic Publishers, Dordrecht, Boston, London

Hammonds KD, Deng H, Heine V, Dove MT (1997) How floppy modes give rise to adsorption sites in Zeolites. Phys Rev Lett 78:3701-3704

Hammonds KD, Dove MT, Giddy AP, Heine V (1994) Crush: A Fortran program for the analysis of the rigid-unit mode spectrum of a framework structure. Am Mineral 79:1207-1209

Hammonds KD, Dove MT, Giddy AP, Heine V, Winkler B (1996) Rigid unit phonon modes and structural phase transitions in framework silicates. Am Mineral 81:1057-1079

Hassan I, Grund HD (1984) The crystal structures of sodalite-group of minerals. Acta Crystallogr B40:6-13

Hazen RM, Finger LM (1982) Comparative Crystal Chemistry - Temperature, Pressure, Composition and the Variation of Crystal Structures. John Wiley & Sons, Chichester

Hazen RM, Sharp ZD (1988) Compressibility of sodalite and scapolite. Am Mineral 73:1120-1122

Henderson CMB, Taylor D (1978) The thermal expansion of synthetic aluminosilicate-sodalites, $M_8(Al_6Si_6O_{24})X_2$. Phys Chem Miner 2:337-347

Henmi C, Kusachi I, Henmi K, Sabine PA, Young BR (1973) A new mineral bicchulite, the natural analogue of gehlenite hydrate, from Tuka, Okoyama prefecture, Japan, and Carneal, County Antrim, Northern Ireland, Mineral J (Tokyo) 7:243–251, Engl Abstr (1974) Am Mineral 59:1330

Homeyer J, Bode O, Rüscher CH, Buhl J-Ch (2001) Temperature-dependent IR spectroscopy on NaCl-sodalite: observation of the exchange of cage ions. Z Krist Suppl 18:110

Hyde BG, Andersson S (1989) Inorganic Crystal Structures. Wiley, New York

Hyde ST, Andersson S, von Schnering HG (1984) The intrinsic curvature of solids. Z Kristallogr 168:1-17

Kahlenberg V (2001) Kristallchemische und strukturelle Untersuchungen an Aluminaten, Gallaten und Ferraten der Erdalkalien(Ca, Sr, Ba). Habilitationsschrift, Universität Bremen

Kanatzidis MG (1997) New directions in synthetic solid state chemistry: chalcophosphate salt fluxes for discovery of new meltinary solids. Curr Opin Solid State Mater Sci 2:139-149

Keijsper J, Ouden CJJ, Post MFM (1989) Synthesis of high-silica sodalite from aqueous systems, a combined experimental and model-based approach. *In*: Zeolites: Facts, Figures, Future. Jacobs PA, van Santen RA (eds) Elsevier Science Publishers, Amsterdam, p 237-247

Knorr K, Winkler B, Milman V (2001) Compression mechanism of cubic silica sodalite [$Si_{12}O_{24}$]: a first principles study of the Im-3m to I-43m phase transition. Z Kristallogr 216:495-500

Koch E (1984) A geometrical classification of cubic point configurations. Z Kristallogr 166:23-52

Koch E (1985) The geometrical characteristics of the alpha-$ThSi_2$ structure type and of its parameter field. Z Kristallogr 173:205-224

Koch E (2000a) Minimal surfaces with self-intersections along straight lines. II. Surfaces forming three-periodic labyrinths. Acta Crystallogr A56:15-23

Koch E (2000b) Self-intersecting three-periodic minimal surfaces forming two-periodic (flat) labyrinths. Z Kristallogr 215:386-392

Koch E (2001) Self-intersecting three-periodic minimal surfaces forming one-periodic tubes or finite polyhedra Z Kristallogr 216:430-437

Koch E, Fischer W (1972) Wirkungsbereichstypen einer verzerrten Diamantkonfiguration mit Kugelpackungscharakter. Z Kristallogr 135:73-92

Koch E, Fischer W (1978) Types of sphere packings for crystallographic point groups, rod groups and layer groups. Z Kristallogr 148:107-152

Koch E, Fischer W (1988) On 3-periodic minimal surfaces with non-cubic symmetry. Z Kristallogr 183:129-152

Koch E, Fischer W (1989a) New surface patches for minimal balance surfaces. II. Multiple catenoids. Acta Crystallogr A45:169-174

Koch E, Fischer W (1989b) New surface patches for minimal balance surfaces. IV. Catenoids with spout-like attachments. Acta Crystallogr A45:558-563

Koch E, Fischer W (1990) Flat points of minimal balance surfaces. Acta Crystallogr A46:33-40

Koch E, Fischer W (1993a) Triply periodic minimal balance surfaces: a correction. Acta Crystallogr A 49: 209-210

Koch E, Fischer W (1993b) A crystallographic approach to 3-periodic minimal surfaces. *In*: Statistical Thermodynamics and Differential Geometry of Microstructured Materials, The IMA Volumes in Mathematics and its Applications, Vol. 51. Davis HT, Nitsche JCC (eds), Springer, New York, p 15-48

Koch E, Fischer W (1995) Sphere packings with three contacts per sphere and the problem of the least dense sphere packing. Z Kristallogr 210:407-414

Koch E, Fischer W (1999a) Sphere packings and packings of ellipsoids. *In*: International Tables for Crystallography, Vol. C (Second revised edition). Wilson AJC (ed) Kluwer Academic Publishers, Dordrecht, Boston, London, p 738-743

Koch E, Fischer W (1999b) Minimal surfaces with self-intersections along straight lines. I. Derivation and properties. Acta Crystallogr A55:58-64

Koch E, Fischer W, Müller U (2002) Normalizers of space groups and their use in crystallography. *In*: International Tables for Crystallography, Bd. A: Space-group symmetry. 5th edition. Hahn T (ed) Kluwer Academic Publishers, Dordrecht, p 877-899

Koch E, Hellner E (1981) The frameworks of sodalite-like structures and of tetrahedrite-like structures. Z Kristallogr 154:95-114

Koch E, Sowa H (2004) Exceptional properties of some sphere packings in the general position of $P6_222$. Acta Crystallogr A60:239-245

Kudoh Y, Takéuchi Y (1985) The effect of pressure on helvite $Mn_8S_2[Be_6Si_6O_{24}]$, Z Kristallogr 173:305-312

Li Z, Nevitt MV, Ghose S (1989) Elastic constants of sodalite $Na_4Al_3Si_3O_{12}Cl$. Appl Phys Lett 55:1730-1731

Liebau F (1985) Structural Chemistry of Silicates - Structure, Bonding, and Classification. Springer, Berlin

Liebau F (2003) Ordered microporous and mesoporous materials with inorganic hosts: definitions of terms, formula notation, and systematic classification. Micropor Mesopor Mater 58:15-72, Erratum (2003). Micropor Mesopor Mater 66:363

Liebau F (2004) Microporous materials of the nth-order: new classes of poroates. Micropor Mesopor Mater 70:103-108

Loewenstein W (1954) The distribution of aluminium in the tetrahedra of silicates and aluminates. Am Mineral 39:92–98

Löns J (1969) Kristallchemische und strukturelle Untersuchungen in der Sodalithgruppe. PhD Dissertation, Universität Hamburg

Löns J, Schulz H (1967) Strukturverfeinerung von Sodalith, $Na_8Al_6Si_6O_{24}Cl_2$. Acta Crystallogr 23:434-435

Martin JD, Greenwood KB (1997) Halo-zeotypes: a new generation of zeolite-type materials. Angew Chem Int Ed 36:2072-75

McCusker LB (2005) IUPAC nomenclature for ordered microporous and mesoporous materials and its application to non-zeolite microporous mineral phases. Rev Mineral Geochem 57:1-16

McCusker LB, Liebau F, Engelhardt G (2001) Nomenclature of structural and compositional characteristics of ordered microporous and mesoporous materials with inorganic hosts (IUPAC Recommendations 2001). Pure Appl Chem 73:381-394

McMullan RK, Ghose S, Haga N, Schomaker V (1996) Sodalite, $Na_4Si_3Al_3O_{12}Cl$: structure and ionic mobility at high temperatures by neutron diffraction. Acta Crystallogr B52:616-627

McMullan RK, Mak TCW, Jeffrey GA (1966) Polyhedral clathrate hydrates. XI. Structure of tetramethylammonium hydroxide pentahydrate. J Chem Phys 44:2338-2345

Megaw HD (1973) Crystal Structures: A Working Approach. W. B. Saunders, Philadelphia, London, Toronto

Melzer R, Depmeier W, Vogt T, Gering E (1995) Neutron and synchrotron radiation high pressure experiments on aluminate sodalite $Sr_8[Al_{12}O_{24}](CrO_4)_2$. Crystal Res Technol 30:767-773

Mondal P, Jeffery JW (1975) The crystal structure of tricalcium aluminate, $Ca_3Al_2O_6$. Acta Crystallogr B31: 689-697

Müller U (2004) Relations between Wyckoff positions. *In*: International Tables for Crystallography, Bd. A1: Group-subgroup relations of space groups. Wondratschek H, Müller U (eds) Kluwer Academic Publishers, Dordrecht, p 428-727

Nishi F, Takéuchi Y (1975) The Al_6O_{18} rings of tetrahedra in the structure of $Ca_{8.5}NaAl_6O_{18}$. Acta Crystallogr B31:1169-1173
Nyman H, Hyde BG (1981) The related structures of α-Mn, sodalite, Sb_2Tl_7, etc. Acta Crystallogr A 37:11-17
Pauling L (1930) The structure of sodalite and helvite. Z Kristallogr 74:213-225
Pentinghaus H, Göttlicher J, Bernotat W (1990) Neue Glieder der Sodalith-Strukturfamile: $Se_4Al_8Si_4O_{24}(MoO_4)$ 2. Ergebnisbericht über Forschungs- und Entwicklungsarbeiten 1990, KfK Karlsruhe, p 12-13
Roth G, Pentinghaus H, Wanklyn BM (1989) Eine neue Variante in der Sodalith-Strukturfamilie: $DyAl_4Si_2O_{12} \cdot MoO_4$. Z Kristallogr 186:251-252
Rouquerol J, Avnir D, Fairbridge CW, Everett DH, Haynes JH, Pernicone N, Ramsay JD, Sing KSW, Unger KK (1994) IUPAC "Recommendations for the Characterization of Porous Solids". Pure Appl Chem 66: 1739-1758
Sahl K (1980) Refinement of the crystal structure of bicchulite, $Ca_2[Al_2SiO_6](OH)_2$. Z Kristallogr 152:13-21
Sahl K, Chatterjee ND (1977) The crystal structure of bicchulite, $Ca_2[Al_2SiO_6](OH)_2$. Z Kristallogr 146:35-41
Samoson A, Lippmaa E, Pines A (1988) High resolution solid state NMR - Averaging of second-order effects by means of a double-rotor. Molecular Physics 65 – 4:1013-1018
Sauer J, Windiks R (2003) Density functional studies of host-guest interactions in sodalites. In: Host – Guest Systems Based on Nanoporous Crystals. Laeri F, Schüth F, Simon U, Wark M (eds), Wiley-VCH, Weinheim p 410-423
Scheikowski M, Müller-Buschbaum HK (1993) Zur Kristallchemie der Blei-Lanthanoid-Oxoaluminate. Zur Kenntnis von $Pb_2HoAl_3O_8$ und $Pb_2LuAl_3O_8$. Z Anorg Allg Chem 619:1755-1758
Schnick W, Lücke J (1992) $Zn_7[P_{12}N_{24}]Cl_2$ – a sodalite with a phosphorus-nitrogen framework. Angew Chem Int Ed 31:213-215
Smith JV (2000) Tetrahedral frameworks of zeolites, clathrates and related materials. In: Microporous and Other Framework Materials with Zeolite-type Structures, Landolt-Börnstein/New Series IV/14, Vol. A. Baur WH, Fischer RX (eds) Springer, Berlin
Sokol AA, Catlow CRA (1996) Electronic structure of sodalite: a computational study. RAU Scientific Rep Solid State Electron Technol 1:44-49
Sokolova EV, Rybakov VB, Pautov LA, Pushcharovskii DY (1993) Structural transitions of tsaregorodtsevite. Dokl Akad Nauk 332:309-311
Sowa H, Koch E (1999) Sphere configurations with the symmetry R-$3m$ $18(h)$ .m. Z Kristallogr 214:316-323
Sowa H, Koch E (2001) A proposal for a transition mechanism from the diamond to the lonsdaleite type. Acta Crystallogr A57:406-413
Sowa H, Koch E (2002) Group-theoretical and geometrical considerations of the phase transition between the high-temperature polymorphs of quartz and tridymite. Acta Crystallogr A58:327-333
Sowa H, Koch E (2004a) Quantification of the deviations from closest sphere packings. Eur J Mineral 16: 255-260
Sowa H, Koch E (2004b) Hexagonal and trigonal sphere packings. II. Bivariant lattice complexes. Acta Crystallogr A60:158-166
Sowa H, Koch E, Fischer W (2003) Hexagonal and trigonal sphere packings. I. Invariant and univariant lattice complexes. Acta Crystallogr A 59:317-326
Stock N, Irran E, Schnick W (1998) Phosphorus oxonitridosodalites: synthesis using a molecular precursor and structural investigation by x-ray and neutron powder diffraction and ^{31}P MAS NMR spectroscopy. Chemistry 4:1822-1828
Stoll P, Dürichen P, Näther Ch, Bensch W (1998) Synthesis and crystal structure of $KCuGd_2S_4$. Z Anorg Allg Chem 624:1807-1810
Szpiro G (2003) Kepler's Conjecture: How Some of the Greatest Minds in History Helped Solve One of the Oldest Math Problems in the World. John Wiley & Sons Inc, New York
Takéuchi Y, Nishi F, Maki, I (1980) Crystal-chemical characterization of the 3 $CaO \cdot Al_2O_3$ – Na_2O solid-solution series. Z Kristallogr 152:259-307
Tarling SE, Barnes,P, Klinowski J (1988) The structure and Si,Al distribution of the ultramarines. Acta Crystallogr B 44:128-135
Taylor D (1972) The thermal expansion of the framework silicates. Mineral Mag 38:593-604
Thayaparam S, Dove MT, Heine V (1994) A computer simulation study of Al/Si ordering in gehlenite and the paradox of the low transition temperature. Phys Chem Mineral 21:110-116
Többens DM (1998) Untersuchungen zu Struktur und Phasenumwandlungen von Kristallen der Aluminatsodalithgruppe. PhD Dissertation, Universität Kiel
Többens DM, Depmeier W (1998) Intermediate Phases in the Ca-rich part of the System $(Ca_{1-x}Sr_x)8[Al_{12}O_{24}]$ $(WO_4)_2$. Z Kristallogr 213:522-531
Többens DM, Depmeier W, Griewatsch C, Peun T (1995) High pressure behavior of some aluminate sodalites, Jahresbericht HASYLAB, 469-470

Uchida E, Iiyama JT (1981) On kamaishilite, $Ca_2Al_2SiO_6(OH)_2$, a new mineral (tetragonal), dimorphous with bicchulite, from Kamaishi Mine. Japan Proc Japan Acad 57B:239-243; Abstract: Am Mineral 67:855 (1982)

van de Goor G, Behrens P, Felsche J (1994) $(C_3H_6O_2)_2(Si_6O_{12})_2$, a new silica sodalite synthesized, using 1,3-dioxolane as template. Microporous Mesoporous Mater 2:493-500

von Schnering HG, Nesper R (1987) Die natürliche Anpassung von chemischen Strukturen an gekrümmte Flächen. Angew Chem Int Ed 99:1097-1119

Weller MT (2000) Where zeolites and oxides merge: semi-condensed tetrahedral frameworks. J Chem Soc Dalton Trans 4227 – 4240

Wells AF (1977) Three-Dimensional Nets, Polyhedra. John Wiley & Sons, New York

Wells AF (1984) Structural Inorganic Chemistry. Clarendon Press, Oxford

Werner S, Barth S, Jordan R, Schulz H (1996) Single crystal study of sodalite at high pressure. Z Kristallogr 211:158-162

Werner S, Plech A (1995) Compressibility of tugtupite at high pressure. Z Kristallogr 210:418-420

Weyl H (1952) Symmetry. Princeton University Press, Princeton

Windiks R, Sauer J (1999) Sodium doped sodium sodalite: magnetic coupling between F centers and hyperfine interactions with framework atoms. Phys Chem Chem Phys 1:4505-4513

Wondratschek H, Müller U (eds) (2004) International Tables for Crystallography. Volume A1. Symmetry relations between spacegroups. Kluwer Academic Publishers, Dordrecht, Boston, London

Wuensch BJ (1964) The crystal structure of tetrahedrite, $Cu_{12}Sb_4S_{13}$. Z Kristallogr 119:437-453

Modular Microporous Minerals: Cancrinite-Davyne Group and C-S-H Phases

Elena Bonaccorsi and Stefano Merlino

Dipartimento di Scienze della Terra
Università di Pisa
Pisa I-56126, Italy
elena@dst.unipi.it   merlino@dst.unipi.it

In this chapter, we illustrate and discuss two distinct groups of microporous phases: the cancrinite group and the C-S-H compounds of the tobermorite and gyrolite families. The compounds in the first group present a three-dimensional purely tetrahedral framework with, apart from a single exception, Si:Al ratio equal to 1; in the mineralogical classifications they are included among feldspathoids and are generally "regarded …… distinct from zeolites, in part, at least, because of the presence of large volatile anions" (Coombs et al. 1998). The members of the second group are characterized by mixed frameworks built up by silicon (and aluminum) tetrahedra and calcium polyhedra. A common feature of both groups is the modular character of their frameworks, which are built up through various stacking ways of a single module (as in the minerals of the cancrinite-davyne family) and two or more modules as in the case of the C-S-H phases.

CANCRINITE-DAVYNE GROUP

Structural aspects

The minerals belonging to the cancrinite group (Merlino 1984; Deer et al. 2004) are feldspathoids with a Si:Al ratio equal to 1, with the only exception of cancrisilite, which has Si:Al = 7:5 (Khomyakov et al. 1991a,b). The available structural data for the phases with Si:Al = 1 indicate that silicon and aluminum regularly alternate in the tetrahedral sites, in accordance with the Loewenstein rule. The structural cavities host alkaline and earth-alkaline cations, and a wide variety of extra-framework anions, as well as H_2O molecules. Their framework is characterized by layers containing six-membered rings of tetrahedra (Fig. 1). Every ring is linked to three similar rings in the preceding layer and to three rings in the succeeding one. If the position of the rings in the first layer is called "A," and the two possible alternative positions in the adjacent layers are "B" and

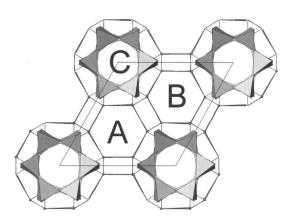

Figure 1. Schematic drawing of the layer of six-membered rings which builds up the minerals of the cancrinite group; the positions A, B and C are indicated.

"C," following the notation of the closest-packed structures, the resulting framework can be described by a sequence of A, B, C symbols, without consecutive repetition of letters. Another group of phases exists, in which six-membered rings may overlap, forming a double ring, or a hexagonal prism. In this case, the stacking sequence symbol contains pairs of A, B and/or C letters, as in the well-known offretite (AAB... stacking sequence), gmelinite (AABB...), and chabazite (AABBCC...) zeolites. Formally, all these natural or synthetic phases as well as the members of the cancrinite group belong to the so-called ABC-6 family of crystal structures (Gies et al. 1999). The natural phases of this wide group are listed in Table 1.

By examining Table 1, it is evident that the minerals traditionally included in the cancrinite group could be divided in two subgroups on the basis of the stacking sequence of layers. The simplest sequence AB... characterizes many natural and synthetic phases including cancrinite, vishnevite, and davyne. On the contrary, other trigonal or hexagonal phases show more complex stacking sequences of layers, which give rise to the occurrence of different cages. The known minerals with complex sequences have 4, 6, 8, 10, 12, 14, 16, and 28 layers for c translation; moreover, domains showing 14, 18 and 24 layer sequences were observed in TEM images (Rinaldi and Wenk 1979; Rinaldi 1982). All these phases with complex sequences were originally named "cancrinite-like" minerals by Leoni et al. (1979), and, later on, they were usually grouped together within the "cancrinite group."

The phases with AB... stacking sequence. There are nine natural phases which display the same kind of framework, built up by the AB... sequence of six-membered rings of tetrahedra. Eight phases are aluminosilicates; tiptopite, instead is a beryllophosphate with formula $[(Li_{2.9}Na_{1.7}Ca_{0.7})(OH)_2(H_2O)_{1.3}](K_2)(Be_6P_6O_{24})$ (Peacor et al. 1987).

The framework of the nine phases is characterized by channels delimited by 12-membered rings of tetrahedra running along [001], denoted as $[6^6 12^{2/2}]$ in the IUPAC nomenclature (McCusker et al. 2001), and by columns of base-sharing cancrinite cages, denoted as $[4^6 6^5]$ but also known as ε-cages or undecahedral cages (Fig. 2).

While the extra-framework content of the channel varies to a large extent in the different minerals, in the small cancrinite cages only two different situations have been observed: either water molecules bonded to sodium cations or chlorine anions bonded to calcium cations occur (see below). Consequently, the minerals will be grouped just taking into account this crystal chemical difference.

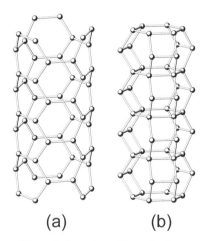

The cancrinite-vishnevite series. On the basis of crystal chemical considerations Pauling (1930) put forward the first hypothesis on the crystal structure of cancrinite; and it was confirmed and refined by Jarchow (1965). Later on, many other structural refinements of natural and synthetic cancrinites have been performed, and many other chemical data have been collected, allowing for a sound definition of its compositional range. The crystal chemical formula of cancrinite may be written as:

$[(Na,Ca)_{5-6}(CO_3)_{1.4-1.7}][Na_2(H_2O)_2][Si_6Al_6O_{24}]$

where the first part of the formula refers to the content of the large channel, the second to the cancrinite cages, and the last one represents the framework

Figure 2. Open channel (a) and column of base-sharing cancrinite cages (b), running along the [001] direction in the AB... framework. Only the tetrahedral sites are indicated.

composition. In the large channel, the amount of allowable carbonate groups is limited by the occurrence of short contacts between two adjacent carbonate groups. By assuming the distance between two carbonate groups in aragonite [2.87 Å] as the lowest possible C-C distance in cancrinite, a maximum content of 1.78 CO_3 groups may be calculated. This value is in agreement with the published chemical analyses (cf. Fig. 3). The ordering of the carbonate groups and vacancies, possibly related to the ordering of neighboring sodium and calcium cations, can give rise to the occurrence of satellite reflections (Jarchow 1965; Brown and Cesbron 1973; Foit et al. 1973; Grundy and Hassan 1982; Hassan and Buseck 1992) pointing to commensurate superstructures with multiple c values, as well as to incommensurate structures.

Actually, almost all the natural cancrinites contain also significant amounts of SO_4 anions (Fig. 3), substituting the carbonate groups. The end-member of this substitutional series is the mineral vishnevite, of chemical formula

$$[(Na,Ca)_{6-x}K_x(SO_4)][Na_2(H_2O)_2][Si_6Al_6O_{24}]$$

The presence of one sulfate group per unit cell, instead of 1.4÷1.7 carbonate groups, allows for the introduction of significant amounts x of potassium cations within the channel. In vishnevite of the type locality $x = 1$ (Hassan and Grundy 1984), and the potassium cation is statistically distributed in the Na1 cation site. However, up to three K cations could occupy three symmetry related sites in the channel, which regularly alternate with three (Na,Ca) sites surrounding the sulfate groups (Pushcharovskii et al. 1989). The apparently disordered distribution of these atoms inside the channel, with two split cation sites and two statistically occupied sulfur sites, depends on the lack of long-range correlation of this ordered sequence in the different channels; actually, a (Na,Ca):K ratio close to 1 greatly favors a long range ordering of sulfate groups and extra-framework cations also within adjacent channels; consequently, superstructure reflections requiring $a_{sup} = \sqrt{3}\ a$ occur. In agreement with this statement, the "high-potassium vishnevite" from Synnyrksii, Russia (Pushcharovskii et al. 1989), which contains about 3 K$^+$ cations per unit cell, has cell parameter $a = 22.24$ Å (Bonaccorsi 1992). A potassium rich phase with $a = \sqrt{3}\ a_{vish} = 22.1$ Å was originally found at Pitigliano (Italy) and named pitiglianoite (Merlino et al. 1991).

In all these phases – cancrinite, vishnevite, pitiglianoite – the cancrinite cages contain sodium cations and water molecules. The sodium cations are strongly bonded to three framework oxygen atoms and to one water molecule (bond distances of 2.30–2.45 Å), in

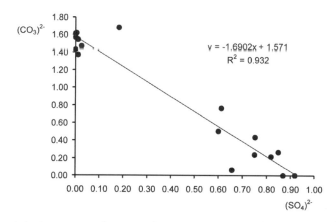

Figure 3. Correlation between $(CO_3)^{2-}$ and $(SO_4)^{2-}$ contents in minerals of the cancrinite-vishnevite series. Modified from Ballirano et al. (1998).

Table 1. The ABC-6 family of minerals. N = number of the layers for unit cell. In italics synthetic phases with no natural counterpart.

Phase	Stacking sequence	N	S.G.	a (Å)	c (Å)	Z	Ideal chemical formula (on the basis of 12 tetrahedral cations)
Single six-membered rings							
sodalite	(ABC)	3	$P\bar{4}3n$	8.882		1	$(Na_8Cl_2)(Si_6Al_6O_{24})$
haüyne	(ABC)	3	$P\bar{4}3n$ (P23)	9.082		1	$[Na_6Ca_2(SO_4)_2](Si_6Al_6O_{24})$
nosean	(ABC)	3	$P\bar{4}3n$	9.090		1	$[Na_8(SO_4)(H_2O)](Si_6Al_6O_{24})$
lazurite	(ABC)	3	$P\bar{4}3n$	9.074		1	$\{(Na,Ca)_8[(SO_4),S,Cl,(OH)]_2\}(Si_6Al_6O_{24})$
helvite	(ABC)	3	$P\bar{4}3n$	8.291		1	$(Mn_8S_2)(Be_6Si_6O_{24})$
genthelvite	(ABC)	3	$P\bar{4}3n$	8.120		1	$(Zn_8S_2)(Be_6Si_6O_{24})$
danalite	(ABC)	3	$P\bar{4}3n$	8.213		1	$(Fe_8S_2)(Be_6Si_6O_{24})$
bicchulite	(ABC)	3	$P\bar{4}3n$	8.829		1	$[Ca_8(OH)_8](Al_8Si_4O_{24})$
tugtupite	(ABC)	3	$I\bar{4}$	8.640	$c = 8.874$	1	$(Na_8Cl_2)(Al_2Be_2Si_8O_{24})$
tsaregorodsevite	(ABC)	3	$I222$	8.984	$b = 8.937$, $c = 8.927$	1	$[N(CH_3)_4]_2(Si_{10}Al_2O_{24})$
cancrinite	(AB)	2	$P6_3$	12.615	5.127	1	$[(Ca,Na)_6(CO_3)_{1-7}][Na_2(H_2O_2)](Si_6Al_6O_{24})$
vishnevite	(AB)	2	$P6_3$	12.685	5.179	1	$[(Na_6(SO_4)][Na_2(H_2O)_2](Si_6Al_6O_{24})$
hydroxycancrinite	(AB)	2	$P3$	12.740	5.182	1	$[Na_6(OH)_2][Na_2(H_2O)_2](Si_6Al_6O_{24})$
cancrisilite	(AB)	2	$P6_3mc$	12.575	5.105	1	$Na_7Al_5Si_7O_{24}(CO_3)\cdot3H_2O$
pitiglianoite	(AB)	2	$P6_3$	22.121	5.221	1	$[(Na_4K_2)(SO_4)][Na_3(H_2O)_2](Si_6Al_6O_{24})$
davyne	(AB)	2	$P6_3/m$ or $P6_3$	12.705	5.368	1	$[(Na_1K)_6(SO_4)_{0.5-1.0}](Ca_2Cl_2)(Si_6Al_6O_{24})$
microsommite	(AB)	2	$P6_3$	22.142	5.345	3	$[Na_4K_2(SO_4)](Ca_2Cl_2)(Si_6Al_6O_{24})$
quadridavyne	(AB)	2	$P6_3/m$	25.771	5.371	4	$[(Na_1K)_6Cl_2](Ca_2Cl_2)(Si_6Al_6O_{24})$
tiptopite	(AB)	2	$P6_3$	11.655	4.692	1	$[(Li_{2.9}Na_{1.7}Ca_{0.7})(OH)_2(H_2O)_{1.3}](K_2)(Be_6P_6O_{24})$
bystrite	(ABAC)	4	$P31c$	12.855	10.700	2	$[(Na_1K)_7Ca](Si_6Al_6O_{24})(S^{2-})_{1.5}\cdot H_2O$
liottite	(ABABAC)	6	$P\bar{6}$	12.87	16.096	3	$[(Na_1K)_{5.3}Ca_{2.7}](Si_6Al_6O_{24})(SO_4)_{1.7}Cl_{1.3}$

afghanite	(ABABACAC)	8	$P31c$	12.801	21.412	4	$[(Na,K)_{5.5}Ca_{2.5}](Si_6Al_6O_{24})(SO_4)_{1.5}Cl_{1.5}$
franzinite	(ABCABACABC)	10	$P321$	12.904	26.514	5	$[(Na,K)_6Ca_2](Si_6Al_6O_{24})(SO_4)_2 \cdot 0.4H_2O$
tounkite	(ABABACACABAC)	12	$P3$	12.755	32.218	6	$[(Na,K)_5Ca_3](Si_6Al_6O_{24})(SO_4)_{1.7}Cl_{1.3}$
marinellite	(ABCBCBACBCBC)	12	$P\bar{6}2c$ or $P31c$	12.88	31.761	6	$[(Na,K)_7Ca](Si_6Al_6O_{24})(SO_4)_{1.3}Cl_{0.3}\cdot H_2O$
farneseite	(ABCABABACBACAC)	14	$P6_3/m$	12.878	37.007	7	$[(Na,Ca,K)_8](Si_6Al_6O_{24})(SO_4)_{1.7}(Cl,H_2O)_{0.9}$
giuseppettite	(ABABABACBABABABC)	16	$P31c$	12.858	42.306	8	$[(Na,K)_{7.25}Ca_{0.75}](Si_6Al_6O_{24})(SO_4)_{1.25}Cl_{0.25}\cdot H_2O$
sacrofanite	(ABCABACACABACBACBACABABACABC)	28	$P\bar{6}2c$	12.865	74.24	14	$[Na,K,Ca]_8(Si_6Al_6O_{24})(SO_4)_{1.86}Cl_{0.14}\cdot 0.57H_2O$

Double six-membered rings

gmelinite	(AABB)	4	$P6_3/mmc$	13.756	10.048	1	$(Na_2,Ca)_4(Al_8Si_{16}O_{48})\cdot 24H_2O$
chabazite	(AABBCC)	6	$R\bar{3}m$	13.803	15.075	1	$Ca_6Al_{12}Si_{24}O_{72}\cdot 36H_2O$
willhendersonite	(AABBCC)	6	$P\bar{1}$	$a = 9.206$ $b = 9.216$ $c = 9.500$	$\alpha = 92.3$ $\beta = 92.7$ $\gamma = 90.1$	1	$Ca_6K_6Al_{18}Si_{18}O_{72}\cdot 30H_2O$
SAPO-56	(AABBCCBB)	8	$P\bar{3}1c$	13.757	19.936	1	$TMHD_{3.26}(Al_{23.5}P_{19.9}Si_{4.9})O_{96}$
AlPO4-52	(AABBCCAACCBB)	12	$P\bar{3}1c$	13.73	28.95	1	$7.2[(TEA^+)(H2PO_4^-)]\cdot [Al_{36}P_{36}]O_{144}$

Alternated single and double six-membered rings

offretite	(AAB)	3	$P\bar{6}m2$	13.261	7.347	1	$KCaMg[Al_5Si_{13}O_{36}]\cdot 16H_2O$
erionite	(AAABAAC)	6	$P6_3/mmc$	13.15	15.05	1	$Na_2K_2Ca_3[Al_{10}Si_{26}O_{72}]\cdot 30H_2O$
bellbergite	(ABBACC)	6	$P6_3/mmc$, $P6_3mc$, or $P\bar{6}2c$	13.244	15.988	1	$(K,Na,Sr)_2Sr_2Ca_2(Ca,Na)_4Al_{18}Si_{18}O_{72}\cdot 30H_2O$
levyne	(AABCCABBC)	9	$R\bar{3}m$	13.338	23.014	1	$Ca_8Na_{2.1}K_{0.9}[Al_{19}Si_{35}O_{108}]\cdot 50H_2O$

One double and two single rings regularly alternated

STA-2	(ABBCBCCACAAB)	12	$R\bar{3}$	12.726	30.939	1	$(Mg_6Al_{30})P_{36}O_{144}\cdot 3R^{2+}\cdot 22H_2O$

a substantially tetrahedral coordination, whereas three other oxygen atoms and the water molecule on the other side form weaker bonds (bond distances greater than 2.8 Å). The column of cancrinite cages contains Na-H_2O···Na-H_2O··· sequences; this atomic distribution excludes the occurrence of a horizontal symmetry plane (the space group of all these phases is $P6_3$), and results in a generally short c parameter (5.1 ÷ 5.2 Å).

Davyne, microsommite, quadridavyne. Davyne, microsommite and quadridavyne host Ca^{2+} and Cl^- in the cancrinite cages, instead of Na^+ and H_2O. The calcium cation lies on the base of the cancrinite cages, and has a bi-pyramidal coordination, forming six bonds with the framework oxygen atoms in the base of the cage, and two equal bonds with the chlorine anions which occupy the center of two base-sharing cancrinite cages. The resulting Ca-Cl-Ca-Cl-... chain along [001] is compatible with the $P6_3/m$ space group, which was actually found in davyne and quadridavyne (Bonaccorsi et al. 1990; Bonaccorsi et al. 1994). Moreover, the c parameter (5.36 Å) of davyne, microsommite and quadridavyne is significantly higher than that (5.20 Å) of the cancrinite-vishnevite phases.

These minerals differ as regards the kind of the extra-framework anions which are placed in the large channel, microsommite being the sulfate-rich and quadridavyne the chloride-rich end-member of the substitutional series $(Na,K)_6(SO_4)Ca_2Cl_2Si_6Al_6O_{24}$ – $(Na,K)_6(Cl_2)Ca_2Cl_2Si_6Al_6O_{24}$, whereas davyne has an intermediate composition. Moreover, davyne may also contain significant amounts of carbonate groups (Ballirano et al. 1998; Binon et al. 2004). The peculiarity of this series is the different ordering which takes place in the two end-members, resulting in different superstructures. The relationships among the unit cells of the three phases are sketched in Figure 4.

The structural results obtained for davyne from Vesuvius (Bonaccorsi et al. 1990; Hassan and Grundy 1990), indicate that the sodium and potassium cations located in the channel occupy two different sites, external and internal respectively. Up to one sulfate can be placed in the large channel, with the S atom occupying one of two symmetry-related sites, ½ c apart. As only the external cation sites may be occupied at the same level of the sulfate group, to prevent too short contact distances, an ordered distribution of two distinct clusters, the former built up by a (SO_4) anion surrounded by three Na cations, the latter built up by three (K,Na) cations, may be hypothesized. The observed disordered distribution of the two groups is the consequence of two factors, both occurring in davyne: (i) the ordering of these groups in the channels can be realized in two geometrically equivalent ways, and adjacent channels display them statistically; and (ii) the substitution of one sulfate group by two chlorine anions, as well as by carbonate groups, could break the ordered sequence within a channel. In microsommite, where such substitutions are not present, the channels are internally ordered and the scheme of ordering

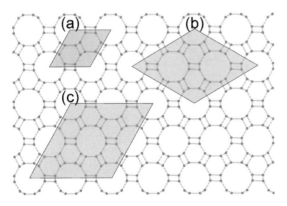

Figure 4. Relationships among the unit cell of (a) davyne, (b) microsommite and (c) quadridavyne, as seen along [001].

is correlated in adjacent channels, so that they are no longer equivalent by translation, and a superstructure develops. Such long-range ordering depends on the temperature, and a davyne-like phase can be obtained by heating microsommite up to 750°C (Bonaccorsi et al. 2001).

A similar thermal behavior has been observed in quadridavyne, the sulfate-free end member of the series. The crystal structure of quadridavyne is not known in detail, as the superstructure reflections are generally very weak. Preliminary structural data were published together with the description of the mineral (Bonaccorsi et al. 1994), indicating a possible ordering scheme for the chlorine anions and the alkali cations inside the channels.

Cancrisilite, hydroxycancrinite. These two phases are substantially similar to the minerals of the cancrinite-vishnevite series as regards the content of the cancrinite cages (Na^+ and H_2O), but they differ for the other chemical components.

Hydroxycancrinite (Nadezhina et al. 1991; Khomyakov et al. 1992) is the natural counterpart of the widely synthesized "basic cancrinite," containing sodium and (OH) groups within the large channel.

Cancrisilite (Khomyakov et al. 1991a,b) is the only member of the group which shows a Si:Al ratio different from 1. Its simplified crystal formula is $Na_7[Al_5Si_7O_{24}](CO_3)\cdot 3H_2O$, cell parameters $a = 12.573$, $c = 5.105$ Å and space group $P6_3mc$.

The phases with complex sequences of layers. Nine minerals were found which may be described as a more or less complex sequence of six-membered single rings (Table 1). Their c parameters correspond to 4, 6, 8, 10, 12, 14, 16, 28 layers. Generally, only one mineral exists for a given number of layers, except the case of marinellite and tounkite, which are structurally different phases presenting the same number of layers (twelve).

The complex-sequence phases may be also conveniently described by examining the different cages stacked along the [001] direction in three distinct columns, corresponding to the A, B and C positions, respectively. In Table 2, the sequences of cages occurring in the different structures are listed, whereas a schematic drawing of the existing structures is reported in Figure 5. The common feature of all these cages is that they are delimited by six-membered and four-membered rings only, at variance with the cages that occur in other phases of the ABC-6 family, which contain also eight-member rings.

The thickness of the cages along c correspond to 2, 3, 4, 6, 8 layers of six-member rings, respectively. Cages corresponding to 5 or 7 layers were never found, probably because they are unsuitable to host an integer number of sulfate groups. As each kind of cage has similar features in the different minerals, a short description of the cages and of their chemical content in these minerals will be presented hereafter.

Cancrinite cage. The cancrinite cage, [$4^6 6^5$] according to the IUPAC rules, has a thickness along c corresponding to two layers. All the minerals with complex sequences of layers contain an even number of cancrinite cages, which, as already seen in the phases with AB... stacking sequence, may host either water molecules or chlorine anions (Fig. 6a and 6b, respectively). In the former case (franzinite, marinellite, giuseppettite, and partially in sacrofanite), the site in the center of the cage may be partially occupied, pointing to a water content lower than the maximum allowable. In the latter case (bystrite, liottite, afghanite, tounkite), the site is fully occupied. A minor substitution of chlorine by fluorine anions has been observed in liottite and afghanite.

Sodalite cage. The sodalite cage, [$4^6 6^8$] in the IUPAC nomenclature, is well known, as it constitutes the building unit of the sodalite group minerals (see Depmeier 2005), as well as of other zeolitic phases, where it can host many types of extra-framework atoms. In the phases

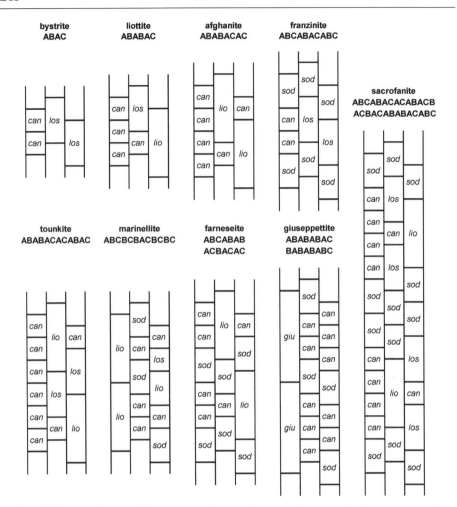

Figure 5. Schematic drawing of the sequences of cages in the cancrinite group minerals showing complex sequences of layers.

of the cancrinite group this kind of cage hosts generally one sulfate group in the center of the cage, whereas (Na,Ca) cations are distributed in split sites near the center of the eight six-membered faces of the cage (Fig. 7a). Typically, the atom distribution within this kind of cage is highly disordered, from both the chemical and geometrical points of view. In marinellite and giuseppettite, the sodalite cages host both sulfate and chlorine anions (Fig. 7a and 7b, respectively). In those cases, the ordering of sulfate and chlorine anions in different sodalite cages causes a decreasing of symmetry from $P\bar{6}2c$ to $P31c$.

Losod cage. The losod cage was first described by Sieber and Meier (1974) in the synthetic compound "Losod," which is a four-layer phase isostructural with bystrite. It has symbol [$4^6 6^{11}$] and has a thickness of four layers. The minerals which contain losod cages (Fig. 8a) are bystrite itself, liottite, franzinite, tounkite and sacrofanite. Apart from bystrite, which probably contains a significant amount of S^{2-} anions, in the other minerals this kind of cage is able to host two sulfate groups. They are surrounded by three (Ca,Na) cations, which may be shared by other neighboring cages, and are separated by groups of three cations.

Table 2. Structures of the complex sequence phases. The sequences of cages along the three columns are indicated by a sequences of number, corresponding to the width of the cages in terms of layers (2 = cancrinite cage, 3 = sodalite cage and so on).

Phase	N	Zdhanov symbol	number and kind of cages in the unit cell	0, 0, z	1/3, 2/3, z	2/3, 1/3, z	Ref.
bystrite	4	|(2)(2)|	2 los, 2 can	22	4	4	(1)
liottite	6	|21|12|	1 lio, 1 los, 4 can	222	24	6	(2)
afghanite	8	|1(2)1|1(2)1|	2 lio, 6 can	2222	26	62	(3)
franzinite	10	82	2 los, 6 sod, 2 can	3223	343	433	(4)
tounkite	12	(2)211(2)112	2 lio, 2 los, 8 can	222222	264	624	(5)
marinellite	12	|1(4)1|1(4)1|	2 lio, 4 sod, 6 can	66	23223	22323	(6)
farneseite	14	|1(5)1|1(5)1|	2 lio, 6 sod, 6 can	322322	6323	3236	(7)
giuseppettite	16	|11(4)11|11(4)11|	2 giu, 4 sod, 10 can	88	2232223	2223223	(8)
sacrofanite	28	|12(8)21|12(8)21|	2 lio, 4 los, 12 sod, 10 can	322223232223	34243363	36334243	(9), (10)

References: (1) Pobedimskaya et al. 1991; (2) Ballirano et al. 1996b; (3) Ballirano et al. 1997; (4) Ballirano et al. 2000; (5) Rozenberg et al. 2004; (6) Bonaccorsi and Orlandi 2004; (7) Camara et al. 2004; (8) Bonaccorsi 2004; (9) Ballirano 1994; (10) Ballirano and Bonaccorsi (*pers. commun.*).

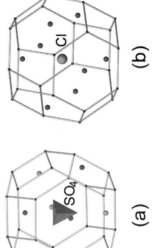

Figure 6. Cancrinite cages hosting Na and H$_2$O (a) and Ca and Cl (b).

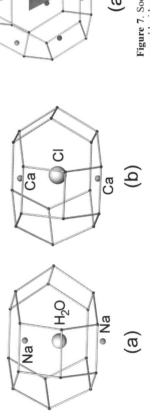

Figure 7. Sodalite cages hosting (a) one sulfate group and (b) one chloride anion in the center of the cage. (Na,Ca) cations are represented as small grey circles.

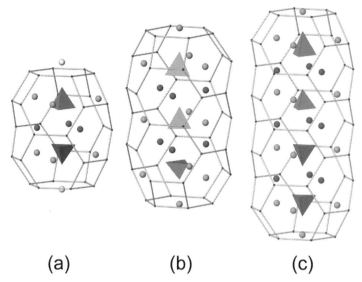

Figure 8. (a) Losod cage, containing two SO$_4$ groups; (b) liottite cage, with three SO$_4$ groups; (c) giuseppettite cage, 21 Å long; it contains four SO$_4$ groups. In the various cages, triples of cations (filled small circles) follow each other along [001].

Liottite cage. This 6-layer cage, [$4^6 6^{17}$] is present in liottite, afghanite, tounkite, marinellite, sacrofanite, as well as in a 14-layer new phase (Camara et al. 2004) [the new mineral and its name, farneseite, have been recently approved by the IMA Commission on New Minerals and Mineral Names, IMA No. 2004-043]. The liottite cage contains three sulfate groups (Fig. 8b).

Giuseppettite cage. This 8-layer long cage, with symbol [$4^6 6^{23}$], was found so far only in giuseppettite, where it contains four sulfate groups (Fig. 8c).

Relationships between chemistry and stacking sequences. The occurrence of so numerous phases showing different stacking sequences and number of layers in the unit cell raises the problem of the stability fields, and of the chemical and physical factors which influence the crystallization.

An earlier trial to set some geometrical constraints on the chemical content in the different cages was attempted by Ballirano et al. (1996a). Those authors recognized that phases showing different stacking sequences display also different chemical composition, especially as regards the extra-framework anion content. In their model, the small cancrinite cages host a chloride anion, whereas the larger cages host sulfate groups (1, 2, 3 and 4 in sodalite, losod, liottite and giuseppettite cages, respectively). According to those authors the SO$_4$ group seems to play a major role, as "…it tends to fill completely the available voids within the framework."

More recently, Sapozhnikov et al. (2004) built a reasonable structural model of tounkite, one of the two 12-layer members of the group, just on the basis of similar crystal chemical considerations. They observed that the phases containing more than one Cl$^-$ anion per 12 (Si+Al) tetrahedral cations, namely liottite and afghanite, display uninterrupted columns of cancrinite cages in their framework. Moreover, their normalized c parameter, namely their actual c parameter divided by $N/2$ (N is defined in Table 1) is higher than the normalized c parameter of the phases with a lower chlorine content (for example franzinite, marinellite, giuseppettite). The authors emphasized the role of the chloride anions in determining the stacking sequence of the different phases. As tounkite shows a high chlorine content, as well as large normalized c parameter, they assumed and afterwards successfully confirmed (Rozenberg et al. 2004) that a continuous column of cancrinite cages should be present in its structure.

In both the mentioned approaches, the occurrence of a well-defined stacking sequence is related to the presence of different amounts of extra-framework anions. In fact, while the sum of the extra-framework cations is equal to 8 per 12 tetrahedral cations in all the idealized formulas of the cancrinite group phases, the sum of the extra-framework anions and of the H_2O molecules is variable, ranging from 2 to 3 in the phases with complex sequences of layers. This variability does not depend on the possible partial occupancy of the anion sites, but is a consequence of the set of different types of cages present in the various structures. In Figure 9, the extra-framework chemical content of the minerals with complex sequence of layers is plotted in a (SO_4) vs. ($Cl + H_2O$) diagram. As additional information, also the anion contents of several sodalite group minerals (ABC... stacking sequence) and cancrinite group minerals (AB... stacking sequence) are plotted in the same diagram.

The minerals with the simple AB... stacking sequence may host the greatest amount of extra-framework anions and/or water molecules. As reported above, for example, microsommite, davyne and quadridavyne, respectively, all contain two chlorine anions within the cancrinite cages, whereas in the large channel one sulfate group (occurring in microsommite) may be partially or completely substituted by two chlorine anions (in davyne and quadridavyne). Similarly, vishnevite and pitiglianoite contain two H_2O molecules in the cancrinite cages and one SO_4 group in the channel. On the contrary, the sum of the extra-framework anions and water molecules in minerals with ABC... stacking sequence (sodalite group) is ideally equal to 2.

The phases with more complex sequence of layers occupy a different field in the plot of Figure 9. They are able to host more anions than the minerals of the sodalite group and more sulfate groups than the phases with AB... stacking sequence. For example, the framework of afghanite, 8-layer member with stacking sequence ABABACAC..., is formed by six cancrinite cages, containing 6 Cl⁻ anions, and two liottite cages, each one hosting three $(SO_4)^{2-}$ group. The formula of afghanite, on the basis of 12 (Si+Al) tetrahedral cations, may be written as $[(Na,K)_{5.5}Ca_{2.5}](Si_6Al_6O_{24})(SO_4)_{1.5}Cl_{1.5}$, where the sum of the extra-framework anions is

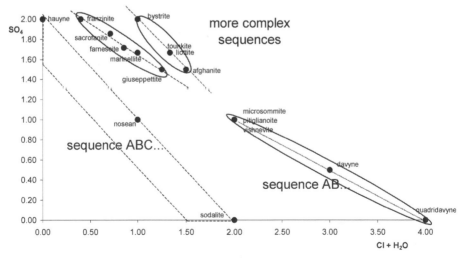

Figure 9. Different compositional fields of the phases formed by the stacking of layers with six-membered single rings.

exactly 3. On the other hand, the framework of franzinite, 10-layer member with stacking sequence ABCABACABC..., is formed by two cancrinite, six sodalite and two losod cages. The formula of franzinite is $[(Na,K)_6Ca_2](Si_6Al_6O_{24})(SO_4)_2 \cdot 0.4H_2O$, where the sum of the extra-framework anions and water molecules is 2.4. As a rule, the occurrence of sodalite cages reduces the amount of extra-framework anions which can be hosted in a structure.

By looking at the diagram of Figure 9, it can also be observed that the points corresponding to the complex-sequence phases present two different trends, suggesting a possible regularity in the stacking sequence of layers on the basis of the $(SO_4):(Cl+H_2O)$ ratio. One trend comprises bystrite, liottite, afghanite and tounkite. These phases are characterized by the absence of sodalite cages, a high chlorine content, a great normalized c parameter around 5.36 Å (Sapozhnikov 2004) and a sum of extra-framework anions ideally equal to 3. The second trend includes franzinite, marinellite, farneseite, giuseppettite and sacrofanite. Their frameworks are characterized by the occurrence of sodalite cages; moreover, they host a lower amount of chlorine anions, and a sum of extra-framework anions and water molecules ranging from 2.3 to 2.7 per formula unit. Additionally, their normalized c parameter is always below 5.30 Å.

These considerations, together with the relationships between the size of the cages and their chemical content discussed in a preceding section, could be useful both for descriptive and structure modeling purposes. For example, on the basis of a good chemical analysis giving the exact amount of the extra-framework anions and water molecules, and knowing the number of the layers from the cell parameter c, the number of the possible cages in the structure could be evaluated.

Genesis of natural compounds

The genesis of the cancrinite-group minerals and the main localities of occurrence are exhaustively accounted for in Deer et al. (2004), and also the recently found minerals of the group (marinellite and farneseite) have occurrences similar to those there described for other complex-sequence minerals.

The most common occurrence of cancrinite as primary phase is in nepheline syenite intrusions, where it crystallizes in the late stage as hydrothermal product in the presence of fluids containing "volatile" components such as carbonate, sulfate and/or chloride anions. Cancrinite occurs also as replacement of nepheline or minerals of the sodalite group. Similar occurrences, even if less common, were reported for vishnevite.

Cancrinite and cancrinite-group minerals formed also in the contact zone between alkaline rocks and limestone. Several minerals with complex sequences of layers are found in ejected skarn blocks in Italian volcanic areas (afghanite, in the Somma-Vesuvius area; liottite, afghanite, franzinite, and farneseite in Monti Vulsini volcanic field, afghanite, franzinite, marinellite, giuseppettite, sacrofanite in the Monti Sabatini volcanic area). Few of these minerals were found also in other localities. Afghanite was first found in Sar-e-Sang lapislazuli mine, Afghanistan, and later in several other localities in contact metamorphic deposits (Ivanov and Sapozhnikov 1975; Hogarth 1979), including Zabargad Island, Egypt (Bonaccorsi et al. 1992). Giuseppettite was also claimed to occur in the Oslo Rift, Norway (Jamtveit et al. 1997), but it was identified only by means of a microprobe analysis.

Thermal behavior

Few studies were performed to characterize the structural modification of cancrinite and related phases by increasing temperature. The structural modifications of carbonate cancrinite upon heating are related to the dehydration and to the consequent movement of the sodium cations in the cancrinite cages (Ballirano et al. 1995; Hassan et al. 2005). The non-linear increase of the cell parameters, and the discontinuity observed at about 500°C were related (Hassan et al. 2005) to the different thermal mechanisms operating in cancrinite. A different behavior upon heating was observed in pitiglianoite (Merlino and Bonaccorsi, unpublished results), where the dehydration process, in the temperature range 200–350°C, corresponds to a significant decrease of the cell volume (Fig. 10).

The linear thermal expansion coefficients up to 650°C were obtained for davyne from two localities and for microsommite (Bonaccorsi et al. 1995). The observed discontinuity in the thermal expansion of the c parameter of microsommite was related to the tilting of the tetrahedra connected along [001] and to the occurrence of a purely displacive phase transition with symmetry change from $P6_3$ to $P6_3/m$ at about 200°C.

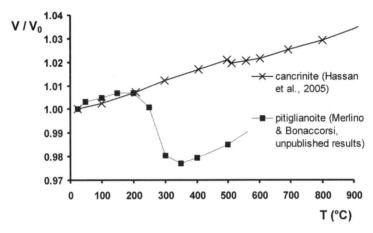

Figure 10. Relative variations of the unit cell volume with increasing temperature for cancrinite (crosses) and pitiglianoite (filled squares).

Synthetic cancrinites

Many synthetic compounds were obtained, showing the same framework topology of cancrinite, and with a wide variability as regards both the extra-framework species and the tetrahedral cations. A list of these compounds may be found in the paper by Sirbescu and Jenkins (1999); additional entries are reported in Table 3. Phases with chemical compositions very similar to those of natural cancrinite can be readily synthesized by hydrothermal reaction of oxides and carbonates, or from gels and carbonates; cancrinite was also obtained by hydrothermal methods starting from a mixture of kaolinite, NaOH and $NaHCO_3$.

After the first approach of Eitel (1922), several other authors studied synthetic CO_3-bearing cancrinite either from a structural point of view (Nithollon and Vernotte 1955; Smolin et al. 1981; Emiraliev and Yamzin 1982; Buhl 1991; Hackbarth et al. 1999) or in order to define its stability field (Barrer and White 1952; Edgar and Burley 1963; Edgar 1964; Zyryanov 1982; Hackbarth et al. 1999; Sirbescu and Jenkins 1999).

As far as structural aspects are concerned, no significant differences were observed with respect to the natural samples (Grundy and Hassan 1982; Ballirano and Maras 2004). The refined occupancy factors for the carbon sites within the channel appears to be lower than in natural cancrinites, pointing to only one carbonate group per formula unit instead of 1.4–1.6 (see also Fig. 3). Moreover, no superstructure reflections were observed in these synthetic CO_3-bearing samples, in agreement with their lower CO_3 content; in fact, the possible ordering

Table 3. Synthetic cancrinites which are not already listed in Sirbescu and Jenkins (1999).

Crystal-chemical formula			a	c	S.G.	Ref.
Channel	Cage	Framework				
$Na_6H_{0.88}(CO_3)_{1.44}$	$Na_2(H_2O)_2$	$[Al_6Si_6O_{24}]$	12.644	5.146	$P6_3$	(1)
$Na_{5.3}CO_3$	Na_2	$[Al_6Si_6O_{24}]$	12.659	5.153	$P6_3$	(2)
$Na_6CO_3 \cdot 1.4(H_2O)$	$Na_2(H_2O)_2$	$[Al_6Si_6O_{24}]$	12.713	5.186	$P6_3$	(3)
$Na_{5.5}(OH)_{1.5} \cdot 5.6H_2O$	$Na_2(H_2O)_2$	$[Al_6Si_6O_{24}]$	12.756	5.198	$P6_3$	(4)
$Na_{5.6}(HCO_3)_{1.2}(CO_3)_{0.2}$	$Na_2(H_2O)_2$	$[Al_6Si_6O_{24}]$	12.725	5.177	$P6_3$	(5)
$Na_6(OH)_2$	$Na_2(H_2O)_2$	$[Al_6Si_6O_{24}]$	12.735	5.182	$P6_3$	(6)
$Na_{5.6}(NO_3)_{1.6}$	$Na_2(H_2O)_2$	$[Al_6Si_6O_{24}]$	12.668	5.166	$P6_3$	(7)
$Na_6(NO_3)_2 \cdot 2H_2O$	$Na_2(H_2O)_2$	$[Al_6Si_6O_{24}]$	12.68	5.18	Not rep.	(8)
$Na_6(S_2O_3)$	$Na_2(H_2O)_2$	$[Al_6Si_6O_{24}]$	12.624	5.170	$P3$	(9)
$Na_6(S_2O_3) \cdot H_2O$	$Na_2(H_2O)_2$	$[Al_6Si_6O_{24}]$	12.73	5.02	Not rep.	(8)
$Na_6(SO_4) \cdot H_2O$	$Na_2(H_2O)_2$	$[Al_6Si_6O_{24}]$	12.674	5.173	Not rep.	(8)
$Na_6S \cdot 2H_2O$	$Na_2(H_2O)_2$	$[Al_6Si_6O_{24}]$	12.669	5.187	Not rep.	(8)
$Na_6(OH)_2 \cdot 0.54Se$	Na_2	$[Al_6Si_6O_{24}]$	12.670	5.165	$P6_3$	(6)
$Na_{2.5}(Se_2)^{2-}{}_{0.15}(Se_2)^{-}{}_{0.20}$	$Na_2(H_2O)_2$	$[Al_4Si_8O_{24}]$	12.639	5.157	Not rep.	(10)
$Li_{4.5} \cdot 4.9H_2O$	$Cs_{1.50}$	$[Al_6Si_6O_{24}]$	12.433	4.969	$P6_3$	(11)
$Li_{2.75}Tl_{1.25} \cdot 2.0H_2O$	Tl_2	$[Al_6Si_6O_{24}]$	12.442	4.988	$P6_3$	(11)
$Li_{5.47}(OH)_{1.47} \cdot 7.8(H_2O)$	Cs_2	$[Al_6Si_6O_{24}]$	12.416	4.970	$P6_3$	(12)
$Na_6Ge(OH)_6$	$Na_2(H_2O)_2$	$[Al_6Ge_6O_{24}]$	13.023	5.204	$P6_3$	(13)
$Na_6Ge(OH)_6$	Cs_2	$[Al_6Ge_6O_{24}]$	12.968	5.132	$P6_3$	(14)
$Na_6Ge(OH)_6$	Cs_2	$[Ga_6Ge_6O_{24}]$	12.950	5.117	($P6_3mc$)	(14)
$Na_6(OH)_2 \cdot 5H_2O$	Cs_2	$[Zn_6P_6O_{24}]$	12.794	5.066	$P6_3$	(15)
$Fe^{3+}Na \cdot 6H_2O$	$(Cs,K)_2$	$[Zn_6P_6O_{24}]$	12.492	4.999	$P6_3$	(16)
$Na_6(OH)_2 \cdot 5.3H_2O$	Cs_2	$[Co_6P_6O_{24}]$	12.851	5.047	$P6_3$	(17)

References: (1) Kanepit and Rieder 1995; (2) Burton et al. 1999; (3) Hackbarth et al. 1999; (4) Fechtelkord et al. 2003; (5) Gesing and Buhl 2000; (6) Bogomolov et al. 1992; (7) Buhl et al. 2000; (8) Hund 1984; (9) Lindner et al. 1995; (10) Lindner et al. 1996; (11) Norby et al. 1991; (12) Fechtelkord et al. 2001; (13) Belokoneva et al. 1986; (14) Lee et al. 2000; (15) Bienok et al 1998; (16) Yakubovich et al. 1986; (17) Bienok et al 2004

of one carbonate group per unit cell should simply correspond to a decrease of symmetry from $P6_3$ to $P3$ (Emiraliev and Yamzin 1982).

The stability of cancrinite as a function of T and $X(CO_2)$ of the vapor phase was investigated at constant pressure by Sirbescu and Jenkins (1999). By increasing T and $X(CO_2)$, cancrinite breakdowns to nepheline, calcite and H_2O, according to the reaction

$$Na_6Ca_{1.5}[Al_6Si_6O_{24}](CO_3)_{1.5} \cdot 1.1H_2O = 6\ NaAlSiO_4 + 1.5\ CaCO_3 + 1.1\ H_2O$$
$$\text{cancrinite} \qquad\qquad \text{nepheline} \qquad \text{calcite} \qquad \text{fluid}$$

The thermodynamic analysis of this reaction allowed an evaluation of the enthalpy and entropy of cancrinite (Sirbescu and Jenkins 1999).

A so-called basic cancrinite, which corresponds to the hydroxycancrinite composition (Nadhezina et al. 1991; Khomyakov et al. 1992), has been obtained from a gel of composition Na$_2$O, Al$_2$O$_3$, 2SiO$_2$, xH$_2$O and aqueous NaOH in excess at 390°C (Barrer and White 1952; Barrer et al. 1970). The occurrence of both H$_2$O molecules and OH$^-$ groups, as well as of two sites for the sodium cations, have been found within the structural channel of this sample. Similar structural results, except for the occurrence of only one cation site within the channel, were obtained for other synthetic basic cancrinites (Bresciani Pahor et al. 1982; Hassan and Grundy 1991).

Structural studies are also available for the nitrate cancrinite, with chemical formula Na$_{7.5}$[Al$_6$Si$_6$O$_{24}$](NO$_3$)$_{1.5}$·2H$_2$O. It was synthesized by hydrothermal reaction of kaolinite, NaOH and NaNO$_3$ (Barrer and White 1952) and later obtained in several different experiments (Hund 1984; Buhl et al. 2000). Different conclusions were reached as regards the position of the nitrate group within the cancrinite framework. While the early provisional data of Barrer and White (1952) suggested that NO$_3$ groups could be located mainly in the cancrinite cages, more recent results rule out this possibility, pointing out that NO$_3$ groups are located only in the channels, exactly as it happens for the CO$_3$ groups (Buhl et al. 2000).

A phase with composition K$_3$Na$_5$(Si$_6$Al$_6$O$_{24}$)(SO$_4$)$_{0.67}$(OH)$_{0.67}$·2.3–2.7H$_2$O and supercell parameters corresponding to pitiglianoite and microsommite was synthesized by Klaska and Jarchow (1977), who named it sulfate-hydrocancrinite.

Among the synthetic products containing more "exotic" guest species, the Cs-bearing cancrinites are particularly interesting as they were shown to contain the cesium cations entrapped in the center of the cancrinite cages (Norby et al. 1991; Fechtelkord et al. 2001; Bieniok et al. 2004). Here, the Cs$^+$ cation is strongly bonded to 12÷15 framework oxygen atoms with bond distances ranging from 3.2 to 3.7 Å, and is virtually not exchangeable. These properties, and the observation that cancrinite crystallizes as a secondary phase in the alteration process of kaolinite under particular conditions of high alkalinity, were the basis for environmental studies in the sedimentary soils near waste storage tanks (Chorover et al. 2003; Mashal et al. 2004). While the extensive uptake of Cs$^+$ into the crystal structure of newly formed nitrate cancrinite could be a promising way for immobilizing this radionuclide, the colloidal nature of the alteration products including cancrinite may actually facilitate the movement of Cs$^+$ in the soils (Zhuang et al. 2003).

More recently, several efforts have been made to synthesize cancrinite from organic solvents, in order to obtain a free channel system after removing, by heating, the small organic precursor molecules. Organic solvents successfully used to crystallize cancrinite are 1,3 and 1,4 butane-diol (Liu et al. 1993; Milestone et al. 1995; Burton et al. 1999; Fechtelkord et al. 2003).

ECR-5 (Vaughan 1986) is a cancrinite structure-type material obtained under application of ammonia and aqueous ammonia solutions; it has a high silica content and an increased adsorption capacity compared to conventional cancrinite. The same silica-rich phase was successively obtained in ammonia-free solutions (Vaughan 1991), disposing of a major pollution problem during the synthesis process.

An additional method successfully used to obtain synthetic cancrinite, together with sodalite and other zeolitic phases, makes use of molten salts (Park et al. 2000).

Other synthetic phases. While a wide literature exists on synthetic compounds isostructural with cancrinite, few synthetic products are known with more complex stacking sequence. Actually, they were obtained only for the stacking sequence ABAC..., corresponding to the natural phase bystrite.

The compound "Losod" was synthesized and its crystal structure was proposed (Sieber and Meier 1974) before the natural counterpart was characterized by Pobedimskaya et al. (1991), who also refined the crystal structure. Losod has chemical formula $Na_6Si_6Al_6O_{24} \cdot 9H_2O$, and cell parameter a = 12.906, c = 10.541 Å. The proposed space group for Losod was $P6_3/mmc$ (Sieber and Meier 1974), but the probable ordering of Si and Al in the tetrahedral sites could lower the symmetry to either $P\bar{6}2c$ or $P31c$ (Baur 1991).

Later on, two other synthetic compounds with ABAC... framework type (LOS) were crystallized, a sodium aluminogermanate containing carbonate groups and water molecules in the cages (Sokolov et al. 1978, 1981; Baur 1991) and a lithium beryllophosphate, containing (HPO_4) and water molecules in the cages (Harrison et al. 1993).

So far, no phases have been synthesized with a framework corresponding to the more complex sequence of layers listed in Table 2. On the other hand, many synthetic compounds were obtained and characterized as formed by layers of six-membered rings with more or less complex stacking sequences but containing double six-membered rings of tetrahedra together with single ones. For example synthetic phases have been found to be isostructural with gmelinite, chabazite, offretite, erionite, bellbergite, and levyne, as reported in the Database of Zeolite Structures (*http://www.iza-structure.org/databases/*; website maintained by Baerlocher and McCusker). Moreover, other synthetic phases belonging to the ABC-6 family with double rings exist, which do not correspond to any natural counterpart (SAPO-56, AlPO4-52, STA-1; their stacking sequences are reported in Table 1).

Ionic exchanges

The cancrinite framework is composed by a large channel running along [001], delimited by 12-membered rings of tetrahedra and with a free diameter of about 6 Å, and by columns of cancrinite cages, delimited by 6- and 4-membered rings of tetrahedra (Fig. 2). As a consequence of this structural arrangement, it could be deduced that the extra-framework ions located within the channel are exchangeable and, on the contrary, the small openings of the cancrinite cages prevent the substitution of the guest species, at least at low temperature and for some cations. The experimental data partially confirm these hypotheses. However, the occurrence of stacking faults in the AB... sequence, with the insertion of a 6-membered ring in position C, could dramatically reduce the ionic exchange capacity of cancrinite. Another obstacle to the cation diffusion within the channel is the presence of other ionic species, which may block the structural micropores.

Complete exchange reactions Na ↔ Li and Na ↔ Ag occur between fused lithium and silver nitrates, respectively, and a synthetic basic-cancrinite (Barrer and Falconer 1956) to give completely exchanged Li- and Ag-cancrinites. These samples were used in successive cation exchanges in solutions containing salts of the ions Li, Na, K, Rb, Cs, Ag and Tl. While 100% exchanged cancrinites were obtained with Li, Na, and Ag, only partial exchanges were realized with K, Tl and Rb, and no exchange at all with Cs. Moreover, the complete and reversible exchange reaction between Li^+ and Na^+ was studied in solution at different temperatures; the calculated Arrhenius energy of activation of the process was estimated to be 13.3 kcal/mol, and the activity coefficient of the reaction was 1.85 (Barrer and Falconer 1956).

The cation exchange reactions between the Na-rich davyne from Zabargad (Bonaccorsi et al. 1992) and different molten salts was studied by the present authors (Bonaccorsi 1992; Merlino and Bonaccorsi, unpublished results), in order to investigate the role of the large cations in the order-disorder transition between davyne and microsommite. Single crystals of davyne with chemical composition $Na_{5.5}Ca_{2.5}[Si_6Al_6O_{24}](SO_4)Cl_{2.5}$ and cell parameters a = 12.74, c = 5.34 Å were kept in fused KI at 680–710°C for 24 hours. The reaction which took place is represented by the following equation, in which the contents of the channel, cage and tetrahedral framework are between square parentheses for the solid phases:

$$[Na_{5.5}Ca_{0.5}(SO_4)Cl_{0.5}][Ca_{2.0}Cl_{2.0}][Si_6Al_6O_{24}] + 3KI \leftrightarrow$$
davyne   liquid

$$[K_{3.0}Na_{2.0}Ca_{0.5}(SO_4)][Ca_{2.0}Cl_{2.0}][Si_6Al_6O_{24}] + 3\,NaI + 0.5NaCl$$
microsommite   liquids

The main conclusions of these experiments were: (i) only the cations located in the channel are involved in the exchange process; (ii) when sufficient amounts of large cations substitute the small Na cations within the channel, a long range microsommite-like ordering takes place, and a supercell with $a = \sqrt{3}\, a_{dav}$ is observed. Very similar results were obtained with NaTlSO$_4$ at 750°C, whereas the exchange reaction with NaKSO$_4$ at 950°C could not be studied: the resulting phase had a different framework, possibly of the sodalite type.

C-S-H COMPOUNDS

The calcium silicate hydrates system, with about thirty stable crystalline phases, besides to ill-crystallized materials, is well more complex than the MgO(FeO)-SiO$_2$-H$_2$O system. Most of the crystalline phases occur as minerals and display OD features and polytypic forms. According to Taylor (1964), "one reason for the greater complexity of the CaO-SiO$_2$-H$_2$O system perhaps lies in the greater ionic radius and in the more electropositive character of calcium, which permit a number of different types of coordination with oxygen. Mg^{2+} and Fe^{2+}, in contrast, are nearly always octahedrally coordinated."

The terminology used in dealing with C-S-H phases, including the expression C-S-H phases itself, has been introduced and developed by cement chemists. In fact the relevance of part of these compounds in the hydration processes of cements, and of Portland cement in particular, stimulated their interest and produced a wide number of research programs aimed at understanding the structural arrangements, the crystal chemical features and the formation processes of the calcium silicate hydrates.

C-S-H compounds are generally obtained in the laboratory through hydrothermal treatment of amorphous or semicrystalline products showing different CaO:SiO$_2$ proportions. This ratio (C:S in the cement chemistry terminology) and the temperature are the main—although not unique—parameters controlling the formation of the different crystalline phases. Figure 11 presents a schematic stability diagram, in the C:S range 0.6 to 1.0, and temperature range up to 300°C, showing the relative stabilities of part of C-S-H phases (Shaw et al. 2000). In this diagram C-S-H(I) indicates poorly crystalline materials formed during the hydration of Portland cement and corresponding, according to Taylor (1986, 1992), to tobermorite-like structures.

Our attention will be particularly devoted to the minerals, and corresponding

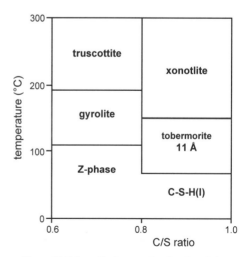

Figure 11. Schematic diagram showing the relative stability of part of the C-S-H compounds prepared under hydrothermal conditions (modified from Fig. 1 in Shaw et al. 2000).

synthetic phases, of the tobermorite and gyrolite groups. Actually not all of them are microporous. In the tobermorite group microporosity characterizes tobermorite 11 Å (in both normal and anomalous forms) and clinotobermorite, whereas their dehydration products do not present pores of suitable dimensions, and the hydration product of tobermorite 11 Å, namely tobermorite 14 Å, is a layer structure, although a peculiar one, as it will be shown in the following, easily transformed into the structure-type of tobermorite 11 Å by heating at 80–100°C. In the group of gyrolite, besides microporous compounds (truscottite, reyerite, fedorite, K-phase) there are layer structures, gyrolite and Z-phase, which, however, may transform, by dehydration, into microporous phases. The close structural relationships among the various phases in both groups, as well as the relatively easy conversion of the layer structures to microporous structures by dehydration, suggest to present and discuss in this chapter all the phases occurring in both structural groups.

NATURAL AND SYNTHETIC COMPOUNDS OF THE TOBERMORITE GROUP

Historical outlook

The particular interest in the structure and crystal chemistry of the tobermorite compounds stemmed not only from their close relationships with the C-S-H phases formed during the hydration processes of Portland cement, but also from their properties as cation exchangers (at least for tobermorite 11 Å) and potential applications in nuclear and hazardous waste disposal. That has stimulated a broad series of studies to obtain a deep knowledge of the structural aspects of these minerals, with special consideration of tobermorite 11 Å for its central position in the family and the ambiguity of its behavior in dehydration processes.

In 1880 Heddle described a hydrate calcium silicate found in three localities of Scotland, two near Tobermory, Mull island, and one at Dunvegan, Skye island; the name tobermorite was given to that compound. Afterwards tobermorite has been discovered in two other Scottish localities by Heddle (1893) and Currie (1905) and subsequently in various localities all over the world, generally occurring as alteration product of calcium carbonate rocks and as vesicle fillings in basalts.

Claringbull and Hey (1952) re-examined the specimens of Heddle; they confirmed that tobermorite was a valid species and observed a close similarity between the diffraction patterns of tobermorite and that of C-S-H-I (calcium silicate hydrate) compounds synthesized and studied by Taylor (1950) and Heller and Taylor (1951, 1952). Moreover they suggested some similarity with crestmoreite, a supposed phase described by Eakle (1917) as occurring in Crestmore (California, USA), together with another phase, riversideite, which would differ from crestmoreite only for the lower water content. Flint and coworkers (1938), by investigating specimens of riversideite and crestmoreite concluded that they are the same and proposed to drop out the name riversideite. Subsequently Taylor (1953a) demonstrated that crestmoreite (and riversideite as well) is an association, at submicroscopic scale, of tobermorite "with different hydration states" and wilkeite, a presently discredited species corresponding to phosphatian fluorellestadite.

Parallel to the researches carried out on natural phases of the tobermorite group, chemists interested in the nature of the compounds produced in the hydration processes of Portland cement studied C-S-H-I (calcium silicate hydrate) phases and found that they occur in three hydration states, called tobermorite 14 Å, tobermorite 11 Å and tobermorite 9 Å (Taylor 1953b). The notations 9 Å, 11 Å and 14 Å refer to the characteristic basal spacings of 9.3, 11.3 and 14.0 Å which these phases present in their X-ray powder diffraction patterns.

An important contribution in understanding the relationships among the various phases and in assessing the nomenclature has been given by McConnell (1954). In samples from Ballycraigy, County Antrim, North Ireland, he found the pure 11 Å phase, as well as the mixed 11 Å –14 Å hydrates. He carried out a careful study of the behavior of the mixed material on heating, indicating that tobermorite 14 Å transforms to tobermorite 11 Å (5CaO·6SiO$_2$·5H$_2$O) in the temperature range 80-100°C; then tobermorite 11 Å transforms to tobermorite 9 Å (5CaO·6SiO$_2$·1H$_2$O) at 300°C. McConnell assumed that the product studied by Eakle (1917) (riversideite – crestmoreite) was actually the 9 Å phase, transformed into the more hydrated terms due to the lack of proper preservation. Finally he observed that the composition of the more hydrated term, which appears at Ballycraigy as a natural gel, was identical to that of plombierite, a mineral with gelatinous character defined by Daubrée (1858), who found it at Plombières, Vosges, France. In conclusion McConnell proposed to assign the following names: *riversideite* to the 9 Å phase, *tobermorite* to the 11 Å phase, *plombierite* to the 14 Å phase.

More recently clinotorbermorite, another member of the group, has been found at Fuka, Japan, by Henmi and Kusachi (1989, 1992); it was subsequently found at Wessels mine, South Africa, and studied by Hoffmann and Armbruster (1997). Also clinotobermorite, which has a basal spacing of ~11 Å, does shrink, as "normal" tobermorite (actually it was found that some specimens of tobermorite 11 Å do not shrink on dehydration and are referred to as "anomalous") upon heating at 300°C, with formation of a compound characterized by a 9 Å spacing.

Note that, except clinotorbermotite, none of the mineral names and species quoted above are officially approved by the Commission on New Minerals and Mineral Names (CNMMN) of the International Mineralogical Association (IMA; cf. the website *www.geo.vu.nl/users/ ima-cnmmn/*). On the other side, the complex situation affecting the natural C-S-H phases of the tobermorite group suggests caution in proposing an official nomenclature.

OD character of the phases in the tobermorite group

The various compounds in the tobermorite group present OD character (Dornberger-Schiff 1956, 1964, 1966; Ďurovič 1997; Merlino 1997; Ferraris et al. 2004) clearly manifested by their diffraction patterns displaying streaks, diffuse reflections and unusual systematic absences rules. The OD character depends on their peculiar crystal chemistry (which will be thoroughly discussed in following sections), in particular on the metrical relationships between the calcium polyhedral module, with the repeat of 3.65 Å, and the tetrahedral chains with the typical repeat of 7.3 Å. Figure 12 shows that the tetrahedral chain may be connected to the calcium ribbons in two distinct but equivalent positions shifted by 3.65 Å in the **b** direction. Consequently the various compounds in the tobermorite group may be described in terms of OD layers, fully ordered in two dimensions, with **a** and **b** translation vectors ($a = 11.3$, $b = 7.4$ Å, $\gamma = 90°$ in all the tobermorite phases), which – thanks to the ambiguity of the connection between the silicon tetrahedral and calcium polyhedral modules – may stack according to two distinct ways along the **c*** direction, giving rise to a whole family of disordered or ordered sequences (polytypes): in all the possible sequences pairs of adjacent layers are geometrically equivalent, no matter whether taken in one member or in different members of the family (principle of OD structures).

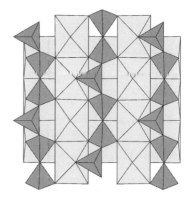

Figure 12. Connection between the silicon tetrahedral chain (dark grey) and the calcium polyhedral module (light grey) in the compounds of the tobermorite group.

The possible members of each family present diffraction patterns characterized by a set of common reflections (those with $k = 2n$ in all the tobermorite compounds), which are always sharp and have the same position in the reciprocal space and the same intensity in all the members of the family: they are called family reflections and define the "family cell," subcell in short, and consent to determine the so called "average structure." Table 4 compares the subcells of the various compounds.

The various members of each family may be distinguished on the basis of the characteristic reflections, namely those with $k = 2n+1$, more or less diffuse (sometimes continuous streaks) along $\mathbf{c}^*$; they present different positions and intensities in the different members of the family. Two main polytypes exist in each tobermorite family and OD theory indicates how to single them out. In fact they correspond to the MDO (Maximum Degree of Order) structures, in which not only pairs but also triples (quadruples,...n-tuples) of OD layers are geometrically equivalent (principle of MDO structures). A detailed treatment of the OD features for the various compounds in the tobermorite group, including the derivation of the MDO polytypes, may be found in the papers devoted to their structure determinations (Merlino et al. 1999, 2000, 2001; Bonaccorsi et al. 2005). The cell parameters and space groups of the various MDO structures are collected in Table 5.

Table 4. Composition and subcell parameters (in Å and °) of the phases in the tobermorite group.

Composition	Phase	Subcell	S.G.	a_S	b_S	c_S	β_S
$Ca_5Si_6O_{16}(OH)_2$	'clinotobermorite 9Å'	monoclinic	$A2/m$	5.58	3.65	18.78	92.8
	riverseideite (tobermorite 9Å)	orthorhombic	$Pnmm$	5.58	3.65	18.78	
$Ca_5Si_6O_{17}.5H_2O$	clinotobermorite	monoclinic	$A2/m$	5.638	3.671	22.642	97.3
	tobermorite 11Å	orthorhombic	$I2mm$	5.632	3.692	22.487	
$Ca_5Si_6O_{16}(OH)_2.7H_2O$	plombierite (tobermorite 14Å)	orthorhombic	$I2mm$	5.63	3.71	27.99	

Table 5. The two main polytypes (MDO structures) for each compound of the tobermorite group, and their crystallographic data (in Å and °).

	Polytype	S.G.	a	b	c	α	β	γ
Clinotobermorite	MDO_1	Cc	11.276	7.343	22.642		97.28	
	MDO_2	$C1$	11.274	7.344	11.468	99.18	97.19	90.02
'Clinotobermorite 9Å'	MDO_1	$C2/c$	11.161	7.303	18.771		92.91	
	MDO_2	$C\bar{1}$	11.156	7.303	9.566	101.08	92.83	89.98
Tobermorite 11Å	MDO_1	$F2dd$	11.265	7.386	44.97			
	MDO_2	$B11m$	6.735	7.385	22.487			123.25
Riversideite (Tobermorite 9Å)	MDO_1	$Fd2d$	11.16	7.32	37.40			
	MDO_2	$P112_1/a$	6.7	7.32	18.70			123.5
Plombierite (Tobermorite 14Å)	MDO_1	$F2dd$	11.2	7.3	56.0			
	MDO_2	$B11b$	6.735	7.425	27.987			123.25

Genesis and parageneses of the natural compounds

The minerals of the tobermorite group (tobermorite 11 Å, sometimes together with plombierite and, in a pair of cases, with riversideite) form through the action of hydrothermal fluids mainly at late stages of the evolution of different geological environments.

One of the most recurrent occurrences is in vugs, fissures or veins in basalts, frequently in association with zeolites and gyrolite. Of this type were the first findings, in Scotland, of tobermorite as red material "totally filling small druses in the cliffs of the shore immediately to the north of the pier of Tobermory in the Island of Mull" and in a stone quarry near the pier of Dunvegan, Isle of Skye (Heddle 1880) and at Loch Eynort, Isle of Skye (Heddle 1893); as well as the subsequent findings in Scotland: in the basalt of Ardtornish Bay at Morvern, Argylleshire, within druses which are generally filled by gyrolite (Currie 1905); 1 km north of Portree, Skye, on the "Staffin Road," in a olivine-dolerite, with xonotlite, gyrolite and zeolites (Sweet et al. 1961); in the vugs of a basaltic volcanic plug, at Castle Hill, near Kilbirnie in Ayrshire (Webb 1971); in Italy: at Prà de la Stua (province of Trento), near Malga Cola, in the cavities of an olivine basalt, together with natrolite, analcite, apophyllite and phillipsite (Gottardi and Passaglia 1965); as an inclusion in an olivine basalt on the eastern slope of Mount Biaena (province of Trento), together with gyrolite (Gottardi and Passaglia 1966); in the quarry Campomorto, Montalto di Castro (Viterbo), where it is the most widespread mineral in the cavities of the phonolite (Passaglia and Turconi 1982); in peralkaline pegmatites of the Khibiny massif, Kola peninsula, with thaumasite, apophyllite, saponite and Kovdor massif, Kola peninsula, with calcite and tacharanite; at Mokraya Synya river (Voykaro-Synninsky ultrabasic massif, North Urals), where tobermorite forms, together with gyrolite, late hydrothermal veinlets in the gabbro. In Germany tobermorite has been found by Walenta (1980) in melilite-nephelinite of Howenegg in Hegau, together with phillipsite, harmotome and mountainite; at Zeilberg quarry, Maroldsweisach, Bavaria; in the Eifel, at Arensberg near Zilsdorf (Hentschel 1973).

Tobermorite occupies the core of amygdales, accompanied by small amounts of tacharanite, in alkaline basalts at Puyuhuapi, in Chilean Patagonia (Aguirre et al. 1998). At Goldfield, Nevada, tobermorite has been produced by hydrothermal alteration of dacitic rocks and it is found associated with alunite as pseudomorphs after plagioclase phenocrysts in the most intense zone of alteration (Harvey and Beck 1962). In the Island of Surtsey it was found among the products of the hydrothermal alteration of the tephra (Jakobsson and Moore 1986). At Heguri, Chiba Prefecture, Japan, both tobermorite 11 Å and barium-bearing tobermorite 11 Å occur in veinlets cutting a basic tuff suffering metamorphism of zeolite facies, the first being accompanied by small amounts of calcite and poorly crystallized hydrated magnesium silicate (Mitsuda 1973), the second by thomsonite (Kato et al. 1984).

The minerals of the tobermorite groups may also be found in connection with the occurrence of hydrothermal conditions at the contact between limestones and dolerites or granodiorites, or more generally through the action of hydrothermal fluids on calcium silicate minerals. Of this type was the well-known finding of tobermorite phases at Crestmore, Riverside County, California, in a crystalline limestone at the contact with granodiorite (Eakle 1917), where also riversideite has been identified. Tobermorite minerals also occur in vugs of larnite rocks at the dolerite-chalk contact at Ballycraigy, Larne, County Antrim, Ireland, together with scawtite (McConnell 1954); at Fuka, Okayama, Japan, in a vein (10 mm width) of the altered rock at the contact between limestones and intrusive igneous rocks (Maeshima et al. 2003); in the Wessels mine (north of the town of Kuruman in the Kalahari Manganese Field, South Africa), where tobermorite was found near a dike and had crystallized in altered portions of the wall rock that had been in contact with the intrusion (Gutzmer and Cairncross 1993); in skarns at Cornet Hill, Apuseni Mountains, Romania (Marincea et al. 2001), which have undergone a late metasomatic event and subsequent hydrothermal events, the late of which resulted in the formation of 11 Å tobermorite, riversideite, as well as thomsonite, gismondine, aragonite and calcite, while the subsequent weathering gave plombierite, portlandite and allophane. Tobermorite 11 Å occurs in garnet-pyroxene skarns at Okur-tau, Uzbekistan, and in garnet-wollastonite skarns at Arimao-Norte, Cuba, with plombierite, riversideite, scawtite (Zadov et al. 1995); in xenoliths of calcite-bearing rocks in basic magmatic rocks, as at Vechec, Eastern Slovakia, and at Ozersky massif, near Baikal lake, together with calcite, plombierite, kilchoanite.

Tobermorite 11 Å has been found (Němec 1982) in the basic Ransko Massif, in the contact parts of the massif or in the xenolith-containing zone, closely associated with apophyllite. Tobermorite 11 Å and plombierite have been found in the rodingites of the Bazhenovskoe deposit of chrysotile, Urals, Russia (Zadov et al. 1995). At Bingham, Utah, a suite of mineral species including, together plombierite, also thaumasite, apophyllite, gyrolite, okenite, stilbite, thomsonite "has formed as hydrothermal alteration product of silicated limestone adjacent to the disseminated copper deposit" (Stephens and Bray 1973).

In one of the most widespread rock types, mainly composed of calcite and spurrite, of the so-called "Mottled Zone," in the Hatrurim region and at the Beersheba Valley in the northern Negev, near Ramleh in the coastal plain (Israel) and at Maaleh Adumin in the Jordanian part of the Judean Desert, minerals of the tobermorite group have been identified in small veinlets, together with a number of other uncommon minerals which include vaterite, portlandite, bayerite, thaumasite, minor ettringite and possibly jennite (Bentor et al. 1963).

In the Maqarin area (North Jordan) tobermorite 11 Å and plombierite were found as secondary minerals filling cavities of weathered marbles, together with ettringite, calcite, jennite and thaumasite, and were probably "precipitated during the weathering of the high grade metamorphic minerals directly from the high pH-water" (Abdul-Jaber and Khoury 1998).

Plombierite has also been collected at Carneal, Co. Antrim, Ireland, in veinlets and cavities in larnite-rich rocks at the chalk-dolerite contact (Nawaz 1977).

Clinotobermorite was firstly described as a new mineral from Fuka, Japan (Henmi and Kusachi 1992), in ghelenite-spurrite-bearing skarns occurring at the contact between quartz monzonite dykes and limestones. Later, it was found in fissures of the Wessels mine, in the Kalahari Manganese zone, South Africa, associated with xonotlite and datolite (Hoffmann and Armbruster 1997). More recently, it was also detected in samples from Bazhenov, Urals (Garbev 2004), were it occurs in association with tobermorite 11 Å, plombierite and diopside.

Synthetic counterparts

According to Taylor (1964) the first definite synthesis of crystalline tobermorite 11 Å was by Flint et al. (1938), who obtained it by hydrothermal treatment of "amorphous hydrate with $CaO:SiO_2$ molar ratio 0.80 at temperatures between 150°C and 275°C, assigned the formula $4CaO \cdot 5SiO_2 \cdot 5H_2O$ to their product and pointed to its possible identity with the natural phase. It was subsequently prepared by Heller and Taylor (1951), Kalousek (1955) and Assarsson (1958). Heller and Taylor (1951) synthesized tobermorite 11 Å by hydrothermal treatment, in the temperature range 110-200°C, of mixtures of $Ca(OH)_2$ and silica gel, as well as of ill-crystallized C-S-H preparations (Taylor 1950), in both cases with $CaO:SiO_2$ ratio approximately 1:1, and interpreted the X-ray diffraction pattern on the basis of an orthorhombic cell with $a = 5.62$, $b = 3.66$, $c = 11.0$ Å, corresponding, apart from the halving of c, to the parameters of the subcell of the natural tobermorite 11 Å, as reported in Table 4. The identity of the synthetic product with the natural phase has been indicated by Claringbull and Hey (1952) and definitely established by McConnell (1954). Although tobermorite 11 Å has been prepared from a variety of starting materials, the synthesis procedures most widely used are either hydrothermal reactions of mixtures of CaO or Ca(OH)$_2$ and finely ground quartz (Heller and Taylor 1951; Sasaki et al. 1996) or autoclaving amorphous C-S-H preparations (Hong and Glasser 2003; Garbev 2004) of appropriate CaO:SiO$_2$ ratio. Actually it has been shown by El Hemaly et al. (1977) that the formation of semi-crystalline C-S-H is the first step in the hydrothermal reaction starting from CaO and finely ground quartz, with subsequent formation of normal and afterwards anomalous tobermorite 11 Å, abutting, for temperature higher than 140°C, to xonotlite. Those authors have also defined the factors (short reaction times, high Ca:Si ratio, presence of Al^{3+} in absence of alkali, low temperature) which tend to stop the reaction at the step of normal tobermorite (El Hemaly et al. 1977).

The possibility to accommodate Al in substitution for Si in tobermorite 11 Å has been firstly shown by Kalousek (1957). The tetrahedral coordination of the aluminum cations has been confirmed by X-ray diffraction (Diamond et al. 1966) and by ^{27}Al and ^{29}Si MASNMR spectroscopy (Komarneni et al. 1985); the upper limit of Al substitution was indicated as Al:(Al+Si) = 0.13–0.14 by Sakiyama et al. (2000).

Normal tobermorite 11 Å may also be obtained by heating tobermorite 14 Å at 55–120°C. The first synthesis of tobermorite 14 Å has been made by Kalousek and Roy (1957) by processing lime and silicic acid, Ca:Si ratio 1:1, at 60°C for six months. Afterwards it was prepared by Hara et al. (1978) and Hara and Inoue (1980) by autoclaving mixtures of lime and amorphous silica (Ca:Si ratios 0.8, 0.9, 1.0) at 140°C for 20 hours and then keeping the product at 60°C for ten months; for some preparations alumina was added, with Al:(Al+Si) = 0.1, and in these cases tobermorite 11 Å was obtained. It is proper to recall that El Hemaly et al. (1977), in one single experiment of their study, obtained tobermorite 14 Å by hydrothermal reaction of lime-quartz mixtures at 105°C. It has been concluded that "14 Å tobermorite is a stable phase below 100°C in the absence of Al" (Hara et al. 1978).

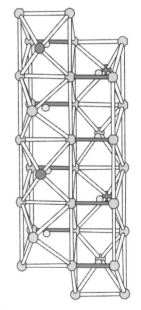

General structural aspects

All the phases of the tobermorite group are clearly distinguished from the other calcium silicate hydrate minerals by the presence of the common structural "complex module," built up by a calcium polyhedral layer (a continuous layer—in the (001) plane —of sevenfold coordinated calcium cations), with tetrahedral chains of wollastonite-type grasped on both sides of it. The "complex module" is C centered with periods $a \approx$ 11.2 Å, $b \approx 7.3$ Å and width $c_0 \approx 11.2$ Å (Merlino et al. 1999, 2000, 2001) and is represented in Figure 12.

The coordination polyhedron of the cations in the calcium layer may be described as a monocapped trigonal prism or, alternatively, as consisting of a pyramidal part on one side and a domatic part on the other side. The polyhedra are connected through edge sharing to build columns running along **b**. As illustrated in Figure 13, there are two types of polyhedra, which alternate along **b**: in one, the pyramidal apical site is occupied by a

Figure 13. Columns of CaO$_7$ polyhedra. The short dome edges are drawn in dark grey. The pyramidal apical sites occupied by O^{2-} or OH$^-$ anion are represented as light grey circles, whereas those occupied by H$_2$O molecules are represented as dark grey circles. The calcium cations are represented as small circles.

water molecule, in the other the apical site is occupied by an O^{2-} or OH^- anion and the corresponding Ca-O bond is significantly displaced from the normal position; in both types of calcium polyhedra the dome edge is substantially shorter (about 2.6 Å) than the two edges of the pyramidal basis parallel to it (nearly 3.0 Å). The polyhedral columns are joined, again through edge sharing, along the **a** direction to build infinite layers. The important aspect of the polyhedral connection is that adjacent columns present the apical ligands (the capping ligands in the "monocapped trigonal prism" description) on opposite sides of the layer and, similarly, the short "dome" edges are placed on opposite sides in adjacent columns, with relative displacement of **b**/4 (Merlino et al. 2000).

Infinite silicate chains of the wollastonite-type ("Dreierketten") run along **b** on both sides of the calcium polyhedral layers. They are built up by "paired" tetrahedra connected by a "bridging" tetrahedron (the terminology of the cement chemists is here used). The chains are linked to the calcium layer with the paired tetrahedra attached to it by sharing the "dome" edges and the bridging tetrahedron sharing the pyramidal O^{2-} (or OH^-) apices. This kind of connection explains well both the shortness of the dome edges, which must conform to the length of the SiO_4 tetrahedral edge, and the deviation from the normal position of the Ca-O_{ap} bond involving the apical oxygen atom, a deviation which is necessary to form Si-O_{ap} bonds of proper distance in the bridging tetrahedra.

It is important to stress that there are two geometrically distinct ways to place the tetrahedral chains on the two sides of the calcium sheet, thus giving rise to two distinct types of "complex module": in the first one, the bridging tetrahedra are placed at right on one side and at left on the other side (or *vice versa*), with respect to the disilicate groups of the corresponding chain, so that the (010) projection of the chains are inclined in the same direction on both sides of the central calcium sheet (complex layer of type A, Fig. 14a); in the second one, the bridging tetrahedra on both sides are all placed at left (or all placed at right) with respect to the corresponding disilicate groups, so that the (010) projection of the chains are inclined in nearly orthogonal directions (complex layer of type B, Fig. 14b).

We shall now examine the occurrence of the complex layers in the various phases of the tobermorite group and how the distinct ways of assembling the layers give rise to the whole series of tobermorite phases. We shall not always specify which kind of complex layer (A or B) we are dealing with, but it is easy to check that types A and B occur in the phases with monoclinic and orthorhombic subcells (Table 4), respectively.

There is a second aspect of modularity which has been briefly discussed previously. In fact the structures of the various compounds, plombierite (tobermorite 14 Å), tobermorite 11 Å, riversideite (tobermorite 9 Å), clinotobermorite and its dehydration product "clinotobermorite 9 Å," occur in two main polytypic variants. In the following, we shall compare the structural

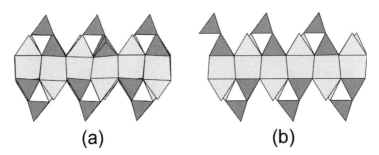

Figure 14. Complex layers of type A (a) and B (b), as seen down **b** (**a** horizontal).

arrangements of the various compounds considering only one polytype for each of them, as the crystal chemical properties and thermal behavior are largely independent on the polytypic form.

11 Å phases: clinotobermorite and tobermorite 11 Å

Clinotobermorite. In the crystal structure of clinotobermorite the complex layers are stacked in the direction normal to them and the wollastonite-type chains are condensed through apical oxygens of the bridging tetrahedra, thus giving rise to double chains of $[Si_6O_{17}]^{10-}$ composition. These hold firmly adjacent calcium layers together, thus building a strong scaffolding hosting additional calcium cations and water molecules (Ca2, W5, W7 and W8, according to the notations used in Merlino et al. 2000) in its cavities. W5 is placed in the center of the eight-membered ring of the silicate double chains; Ca2, W7 and W8 are located in the central zone of the channels running along **b** and limited by adjacent tetrahedral chains.

Figure 15 shows the position of these "zeolitic" components, with indication of the strong bonds (average length 2.35 Å) between Ca2 and the three "zeolitic" water molecules and two oxygen atoms of the polyhedral framework. Through careful consideration of the hydrogen-bond system involving the "zeolitic" water molecules and the actual geometry of the calcium polyhedra, it was concluded that O6 and O6A are O^{2-} anions and the composition of the polyhedral framework has been derived as $[Ca_4(Si_6O_{17})\cdot 2H_2O]^{2-}$, the negative charge being compensated by the corresponding positive charge of the "zeolitic" part $[Ca\cdot 3H_2O]^{2+}$. The resulting crystal chemical formula of clinotobermorite is $Ca_5Si_6O_{17}\cdot 5H_2O$.

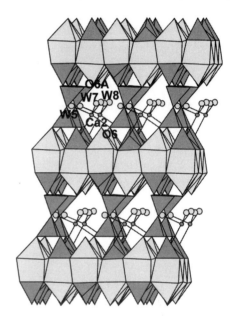

Figure 15. Crystal structure of clinotobermorite (triclinic polytype MDO$_2$ in Table 5), as seen along **b**, **a** horizontal. The projection axis is inclined by 7° to favor a better appreciation of the structural arrangement.

Tobermorite 11 Å. Also in tobermorite 11 Å the complex layers are connected through condensation of the wollastonite chains to build Si_6O_{17} double chains. However the double chains present different shapes and symmetries (2/*m* in clinotobermorite, 2*mm* in tobermorite 11 Å) in the two compounds (Fig. 16).

As we previously recalled, two varieties of tobermorite 11 Å are known, normal and anomalous. The crystal structures of the anomalous (specimen from Wessels Mine, South Africa) and normal (specimen from Bazhenovskoe

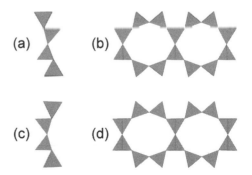

Figure 16. Silicate double chains in clinotobermorite, (a) and (b), displaying 2/*m* symmetry, and in tobermorite 11 Å, (c) and (d), displaying 2*mm* symmetry.

deposit, Urals, Russia) tobermorite 11 Å (Merlino et al. 2001) are compared in Figure 17. In both cases the polytype MDO_2 of Table 5 is illustrated.

In the anomalous variety only water molecules are located in the cavities of the structure, in positions corresponding to those of the water molecules in clinotobermorite: W2 in the center of the eight-membered rings of the silicate double chains, as W5 in clinotobermorite; W1 and W3 in the central zone of the channels, as W7 and W8 in clinotobermorite; the three molecules lie on the symmetry plane. Due to the absence of the "zeolitic" cation and the weakness of the hydrogen bonding of the water molecules with the oxygen atoms of the framework, it was concluded that O6 is a hydroxyl anion, the composition of the polyhedral framework is $[Ca_4(Si_6O_{15})(OH)_2 \cdot 2H_2O]$ and the crystal chemical formula of the anomalous tobermorite from Wessels Mine is $Ca_4Si_6O_{15}(OH)_2 \cdot 5H_2O$.

As regards the normal tobermorite from Urals, the results of the structural studies point to the presence of "zeolitic" calcium cations Ca2 in general position with occupancy 1/4, in addition to water molecules W2, with full occupancy and in position similar to that of the corresponding molecule in the anomalous variety, and W1, W3 with half occupancy and located in positions not too different from those occupied by the corresponding molecules in anomalous tobermorite 11 Å, apart a small (0.4 Å) displacement from the symmetry plane. Because of the particular composition of the normal tobermorite from Urals with 0.5 "zeolitic" cations per unit formula (Ca:Si = 4.5:6), two situations occur: *a*) the "zeolitic" calcium cation Ca2 is located on one side of the reflection plane, whereas W1 and W3 are located on the other side (Fig. 18); exactly as in the case of clinotobermorite, the O6 oxygen atom on one side of the mirror plane is strongly linked to Ca2, whereas the O6 atom on the other side is hydrogen bonded to W1 and W3 water molecules; consequently both O6 sites are occupied by O^{2-} anions; *b*) Ca2 is absent and possibly the W1 and W3 molecules shift their position

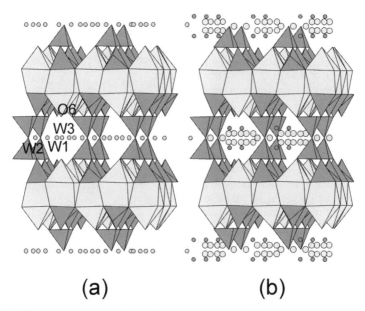

Figure 17. (a) The crystal structures of anomalous tobermorite from Wessels mine (only water molecules in the cavities). (b) The crystal structure of normal tobermorite from Urals (water molecules, large circles, and calcium cations, small dark grey circles, in the cavities). In both cases the monoclinic polytype MDO_2 of Table 5 is represented, as seen along **b**, **c** vertical, with an inclination (10°) of the projection axis to favor a better appreciation of the structural arrangement.

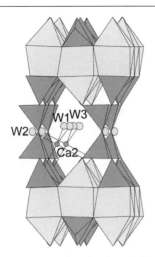

Figure 18. The situation a) in normal tobermorite from Urals, discussed in the text, is here illustrated: Ca2 cation (small dark grey circle) is located on one side and W1 and W3 are located on the opposite side of the plane passing through the bridging oxygen atoms which connect the wollastonite chains into double chains, as well as through the W2 water molecules.

toward the mirror plane (as it happens in the anomalous variety) and both O6 sites are occupied by hydroxyl anions. As the two situations occur with the same frequency, the composition of the polyhedral framework is $[Ca_4Si_6O_{16}(OH)\cdot 2H_2O]^{1-}$, the negative charge being compensated by the corresponding positive charge of the "zeolitic" part $[Ca_{0.5}\cdot 3H_2O]^{1+}$. The resulting crystal chemical formula of the normal tobermorite from Urals is $Ca_{4.5}Si_6O_{16}(OH)\cdot 5H_2O$.

Tobermorite specimens from different localities present different amounts of "zeolitic" calcium cations: their increasing causes an increasing occurrence of the situation *a*) described above, till the attainment of the extreme composition $Ca_5Si_6O_{17}\cdot 5H_2O$. At that composition, exactly corresponding to that of clinotobermorite, only the situation *a*) occurs and the distribution of calcium cations and water molecules in the cavities is substantially similar to that realized in clinotobermorite and described by Figure 18.

The whole compositional range for tobermorite 11 Å, excluding substitution of silicon by aluminum in the tetrahedra, spans between $Ca_4Si_6O_{15}(OH)_2\cdot 5H_2O$ and $Ca_5Si_6O_{17}\cdot 5H_2O$. Both in natural and in synthetic compounds, silicon may be partly substituted by aluminum [up to Al:(Al+Si) = 0.13–0.14, according to Sakiyama et al. (2000)], a substitution which may be balanced by a higher Ca content and/or higher O^{2-} by OH^- substitution. That has been confirmed by the structural study carried on a synthetic specimen of tobermorite 11 Å of composition near to $Ca_5Si_{5.5}Al_{0.5}O_{16}(OH)\cdot 5H_2O$, through Rietveld refinement of synchrotron radiation powder diffraction data (Yamazaki and Toraya 2001).

9 Å phases: "clinotobermorite 9 Å" and riversideite

"Clinotobermorite 9 Å" and riversideite, are obtained by heating, at 300°C, clinotobermorite and tobermorite 11 Å, respectively. In "clinotobermorite 9 Å" (Fig. 19) adjacent complex layers are now wedged together, the ridges of one fitting into the hollows of the other: the apical oxygen atoms of each single tetrahedral chain on one complex layer are connected to the calcium sheet of the adjacent complex layer just at the positions which in clinotobermorite are occupied by water molecules (and which are now occupied by hydroxyl anions). All the water molecules are lost in the dehydration process and Ca2 is now firmly linked to six oxygen atoms of the framework, with four shorter (2.32–2.36 Å) and two longer (2.62–2.67 Å) bonds. A careful examination of the bond lengths and calculation of the bond-valence balance indicate that all the oxygen sites correspond to O^{2-} anions, except the hydroxyl anion previously mentioned; thus, the crystal chemical formula is $Ca_5[Si_3O_8(OH)]_2$.

The comparison between the crystal structures of clinotobermorite and that of its dehydration product lets guess the following path of the dehydration: a) decondensation of the double chains through action of water molecules, namely $[Si_6O_{17}]^{10-} + H_2O = 2[Si_3O_8(OH)]^{5-}$, and consequent disconnection of adjacent complex layers; b) loss of water molecules and tight reconnection of the layers, after parallel shifting of alternate layers by **b**/2. The whole transformation does not require rearrangements inside the complex layers and proceeds topotactically (Merlino et al. 2000).

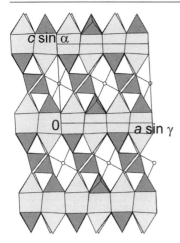

Figure 19. Crystal structure of "clino-tobermorite 9 Å" (triclinic polytype MDO$_2$ of Table 5) as seen along **b**. The four shortest Ca2-O bonds are indicated.

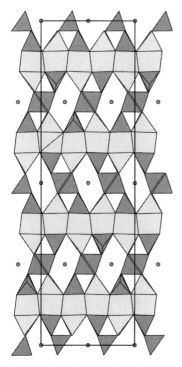

Figure 20. Structural model of riversideite (orthorhombic polytype MDO$_1$ of Table 5) as seen along **b**, **a** horizontal.

The structure of tobermorite 11 Å is built up by complex modules of type B and after the decondensation of the double chains, which constitutes the first step of the dehydration process, the complex layers facing each other cannot reconnect without a rearrangement of the tetrahedral chains, driven by the "zeolitic" water molecules. An ordered rearrangement occurring on both sides of each second complex layer, with subsequent reconnection of alternating unaltered and rearranged layers gives rise to the structure of riversideite (tobermorite 9 Å) as represented in Figure 20.

The structural details of riversideite are still unknown; in fact only the set of reflections with $k = 2n$, corresponding to the "subcell" structure, may be measured, the reflections with $k = 2n+1$ appearing as continuous diffuse streaks, indicative of high disorder in the sequence of the OD layers. Moreover some diffuseness observed also in the "subcell" reflections indicates that the ordered rearrangement we have sketched in Figure 20 is sometimes interrupted by alternate rearrangements.

14 Å phase: plombierite

The crystal structure of plombierite (tobermorite 14 Å) is built up by the same complex module of type B which characterizes tobermorite 11 Å. However the wollastonite-type chains facing each other on adjacent layers are not condensed into double chains; moreover they show a relative shift of **b**/2. The unconnected complex modules are moved apart and the space in between contains calcium cations and a larger amount of water molecules with respect to the 11 Å phase. The monoclinic MDO$_2$ polytype of plombierite (Table 5) is represented in Figure 21.

The water molecules in the interlayer region (W2, W3 and W4) as well as Ca2 calcium cations have half occupancies, whereas W1 is fully occupied; two W1, one W2, one W3 water molecules and two O5 anions (the apical oxygen atoms of the bridging silicon tetrahedra on the wollastonite chains) complete a nearly regular octahedron around Ca2. Two distinct ordering schemes may be sketched, in which Ca(H$_2$O)$_4$O$_2$ octahedra and W4 water molecules regularly alternate in columns along [010]. Each one may be realized in the space group B1 (Fig. 22). The random distribution of the two ordering schemes in the crystal restores the B11b symmetry. The crystal chemical formula indicated by the structural study (Bonaccorsi et al. 2005) is Ca$_5$Si$_6$O$_{16}$(OH)$_2$·7H$_2$O in agreement with chemical (Mitsuda and Taylor 1978; Maeshima et al. 2003),

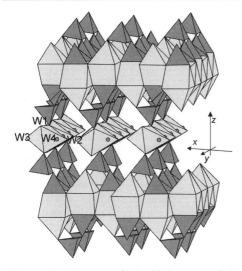

Figure 21. Structure of plombierite, monoclinic polytype MDO$_2$. In the interlayer sheet the half occupied CaO$_2$(H$_2$O)$_4$ octahedron drawn in light grey; the half occupied W4 molecules are represented by dark grey circles.

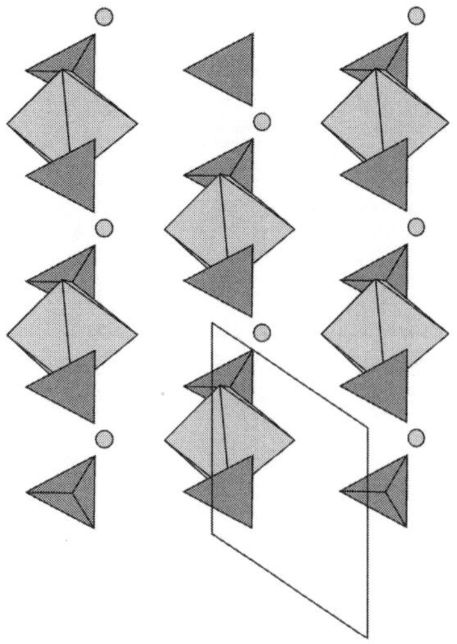

Figure 22. Ordering of interlayer material, in the structure of plombierite (monoclinic polytype MDO$_2$) as seen along **c**, **b** vertical.

thermal (Farmer et al. 1966; Maeshima et al. 2003) and MASNMR data (Wieker et al. 1982; Komarneni et al. 1987a; Cong and Kirkpatrick 1996; Maeshima et al. 2003).

Thermal behavior

The thermal study carried on tobermorite 14 Å, Ca$_5$Si$_6$O$_{16}$(OH)$_2$·7H$_2$O, from Crestmore (Farmer et al. 1966) points to a first weight loss at 55°C, when the 11 Å phase, Ca$_5$Si$_6$O$_{17}$·5H$_2$O, forms. The new phase persists, with continuous loss of water, till 250°C, at which temperature the 9 Å phase, Ca$_5$Si$_6$O$_{16}$(OH)$_2$, appears. This phase persists till 450°C; at 450 to 650°C the hydroxyl groups are lost through a gradual process leading to a phase with increased layer thickness (9.7 Å), at least in the Crestmore specimen. Then, between 850 and 900°C, wollastonite, CaSiO$_3$, forms, after an intermediate (730 to 775°C) wollastonite-like phase. The sketched sequence of steps, the water losses and the consequent structural rearrangements are in keeping with the structural aspects presented for the various phases in the preceding chapter.

Two relevant points deserve careful consideration and comment. The formation of the 9 Å phase from tobermorite 11 Å (*normal behavior*) in the dehydration process, is not straightforward; in fact, some natural and synthetic specimens do dehydrate at 250-300°C without shrinking, namely preserving the basal spacing of 11.3 Å (*anomalous behavior*). This distinct behavior had been previously explained with the presence of "interlayer Si-O-Si linkages" (double chains) in "anomalous" tobermorites and their absence (single chains) in "normal" tobermorites, an absence consenting an approach of adjacent calcium layers on dehydration, with corresponding decrease of the basal spacing. This explanation is obviously untenable since the structural studies have shown that double chains occur in both "normal" and "anomalous" tobermorites, as well as in clinotobermorite which, similarly, transforms to a 9 Å phase on dehydration at 300°C. In keeping with the present structural knowledge, an alternative explanation may be put forward. In clinotobermorite, Ca$_5$Si$_6$O$_{17}$·5H$_2$O, "zeolitic" calcium cations in the cavities of the structure are strongly bonded to oxygen atoms of the "complex layers" and to "zeolitic" water molecules. By heating at

300°C the water molecules are lost and a wide rearrangement is required to properly complete the calcium coordination: as the missing ligands may be offered only by atoms of the complex layers, chain decondensation occurs to consent the proper approach of adjacent layers.

As regards tobermorite 11 Å, we have already said that the composition spans between $Ca_4Si_6O_{15}(OH)_2 \cdot 5H_2O$ (no "zeolitic" cations) and $Ca_5Si_6O_{17} \cdot 5H_2O$ (one "zeolitic" calcium cation per unit formula, as in clinotobermorite). In the latter case, as well as in the general case of substantial "zeolitic" calcium content, the loss of water upon heating has the same consequences just described for clinotobermorite, with formation of tobermorite 9 Å. On the contrary, when no (or very few) "zeolitic" calcium cations occur in the cavities of tobermorite 11 Å, no requirement of proper calcium coordination is caused by the dehydration, no chain decondensation takes place and the specimen preserves the basal spacing of 11.3 Å.

The preceding explanation may be generalized in the following way: the normal-anomalous behavior depends on the nature and amount of "zeolitic" cations. Ferreira et al. (2003) prepared novel microporous sodium lanthanide silicates with the structure type of tobermorite 11 Å and with idealized crystal chemical formula $KNa_2Ln_2Si_6O_{16}(OH) \cdot 4H_2O$, actually $KNa_{0.44}(Na_{1.72} Ln_{2.28})Si_6O_{17} \cdot 3.56H_2O$, ($Ln$ = Eu, Tb, Sm, Ce, Gd). These compounds display the structure type of tobermorite 11 Å, with layers of alternating Ln^{3+} and Na^+ polyhedra substituting for the calcium layers and with K^+ "zeolitic" cations located in the cavities of the structure together with water molecules. They have anomalous behavior, as do not shrink on dehydration. This may be explained with the presence of K^+ instead than Ca^{2+} as extra-framework cation; K^+ cation, with a sensibly larger ionic radius, has satisfactory coordination also without extra-framework water molecules; moreover, and more important, its radius is too large for a proper positioning of K^+ in a shrunk 9 Å phase (the Ca2 calcium cation in "clinotobermorite 9 Å" has four Ca-O distances at 2.3 Å and two longer distances at 2.6 Å).

Technological properties

The cation exchange and selectivity properties of tobermorite 11 Å have been firstly reported by Komarneni et al. (1982) and Komarneni and Roy (1983). These authors observed that an exchange capability is mainly exhibited by (Al+alkali)-substituted tobermorite 11 Å, which, moreover, shows high selectivity for cesium cations.

The exchange behavior of Al-substituted tobermorite was the subject of several studies, for example by Shrivastava et al. (1991), Al-Wakeel et al. (2001), Šiačiūnas et al. (2002). These last authors have determined the sorption capacity of the compound for Co^{2+}, Ni^{2+}, Cu^{2+} and Zn^{2+} under static and dynamic conditions, and suggested its use for the elimination of heavy metals from waste water.

The high selectivity of (Al+Na)-substituted tobermorite for cesium has been confirmed by Komarneni et al. (1987b), Komarneni and Guggenheim (1988), Tsuji and Komarneni (1989), Miyake et al. (1989), who defined the detailed ion-exchange characteristics of (Al+Na)- and (Al+K)-substituted tobermorites for K^+ and Cs^+ cations, respectively, as regards their kinetic, equilibrium, and thermodynamic aspects. Tsuji et al. (1991) determined the exchange properties of (Al+Na)-substituted tobermorites for Cs^+ and Li^+ cations, as well as their water sorption capacity as a function of the aluminum for silicon substitution.

The high selectivity of Al-substituted tobermorites may find very useful application "in decontaminating circulation water in nuclear reactors and radioactive waste solutions" (Komarneni et al. 1987b). Moreover these compounds are stable in cement and therefore "may be incorporated in cement or concrete for use in nuclear as well as hazardous waste disposal" (Komarneni and Roy 1983).

The exchange, selectivity, sequestration features of (Al+alkali)-substituted tobermorite have been so far interpreted on the basis of incompletely known structural models. According to the present knowledge of the real structure of tobermorite 11 Å, in both its normal and anomalous forms, the previous interpretations require substantial revisions and new experimental studies are highly recommended.

The interesting photoluminescence properties presented by the sodium lanthanide silicates with the structure-type of tobermorite 11 Å (Ferreira et al. 2003) discussed above, are considered in the chapter by Rocha and Lin (2005) in this volume.

NATURAL AND SYNTHETIC COMPOUNDS OF THE GYROLITE GROUP

The known natural and synthetic phases in the gyrolite group are presented in Tables 6 and 7, with indication of their chemical compositions and crystallographic properties. It may be observed that the synthetic phases are C-S-H compounds *s.s.*, whereas in the natural

Table 6. Crystal chemical formulae of the natural and synthetic compounds of the gyrolite group and related mineral phases.

Mineral phases		*Synthetic compounds*	
gyrolite	$NaCa_{16}Si_{23}AlO_{60}(OH)_8 \cdot 14H_2O^{(1)}$	gyrolite	$Ca_{16}Si_{24}O_{60}(OH)_8 \cdot (14+x)H_2O$
truscottite	$Ca_{14}Si_{24}O_{58}(OH)_8 \cdot 2H_2O^{(2)}$	truscottite	$Ca_{14}Si_{24}O_{58}(OH)_8 \cdot 2H_2O^{(2)}$
reyerite	$(Na,K)_2Ca_{14}Si_{22}Al_2O_{58}(OH)_8 \cdot 6H_2O^{(3)}$	K-phase	$Ca_7Si_{16}O_{38}(OH)_2^{(5)}$
fedorite	$(Na,K)_2(Ca,Na)_7(Si,Al)_{16}O_{38}(F,OH)_2 \cdot 3.5H_2O^{(4)}$	Z-phase	$Ca_9Si_{16}O_{40}(OH)_2 \cdot 14H_2O^{(6)}$

Related mineral phases	
minehillite	$(K,Na)_2Ca_{28}Zn_5Al_4Si_{40}O_{112}(OH)_{16}^{(7)}$
tungusite	$Ca_{14}Fe_9Si_{24}O_{60}(OH)_{22}^{(8)}$
martinite	$(Ca,Na)_6Na_7(Si,S)_{14}B_2O_{38}(OH,F,Cl)_4 \cdot 6H_2O^{(9)}$
orlymanite	$Ca_4Mn_3Si_8O_{20}(OH)_6 \cdot 2H_2O^{(10)}$

[1] The composition refers to the specimen from Qarusait; most gyrolite samples are quite well accounted for by the simpler formula $Ca_{16}Si_{24}O_{60}(OH)_8 \cdot (14+x)H_2O$, with minor substitutions of sodium and aluminum in calcium and silicon sites respectively and water in the hydroxyl sites, and $0 \leq x \leq 3$ (Merlino 1988a).
[2] Lachowski et al. 1979
[3] Merlino 1972, 1988b
[4] Mitchell and Burns 2001
[5] Gard et al. 1981a,b
[6] Merlino 1988a
[7] Dai et al. 1995
[8] Ferraris et al. 1995
[9] McDonald and Chao 2002
[10] Peacor et al. 1990

Table 7. Crystallographic data of the phases of the gyrolite group (Å and °).

	S.G.	a	b	c	α	β	γ	References
gyrolite	$P\bar{1}$	9.74	9.74	22.40	95.7	91.5	120.0	Merlino 1988a
reyerite	$P\bar{3}$	9.765		19.067				Merlino 1988b
truscottite	$P\bar{3}$	9.731		18.84				Lachowski et al. 1979
fedorite	$P\bar{1}$	9.630	9.639	12.612	102.42	96.23	119.89	Mitchell and Burns 2001
K-phase[1]	$P\bar{1}$	9.70	9.70	12.25	101.9	96.5	120.0	Gard et al. 1981a,b
Z-phase[2]	$P\bar{1}$	9.70	9.70	15.24	90.94	93.19	120.0	

[1] The transformation matrix [110/$\bar{1}$00/001] has been applied.
[2] Structural model discussed in this paper.

phases limited Al for Si substitutions occur, being the charge compensation obtained through introduction of alkali cations. Table 6 presents also some minerals which display a more complex composition but deserve consideration because of their close structural and crystal chemical relationships with the other phases, as it will be shown below.

Natural phases: occurrences and composition

Gyrolite. Gyrolite has long been of interest to mineralogists and chemists, in particular those concerned with the chemistry of cement, who tried to define its crystal chemistry and establish its relationships with reyerite, truscottite and Z-phase of Assarsson. It was first found in Skye, Scotland, by Anderson (1851), who indicated its approximate composition and was subsequently identified in several other localities, generally in association with calcite, zeolites and other calcium silicate hydrates, as okenite, tacharanite, tobermorite, xonotlite. It mainly occurs in vugs and amygdules related to the latest crystallization stages of basaltic rocks; but it was also found within andesitic tuffs (Kobayashi and Kawai 1974); in drill cores from Yellowstone Park, USA, within hydrothermally altered sediments, pyroclastites and rhyolitic flows (White et al. 1975; Bargar et al. 1981) together with zeolites and calcium silicate hydrates; as very uncommon mineral in some ore deposits, as in the iron sulfides ore deposit of Ortano, Island of Elba, Italy (Garavelli and Vurro 1984) and in hydrothermally metamorphosed limestone at Bingham copper mine, Utah, USA (Stephens and Bray 1973).

Indication of gyrolite occurrences, together with the chemical data of the specimens, may be found in Merlino (1988a) and in general compilations, as Handbook of Mineralogy (Anthony et al. 1995) and Dana's New Mineralogy (Gaines et al. 1997).

On the basis of X-ray diffraction data, chemical analysis and dehydration study of samples from Bombay, India, Mackay and Taylor (1953) indicated that gyrolite is a sheet silicate with a structure based on hexagonal or pseudo-hexagonal structural modules, with $a = 9.72$, $c = 22.1$ Å, and chemical composition $Ca_{16}Si_{24}O_{60}(OH)_8 \cdot 12H_2O$. Various structural arrangements were hypothesized: Mackay and Taylor (1953) assumed that the hexagonal or pseudo-hexagonal modules are stacked with successive angular displacements of 60°. Strunz and Micheelsen (1958) maintained they found a trigonal one-layer crystal of gyrolite, with $a = 9.80$, $c = 22.08$ Å and chemical composition $Ca_{18}Si_{24}O_{60}(OH)_{12} \cdot 12H_2O$, and Cann (1965), who studied gyrolite closely associated with reyerite from Mull, Scotland, suggested that "the structure was made up of three layers of relatively low symmetry to give a trigonal structure with $a = 9.76$, $c = 67.0$ Å," and proposed the ideal formula $Ca_{16}Si_{24}O_{60}(OH)_8 \cdot 14H_2O$. A different structural arrangement has been hypothesized by Mamedov and Belov (1958), who sketched for gyrolite and truscottite a succession of octahedral calcium layers and tetrahedral silicon layers, these last built up through connections of parallel wollastonite chains building up alternating rows of pentagonal and octagonal rings.

The crystal structure of gyrolite, its crystal chemical features and relationships with the other natural and synthetic members of the group were elucidated by Merlino (1988a) who found a triclinic cell, space group $P\bar{1}$, $a = 9.74$, $b = 9.74$, $c = 22.40$ Å, $\alpha = 95.7$, $\beta = 91.5$, $\gamma = 120.0°$, crystal chemical formula $NaCa_{16}Si_{23}AlO_{60}(OH)_8 \cdot 14H_2O$ for a specimen from Qarusait, Greenland; moreover, on the basis of the best analytical data reported in the literature for gyrolites from different localities, the following general formula has been proposed: $(Ca,Na,Mg,Fe,\square)_{17}(Si,Al)_{24}O_{60}(OH,H_2O)_8 \cdot (14+x)H_2O$.

Reyerite. This rare mineral has been firstly found, according to Bøggild (1908), by Giesecke in 1811 at Niakornak, Greenland and was studied by Bøggild himself and Cornu and Himmelbauer (1906). Its troubled history as a mineral species, identified from time to time with gyrolite and truscottite, has been reported by Chalmers et al. (1964) in their comprehensive study of reyerite from the type locality, presenting chemical analysis, infrared absorption, X-ray powder and single crystal diffraction data and thermal weight loss study. They indicated a trigonal symmetry, space groups $P3$ or $P\bar{3}$, with $a = 9.74$, $c = 19.04$ Å, and a cell content $KCa_{14}Si_{24}O_{60}(OH)_5 \cdot 5H_2O$, with minor substitution of Si by Na+Al.

Reyerite has been subsequently found in few other places. At 'S Airde Beinn, Isle of Mull, Scotland, it was found in amygdales, together with analcime, thomsonite, natrolite and gyrolite, just within the metamorphic aureole of the volcanic plug of the plateau basalt (Cann 1965). According to Cann (1965) it was formed by mild metamorphism of gyrolite and it was suggested that the formation in the type locality was similar. Reyerite was later found in "chlorite-containing amygdales in a diabase dike in Rawlings quarry, Brunswick County, Virginia, that has been subjected to an episode of low-grade regional metamorphism" (Clement and Ribbe 1973), who confirmed the chemical and crystallographic data of Chalmers et al. (1964). Reyerite was found, together with tobermorite 11 Å, with minor chlorite, calcite, pectolite and thomsonite, in central zones within amygdales in olivine basalt at Allt Coir' a' Ghobhainn, near Drynoch, Isle of Skye, Scotland, by Livingstone (1988), who maintains it formed through the action of a Ca- and Si-rich fluid within the deepest and hottest zones of the lava pile during zeolitization.

Merlino (1972, 1988b) determined the crystal structure of reyerite on a specimen of the type locality, indicating the space group $P\bar{3}$, with $a = 9.765$, $c = 19.067$ Å and cell content (Na, K)$_2$Ca$_{14}$Si$_{22}$Al$_2$O$_{58}$(OH)$_8 \cdot 6H_2O$, and stressing the distinctions between reyerite and truscottite.

Truscottite. Truscottite was firstly found by Hövig (1914) in the Lebong Donok mine, Benkulen, Sumatra. The specimen was studied by Mackay and Taylor (1954) who found a Laue symmetry $\bar{3}$ or $\bar{3}m$ with $a = 9.72$, $c = 18.71$ Å, and pointed to an ideal chemical content $12CaO \cdot 24SiO_2 \cdot 6H_2O$.

It was subsequently found in one of the auriferous quartz veins cutting Miocene pyroclastic rocks at Toi mine, Shizuoka prefecture, Japan, by Minato and Kato (1967), who, taking into consideration the experimentally established stability range under low pressure (see succeeding section), supposed that the vein including truscottite had been formed under temperatures above 180°C and possibly below 355°C, and probably in the lower part of the stability range (Minato and Kato 1967). Truscottite was also found in drill cores from hot spring and geyser areas of Yellowstone National Park (Bargar et al. 1981); at Hishikani deposit, Kagoshima Prefecture, Japan (Izawa and Yamashita 1995); in deep drill hole at Kilauea Volcano (Grose and Keller 1976).

Lachowski et al. (1979) carried out a wide chemical and crystallographic study on specimens of natural and synthetic truscottite and, taking into account the results of the structural study of reyerite (Merlino 1972), definitely established the crystal chemical formula $Ca_{14}Si_{24}O_{58}(OH)_8 \cdot 2H_2O$.

Fedorite. Fedorite has been firstly found by Khukarenko et al. (1965) in fenitized sandstone of the Turiy Peninsula, Kola, Russia, as fine veinlets, in association with narsarsukite, and partly replaced by quartz and apophyllite. It was later found also in metasomatic-hydrothermal charoite-carbonatite rocks of the Little Murun potassic complex, Siberia, Russia (Konyev et al. 1993).

The structure of fedorite has been determined by Sokolova et al. (1983) and subsequently refined with X-ray and neutron diffraction data by Joswig et al. (1988); the studies were carried out with a specimen from Turiy in the space group $C\bar{1}$, with $a = 9.650$, $b = 16.706$, $c = 13.153$ Å, $\alpha = 93.42$, $\beta = 114.92$, $\gamma = 90.0°$. Mitchell and Burns (2001) refined the crystal structures of specimens from both known localities, in the space group $P\bar{1}$, with $a = 9.630$, $b = 9.639$, $c = 12.612$ Å, $\alpha = 102.42$, $\beta = 96.23$, $\gamma = 119.89°$ (specimen from Turiy); the unit cell given by Mitchell and Burns (2001) may be obtained from the cell of Sokolova et al. (1983) and Joswig et al. (1988) through the transformation matrix [−½ −½ 0 / −½ ½ 0 / 1 0 1].

Synthetic phases: preparation and composition

Gyrolite. Gyrolite was firstly synthesized by Flint et al. (1938) from glasses and gels characterized by Ca:Si ratios 0.50–0.66 in the temperature range 150–400°C; actually the product they obtained at the higher temperatures was most probably truscottite (Taylor 1964). It was subsequently prepared by Mackay and Taylor (1954) by autoclaving mixtures of calcium hydroxide and moist silica gel (Ca:Si ratio 0.66) at 150°C for 76 days, and afterwards by several other research groups, who defined the temperature conditions and the preparation procedures to obtain well crystallized products, as well as the stability field with respect to truscottite and xonotlite.

Recently the formation of gyrolite has been the subject of wide and relevant investigations (Shaw et al. 2002; Garbev 2004; Šiačiūnas and Baltakys 2004). Garbev (2004) carried out a wide study on the hydrothermal syntheses of C-S-H phases from mechano-chemically prepared gels with various C:S ratios. Gyrolite has been obtained by hydrothermal treatment from 110 to 220°C of preparations with C:S ratios 0.5 and 0.66. It was clearly shown how the crystallization of gyrolite was preceded by the formation of the Z-phase and a reliable mechanism for its formation was presented. Similar results were obtained by Šiačiūnas and Baltakys (2004) and by Shaw et al. (2002) who studied dynamically, at 190-240°C, by energy dispersive powder diffraction, with X-rays from a synchrotron source, the hydrothermal crystallization of gyrolite. This is a continuous process starting with the formation of a C-S-H gel with well ordered Ca(O,OH)$_2$ layers in the (001) plane, proceeding with progressive ordering in the **c** direction and formation of Z-phase which finally transforms to gyrolite.

Truscottite. As remarked by Taylor (1964) "the earliest report of the synthesis of truscottite, as distinct from gyrolite, was by Buckner et al. (1960), who obtained it together with other products from a hydrothermal run at 295°C and about 2000 kg/cm^2," although it had probably already been obtained by Jander and Franke (1941) and by Funk and Thilo (1955). Subsequently it was prepared by Funk (1961) and Meyer and Jaunarais (1961) who indicated the composition 2CaO·4SiO$_2$·H$_2$O, whereas Harker (1964) proposed the composition 3CaO·5SiO$_2$·1.5H$_2$O. Lachowski et al. (1979) presented the results of hydrothermal syntheses carried out for 7 days at saturated steam pressure and temperatures between 300 and 345°C, with different starting mixtures at C:S ratio *ca.* 0.6; some preparations had Al(OH)$_3$ or KOH as additional components. By careful examination of the products through analytical electron microscopy and X-ray powder diffraction, and by taking into account the results of the structural study of reyerite (Merlino 1972, 1988b), the crystal chemical formula Ca$_{14}$Si$_{24}$O$_{58}$(OH)$_8$·2H$_2$O was established. It was also shown that truscottite can accommodate aluminum, up to ~1.5 atoms per formula unit, and potassium, up to 0.5 atoms per formula unit. With some preparation, in particular when mixtures of CaO and silicic acid are reacted, gyrolite-truscottite intergrowths are formed as intermediate products, as already observed by Meyer and Jaunarais (1961) and Harker (1964).

Z-phase. This compound, with composition assumed as CaO·2SiO$_2$·2H$_2$O, was first obtained by Funk and Thilo (1955) by autoclaving a C-S-H gel at 180°C. Subsequently Assarsson (1957, 1958) synthesized a similar product by hydrothermal treatment of mixtures of CaO and amorphous SiO$_2$ at 140–240°C and named it Z-phase. The identity of the preparation by Funk and Thilo and Z-phase of Assarson was indicated by Taylor (1962) and confirmed by Funk (1961), who carried out new preparations and, on the basis of a DTA study, concluded that the composition was 2CaO·4SiO$_2$·3H$_2$O. Z-phase was also obtained by Harker (1964) by autoclaving a mixture of lime and silicic acid at 195°C. It has been recalled in a preceding section that Z-phase occurs as an intermediate in the formation of gyrolite. A wide study of the Z-phase obtained hydrothermally at 120°C by decomposition of Al-substituted tobermorite was carried out by Gard et al. (1975), who studied the product through X-ray and electron diffraction, thermal and infra-red investigations. They indicated that the phase is built up by hexagonal structural modules with $a = 9.65$, $c = 15.3$ Å, suggested structural relationships with gyrolite and truscottite and proposed for the structural module a composition between $8CaO \cdot 16SiO_2 \cdot 14H_2O$ and $8CaO \cdot 16SiO_2 \cdot 16H_2O$. On the basis of the careful analytical data of Gard et al. (1975) and assuming a close structural relationships of Z-phase with gyrolite, Merlino (1988a) proposed the crystal chemical formula $Ca_9Si_{16}O_{40}(OH)_2 \cdot (14+x)H_2O$, with $0 \leq x \leq 3$. As it will be discussed in another section, a structural model may be sketched for Z-phase in the space group $P\bar{1}$, with $a = 9.70$, $b = 9.70$, $c = 15.24$ Å, $\alpha = 90.94$, $\beta = 93.19$, $\gamma = 120.0°$.

K-phase. It was hydrothermally synthesized by Gard et al. (1981a) from a mixture of CaO and very finely divided reactive silicic acid (Ca:Si ratios 0.2 to 0.6) at temperatures 300 to 400°C. K-phase, if formed, was always mixed with other phases. These almost always included either truscottite, or in short runs, truscottite randomly interstratified with gyrolite. The optimum time for preparing K-phase was about 5 d at 350-375°C, or 5 h at 400°C (Gard et al. 1981a). Through microprobe analyses, transmission electron microscopy, as well as electron and X-ray diffraction studies, Gard et al. (1981a, b) proposed a structural model of K-phase in the space group $P\bar{1}$, with $a = 9.70$, $b = 9.70$, $c = 12.25$ Å, $\alpha = 108.6$, $\beta = 78.1$, $\gamma = 120°$ (by applying the transformation matrix [110/-100/001] we obtain a unit cell with $a = 9.70$, $b = 9.70$, $c = 12.25$ Å, $\alpha = 101.9$, $\beta = 96.5$, $\gamma = 120°$, in agreement with the cell parameters of fedorite, as given in Table 7), and defined the crystal chemical formula $Ca_7Si_{16}O_{38}(OH)_2$. It seems proper to recall here that recently Garbev (2004) has shown that the powder diffraction pattern of the product obtained by Gard et al. (1975) by dehydration of the Z-phase corresponds to that of K-phase, which indicates, according to Garbev (2004) that Gard et al. (1975) had already obtained the K-phase six years before they synthesized it through hydrothermal treatment.

STRUCTURAL ASPECTS

The various phases of the gyrolite group are built up by the stacking of all (as in gyrolite) or some of the following structural modules: S_1 and S_2 tetrahedral sheets, O octahedral and X interlayer sheets, all presenting trigonal or hexagonal symmetry, with $a = b \approx 9.7$ Å, $\gamma = 120°$.

The tetrahedral sheet S_1 (Fig. 23) may be described as made up by two-dimensional connection of groups of four tetrahedra, characterized by a central tetrahedron and three peripheral tetrahedra, all pointing in one direction; each group of four up-pointing tetrahedra is connected to three groups of four down-pointing tetrahedra (and *vice versa*) to build an infinite sheet with composition Si_8O_{20} and characterized by the presence of two kinds of six-membered rings; the first presents alternatively up- and down-pointing tetrahedra and is denoted as 1,3,5-ring; the second, with oval shape, has three down-pointing tetrahedra, followed by three up-pointing tetrahedra and is denoted as 1,2,3-ring.

Also the tetrahedral sheet S_2 (Fig. 24) may be described as built up by connection of groups of four tetrahedra, although in this case the central tetrahedron and the peripheral tetrahedra have opposite directions; the resulting infinite sheet with composition Si_8O_{20} presents two types of six-membered rings: the first is an almost hexagonal ring of tetrahedra pointing in one direction (up pointing in Fig. 24), the second (1,4-ring), with oval shape, is composed of two separated pairs of up-pointing tetrahedra, connected by two single down-pointing tetrahedra.

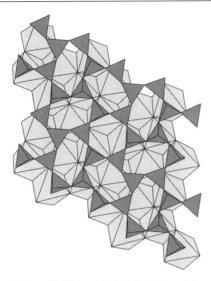 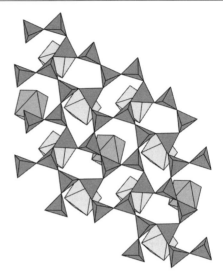

Figure 23. The tetrahedral sheet S_1 and its connection with the octahedral module O, as seen normal to (001) of gyrolite, **b** vertical.

Figure 24. The tetrahedral sheet S_2 and its connection with the module X, as seen normal to (001) of gyrolite, **b** vertical. The module X is built up, in gyrolite from Qarusait, by calcium (light grey) and sodium (dark grey) octahedra.

The S_1 and S_2 sheets may also be described as formed by the connection of four-repeat tetrahedral chains, the translational unit (9.7 Å long) presenting three tetrahedra pointing in one direction and one tetrahedron pointing in the opposite direction: in the S_1 sheet inversion-equivalent chains follow each other, whereas in the S_2 sheet similar chains follow each other.

Calcium octahedra are connected by edge sharing to build infinite O sheets (Fig. 23) with seven octahedra within the unit net and chemical composition $Ca_7(O,OH)_{14}$. The nature of the ligands depends on the kind of connection along the stacking direction: the oxygen atoms of the octahedral sheet which are coordinated also to silicon cations of an adjacent tetrahedral sheet correspond to oxide anions, whereas the remaining oxygen atoms correspond to hydroxyl anions.

The interlayer X sheet, which characterizes the structures of gyrolite and Z-phase, is also called "calcium-water layer" and its actual composition may slightly vary in gyrolites from different localities or in different synthetic preparations. Figure 24 represents the particular X layer in gyrolite from Qarusait (Merlino 1988a), with one sodium and two calcium octahedra in the unit net. The calcium cations are coordinated by two oxide anions, the apical oxygen atoms in the SiO_4 tetrahedra of the S_2 and $\bar{S}_2$ sheets sandwiching the X layer, and four water molecules, whereas the sodium cation is coordinated by six water molecules; the resulting composition of the X layer is $[Ca_2Na \cdot 14H_2O]^{+5}$. In most natural and synthetic samples the X layer presents only two calcium cations and the corresponding composition may be expressed as $[2Ca \cdot (14+x)H_2O]^{+4}$.

It is now possible to describe (Table 8) the structures of the various natural and synthetic phases of the gyrolite group just indicating the stacking of the constituting modules in the [001] direction (cf. Ferraris et al. 2004) and presenting, where necessary, the additional features which characterize the phase under consideration. The polytypic aspects are not considered here, but will be shortly discussed in a successive section.

Table 8. Modular schemes, crystal chemical formulae, space group symmetry and d_{001} distances (Å) for the various natural and synthetic phases of the gyrolite group.

Modular scheme	Phase	Crystal chemical formula	S.G.	d_{001}
$S_1 O S_2 \bar{S}_2 \bar{O} S_1$	Reyerite	$(Na,K)_2Ca_{14}Si_{22}Al_2O_{58}(OH)_8 \cdot 6H_2O$	$P\bar{3}$	19.07
	Truscottite	$Ca_{14}Si_{24}O_{58}(OH)_8 \cdot 2H_2O$	$P\bar{3}$	18.84
$S_1 O S_2 X \bar{S}_2 \bar{O} S_1$	Gyrolite	$Ca_{16}Si_{24}O_{60}(OH)_8 \cdot (14+x)H_2O$	$P\bar{1}$	22.20
$O S_2 \bar{S}_2 O$	Fedorite	$(Na,K)_2(Ca,Na)_7(Si,Al)_{16}O_{38}(F,OH)_2 \cdot 3.5H_2O$	$P\bar{1}$	11.91
	K-phase	$Ca_7Si_{16}O_{38}(OH)_2$	$P\bar{1}$	11.59
$O S_2 X \bar{S}_2 O$	Z-phase	$Ca_9Si_{16}O_{40}(OH)_2 \cdot (14+x)H_2O$	$P\bar{1}$	15.30

Gyrolite. The crystal structure of gyrolite (Merlino 1988a) is represented in Figure 25: two centrosymmetrically related octahedral sheets O and $\bar{O}$ are present in the unit cell, both sandwiched between two tetrahedral sheets of different kind, building up a complex layer $\bar{S}_2 \bar{O} S_1 O S_2$, with ideal composition $[Ca_{14}Si_{24}O_{60}(OH)_8]^{-4}$ and trigonal symmetry. This symmetry is not preserved in the whole structure in which the complex layer and the X interlayer sheet, $[2Ca \cdot 14H_2O]^{+4}$, regularly alternate giving rise to a triclinic arrangement $S_1 O S_2 X \bar{S}_2 \bar{O} S_1$, with composition $Ca_{16}Si_{24}O_{60}(OH)_8 \cdot 14H_2O$. When Al for Si substitutions occur, the charge balance may be restored by introduction of sodium cations in the X interlayer sheet.

Reyerite and truscottite. The crystal structure of reyerite (Merlino 1972, 1988b) is represented in Figure 26 and may be described by the scheme $S_1 O S_2 \bar{S}_2 \bar{O} S_1$, which is valid also for truscottite: two inversion related octahedral sheets, O and $\bar{O}$, are present in the unit cell, both sandwiched between a single S_1 and a double $S_2 \bar{S}_2$ tetrahedral sheets. In reyerite, as indicated in Figure 26, two aluminum cations per unit cell are perfectly ordered in the $S_2 \bar{S}_2$ double sheet; the whole tetrahedral-octahedral scaffolding has composition $[Ca_{14}Si_{22}Al_2O_{58}(OH)_8]^{2-}$ and the charge balance is restored by "zeolitic" alkali cations which are placed, together with water molecules, in the cavities of the structure at the level of the double tetrahedral sheet; the resulting composition is $(Na,K)_2 Ca_{14}Si_{22}Al_2O_{58}(OH)_8 \cdot 6H_2O$.

Notwithstanding the common structural scheme, reyerite and truscottite present significant differences in their infrared absorption spectra, in the region at 600–850 cm^{-1}, associated with Si-O-Si linkages, and in the band at 1640 cm^{-1}, attributed to molecular

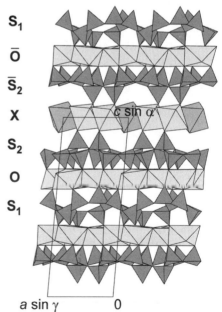

Figure 25. Crystal structure of gyrolite from Qarusait, as seen along **b**. The sequence of the modules is indicated. In the X sheet the calcium and sodium octahedra are drawn light and dark grey, respectively.

water (Chalmers et al. 1964), and in chemical composition, with higher water, alkali and aluminum contents in reyerite, and also in the length of their c parameters. All the differences in chemistry, infrared spectra, as well as in the c parameter, may be easily explained assuming that reyerite is characterized, in comparison with natural and synthetic truscottite, by the presence of two ordered Al cations in the unit cell. This clearly explains the chemical features of reyerite: the Si by Al substitution requires the introduction of alkali cations in the cavities of the $S_2 \bar{S}_2$ sheet, to restore the charge balance, and also the introduction of water molecules to complete their coordination. The differences in the band at 1640 cm^{-1} are dependent on the different water content in the two phases, whereas the differences in the region 600-850 cm^{-1} are related to the presence of Si-O-Al linkages in reyerite.

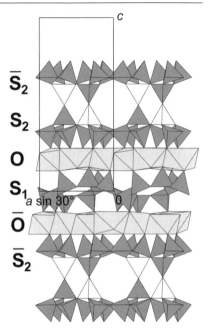

Figure 26. Crystal structure of reyerite, as seen along **b**. The sequence of the modules is indicated. In the S_2 sheet the AlO$_4$ tetrahedra are drawn white. The disordered distribution of the "zeolitic" alkali cations and water molecules is not indicated.

Fedorite and K-phase. The crystal structure of fedorite (Sokolova et al. 1983; Joswig et al. 1988; Mitchell and Burns 2001) is represented in Figure 27 and may be described by the scheme O $S_2 \bar{S}_2$ O, which is valid also for the K-phase (Gard et al. 1981a, b). There are between fedorite and K-phase the same relationships already described in the case of reyerite and truscottite: Si by Al substitution in the tetrahedral sheet and Ca by Na substitution in the octahedral sheet are compensated by the introduction of "zeolitic" alkali cations in the cavities of the $S_2 \bar{S}_2$ sheet, together with water molecules to complete their coordination.

Z-phase. Although its structure has not yet definitely established, the structural scheme O S_2 X $\bar{S}_2$ O has been proposed (Gard et al. 1975; Merlino 1988a) and a reliable model may be derived on the basis of our knowledge of the structures of K-phase and gyrolite and the available crystallographic information collected by Gard et al. (1975) through electron diffraction investigations. They obtained an electron diffraction pattern that could be indexed on a pseudo-hexagonal unit cell with $a = 9.65$ Å and estimated a c period between 15 and 16 Å, with an angle between **c** and **c*** of 4±1°. In building

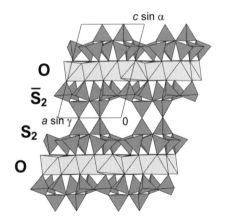

Figure 27. Structure of fedorite, as seen along **b**. The alkali cations and water molecules distributed inside the cavities are not represented.

the structural model for Z-phase we find, in the sequence of the modules, two points of possible ambiguity, namely in fixing the relative position of $\bar{S}_2$ and S_2 layers on both sides of the O module and of the X module. As regards the first point we may assume the same positioning as found in K-phase (and fedorite); as regards the second point there are six possible stacking ways equivalent to that realized in gyrolite, but only one gives an angle between **c** and **c*** (4.3°) corresponding to the value obtained by Gard et al. (1975). For the structure so derived, represented in Figure 28, the calculated intensities are in satisfactory agreement with those of the $hk0$ electron diffraction pattern presented by Gard et al. (1975).

Thermal behavior

The thermal studies carried on the various phases may be easily interpreted on the basis of the structural arrangements just discussed. By weight loss studies of reyerite, Chalmers et al. (1964) determined that water is lost in two steps, the first, corresponding to the loss of 5.8 water molecules is completed at 400°C, the second one, which begins at 650°C, corresponds to the loss of 3.6 water molecules; they attribute the two steps to the loss of water molecules and hydroxyl anions respectively, which is in good agreement with the six water molecules and the eight hydroxyl anions indicated by the structure analysis (Merlino 1988b).

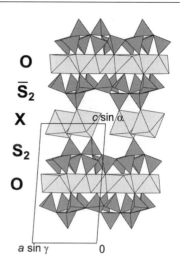

Figure 28. The structural model of Z-phase, as seen along **b**. The interlayer X module is built up by $CaO_2(H_2O)_4$ octahedra.

Similarly the dehydration curve presented by Mackay and Taylor (1954) for truscottite indicates a total weight loss of 4.6%, 1.5% in the temperature range 110 to 350°C, corresponding to the two water molecules in the crystal chemical formula, and 3.1% from 350 to 650°C, corresponding to the eight hydroxyl anions in the formula (Table 6).

The thermal behavior of gyrolite was firstly described for a specimen from Bombay by Mackay and Taylor (1953): also in this case the result of the study clearly shows that the water is lost in two stages, the first completed at 450°C, the second completed by 850°C; the weight losses in the two stages closely correspond to the water molecules and hydroxyl anions in the crystal chemical formula. A careful thermal study of gyrolite from Ortano (Italy) was carried out by Garavelli and Vurro (1984) and the results were carefully discussed and convincingly explained on the basis of the structural arrangement.

A thermal study of the Z-phase has been presented by Garbev (2004) who maintains that at 500°C the phase has lost all the molecular water and the transformation into the K-phase (see the discussion in the following paragraph) has been completed.

Very interesting results were obtained by following the dehydration processes of gyrolite and Z-phase through X-ray powder diffraction. Gard et al. (1975) found that by heating Z-phase at 500°C the basal (001) spacing, 15.3 Å, shortens to 11.8 Å; they also collected the powder diffraction pattern of the heated material, which, as we have already said, was subsequently shown by Garbev (2004) to correspond to the pattern of the K-phase. The Z-phase to K-phase transformation on dehydration at 400°C has been definitely proved and discussed by Garbev et al. (2004). They have also shown that, similarly, synthetic gyrolite transforms to truscottite at 400°C. In both transformations the water content of the X layer is lost and the separated S_2 and $\bar{S}_2$ sheets are condensed to form the characteristic double sheet of K-phase and truscottite. They also discuss about the possible location, in the dehydrated phases, of the two calcium cations of the interlayer X sheet, indicating that most probably they behave as the "zeolitic" alkali cations in fedorite and reyerite and that the charge balance may be restored through combination of the "redundant" oxide anions resulting from the condensation process and water molecules to give hydroxyl anions: $2\ O^{2-} + 2\ H_2O = 4\ OH^-$.

The dehydration process of a crystal of gyrolite from Qarusait has been also studied on a single crystal X-ray diffractometer equipped with heating device (Bonaccorsi and Merlino 2004). Structural refinements were carried out with data collected from the same crystal at room temperature and at 250°C, then again at room temperature after placing the crystal in water for three weeks. The crystal heated at 250°C has the structure-type of reyerite, with trigonal symmetry and $c = 19.1$ Å, with calcium and sodium cations distributed in the cavities of the $S_2 \overline{S}_2$ double sheet; after wetting, partial rehydration occurs, but the structure-type does not change. Gyrolite specimens from other localities (for example Antrim, Ireland; Ortano, Italy; Poona, India) present a different behavior on heating: their basal spacing remain unchanged at ≈ 22 Å on dehydration; it is possible that small differences in composition—we guess in the aluminum content—may explain the distinct behavior (Bonaccorsi and Merlino 2004).

Related mineral phases

Four other natural phases are known which may be classified as members of the gyrolite group: minehillite (related to reyerite), tungusite (related to gyrolite), martinite (related to Z-phase), and orlymanite, whose structure is still unknown.

Minehillite. It was described from Franklin, New Jersey, by Dunn et al. (1984) who suggested it is "a secondary, relatively low-temperature, hydrothermal, replacement mineral" with close relationships with reyerite. The results of the structural study, carried on by Dai et al. (1995) in the space group $P\overline{3}c1$, with $a = 9.777$, $c = 33.293$ Å, lead to the crystal chemical formula $(K,Na)_2Ca_{28}Zn_5Al_4Si_{40}O_{112}(OH)_{16}$. The structure is illustrated in Figure 29 and its relationships with that of reyerite may be understood on the basis of the following scheme:

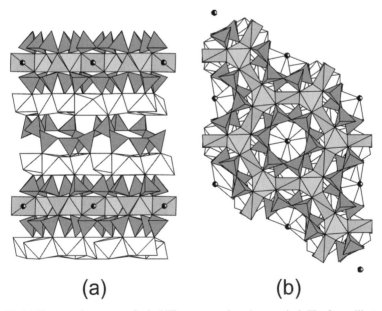

Figure 29. (a) The crystal structure of minehillite as seen along **b**, **c** vertical. The figure illustrates the sequence of the layers. (b) The figure represents the complex layer built up by sheets of hexagonal silicate rings, with the sheet of Al octahedra and Zn tetrahedra in between, as seen along **c**. The small circles in both (a) and (b) represent the K^+ cations located at 0, 0, 0.

(K,Na)	Ca$_{14}$	Zn$_{2.5}$□$_{0.5}$Al$_2$	Si$_{20}$O$_{56}$(OH)$_8$	minehillite
(Na,K)$_2$	Ca$_{14}$	Si$_2$Al$_2$O$_2$	Si$_{20}$O$_{56}$(OH)$_8$·6H$_2$O	reyerite

The structure of minehillite may be derived from that of reyerite by substituting, in the $S_2\bar{S}_2$ double layers, the pairs of inverted tetrahedra sharing common apices (Si-O-Al), with Al octahedra which build up an infinite layer through interconnection, by edge sharing, with Zn tetrahedra. The alkali cations are located in the centers of the rings of Al and Zn polyhedra. As clearly shown by Figure 29b, the slab built up by the sheets of silicate hexagonal rings and the sheet of Al and Zn polyhedra in between, is closely similar to that occurring in beryl, with zinc substituting for beryllium, or in the minerals of the osumilite group, with various cations substituting for Al and Zn. The thickness of this slab is drastically smaller than the thickness of the double tetrahedral layer $S_2\bar{S}_2$ in reyerite, thus reducing the size of the cavities; this explains why only a single alkali cation is present and the water molecules are absent. The presence of zinc in the hydrothermal solutions from which the mineral formed was probably responsible for the crystallization of minehillite instead than reyerite.

Tungusite. It was firstly found (Kudriashova 1966) in pillow lavas on the right-hand bank of the Lower Tunguska river, Tura, East Siberia, as radially fibrous flakes on the walls of amygdales, with intergrown reyerite and gyrolite. New findings were reported by Anastasenko (1978) in basalts of the North-West Siberian platform (basins of rivers Eramchino, Tutonchana and Kureyka). On the basis of chemical data, and electron and X-ray powder diffraction studies, a structural model closely related to the structure-type of gyrolite and the ideal crystal chemical formula Ca$_{14}$Fe$_9$Si$_{24}$O$_{60}$(OH)$_{22}$ have been proposed by Ferraris et al. (1995), who assumed the space group symmetry $P\bar{1}$, with a = 9.714, b = 9.721, c = 22.09 Å, α = 90.13, β = 98.3, γ = 120.0°. The structural relationships between tungusite (Fig. 30) and gyrolite are described by the following scheme:

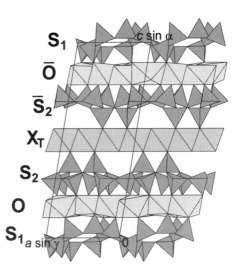

Figure 30. Crystal structural model of tungusite, as seen along **b**. The sequence of building modules is indicated.

Ca$_{14}$Si$_{24}$O$_{60}$(OH)$_8$	Fe$_9$(OH)$_{14}$	tungusite
Ca$_{14}$Si$_{24}$O$_{60}$(OH)$_8$	Ca$_2$·(14+x)H$_2$O	gyrolite

the X layer of gyrolite being substituted in tungusite by a continuous sheet X_T of edge sharing iron octahedra.

Martinite. It was discovered as last-stage phase in sodalite syenite xenoliths at the Poudrette quarry, Mont St. Hilaire, Quebec, by McDonald and Chao (2002), who determined and refined its crystal structure in the space group $P\bar{1}$, a = 9.544, b = 14.027, c = 9.535 Å, α = 71.06, β = 119.79, γ = 105.85° and defined its crystal chemical formula. The structure of martinite is related to that of the Z-phase and their relationships are described by the scheme:

(Ca,Na)$_7$	(Si,S)$_{14}$B$_2$O$_{38}$(OH,F,Cl)$_4$	Na$_6$·6H$_2$O	martinite
Ca$_7$	Si$_{16}$O$_{40}$(OH)$_2$	Ca$_2$·14H$_2$O	Z-phase

The S_2 and O modules present in martinite a more complex composition than in Z-phase, whereas the X layer of Z-phase is substituted by a layer of sodium cations and water molecules.

Orlymanite. This mineral has been found at the Wessels Mine, South Africa, as dark brown spherules in association with inesite and minor calcite. The species was defined by Peacor et al. (1990) who proposed the crystal chemical formula, $Ca_4Mn_3Si_8O_{20}(OH)_6 \cdot 2H_2O$, and determined its unit cell parameters $a = 9.60$, $c = 35.92$ Å, space group $P3$ or $P\bar{3}$. The structure has not been solved, but the cell parameters and the chemical composition point to a close relationships with the phases of the gyrolite group.

Ambiguities in the layer stacking and polytypism

Zvyagin (1997) carried on a study of the possible polytypes in the phases of the gyrolite group. In reyerite (and truscottite as well) the so called complex modules R = $\bar{S}_2$ O S_1 O S_2 are directly connected through the double sheet S_2 $\bar{S}_2$. The different periodicity of the O and S sheets (both S_1 and S_2 unit nets contain seven unit nets of the O sheet) is the cause of the possible polytypism in reyerite. Four distinct polytypes are derived, all presenting the common "complex layer" R; the R modules follow each other in different ways in the different polytypes, giving rise to the known reyerite structure with space group $P\bar{3}$ and $c = 19$ Å, and to three other possible arrangements, presenting two R modules in the unit translation $c = 38$ Å and space group $P6_3$, $P\bar{3}1c$, $P\bar{3}c1$. This last structure is closely related to that of minehillite through substitution of the S_2 $\bar{S}_2$ double sheet by the aluminum-zinc module which characterizes the crystal structure of minehillite.

In gyrolite the X module may have six different azimuthal orientations around the symmetry axis of the R complex layer. A careful analysis by Zvyagin (1997) has derived the possible homogeneous polytypes which comprise, besides the one-layer triclinic structure we have described in a previous section and represented in Figure 25, one trigonal three-layer ($c = 66.5$ Å) and one hexagonal six-layer ($c = 135.0$ Å) polytype, as well as several other triclinic and monoclinic two-layer polytypes.

In fedorite and in K phase, ...O S_2 $\bar{S}_2$ O S_2..., there are three distinct ways for positioning S_2 and $\bar{S}_2$ sheets on both sides of O: one of them is realized in the actual structures of fedorite and K-phase. In the more complex structure of Z-phase, ... $\bar{S}_2$ O S_2 X $\bar{S}_2$ O S_2..., if we assume that the S_2 and $\bar{S}_2$ sheets on both sides of the O sheet are placed as in the actual structure of fedorite, the problem is similar to that already met in gyrolite. As we have already said in a previous section, it was possible to derive the correct azimuthal orientation of the X module by comparing the calculated crystallographic parameters for each case with the crystallographic information by Gard et al. (1975).

ACKNOWLEDGMENTS

The authors are grateful to K. Garbev for his careful reading of the text and his valuable suggestions. The research was supported by MIUR (Rome) through the FIRB project "Properties and technological applications of minerals and their synthetic analogues" and the PRIN project "Microstructural and modular aspects in minerals: analyses and applications."

REFERENCES

Abdul-Jaber QH, Khoury H (1998) Unusual mineralisation in the Maqarin Area (North Jordan) and the occurrence of some rare minerals in the marbles and the weathered rocks. N Jb Geol Paläont Abh 208: 603-629

Aguirre L, Dominguez-Bella S, Morata D, Wittke O (1998) An occurrence of tobermorite in tertiary basalts from Patagonia, Chile. Can Mineral 36:1149-1155

Al-Wakeel EI, El-Korashy SA, El-Hemaly SA, Rizk MA (2001) Divalent ion uptake of heavy metal cations by (aluminum + alkali metals)-substituted synthetic 1.1nm-tobermorites. J Mater Sci 36:2405-2415

Anastasenko GF (1978) Boron-bearing traps of the North-West Siberian platform. Leningrad University Press. Leningrad (in Russian)

Anderson T (1851) Description and analysis of gyrolite. Philos Mag Sect IV 1:111-115

Anthony JW, Bideaux RA, Bladh KW, Nichols MC (1995) Handbook of Mineralogy. Vol. II Silica, silicates. Mineral Data Publishing, Tucson, Arizona

Assarsson GO (1957) Hydrothermal reactions between calcium hydroxide and amorphous silica; the reactions between 180 and 220°C. J Phys Chem 61:473-479

Assarsson GO (1958) Hydrothermal reactions between calcium hydroxide and amorphous silica; the reactions between 120 and 160°C. J Phys Chem 62:223-228

Baerlocher Ch, McCusker LB Database of Zeolite Structures: *http://www.iza-structure.org/databases/*

Ballirano P (1994) Crystal chemistry of cancrinites. Plinius 11:81-86

Ballirano P, Bonaccorsi E, Maras A, Merlino S (1997) Crystal structure of afghanite, the eight-layer member of the cancrinite group: evidence for long-range Si,Al ordering. Eur J Mineral 9:21-30

Ballirano P, Bonaccorsi E, Maras A, Merlino S (2000) The crystal structure of franzinite, the ten-layer mineral of the cancrinite group. Can Mineral 38:657-668

Ballirano P, Bonaccorsi E, Merlino S, Maras A (1998) Carbonate groups in davyne: structural and crystal-chemical considerations. Can Mineral 36:1285-1292

Ballirano P, Maras A (2004) The crystal structure of a "disordered" cancrinite. Eur J Mineral 16:135-141

Ballirano P, Maras A, Buseck PR (1996a) Crystal chemistry and IR spectroscopy of Cl- and SO_4-bearing cancrinite-like minerals. Am Mineral 81:1003-1012

Ballirano P, Maras A, Caminiti R, Sadun C (1995) Carbonate-cancrinite: In situ real time thermal processes studied by means of energy-dispersive X-ray powder diffractometry. Powder Diffr 10:173-177

Ballirano P, Merlino S, Bonaccorsi E, Maras A (1996b) The crystal structure of liottite, a six-layer member of the cancrinite group. Can Mineral 34:1021-1030

Bargar KE, Beeson MH, Keith TEC (1981) Zeolites in Yellowstone National Park. Min Record 12:29-38

Barrer RM, Cole JF, Villiger H (1970) Chemistry of soil minerals. Part VII. Synthesis, properties, and crystal structures of salt-filled cancrinites. J Chem Soc A:1523-1530

Barrer RM, Falconer JD (1956) Ion exchange in feldspathoids as a solid-state reaction. Proc Royal Soc 236: 227-249

Barrer RM, White EAD (1952) The hydrothermal chemistry of silicates. Part II. Synthetic crystalline sodium aluminosilicates. J Chem Soc (London):1561-1571

Baur WH (1991) The framework of $Na_8Al_6Ge_6O_{24} \cdot CO_3 \cdot 2H_2O$ has the LOS topology. Zeolites 11:639

Belokoneva EL, Uvarova TG, Dem'yanets LN (1986) Crystal structure of synthetic Ge-cancrinite $Na_8[Al_6Ge_6 O_{24}]Ge(OH)_6 \cdot 2H_2O$. Sov Phys Crystallogr 31:516-519

Bentor YK, Gross S, Heller L (1963) Some unusual minerals from the "mottled zone" complex, Israel. Am Mineral 48:924-930

Bieniok A, Brendel U, Amthauer G (2004) Synthesis and characterisation of a microporous cobaltphosphate with the cancrinite framework structure. Micro- and Mesoporous Mineral Phases (Accad Lincei, Roma). Volume of Abstracts, p 63-64

Bieniok A, Brendel U, Paulus E (1998) Synthese und Struktur eines Zinkphosphat-Cancrinits. Z Kristallogr Suppl. Issue 15:26.

Binon J, Bonaccorsi E, Bernhardt HJ, Fransolet AM (2004) The mineralogical status of "cavolinite" from Vesuvius, Italy, and crystallochemical data on the davyne subgroup. Eur J Mineral 16:511-520

Bøggild OB (1908) On Gyrolite from Greenland. Meddelelser om Grønland 34:91

Bogomolov VN, Efimov AN, Ivanova MS, Poborchii VV, Romanov SG, Smolin YuI, Shepelev YuF (1992) Structure and optical properties of a one-dimensional chain of selenium atoms in a cancrinite channel. Sov Phys Solid State 34:916-919

Bonaccorsi E (1992) Feldspatoidi del gruppo della davyna: cristallochimica, trasformazioni di fase, scambi ionici. PhD Dissertation, Dipartimento di Scienze della Terra, Università di Pisa

Bonaccorsi E (2004) The crystal structure of giuseppettite, the 16-layer member of the cancrinite-sodalite group. Microporous Mesoporous Mater 73:129-136

Bonaccorsi E, Comodi P, Merlino S (1995) Thermal behavior of davyne-group minerals. Phys Chem Minerals 22:367-374

Bonaccorsi E, Merlino S (2004) Calcium silicate hydrate (C-S-H) minerals: structures and transformations. 32nd Int Geol Congr Abs 1:215

Bonaccorsi E, Merlino S, Kampf AR (2005) The crystal structure of tobermorite 14 Å (plombierite), a C-S-H phase. J Am Ceram Soc 88:505-512

Bonaccorsi E, Merlino S, Orlandi P, Pasero M, Vezzalini G (1994) Quadridavyne, [(Na,K)$_6$Cl$_2$][Ca$_2$Cl$_2$][Si$_6$Al$_6$O$_{24}$], a new feldspathoid mineral from Vesuvius area. Eur J Mineral 6:481-487

Bonaccorsi E, Merlino S, Pasero M (1990) Davyne: its structural relationships with cancrinite and vishnevite. N Jb Mineral Mh 1990:97-112

Bonaccorsi E, Merlino S, Pasero M (1992) Davyne from Zabargad (St. John's) Island: peculiar chemical and structural features. Acta Vulcanol 2 Marinelli Volume:55-63

Bonaccorsi E, Merlino S, Pasero M, Macedonio G (2001) Microsommite: crystal chemistry, phase transitions, Ising model and Monte Carlo simulations. Phys Chem Miner 28:509-522

Bonaccorsi E, Orlandi P (2003) Marinellite, a new feldspathoid of the cancrinite-sodalite group. Eur J Mineral 15:1019-1027

Bresciani Pahor N, Calligaris M, Nardin G, Randaccio L (1982) Structure of a basic cancrinite. Acta Crystallogr B38:893-895

Brown L, Cesbron F (1973) Sur les surstructures des cancrinites. C R Acad Sc Paris Sèrie D 276:1-4

Buckner DA, Roy DM, Roy R (1960) Studies in the system CaO-Al$_2$O$_3$-SiO$_2$-H$_2$O. II: The system CaSiO$_3$-H$_2$O. Am J Science 258:132-147

Buhl J-Ch (1991) Synthesis and characterization of the basic and non-basic members of the cancrinite-natrodavyne family. Thermochim Acta 178:19-31

Buhl J-Ch, Stief F, Fechtelkord M, Gesing TM, Taphorn U, Taake C (2000) Synthesis, X-ray diffraction and MAS NMR characteristics of nitrate cancrinite Na$_{7.6}$[AlSiO$_4$]$_6$(NO$_3$)$_{1.6}$(H$_2$O)$_2$. J Alloys Compd 305:93-102

Burton A, Feuerstein M, Lobo RF, Chan JCC (1999) Characterization of cancrinite synthesized in 1,3-butanediol by Rietveld analysis of powder neutron diffraction data and solid-state ^{23}Na NMR spectroscopy. Microporous Mesoporous Mater 30:293-305

Camara F, Bellatreccia F, della Ventura G, Mottana A (2004) A new member of the cancrinite-sodalite group with a 14 layers stacking sequence. Micro- and Mesoporous Mineral Phases (Accad Lincei, Roma) Volume of Abstracts, p 186-187

Cann JR (1965) Gyrolite and reyerite from 'S Airde Beinn, northern Mull. Mineral Mag 35:1-4

Chalmers RA, Farmer VC, Harker RI, Kelly S, Taylor HFW (1964) Reyerite. Mineral Mag 33:821-840

Chorover J, Choi S, Amistadi MK, Karthikeyan KG, Crosson G, Mueller KT (2003) Linking Cesium and Strontium Uptake to Kaolinite Weathering in Simulated Tank Waste Leachate. Environ Sci Technol 37: 2200-2208

Claringbull GF, Hey MH (1952) A re-examination of tobermorite. Mineral Mag 29:960-962

Clement SC, Ribbe PH (1973) New locality, formula, and proposed structure for reyerite. Am Mineral 58: 517-522

Cong X, Kirkpatrick RJ (1996) ^{29}Si and ^{17}O NMR investigation of the structure of some crystalline calcium silicate hydrates. Adv Cem Bas Mater 3:133-143

Coombs DS, Alberti A, Armbruster T, Artioli G, Colella C, Galli E, Grice JD, Liebau F, Mandarino JA, Minato H, Nickel EH, Passaglia E, Peacor DR, Quartieri S, Rinaldi R, Ross M, Sheppard RA, Tillmanns E, Vezzalini G (1998) Recommended nomenclature for zeolite minerals: Report of the Subcommittee on Zeolites of the International Mineralogical Association, Commission on New Minerals and Mineral Names. Eur J Mineral 10:1037-1081

Cornu P, Himmelbauer A (1906) Reyerit. Tsch Mineral Petr Mitt 25:519-520

Currie J (1905) Note on some new localities for Gyrolite and Tobermorite. Mineral Mag 14:93-95

Dai Y, Post JE, Appleman DE (1995) Crystal structure of minehillite: twinning and structural relationships to reyerite. Am Mineral 80:173-178

Daubrée GA (1858) Sur la relation des sources thermales de Plombières avec les filons métallifères et sur la formation contemporaine des zéolithes. Annales des Mines ser. 5 13:227-256

Deer WA, Howie RS, Wise WS, Zussman J (2004) Rock-forming minerals. Vol. 4B Second Edition. Framework silicates: silica minerals, feldspathoids and the zeolites. The Geological Society, London

Depmeier W (2005) The sodalite family – a simple but versatile framework structure. Rev Mineral Geochem 57:203-240

Diamond S, White JL, Dolch WL (1966) Effects of isomorphous substitutions in hydrothermally-synthesized tobermorite. Am Mineral 51:388-401

Dornberger Schiff K (1956) On the order-disorder (OD-structures). Acta Crystallogr 9:593-601

Dornberger Schiff K (1964) Grundzüge einer Theorie von OD-Strukturen aus Schichten, Abh Deutschen Akad Wiss Berlin, Klasse für Chem Geol Biol 3:1-107

Dornberger-Schiff K (1966) Lehrgang über OD-Strukturen. Akademie-Verlag, Berlin

Dunn PJ, Peacor DR, Leavens PB, Wicks FJ (1984) Minehillite: A new layer silicate from Franklin, New Jersey, related to reyerite and truscottite. Am Mineral 69:1150-1155

Ďurovič S (1997) Fundamentals of the OD theory. EMU Notes in Mineralogy 1:3-28

Eakle AS (1917) Minerals associated with the crystalline limestone at Crestmore, Riverside County, California. Bull Dept Geol Univ Calif 10/19:327-330

Edgar AD (1964) Studies on cancrinites: II – Stability fields and cell dimensions of calcium and potassium-rich cancrinites. Can Mineral 8:53-67

Edgar AD, Burley BJ (1963) Studies on cancrinites I –Polymorphism in sodium carbonate rich cancrinite-natrodavyne. Can Mineral 7:631-642

Eitel W (1922) Über das System $CaCO_3$-$NaAlSiO_4$ (Calcit-Nephelin) und den Cancrinit. N Jb Mineral 2: 45-61

El Hemaly SAS, Mitsuda T, Taylor HFW (1977) Synthesis of normal and anomalous tobermorites. Cem Concr Res 7:429-438

Emiraliev A, Yamzin II (1982) Neutron-diffraction refinement of the structure of a carbonate-rich cancrinite. Sov Phys Crystallogr 27:27-30

Farmer VC, Jeevaratnam J, Speakman K, Taylor HFW (1966) Thermal decomposition of 14 Å tobermorite from Crestmore. In: Proc. Symp. 'Structure of Portland cement paste and concrete' Sp. Report 90, Highway Res. Board. Washington DC, p 291-299

Fechtelkord M, Posnatzki B, Buhl J-Ch (2003) Characterization of basic cancrinite synthesized in a butanediol-water system. Eur J Mineral 15:589-598

Fechtelkord M, Posnatzki B, Buhl J-Ch, Fyfe CA, Groat LA, Raudsepp M (2001) Characterization of synthetic Cs-Li cancrinite grown in a butanediol-water system: An NMR spectroscopic and Rietveld refinement study. Am Mineral 86:881-888

Ferraris G, Makovicky E, Merlino S (2004) Crystallography of Modular Materials. IUCr Monographs in Crystallography, Oxford University Press, Oxford

Ferraris G, Pavese A, Soboleva S (1995) Tungusite: new data, relationships with gyrolite and structural model. Mineral Mag 59:535-543

Ferreira A, Ananias D, Carlos LD, Morais CM, Rocha J (2003) Novel microporous lanthanide silicates with tobermorite-like structure. J Am Chem Soc 125:14573-14579

Flint EP, McMurdie HF, Wells LS (1938) Formation of hydrated calcium silicate at elevated temperatures and pressures. J Res Natl Bur Standards 21:617-638

Foit FFJr, Peacor DR, Heinrich EW (1973) Cancrinite with a new superstructure from Bancroft, Ontario. Can Mineral 11:940-951

Funk H (1961) Chemische Untersuchungen von Silicaten. XXVI. Über Calciumsilicathydrate mit der Zusammensetzung $CaO·2SiO_2·0.5$-$2H_2O$ und die Synthese des Reyerit (=Truscottit) ($CaO·2SiO_2·0.5H_2O$). Z anorg allg Chem 313:1-11

Funk H, Thilo E (1955) Über Hydrogensilicate. IV. Das Calcium-trihydrogenmonosilikat $Ca[OSi(OH)_3]_2$ und seine Umwandlung in das Calciumtetrahydrogendisilikat $Ca[Si_2O_3(OH)_4]$. Z anorg allg Chem 278:237-248

Gaines RV, Skinner HCW, Foord EE, Mason B, Rosenzweig A (1997) Dana's New Mineralogy, 8th Ed. John Wiley & Sons, New York

Garavelli CL, Vurro F (1984) Gyrolite from Ortano (Island of Elba). Rend Soc Ital Mineral Petrol 39:695-704

Garbev K (2004) Structure, properties and quantitative Rietveld analysis of calcium silicate hydrates (C-S-H-phases) crystallized under hydrothermal conditions. PhD Dissertation. Ruprecht-Karls-University, Heidelberg

Garbev K, Black L, Stumm A, Stemmermann P, Gasharova B (2004) Polymerization reactions by thermal treatment of gyrolite-group minerals – an IR spectroscopic and X-ray diffraction study based on synchrotron radiation. In: Applied Mineralogy. Pecchio et al (eds) 2004 ICAM-BI, São Paulo, p 245-248

Gard JA, Luke K, Taylor HFW (1981a) $Ca_7Si_{16}O_{40}H_2$, a new calcium silicate hydrate phase of the truscottite group. Cem Concr Res 11:659-664

Gard JA, Luke K, Taylor HFW (1981b) Crystal structure of K-phase, $Ca_7Si_{16}O_{40}H_2$. Sov Phys Crystallogr 26: 691-695

Gard JA, Mitsuda T, Taylor HFW (1975) Some observations on Assarsson's Z-phase and its structure relations to gyrolite, truscottite and reyerite. Mineral Mag 43:325-332

Gesing ThM, Buhl J-Ch (2000) Structure and spectroscopic properties of hydrogencarbonate containing alumosilicate sodalite and cancrinite. Z Kristallogr 215:413-418

Gies H, Kirchner R, van Koningsveld H, Treacey MMJ (1999) Faulted Zeolite Framework Structures. In: Proc. 12th International Zeolite Conference, Baltimore, Maryland, USA, July 5-11, 1998, Treacey MMJ, Marcus BK, Bisher ME, Higgins JB (Eds) Materials Research Society, MRS, Warrendale, Pennsylvania, p 2999-3029

Gottardi G, Passaglia E (1965) Tobermorite "non espandibile" di Prà de la Stua (Trento). Period Mineral 35: 197-204

Gottardi G, Passaglia E (1966) Tobermorite "non espandibile" e Gyrolite del Monte Biaena (Trento). Period Mineral 36:1079-1083

Grose LT, Keller GV (1976) Petrology of deep drill hole, Kilauea Volcano. Am Geophys Union Trans 57: 1017

Grundy HD, Hassan I (1982) The crystal structure of a carbonate-rich cancrinite. Can Mineral 20:239-251

Gutzmer J, Cairncross B (1993) Recent discoveries from the Wessels Mine, South Africa. Min Record 24: 365-368

Hackbarth K, Gesing ThM, Fechtelkord M, Stief F, Buhl J-Ch (1999) Synthesis and crystal structure of carbonate cancrinite $Na_8[AlSiO_4]_6CO_3(H_2O)_{3.4}$, grown under low-temperature hydrothermal conditions. Microporous Mesoporous Mater 30:347-358

Hara N, Chan CF, Mitsuda T (1978) Formation of 14 Å tobermorite. Cem Concr Res 8:113-116

Hara N, Inoue N (1980) Thermal behavior of 11 Å tobermorite and its lattice parameters. Cem Concr Res 10: 53-60

Harker RI (1964) Dehydration series in the system $CaO-SiO_2-H_2O$. J Am Ceram Soc 47:521-529

Harrison WTA, Gier TE, Stucky GD (1993) Synthesis and structure of $Li_8(HPO_4)(BePO_4)_6 \cdot H_2O$ – a new zeolite, LOS-type beryllophosphate molecular sieve. Zeolites 13:242-248

Harvey RD, Beck CW (1962) Hydrothermal regularly interstratified chlorite-vermiculite and tobermorite in alteration zones at Goldfield, Nevada. Clays Clay Miner 9:343-354

Hassan I, Antao SM, Parise JB (2005) Cancrinite: structures, phase transition, dehydration, and Na mobilization at high temperatures. Eur J Mineral (accepted)

Hassan I, Buseck PR (1992) The origin of the superstructure and modulations in cancrinite. Can Mineral 30: 49-59

Hassan I, Grundy HD (1984) The character of the cancrinite-vishnevite solid-solution series. Can Mineral 22: 333-340

Hassan I, Grundy HD (1990) Structure of davyne and implications for stacking faults. Can Mineral 28:341-349

Hassan I, Grundy HD (1991) Crystal structure of basic cancrinite, ideally $Na_8[Al_6Si_6O_{24}](OH)_2 \cdot 3H_2O$. Can Mineral 29:377-383

Heddle MF (1880) Preliminary notice of substances which may prove to be new minerals. Mineral Mag 4: 117-123

Heddle MF (1893) On pectolite and okenite from new localities: the former with new appearances. Trans Geol Soc Glasgow 9:241-255

Heller L, Taylor HFW (1951) Hydrated calcium silicates. Part II. Hydrothermal reactions: lime:silica ratio 1: 1. J Chem Soc 1951:2397-2401

Heller L, Taylor HFW (1952) Hydrated calcium silicates. Part III. Hydrothermal reactions of lime:silica molar ratio 3:2. J Chem Soc 1952:1018-1020

Henmi C, Kusachi I (1989) Monoclinic tobermorite from Fuka, Bitchu-cho, Okayama Prefecture, Japan. J Min Petr Econ Geol 84:374-379 (in Japanese)

Henmi C, Kusachi I (1992) Clinotobermorite $Ca_5Si_6(O,OH)_{18} \cdot 5H_2O$, a new mineral from Fuka, Okayama Prefecture, Japan. Mineral Mag 56:353-358

Hentschel G (1973) Begleitmineralien des Basalts vom Arensberg bei Zilsdorf/Eifel. Notizbl Hess Landesamt Bodenforsch 101:310-316

Hoffmann C, Armbruster T (1997) Clinotobermorite, $Ca_5[Si_3O_8(OH)]_2 \cdot 4H_2O$ – $Ca_5[Si_6O_{17}] \cdot 5H_2O$, a natural C-S-H(I) type cement mineral: determination of the substructure. Z Kristallogr 212:864-873

Hogarth DD (1979) Afghanite: new occurrences and chemical composition. Can Mineral 17:47-52

Hong S-Y, Glasser FP (2003) Phase relations in the $CaO-SiO_2-H_2O$ system to 200°C at saturated steam pressure. Cem Concr Res 34:1529-1534

Hövig P (1914) Truscottiet. Jaarb Mijnwezen Ned Oost-Indië Batavia 41 (for 1912):202

Hund F (1984) Nitrat-, Thiosulfat-, Sulfat- und Sulfid-Cancrinit. Z Anorg Allg Chem 509:153-160

Ivanov VG, Sapozhnikov AN (1975) The first find of afghanite in the U.S.S.R.. Zap Vses Mineral Obs 104: 3328-3331 (in Russian)

Izawa E, Yamashita M (1995) Truscottite from the Hishikari deposit, Kagoshima Prefecture. J Soc Resource Geology 45:251-252

Jakobsson S, Moore JG (1986) Hydrothermal minerals and alteration rates at Surtsey volcano, Iceland. Geol Soc Am Bull 97:648-659

Jamtveit B, Dahlgren S, Austrheim H (1997) High-grade contact metamorphism of calcareous rocks from the Oslo Rift, Southern Norway. Am Mineral 82:1241-1254

Jander W, Franke B (1941) Die Bildung von Calciumhydrosilikaten aus Calciumoxyd und Kieselsauregel bei 300° und 350° und hohen Drucken. Z Anorg Allg Chem 247:161-179

Jarchow O (1965) Atomanordnung und Strukturverfeinerung von Cancrinit. Z Kristallogr 122:407-422

Joswig W, Drits VA, Sokolova GV (1988) Refinement of structure of fedorite. Sov Phys Crystallogr 33:763-765
Kalousek GL (1955) Tobermorite and related phases in the system $CaO-SiO_2-H_2O$. J Am Concr Inst 51:989-1011
Kalousek GL (1957) Crystal chemistry of the hydrous calcium silicates: I. Substitution of aluminum in the lattice of tobermorite. J Am Ceram Soc 40:124-132
Kalousek GL, Roy R (1957) Crystal chemistry of hydrous calcium silicates: II. Characterization of interlayer water. J Am Ceram Soc 40:236-239
Kanepit VN, Rieder EÉ (1995) Neutron diffraction study of cancrinite. J Struct Chem 36:694-696
Kato A, Matsubara S, Tiba T, Sakata Y (1984) A Barium-bearing Tobermorite from Heguri, Chiba prefecture, Japan. Bull Natn Sci Mus Tokyo, Ser. C, 1984:10131-10140
Khomyakov AP, Nadezhina TN, Rastsvetaeva RK, Pobedimskaya EA (1992) Hydroxycancrinite $Na_8[Al_6Si_6O_{24}](OH)_2 \cdot 3H_2O$: A new mineral. Zap Veseross Mineral Obs 121(1):100-105 (in Russian)
Khomyakov AP, Pobedimskaya EA, Nadezhina TN, Tereneteva LE, Rastsvetaeva RK (1991a) Structural mineralogy of high-Si cancrinite. Moscow Univ Geol Bull 46:71-75
Khomyakov AP, Semenov EI, Pobedimskaya EA, Nadezhina TN, Rastsvetaeva RK (1991b) Cancrisilite $Na_7[Al_5Si_7O_{24}]CO_3 \cdot 3H_2O$: A new mineral of the cancrinite group. Zap Vses Mineral Obs 120(6):80-84 (in Russian)
Khukarenko AA, Orlova MP, Bulakh AG, Bagdasarov EA, Rimskaya-Korsakov OM, Nefedov EI, Il'inskii GA, Sergeev AS, Abakumova NB (1965) Caledonian complex of ultrabasic alkaline rocks and carbonatites of the Kola Peninsula and Northern Karelia. Nedra Press, Leningrad, Russia (in Russian)
Klaska R, Jarchow O (1977) Sinthetischer Sulfat-Hydrocancrinit vom Mikrosommit-typ. Naturwiss 64:93
Kobayashi A, Kawai T (1974) Gyrolite found in the andesitic tuffs, near the Sayama lake, Ueda city, Nagano Prefecture, Japan. Geosci Mag 25:367-370
Komarneni S, Breval E, Miyake M, Roy R (1987b) Cation-exchange properties of (Al+Na)-substituted synthetic tobermorites. Clays Clay Minerals 35:385-390
Komarneni S, Guggenheim S (1988) Comparison of cation exchange in ganophyllite and [Na+Al]-substituted tobermorite: crystal-chemical implications. Mineral Mag 52:371-375
Komarneni S, Roy DM (1983) Tobermorites: a new family of cation exchangers. Science 221:647-648
Komarneni S, Roy DM, Fyfe CA, Kennedy GJ (1987a) Naturally occurring 1.4nm tobermorite and synthetic jennite: characterization by ^{27}Al and ^{29}Si MASNMR spectroscopy and cation exchange properties. Cem Concr Res 17:891-895
Komarneni S, Roy DM, Roy R (1982) Al-substituted tobermorite: shows cation exchange. Cem Concr Res 12:773-780
Komarneni S, Roy R, Roy DM, Fyfe CA, Kennedy GJ, Bothner-By AA, Dadok J, Chesnick AS (1985) ^{27}Al and ^{29}Si magic angle spinning nuclear magnetic resonance spectroscopy of Al-substituted tobermorites. J Mater Science 20:4209-4214
Konyev AA, Vorobyev YI, Bulakh AG (1993) Charoit – der Schmukstein aus Sibirien und seine seltenen Begleit-minerale. Lapis 1993:13-20
Kudriashova VI (1966) Tungusite, a new hydrous silicate of calcium. Dokl Akad Nauk SSSR 171:163-166
Lachowski EE, Murray LW, Taylor HFW (1979) Truscottite: composition and ionic substitutions. Mineral Mag 43:333-336
Lee Y, Parise JB, Tripathi A, Kim SJ, Vogt T (2000) Synthesis and crystal structures of gallium and germanium variants of cancrinite. Microporous Mesoporous Mater 39:445-455
Leoni L, Mellini M, Merlino S, Orlandi P (1979) Cancrinite-like minerals: new data and crystal chemical considerations. Rend Soc Ital Mineral Petrol 35:713-719
Lindner G-G, Hoffmann K, Witke K, Reinen D, Heinemann C, Koch W (1996) Spectroscopic properties of Se_2^{2-} and Se_2^- in cancrinite. J Solid State Chem 126:50-54
Lindner G-G, Massa W, Reinen D (1995) Structure and properties of hydrothermally synthesized thiosulfate cancrinite. J Solid State Chem 117:386-391
Liu C, Li S, Tu, K, Xu R (1993) Synthesis of cancrinite in butane-1,3-diol systems. Chem Comm 1993:1645-1646
Livingstone A (1988) Reyerite, tobermorite, calcian analcime and bytownite from amygdales in a Skye basalt. Mineral Mag 52:711-713
Mackay AL, Taylor HFW (1953) Gyrolite. Mineral Mag 30:80-91
Mackay AL, Taylor HFW (1954) Truscottite. Mineral Mag 30:450-457
Maeshima T, Noma H, Sakiyama M, Mitsuda T (2003) Natural 1.1 and 1.4 nm tobermorites from Fuka, Okayama, Japan: chemical analysis, cell dimensions, ^{29}Si NMR and thermal behavior. Cem Concr Res 33:1515-1523
Mamedov KS, Belov NV (1958) The crystal structure of micaceous Ca-hydrosilicates: okenite, nekoite, truscottite, gyrolite. A new silico-oxygen radical $[Si_6O_{15}]$. Dokl Akad Nauk SSSR 121:720-723

Marincea S, Bilal E, Verkaeren J, Pascal M-L, Fonteilles M (2001) Superposed parageneses in the spurrite-, tilleyite- and gehlenite-bearing skarns from Cornet Hill, Apuseni Mountains, Romania. Can Mineral 39: 1435-1453

Mashal K, Harsh JB, Flury M, Felmy AR, Zhao H (2004) Colloid formation in Hanford sediments reacted with simulated tank waste. Environ Sci Technol 38:5750-5756

McConnell JDC (1954) The hydrated calcium silicates riversideite, tobermorite, and plombierite. Mineral Mag 30:293-305

McCusker LB, Liebau F, Engelhardt G (2001) Nomenclature of structural and compositional characteristics of ordered microporous and mesoporous materials with inorganic hosts. (IUPAC Recommendations 2001). Pure Appl Chem 73:381-394

McDonald AM, Chao GY (2002) Martinite, a new borosilicate mineral from Mont Saint-Hilaire, Quebec, Canada: description and crystal structure determination. 18th Gen Meet IMA 1:139

Merlino S (1972) New tetrahedral sheets in reyerite. Nature Phys Sci 238:124-125

Merlino S (1984) Feldspathoids: their average and real structures. In: Feldspars and Feldspathoids. Brown WL (ed), Riedel Publishing Company, Dordrecht, p 457-470

Merlino S (1988a) Gyrolite: its crystal structure and crystal chemistry. Mineral Mag 52:377-387

Merlino S (1988b) The structure of reyerite, $(Na,K)_2Ca_{14}Si_{22}Al_2O_{58}(OH)_8 \cdot 6H_2O$. Mineral Mag 52:247-256

Merlino S (1997) OD approach in minerals. EMU Notes in Mineralogy 1:29-54

Merlino S, Bonaccorsi E, Armbruster T (1999) Tobermorites: Their real structure and order-disorder (OD) character. Am Mineral 84:1613-1621

Merlino S, Bonaccorsi E, Armbruster T (2000) The real structures of clinotobermorite and tobermorite 9 Å: OD character, polytypes, and structural relationships. Eur J Mineral 12:411-429

Merlino S, Bonaccorsi E, Armbruster T (2001) The real structure of tobermorite 11 Å: normal and anomalous forms, OD character and polytypic modifications. Eur J Mineral, 13:577-590

Merlino S, Mellini M, Bonaccorsi E Pasero M, Leoni L, Orlandi P (1991) Pitiglianoite, a new feldspathoid from southern Tuscany, Italy: chemical composition and crystal structure. Am Mineral 76:2003-2008

Meyer JW, Jaunarais KL (1961) Synthesis and crystal chemistry of gyrolite and reyerite. Am Mineral 46: 913-933

Milestone NB, Hughes SM, Stonestreet PJ (1995) Synthesis of zeolites in anhydrous glycol systems. Stud Surf Sci Catal 98:42-43

Minato H, Kato A (1967) Truscottite from the Toi mine, Shizuoka Prefecture. Mineral J (Japan) 5:144-156

Mitchell RH, Burns PC (2001) The structure of fedorite: a re-appraisal. Can Mineral 39:769-777

Mitsuda T (1973) Paragenesis of 11 Å tobermorite and poorly crystalline hydrated magnesium silicate. Cem Concr Res 3:71-80

Mitsuda T, Taylor HFW (1978) Normal and anomalous tobermorites. Mineral Mag 42:229-235

Miyake M, Komarneni S, Roy R (1989) Kinetics, equilibria and thermodynamics of ion exchange in substituted tobermorites. Mater Res Bull 24:311-320

Nadezhina TN, Rastsvetaeva RK, Pobedimskaya EA, Khomyakov AP (1991) Crystal structure of natural hydroxyl-containing cancrinite. Sov Phys Crystallogr 46:325-327

Nawaz R (1977) A second occurrence of killalaite. Mineral Mag 41:546-548

Němec D (1982) Assemblages of fissure minerals in the basic Ransko Massif. N Jb Mineral Abh 145:256-269

Nithollon P, Vernotte MP (1955) Structure cristalline de la cancrinite. Publ Sci Tech Ministère Air, France, Notes Tech 53, 48 pp

Norby P, Krogh Andersen IG, Krogh Andersen E, Colella C, De Gennaro M (1991) Synthesis and structure of lithium cesium and lithium thallium cancrinites. Zeolites 11:248-253

Park M, Choi CL, Lim WT, Kim MC, Choi J, Heo NH (2000) Molten-salt method for the synthesis of zeolitic materials II. Characterization of zeolitic materials. Microporous Mesoporous Mater 37:91-98

Passaglia E, Turconi B (1982) Silicati ed altri minerali di Montalto di Castro (Viterbo). Rivista Mineralogica Italiana 1982:97-110

Pauling L (1930) The structure of some sodium and calcium aluminosilicates. Proc Natl Acad Sci 16:453-459

Peacor DR, Dunn PJ, Nelen JA (1990) Orlymanite, $Ca_4Mn_3Si_8O_{20}(OH)_6 \cdot 2H_2O$, a new mineral from South Africa: A link between gyrolite-family and conventional phyllosilicate minerals? Am Mineral 75:923-927

Peacor DR, Rouse RC, Ahn J-Ho (1987) Crystal structure of tiptopite, a framework beryllophosphate isotypic with basic cancrinite. Am Mineral 72:816-820

Pobedimskaya YeA, Terent'eva LYe, Rastsvetaeva RK, Sapozhnikov AN, Kashaev AA, Dorokhova GI (1991) Kristallicheskaya struktura bystrita. Dokl Akad Nauk SSSR 319:873-878 (in Russian)

Pushcharovskii DYu, Yamnova NA, Khomyakov AP (1989) Crystal structure of high-potassium vishnevite. Sov Phys Crystallogr 34:37-39

Rinaldi R (1982) More stacking variations in cancrinite-related minerals; how many more new minerals? J Microsc Spectrosc Electron 7:76a-77a

Rinaldi R, Wenk H-R (1979) Stacking variations in cancrinite minerals. Acta Cryst A35:825-828
Rocha J, Lin Z (2005) Microporous mixed octahedral-pentahedral-tetrahedral framework silicates. Rev Mineral Geochem 57:173-202
Rozenberg KA, Sapozhnikov AN, Rastsvetaeva RK, Bolotina NB, Kashaev AA (2004) Crystal structure of a new representative of the cancrinite group with a 12-layer stacking sequence of tetrahedral rings. Crystallogr Rep 49:635-642
Sakiyama M, Maeshima T, Mitsuda T (2000) Synthesis and crystal chemistry of Al-substituted 11 Å tobermorite. J Soc Inorg Mater Jpn 7:413-419
Sapozhnikov AN (2004) Influence of the chemical composition on the framework configuration of cancrinite-like minerals. Micro- and Mesoporous Mineral Phases (Accad Lincei, Roma). Volume of Abstracts, p 290-293
Sapozhnikov AN, Levitsky VI, Cherepanov DI, Suvorova LF, Bogdanova LA (2004) About influence of chlorine upon the framework configuration in cancrinite-like minerals. Zap Veseross Mineral Obs 133(5): 93-102 (in Russian)
Sasaki K, Masuda T, Ispida H, Mitsuda T (1996) Structural degradation of tobermorite during vibratory milling. J Am Ceram Soc 79:1569-1574
Shaw S, Clark SM, Henderson CMB (2000) Hydrothermal formation of the calcium silicate hydrates, tobermorite [$Ca_5Si_6O_{16}(OH)_2 \cdot 4H_2O$] and xonotlite [$Ca_6Si_6O_{17}(OH)_2$] an in-situ synchrotron study. Chem Geol 167:129-140
Shaw S, Henderson CMB, Clark SM (2002) In-situ synchrotron study of the kinetics, thermodynamics, and reaction mechanism of the hydrothermal crystallization of gyrolite $Ca_{16}Si_{24}O_{60}(OH)_8 \cdot 14H_2O$. Am Mineral 87:533-541
Shrivastava O P, Komarneni S, Breval E (1991) Mg^{2+} uptake by synthetic tobermorite and xonotlite. Cem Concr Res 21:83-90
Šiaučiūnas R, Baltakys K (2004) Formation of gyrolite during hydrothermal synthesis in the mixtures of CaO and amorphous SiO_2 or quartz. Cem Concr Res 34:2029-2036
Šiaučiūnas R, Palubinskaite D, Ivaunaskas R (2002) Elimination of heavy metals from water by modified tobermorite. Envir Res Engin Manag 3(21):61-66
Sieber W, Meier WM (1974) Formation and properties of Losod, a new sodium zeolite. Helv Chim Acta 57: 1533-1549
Sirbescu M, Jenkins DM (1999) Experiments on the stability of cancrinite in the system Na_2O-CaO-Al_2O_3-SiO_2-CO_2-H_2O. Am Mineral 84:1850-1860
Smolin YuI, Shepelev YuF, Butikova IK, Kobyakov IB (1981) Crystal structure of cancrinite. Sov Phys Crystallogr 26:33-35
Sokolov YuA, Maksimov BA, Galiulin RV, Ilyukhin VV, Belov NV (1981) Determination of crystal structure of cancrinite-like $Na_8Al_6Ge_6O_{24}(CO_3) \cdot 2H_2O$. Patterson and diffraction pseudosymmetry. Sov Phys Crystallogr 26:161-164
Sokolov YuA, Maksimov BA, Ilyukhin VV, Belov NV (1978) Low-temperature investigation of the crystal structure of sodium aluminogermanate $Na_8Al_6Ge_6O_{24}(CO_3) \cdot 3H_2O$. Sov Phys Dokl 23:789-791
Sokolova GV, Kashaev AA, Drits VA, Ilyukhin VV (1983) The crystal structure of fedorite. Sov Phys Crystallogr 28:95-96
Stephens DJ, Bray E (1973) Occurrence and infrared analysis of unusual zeolitic minerals from Bingham, Utah. Min Record 4:67-72
Strunz H, Micheelsen H (1958) Calcium phyllosilicates. Naturwiss 45:515
Sweet JM, Bothwell DI, Williams DL (1961) Tacharanite and other hydrated calcium silicates from Portree, Isle of Skye. Mineral Mag 32:745-753
Taylor HFW (1950) Hydrated calcium silicates. Part I. Compound formation at ordinary temperatures. J Chem Soc 1950:3682-3690
Taylor HFW (1953a) Crestmoreite and riversideite. Mineral Mag 30:155-165
Taylor HFW (1953b) Hydrated calcium silicates. Part V. The water content of calcium silicate hydrate (I). J Chem Soc 1953:163-171
Taylor HFW (1962) Hydrothermal reactions in the systems CaO-SiO_2-H_2O and the steam curing of cement and cement-silica products. *In:* Chemistry of Cement, Proceedings of the Fourth International Symposium. National Bureau of Standards Monograph 43. U.S. Department of Commerce, Washington DC, p 167-204
Taylor HFW (1964) The calcium silicate hydrates. *In:* Chemistry of Cements. Vol. 1. Taylor HFW (Ed) Academic Press, London, p 167-232
Taylor HFW (1986) Proposed structure for calcium silicate hydrate gel. J Am Ceram Soc 69:464-467
Taylor HFW (1992) Tobermorite, jennite, and cement gel. Z Kristallogr 202:41-50
Tsuji M, Komarneni S (1989) Alkali metal ion selectivity of Al-substituted tobermorite. J Mater Res 4:698-703

Tsuji M, Komarneni S, Malla P (1991) Substituted tobermorites: ^{27}Al and ^{29}Si MASNMR, cation exchange, and water sorption studies. J Am Ceram Soc 74:274-279

Vaughan DEW (1986) A crystalline zeolite composition having a cancrinite-like structure and a process for its preparation. Eur Patent 0 190 903 A3

Vaughan DEW (1991) Process for preparation of an ECR-5 crystalline zeolite composition. US Patent 5,015,454

Walenta K (1980) Zeolithparagenesen aus dem Melilith-Nephelinit des Howenegg im Hegau. Aufschluss 25: 613-626

Webb ABSJ (1971) Tobermorite from Castle Hill near Kilbirnie, Ayrshire. Mineral Mag 38:253

White DE, Fournier RO, Muffler LJP, Truesdell AH (1975) Physical results of research drilling in thermal areas of Yellowston National Park, Wyoming. U S Geol Survey Prof Paper 892

Wieker W, Grimmer A-R, Winkler A, Mägi M, Tarmak M, Lippmaa E (1982) Solid-state high-resolution ^{29}Si NMR spectroscopy of synthetic 14 Å, 11 Å and 9 Å tobermorites. Cem Concr Res 12:333-339

Yakubovich OV, Karimova OV, Mel'nikov OK (1994) A new representative of the cancrinite family $(Cs,\bar{K})_{0.33}$ $[Na_{0.18}Fe_{0.16}(H_2O)_{1.05}]\{ZnPO_4\}$: preparation and crystal structure. Crystallogr Rep 39:564-568

Yamazaki S, Toraya H (2001) Determination of positions of zeolitic calcium atoms and water molecules in hydrothermally formed aluminum-substituted tobermorite-1.1nm using synchrotron radiation powder data. J Am Ceram Soc 84:2685-2690

Zadov AE, Chukanov NV, Organova NI, Belakovsky DI, Fedorov AV, Kartashov PM, Kuzmina OV, Litzarev MA, Moknov AV, Loskutov AB, Finko VI (1995) New results: the investigations on the minerals of the tobermorite group. Proc Russ Miner Soc 124:36-54

Zhuang J, Flury M, Jin Y (2003) Colloid-facilitated Cs transport through water-saturated Hanford sediment and Ottawa sand. Environ Sci Technol 37:4905-4911

Zvyagin BB (1997) Modular analysis of crystals. EMU Notes in Mineralogy 1:345-372

Zyryanov VN (1982) Cancrinite equilibria in the system Can-Ne-Fsp-(K,Na),CO_3^{aq}. Internat Geol Rev 24: 671-676

A Short Outline of the Tunnel Oxides

Marco Pasero

Dipartimento di Scienze della Terra
Università di Pisa
Via S. Maria 53
I-56126 Pisa, Italy
pasero@dst.unipi.it

INTRODUCTION

Within the large and comprehensive group of oxide minerals (class 4 in the Strunz classification; Strunz and Nickel 2001), the most relevant subgroup in the frame of microporous materials is that of the so-called "tunnel oxides." This rather generic term historically refers to a number of minerals which, from a chemical point of view, are (mainly) manganese oxides. In nature manganese occurs in three different oxidation states—Mn^{2+}, Mn^{3+} and Mn^{4+}—with the latter being the dominant form in tunnel oxides. Tetravalent manganese typically has octahedral coordination, and using only $[Mn^{4+}O_6]$ building modules linked together *via* corner- and edge-sharing it is feasible to construct many framework structures. Besides, a set of titanate minerals display the same feature, as might be expected due to the similar crystal-chemical behaviour of Mn^{4+} and Ti^{4+} cations. Therefore, this note is devoted to describing the structural principles and arrangements of minerals—and a number of synthetic compounds as well—in which the dominant cations are Mn^{4+} and Ti^{4+}.

Starting from the basic formula of manganese dioxide, $Mn^{4+}O_2$, incorporation of mono- and divalent cations (primarily alkali and alkali earths) within the tunnels of the structures, can be accommodated by partial reduction of manganese. Until recently, there was a considerable uncertainty and lively debate (e.g., Burns et al. 1983; Burns et al. 1985; Giovanoli 1985) concerning the valence of the reduced species and whether Mn^{2+} or Mn^{3+} was present. It is now commonly accepted on the basis of several high-quality structural studies that Mn^{3+} replaces Mn^{4+}. In titaniferous phases, charge balance accompanying the inclusion of tunnel cations is not adjusted by reduction to Ti^{3+}, but through incorporation of Fe^{3+}, V^{3+} or Cr^{3+} substituting for Ti^{4+}. The generic substitutions can be summarized in the generic formula:

$$A^+_x(M^{3+}_x M^{4+}_{1-x})O_2 \text{ or } A^{2+}_x(M^{3+}_{2x} M^{4+}_{1-2x})O_2$$

where $A = Na^+$, K^+, Rb^+, Mg^{2+}, Ba^{2+}, Pb^{2+}; $M^{3+} = Mn$, Fe, V, Cr, Al; and $M^{4+} = Mn$ or Ti. Water molecules can also enter the tunnels.

BASIC STRUCTURAL FEATURES

In all these compounds $[Mn^{4+}O_6]$ or $[Ti^{4+}O_6]$ octahedra are arranged in edge-sharing columns, which in turn link together, again by edge-sharing, to construct ribbons, with widths potentially ranging from 1 to ∞. Cross-linking by corner-sharing of columns or ribbons in near perpendicular directions, gives rise to a number of different tunnel structures that may be square or rectangular, depending on the dimensions of the ribbons. The ideal topological symmetry of the frameworks is either tetragonal (for structures with the same dimensionality in the two directions, e.g., 1×1, 2×2,...) or orthorhombic. In most cases the real symmetry is lower, the deviation arising from minor distortions of the framework and/or ordered distributions of cations. In a number of examples substitution of M^{4+} by trivalent (or even divalent) cations can be balanced by the insertion of large mono- or divalent cations within the tunnels. Typically, the unit cell axis along the direction in which the octahedral columns or ribbons run is 2.8–2.9 Å, corresponding to a single octahedron repeat. Clearly, the two unit cell parameters in the orthogonal plane depend on the width of the ribbons. What follows is a short outline of all the known tunnel structure-types belonging to this group in order of increasing complexity. Throughout the paper we will speak of "tunnels" to conform to the historical naming of those structures and the long-time custom, although we are aware that, according to the IUPAC recommendation, it should be more correct to denote them as "channels" (McCusker 2005).

1×1 tunnels

The basic 1 × 1 structure is adopted by the simplest and most common form of manganese dioxide, β-MnO_2 or pyrolusite. This mineral (tetragonal, $P4_2/mnm$, $a=4.3983$, $c=2.8730$ Å; Baur 1976) is isostructural with rutile, the commonest polymorph of TiO_2. Many other $M^{4+}O_2$ compounds isostructural with pyrolusite and rutile are known, including cassiterite ($M=Sn^{4+}$), plattnerite ($M=Pb^{4+}$) and argutite ($M=Ge^{4+}$), as well as stishovite, the high pressure polymorph of silica. According to De Wolff (1959), another form of pyrolusite exists—it has orthorhombic symmetry and is thought to be an alteration product of manganite, γ-MnO(OH). Rietveld refinement of orthorhombic pyrolusite has been completed (Yoshino et al. 1992, 1993) and showed that Mn^{4+} is partially substituted by Mn^{3+} (with charge compensation by H^+) in keeping with the general formula $Mn^{4+}_{1-x}Mn^{3+}_xO_{2-x}(OH)_x$. According to Kikuchi et al. (1994) only pure pyrolusite ($x=0$ in the above chemical formula) has tetragonal symmetry, with the following unit cell parameters derived by means of lattice energy calculations: $a=4.4424$, $c=2.8359$ Å (tetragonal, $x=0$); $a=4.4609$, $b=4.6113$, $c=2.7461$ Å (orthorhombic, $x=0.1$). The pure Mn^{3+} end member ($x=1$) is manganite *stricto sensu*, γ-MnO(OH), whose crystal structure has been studied by Dachs (1963). In pyrolusite and related phases the channels of the 1 × 1 framework are too small for incorporation of large cations.

1×2 tunnels

The crystal structure of ramsdellite was first determined by Byström (1949) as orthorhombic, space group $Pbnm$, $a=4.533$, $b=9.27$, $c=2.866$ Å. More recently, refinements of natural (Miura et al. 1990) and synthetic γ-MnO_2 (Fong et al. 1994) were presented that confirmed the essential features of the earlier work.

The hydroxide form of ramsdellite in which $Mn^{4+} \rightarrow Mn^{3+} + H^+$ is groutite, $Mn^{3+}O(OH)$, which corresponds to synthetic α-MnO(OH), and initially reported to be isostructural with ramsdellite (Dent Glasser and Ingram 1968). Lately, the structural relationship between ramsdellite and groutite has been examined in detail by Post et al. (2001), who discovered that most natural ramsdellites contain domains of a second isostructural phase with larger cell parameters, intermediate between those of ramsdellite and groutite. The intermediate phase, called "groutellite," has a longer <Mn–O> distance and increased Jahn-Teller distortion than pure ramsdellite, indicating a partial substitution of Mn^{4+} by Mn^{3+}. The crystal structure of "groutellite" has been refined by Post and Heaney (2004).

1×3 tunnels

Presently, no 1 × 3 tunnel oxide has been reported among minerals or synthetic compounds although unit cell scale occurrences are known in nsutite. This mineral which is polymorphic with synthetic γ-MnO_2 was defined as a new mineral species by Zwicker et al. (1962) and is named after the type locality (Nsuta, Ghana). Comprehensive high resolution transmission electron microscopy (HRTEM) of nsutite from the type locality and Piedras Negras, Mexico

(Turner and Buseck 1983) revealed at both occurrences a fine-scale intergrowth of domains of pyrolusite and ramsdellite, or more complex and faulted sequences. As such, maybe nsutite should not deserve the status of mineral, although it is still considered a valid species (Gaines et al. 1997). In the HRTEM image (Fig. 1) 1×3 cavities alternate with 1×2 cavities giving rise to a complex superstructure, in a narrow (*ca.* 100 Å wide) domain. So, while 1×3 cavities may be formed, it could not be energetically favorable for them to exist as extended structures. In this respect, however, it is worth noting that the relative stability of the various polymorphs of manganese dioxide seems little affected by their "openness" (Fritsch et al. 1997).

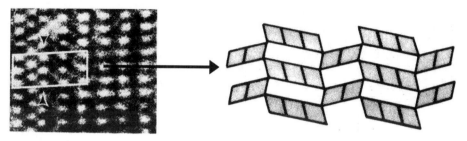

Figure 1. On the left side, HRTEM photo of nsutite from Piedra Negras, Mexico, showing a regular alternation of 1×2 and 1×3 tunnels. The structural arrangement of ribbons in the region within the B box is sketched on the right side [Used by permission of Nature Publishing Group, from Turner and Buseck (1983), Nature, Vol. 304, Fig. 2, p. 144].

2×2 tunnels

The 2×2 tunnel structure is most common, being displayed in several minerals and synthetic compounds. The best known example is hollandite which can be represented by the simplified formula $BaMn^{4+}_6Mn^{3+}_2(O,OH)_{16}$, although wide compositional variability has been reported involving substitutions of Mn^{3+} by Fe^{3+} or Al^{3+}. The crystal structure corresponds to the α-MnO_2 type and was solved by Byström and Byström (1950, 1951) from Weissenberg films. While the original description was made in the $I4/m$ space group, i.e., assuming the ideal topological symmetry, a more precise refinement (Post et al. 1982) showed the true structure to be monoclinic $I2/m$, with $a = 10.026$, $b = 2.8782$, $c = 9.729$ Å, $\beta = 91.03°$.

Cryptomelane is analogous to hollandite, the difference lying only in the A cation where potassium replaces barium. Its structure has been refined by Post et al. (1982) in the $I2/m$ space group with $a = 9.956$, $b = 2.8705$, $c = 9.706$ Å, $\beta = 90.95°$. A monoclinic-tetragonal transformation of cryptomelane at high temperature has been observed by Kudo et al. (1990). Moreover, Vicat et al. (1986) refined, at room temperature, the structure of a synthetic cryptomelane in the tetragonal space group $I4/m$, with $a = 9.866$, $c = 2.872$ Å.

Coronadite is the plumbous analogue of hollandite and cryptomelane with its crystal structure refined (Post and Bish 1989) using two different samples (from Bou Tazzoult, Morocco and from Broken Hill, Australia) in the space group $I2/m$, with $a = 9.938$, $b = 2.8678$, $c = 9.834$ Å, $\beta = 90.39°$ (unit cell from the Moroccan sample).

Among titanium oxides, priderite displays the 2×2 topology. The simplified chemical formula of priderite is $(K,Ba)_{1-2}(Ti^{4+},Fe^{3+})_8O_{16}$. However, in this case the symmetry is tetragonal rather than monoclinic as proven by two independent and near simultaneous studies (Sinclair and McLaughlin 1982; Post et al. 1982), the latter authors determination pointing to $I4/m$, $a = 10.139$, $c = 2.9664$ Å. Priderite with octahedral Fe^{2+} in place of Fe^{3+} has been synthesized at high pressure and is thought to occur in the mantle (Foley et al. 1994)

The crystal structures of four other titanate minerals belonging to the 2×2 family are reported. The crystal structure of mannardite, ideally $BaTi_6V_2O_{16} \cdot H_2O$, was solved by Szymański (1986) ($I4_1/a$, with $a=14.357$, $c=5.908$ Å), and although topologically identical to hollandite, has been described in a larger cell, with c parameter multiplied by 2 and a parameter multiplied by $\sqrt{2}$ presumably as a result of intertunnel ordering of barium. A similar behavior was also recorded on a sample of mannardite from Kyrgyzstan, which gave a quintuple c parameter ($a=10.071$, $c=14.810$ Å; Bolotina et al. 1992). In the closely related mineral ankangite, a supercell with a c parameter 14 times as large as the subcell was reported (Wu and Li 1990; Shi et al. 1991). The crystal structure of redledgeite $BaTi_6Cr_2O_{16} \cdot H_2O$ was solved by Gatehouse et al. (1986) in the monoclinic space group $I2/m$, and then redetermined by Foley et al. (1997) in the tetragonal space group $I4/m$ (with $a=10.150$, $c=2.9520$ Å). The latter authors also proposed a plausible occupancy model for Ba cations within the tunnels, in which Ba-Ba distances shorter than 4.1 Å were avoided. According to the proposed model, the general chemical formula of redledgeite can be written $Ba_x(M^{3+}_{2x}Ti^{4+}_{8-2x})O_{16}$, where $x \leq 1.33$. Henrymeyerite is a Ba-Fe titanate with ideal chemical formula $BaFe^{2+}Ti_7O_{16}$ found in the Kovdor complex, Kola peninsula, Russia (Mitchell et al. 2000). Henrymeyerite, a rare example where the tetravalent framework cation is replaced by a divalent species, is tetragonal ($I4/m$, $a=10.219$, $c=2.963$ Å). A phase with the same composition was also synthesized.

The structural distortion which lowers the symmetry from tetragonal to monoclinic in hollandite-type minerals has been discussed by Post et al. (1982) and Zhang and Burnham (1994), who studied the dependence of the symmetry on the radii of A (tunnel) and B (octahedral) cations.

A number of other minerals have been described, whose structures remain unknown, but which are likely to have the same 2×2 topology. Manjiroite is the sodium-rich analogue of hollandite with formula $(Na,K)(Mn^{4+},Mn^{3+})_8O_{16}$, and with unit cell parameters refined in the tetragonal space group $I4/m$ to $a=9.916$, $c=2.864$ Å (Nambu and Tanida 1967). Like priderite, it is possible that manjiroite may display the ideal topological symmetry. On the other hand strontiomelane, the Sr-dominant analogue of cryptomelane, has the ideal chemical formula $SrMn^{4+}_6Mn^{3+}_2O_{16}$ and $P2_1/n$ symmetry, which is a subgroup of $I2/m$, the space group of the majority of the hollandite-group minerals. The lowering of symmetry in strontiomelane is related to the doubling of the b axis (unit cell: $a=10.00$, $b=5.758$, $c=9.88$ Å, $\beta=90.64°$; Meisser et al. 1999).

A lead titanate analogue of coronadite has yet to be found, although plumbous henrymeyerite was recently described from the Murun alkaline complex, Yakutia, Russia (Reguir et al. 2003). In that sample, Pb reaches 0.45 apfu although not enough to be considered a new mineral species. Another feature of henrymeyerite from this latter occurrence, besides the solid-solution between Ba and Pb end members, is that iron is mainly in its trivalent state and its chemical formula may be best described as $(Ba,Pb)(Fe^{3+}_2Ti_6)O_{16}$. Moreover, Reguir et al. (2003) also synthesized the ideal Pb end member with composition $Pb(Fe^{3+}_2Ti_6)O_{16}$, and a number of intermediate compounds. For one of them, with composition $(Ba_{0.58}Pb_{0.51})_{\Sigma 1.09}(Ti,Fe)_8O_{16}$ a Rietveld refinement was also carried out.

Thus, the 2×2 tunnel structure allows considerable flexibility in chemical composition, as a result not only of tunnel cation exchange, but also of octahedral framework replacements. Indeed, besides the manganate and titanate minerals, many synthetic compounds are known to assume this framework topology. In early work, Bayer and Hoffman (1966) synthesized several compounds with the general formulae $A_2(Ti^{4+}_6M^{3+}_2)O_{16}$ (where A=K, Rb and M^{3+}=Al, Ti, Cr, Fe, Ga), or $A_2(Ti^{4+}_7M^{2+})O_{16}$ (where A=K, Rb and M^{2+}=Mg, Co, Ni, Cu, Zn). A compound with ideal chemical formula $BaAl_2Ti_6O_{16}$ was studied by Sinclair et al. (1980).

High pressure transformations of a number of alkali aluminosilicate and aluminogermanates ($KAlSi_3O_8$, $KAlGe_3O_8$, $NaAlGe_3O_8$, $RbAlGe_3O_8$) which assume the hollandite structure have been discussed by Ringwood et al (1967b). Moreover, Ringwood et al (1967a) described in detail a high-pressure (9 GPa) hollandite-type modification of K-feldspar ($KAlSi_3O_8$) with octahedrally coordinated Si^{4+}. This represented, at that time, the second known occurrence of a compound with $^{[6]}Si$ after stishovite. The crystal structure of high-pressure $KAlSi_3O_8$ was refined by powder XRD (Yamada et al. 1984) and single-crystal methods (Zhang et al. 1993). Since then, other very high pressure aluminosilicate phases were reported with the hollandite-type structure, e.g., $(Ca_{0.5}Mg_{0.5})Al_2Si_2O_8$ at 50 GPa (Madon et al. 1989), or $Pb_{0.8}Al_{1.6}Si_{2.4}O_8$ at 16.5 GPa (Downs et al. 1995).

2×3 tunnels

Romanechite, $Ba(Mn^{4+},Mn^{3+})_5O_{10}\cdot H_2O$, has been known since the earliest days of mineralogy and it is often referred to as psilomelane. The latter name is now discredited, being used for any poorly defined hard black manganese oxides. The crystal structure of romanechite was solved by Wadsley (1953) and re-determined by Turner and Post (1988). Romanechite is monoclinic, $C2/m$, $a=13.929$, $b=2.8459$, $c=9.678$ Å, $\beta=92.39°$. Structurally, it is a manganese oxide with 2x3 tunnels. Tunnels are occupied by barium cations and water molecules; the latter take part in the coordination polyhedra of barium, completed by oxygen atoms belonging to the walls of the tunnel. Turner and Post (1988) also refined a superstructure of romanechite, having a triple b axis, as a result of ordering of Ba and H_2O along the tunnel length; $3b$ (and, besides, $2a$) multiples were already detected by Chukhrov et al. (1983) by electron diffraction, and were ascribed to the same phenomenon, although this not supported by HRTEM.

A HRTEM study carried out by Turner and Buseck (1979) on samples from Rattlesnake mine, Socorro Co., New Mexico, USA showed the conspicuous occurrence of romanechite-hollandite intergrowths; besides, some unusual insulated tunnels (2×4 and 2×7) were also observed. Recently, a compound structurally related to romanechite with Na+H_2O within the 2×3 tunnels has been synthesized by Shen et al. (2004).

2×4 and 2×5 tunnels

Even larger tunnels have been observed outside the mineral kingdom. 2×4 tunnels were found in compounds with ideal formula close to $Rb_{0.25}(Mn^{4+},Mn^{3+})O_2$ (Rziha et al. 1996) and $Na_{0.33}(Mn^{4+},Mn^{3+})O_2\cdot xH_2O$ (Xia et al. 2001; Liu and Ooi 2003). Another synthetic compound with composition $Rb_{0.27}(Mn^{4+},Mn^{3+})O_2$ is based upon 2×5 tunnels (Tamada and Yamamoto 1986). This suggests that the dimensions of the tunnels are related to the extent of the substitution $Mn^{4+} \rightarrow Mn^{3+} + A^+$ and that the ionic radius of the A^+ cation can play some role in the formation of large tunnels. It is reasonable to expect that more complex tunnel structures can be tailored in controlled chemical environments.

3×3 tunnels

The first insight that todorokite displays a tunnel structure based on a new kind of net, 3×3 octahedra wide, was gained through electron microscopy. Turner and Buseck (1981) proposed this new structural model for todorokite, today well accepted, by means of a careful HRTEM study carried out on samples from Chargo Redondo (Cuba) and Bombay (India). Besides unravelling the basic structural features of the mineral, that study also revealed that at the atomic scale, tunnels with different dimensions may be intergrown. In particular, locally, 3×4, 3×5, 3×6, and 3×7 tunnels were observed. Normally, such tunnels are isolated within dominant 3×3 tunnels, with few exceptions (for instance, a sequence of three 3×4 tunnels is shown). Faulted sequences develop along only one direction: in todorokite, the walls along the other direction are invariably 3 octahedra wide. Chukhrov et al. (1980, 1985) also recorded a number of selected area electron diffraction patterns in todorokites from Bakal (Russia),

Sterling Hill (USA) and Takhta-Karacha (Central Asia) that were interpreted as arising from the occurrence of 3×9 wide tunnels. Although this hypothesis is supported by SAED, rather than high resolution lattice imaging, it is strong evidence of substantial inhomogeneity at the unit cell scale. A typical feature of todorokite, revealed by both optical and electron microscopy, was the occurrence of trilling (three twin individuals at 120°) a feature also observed in a synthetic compound with the same topology as todorokite (Golden et al. 1986). Finally, the crystal structure of todorokite was solved by Post and Bish (1988) by powder XRD since the mineral typically occurs as fine-grained crystals. A more precise Rietveld refinement using synchrotron radiation was presented recently (Post et al. 2003b).

3×4 tunnels

The prototype woodruffite occurs at Sterling Hill, NJ, USA (Frondel 1953) and was earlier considered as a Zn-rich variety of todorokite. At the type locality, woodruffite invariably occurs as fine-grained masses, which are unsuitable for single-crystal XRD analysis. Another occurrence (Mapimi, Durango, Mexico) gave tiny, needle-like crystals that proved suitable for investigation using a high intensity synchrotron X-ray source (Post et al. 2003a) that lead to a structure solution which confirmed an opening of 3×4 octahedral units, the largest so far described in either natural or synthetic systems.

A comprehensive list of known natural and synthetic tunnel oxides, including the basic chemical formula and the unit cell parameters, is reported in Table 1.

STRUCTURAL DETAILS

Octahedral distortion

By considering the complete set of structural data for the tunnel oxides, some recurring features concerning octahedral distortion can be rationalised. For testing purposes, only good-quality structural data for each type of tunnel structure were selected, and the octahedral distortion in each independent octahedron was calculated using the well-known equation $\Delta = 1/6 \; \Sigma[(R_i - R)/R]^2$, where R_i is an individual M–O bond length and R is the average bond length in each octahedron. The results are presented in Table 2.

Overall, it can be seen that octahedral distortion is related to increasing structural complexity, the lowest distortion occurring in the 1×1 structure and increasing for larger tunnels. In pyrolusite, the [MnO$_6$] octahedron shows moderate distortion, with four equatorial Mn–O bonds slightly shorter (by 0.015 Å on average) than two apical Mn–O bonds. In ramsdellite, with a 1×2 tunnel, Mn–O distances are in the range 1.81–1.97 Å and octahedral distortion is greater than in pyrolusite. Such a trend is confirmed by the structures with larger tunnels.

The relationship between the dimensions of the tunnels and the degree of distortion can be related to the greater extent, in structures with large tunnels, of the substitution $^{VI}M^{4+} = {}^{VI}M^{3+} + A^+$ (or $^{VI}M^{4+} = 2{}^{VI}M^{3+} + A^{2+}$). The distortion is more evident when the dominant trivalent cation is Mn^{3+}, which is known to display Jahn-Teller effects. In addition, in tunnel oxides with a square outline (e.g., 2×2, 3×3) distortion is less pronounced than expected. It is also noted that in cases of wide ribbons, the distortion is most evident in external octahedra due to the tendency for trivalent cations to concentrate in those sites.

Extra-framework positions

The capacity of the tunnel oxides to incorporate larger cations is obviously related to channel dimension, and consequently the 1×1 tunnel structures (pyrolusite- and rutile-type) circumscribe an interstitial that is too small to accommodate a cation. Similarly, the 1×2 the tunnels do not show significant cation incorporation, although Potter and Rossman (1979)

Table 1. A selection of unit cell parameters for tunnel oxide compounds.

Name and chemical formula	S.G.	Unit cell	Ref.
1×1			
Synthetic α-MnO$_2$	$P4_2/mnm$	a 4.3983, c 2.8730 Å	[1]
Synthetic TiO$_2$	$P4_2/mnm$	a 4.593, c 2.959 Å	[2]
Synthetic GeO$_2$	$P4_2/mnm$	a 4.3975, c 2.8625 Å	[3]
Synthetic PbO$_2$	$P4_2/mnm$	a 4.9578, c 3.3878 Å	[4]
Cassiterite, SnO$_2$	$P4_2/mnm$	a 4.737, c 3.185 Å	[5]
Stishovite, SiO$_2$	$P4_2/mnm$	a 4.1790, c 2.6649 Å	[3]
1×2			
Ramsdellite, MnO$_2$	$Pbnm$	a 4.533, b 9.27, c 2.866 Å	[6]
Synthetic γ-MnO$_2$	$Pnam$	a 9.3229, b 4.4533, c 2.8482 Å	[7]
Groutite, MnO(OH)	$Pbnm$	a 4.560, b 10.700, c 2.870 Å	[8]
Goethite, FeO(OH)	$Pbnm$	a 4.62, b 9.95, c 3.01 Å	[9]
Diaspore, AlO(OH)	$Pbnm$	a 4.4007, b 9.4253, c 2.8452 Å	[10]
2×2			
Hollandite, Ba(Mn^{4+},Mn^{2+})O$_{16}$	$I4/m$	a 9.96, c 2.86 Å	[11]
Hollandite, Ba(Mn^{4+},Mn^{3+},Fe^{3+})(O,OH)$_{16}$	$I2/m$	a 10.026, b 2.878, c 9.729 Å, β 91.03°	[12]
Cryptomelane, K(Mn^{4+},Mn^{2+})O$_{16}$	$I2/m$	a 9.79, b 2.88, c 9.94 Å, β 90.62°	[13]
Cryptomelane, K(Mn^{4+},Mn^{3+},)(O,OH)$_{16}$	$I2/m$	a 9.956, b 2.870, c 9.706 Å, β 90.95°	[12]
Coronadite, Pb(Mn^{4+},V^{3+})O$_{16}$	$I2/m$	a 9.938, b 2.868, c 9.834 Å, β 90.39°	[14]
Priderite, K(Ti,Fe^{3+})O$_{16}$	$I4/m$	a 10.139, c 2.966 Å	[12]
Akaganéite, FeO(OH)	$I2/m$	a 10.587, b 3.031, c 10.515 Å, β 90.03°	[15]
Manjiroite, (Na,K)(Mn^{4+},Mn^{3+})$_8$O$_{16}$	$I4/m$	a 9.916, c 2.864 Å	[16]
Strontiomelane, SrMn$^{4+}{}_6$Mn$^{3+}{}_2$O$_{16}$	$P2_1/n$	a 10.00, b 5.758, c 9.88 Å, β 90.64°	[17]
Ankangite, Ba(Ti,V)$_8$O$_{16}$	$I4/m$	a 10.139, c 2.961 Å	[18]
Mannardite, BaTi$_6$V$_2$O$_{16}$·H$_2$O	$I4_1/a$	a 14.357, c 5.908 Å	[19]
Mannardite, Ba(Ti,V,Cr)$_8$(O,OH)$_{16}$	$I4/m$	a 10.071, c 14.810 Å	[20]
Redledgeite, BaTi$_6$Cr$_2$O$_{16}$·H$_2$O	$I2/m$	a 10.129, b 2.95, c 10.135 Å, β 90.05°	[21]
Redledgeite, BaTi$_6$(Cr,Fe,V)$_2$O$_{16}$	$I4/m$	a 10.1500, c 2.9520 Å	[22]
Henrymeyerite, BaFe^{2+}Ti$_7$O$_{16}$	$I4/m$	a 10.219, c 2.963 Å	[23]
Synthetic (Ba$_{0.58}$Pb$_{0.51}$)(Ti,Fe^{3+})$_8$O$_{16}$	$I4/m$	a 10.1124, c 2.9714 Å	[24]
Synthetic Ba(Ti,Al,Ni)$_8$O$_{16}$	$I4/m$	a 10.039, c 2.943 Å	[25]
Synthetic KAlSi$_3$O$_8$	$I4/m$	a 9.38, c 2.74 Å	[26]
Synthetic KAlSi$_3$O$_8$	$I4/m$	a 9.315, c 2.723 Å	[27]
Synthetic KAlSi$_3$O$_8$ (at 4.47 Gpa)	$I4/m$	a 9.237, c 2.706 Å	[27]
Synthetic (Ca$_{0.5}$Mg$_{0.5}$)Al$_2$Si$_2$O$_8$	$I2/m$	a 9.384, b 8.148, c 9.258 Å, β 90.47°	[28]
Synthetic Pb$_{0.8}$Al$_{1.6}$Si$_{2.4}$O$_8$	$I4/m$	a 9.414, c 2.750 Å	[29]
2×3			
Romanechite, Ba(Mn^{4+},Mn^{3+})$_5$O$_{10}$·H$_2$O	$C2/m$	a 13.929, b 2.8459, c 9.678 Å, β 92.39°	[30]
2×4			
Synthetic Rb$_{0.25}$(Mn^{4+},Mn^{3+})O$_2$	$C2/m$	a 14.191, b 2.851, c 24.343 Å, β 91.29°	[31]
Synthetic Na$_{0.33}$(Mn^{4+},Mn^{3+})O$_2$·nH$_2$O	$C2/m$	a 14.434, b 2.849, c 23.976 Å, β 98.18°	[32]
2×5			
Synthetic Rb$_{0.27}$(Mn^{4+},Mn^{3+})O$_2$	$A2/m$	a 15.04, b 2.886, c 14.64 Å, β 92.4°	[33]
3×3			
Todorokite, (Na,Ca)(Mn^{4+},Mg)$_6$O$_{12}$·5H$_2$O	$P2/m$	a 9.769, b 2.8512, c 9.560 Å, β 94.47°	[34]
3×4			
Woodruffite, Zn$_{0.2}$(Mn^{4+},Mn^{3+})·0.7H$_2$O	$C2/m$	a 24.810, b 2.8503, c 9.581 Å, β 93.845°	[35]

References: [1] Baur 1976; [2] Meagher and Lager 1979; [3] Baur and Khan 1971; [4] D'Antonio and Santoro 1980; [5] Baur 1956; [6] Byström 1949; [7] Fong et al. 1994; [8] Dent Glasser and Ingram 1968; [9] Hoppe 1941; [10] Hill 1979; [11] Byström and Byström 1950; [12] Post et al. 1982; [13] Mathieson and Wadsley 1950; [14] Post and Bish 1989; [15] Post et al. 2003c; [16] Nambu and Tanida 1967; [17] Meisser et al. 1999; [18] Shi et al. 1991; [19] Szymański 1986; [20] Bolotina et al. 1992; [21] Gatehouse et al. 1986; [22] Foley et al. 1997; [23] Mitchell et al. 2000; [24] Reguir et al. 2003; [25] Sinclair et al. 1980; [26] Ringwood et al. 1967a; [27] Zhang et al. 1993; [28] Madon et al. 1989; [29] Downs et al. 1995; [30] Turner and Post 1988; [31] Rziha et al. 1996; [32] Xia et al. 2001; [33] Tamada and Yamamoto 1986; [34] Post et al. 2003b; [35] Post et al. 2003a

Table 2. Octahedral distortion index Δ ($= 1/6 \; \Sigma[(R_i-R)/R]^2$), computed for all independent octahedra in the given type tunnel structures. In column 2, the number between square brackets refers to the number of equivalent octahedra of the same kind within each ribbon.

Structure type	Octahedron	Δ ($\times 10^3$)	Ref.
1 × 1 pyrolusite	M1	0.022	[1]
1 × 2 ramsdellite	M1 [×2]	0.716	[2]
2 × 2 priderite	M1 [×2]	0.228	[3]
2 × 3 romanechite	M1 (central in the 3-ribbon)	0.012	[4]
	M2 (external in the 3-ribbon) [×2]	1.520	
	M3 (2-ribbon) [×2]	0.198	
2 × 4 RUB-7 (synth.)	M1 (external in the 4-ribbon)	6.531	[5]
	M2 (internal in the 4-ribbon)	1.547	
	M3 (internal in the 4-ribbon)	1.627	
	M4 (external in the 4-ribbon)	5.522	
	M5 (2-ribbon)	3.434	
	M6 (2-ribbon)	1.379	
2 × 5 $Rb_{0.27}MnO_2$ (synth.)	M1 (central in the 5-ribbon)	0.012	[6]
	M2 (mid-position in the 5-ribbon) [×2]	0.840	
	M3 (external in the 5-ribbon) [×2]	8.451	
	M4 (2-ribbon) [×2]	0.459	
3 × 3 todorokite	M1 (central in the 3-ribbon)	0.010	[7]
	M3 (central in the 3-ribbon)	0.388	
	M2 (external in the 3-ribbon) [×2]	0.539	
	M4 (external in the 3-ribbon) [×2]	0.237	
3 × 4 woodruffite	M1 (internal in the 4-ribbon) [×2]	0.184	[8]
	M2 (external in the 4-ribbon) [×2]	5.294	
	M3 (external in the 3-ribbon) [×2]	0.849	
	M4 (central in the 3-ribbon)	0.030	

References: [1] Baur 1976; [2] Fong et al. 1994; [3] Sinclair and McLaughlin 1982; [4] Turner and Post 1988; [5] Rziha et al. 1996; [6] Tamada and Yamamoto 1986; [7] Post et al. 2003b; [8] Post et al. 2003a

using IR spectroscopy suggest the constant occurrence of minor water in ramsdellite, which should be concentrated in a well-defined crystallographic site within the tunnel.

In the ideal 2×2 tunnel structure, the basic tunnel site is the inversion center at $(0,0,0)$ and a cation placed in this position is coordinated by 4+4 oxygen atoms, four placed at $z = \frac{1}{2}$, four at $z = -\frac{1}{2}$, with A–O distances in the range 2.8–3.0 Å, in the disposition of a distorted tetragonal prism. Moreover, four additional oxygens at distances of ca. 3.5 Å, and at the same z level as the tunnel cation, can be considered as non-coordinating capping anions, one for each of the four vertical faces of the prism (Fig. 2). In several compounds, however, the tunnel cation (K^+, Na^+, Ba^{2+}, Pb^{2+}), is displaced away from the central position. This depends on the high "rigidity" of the cavities, in which the two square planes are forced to be separated by a c translation. Therefore, each tunnel cation, as a function of its ionic radius and of the local geometry of the octahedral framework, may shift closer to either of the two bases of the prism, and possibly may complete its coordination with some of the capping anions. In the case of Pb^{2+} this is favored by the occurrence of a lone pair of electrons. Another reason

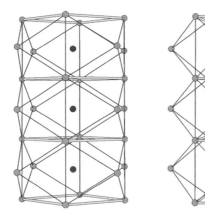

Figure 2. Two perspective views of the tunnel sites in the idealized structure of hollandite-type compound (2×2 tunnels). The tunnel cation is ideally coordinated by 8+4 oxygens (tetracapped tetragonal prism).

for the displacement of the cations from the $(0,0,0)$ position is related to the implausibility of placing neighboring tunnel cations on consecutive sites along the c axis, which is shorter than 3 Å. This, in some cases, results in several, partly occupied, cation sites along the tunnel axis, which can be described as an incommensurate stacking of tunnel cations with respect to the basic c parameter, or alternatively, as superstructures having multiple c parameters. Some examples have been mentioned above.

Structures with larger tunnels can incorporate rather more complex sets of cations and/or water molecules. For instance, in romanechite (2×3) the tunnels host two equivalent, IX-coordinated, cation sites, which are occupied by barium. In todorokite (3×3), the tunnels host three independent water molecules, linked by hydrogen bonds between each other and with the oxygen of the octahedral framework. Moreover, a Mg^{2+} cation is placed in the very center of the tunnel, and is octahedrally coordinated by 6 oxygens belonging to the water molecules. In woodruffite (3×4), the tunnels contain four independent water molecules connected to each other and framework oxygens through hydrogen bonding. Moreover, a Zn^{2+} cation is coordinated by 4 oxygens belonging to the water molecules to form a distorted tetrahedron. In synthetic 2×4 and 2×5 compounds, the large cavities are filled by Rb^+ cations distributed over four partially occupied independent sites and variably coordinated by 8 to 10 oxygen atoms.

Besides pyrolusite, 1×1 tunnels also occur in all structures with larger tunnels where ribbons cross. Some examples of partially occupied 1×1 tunnels have been reported for a Li-bearing todorokite (Duncan et al. 1998) and for woodruffite (Post et al. 2003a). In the latter case, a small Zn occupancy within the 1×1 tunnels is tentatively coupled with a corresponding cation deficiency in the neighbor octahedral sites.

POLYSOMATIC RELATIONSHIPS AMONG COMPOUNDS WITHIN THE PYROLUSITE-"BUSERITE" FAMILY

For these tunnel structures a polysomatic family can be described, that is referred to as the "pyrolusite-'buserite' family," as these structures display the smallest and the largest tunnels so far reported. This family was already sketched by Veblen (1991), who proposed a nomenclature similar to that used for biopyriboles, using P (for pyrolusite) and B (for birnessite). However, unlike biopyriboles, different stacking of P and B can occur along two directions, therefore the symbol for denoting any member of the family becomes quite long in cases of complex sequences. For instance, the largest tunnel structure so far described, woodruffite, should have the symbol (PBB)×(PBBB). It is more convenient to denote these structures simply by the numeric symbol which defined the openness of the tunnels, which is self-explanatory. In general, it will be a symbol of the kind M×N, with both M and N potentially ranging between 1 and ∞, being M ≤ N (this is purely formal, since of course the structures, e.g., 2×3 and 3×2 are topologically identical between each other, and should not be considered twice). To date, several polysomes have been reported. These have been collected in Table 3, where equidimensional ribbons along the two orthogonal directions are put in the same rows and columns. The values given in Table 3 for each polysome are indicative of the dimensions of a single M×N tunnel, and have been extrapolated from the rough unit cell parameters by subtracting the thickness of the ribbons (octahedral walls). This aims at showing the internal consistency in the lengths and widths of tunnels with the same dimensionality, including the open tunnel of the last column, and the soundness of the polysomatic approach. A schematic drawing of the various tunnel structures is reported in Figure 3.

Table 3. Estimated dimensions of tunnels (in Å) in selected members of the pyrolusite-"buserite" polysomatic family.

1×1 pyrolusite 2.2 × 2.2	[…]	[…]	[…]	[…]	[…]	1×∞ lithiophorite 2.3
	2×2 hollandite 4.7 × 4.7	2×3 romanechite 4.7× 7.3	2×4 RUB-7 4.7 ×9.8	2×5 $Rb_{0.27}MnO_2$ 4.9 × 12.5	[…]	2×∞ birnessite 4.7
		3×3 todorokite 7.3×7.3	3×4 woodruffite 7.2×9.9	[…]	[…]	3×∞ 'buserite' 7.4

Polysomes with infinite ribbons

In the last column of Table 3 some additional minerals have been included that correspond to those M×N polysomes in which N=∞, although these structures, being formed by infinite ribbons (or in other words by octahedral layers) do not have closed tunnels. So far, 1×∞ (lithiophorite), 2×∞ (birnessite), and 3×∞ ('buserite') polysomes are known to occur. The idealized chemical formula of lithiophorite may be written as $(LiAl_2)(Mn^{4+}_2Mn^{3+})O_6(OH)_6$; its crystal structure, which was determined by Wadsley (1952) and then refined by Pauling and Kamb (1982) and by Post and Appleman (1994), consist of regularly alternating $(LiAl_2)$ and $(Mn^{4+}_2Mn^{3+})$ octahedral layers linked by hydrogen bonds. The distance between neighboring layers is ca. 4.7 Å. Birnessite, a manganese oxide mineral originally described from Scotland with formula $(Na,Ca)Mn_7O_{14} \cdot 2.8H_2O$ (Jones and Milne 1956), is poorly characterized due to its small crystal dimensions and disordered nature. However, synthetic birnessite is amenable to crystal structure solution (Post and Veblen 1990) and has been determined to be a phyllomanganate in which Mn^{4+}-centered octahedral layers are interleaved with cations (Na, Ca, Mg) and water molecules, with charge balance achieved by Mn^{4+}-Mn^{3+} substitution. The distance between neighbor layers is ca. 7.1 Å. "Buserite" is not a valid mineral species, although the name has been extensively used in reference to a disordered phyllomanganate with a dominant 10 Å distance between octahedral layers. Such compounds have been synthesized (Kuma et al. 1994) in which the interlayer cations are Na^+, Mg^{2+}, Ca^{2+}. Typically, "buserite" is obtained as intermediate product during the synthesis of birnessite (e.g., Post 1999; Feng et al. 2004).

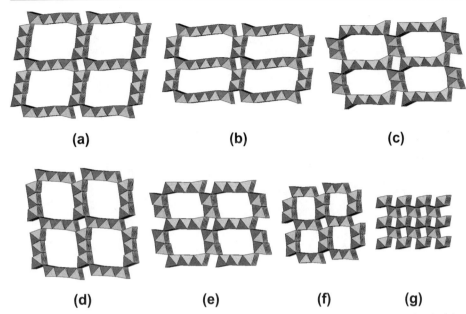

Figure 3. Tunnel structures: (a) woodruffite, 3×4; (b) $Rb_{0.27}MnO_2$ (synthetic), 2×5; (c) RUB-7 (synthetic), 2×4; (d) todorokite, 3×3; (e) romanechite, 2×3; (f) hollandite, 2×2; (g) pyrolusite, 1×1.

True 1×2 polysome

Historically, ramsdellite has been considered a 1×2 tunnel structure and included together with other members of the tunnel oxides, but in strict polysomatic terms this is incorrect. The common feature in tunnel oxides is the occurrence of two sets of ribbons formed by edge-sharing octahedral chains with dimensions ranging from 1 to ∞ that develop in orthogonal directions. This is not the case for ramsdellite where the ribbons (with width corresponding to two octahedral chains) are parallel. Confusingly, in ramsdellite the notation "1×2" refers to the opening of the tunnels but not to the width of the ribbons. Consequently 1×1 tunnels found in all other members of the polysomatic family are absent (Fig. 4). The notation "1×2," whenever referred to the polysomatic description and not to the tunnels, denotes a different structure, which has not been yet reported for any mineral although it has been recognized as local domain within nsutite by HRTEM (Fig. 5).

APPLICATIONS

The technological importance of tunnel oxides arises from three features:

1. Tunnel oxides are very common in nature. A potentially important resource is provided by those manganese oxides that occur in large deposits on the ocean floor and products of mid-ocean ridges (Menard and Shipek 1958).

2. The open structure of tunnel oxides makes them possibly important as cation exchange materials for the sorption of heavy metals. In synthetic analogues these properties may be tailored through controlling the tunnel dimensions.

3. These minerals typically occur as very fine scale aggregates and as coatings with large surface areas that may further enhances their exchange capacity.

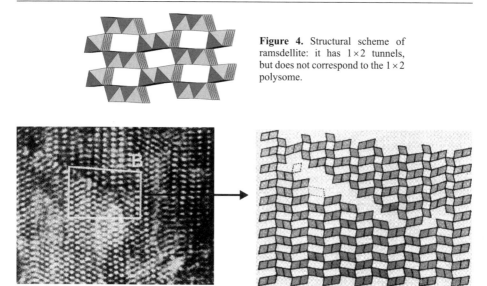

Figure 4. Structural scheme of ramsdellite: it has 1×2 tunnels, but does not correspond to the 1×2 polysome.

Figure 5. On the left side, HRTEM photo of nsutite from Piedra Negras, Mexico, showing several defects. The structural arrangement of ribbons in the region within the B box is sketched on the right side; the upper right region corresponds to the "true" 1×2 polysome, whereas the lower left region corresponds to ramsdellite, the "false" 1×2 polysome [Used by permission of Nature Publishing Group, from Turner and Buseck (1983), Nature, Vol. 304, Fig. 4, p. 145].

These three features are common with zeolites, although the first is perhaps limited due to the difficulty of mining deep ocean deposits. Generally however, technological applications do not require comprehensive characterization of fine-grained and poorly crystalline phases, as their useful industrial properties are apparently independent of crystal chemistry, at least as the first approximation. According to Post (1999), the most common manganese oxide minerals in soils are lithiophorite, hollandite, and birnessite.

Already manganese tunnel oxides are exploited due to their microporous features. Specifically, tunnel oxides are often referred to as Octahedral Molecular Sieves (OMS), a terminology that mirrors the acronym ZSM that is applied to zeolite-like molecular sieves based on a tetrahedral framework. The wide range of applications of tunnel oxides include:

i. The inclusion in a high level radioactive waste of a compound with composition $BaAl_2Ti_6O_{16}$ and structurally related to hollandite (2×2) called SYNROC (Ringwood et al. 1979) specifically for the immobilisation of radioactive cesium.

ii. Amelioration of pollution using manganese oxides that may act as natural sinks for heavy metals in contaminated waters from mines and other industrial activities (Whitney 1975; Lind and Hem 1993). For instance, the capability of cryptomelane to sorb cations such as Co^{2+}, Zn^{2+}, and Cd^{2+} has been discussed by Ghoneimy (1997) and Randall et al. (1998).

iii. Photocatalytic oxidation (e.g., birnessite, todorokite) for environmental remediation and modifying soil chemistry (Oscarson et al. 1981; Shen et al. 1993).

iv. Ion exchange media in a variety of uses. Synthetic todorokite (3×3) obtained in a two-step procedure after synthetic birnessite (2×∞), shows significant cation exchange properties similar to zeolites (Golden et al. 1986). This may prove to be true for other oxides with large tunnels (3×4, 2×5).

Finally, tunnel oxides are of considerable scientific interest in pure and applied geosciences. As noted earlier, feldspars assume hollandite-type structures at very high pressure and may represent one of the major phases in the Earth's mantle.

ACKNOWLEDGMENTS

Marco Bellezza assisted with the art work. The paper benefited from reviews by Giovanni Ferraris, Sergey Krivovichev, Emil Mackovicky, and Tim White.

REFERENCES

Baur WH (1956) Über der Verfeinerung der Kristallstrukturbestimmung einiger Vertreter des Rutiltyps: TiO_2, SnO_2, GeO_2 und MgF_2. Acta Crystallogr 9:515-520

Baur WH (1976) Rutile-type compounds. V. Refinement of MnO_2 and MgF_2. Acta Crystallogr B32:2200-2204

Baur WH, Khan AA (1971) Rutile-type compounds. IV. SiO_2, GeO_2 and a comparison with other rutile-type structures. Acta Crystallogr B27:2133-2139

Bayer G, Hoffman W (1966) Complex alkali titanium oxides $A_xB_yTi_{8-y}O_{16}$ of the α-MnO_2 structure-type. Am Mineral 51:511-516

Bolotina NB, Dmitrieva MT, Rastsvetaeva RK (1992) Modulated structures of a new natural representative of the hollandite series. Sov Phys Crystallogr 37:311-315

Burns RG, Burns VM, Stockman HW (1983) A review of the todorokite-buserite problem: implications to the mineralogy of marine manganese nodules. Am Mineral 68:972-980

Burns RG, Burns VM, Stockman HW (1985) The todorokite-buserite problem: further considerations. Am Mineral 70:205-208

Byström A, Byström AM (1950) The crystal structure of hollandite, the related manganese oxide minerals and α-MnO_2. Acta Crystallogr 3:146-154

Byström A, Byström AM (1951) The positions of the barium atoms in hollandite. Acta Crystallogr 4:469

Byström AM (1949) The crystal structure of ramsdellite, an orthorhombic modification of MnO_2. Acta Chem Scand 3:163-173

Chukhrov FV, Gorshkov AI, Dmitrieva MT, Sivtsov AV (1983) Crystallochemistry of romanechite. Int Geol Rev 25:517-525

Chukhrov FV, Gorshkov AI, Drits VA, Dikov YP (1985) Structural varieties of todorokite. Int Geol Rev 27:1481-1491

Chukhrov FV, Gorshkov AI, Sivtsov AV, Berezovskaya VV (1980) Structural varieties of todorokite. Int Geol Rev 22:75-83

D'Antonio P, Santoro A (1980) Powder neutron diffraction study of chemically prepared β-lead dioxide. Acta Crystallogr B36:2394-2397

Dachs H (1963) Neutronen- und Röntgenuntersuchungen an Manganit, MnOOH. Z Kristallogr 118:303-326

De Wolff PM (1959) Interpretation of some γ-MnO_2 diffraction patterns. Acta Crystallogr 12:341-345

Dent Glasser LS, Ingram L (1968) Refinement of the crystal structure of groutite, α–MnOOH. Acta Crystallogr B24:1233-1236

Downs RT, Hazen RM, Finger LW (1995) Crystal chemistry of lead aluminosilicate hollandite: a new high-pressure synthetic phase with octahedral Si. Am Mineral 80:937-940

Duncan MJ, Leroux F, Corbett JM, Nazar LF (1998) Todorokites as a Li insertion cathode. J Electrochem Soc 145:3746-3757

Feng XH, Liu F, Tan WF, Liu XW (2004) Synthesis of birnessite from the oxidation of Mn^{2+} by O_2 in alkali medium: effects of synthesis conditions. Clays Clay Mineral 52:240-250

Foley JA, Hughes JM, Drexler JW (1997) Redledgeite, $Ba_x([Cr,Fe,V]^{3+}{}_{2x}Ti_{8-2x})O_{16}$, the $I4/m$ structure and elucidation of the sequence of tunnel Ba cations. Can Mineral 35:1531-1534

Foley S, Hoefer H, Brey G (1994) High-pressure synthesis of priderite and members of the lindsleyite-mathiasite and hawthorneite-yimengite series. Contrib Mineral Petrol 117:164-174.

Fong C, Kennedy BJ, Elcombe MM (1994) A powder neutron diffraction study of λ and γ manganese dioxide and of $LiMn_2O_4$. Z Kristallogr 209:941-945

Fritsch S, Post JE, Navrotsky A (1997) Energetics of low-temperature polymorphs of manganese dioxide and oxyhydroxide. Geochim Cosmochim Acta 61:2613-2616

Frondel C (1953) New manganese oxides: hydrohausmannite and woodruffite. Am Mineral 38:761-769

Gaines RV, Skinner HCW, Foord EE, Mason B, Rosenzweig A (1997) Dana's New Mineralogy, 8th ed. Wiley, New York

Gatehouse BM, Jones GC, Pring A, Symes RF (1986) The chemistry and structure of redledgeite. Mineral Mag 50:709-715

Ghoneimy HF (1997) Adsorption of Co^{2+} and Zn^{2+} on cryptomelane-type hydrous manganese dioxide. J Radioanal Nucl Chem 223:61-65

Giovanoli R (1985) A review of the todorokite-buserite problem: implications to the mineralogy of marine manganese nodules: discussion. Am Mineral 70:202-204

Golden DC, Chen CC, Dixon JB (1986) Synthesis of todorokite. Science 231:717-719

Hill R (1979) Crystal structure refinement and electron density distribution in diaspore. Phys Chem Minerals 5:179-200

Hoppe W (1941) Über die Kristallstruktur von α-AlOOH (Diaspor) und α-FeOOH (Nadeleisenerz). Z Kristallogr 103:73-89

Jones LHP, Milne AA (1956) Birnessite, a new manganese oxide mineral from Aberdeenshire, Scotland. Mineral Mag 31:283-288

Kikuchi T, Yoshino T, Miura H (1994) Orthorhombic distortion in pyrolusites. 16th Gen Meet Int Mineral Ass, Pisa, 4-9 Sep, abstr, 203-204

Kudo H, Miura H, Hariya Y (1990) Tetragonal-monoclinic transformation of cryptomelane at high temperature. Mineral J 15:50-63

Kuma K, Usui A, Paplawsky W, Gedulin B, Arrhenius G (1994) Crystal structures of synthetic 7 Å and 10 Å manganates substituted by mono- and divalent cations. Mineral Mag 58:425-447

Lind CJ, Hem JD (1993) Manganese minerals and associated fine particulates in the streambed of Pinal Creek, Arizona, U.S.A.: a mining-related acid drainage problem. Appl Geochem 8:67-80

Liu ZH, Ooi K (2003) Preparation and alkali-metal ion extraction/insertion reactions with nanofibrous manganese oxide having 2×4 tunnel structure. Chem Mater 15:3696-3703

Madon M, Castex J, Peyronneau J (1989) A new aluminosilicate high-pressure phase as a possible host of calcium and aluminum in the lower mantle. Nature 342:422-425

Mathieson AM, Wadsley AD (1950) The crystal structure of cryptomelane. Am Mineral 35:99-101

McCusker LB (2005) IUPAC nomenclature for ordered microporous and mesoporous materials and its application to non-zeolite microporous mineral phases. Rev Mineral Geochem 57:1-16

Meagher EP, Lager GA (1979) Polyhedral thermal expansion in the TiO_2 polymorphs: refinement of the crystal structures of rutile and brookite at high temperature. Can Mineral 17:77-85

Meisser N, Perseil EA, Brugger J, Chiappero PJ (1999) Strontiomelane, $SrMn^{4+}_6Mn^{3+}_2O_{16}$, a new mineral species of the cryptomelane group from St. Marcel-Praborna, Aosta Valley, Italy. Can Mineral 37:673-678

Menard HW, Shipek CJ (1958) Surface concentrations of manganese nodules. Nature 182:1156-1158

Mitchell RH, Yakovenchuk VN, Chakhmouradian AR, Burns PC, Pakhomovsky YA (2000) Henrymeyerite, a new hollandite-type Ba-Fe titanate from the Kovdor Complex, Russia. Can Mineral 38:617-626

Miura H, Kudou H, Choi JH, Hariya Y (1990) The crystal structure of ramsdellite from Pirika Mine. J Fac Sci Hokkaido Univ, Ser 4, Geol Mineral 22:611-617

Nambu M, Tanida K (1967) Manjiroite, a new manganese dioxide mineral, from Kohare Mine, Iwate prefecture, Japan. J Japan Ass Mineral Petrol Econ Geol 58:39-54

Oscarson DW, Huang PM, Liaw WK (1981) The role of manganese in the oxidation of arsenite by freshwater lake sediments. Clays Clay Minerals 29:219-225

Pauling L, Kamb B (1982) The crystal structure of lithiophorite. Am Mineral 67:817-821

Post JE (1999) Manganese oxide minerals: crystal structures and economic and environmental significance. Proc Nat Acad Sci USA 96:3447-3454

Post JE, Appleman DE (1994) Crystal structure refinement of lithiophorite. Am Mineral 79:370-374

Post JE, Bish DL (1988) Rietveld refinement of the todorokite structure. Am Mineral 73:861-869

Post JE, Bish DL (1989) Rietveld refinement of the coronadite structure. Am Mineral 74:913-917

Post JE, Heaney PJ (2004) Neutron and synchrotron X-ray diffraction study of the structures and dehydration behaviors of ramsdellite and "groutellite." Am Mineral 89:969-975

Post JE, Heaney PJ, Cahill CL, Finger LW (2003a) Woodruffite: a new Mn oxide structure with 3×4 tunnels. Am Mineral 88:1697-1702

Post JE, Heaney PJ, Hanson J (2003b) Synchrotron X-ray diffraction study of the structure and dehydration behavior of todorokite. Am Mineral 88:142-150

Post JE, Heaney PJ, Hanson JC (2001) Temperature-resolved synchrotron X-ray diffraction study of ramsdellite and groutite. Geol Soc Am Ann Meet, Boston, abstr, 33:362-363

Post JE, Heaney PJ, Von Dreele RB, Hanson JC (2003c) Neutron and temperature-resolved synchrotron X-ray powder diffraction study of akaganéite. Am Mineral 88:782-788

Post JE, Veblen DR (1990) Crystal structure determination of synthetic sodium, magnesium, and potassium birnessite using TEM and the Rietveld method. Am Mineral 75:477-489

Post JE, Von Dreele RB, Buseck PR (1982) Symmetry and cation displacements in hollandites: structure refinements of hollandite, cryptomelane and priderite. Acta Crystallogr B38:1056-1065

Potter RM, Rossman GR (1979) The tetravalent manganese oxides: identification, hydration, and structural relationships by infrared spectroscopy. Am Mineral 64:1199-1218

Randall SR, Sherman DM, Ragnarsdottir KV (1998) An extended X-ray absorption fine structure spectroscopy investigation of cadmium sorption on cryptomelane (KMn_8O_{16}). Chem Geol 151:95-106

Reguir EP, Chakhmouradian AR, Mitchell RH (2003) Pb-bearing hollandite-type titanates: a first natural occurrence and reconnaissance synthesis study. Mineral Mag 67:957-965

Ringwood AE, Kesson SE, Ware NG, Hibberson W, Major A (1979) Immobilization of high level nuclear reactor wastes in SYNROC. Nature 278: 219-223

Ringwood AE, Reid AF, Wadsley AD (1967a) High pressure $KAlSi_3O_8$, an aluminosilicate with sixfold coordination. Acta Crystallogr 23:1093-1095

Ringwood AE, Reid AF, Wadsley AD (1967b) High pressure transformation of alkali aluminosilicates and aluminogermanates. Earth Planet Sci Letters 3:38-40

Rziha T, Gies H, Rius J (1996) RUB-7, a new synthetic manganese oxide structure type with a 2×4 tunnel. Eur J Mineral 8:675-686

Shen X, Ding Y, Liu J, Laubernds K, Zerger RP, Polverejan M, Son YC, Aindow M, Suib SL (2004) Synthesis, characterization and catalytic applications of manganese oxide octahedral molecular sieve (OMS) nanowires with a 2×3 tunnel structure. Chem Mater 16:5327-5335

Shen YF, Zerger RP, DeGuzmaz RN, Suib SL, McCurdy L, Potter DI, O'Young CL (1993) Manganese oxide octahedral molecular sieves: preparation, characterization, and applications. Science 260:511-515

Shi N, Ma Z, Liu W (1991) Crystal structure determination of ankangite with one-dimensional incommensurate modulation. Acta Petrol Mineral 10:233-245 (in Chinese)

Sinclair W, McLaughlin GM (1982) Structure refinement of priderite. Acta Crystallogr B38:245-256

Sinclair W, McLaughlin GM, Ringwood AE (1980) The structure and chemistry of a barium titanate hollandite-type phase. Acta Crystallogr 36:2913-2918

Strunz H, Nickel EH (2001) Strunz Mineralogical Tables, 9th ed. Schweizerbart'sche, Stuttgart

Szymański JT (1986) The crystal structure of mannardite, a new hydrated cryptomelane-group (hollandite) mineral with a doubled short axis. Can Mineral 24:67-78

Tamada O, Yamamoto N (1986) The crystal structure of a new manganese dioxide ($Rb_{0.27}MnO_2$) with a giant tunnel. Mineral J 13:130-140

Turner S, Buseck PR (1979) Manganese oxide tunnel structures and their intergrowths. Science 203:456-458

Turner S, Buseck PR (1981) Todorokites: a new family of naturally occurring manganese oxides. Science 212:1024-1027

Turner S, Buseck PR (1983) Defects in nsutite (γ-MnO_2) and dry-cell battery efficiency. Nature 304:143-146

Turner S, Post JE (1988) Refinement of the substructure and superstructure of romanechite. Am Mineral 73:1155-1161

Veblen DR (1991) Polysomatism and polysomatic series: a review and applications. Am Mineral 76:801-826

Vicat J, Fanchon E, Strobel P, Qui DT (1986) The structure of $K_{1.33}Mn_8O_{16}$ and cation ordering in hollandite-type structures. Acta Crystallogr B42:162-167

Wadsley AD (1952) The structure of lithiophorite, $(Al,Li)MnO_2(OH)_2$. Acta Crystallogr 5:676-680

Wadsley AD (1953) The crystal structure of psilomelane, $(Ba,H_2O)Mn_5O_{10}$. Acta Crystallogr 6:433-438

Whitney PR (1975) Relationship of manganese-iron oxides and associated heavy metals to grain size in stream sediments. J Geochem Expl 4:251-263

Wu XJ, Li FH (1990) Electron microscopy study of incommensurately modulated structure of ankangite. Acta Crystallogr 46:111-117

Xia GG, Tong W, Tolentino EN, Duan NG, Brock SL, Wang JY, Suib SL, Ressler T (2001) Synthesis and characterization of nanofibrous sodium manganese oxide with a 2×4 tunnel structure. Chem Mater 13:1585-1592

Yamada H, Matsui Y, Ito E (1984) Crystal-chemical characterization of $KAlSi_3O_8$ with the hollandite structure. Mineral J 12:29.34

Yoshino T, Miura H, Hariya Y (1992) Crystal structure of orthorhombic pyrolusite. 29th Int Geol Congr, Kyoto, Aug 24 - Sep 3, abstr, 216

Yoshino T, Miura H, Hariya Y (1993) Crystal structure of orthorhombic pyrolusite from Imini Mine, Marrakesh, Morocco. In: Mineral resources symposia. S Ishihara, T Urabe, H Ohmoto (eds) Society of Resource Geologists of Japan, Tokyo, p. 62-65

Zhang J, Burnham CW (1994) Hollandite-type phase: geometric consideration of unit-cell size and symmetry. Am Mineral 79:168-174

Zhang J, Ko J, Hazen RM, Prewitt CT (1993) High-pressure crystal chemistry of $KAlSi_3O_8$ hollandite. Am Mineral 78:493-499

Zwicker WK, Meijer WOJG, Jaffe HW (1962) Nsutite, a widespread manganese oxide mineral. Am Mineral 47:246-266

Apatite – An Adaptive Framework Structure

Tim White, Cristiano Ferraris, Jean Kim and Srinivasan Madhavi

School of Materials Science and Engineering
Nanyang Technological University
N 4.1-01-30, Nanyang Avenue
Singapore 639798

This chapter was written in commemoration of the 60th anniversary of the publication "The atomic structure of fluor-apatite and its relation to that of tooth and bone material." by CA Beevers and DB McIntyre (1946) Mineralogical Magazine 27:254-257.

INTRODUCTION

Apatite, the most common phosphate mineral, is generally described by the formula $Ca_5(PO_4)_3(OH,F,Cl)$ or, more completely with regard to its usual description in $P6_3/m$ symmetry, by the unit cell content $[Ca_4][Ca_6][(PO_4)_6][OH,F,Cl]_2$. An earlier volume of the *Reviews* series (Kohn et al. 2002) has dealt with the mineralogy and crystallography of apatite *sensu stricto* (Hughes and Rakovan 2002) and the diverse compounds that adopt apatite or apatite-like structures (Pan and Fleet 2002; Huminicki and Hawthorne 2002). In addition, apatite compilations have appeared at regular intervals (Wychoff 1965; McConnell 1973; Nriagu and Moore 1984; Brown and Constantz 1994; Elliott 1994) as the breadth of apatite chemistry has expanded and the level of understanding of its importance with respect to the mineral, materials, environmental, and biological sciences increased. It is not therefore, the purpose of this chapter to restate these excellent reviews, but rather, to focus on microporosity in apatites, a feature which allows ion conduction and exchange, and that is proving to be an important consideration for fashioning synthetic analogues of these minerals in technologically advantageous ways.

The notion of apatite as an industrially significant microporous mineral is not new. In 1944, V. M. Goldschmidt who had studied apatite deposits in Scandinavia and at that time had found refuge from war-ravaged Europe at the Macaulay Institute of Soil Research, Aberdeen (Kauffman 1997; McIntyre 2004) persuaded C. A. Beevers to undertake a new refinement of fluorapatite. The results collected in the seminal paper of Beevers and McIntyre (1946), lead not only to the most accurate crystallographic data available at that time, but also provided the first overview of $[Ca_4][Ca_6][(PO_4)_6][F]_2$, noting that it was composed of CaO_6 columns linked together with PO_4 groups to form "a hexagonal network like a honeycomb with channels extending right through the structure in the c direction" (Fig. 1). More generally, it was recognized that the one-dimensional tunnel structure was determined primarily by "the calcium and phosphate arrangement, which is likely to be a strong and stable one." Into these tunnels were inserted the remaining calcium and fluorine. In comparing the hydroxyl and fluorapatites it was observed that OH^-, being slightly larger than F^-, leads to the former structure being expanded compared to the latter. Thus, Beevers and McIntyre identified the three critical elements of the microporous description of apatites namely: (1) the structure can be considered a tunnel structure with walls composed of corner-connected CaO_6 and PO_4 polyhedra as relatively invariant units; (2) filling of these tunnels by Ca and anions (OH, F) leads to characteristic adjustments that best satisfy bond-length requirements; and (3) even slight changes in the ionic radii of the tunnel atoms lead to expansion or contraction of the tunnel. On this basis, it was surmised that the "very critical fit" of the fluorine and hydroxyl ions was responsible for the greater stability of fluorapatite, consistent with the observation that bone could take up fluorine selectively even from dilute solutions. The possibility of reducing dental caries by increasing fluorine content was thus established and provided the fundamental underpinning for water fluorination technology.

While the microporous description of apatites pioneered by Beevers and McIntyre was largely overlooked for many years, it has become increasingly significant as apatites of various chemistries are investigated for assorted applications in chemical synthesis, clean energy and environmental remediation. For example, the so-called "lacunary" apatites are prospective fuel cell electrolytes, while lead apatites are potential photocatalysts, and radiation resistant phosphate apatites may be useful for retaining nuclear wastes. Although apatites have one-dimensional channels, as distinct from the three-dimensional channels in classic zeolites, they do display several zeolitic features including: a framework which can be tuned to accommodate different tunnel contents; an ability to accept large cations of different valance through the introduction of framework counter ions; and reversible ion exchange for some anions and cations. Fluorapatite can be described formally using microporous nomenclature (McCusker et al. 2001) as

$$| Ca^{II[6]}F^{I[3]} | [Ca^{II[7]}P^{V[4]}O^{II[4]}]_p \{1[4^{12}12^2]\}$$

Unlike zeolites however, completely empty channels have not yet been reported. Most recently, it has been recognized, in both natural and synthetic materials, that intergrowth of tunnels of different size at the nanoscale is possible, a feature with important technology performance implications. This article will focus on the description of apatites as microporous compounds, systematize the correlation of chemistry and tunnel geometry, and summarizes the crystal chemical foundation of several contemporary apatite-based technologies.

DESCRIPTIVE CRYSTALLOGRAPHY

While the general formula for apatites can be written as $[A_4][A_6][(BO_4)_6][X]_2$ the variants $[A_4][A_6][(BO_3)_6][X]_2$ and $[A_4][A_6][(BO_5)_6][X]_2$ occur less commonly (White and Dong 2003). Historically, apatite has generally been regarded as a "difficult" structure from the perspective of descriptive crystallography as, with the exception of the $BO_3/BO_4/BO_5$ units, its polyhedra appear irregular and relationships to other structures including glaserite (Moore and Araki 1977) and perovskite (Dong et al. 2005b) are not obvious. In addition, while many apatites are hexagonal, there is increasing recognition that monoclinic varieties are not unusual (Elliott et al. 1973). Nonetheless, it is well known that in comparing structures and for relating structure to properties, reference to idealized models is invaluable (Andersson 1978). For this reason, several alternate descriptions of apatite have appeared in an effort to make the structure accessible to comparative crystal chemistry.

Stuffed alloy representation

The description of many crystal structures can be simplified by considering the atomic arrangements as anion-stuffed alloy structures (O'Keeffe and Hyde 1985). Apatite is no exception, and the correspondence between the topology of the Ca_5P_3 cation array of $Ca_5(PO_4)_3F$ with the Mn_5Si_3 (D 8_8) alloy type (Wondratschek et al. 1964; Vegas et al. 1991) or Ca_5P_3F with Mo_5Si_3C (Schriewer and Jeitschko 1993) is well known. An illustration of Mn_5Si_3 and fluorapatite $[Ca(1)_4][Ca(2)_6][(PO_4)_6][F]_2$ emphasizing the cation arrangement is shown in Figure 2. The core unit is a $Ca(2)_6$ octahedron which is capped on every edge by

Apatite – An Adaptive Framework Structure 309

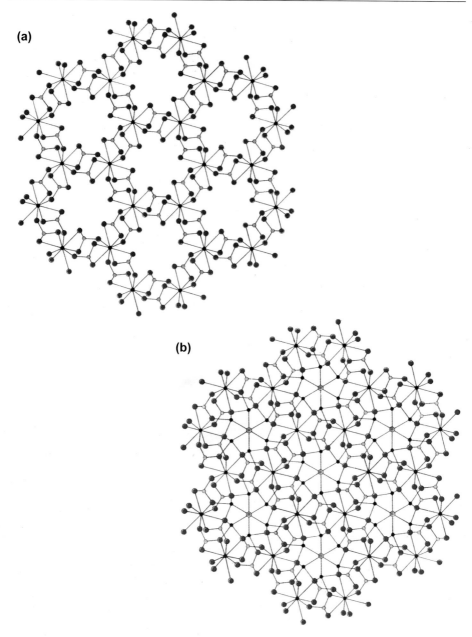

Figure 1. The original color illustrations of fluorapatite as presented by Beevers and McIntyre (1946). In their own words: (a) "[The] linkage of the Ca-O columns by the PO$_4$ groups produces a hexagonal network like a honeycomb with channels extending right through the structure in the c-direction," and (b) "The walls of these channels are lined with oxygen atoms, the arrangement being such that in the wall of each channel there are six 'caves' per unit length. Into these caves calcium atoms will just fit, one Ca going about half-way into each. Thus, about half the area of each Ca is left exposed on the inside of the channels and the F atoms just fit between the group of the three Ca's at one level." [Reproduced with permission of the Mineralogical Society, from Beevers and McIntyre (1946), *Mineralogical Magazine*, Vol. XXVII, Fig. 4 Plate XVII and Fig. 5 Plate XVIII, p. 254-257.]

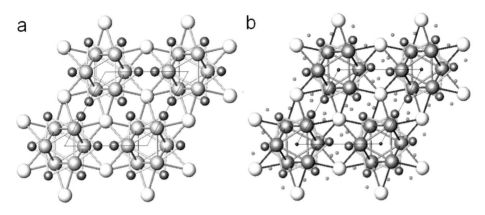

Figure 2. (a) The structure of Mn_5Si_3 emphasizing the Mn_6 octahedra that are capped on every edge by Mn or Si. (b) Fluorapatite represented similarly showing $Ca(1)_4(Ca(2)_6P_6$ capped octahedra with the interstices occupied by oxygen and fluorine.

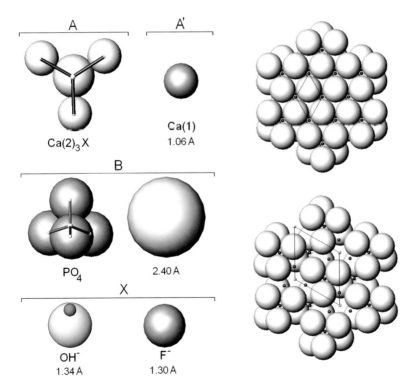

Figure 4. Representation of hydroxyapatite as close packed phosphate "spheres." The PO_4 tetrahedra are simplified as spheres of approximate radius 2.40 Å. Perfect close packing (upper right) is disrupted by the introduction of $Ca(2)_3OH$ units that cause rotation of the spheres (lower right), but the Ca(1) atoms are a better fit to the octahedral phosphate interstices.

Apatite – An Adaptive Framework Structure 311

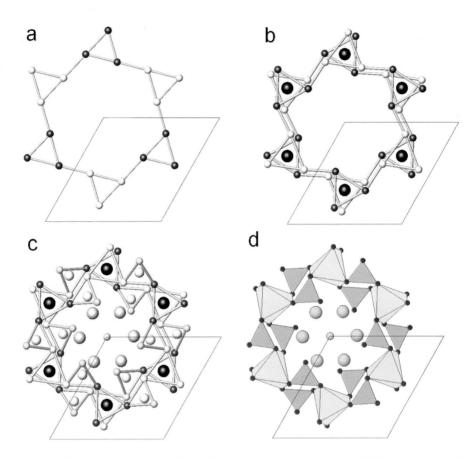

Figure 6. Fluorapatite can be derived by stacking triangular anion nets along c. (a) The prototype net of O(1) and O(2) triangles connected through their apices. (b) In fluorapatite alternate triangles are twisted about [001] to create metaprisms that contain Ca(1) atoms. (c) The O(3) atoms are inserted between the O(1)-(2) nets to create the tetrahedral interstices for P. The apatite framework is now complete. Into the tunnel framework Ca(2) and F atoms are introduced, with the twisting between the O(1)$_3$ and O(2)$_3$ triangles adjusted to best accommodate the Ca(2)O$_6$F co-ordination sphere. (d) The structure, in polyhedral representation, emphasizing the Ca(1)O$_6$ metaprisms and PO$_4$ tetrahedra.

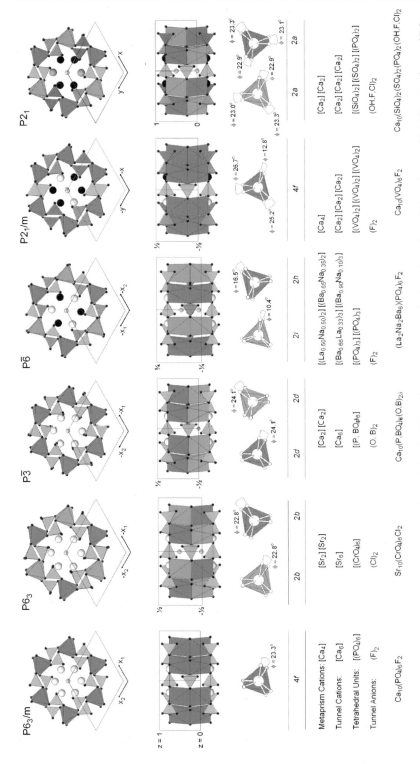

Figure 14. Polyhedral drawings of representative apatites in the six principle apatite space groups emphasizing the cation ordering schemes of the AO_6 metaprisms and BO_4 tetrahedra. In passing from $P6_3/m$ to $P2_1$, the number of unique cation acceptors sites increases to accommodate more complex chemistries or less symmetrical polyhedral distortions. For all symmetries, the metaprism twist angle can be used as a measure of tunnel expansion or collapse as a function of its content. In the case of the monoclinic apatites, the metaprisms have three distinct twist angles; however, their average is a suitable proxy for the single values observed at higher symmetries when investigating crystal chemical trends.

Ca(1) or P (Nyman and Andersson 1979). The entire construction is completed by joining these $Ca(1)_4Ca(2)_6P_6$ capped octahedra through their faces along [001] and by the Ca(1) apices in (001). The anions (oxygen and fluorine) are then introduced into the interstices between the cations. While the positions of the anions are variable, those of the cations are essentially fixed, although there is slight rotation of P as compared to Si about [001].

The equivalence of alloy cation arrays and apatite oxides has been explored in detail by Vegas and Jansen (2002) who compiled a list of 14 corresponding intermetallic and apatite structures (Table 1). A further advantage of the stuffed alloy description is that the $D\,8_8$ alloy structure having $P6_3/mcm$ symmetry may be taken as the aristotype of the apatite family, with all other members adopting maximal non-isomorphic subgroups as shown in Figure 3.*

Table 1. Equivalency between apatite cation lattices and known alloys [after Vegas and Jansen 2002)].

Apatite-like Compounds	Cation Array	Alloy Equivalent
$Ca_5(AsO_4)_3Cl$	Ca_5As_3	Ca_5As_3
$Sr_5(AsO_4)_3Cl$	Sr_5As_3	Sr_5As_3
$Ba_5(AsO_4)_3Cl$	Ba_5As_3	Ba_5As_3
$Ba_5(AsO_4)_2(SO_4)S$	Ba_5As_2S	Ba_5As_3
$Y_5(SiO_4)_3N$	Y_5Si_3	Y_5Si_3
$CaNd_4(SiO_4)_3O$	$CaNd_4Si_3$	Nd_5Si_3
$CdNd_4(SiO_4)_3O$	$CdNd_4Si_3$	Nd_5Si_3
$La_{4.67}(SiO_4)_3O$	$La_{4.67}Si_3$	La_5Si_3
$CaLa_4(SiO_4)_3O$	$CaLa_4Si_3$	La_5Si_3
$Ce_{4.67}(SiO_4)_3O$	$Ce_{4.67}Si_3$	Ce_5Si_3
$Sm_{4.67}(SiO_4)_3O$	$Sm_{4.67}Si_3$	Sm_5Si_3
$MnSm_4(SiO_4)_3O$	$MnSm_4Si_3$	Sm_5Si_3
$Gd_{4.67}(SiO_4)_3O$	$Gd_{4.67}Si_3$	Gd_5Si_3
$Dy_{4.67}(SiO_4)_3O$	$Dy_{4.67}Si_3$	Dy_5Si_3

Close-packed metalloid units

Another simplification was developed by Elliott (1973) who considered that, to a first approximation, the structure of hydroxyapatite could be conveyed by considering the PO_4^{3-} radicals as spheres (diameter ~2.40 Å) arranged in hexagonal close packing (*hcp*), with interstitial insertion of the remaining cations and anions (Fig. 4: color figure on page 310). The basis for this approach was that in *hcp* continuous channels are formed along [001] (as in apatite) into which Ca(1) may be inserted to form the columns noted earlier, while OH and Ca(2) jointly occupy the remaining channels. Thus, in an ideal *hcp* compound $A_3B_3 \equiv [A][A']_2[B]_3$ where A and A' are interstitials and B are the close-packed components, the equivalent description for $Ca_5(PO_4)_3(OH)$ hydroxyapatite will be $[Ca(2)_3(OH)][Ca(1)]_2[PO_4]_3$.

Introduction of ions [Ca(1)] and ionic entities $[Ca(2)_3(OH)]$ in the channels leads to the phosphate ions no longer being close packed. However, while Ca(1) is a near fit to the

---

* It should be noted that the non-isomorphic subgroups do not include $P2_1/b$ that has been assigned to chlorapatite (Mackie et al. 1972) amongst others. This is because $P6_3/m$ symmetry is broken by the need to accommodate order correlation of chlorine and X-anion vacancies between apatite channels. Such structures are perhaps best considered intergrowths of commensurate and incommensurate lattices (Alberius-Henning et al. 1999b).

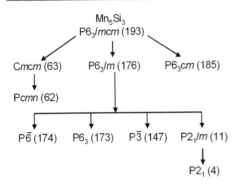

Figure 3. Symmetry relationships of apatite hettotypes derived from the Mn₅Si₃ aristotype. [Used with permission of the International Union of Crystallography (http://journals.iucr.org/) from White and Dong (2003) *Acta Crystallographica*, Vol. B59, Fig. 3, p. 3.]

octahedral phosphate interstice, Ca(2)₃OH can only be accommodated through rotation of the Ca(1)(PO₄)₆ columns (Fig. 4). It then becomes apparent that the a lattice constant will be determined by the extent of rotation of the columns, but the c constant will be determined by the size of the A interstitial. Consequently, and as observed experimentally, variation in the c/a ratio is primarily controlled by contraction and dilation of a in response to the effective size of the A component. Following this logic it is possible to use the cell constants of cadmium, calcium, strontium and lead fluor-, hydroxy, chlor- and brom-apatites to extrapolate to the c/a ratio of a hypothetical material containing an X anion of zero radius as shown in Figure 5, leading to values of 0.806–0.831, which bracket the ideal geometrical ratio of $1.633/2 = 0.817$. Because the X anion is not the only disrupting influence of perfect phosphate close-packing the approach to *hcp* by back-extrapolation is also limited by the size of the A(2) cation, as evidenced by the reduction in slope, and poorer packing in the X = 0 limit, in passing from smaller cadmium to larger lead. In addition, the recognition that phosphate (or indeed any BO₄ unit) is an intrinsically stable entity in apatite allows other crystal chemical observations to be readily explained, including the fact that non-disruptive X ion exchange in single crystal material is commonly observed.

Derivation from triangular anion nets

While many workers have noted the "irregularity" of the anion arrangement in apatites and the difficulty in describing the structure using cation-centered polyhedra, it is feasible, following from the idealized representation of Povarennykh (1972) to derive a prototype from a triangular network of oxygen. In this method, the starting point are regular anion triangles [O(1) and O(2) in $P6_3/m$ fluorapatite] connected through a single vertex as shown in Figure 6a (color figure on page 311). These layers occur at $z = ¼$ and $¾$ but in real apatites do not superimpose in the [001] projection precisely. Rather, in each layer, alternate oxygen triangles twist slightly in (001) to produce interstices in which A(1) atoms reside at $z = 0$ and $½$ (Fig. 6b). Tetrahedral interstices are formed by the introduction of O(3) atoms above and below the O(1) and O(2) layers that accommodate B cations at $z = ¼$ and $¾$. The tunnel structure is now evident. Finally, A(2) and X atoms are introduced (Fig. 6c). Depiction of this model in polyhedral terms (Fig. 6d) highlights the corner connectedness of the metaprisms and tetrahedra, and the location of relatively more mobile and reactive tunnel species, reminiscent of the description by Beevers and McIntyre (1946). Several corollaries follow from this derivation.

First, if apatites are derived from triangular networks through twisting the O(1) and O(2) triangles to produce metaprisms, it is then possible to conceive of two possible prototypes. In one, the rotation between triangles in successive layers is 0° yielding an idealized structure containing A(1)O₆ trigonal prisms, while in the second, the rotation is set at 60° that generates A(1)O₆ octahedra (Fig. 7). The intermediate structure with a rotation of 30° generates an A(1)O₆ metaprism (a polyhedron intermediate between a trigonal prism and an octahedron) and a structure clearly resembling fluorapatite (Fig. 6d). Second, as the tunnel walls are defined by the shortest, and therefore strongest, metal-oxygen bond lengths their integrity is maintained throughout. It is also apparent that the tunnel volume is largest and smallest when

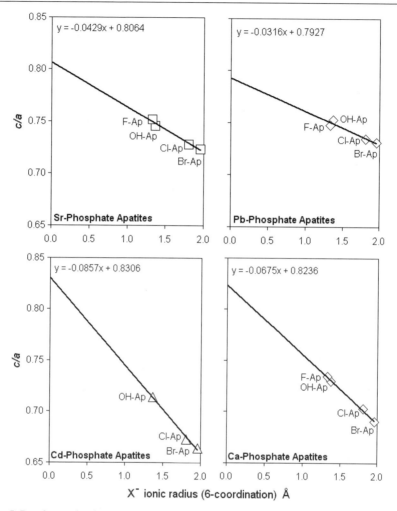

Figure 5. Based upon the closest packing phosphate model, back extrapolation to a hypothetical apatite containing no X anions leads to c/a ratios close to the theoretical value of 0.817. Apatites with larger A cations such as lead remain slightly expanded even with complete "removal" of the X atoms.

the rotation angles are 0° and 60° respectively. Finally, because metaprism rotation controls tunnel diameter, adjustments can be made spontaneously in response to changes in chemistry and tunnel filling.

Formally, the deviation of apatites from the prototypes can be described by a single parameter—the metaprism twist angle (ϕ)—which is defined as the (001) projected angle of O(layer 1)-A-O(layer 2). The degree of twist is controlled by the relative sizes of the wall and tunnel atoms, which in turn is related to chemistry. For example, in the simple chemical series $Cd_{10}(PO_4)_6(OH)_2$, $Cd_{10}(PO_4)_6Cl_2$ and $Cd_{10}(PO_4)_6Br_2$, ϕ becomes progressively more acute as the tunnel expands to accommodate the larger X anion (Fig. 8). While this trend illustrates the essential capacity of the twist angle to monitor apatite crystal chemistry, it is also evident that even in compositionally complex apatites ϕ can be an exquisitely sensitive probe for the detection of unexpected departures from stoichiometry or thermodynamic equilibrium.

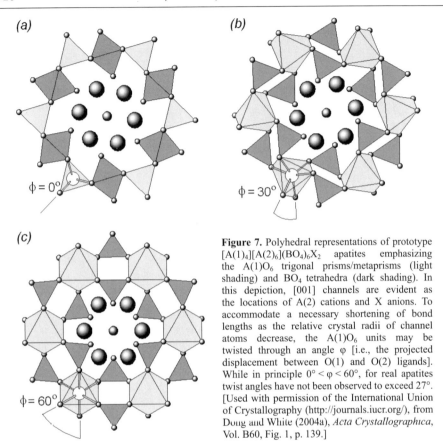

Figure 7. Polyhedral representations of prototype $[A(1)_4][A(2)_6](BO_4)_6X_2$ apatites emphasizing the $A(1)O_6$ trigonal prisms/metaprisms (light shading) and BO_4 tetrahedra (dark shading). In this depiction, [001] channels are evident as the locations of A(2) cations and X anions. To accommodate a necessary shortening of bond lengths as the relative crystal radii of channel atoms decrease, the $A(1)O_6$ units may be twisted through an angle φ [i.e., the projected displacement between O(1) and O(2) ligands]. While in principle $0° < \varphi < 60°$, for real apatites twist angles have not been observed to exceed 27°. [Used with permission of the International Union of Crystallography (http://journals.iucr.org/), from Dong and White (2004a), *Acta Crystallographica*, Vol. B60, Fig. 1, p. 139.]

CRYSTAL CHEMICAL SYSTEMATICS

It has been long known that attempting a systematic consideration of apatite crystal chemistry from literature data is problematical due to incomplete crystallographic and compositional characterization of materials. In this regard, the critical remarks of Felsche (1972) and McConnell (1973) remain germane to the present day. Essentially the difficulty of seeking a methodical arrangement of data arises from three potential sources of error:

1. *Composition*: It is generally true that even in contemporary crystallographic determinations and refinements independent verification of crystal composition is not undertaken. While misinterpretation is less likely in single crystal studies, modern research is often reliant on powder methods, particularly Rietveld analysis, as large single crystals are unavailable. However, poor agreement has been observed where chemical analyses and refined occupancies are compared (e.g., Wilson et al. 2003; Dong et al. 2005a).

2. *Equilibration*: There are often quite substantial discrepancies in the unit cell constants of apatites with the same nominal composition. Aside from the issue of bulk composition, the cell parameters will be influenced by cation partitioning over the available crystallographic sites. Moreover, even in moderately complex apatites the approach to equilibrium may require several weeks annealing close to the solidus (Dong and White 2004a).

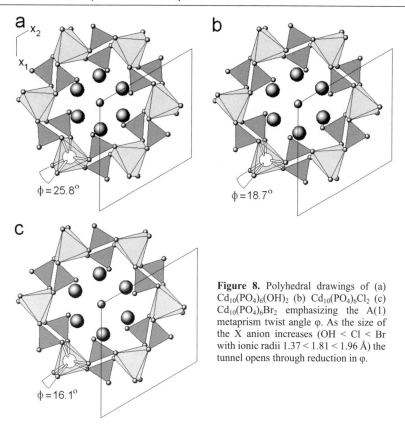

Figure 8. Polyhedral drawings of (a) $Cd_{10}(PO_4)_6(OH)_2$ (b) $Cd_{10}(PO_4)_6Cl_2$ (c) $Cd_{10}(PO_4)_6Br_2$ emphasizing the A(1) metaprism twist angle φ. As the size of the X anion increases (OH < Cl < Br with ionic radii 1.37 < 1.81 < 1.96 Å) the tunnel opens through reduction in φ.

3. *Partitioning*: Without the benefit of complete structure refinement, neither the distribution of atoms over cation acceptor sites, nor the confirmation of space group is possible. While the majority of apatites have been assigned $P6_3/m$, reports of lower symmetry analogues whose formation is intimately linked to composition and equilibration are becoming more common.

It is with these caveats that the following discussion should be considered. While a comprehensive data set has been assembled (Appendix A), the quality of individual entries will vary, however it is anticipated that the general trends synthesized below will nonetheless remain valid. In certain instances, materials subsets are considered in detail to demonstrate methods for detecting less reliable crystallographic refinements.

X anion ordering

The structural feature of apatites that has been most thoroughly, and reliably, studied has been the location of the X anion as a function of its size with respect to the surrounding A(2) cations (e.g., Mackie et al. 1972; Hata et al. 1979; Hashimoto and Matsumoto 1998; Kim et al. 2000). In the case of $Ca_{10}(PO_4)_6F_2$ the fluorine neatly fits within a $Ca(2)_3$ triangle, but substitution by OH^-, Cl^- and Br^- leads to consecutively greater displacements out of the triangular plane. In $P6_3/m$ this displacement leads to degeneracy of the $2a$ (0, 0, ¼) position as it splits into $4e$ (0, 0, z) symmetry to accommodate filled and vacant interstices. While the distribution of larger anions is often regarded as statistical, it has been observed that equilibrated materials or minerals can exhibit ordering both along and between anion-vacancy strings in the tunnels to yield superstructures. For example, Bauer and Klee (1993) have shown that chlorapatite $Ca_{10}(PO_4)_6Cl_2$ can adopt $P2_1/b$ symmetry and a doubled b-axis ($a = 9.643$ Å, $b = 19.279$ Å, $c = 6.766$ Å, $\gamma = 120.01°$). More complex incommensurate ordering schemes have also been identified. In $Cd_{10}(PO_4)_6Br_2$, an incommensurate superstructure (R: P-3(00γ) $a = 16.932$ Å, $b = 16.932$ Å, $c = 6.451$ Å) has been reported with modulation wave vector $\mathbf{q} = 0.778\mathbf{c}^*$ (Alberius-Henning et al. 2000). The germanate $La_{9.33}(GeO_4)_6O_2$ has $\mathbf{q} = 1.6$-$1.7\mathbf{c}^*$ (Berastegui et al. 2002), and the instances of long range X-site ordering are being reported more frequently.

Correlation of cell parameters and ionic radii

A general classification of the fluor- and chlorapatites and their stability fields as a function of A and B cation radii was devised by Kreidler and Hummel (1970). The principles set out remain essentially correct; metalloids can adopt 4+, 5+ and 6+ valence states and have sizes ranging from S^{6+} (0.12 Å) to V^{5+} (0.355 Å) with appropriate charge compensation by A^+/A^{2+} cations with radii from Cd^{2+} (1.03 Å) to Ba^{2+} (1.38 Å). For larger metalloids (i.e., V and As) monoclinic symmetry may be favored. However, end members containing Al^{3+}, Sb^{5+}, W^{6+}, Mo^{6+}, Ta^{5+} and Nb^{5+} are not stable. Since this early work, it has been established that limited substitution of many of these metalloids is possible. The major outstanding question at that time was the possible role of manganese, but it has now been demonstrated unequivocally from single crystal studies that manganese can be accommodated in either the A or B site, and $Mn_{10}(PO_4)_6Cl_{1.8}(OH)_{0.2}$ (Engel et al. 1975b) and $Ba_{10}(MnO_4)_6Cl_2$ (Reinen et al. 1986) have been prepared.

Ito (1968) conducted an investigation of remarkable scope for the $A_4REE_6(SiO_4)_6(OH)_2$ hydroxy- and $A_4REE_6(SiO_4)_6O_2$ oxy-silicate apatites with A = Mn, Ca, Sr and Pb, the results of which are summarized in Figure 9. An underlying assumption of these formulations is that the smaller A(1) sites will be filled almost entirely by the divalent cation, or at least, the A(1)/A(2) partitioning remains constant across REE compositional series. While this is certainly true for larger REEs such as La, Ce and Nd (Schroeder and Matthew 1978; Fahey et al. 1985; Skakle et al. 2000) it is likely that smaller REEs will partition more strongly to the A(1) site. Consequently, while for the most part the apatite lattices expanded linearly as a function of ionic radii, it is evident that for the $Ca_4REE_6(SiO_4)_6(OH)_2$ and $Mn_4REE_6(SiO_4)_6O_2$ series two distinct linear segments are present, with the discontinuity appearing for REE < Dy, possibly arising because smaller REEs are similar in size to Ca and Mn, leading to mixing over the A(1) and A(2) sites, and therefore a change in the cell constant.

An alternative method to systematize the lattice constants of hexagonal apatites is to examine the cell volume versus the c/a ratio when the A cation is fixed and only (non-silicate) BO_4 and X components are varied (Fig. 10). While there is a considerable spread of data for the calcium apatites, clear trends appear for the strontium, lead and barium apatites. In these cases, gradual incorporation of larger metalloids and halides leads first to an increase in c/a while at larger cell volumes this ratio decreases.

Classification by metaprism twist angle (φ)

The derivation of apatites from a prototype, and in particular the relationship between metaprism twist angle (ϕ) and composition, provides the basis for examining crystal chemical systematics in a more general sense. It has been shown previously that for a fixed A cation, ϕ varies inversely with the average effective ionic radius of the unit cell (White and Dong 2003) because filling of the microporous channels with larger atoms is accommodated by a reduction of twist angle to increase its volume (Table 2, Fig. 11). The rate of change of ϕ is greatest for the smallest A cation (Cd^{2+}) and slowest for the largest (Ba^{2+}) as the channel diameter is already enlarged in latter instance. Twist angle outliers are evident for the Pb- and Ba-apatites, and this data may indicate that these compounds deviate from the reported composition.

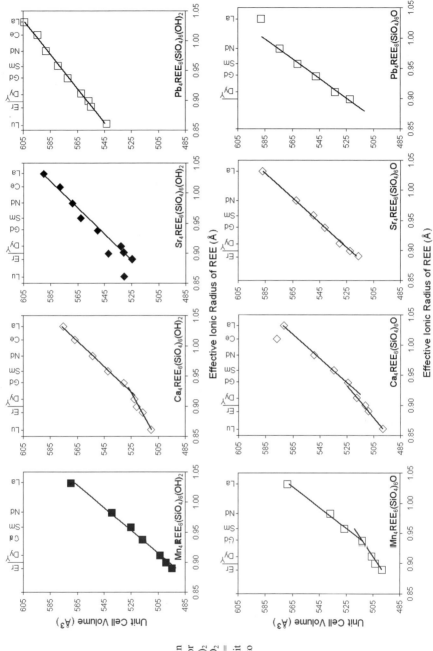

Figure 9. Trends in unit cell volume for $A_4REE_6(SiO_4)_6(OH)_2$ and $A_4REE_6(SiO_4)_6O_2$ apatites with A = Mn, Ca, Sr, Pb [unit cell data from Ito (1968)].

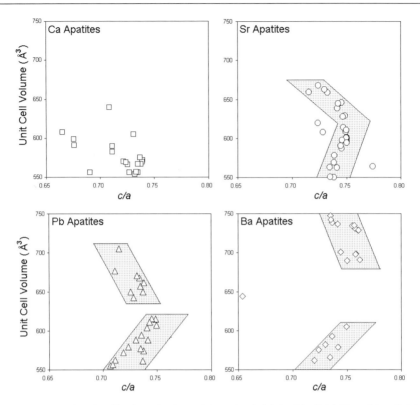

Figure 10. Trends in unit cell volume for $Ca_{10}(BO_4)_6X_2$, $Sr_{10}(BO_4)_6X_2$, $Pb_{10}(BO_4)_6X_2$, and $Ba_{10}(BO_4)_6X_2$ apatites using data for $P6_3/m$ phases extracted from Appendix A.

In apatite solid solution series, trends in twist angle can be used to establish the reliability of crystallographic data, which may be compromised due to disequilibrium effects, or in the case of powder diffraction data, refinement to false minima. For example, the compounds $Ca_{10-x}Pb_x(PO_4)_6Br_2$ were prepared by solid state reaction of the calcium and lead members by heating for 15 h at 850°C (Kim 2001), and the products characterized by Rietveld co-refinement of X-ray and neutron data (Table 3). It would be expected that in passing from the calcium to lead end member ϕ should decrease, with larger Pb^{2+} occupying the A(2) tunnels sites partially accommodated through a reduction in metaprism twist angle. However, while there is a slight downward trend in ϕ from $x = 0$ to 9.53, the values are quite variable, with ϕ for $x = 10$ increasing dramatically to 20.6°. These data are explicable in terms of disequilibrium and compositional changes. Assuming that the single crystal determinations for $Ca_{10}(PO_4)_6Br_2$ (Elliott et al. 1981), $Pb_{10}(PO_4)_6(OH)_2$ (Barinova et al. 1998), and $Ca_{10}(PO_4)_6(OH)_2$ (Kay et al. 1964) are reliable, trend lines can be drawn as shown in Figure 12 that relate ϕ to composition (Table 4). The data of Kim (2001) may be superimposed on this and interpreted in three segments.

For $x < 3$, ϕ lies above the $Ca_{10}(PO_4)_6Br_2$ - $Pb_{10}(PO_4)_6Br_2$ trend line consistent with an enrichment of Pb in the A(1) position and the loss of bromine and its replacement by oxygen.

For $3 < x < 9$, ϕ's lie between the $Ca_{10}(PO_4)_6Br_2$ - $Pb_{10}(PO_4)_6Br_2$ and $Ca_{10}(PO_4)_6Br_2$ - $Pb_{10}(PO_4)_6(OH)_2$ trends consistent with partial oxidation of the bromo-apatites, and as shown by the lead partitioning coefficient, incomplete equilibration of Ca/Pb separation over the A(1) and A(2) sites.

Table 2. Correlation of metaprism twist angle and average effective ionic radius.

Apatite	Average Radius (Å)	Twist Angle (°)
$Ca_{10}(PO_4)_2F_2$	1.143	23.3
$Ca_{10}(PO_4)_6(OH)_2$	1.146	23.2
$Ca_{10}(PO_4)_6Cl_2$	1.166	19.1
$Ca_{10}(CrO_4)_6(OH)_2$	1.171	17.8
$Ca_{10}(PO_4)_6Br_2$	1.173	16.3
$Ca_{10}(AsO_4)_6Cl_2$	1.189	13.0
$Ca_4Pb_6(AsO_4)_6Cl_2$	1.214	5.2
$Cd_{10}(PO_4)_6(OH)_2$	1.139	25.8
$Cd_{10}(PO_4)_6Cl_2$	1.158	19.5
$Cd_{10}(PO_4)_6Br_2$	1.165	16.0
$Cd_{10}(AsO_4)_6Br_2$	1.189	11.6
$Cd_{10}(VO_4)_6Br_2$	1.192	8.8
$Cd_{10}(VO_4)_6I_2$	1.203	8.4
$Sr_{10}(PO_4)_6F_2$	1.180	24.3
$Sr_{10}(PO_4)_6(OH)_2$	1.183	23.0
$Sr_{10}(PO_4)_6Cl_2$	1.203	21.1
$Sr_{10}(PO_4)_6Br_2$	1.210	19.6
$Pb_6(PO_4)_2$	1.122	27.3
$Pb_{10}(PO_4)_6(OH)_2$	1.189	26.7
$Pb_{10}(GeO_4)_2(CrO_4)$	1.151	25.8
$Pb_{10}(PO_4)_6F_2$	1.185	23.5
$Pb_{10}(PO_4)_6Cl_2$	1.208	17.6
$Pb_{10}(VO_4)_6Cl_2$	1.235	17.5
$Pb_{10}(VO_4)_6I_2$	1.253	16.7
$Pb_{10}(AsO_4)_6Cl_2$	1.232	5.2
$Ba_{10}(PO_4)_6F_2$	1.219	22.5
$Ba_{10}(MnO_4)_6Cl_2$	1.254	22.3
$Ba_{10}(PO_4)_6(OH)_2$	1.222	22.2
$Ba_{10}(PO_4)_6Cl_2$	1.242	21.0
$Ba_{10}(AsO_4)_4(SO_4)_2S$	1.259	16.2

For $x = 10$, ϕ is unexpectedly high probably as a result of substantial Br loss, and consequently its twist angle is close to that determined for $Pb_{10}(PO_4)_6(OH)_2$. Furthermore, the difference in ϕ for $Pb_{10}(PO_4)_6(OH)_2$ in the determinations of Barinova et al. (1998) and Brückner et al. (1995) may arise from the partial loss of oxygen in the latter material, and its approach to $Pb_9(PO_4)_6$ stoichiometry. Inspection of Brückner et al.'s refined powder X-ray diffraction data reports large isotropic thermal parameters (B = 1.85 Å^2) for lead that could also be consistent with vacancies.

In passing from $Ca_{10}(PO_4)_6(OH)_2$ to $Pb_{10}(PO_4)_6(OH)_2$ ϕ decreases from 23.2° to 22.1°, thus it follows that as the reliable determination for ϕ in $Ca_{10}(PO_4)_6Br_2$ is 14.8°, the stoichiometric end member $Pb_{10}(PO_4)_6Br_2$ is predicted to adopt a metaprism twist angle close to 13.7° (Table 4). In this respect, the near end member of Kim (2001) with $x = 9.5$ is of particular interest as it refines to a nearly ideal $k_{Pb} = 0.62$, and has apparently suffered only a small loss of Br, leading to $\phi = 14°$ for this material.

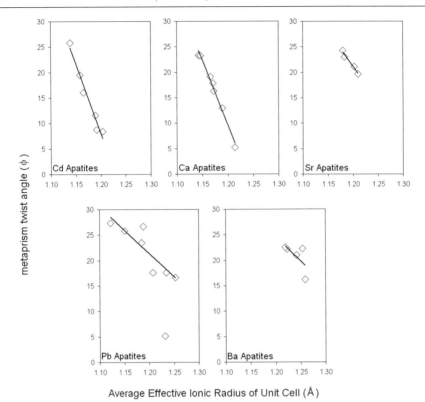

Figure 11. Trends in metaprism twist angle φ for $Cd_{10}(BO_4)_6X_2$, $Ca_{10}(BO_4)_6X_2$, $Sr_{10}(BO_4)_6X_2$, $Pb_{10}(BO_4)_6X_2$ and $Ba_{10}(BO_4)_6X_2$ apatites as a function of average effective ionic radius for the whole unit cell contents using data extracted from Table 2.

Prediction of cell parameters

As apatite topology can be modified by twisting regular triangular nets, and the $A(1)O_6$ and BO_4 polyhedra are corner connected, it is feasible to predict and model the relative changes in unit cell parameters as a function of experimentally derived φ's. From a purely geometrical consideration, and using the construction shown in Figure 13, it has been shown by Dong and White (2004b) that the basal unit cell dimension a is related to φ and the equilateral triangle edge t such that

$$a = t\sqrt{13 - 28\sin^2\left(\frac{\phi}{4}\right) + 16\sin^4\left(\frac{\phi}{4}\right)}$$

where t is a scaling parameter specific to each apatite system. For example, $t = 2.729 + 0.017x$ and x is the stoichiometry in $(Pb_xCa_{10-x})(VO_4)_6(F_{2-2y}O_y)$. The compositional adjustment factor for t is derived from the increase in ionic radii for Ca^{2+} (IR = 1.18 Å) and Pb^{2+} (IR = 1.35 Å) divided by total A-cation formula content, i.e., $(1.35-1.18)/10 = 0.017$ Å.

Similarly, the c cell parameter can be related to φ as

$$c = 2\sqrt{h_{\phi=0}^2 - \frac{4t^2}{3}\sin^2\left(\frac{\phi}{2}\right)}$$

Table 3. Co-refined synchrotron X-ray and neutron data at for $Ca_{10-x}Pb_x(PO_4)_6Br_2$, $0 \leq x \leq 10$ apatites. [from Kim 2001].

			$x = 0$	0.9	2.9	5.0	7.4	9.2	9.5	10
a (Å)			9.7682(6)	9.7823(5)	9.8452(3)	9.9071(3)	9.9708(2)	10.0116(4)	10.0356(2)	10.0618(3)
c (Å)			6.7388(2)	6.7485(3)	6.8596(4)	7.0086(5)	7.1682(3)	7.2662(3)	7.3203(3)	7.3592(1)
Ca(I), Pb(I)†	$4f$ (⅓, ⅔, 0)	z	0.0063(6)	0.0059(4)	0.010(5)	0.0075(3)	0.0066(5)	0.0059(4)	0.0069(2)	0.0067(3)
		B(Å²)	0.3(3)*	0.3(1)*	0.5(3)*	0.5(4)*	0.3(2)*	0.4(5)*	0.3(1)*	0.3(1)*
Ca(II), Pb(II)†	$6h$ (x, y, ¼)	x	0.2600(3)	0.2589(4)	0.2614(2)	0.2634(3)	0.2630(5)	0.2620(3)	0.2615(2)	0.2618(1)
		y	1.0031(4)	0.9832(3)	1.0052(3)	1.0781(4)	1.012(3)	0.8431(4)	0.901(3)	0.0053(2)
		B(Å²)	0.3(3)*	0.3(4)*	0.4(2)*	0.4(4)*	0.25(5)*	0.3(4)*	0.4(4)*	0.3(1)*
P	$6h$ (x, y, ¼)	x	0.4063(4)	0.4072(4)	0.4001(3)	0.4072(5)	0.3946(4)	0.3802(5)	0.3711(4)	0.3497(3)
		y	0.3724(4)	0.3727(3)	0.3832(4)	0.3995(4)	0.4158(3)	0.4012(4)	0.4345(3)	0.4877(3)
		B(Å²)	0.4(2)*	0.4(2)*	0.4(2)*	0.3(1)*	0.24(4)*	0.6(2)*	0.4(4)*	0.3(3)*
O(1)	$6h$ (x, y, ¼)	x	0.3490(8)	0.3495(3)	0.3497(4)	0.3558(4)	0.3462(5)	0.3523(4)	0.3648(3)	0.4128(3)
		y	0.4929(6)	0.4946(3)	0.4940(4)	0.4851(2)	0.4872(3)	0.4953(3)	0.4612(2)	0.3795(4)
		B(Å²)	0.4(5)*	0.4(5)*	0.5(2)*	0.4(4)*	0.4(2)*	0.5(3)*	0.7(4)*	0.5(2)*
O(2)	$6h$ (x, y, ¼)	x	0.5918(6)	0.5924(6)	0.5895(4)	0.5910(4)	0.5986(5)	0.5909(6)	0.5916(5)	0.5904(3)
		y	0.4691(7)	0.4613(5)	0.4689(5)	0.4692(2)	0.4703(4)	0.3622(4)	0.4653(4)	0.4684(3)
		B(Å²)	0.3(7)*	0.4(5)*	0.5(1)*	0.2(1)*	0.4(4)*	0.3(4)*	0.5(3)*	0.4(2)*
O(3)	$12i$ (x, y, z)	x	0.3536(5)	0.3584(4)	0.359(4)	0.3610(6)	0.3633(5)	0.3625(3)	0.3694(4)	0.3643(2)
		y	0.2638(4)	0.2685(4)	0.2692(4)	0.2702(5)	0.2739(4)	0.2703(4)	0.2815(3)	0.2747(2)
		z	0.06445(5)	0.067332(3)	0.0677(3)	0.0701(3)	0.0811(3)	0.0726(5)	0.0823(2)	0.0809(3)
		B(Å²)	0.6(2)*	0.6(1)*	0.5(4)*	0.5(1)*	0.5(3)*	0.5(3)*	0.5(4)*	0.5(2)*
Br	$4e$ (0, 0, z)	B(Å²)	8.3(5)*	1.2(4)*	0.8(3)*	1.1(4)*	0.7(2)*	0.5(2)*	0.4(5)*	0.3(3)*
R_{Bragg}			(5–45)#5.2	6.90	7.04	6.97	7.84	7.46	6.85	6.14
			(45–85)#5.38	7.12	8.95	7.82	7.42	8.11	6.99	7.09
			Neutron 3.74	7.99	9.54	6.52	7.27	6.54	7.71	2.81

† End members have only Ca or Pb sites present.
* Equivalent value derived from anisotropic values $Biso = 8\pi^2(U_{11} + U_{22} + U_{33})/3$ (Å²).
# The values refer to data from the two IPs used.

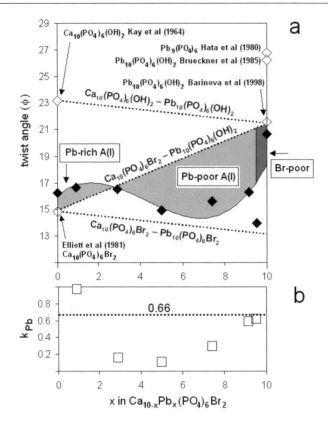

Figure 12. (a) Trends in metaprism twist angle, φ, for the joins $Ca_{10}(PO_4)_6(OH)_2 - Pb_{10}(PO_4)_6(OH)_2$, $Ca_{10}(PO_4)_6Br_2 - Pb_{10}(PO_4)_6(OH)_2$ and $Ca_{10}(PO_4)_6Br_2 - Pb_{10}(PO_4)Br_2$. Open diamonds are derived from single crystal X-ray diffraction determinations. Closed diamonds are derived from the powder neutron diffraction data of Kim (2001) (see also Table 3) for $Ca_{10-x}Pb_x(PO_4)_6Br_2$ solid solution members. Consideration of φ allows the spread of this latter data to be explained in terms of disequilibrium and loss of bromine. (b) Partitioning coefficients for lead $k_{Pb}(A1)/(A2)$ sites is variable due to crystallization times being too short to achieve equilibrium. As the A(2) sites are larger than A(1) preferential entry of lead into A(2) would be expected, except near the lead end member, where $k_{Pb} = 0.66$ for equal partitioning over the A-sites is approached.

where the prism height $h_{\phi=0}$ is the metaprism height in the aristotype. Using again the example of $(Pb_xCa_{10-x})(VO_4)_6(F_{2-2y}O_y)$ $h_{\phi=0} = 3.555 + 0.017x$ at $\phi = 0°$.

While this model contains obvious assumptions and limitations, in particular the projected edges of the metaprisms and tetrahedra are not equal as supposed, it is nonetheless useful for recognizing unexpected departures in Vegard's Law when studying solid solution series.

Symmetry and flexibility

As noted earlier, apatites adopt any of the maximal isomorphic subgroup symmetries of $P6_3/mcm$ (Fig. 3). Broadly, there are two drivers that will impose lower symmetry than the commonly observed $P6_3/m$ on a given apatite—*complex chemistries*, where multiple cation-acceptor sites are required, and *bond strain* induced by "poorly" fitting atoms. As summarized in Table 5, passing from $P6_3/m$ to $P2_1/m$ increases the number of unique Wyckoff positions from 7 to 18 resulting in a concomitant enhancement of crystal chemical flexibility. Polyhedral drawings of representative $A_{10}(BO_4)_6X_2$ apatites in all subgroups are shown in Figure 14 (color

Table 4. Comparison of metaprism twist angle φ in $Ca_{10-x}Pb_x(PO_4)_6(OH,Br)_2$ as a function of composition.

Composition	Method	φ (°)	k_{Pb}	Reference
Refined Composition				
$Ca_{10}(PO_4)_6Br_2$	X-ray & neutron powder	16.2		Kim (2001)
$Ca_{9.1}Pb_{0.9}(PO_4)_6Br_2$		16.6	0.97	
$Ca_{7.1}Pb_{2.9}(PO_4)_6Br_2$		16.5	0.16	
$Ca_{5.0}Pb_{5.0}(PO_4)_6Br_2$		14.9	0.11	
$Ca_{2.6}Pb_{7.4}(PO_4)_6Br_2$		15.6	0.30	
$Ca_{0.8}Pb_{9.2}(PO_4)_6Br_2$		16.3	0.59	
$Ca_{9.1}Pb_{0.9}(PO_4)_6Br_2$		14.0	0.62	
$Pb_{10}(PO_4)_6(OH)_2$		20.6		
Nominal Composition				
$Ca_{10}(PO_4)_6Br_2$	X-ray single crystal	14.8		Elliott et al. (1981)
$Ca_{10}(PO_4)_6(OH)_2$	Neutron single crystal	22.1		Kay et al. (1964)
$Pb_{10}(PO_4)_6Br_2$	—	13.7		predicted
$Pb_{10}(PO_4)_6(OH)_2$	X-ray single crystal	22.1		Barinova et al. (1998)
"$Pb_{10}(PO_4)_6(OH)_2$"	X-ray powder	26.7		Brückner et al. (1995)
$Pb_9(PO_4)_6$	X-ray single crystal	27.3		Hata et al. (1980)

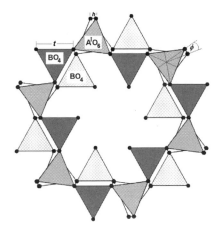

Figure 13. Construction used for the derivation of equations for the calculation of cell parameters from overlapping, at different heights, triangular anion nets. The triangle edge length (t), metaprism height (h) and twist angle (φ) are shown. This idealized representation should be compared with the real structures in Figure 14. [Used with permission of the International Union of Crystallography (http://journals.iucr.org/), from Dong and White (2004a), *Acta Crystallographica*, Vol. B60, Fig. 7, p. 152.]

figure on page 312). The compilation in Appendix A contains approximately 500 distinct compounds with apatites lower in the hierarchical tree less common, or at least, not yet reported as extensively.

P6$_3$/m. In this space group the metaprisms and tetrahedral cations are confined to lattice positions with 4f and 6h symmetry respectively. The prototype mineral $Ca_{10}(PO_4)_6(OH,F)_2$ adopts this space group, as do many binary representatives including vanadates, arsenates, silicates, chromates and germanates. More intricate chemistries such as $Nd_2Ca_6Na_2(PO_4)_6F_2$ have been reported (Mayer et al. 1975), but rarely. For this space group all metaprisms exhibit the same twist angle φ.

P6$_3$ and $P\bar{3}$. These space groups allow ordering of metaprism sites (2b × 2 and 2d × 2) and in addition free the tetrahedrally bonded tunnel oxygen to twist. Therefore, they are observed

Table 5. Atom acceptor sites in apatites adopting different space groups.

Site Types	$P6_3/m$ 176		$P6_3$ 173		Space Group $P\bar{3}$ 147		$P\bar{6}$ 174		$P2_1/m$ 11	
	Wyckoff Number	No. Site Types	Wyckoff Number	No. Site Types	Wyckoff Number	No. Site Types	Wyckoff Number	No. Site Types	Wyckoff Number	No. Site Types
Large A Cations	$6h$	1	$6c$	1	$6g$	1	$3k, 3j$	2	$2a, 2e \times 2$	3
Small A Cation	$4f$	1	$2b \times 2$	2	$2d \times 2$	2	$2i, 2h$	2	$4f$	1
B Cations	$6h$	1	$6c$	1	$6g$	1	$3k, 3j$	2	$2e \times 3$	3
Total No. of Cation Acceptor Sites		3		4		4		6		7
Oxygen anions	$6h \times 2, 12i$	3	$6c \times 4$	4	$6g \times 4$	4	$3k \times 2$, $3j \times 2, 6l \times 2$	6	$2e \times 6$, $4f \times 3$	9
X anions	$2a$ or $2b$ or $4e$	1	$2a$	1	$1a, 1b$	2	$1a, 1b$ or $2g$	1/2	$2a$ or $2e$	2
Total No. of Anion Acceptor Sites		4		5		6		7/8		11
Examples of Cation Occupancy	[Ca$_6$][Ca$_4$][P$_6$]		[K$_6$][K$_2$Sn$_2$][S$_6$]		[Ba$_4$Nd$_1$Na$_1$][Nd$_2$Na$_2$][P$_6$]		[Ba$_5$La$_1$][La$_1$Na$_{1.3}$Ba$_{1.7}$][P$_6$]		[Ca$_6$][Ca$_4$][Si$_3$S$_3$]	
[A$_6$][A$_4$][B$_6$][O$_{24}$][X$_2$]	Ca$_{10}$(PO$_4$)$_6$F$_2$		K$_6$Sn$_4$(SO$_4$)$_6$Cl$_2$		Ba$_4$Nd$_3$Na$_3$(PO$_4$)$_6$F$_2$		Ba$_{6.7}$La$_2$Na$_{1.3}$PO$_4$)$_6$F$_2$		Ca$_{10}$(SiO$_4$)$_3$(SO$_4$)$_3$ (F$_{0.16}$Cl$_{0.48}$(OH)$_{1.36}$)	

where tetrahedral distortion is needed, for example in chromate apatites such as $Sr_{10}(CrO_4)_6Cl_2$, or where two components are incorporated in the tetrahedra, as in $Ca_{10}(P,BO_4)_6(O,B)_{2\delta}$. Where the dual symmetry of the metaprism sites is exploited as in $Sr_{7.3}Ca_{2.7}(PO_4)_6F_2$, the metaprisms have the same twist angle with chemical differences accommodated by oxygen displacements along z.

P$\bar{6}$. For this symmetry, two unique tetrahedral sites are available ($3k$ and $3j$), in addition to two independent metaprism positions ($2i$ and $2h$). As a result, the metaprisms exhibit two twist angles as observed in $(La_2Na_2Ba_6)(PO_4)_6F_2$.

P2_1/m. This space group allows flexibility in the tetrahedral positions ($2e \times 3$), but permits only a single metaprism site ($4f$). This symmetry is particularly useful where fluorine is the X anion, as the tunnel has difficulty in collapsing sufficiently purely by metaprism twisting. It has recently been confirmed that $Ca_{10}(VO_4)_6F_2$ adopts this arrangement (Dong and White 2004b), and $Cd_{10}(PO_4)_6F_2$ and $Ca_{10}(AsO_4)_6F_2$ may be isostructural (Kreidler and Hummel 1970).

P2_1. In their lowest symmetry, apatites have 3 independent tetrahedral sites and 2 independent metaprisms. This structure was first confirmed for ellestadite, ideally $Ca_{10}(SiO_4)_2(SO_4)_2(PO_4)_2(OH,F,Cl)_2$ (Organova et al. 1994). In both monoclinic space groups, the metaprisms can no longer be described uniquely by a single twist angle.

MICROPOROSITY IN NATURAL APATITES

Apatites are not conventionally regarded as microporous. However, as described above, these minerals do possess crystallographic features—an adaptable framework and "channels"—that while not making them zeolitic may predispose them to exhibiting some of the characteristics of zeolites. Indeed, in certain geological settings apatites and zeolites co-exist. Natural phosphate apatite generally forms as well-shaped hexagonal crystals, which may be elongated or stubby; less commonly in tabular plates or columnar forms; as globular masses, acicular, grainy; and earthy aggregates; but most frequently in enormous beds of massive material, from which industrial phosphate is mined (phosphorites).

Besides apatite *sensu stricto* other end members are well known including vanadinite $Pb_{10}(VO_4)_6Cl_2$, pyromorphite $Pb_{10}(PO_4)_6Cl_2$ and mimetite $Pb_{10}(AsO_4)_6Cl_2$. Vanadinite ranges in color from brown through yellow to orange to red, often forming highly aesthetic crystals generally as short hexagonal prisms terminated by a pinacoid, or flat basal face. The high luster and deep red color of vanadinite appeals to mineral collectors. Named for its vanadium content, vanadinite occurs in association with lead deposits and was in the past an important ore of vanadium but the metal is now usually obtained as a by-product in the production of other metals. A secondary mineral also found in the oxidized zones of lead ore deposits is pyromorphite. It is typically found as green to yellow barrel-shaped hexagonal prisms, in clusters or as druses on matrix. This lead chloride phosphate forms a complete series with mimetite, and many specimens are intermediates. The end-member mimetite is generally found in association with vanadinite and pyromorphite and/or in other settings where lead and arsenic occur together. Usually found as small hexagonal prisms, its color is quite variable, ranging from pale yellow to yellowish-brown to orange-yellow to orange-red, white and colorless. A complete list of natural apatite types is given in Table 6, and while some are rarely occurring, they provide valuable insights for the synthesis of new apatites with specific technological features.

Occurrence

As the most abundant natural phosphate, apatite plays a critical role in the global geobiochemical cycle of phosphorous starting with its mobilization at the Earth's surface, its

Table 6. Mineral apatites and related structures

Name	Formula	Reference
alforsite	$Ba_{10}(PO_4)_6Cl_2$	Hata et al. (1979)
belovite-(Ce)	$Sr_6Na_2(Ce, La)_2(PO_4)_6(F,OH)_2$	Klevtsova and Borisov (1964)
belovite-(La)	$Sr_6Na_2(La, Ce)_2(PO_4)_6(F,OH)_2$	Kabalov et al. (1997)
carbonate-fluorapatite	$Ca_{10}(PO_4, CO_3)_6F_2$	Leventouri et al. (2000)
carbonate-hydroxylapatite	$Ca_{10}(PO_4, CO_3)_6(OH)_2$	El Feki et al. (1999)
chlorapatite	$Ca_{10}(PO_4)_6Cl_2$	Kim et al. (2000)
clinomimetite	$Pb_{10}(AsO_4)_6Cl_2$	Dai et al. (1991)
fermorite	$(Ca, Sr)_{10}[(As, P)O_4]_6(OH)_2$	Hughes and Drexler (1991)
fluorapatite	$Ca_{10}(PO_4)_6F_2$	Kim et al. (2000)
hedyphane	$Pb_6Ca_4(AsO_4)_6Cl_2$	Rouse et al. (1984)
hydroxylapatite	$Ca_{10}(PO_4)_6(OH)_2$	Kim et al. (2000)
Kuannersuite-(Ce)	$Na_4Ba_4(Ce,Nd,La)_2(PO_4)_6(F,Cl)_2$	Friis et al. (2004)
johnbaumite	$Ca_{10}(AsO_4)_6(OH)_2$	Dunn et al. (1980)
mimetite	$Pb_{10}(AsO_4)_6Cl_2$	Dai et al. (1991)
morelandite	$(Ba, Ca, Pb)_{10}[(As, P)O_4)]_6Cl_2$	Dunne and Rouse (1978)
pyromorphite	$Pb_{10}(PO_4)_6Cl_2$	Dai and Huges (1989)
strontium-apatite	$(Sr, Ca)_{10}(PO_4)_6(OH,F)_2$	Sudarsanan and Young (1980)
svabite	$Ca_{10}(AsO_4)_6F_2$	Kreidler and Hummel (1970)
turneaureite	$Ca_{10}[(As, P)O_4]_6Cl_2$	Wardojo and Hwu (1996)
vanadinite	$Pb_{10}(VO_4)_6Cl_2$	Dai and Huges (1989)
britholite-(Ce)	$(Ce, Ca)_{10}[(Si, P)O_4)]_6(OH,F)_2$	Genkina et al. (1991)
britholite-(Y)	$(Y, Ca)_{10}[(Si, P)O_4)]_6(OH,F)_2$	Noe et al. (1993)
fluorbritholite-(Ce)	$(Ce, La, Na)_{10}[(Si, P)O_4]_6F_2$	Hughes et al (1992)
chlorellestadite	$Ca_{10}[(Si, S)O_4]_6(Cl,F)_2$	Organova et al. (1994)
fluorellestadite	$Ca_{10}[(Si, P, S)O_4]_6(F,OH,Cl)_2$	Hughes and Drexler (1991)
hydroxylellestadite	$Ca_{10}[(Si, S)O_4]_6(OH,Cl,F)_2$	Sudarsanan (1980)
mattheddleite	$Pb_{20}[SiO_4)_7(SO_4)]_4Cl_4$	Steele et al. (2000)
cesanite	$Na_6Ca_4(SO_4)_6(OH)_2$	Piotrowski et al. (2002b)

transport through the living environment, and ultimately, re-deposition through the formation of new geological apatites by sedimentary processes and/or tectonic recycling.

Igneous. Members of the apatite group are often accessory minerals in almost all igneous rocks from basic to acidic, sometimes amounting for as much as 5% by volume, although 0.1–1% is usual. In igneous rocks, the appearance of apatite is due not only to its low solubility in melts and aqueous solutions, but also because the common rock-forming minerals do not accept phosphorous. Fluorapatite is generally dominant, often with appreciable incorporation of chlorine and hydroxyl. Bromine and iodine can also be present but their concentrations are much lower. Apatite appears in both plutonic and effusive acidic and basic rocks, granitic pegmatites, as well as in hydrothermal veins-cavities (Piccoli and Candela 2002). Carbonatites generally contain appreciable apatite and there are several apatite-rich areas in the Khibina tundra, Kola Peninsula where apatite-nepheline rocks (Pletchov and Sinogeikin 1996) accommodate both crystals and botryoidal apatite (i.e., aggregates resembling bunches of grapes). Apatite forms about 3% of the Palabora shonkinite in the Transvaal and in this locality both apatite-diopside and apatite rocks contain up to 96% phosphate (Birkett and Simandl 1999). Carbonate-apatite also occurs in the calcitic carbonatites of the Alnö alkaline complex (Wilke 1997).

Metamorphic. Apatite appears in both thermally and regionally metamorphosed rocks as well as precipitates from hydrothermal solutions (Spear and Pyle 2002). The compositions of metamorphic apatite typically fall along the F-OH join although apatite with small amounts of chlorine has been reported. Fluorapatite is frequent in metasomatized calc-silicate rocks and impure limestones, whereas chloroapatite is associated with scapolite in rocks that have undergone chlorine metasomatism (Boudreau 1995). As accessory minerals, apatites have become important for research in metamorphic petrology since the partitioning of trace elements in zoned crystals contains detailed information concerning the reaction history of the host metamorphic rock. The characteristic of apatite to accumulate trace and rare elements in its structure is also important for *in situ* dating techniques based on quantification of such elements and their isotope ratios.

Sedimentary. The presence of apatite, but more generally of phosphates, in sedimentary environments is strictly controlled through interaction of the bio- and geospheres (Knudsen and Gunter 2002). Apatites that crystallize in such environments are often carbonate-fluorapatites where the PO_4^{3-} group is partially substituted by CO_3^{2-}. Exploitable phosphorites occur as sediments of marine origin formed by the upwelling of phosphate-rich waters that are characteristic of large delta systems such as the Volga river outflow to the Caspian Sea (Bushinskiy 1964). Other enormous deposits are the Neocene-Holocene phosphorites of Australia, the Miocene phosphorites of Cuba, the deposits in the upper Oligocene of the San Gregorio Formation at San Juan de la Costa, Baja California, the deposits on the Namibian continental shelf, sediments on the South African continental margin, the Moroccan offshore deposits, the Neocene phosphorites of the Sea of Japan, and the submerged phosphate deposit of Mataiva Atoll, French Polynesia (Burnett and Riggs 1990).

Finally, the role of apatites as the primary minerals in bones, teeth, and in general, for all hard tissues of the human body is well known. Their low perishability and persistence make bones and teeth an important resource in palaeontology and archaeology, particularly for the study of palaeoambients (Kohn and Cerling 2002).

Spinodal decomposition

For a great many mineral families unit cell scale inhomogeneities, or phase separation, are well documented for accommodating nonstoichiometry, especially where the approach to equilibrium is slow. It is then perhaps surprising that the apatites, especially those of some chemical complexity, have stood apart as true "solid solutions" having statistical distributions of cations, notwithstanding that long range ordering of the X anions is known. One reason for this may be that the recognition of spontaneous phase separation at fine scales is usually detected by high-resolution and analytical transmission electron microscopy (HRTEM-AEM). However, many apatites are electron beam sensitive making data collection difficult and its interpretation suspect. Nonetheless, recent atomic scale studies of both natural and synthetic apatites are beginning to reveal unexpected nanometric complexity that arises as the tunnel framework adapts to allow intergrowth of channels of different diameters that separate thermodynamically incompatible chemical compositions.

The co-existence of polyphase apatite assemblages need not be obvious. For example, centimeter-sized blue gem-grade crystals from Ipirá, Brazil (*BAp*) (Ferraris et al. 2004) with an average bulk composition of $(Ca_{3.946}Na_{0.052}Y_{0.001})_{\Sigma=4}(Ca_{5.984}Sr_{0.003}Pb_{0.001}REE^{3+}_{0.008}Th_{0.004})_{\Sigma=6}(P_{5.692}Si_{0.182}S_{0.132})_{\Sigma=6}O_{24}(F_{1.541}Cl_{0.122}OH_{0.337})_{\Sigma=2}$ were shown to possess a complex nanostructure (Table 7). Despite the faceted nature of the crystals, Rietveld analysis of powder X-ray data readily refined into a two-phase apatite model with a wt% ratio of approximately 1: 3, and domain sizes of 400 and 250 nm (Fig. 15a,b). These domains could be observed directly by bright-field TEM, and in combination with broad beam AEM analyses, it was established that the two domains are F- and Cl-rich apatites. At finer scale, both domain types were punctuated by bright areas approximately 5-10 nm in diameter that were most clearly evident within the darker chlorapatite areas and which exhibited poorly defined hexagonal facets (Fig. 15c, d). Microchemical analysis revealed partitioning of Si and S into these ellestadite-apatite nanodomains and is consistent with the substitutions $2P^{5+} \leftrightarrow Si^{4+} + S^{6+}$ and $Ca^{2+} + P^{5+} \leftrightarrow$

Na⁺ + S⁶⁺, however these smaller domains were not detected by powder X-ray diffraction. Thus, an apparent single crystal is actually a composite of three apatite phases with distinct chemical compositions.

In detail, Rietveld refinement of the larger apatite domains revealed anisotropic broadening of the diffraction lines where the reflections $hk0$ are broader than the reflections $00l$ (evidence by the separation of the K_α doublet) and, in addition, non-indexed lines at high-angles. These effects can be explained by the presence of two apatite phases with similar cell parameter c but with two different parameters a. High-resolution images of (001) domain boundaries showed coherent interfaces, however darker contrast can be attributed to strain arising from lattice mismatch to preserve the coherency of the surrounding domains.

It is postulated that the nanostructure of these single crystals arises from consecutive, or nested, spinodal decompositions as shown schematically in Figure 16. In these simplified representations, the amplitudes of the compositional modulations are assumed constant, with long wavelength separation of F rich-Ap and Cl rich-Ap arising from the bulk composition. Before, during or after this long wave event decomposition with high frequency leads to the formation of ellestadite. This combination of periodicities may indicate that apatite experienced two significant geothermometric events. The long period modulation would have developed close to the coherent spinodal decomposition temperature (T_s) resulting in the separation of X anions primarily and the formation of fluorine-rich and chlorine rich apatites. The short period modulation occurred at higher temperature (perhaps contact metamorphism), and was diffusion-dependant, leading to smaller ellestadite domains. If decomposition into the long and short period modulations occurred separately, it would be expected that compositionally distinct ellestadites would separate from the F rich-Ap and Cl rich-Ap regions. This was not observed, suggesting that simultaneous phase separation occurred. However, it has not been possible to resolve such subtle compositional variations and the question of whether spinoidal decomposition is the most appropriate description of these observations remains unproven.

Historically, apatite was so named (apatos = deception) because of its frequent misidentification as more precious types of gemstones. This reputation would appear to remain intact today as it is not immediately obvious that large, gem grade apatites should be worthy of detailed microscopic examination, or possess such intricate phase structures.

Table 7. Ipirá apatite chemical analyses for the major elements. The atoms per formula unit (apfu) are calculated based on 26 (O, OH, F, Cl). The estimated standard deviations (esd) are indicated only for oxides analyzed by EMPA on the basis of 30 point analyses.

Oxide	Weight %	esd	cation	apfu
Na_2O	0.16	0.02	Na	0.052
MgO	< D.L. (0.25)	—	Mg	0.000
Al_2O_3	< D.L. (0.20)	—	Al	0.000
SiO_2	1.12	0.13	Si	0.188
P_2O_5	39.88	0.60	P	5.686
SO_3	1.05	0.09	S	0.132
K_2O	< D.L. (0.10)	—	K	0.000
CaO	55.04	0.99	Ca	9.938
TiO_2	< D.L. (0.03)	—	Ti	0.000
SrO	0.03	—	Sr	0.003
Y_2O_3	0.01	—	Y	0.001
La_2O_3	0.03	—	La	0.002
Ce_2O_3	0.06	—	Ce	0.004
Nd_2O_3	0.02	—	Nd	0.002
ThO_2	0.10	—	Th	0.004
PbO	0.01	—	Pb	0.001
MnO	< D.L. (0.03)	—	Mn	0.000
Fe_2O_3	< D.L. (0.10)	—	Fe	0.000
F	2.89	—	F	1.517
Cl	0.43	—	Cl	0.120
H_2O	0.35	—	OH	0.363
	101.18			
O ≡ F, Cl	1.27			
Total	99.91			

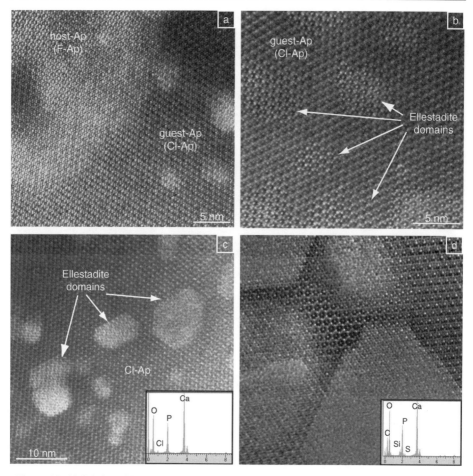

Figure 15. (a) [001] HRTEM image of the interface between *Cl-Ap* (bright contrast) and *F-Ap* (dark contrast) crystals. Within the *Cl-Ap* (b), smaller ellestadite single crystal domains with bright contrast are present. (c) [001] TEM image and (d) [001] HRTEM image showing the development of faceting in ellestadite domains as a result of electron irradiation. Inserts of AEM analyses clearly show enrichment of Si + S in the ellestadite (inset in d) compared to the surrounding *Cl-Ap* (inset in c).

MICROPOROSITY IN SYNTHETIC APATITES

The first known synthesis of an apatite, probably $Ca_{10}(PO_4)_6(OH)_2$ was reported by Daubrée (1851) who obtained it by passing phosphorus trichloride vapor over red hot lime. Since then, a large array of preparative routes have been employed in the synthesis of both phosphate and non-phosphate apatites as tabulated in Appendix B. Broadly, the techniques reported can be divided into three classes.

1. *Solid-state reaction.* High temperature sintering (> 500°C) of stoichiometric mixtures of the starting materials can be used with the phase-forming temperatures deduced from the respective phase diagrams. Depending on the elements involved and their oxidation states, the sintering atmosphere may be oxidizing, reducing or inert although most are carried out in air. Generally, a series of mixing, grinding, pelletizing and heating steps ensure formation of single-phase apatites. This is the

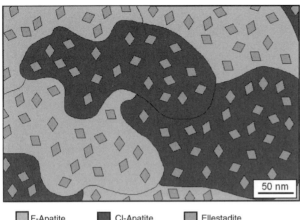
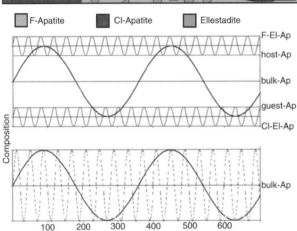

Figure 16. Schematic illustration of the three domain types in Ipirá apatite (upper part) and representation of the spinodal compositional modulations that may have lead to domain formation. The upper schematic illustrates consecutive compositional modulations in which the *ellestadite* domains arise after separation of *F-Ap* and *Cl-Ap*, while the lower portion supposes simultaneous decomposition into *F-Ap*, *Cl-Ap* and ellestadite.

most commonly used method for bulk processing of ceramic apatite powders, and in particular, for the study of phase stability. However, powders prepared in this way usually have irregular external form, large grain sizes, and often exhibit compositional heterogeneity from incomplete reaction owing to the small diffusion coefficients of ions within solids. Non-stoichiometry can also arise due to evaporation of volatile species, particularly the halides and precautions are necessary to prevent such losses.

2. *Hydrothermal reaction.* Under this reaction regime, precursor solutions (especially phosphate apatites) are treated epithermally and hydrothermally. Mass transport is superior to solid-state methods yielding compositionally homogeneous, uniform and easily sinterable powders of good crystallinity. Hydroxyapatites are frequently synthesized by this method (Yoshimura and Suda 1994).

3. *Soft chemical reaction.* This method offers a certain degree of control over the grain size and morphology of apatites. The occurrence of secondary impurity phases is reduced if homogeneous precipitation is achieved. However, powders generated in this way are usually less crystalline as compared to those derived from solid-state reaction.

The largest proportion of apatites have been synthesized by high-temperature solid state reaction. Less frequently used are hydrothermal methods and soft chemical precipitation at relatively low temperatures.

Methods developed to grow large single crystals are used rarely. Gel growth has been applied to calcium phosphate synthesis (McCauley and Roy 1974), while stoichiometric melts were used for the preparation of $Ca_2La_8(SiO_4)_6O_2$ (Ito 1968), $Cd_{10}(PO_4)_6Br_2$ and $Cd_{10}(PO_4)_6Cl_2$ (Sudarsanan et al. 1977), and $Pb_{9.85}(VO_4)_6I_{1.7}$ (Audubert et al. 1999). Melt grown materials are sometimes severely strained as they experience large temperature gradients during crystallization. To overcome this limitation, the flux growth method (Prener 1967), using CaF_2, $CaCl_2$, and $Ca(OH)_2$ mixed with the starting apatite powders, reduce the liquidus temperature and yield crystals with less strain (Yoshimura and Suda 1994). Finally, the sol-gel and polymerization methods (Brendel et al. 1992) have also been developed for processing apatites.

Until comparatively recently, relatively little attention was given to the presence of carbon dioxide in starting materials, particularly the alkali solutions used to control pH, and in distilled water. As a consequence many preparations of "pure" apatites, in particular hydroxyapatites, contain significant carbonate. In fact, it can be difficult to obtain apatites that do not contain carbon dioxide, unless very deliberate methods for exclusion are used (McConnell 1973; Wilson et al. 2003).

Cell constant discontinuities and miscibility gaps

While it may not always be feasible to determine directly the absolute stoichiometry of apatites or the partitioning coefficients of elements over the available cation-acceptor sites, it is in principle possible to follow these effects indirectly by monitoring unit cell constants. However, this may not always be straightforward. Figure 17 shows the distribution of a and c cell constants in hydroxyapatite for 25 data sets gathered by single crystal and powder methods (Table 8). For more than half the materials the ideal Ca/P ratio of 1.667 is assigned, while the remainder are slightly non stoichiometric. It is immediately apparent that while there is a substantial spread of data, well in excess of experimental error, no clear trend is evident suggesting that disequilibrium or compositional effects may be responsible. For the cluster of data near the centre of the graphic, there is obvious variation in a with near constant c, as

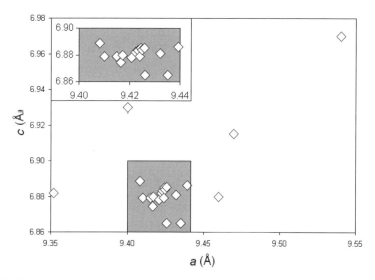

Figure 17. Distribution of lattice constants reported for hydroxyapatite (see also Table 8). The insert shows greater detail surrounding the closest grouped values. The reasons for this distribution may be related to cation order-disorder or compositional variation, especially the uptake of carbonate.

Table 8. Reported parameters for hydroxyapatite.

Reported Composition	Ca/P	a (Å)	c (Å)	c/a	Vol (Å³)	Refs.
$Ca_{10}P_6O_{26}H_2$	1.667	9.3520	6.8820	0.736	521.26	(1)
$Ca_{10}P_6O_{26}H_2$	1.667	9.410	6.879	0.731	527.52	(2)
$Ca_{10}P_6O_{26}H_2$	1.667	9.4166	6.8745	0.730	527.91	(3)
$Ca_{10.042}P_{5.952}O_{26.1}H_{2.292}$	1.687	9.4081	6.8887	0.732	528.05	(4)
$Ca_{9.74}P_6O_{26.08}H_{2.08}$	1.623	9.415	6.879	0.731	528.08	(5)
$Ca_9P_6O_{25.68}H_{1.68}$	1.500	9.426	6.865	0.728	528.23	(5)
$Ca_{10.132}P_{5.958}O_{27.09}H_{3.258}$	1.701	9.4172	6.8799	0.731	528.39	(4)
$Ca_{10}P_6O_{26}H_2$	1.667	9.4207	6.878	0.730	528.64	(6)
$Ca_{10}P_6O_{26}H_2$	1.667	9.424	6.879	0.730	529.09	(7)
$Ca_{10}P_6O_{26}H_2$	1.667	9.4223	6.8818	0.730	529.11	(2)
$Ca_{8.8}P_6O_{25.92}H_{1.92}$	1.467	9.435	6.865	0.728	529.24	(5)
$Ca_{10.084}P_{5.94}O_{27.15}H_{3.39}$	1.698	9.4232	6.8833	0.730	529.33	(4)
$Ca_{10}P_6O_{26}H_2$	1.667	9.4249	6.8838	0.730	529.56	(2)
$Ca_{10}P_6O_{26}H_2$	1.667	9.4244	6.885	0.731	529.59	(2)
$Ca_{10}P_6O_{26}H_2$	1.667	9.4257	6.8853	0.730	529.76	(2)
$Ca_{10}P_6O_{26}H_2$	1.667	9.432	6.881	0.730	530.14	(8)
$Ca_{10}P_6O_{26}H_2$	1.667	9.432	6.881	0.730	530.14	(9)
$Ca_{10}P_6O_{26}H_2$	1.667	9.40	6.93	0.737	530.30	(10)
$Ca_{9.868}P_{5.586}O_{26.35}H_{4.006}$	1.767	9.4394	6.8861	0.730	531.36	(4)
$Ca_{8.86}P_6O_{26}H_4$	1.477	9.46	6.88	0.727	533.21	(11)
$Ca_{10}P_6O_{26}H_2$	1.667	9.470	6.915	0.730	537.06	(2)
$Ca_{10}P_6O_{26}H_2$	1.667	9.54	6.970	0.731	549.36	(2)

References: (1) Tomita et al. 1996 (2) Saenger and Kuhs 1992 (3) Hughes et al. 1989 (4) Wilson et al. 1999 (5) JeanJean et al. 1996a (6) Pritzkow and Rentsch 1985 (7) Sudarsanan and Young 1969 (8) Posner and Diorio 1958 (9) Kay et al. 1964 (10) Hendricks et al. 1932 (11) JeanJean et al. 1994

would be expected if the twist angle (ϕ) was modified (smaller *a* corresponding to larger ϕ) due to replacement of phosphate by carbonate and/or nonstoichiometry of the Ca(2) site.

Correlations can be easier to recognize in solid solution series such as $Ca_{10-x}Pb_x(PO_4)_6(OH)_2$. While early studies (e.g., Narasaraju et al. 1972) appeared to show linear trends in lattice parameters between the lead and calcium end members, subsequent work by Engel et al. (1975a), Verbeeck et al. (1981) and Bigi et al. (1989, 1991) revealed more complex behavior (Fig. 18) due to cation ordering. Not only were discontinuities in lattice parameters as a function of composition observed but also distinct and large changes in lattice parameter were recorded depending on firing temperature. Moreover, both Verbeeck et al. (1981) and Bigi et al. (1989, 1991) found miscibility gaps, *albeit* at slightly different compositions.

Nanodomain intergrowths at disequibrium

While a great range of apatites have been synthesized, the very flexibility which gives rise to such an array of compounds, translates to a need for particular rigor during crystal chemical data collection and interpretation. The recent synthesis of the fluoro-vanadinites $(Pb_xCa_{10-x})(VO_4)_6F_{2\delta}$, $0 < x < 9$, which compared and contrasted the crystal structure and local atomic order in equilibrated and non-equilibrated material, is a case in point (Dong and White,

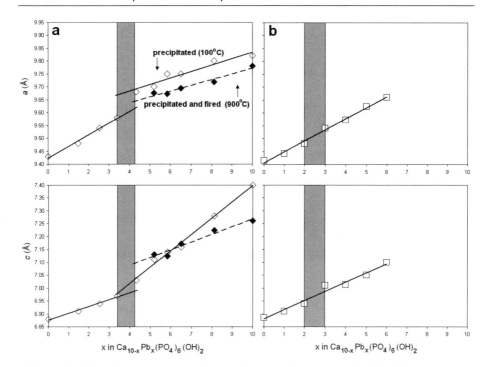

Figure 18. Unit cell parameters as a function of composition in $Ca_{10-x}Pb_x(PO_4)_6(OH)_2$ apatites as determined by (a) Bigi et al. (1991) and (b) Verbeeck et al. (1981). Differences in the trend lines can be attributed to disequilibrium. Both authors reported a miscibility gap (shaded in the drawing) at approximately $2 < x < 4$.

2004a,b). In these experiments, stoichiometric mixtures were annealed 10 h, 1 week, 2 weeks and 7 weeks at 800°C with intermediate grinding. At each stage, the material was assessed by XRD and TEM, and in all cases, single-phase apatite was produced. It was observed that as x increased the unit cell constants dilated as lead (effective ionic radii = 1.29 Å for VIII coordination) replaced calcium (1.12 Å). However, trends in cell constants for 10 h and 4 weeks materials as a function of composition were quite distinct (Fig. 19). Neither equilibrated nor non-equilibrated apatites obeyed Vegard's Law across the entire compositional range from $0 < x < 10$. Rather, the cell constants were best described as two linear segments. This can be understood by considering that vanadinite contains four A(1) sites (the preferred location of Ca) and six A(2) sites (favored by Pb). The change in the slope reflects the changes in the lead partitioning coefficient k_{Pb}.*

The approach to equilibrium is surprisingly slow. For "$Pb_5Ca_5(VO_4)_6F_2$" the c/a lattice constant ratio decreases exponentially as annealing continues such that equilibrium is approached only after 30 days annealing (Fig. 20). Concomitant with this adjustment is the enrichment of lead in the A(2) position and a decrease in the $A(2)O_6$ metaprism twist angle (φ) (Table 9, Fig. 21). Fully equilibrated vanadinite that was heat treated for 7 weeks had a substantially smaller $k_{Pb}[A(1)/A(2)]$ of 0.17 as compared to 0.52 for 10 h material, and φ was reduced to 14.4° from 22.0°.

---

* $k_{Pb}[A(1)/A(2)] = [2-2N(Ca1)]/[3-3N(Ca2)]$ where $N(Ca)$ is the fractional occupancy of the site. For equal partitioning of Pb over A^I and A^{II} k_{Pb} would be 0.66 for all compositions.

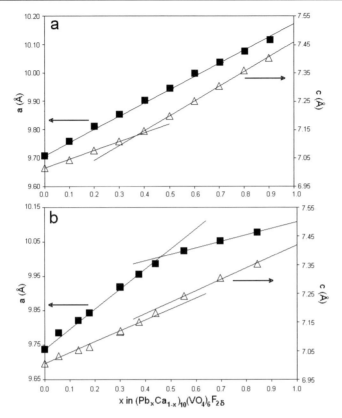

Figure 19. Variation in cell constants for vanadinites of composition $(Pb_xCa_{1-x})_{10}(VO_4)_6F_{2\delta}$ when in disequilibrium after 10 h annealing (a) and in equilibrium after 4 weeks annealing (b).

At the unit cell scale, the structure of the nonequilibrated apatite is complex as shown Figure 22. Nanodomains can be ascribed to local changes in composition, and in particular, to variation in the Ca/Pb ratio. This interpretation is supported by the fact that similar images collected from equilibrated vanadinite failed to display the same complex nanostructure. Within each nanodomain it would be expected that the twist angle will be unique, and the φ derived from Rietveld analysis reflects an average of sorts of these angles. Nonetheless, a combination of powder X-ray diffraction and high resolution transmission electron microscopy have demonstrated that long annealing times of several weeks are necessary to completely equilibrate calcium-lead partitioning in $(Pb_xCa_{10-x})(VO_4)_6F_{2\delta}$ with $0 < x < 9$ vanadinite apatite.

Influence of pressure

As apatites behave as flexible one-dimensional tunnel structures, it follows that the application of pressure will lead to their compression through an increase in φ. This has been observed by Comodi et al. (2001) who used single crystal X-ray diffraction to determine the structure of synthetic $Ca_{10}(PO_4)_6F_2$ up to pressures of 6.89 GPa. Their data, which includes a large set of unit cell constants and complete structure analyses at atmospheric pressure, 3.04 GPa and 4.72 GPa, showed that the least compressible units were the PO_4 tetrahedra and CaO_6 metaprisms, while the larger CaO_6F polyhedra (the tunnel contents) were most compressible, with bulk moduli of 270, 100 and 84 Gpa, respectively (Fig. 23). During compression to 4.72 GPa the twist angle increased 4.2% from 20.8° to 21.7°, while a and c decreased 1.6% and

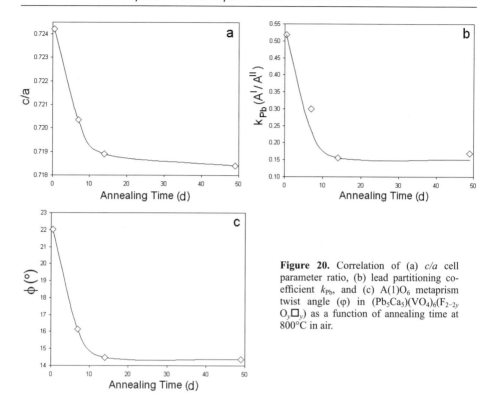

Figure 20. Correlation of (a) c/a cell parameter ratio, (b) lead partitioning coefficient k_{Pb}, and (c) $A(1)O_6$ metaprism twist angle (φ) in $(Pb_5Ca_5)(VO_4)_6(F_{2-2y}O_y\square_y)$ as a function of annealing time at 800°C in air.

1.2%, respectively (Fig. 23). Unlike zeolites, which amorphize under pressure, fluorapatite is stable possibly because structural flexibility is essentially restricted to two-dimensions. It might also be expected that apatites in which the tunnels are more fully extended and therefore have smaller twist angles would be most readily compressed. Brunet et al. (1999) found this trend for hydroxyapatite (ϕ = 23.2°), fluorapatite (23.3°) and chlorapatite (19.1°) with bulk moduli of 98(2), 98(2) and 93(4) GPa respectively. If this phenomena is generally correct it would be predicted that $Ca_{10}(CrO_4)_6(OH)_2$ (17.8°) and $Ca_{10}(AsO_4)_6Cl_2$ (13.0°) should become progressively softer.

ION EXCHANGE PROPERTIES

There are several possible mechanisms for apatite replacement including dissolution-reprecipitation processes, leaching and ion exchange, and a substantial literature has accumulated especially in relation to the treatment of heavy metal wastes (e.g., Reichert and Binner 1996; Sugiyama et al. 2003; Lower et al. 1998; Manecki et al. 2000). However, this discussion is restricted for the most part to pure lattice exchange reactions.

Cadmium. Extensive investigation of cadmium uptake by hydroxyapatites (Jeanjean et al. 1994, 1996a,b; Mandjiny et al. 1998; Fedoroff et al. 1999; McGrellis et al. 2001) has shown unequivocally that cadmium is incorporated in the structure by diffusion and substitution for calcium ions. However, to accelerate the reaction sodium-doped (~ 1 atomic %) apatites were used as the starting materials with probable replacements

$$Ca^{2+} + OH^- \rightarrow Na^+ + H_2O \text{ or } Ca^{2+} + (PO_4)^{3-} \rightarrow Na^+ + (HPO_4)^{2-}.$$

Table 9. Refined crystal and atomic parameters for vanadinites annealed at 800°C for intervals from 10 hours to 7 weeks.

Time/Parameter	10 hours	1 week	2 weeks	7 weeks
Composition	$(Pb_{4.62}Ca_{5.38})(VO_4)_6F_2$	$(Pb_{4.90}Ca_{2.10})(VO_4)_6(F_{0.6}O_{0.7})$	$(Pb_{4.78}Ca_{5.22})(VO_4)_6(F_{0.6}O_{0.7})$	$(Pb_{4.76}Ca_{5.64})(VO_4)_6(F_{0.6}O_{0.7})$
Crystal Formula	$[Pb_{1.56}Ca_{2.44}][Pb_{2.06}Ca_{2.94}]$ $(VO_4)_6F_2$	$[Pb_{1.12}Ca_{2.88}][Pb_{3.78}Ca_{2.22}]$ $(VO_4)_6(F_{0.3}O_{0.7})$	$[Pb_{0.64}Ca_{3.36}][Pb_{4.14}Ca_{1.96}]$ $(VO_4)_6(F_{0.25}O_{0.75})$	$[Pb_{0.68}Ca_{3.32}][Pb_{4.08}Ca_{1.92}]$ $(VO_4)_6(F_{0.3}O_{0.7})$
a (Å)	9.9462(1)	10.0011(1)	10.0025(2)	10.0124(2)
c (Å)	7.1983(1)	7.1993(1)	7.1858(1)	7.1880(1)
Volume (Å^3)	616.7	623.6	622.6	624.0
c/a	0.7237	0.7199	0.7184	0.7179
N(Ca1)	0.61	0.72	0.84	0.83
N(Ca2)	0.49	0.37	0.31	0.32
$k_{Pb}(A^I/A^{II})^*$	0.52	0.30	0.16	0.17
φ (°)	22.0	16.1	14.5	14.4
z(A1)	0.0095(8)	0.0083(8)	0.0062(11)	0.0059(12)
B(A1)	0.6	1.2	0.4	1.4
x(A2)	0.2427(3)	0.2442(2)	0.2418(2)	0.2426(2)
y(A2)	0.0023(4)	0.0028(3)	0.0026(3)	0.0036(3)
B(A2)	0.6	1.1	1.3	1.7
x(V)	0.4020(7)	0.4066(5)	0.4090(6)	0.4093(6)
y(V)	0.3758(6)	0.3770(5)	0.3813(6)	0.3805(6)
B(V)	0.3	0.2	0.2	0.6
x(O1)	0.3195(18)	0.3363(16)	0.3407(17)	0.3393(18)
y(O1)	0.4915(18)	0.5014(16)	0.5073(16)	0.5057(17)
x(O2)	0.6012(20)	0.6046(19)	0.6056(20)	0.6074(20)
y(O2)	0.4755(18)	0.4782(16)	0.4787(17)	0.4807(17)
x(O3)	0.3353(13)	0.3436(10)	0.3545(10)	0.3499(11)
y(O3)	0.2546(13)	0.2594(11)	0.2635(11)	0.2639(12)
z(O3)	0.0628(13)	0.0602(13)	0.0605(13)	0.0583(14)
N(F)	1.0	0.6	0.5	0.6
R_b (%)	4.2	4.4	5.0	4.9

*$k_{Pb}[A(1)/A(2)] = (2-2N(Ca1))/[3-3N(Ca2)]$. For equal partitioning of Pb over A(1) and A(2) $k_{Pb} = 0.66$

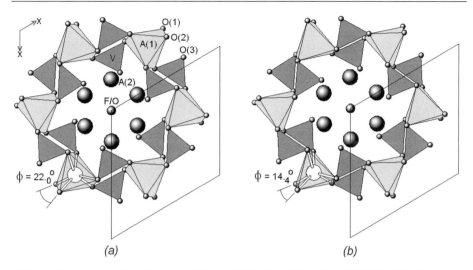

Figure 21. Structures of the fluoro-vanadinites (a) $[Pb_{1.56}Ca_{2.44}][Pb_{2.06}Ca_{2.94}](VO_4)_6F_2$ with refined $x = 4.62$ and (b) $[Pb_{0.68}Ca_{3.32}][Pb_{4.08}Ca_{1.92}](VO_4)_6(F_{0.3}O_{0.7})$ with $x = 4.76$ synthesized from stock powders but annealed respectively for 10 h and 7 weeks at 800°C (see Table 9). The $A(1)O_6$ metaprisms and VO_4 tetrahedra are emphasized in [001] projection. As A(2) becomes progressively lead-rich the channel opens through A-O bond dilation and a reduction of metaprism twist angle (φ). [Reproduced with permission of the International Union of Crystallography (http://journals.iucr.org/) from Dong and White (2004a) *Acta Crystallographica*, Vol. B60, Fig. 4, p. 140.]

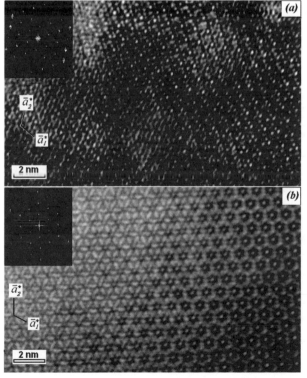

Figure 22. [001] high resolution electron micrographs collected from $(Pb_5Ca_5)(VO_4)_6(F_{1-2y}O_y)_2$ thin crystal wedges annealed for (a) 10 h and (b) 7 weeks. Inserts of the power transforms confirm the orientation. The upper image shows poorly equilibrated microdomains 2 nm in diameter arising from local variation in Ca/Pb content. Conversely, the equilibrated sample shows regular contrast changes in passing from the thin edge (left) to thicker crystal (right). [Used with permission of the International Union of Crystallography (http://journals.iucr.org/), from Dong and White (2004a), *Acta Crystallographica*, Vol. B60, Fig. 6, p. 142.]

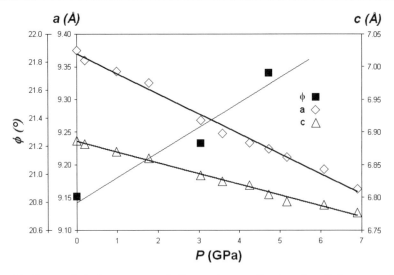

Figure 23. Changes in lattice parameters in fluorapatite as a function of applied pressure. During compression, the tunnel collapses slightly through an increase in metaprism twist angle φ. (Data taken from Comodi et al. 2001).

Crystal structure refinement before exchange were consistent with near full occupancy of the A(1) sites and partial occupancy of the A(2) sites. Ion exchange was monitored over the temperature range 28–75°C using a solution with [Cd] = 3.96 × 10^{-3} mol/L (2 mol of Cd per mole of apatite) and contact times of 1 min to 12 h. For these conditions, the quantity of calcium released varied directly with the quantity of cadmium fixation, however the release of sodium was negligible. At lower concentrations, cadmium partitioned entirely to the A(2) tunnel sites, while for larger quantities (~ 3.9 atomic %) both the A(1) and A(2) sites were occupied. As the external morphology of the apatite crystals was maintained it is reasonable to suspect lattice diffusion occurs, but the precise mechanism by which cadmium enters the metaprisms is not yet understood. As exchange is only partially reversible with ~20% recovery of cadmium it seems possible that it is quite difficult to remove exchanged ions from the A(1) positions.

Selenium. The interaction of selenite (SeO$_3$)$^{2-}$ and biselenite (HSeO$_3$)$^-$ with sodium-bearing hydroxyapatite was studied by Monteil-Rivera et al. (1999). Using solutions with initial selenium concentrations from 10^{-3}–10^{-6} mol/L and ambient reaction temperature, it was shown that the total mole release of phosphorous equaled the uptake of selenium. As neither diffraction nor microscopy indicated the presence of second phase, it was argued that the dominant mechanism for sorption was exchange. Because no oxidation of Se(IV) took place in the exchange solutions possible mechanisms for exchange include

$$2(PO_4)^{3-} + Ca^{2+} \rightarrow 2(SeO_3)^{2-} + \square \quad \text{or} \quad (PO_4)^{3-} + Ca^{2+} \rightarrow (HSeO_3)^- + \square$$

Although the incorporation of BO$_3$ radicals is less usual in apatites, it is not unknown. The mineral finnemanite, Pb$_{10}$(AsO$_3$)$_6$Cl$_2$, shows complete substitution of (As^{V}O$_4$)$^{3-}$ by (As^{III}O$_3$)$^{3-}$ and mimetite, Pb$_{10}$(AsO$_4$)$_6$Cl$_2$, can be produced from arsenite form by direct oxidation as shown in Figure 24 (Effenberger and Pertlik 1977). Therefore, introduction of SeO$_3$ into the tetrahedral phosphate positions may be possible. Structural studies to confirm this point would be invaluable.

Lead. Pyromorphite, Pb$_{10}$(PO$_4$)$_6$(OH,Cl)$_2$, is frequently cited as a useful phase for the fixation of lead as it is sparingly soluble over a wide pH range. Crystallization occurs either

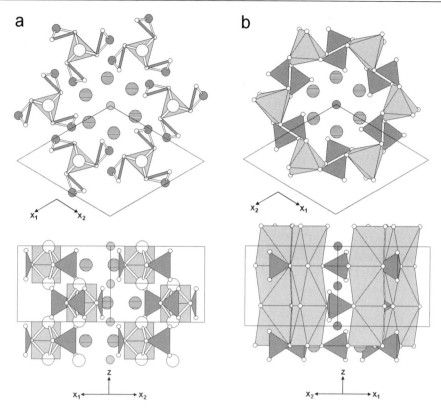

Figure 24. Structure drawings of (a) finnemanite $Pb_{10}(As_{III}O_3)_6Cl_2$ composed of PbO_3 half prisms and $As_{III}O_3$ triangular co-ordination, and its oxidized form (b) mimetite $Pb_{10}(As_VO_4)_6Cl_2$ constructed from the more usual PbO_6 and As_VO_4 polyhedra. The stability of the $As_{III}O_3$ unit suggests that reports of ion exchange of $Se_{IV}O_3$ for PO_4 may be crystallographically feasible.

by replacement of hydroxyapatite or through amendment with various phosphates (Xu and Schwartz 1994). But, there appears to be little evidence for cation exchange (Arnich et al. 2003). Miyake et al. (1986) reportedly conducted Rietveld analysis of lead-exchanged hydroxy-, fluor- and chlorapatites. However, their data do not show preferential entry of lead into the A(2) site suggestive of disequilibrated material, as might be produced by rapid reprecipitation, and the cell parameters are consistent with near pure lead endmembers $Pb_{10}(PO_4)_6(F,OH,Cl)_2$.

While ion exchange in phosphate apatites appears to be established for cadmium and selenium, it is clearly a competitive process with precipitation and in those cases where insoluble metal phosphates form, then lattice exchange will be overwhelmed. It is also apparent that exchange reactions are kinetically slow and as noted by Fedoroff et al. (1999) may take several weeks or longer to equilibrate.

APATITE TECHNOLOGIES

Chemical fertilizers were introduced after a method for production was devised by Lawes in 1843, and by the 1850s, more than a dozen superphosphate works were operating in Britain and Germany (Brock 1993). In 1900, world production was over 4.5 million metric tons per year (Mtpy) with large quantities of sulfuric acid being used in processing. Today three main types of phosphate fertilizers are used—normal super phosphate (20% of the total), ammonium phosphate (65%), and triple super phosphate (10%)—with total consumption amounting to 130 Mtpy (International Fertilizer Industry Association). This traditional use of phosphate rock, including apatite *sensu stricto*, as a raw material for phosphoric acid extraction remains the largest application in terms of volume and investment. However, over the past two decades a number of advanced ceramic technologies based upon apatites have emerged.

Remediation of radioactive wastes

Ceramic options for the disposal of nuclear wastes have been under development in various forms for more than 50 years (Lutze and Ewing 1988) with the first experiments detailing the possible role of apatite presented by Roy and his coworkers (1982). Apatites are appropriate immobilization matrices for three reasons. First, they can accommodate a wide range of fission products, actinides and processing contaminants, often to high concentration. Second, the radiation resistance of apatites in nature is well known (Utsunomiya et al. 2003), and extensive laboratory investigations suggest that durability will not be compromised even under high radiation fluxes as might be experienced shortly after disposal (Ewing et al. 2000; Ewing and Wang 2002). Finally, the synthesis of apatites is relatively straightforward in shielded environments. For certain fission products such as ^{129}I, apatite is one of a limited number of less soluble structures that can condition such waste for storage (Audubert et al. 1997). Most recently, silicate apatites have been studied for the incorporation of plutonium and uranium (Vance et al. 2003).

The intrinsic potential of the apatite structure type for radioactive waste disposal arises directly from crystal chemical adaptation of its framework structure (Dong et al. 2002; Kim et al. 2004), its capacity to be nonstoichiometric, and the ability to increase the number of cation acceptor sites through reduction in symmetry. These properties allow apatites to respond to variations in waste stream composition without intervention to adjust the constitution of stabilization additives.

Catalysis

Several apatites are under investigation as heterogeneous catalysts (Matsumura et al. 1997). For example, hydroxyapatite showed high catalytic activity and selectivity in a Michael addition reaction (Zahouily et al. 2003) involving thiophenol and various chalcones. Interestingly, it is believed that in this instance the catalytic activity of hydroxyapatite does not originate from inside the tunnels, as is the case for some zeolites, but rather at the surface whose acidic character enhanced thiol nucleophilicity. If such catalytic reactions proved viable, they would be environmentally benign compared to basic homogeneous catalysts.

Hydroxyapatite also shows potential in photocatalysis, both as a catalyst in its own right, and as a carrier material for semiconductors such as titania (Nishikawa and Omamiuda 2002). In this case, the highly sorptive capacity of apatite is coupled with its catalytic activity to simultaneously destroy pollutants and capture them for later disposal.

Fuel cell electrolytes

Alumino-silicate (Kahlenberg and Kruger 2004) and germano-silicate (Sansom and Slater 2004) apatites may play a role as electrolytes in solid oxide fuel cells (SOFCs) that will be important components of future clean energy systems. The aim is to achieve high ionic conductivity (σ) at the lowest temperatures possible. Thus, for example, while yttria stablised zirconia (YSZ) is presently the *de facto* standard oxide with $\sigma = 1 \times 10^{-3}$ S cm^{-1} at 500°C, La$_{10}$(SiO$_4$)$_6$O$_3$ is more than four times as conductive with $\sigma = 4.3 \times 10^{-3}$ S cm^{-1} at the same temperature (Nakayama et al. 1995). Reducing SOFC operating temperatures (currently between 950 and 1050°C) could increase their service life by reducing damaging reactions at interfaces and make them much less expensive due to replacement of ceramic interconnectors by less expensive metals.

A recent neutron diffraction investigation by León-Reina et al. (2004) of $La_{10-x}(SiO_4)_6O_{3-1.5x}$ ($9.33 \leq 10-x \leq 9.73$) conducted at room temperature, 500°C and 900°C showed the presence of interstitial oxygen [the "O(5) site"] in the tunnels located in positions almost directly above or below the La(2) cations (Fig. 25a). Using computer modeling techniques Tolchard et al. (2003) demonstrated that local relaxation of the SiO_4 tetrahedra permit O(5)-vacancy conduction through the apatite tunnels by a sinusoidal pathway. It is perhaps relevant that $A_{10}(BO_5)_6X_2$ apatites such as $Sr_{10}(ReO_5)_6Cl_2$ are fully stoichiometric in terms of O(5) resulting in the expansion of the BO_4 tetrahedra to BO_5 square pyramids (Fig. 25b). The synthesis of mixed silico-rhenate apatites could therefore be a useful method for controlling O(5)-vacancy combinations and to optimize ionic conductivity.

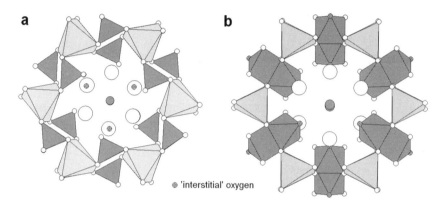

Figure 25. Location of "interstitial" oxygen in (a) $La_{9.55}(SiO_4)_6O_{2.32}$ and (b) $Sr_{10}(ReO_5)_6Cl_2$.

Permeable reaction barriers

The concept of a permeable reaction barrier (PRB) involves the installation of trenches around contaminated sites that will be filled with material of specific chemical reactivity and high permeability to water, to effectively prevent the migration of pollutants to the wider environment. Several designs have been implemented including zeolites, zero valent iron and apatites. In the latter case, a technology known as Apatite II® is best developed and is being trialed at various sites (Wright et al. 2004). This phosphate apatite has been optimized chemically and microstructurally to adsorb and fix a wide range of toxic and radioactive species and has the nominal composition $Ca_{10-x}Na_x(PO_4)_{6-x}(CO_3)_x(OH)_2$. The principle features of this material that make it especially suitable for remediation include low fluorine content, higher substitutions of carbonate and sodium, high purity with few trace elements, X-ray amorphosity and high porosity.

For metal fixation, dissolution-precipitation reactions are the dominant immobilization mechanism rather than lattice ion exchange. In the case of lead (and also uranium, cerium and plutonium) the reaction sequence is

$$Ca_{10-x}Na_x(PO_4)_{6-y}(CO_3)_y(OH)_2 + 14H^+ \rightarrow$$
$$(10-x)Ca^{2+} + xNa^+ + (6-y)[H_2(PO_4)]^- + yH_2CO_3 + 2H_2O$$

$$10Pb^{2+} + 6[H_2(PO_4)]^- + 2OH^- \rightarrow Pb_{10}(PO_4)_6(OH)_2 + 12H^+$$

However, for zinc, cadmium and other transition metals non-apatite phases are precipitated.

CONCLUSION

The crystal chemistry of apatites continues to be a highly active field of study more than 70 years after Hausen (1929) recognized eight apatite types including fluorapatite, alkaliapatite, chlorapatite, carbonate apatite, sulfate apatite, hydroxyapatite, manganese apatite and rare earth bearing apatites. The ongoing pursuit for understanding these materials is driven both by a fundamental interest in exploring the range of natural and synthetic compositions, and technology-focused initiatives to develop new materials with specific properties. This article has attempted to draw together several lines of contemporary thought concerning the underlying principles that lead to the crystallization of such a diverse structural family. However, the topic is vast, and inevitably, important areas have been excluded or dealt with in a cursory fashion. Aside from the voluminous literature concerning biological apatites (Weiner and Wagner 1998; Elliott 2002), other interesting apatite variants are known. For instance, Carrillo-Cabrera and von Schnering (1999) were the first to insert metal-oxygen strings along the apatite channels when they synthesized $Sr_{10}(VO_4)_6(Cu_{0.896}O_{0.95})_2$. More recent work has lead to a number of phosphate analogues including $Ca_{10}(PO_4)_6(Cu_{0.27}O_{0.86}H_y)_2$, $Sr_{10}(PO_4)_6(Cu_{0.33}O_{0.66})_2$ and $Ba_{10}(PO_4)_6(Cu_{0.30}O_{0.86}H_y)_2$ that display brilliant blue colors and may find application as paint pigment (Karpov et al. 2003; Kazin et al. 2003).

Polysomatism is another structural modification that has to date received little attention. While $[A(1)_4][A(2)_6](BO_4)_6X_2$ apatites regularly fill every other BO_4 tetrahedral interstice occurring as corner-connected strings along [001] alternate filling schemes are possible. Accordingly, nasonite with the ideal formula $Pb_6Ca_4(Si_2O_7)_3Cl_2$ contains alternating pairs of filled and empty tetrahedra (Engel 1972; Giuseppetti et al. 1971). It is then possible to construct an apatite-nasonite polysomatic series in which ganomalite is an intermediate member (White and Dong 2003) having the ideal formula $Pb_6Ca_{3.33}Mn_{0.67}(SiO_4)_2(Si_2O_7)_2$ in which Si_2O_7 units alternate with SiO_4 tetrahedra (Carlson et al. 1997). In addition to these minerals, the synthetic analogue of ganomalite $Pb_{10}(GeO_4)_2(Ge_2O_7)_2$ has been studied extensively due to its ferroelectric properties (Newnham et al. 1973; Iwata 1977) and Stemmermann (1992) synthesized the longer period $Pb_{40}(Si_2O_7)_6(Si_4O_{13})_3$. All the members so far reported for the apatite-nasonite family are lead-rich with the role of lone-pair stabilization yet to be fully investigated.

As there are numerous possibilities for structural adaptation and their utility in contemporary technologies is growing, it is anticipated that apatites will continue to surprise in terms of their chemical and physical complexity for some time to come.

ACKNOWLEDGMENTS

The authors are most appreciative of the efforts of Professor Donald B. McIntyre for recovering copies of the plates reproduced in Figure 1, and for providing the historical context of his work with C.A. Beevers and the influence of V.M. Goldschmidt. Reproduction of the color plates was generously funded by the Carnegie Trust for the Universities of Scotland. This work was supported through Singapore Agency for Science, Technology and Research (A*STAR) Grants 012 105 0123, 033 141 01 and M47030104. The authors are grateful to Dr. Dong ZhiLi for access to unpublished data and for the calculation of the twist angles reported throughout the text.

REFERENCES

Akao M, Aoki H, Innami Y, Minamikata S, Yamada T (1989) Flux growth and crystal structure of pyromorphite. Iyo Kiazai Kenkyusho Hokoku, Tokyo Ika Shika Daigu 23:25-29

Akhavan-Niaki AN (1961) Contribution a l'étude des substitutions dans les apatites. Ann Chim (Paris) 6: 51-79
Alberius-Henning P, Landa-Cánovas, Larsson A-K, Lidin S (1999a) Elucidation of the crystal of oxyapatite by high-resolution electron microscopy. Acta Crystallogr B55:170-176
Alberius-Henning P, Lidin S, Petrícek V (1999b) Iodo-oxyapatite, the first example from a new class of modulated apatites. Acta Crystallogr B55:165-169
Alberius-Henning PA, Moustiakimov M, Lidin S (2000) Incommensurately modulated cadmium apatites. J Solid State Chem 150:154-158
Andersson S (1978) An alternative description of the structures of Rh_7Mg_{44} and Mg_6Pd. Acta Crystallogr A34: 833-835
Andres-Verges M, Highes-Rolando FJ, Valenzuela-Calahorro C, Gonzalez-Diaz PF (1983) On the structure of calcium-lead phosphate apatites. Spectrochim Acta A39:1077-1082
Arnich N, Lanhers MC, Laurensot F, Podor R, Montiel A, Burnel D (2003) In vitro and in vivo studies of lead immobilization by synthetic hydroxyapatite. Env Poll 124:139-149
Audubert F, Carpena J, Lacout JL, Tetard F (1997) Elaboration of an iodine-bearing apatite – iodine diffusion into a $Pb_3(VO_4)_2$ matrix. Solid State Ionics 95:113-119
Audubert F, Savariault JM, Lacout JL (1999) Pentaleadtris (vanadate) iodide, a defect vanadinite-type compound. Acta Crystallogr C55:271-273
Badraoui B, Bigi A, Debbabi M, Gazzano M, Roveri N, Thouvenot R (2002a) X-ray powder diffraction and solid-state NMR investigations in cadmium-lead hydroxyapatites. Eur J Inorg Chem 6:1261-1267
Badraoui B, Bigi A, Debbabi M, Gazzano M, Roveri N, Thouvenot R (2002b) Physicochemical properties and structural refinement of strontium-lead hydroxyapatites. Eur J Inorg Chem 7:1864-1870
Badrour L, Sadel A, Zahir M, Kimakh L, El Hajbi (1998) Synthesis and physical and chemical characterization of $Ca_{10-x}Ag_x(PO_4)_6(OH)_{2-x}\Box_x$ apatites. Ann Chim Sci Mat 23:61-64
Banks E, Jaunarajs KL (1965) Chromium analogs of apatite and spodiosite. Inorg Chem 4:78-83
Barinova AV, Bonin M, Pushcharovskii DYu, Rastsvetaeva RK, Schenk K, Dimitrova OV (1998) Crystal structure of synthetic hydroxylpyromorphyte $Pb_5(PO_4)_3(OH)$. Crystallogr Rep 43:189
Baud G, Besse JP, Capestan M, Sueur G, Chevalier R (1980) Etude comparative d'apatites contenant l'ion $(ReO_5)^{3-}$. Structure des fluoro et des carbonatoapatites. Ann Chim Sci Mat 5:575-583
Baud G, Besse JP, Sueur G, Chevalier R (1979) Structure de nouvelles apatites au rhenium contenant des anions volumineux: $Ba_{10}(ReO_5)_6X_2$ (X = Br, I). Mater Res Bull 14:675-682
Bauer M, Klee WE (1993) Induced ferrielectricity in chlorapatite. Z Kristallogr 206:15-24
Beevers CA, McIntyre DB (1946) The atomic structure of fluor-apatite and its relation to that of tooth and bone material. Mineral Mag 27:254-257
Belokoneva EL, Troneva EA, Dem'yanets LN, Duderov NG, Belov NV (1982) Crystal structure of synthetic fluoropyromorphite $Pb_5(PO_4)_3F$. Sov Phys Crystallogr 27:476-477
Benmoussa H, Mikou M, Bensaoud A, Bouhaouss A, Morineaux R (2000) Electrical properties of lanthanum containing vanadocalcic oxyapatite. Mater Res Bull 35:369-375
Berastegui P, Hull S, Garcia Garcia FJ, Grins J (2002) A structural investigation of $La_2(GeO_4)O$ and alkaline-earth-doped $La_{9.33}(GeO_4)_6O_2$. J Solid State Chem 168:294-305
Bertoni E, Bigi A, Cojazzi G, Gandolfi M, Panzavolta S, Roveri N (1998) Nanocrystals of magnesium and fluoride substituted hydroxyapatite. J Inorg Biochem 72:29-35
Besse JP, Baud G, Chevalier R, Zarembowitch J (1980) Mise en évidence de l'ion O^{2-} dans l'apatite au rhenium $Ba_5(ReO_5)_3O_2$. Mater Res Bull 15:1255-1261
Besse JP, Baud G, Levasseur G, Chevalier R (1979) Structure crystalline de $Ba_5(ReO_5)_3Cl$: une nouvelle apatite contenant l'ion $(ReO_5)^{3-}$. Acta Crystallogr B35:1136-1139
Bigi A, Falini G, Foresti E, Gazzano M, Ripamonti A, Roveri N (1996) Rietveld structure refinements of calcium hydroxylapatite containing magnesium. Acta Crystallogr B52:87-92
Bigi A, Gandolfi M, Gazzano M, Ripamonti A, Roveri N, Thomas S (1991) Structural modifications of hydroxyapatite induced by lead substitution for calcium. J Chem Soc Dalton Trans 2883-2886
Bigi A, Ripamonti A, Brückner S, Gazzano M, Roveri N, Thomas SA (1989) Structure refinements of lead-substituted calcium hydroxyapatite by X-ray powder fitting. Acta Crystallogr B45:247-251
Birkett TC, Simandl GJ (1999) Carbonatite-associated deposits: magmatic, replacement and residual. *In:* Selected British Columbia Mineral Deposit Profiles. Vol 3. Simandl GJ, Hora ZD, Lefebure DV (eds) British Columbia Ministry of Energy and Mines
Boechat CB, Eon J-C, Rossi AM, de Castro Perez CA, da Silva San Gil RA (2000) Structure of vanadate in calcium phosphate and vanadate apatite solid solutions. Phys Chem Chem Phys 2:4225-4230
Bondareva OS, Malinovskii YuA (1986) Crystal structure of synthetic Ba hydroxylapatite. Sov Phys Crystallogr 31:136-138
Borodin LS, Kazakova ME (1954) Belovite-(Ce): a new mineral from an alkaline pegmatite. Dokl Akad Nauk SSSR 96:613-616 (in Russian)

Boudreau AE (1995) Formation of chlor- and fluor-apatite in layered intrusions: a comment. Mineral Mag 59: 757-760
Boyer L, Carpena J, Lacout JL (1997) Synthesis of phosphate-silicate apatites at atmospheric pressure. Solid State Ionics 95:121-129
Boyer L, Savariault JM, Carpena J, Lacout JL (1998) A neodymium-substituted britholite compound. Acta Crystallogr C54:1057-1059
Brégiroux D, Audubert F, Champion E, Bernache-Assollant D (2003) Mechanical and thermal properties of hot pressed neodymium-substituted britholite $Ca_9Nd(PO_4)_5(SiO_4)F_2$. Mater Lett 57:3526-3531
Brendel T, Engel A, Russel C (1992) Hydroxyapatite coating by polymeric route. J Mater Sci Mater Med 3: 175-179
Brock WH (1993) The Norton history of chemistry. WW Norton & Company, New York (p 285)
Brown PW, Constantz B (1994) Hydroxyapatite and related materials. CRC Press, Boca Raton
Brückner S, Lusvardi G, Menabue L, Saladini M (1995) Crystal structure of lead hydroxyapatite from powder X-ray diffraction data. Inorg Chim Acta 236:209-212
Brunet F, Allan DR, Redfern SA, Angel RJ, Miletich R, Reichmann HJ, Sergent J, Hanfland M (1999) Compressibility and thermal expansivity of synthetic apatites $Ca_5(PO_4)_3X$ with X = OH, F and Cl. Eur J Mineral 11:1023-1035
Burnett WC, Riggs SR (1990) Phosphate deposits of the world, Neogene to modern phosphorites. Vol 3 Burnett, WC Riggs SR (eds). Cabridge University Press, New York
Bushinskiy GI (1964) On shallow water origin of phosphorite sediments, deltaic and shallow marine deposits. Develop Sediment 1:62-70
Buvaneswari G, Varadaraju UV (2000) Synthesis and characterization of new apatite-related phosphates. J Solid State Chem 149:133-136
Calos NJ, Kennard CHL (1990) Crystal structure of mimetite, $Pb_5(AsO_4)_3Cl$. Z Kristallogr 191:125-129
Calvo C, Faggiani R, Krishnamachari N (1975) Crystal structure of $Sr_{9.402}Na_{0.209}(PO_4)_6B_{0.996}O_2$ - a deviant apatite. Acta Crystallogr B31:188-192
Carlson S, Norrestam R, Holstam D, Spengler R (1997) The crystal structure of ganomalite, $Pb_9Ca_{5.44}Mn_{0.56}Si_9O_{33}$. Z Kristallogr 212:208-212
Carrillo-Cabrera W, Von Schnering HG (1999) Pentastrontium tris[tetraoxovanadate(V)] catena-monoxocuprate(I), $Sr_5(VO_4)_3(CuO)$ - an apatite derivative with inserted linear $[CuO]^{1-}$ chains. Z Anorg Allg Chem 625:183-185
Cockbain AG, Smith GV (1967) Alkaline-earth-rare-earth silicate and germanate apatites. Mineral Mag 36: 411-421
Collin RL (1959) Strontium-calcium hydroxyapatite solid solutions: preparation and lattice constant measurements. J Am Chem Soc 81:5275-5278
Comodi P, Liu Y, Zanazzi PF, Montagnoli M (2001) Structural and vibrational behaviour of fluorapatite with pressure. Part I: in situ single-crystal X-ray diffraction investigation. Phys Chem Mineral 28:219-224
Corker DL, Chai BHT, Nicholls J, Loutts GB (1995) Neodymium-doped $Sr_5(PO_4)_3F$ and $Sr_5(VO_4)_3F$. Acta Crystallogr C51:549-551
Dai YS, Hughes JM (1989) Crystal structure refinements of vanadinite and pyromorphite. Can Mineral 27: 189-192
Dai YS, Hughes JM, Moore PB (1991) The crystal structures of mimetite and clinomimetite, $Pb_5(AsO_4)_3Cl$. Can Mineral 29:369-376
Dardenne K, Vivien D, Huguenin D (1999) Color of Mn(V)-substituted apatites $A_{10}[(B,Mn)O_4]_6F_2$, A = Ba, Sr, Ca; B = P, V. J Solid State Chem 146:464-472
Daubrée A (1851) Expériences sur la production artificielle de l'apatite, de la topaze, et de quelques autres métaux fluoriféres. Compt Rend Acad Sci Paris 32:625
De Boer BG, Sakthivel A, Cagle JR, Young RA (1991) Determination of the antimony substitution site in calcium fluorapatite from powder x-ray diffraction data. Acta Crystallogr B47:683-692
Donaldson JD, Grimes SM (1984) Novel tin(II) sites in X-ray crystal structures of the tin(II) halide sulphates $K_3Sn_2(SO_4)_3X$ (X = Br or Cl). J Chem Soc Dalton Trans 1301-1305
Dong ZL, Kim J, White TJ (2005a) Crystal chemical systematics in $[Ca]_{10}[(P_{10-x}V_x)O_4]_6[F]_2$ and $[Ca]_{10}[(P_{10-x}V_x)O_4]_6[Cl]_2$ apatites. In preparation
Dong ZL, Sun K, Wang LM, White TJ, Ewing RC (2005b) Electron irradiation induced transformation of $(Pb_5Ca_5)(VO_4)_6F_2$ apatite to $CaVO_3$ perovskite. J Am Ceram Soc 88:184-190
Dong ZL, White TJ (2004a) Calcium-lead fluoro-vanadinite apatites. I. Disequilibrium structures. Acta Crystallogr B60:138-145
Dong ZL, White TJ (2004b) Calcium-lead fluoro-vanadinite apatites. II. Equilibrium structures. Acta Crystallogr B60:146-154
Dong ZL, White TJ, Wei B, Laursen, K. (2002) Model apatite systems for the stabilization of toxic metals: I. calcium lead vanadate. J Am Ceram Soc 85:2515-2522

Dunn PJ, Peacor DR, Newberry N (1980) Johnbaumite, a new member of the apatite group from Franklin, New Jersey. Am Mineral 65:1143-1145

Dunn PJ, Rouse RC (1978) Morelandite, a new barium arsenate chloride member of the apatite group. Can Mineral 16:601-604

Effenberger H, Pertlik F (1977) Anzeiger der oesterreichischen Akademie der Wissenschaften, mathematisch-naturwissenschaftliche Klasse 114:209-211

El Feki H, Savariault JM, Ben Salah A (1999) Structure refinements by the Rietveld method of partially substituted hydroxyapatite: $Ca_9Na_{0.5}(PO_4)_{4.5}(CO_3)_{1.5}(OH)_2$. J All Comp 287:114-120

El Koumiri M, Oishi S, Sato S, El Ammari L, Elouadi B (2000) The crystal structure of the lacunar apatite $NaPb_4(PO_4)_3$. Mater Res Bull 35:503-513

Elliott JC (1973) Problems of composition and structure of mineral components of hard tissues. Clinic Orthopaedics Related Res 93:313-345

Elliott JC (1994) Structure and chemistry of the apatites and other calcium orthophosphates. Elsevier, Amsterdam

Elliott JC (2002) Calcium phosphate biominerals. Rev Mineral Geochem 48:427-453

Elliott JC, Bonel G, Trombe JC (1980) Space group and lattice constants of $Ca_{10}(PO_4)_6CO_3$. J Appl Crystallogr 13:618-621

Elliott JC, Dykes E, Mackie PE (1981) Structure of bromapatite, $Ca_5(PO_4)_3Br$, and the radius of the bromide ion. Acta Crystallogr B37:435-438

Elliott JC, Mackie PE, Young RA (1973) Monoclinic hydroxyapatite. Science 180:1055-1057

El Ouenzerfi R, Goutaudier C, Panczer G, Moine B, Cohen-Adad MT, Trabelsi-Ayedi M, Kbir-Ariguib N (2003) Investigation of the $CaO-La_2O_3-SiO_2-P_2O_5$ quaternary diagram: synthesis, existence domain, and characterization of apatitic phosphosilicates. Solid State Ionics 156:209-222

Engel G (1965) Untersuchungen zur Kristallchemie von Cadmiumphosphaten,-arsenaten und-vanadaten mit Apatitstruktur. Diplomarbeit Tech Hochsch Karlsruhe

Engel G (1970) Hydrothermalsynthese von Bleihydrooxylapatiten $Pb_5(XO_4)_3OH$ mit X = P, As, V. Naturwissenschaften 57:355

Engel G (1972) Ganomalite an intermediate between the nasonite and apatite types. Naturwissenschaften 59:121-122

Engel G (1978) Fluoroberyllate mit apatitstruktur und ihre beziehungen zu sulfaten und silicaten. Mater Res Bull 13:43-48

Engel G, Deppisch B (1988) Die Kristallstruktur von $Pb_5(GeO_4)_2SO_4$ und $Pb_5(Ge\ O_4)_2\ CrO_4$, zweier Bleiapatite mit unbesetzten Halogenlagen. Z Anorg Allg Chem 562:131-140

Engel G, Fischer U (1990) Die Kristallstruktur von $Na_3Pb_2(BeF_4)_3F$, einem Fluoroberyllat mit Apatitstruktur. J Less-Common Metals 158:123-130

Engel G, Krieg, Reif G (1975a) Mischkristallbildung und Kationenordnung im System Bleihydroxylapatit-Calciumhydroxylapatit. J Solid State Chem 15:117-126

Engel G, Pretzsch J, Gramlich V, Baur WH (1975b) The crystal structure of hydrothermally grown manganese chlorapatite, $Mn_5(PO_4)_3Cl_{0.9}(OH)_{0.1}$. Acta Crystallogr B31:1854-1860

Engin NO, Tas AC (2000) Preparation of porous $Ca_{10}(PO_4)_6(OH)_2$ and $\beta-Ca_3(PO_4)_2$ bioceramics. J Am Ceram Soc 83:1581-1584

Ewing RC, Meldrum A, Wang LM, Wang SX (2000) Radiation-induced amorphization. Rev Mineral Geochem 39:319-354

Ewing RC, Wang LM (2002) Phosphates as Nuclear Waste Forms. Rev Mineral Geochem 48:673-699

Fahey JA, Weber WJ, Rotella FJ (1985) An X-ray and neutron powder diffraction study of the $Ca_{2+x}Nd_{8-x}(SiO_4)_6O_{2-0.5x}$ system. J Solid State Chem 60:145-158

Fayos J, Watkin DJ, Pérez-Méndez M (1987) Crystal structure of apatite-like compound $K_3Ca_2(SO_4)_3F$. Am Mineral 72:209-212

Fedoroff M, Jeanjean J, Rouchaud JC, Mazerolles I, Trocellier P, Maireles-Torres P, Jones DJ (1999) Sorption kinetics and diffusion of cadmium in calcium hydroxyapatites. Solid State Sci 1:71-83

Felsche J (1972) Rare earth silicates with the apatite structure. J Solid State Chem 5:266-275

Ferraris C, White TJ, Plévert J, Wegner R (2004) First evidence for nano-scale exsolution in apatite. Submitted

Fleet ME, Liu X (2003) Carbonate apatite type A synthesized at high pressure: new space group ($P\overline{3}$) and orientation of channel carbonate ion. J Solid State Chem 174:412-417

Friis H, Balić-Žunić T, Pekov IV, Petersen OV (2004) Kuannersuite-(Ce), $Ba_6Na_2REE_2(PO_4)_6FCl$, a new member of the apatite group, from the ilímaussaq alkaline complex, South Greenland: description and crystal chemistry. Can Mineral 42:95-106

Gaude J, L'Haridon P, Hamon C, Marchand R, Laurent Y (1975) Composés à structure apatite. I. Structure de l' oxynitrure $Sm_{10}Si_6N_2O_{24}$. Bull Soc Fr Mineral Crist 98:214-217

Genkina EA, Malinovskii YuA, Khomyakov AP (1991) Crystal structure of Sr-containing britholith. Sov Phys Crystallogr 36:19-21

Giuseppetti G, Rossi G, Tadini C (1971) The crystal structure of nasonite. Am Mineral 56:1174-1179
Grisafe DA, Hummel FA (1970a) Pentavalent ion substitutions in the apatite structure Part A. Crystal chemistry. J Solid State Chem 2:160-166
Grisafe DA, Hummel FA (1970b) Crystal chemistry and color in apatites containing cobalt, nickel and rare-earth ions. Am Mineral 55:1131-1145
Gunawardane RP, Howie RA, Glasser FP (1982) Structure of the oxyapatite $NaY_9(SiO_4)_6O_2$. Acta Crystallogr B38:1564-1566
Habelitz S, Pascual L, Durán A (1999) Nitrogen-containing apatite. J Eur Ceram Soc 19:2685-2694
Hashimoto H, Matsumoto T (1998) Structure refinements of two natural pyromorphites, $Pb_5(PO_4)_3Cl$, and crystal chemistry of chloroapatite group, $M_5(PO_4)_3Cl$. Z Kristallogr 213:585-590
Hata M, Marumo F (1983) Syntheses and superstructures of $(Cd, M)_5(PO_4)_3OH$ (M = Mn, Fe, Co, Ni, Cu, Hg). Mineral J 11:317-330
Hata M, Marumo F, Iwai S (1979) Structure of barium chlorapatite. Acta Crystallogr B35:2382-2384
Hata M, Marumo F, Iwai S, Aoki H (1980) Structure of a lead apatite $Pb_9(PO_4)_6$. Acta Crystallogr B36:2128-2130
Hata M, Okada K, Iwai S (1978) Cadmium hydroxyapatite. Acta Crystallogr B34:3062-3064
Hausen H (1929) Die Apatite, deren chemische Zusammensetzung und ihr Verhältnis zu den physikalischen und morphologischen Eigenschaften. Acta Acad Åbo Math-Phys 5
Hendricks SB, Jefferson ME, Mosley VM (1932) The crystal structures of some natural and synthetic apatite-like substances. Z Kristallogr Kristallogeom Kristallphys Kristallchem 81:352-369
Herdtweck E (1991) Structure of decastrontium hexachromate(V) difluoride. Acta Crystallogr C47:1711-1712
Higuchi M, Katase H, Kodaira K, Nakayama S (2000) Float zone growth and characterization of $Pr_{9.33}(SiO_4)_6O_2$ and $Sm_{9.33}(SiO_4)_6O_2$ single crystals with an apatite structure. J Crystal Growth 218:282-286
Hitmi N, LaCabanne C, Bonel G, Roux P, Young RA (1986) Dipole co-operative motions in an A-type carbonated apatite, $Sr_{10}(AsO_4)_6CO_3$. J Phys Chem Solids 47:507-515
Howie RA, Moser W, Starks RG, Woodhams FWD, Parker W (1973) Potassium tin(II) sulphate and related tin apatites: Mossbauer and X-ray studies. J Chem Soc Dalton Trans 14:1478-1484
Huang J, Sleight AW (1993) The apatite structure without an inversion center in a new bismuth calcium vanadium oxide $BiCa_4V_3O_{13}$. J Solid State Chem 104:52-58
Hughes JM, Cameron M, Crowley KD (1989) Structural variations in natural F, OH, and Cl apatites. Am Mineral 74:870-876
Hughes JM, Cameron M, Crowley KD (1991) Ordering of divalent cations in the apatite structure: Crystal structure refinements of natural Mn- and Sr-bearing apatite. Am Mineral 76:1857-1862
Hughes JM, Drexler JW (1991) Cation substitution in the apatite tetrahedral site: crystal structures of type hydrxylellstadite and type fermorite. N Jahrb Mineral Monatsh 1991:327-336
Hughes JM, Mariano AN, Drexler W (1992) Crystal structures of synthetic Na-REE-Si oxyapatites, synthetic monoclinic britholite. N Jahrb Mineral Monatsh 1992:311-319
Hughes JM, Rakovan J (2002) The crystal structure of apatite $Ca_5(PO_4)_3(F, OH, Cl)$. Rev Mineral Geochem 48:1-12
Huminicki DMC, Hawthorne FC (2002) The crystal chemistry of the phosphate minerals. Rev Mineral Geochem 48: 123-254
Ikoma T, Yamazaki A, Nakamura S, Akao M (1999) Preparation and structure refinement of monoclinic hydroxyapatite. J Solid State Chem 144:272-276
Ito J (1968) Silicate apatites and oxyapatites. Am Mineral 53:890-907
Ivanov SA (1990) Refinement of the crystal structure of $Pb_5(GeO_4)(VO_4)_2$ relative to the powder diffraction patterns profile. J Struct Chem 31:80-84
Ivanov SA, Zavodnik VE (1989) Crystal structure of $Pb_5GeV_2O_{12}$. Sov Phys Crystallogr 34:493-496
Ivanova TI, Frank-Kamenetskaya OV, Kol'tsov AB, Ugolkov VL (2001) Crystal structure of calcium-deficient carbonated hydroxyapatite - thermal decomposition. J Solid State Chem 160:340-349
Iwata Y (1977) Neutron diffraction study of the structure of paraelectric $Pb_5Ge_3O_{11}$. J Phys Soc Japan 43: 961-967
Jeanjean J, McGrellis S, Rouchaud JC, Fedoroff M, Rondeau A, Perocheau A, Dubis A (1996a) A crystallographic study of the sorption of cadmium on calcium hydroxyapatites: incidence of cationic vacancies. J Solid State Chem 126:195-201
Jeanjean J, Vincent U, Fedorof M (1994) Structural modification of calcium hydroxyapatite induced by sorption of cadmium ions. J Solid State Chem 108:68-72
Jeanjean J, Vincent U, Fedorov M (1996b) Structural modification of calcium hydroxyapatite induced by sorption of cadmium ions. J Solid State Chem 108:68-72
Kabalov YK, Sokolova EV, Pekov IV (1997) Crystal structure of belovite-(La) Dokl Akad Nauk 355:182-185

Kahlenberg V, Krüger H (2004) LaAlSiO$_5$ and apatite-like La$_{9.71}$(Si$_{0.81}$Al$_{0.19}$O$_4$)$_6$O$_2$ – the crystal structures of two synthetic lanthanum aluminosilicates. Solid State Sci 6:553-560

Kalsbeek N, Larsen S, Ronsbo JG (1990) Crystal structures of rare elements rich apatite analogues. Z Kristallogr 191:249-263

Karpov AS, Nuss J, Jensen M (2003) Synthesis, crystal structure and properties of calcium and barium hydroxyapatites containing copper ions in hexagonal channels. Solid State Sci 5:1277-1283

Kauffman GB (1997) Victor Moritz Goldschmidt (188-1947): A tribute to the founder of modern geochemistry on the fiftieth anniversary of his death. Chem Educator 2:S1430-4171.

Kay MI, Young RA, Posner AS (1964) Crystal structure of hydroxyapatite. Nature 204:1050-1052

Kazin PE, Karpov AS, Jansen M, Nuss J, Tretyakov YD (2003) Crystal structure and properties of strontium phosphate apatite with oxocuprate ions in hexagonal channels. Z Anorg Allg Chem 629:344-352

Khomyakov AP, Kulikova IM, Rastsvetaeva RK (1997) Fluorcaphite, Ca(Sr,Na,Ca)(Ca,Sr,Ce)$_3$(PO$_4$)$_3$F - a new mineral with the apatite structural motif. Zap Vser Mineral Obshch 126:87-97 (in Russian)

Khomyakov AP, Lisitsin DV, Kulikova IM, Rastsvetaeva RK (1996) Deloneite-(Ce) NaCa$_2$SrCe(PO$_4$)$_3$F - a new mineral with a belovite-like structure. Zap Vser Mineral Obshch 125:83-94 (in Russian)

Kikuchi M, Yamazaki A, Otsuka R, Akao M, Aoki H (1994) Crystal structure of Sr-substituted hydroxyapatite synthesized by hydrothermal method. J Solid State Chem 113:373-378

Kim JY (2001) Structural studies of lead containing apatites. PhD dissertation, School of Chemistry – The University of Sydney

Kim JY, Dong ZL, White, TJ (2005) Model apatite systems for the stabilization of toxic metals: II netalloid substitutions. J Am Ceram Soc, in press

Kim JY, Fenton RR, Hunter BA, Kennedy BJ (2000) Powder diffraction studies of calcium and lead apatites. Aust J Chem 53:679-686

Klement R, Dihn P (1941) Isomorphe apatitarten. Naturwissenschaften 29:301

Klement R, Harth R (1961) Das verhalten von tertiären erdalkaliphosphaten,-arsenaten und-vanadaten in geschmolzenen halogeniden. Chem Ber 94:1452-1456

Klement R, Haselbeck H (1965) Apatite and wagnerite zweiwertige metalle. Z Anorg Allgem Chem 336: 113-128

Klevtsova RF (1964) About the crystal structure of strontiumapatite. Zh Strukt Khim 5:318-320 (in Russian)

Klevtsova RF, Borisov SV (1964) The crystal structure of belovite. Zh Strukt Khim 5:151-153 (in Russian)

Kluver E, Mueller-Buschbaum (1995) Uber einen lanthanoid-mangan-apatit: Nd$_4$Mn(SiO$_4$)$_3$O. Z Naturforsch Teil B50:61-65

Knudsen AC, Gunter ME (2002) Sedimentary phosporites – An example: phosphoria formation, southeastern Idaho, U.S.A. Rev Mineral Geochem 48:363-390

Kohn MJ, Cerling TE (2002) Stable isotope compositions of biological apatite. Rev Mineral Geochem 48: 455-488

Kohn MJ, Rakovan J, Hughes JM (eds) (2002) Phosphates – geochemical, geobiological, and materials importance. Mineralogical Society of America and Geochemical Society, Washington DC

Kottaisamy M, Jagannathan R, Jeyagopal P, Rao RP, Narayanan (1994) Eu^{2+} luminescence in M$_5$(PO$_4$)$_3$X apatites, where M is Ca^{2+}, Sr^{2+} and Ba^{2+}, and X is F$^-$, Cl$^-$, Br$^-$ and OH$^-$. J Phys D: Appl Phys 27:2210-2215

Kreidler ER, Hummel FA (1970) The crystal chemistry of apatite: structure fields of fluor- and chlorapatite. Am Mineral 55:170-184

Kutoglu A Von (1974) Structure refinement of the apatite Ca$_5$(VO$_4$)$_3$(OH). N Jahrb Mineral Monatsh 1974: 210-218

Kutoglu A Von, Schulien S (1972) Synthese und Kristallstruktur eines Calcium-vanadates. Naturwiss 59:36

Kuzmin EA, Belov NV (1965) Crystalline structure of simple silicates, lanthanum and samarium. Dokl Akad Nauk SSSR 165:88-90

León-Reina L, Losilla ER, Martínez-Lara M, Bruque S, Aranda MAG (2004) Interstitial oxygen conduction in lanthanum oxy-apatite electrolytes. J Mat Chem 14:1142-1149

Leventouri Th, Chakoumakos BC, Moghaddam HY, Perikatsis V (2000) Powder neutron diffraction studies of a carbonate fluorapatite. J Mater Res 15:511-517

Liu DM, Quanzu Yang Q, Troczynski T, Tseng WJ (2002) Structural evolution of sol-gel-derived hydroxyapatite. Biomaterials 23:1679-1687

Liou S-C, Chen S-Y, Lee H-Y, Bow J-S (2004) Structural characterization of nano-sized calcium deficient apatite powders. Biomaterials 25:189-196

Lower SK, Maurice PA, Traina SJ, Carlson EH (1998) Aqueous Pb sorption by hydroxylapatite: application of atomic force microscopy to dissolution, nucleation, and growth studies. Am Mineral 83:147-158

Lutze W, Ewing RC (1988) Radioactive waste forms for the future. North-Holland, Amsterdam

Mackie PE, Elliott JC, Young RA (1972) Monoclinic structure of synthetic Ca$_5$(PO$_4$)$_3$Cl chlorapatite. Acta Crystallogr B28:1840-1848

Malinovskii YuA, Genkina EA, Dimitrova OV (1990) TR-ordering in crystal structure of $La_3Nd_{11}(SiO_4)_9O_3$. Sov Phys Crystallogr 35:184-186

Malinovskii YuA, Pobedimskaya EA, Belov NV (1975) Synthesis and crystal structure of the germanate-carbonate apatite, $Ba_5(Ge,C)(O,OH)_4)_3(OH)$. Sov Phys Crystallogr 20:395-396

Mandjiny S, Matis KA, Zouboulis I, Fedoroff M, Jeanjean J, Rouchaud JC, Toulhoat N, Potocek V, Loos-Neskovic C, Maireles-Torres P, Jones D (1998) Calcium hydroxyapatites: evaluation of sorption properties for cadmium ions in aqueous solution. J Mater Sci 33:5433-5439

Manecki M, Maurice PA, Traina SJ (2000) Uptake of aqueous Pb by Cl⁻, F⁻, and OH⁻ apatites: mineralogic evidence for nucleation mechanism. Am Mineral 85:932-942

Manjubala I, Sivakumar M, Najma Nikkath S (2001) Synthesis and characterisation of hydroxyl/fluoroapatite solid solution. J Mater Sci 36:5481-5486

Mathew M, Brown WE, Austin M, Negas T (1980) Lead alkali apatites without hexad anion: the crystal structure of $Pb_8K_2(PO_4)_6$. J Solid State Chem 35:69-76

Mathew M, Mayer I, Dickens B, Schroeder LW (1979) Substitution in barium-fluoride apatite: the crystal structures of $Ba_{10}(PO_4)_6F_2$, $Ba_6La_2Na_2(PO_4)_6F_2$ and $Ba_4La_3Na_3(PO_4)_6F_2$. J Solid State Chem 28:79-95

Matsumura Y, Kanai H, Moffat JB (1997) Catalytic oxidation of carbon monoxide over stoichiometric and non-stoichiometric hydroxyapatites. J Chem Soc, Faraday Trans 93:4383-4387

Mattausch Von HJ, Mueller-Buschbaum HK (1973) The crystal structure of $Ba_5(CrO_4)_3OH$. Z Anorg Allg Chem 400:1-9

Maunaye M, Hamon C, L'Haridob P, Laurent Y (1976) Composés à structure apatite. IV. Étude structurale de l'oxynitrure $Sm_8Cr_2Si_6N_2O_{24}$. Bull Soc Mineral Crist Fr 99:203-205

Mayer I, Cohen S (1983) The crystal structure of $Ca_6Eu_2Na_2(PO_4)_6F_2$. J Solid State Chem 48:17-20

Mayer I, Fischbein E, Cohen S (1975) Apatites of divalent europium. J Solid State Chem 14:307-312

Mayer I, Roth RS, Brown WE (1974) Rare earth substituted fluoride-phosphate apatites. J Solid State Chem 11:33-37

Mayer I, Semadja A (1983) Bismuth substituted calcium, strontium, and lead apatites. J Solid State Chem 46:363-366

Mazelsky R, Ohlmann RC, Steinnbruegge KB (1968) Crystal growth of a new laser material, fluorapatite. J Electrochem Soc 115:68-70

Mazza D, Tribaudino M, Delmastro A, Lebech B (2000) Synthesis and neutron structure of $La_5Si_2BO_{13}$, an analogue of the apatite mineral. J Solid State Chem 155:389-393

McCarthy G, White WB, Roy R (1967) Preparation of $Sm_4(SiO_4)_3$. J Inorg Nucl Chem 29:253-254

McCauley JW, Roy R (1974) Controlled nucleation and crystal growth of various $CaCO_3$ phases by the silica gel technique. Am Miner 59:947-963

McConnell D (1973) Apatite. Its Crystal Chemistry, Mineralogy, Utilization and Geologic and Biologic Occurrences. Springer, New York

McCusker LB, Liebau F, Engelhardt G (2001) Nomenclature of structural and compositional characteristics of ordered microporous and mesoporous materials with inorganic hosts. Pure Appl Chem 73:381-394

McGrellis S, Serafini JN, Jeanjean J, Pastol JL, Fedorof M (2001) Influence of the sorption protocol on the uptake of cadmium ions in calcium hydroxyapatite. Sep Pur Tech 24:129-138

McIntyre DB (2004) Personal Communication: *"...it was Goldschmidt - a great crystallographer recognized today as "father of geochemistry" - who persuaded Beevers to refine the structure of apatite. Goldschmidt led the field in the determination of ionic radii, and I was privileged to hear him on how appropriate ions are captured by crystal structures acting as 3-dimensional fishing nets.*

After his escape from the Nazis, Goldschmidt found refuge at the Macaulay Institute of Soil Research. He had already studied apatite deposits in Scandinavia and he knew well the importance of apatite as the source of P in soil. I believe he was confident that F entered apatite because it fitted so well into the Ca-P-O structure, and that this was why in 1944-45 Beevers took time out from structures like strychnine and sucrose to refine the structure of apatite."

Merker L, Wondratschek H (1959) Bleiverbindungen mit apatitstruktur, insbesondere blei-jod-und blei-brom-apatite. Z Anorg Allg Chem 300:41-50

Miyake M, Ishigaki K, Suzuki T (1986) Structure refinements of Pb^{2+} ion-exchanged apatites by X-ray powder patter-fitting. J Solid State Chem 61:230-235

Mohseni-Koutchesfehani S (1961) Contribution a l'étude des apatites barytiques. Ann Chim 463-479

Monteil-Rivera F, Masset S, Dumonceau J (1999) Sorption of selenite ions on hydroxyapatite. J Mat Sci Lett 18:1143-1145

Moore PB, Araki T (1977) Samuelsonite: its crystal structure and relation to apatite and octacalcium phosphate. Am Mineral 62:229-245

Morgan MG, Wang M, Mar A (2002) Samarium orthosilicate oxyapatite, $Sm_5(SiO_4)_3O$. Acta Crystallogr E58:i70-i71

Müller-Buschbaum HK, Sander K (1978) Zur Kristallstruktur von $Sr_5(CrO_4)_3Cl$. Z Naturforsch 33b:708-710

Nadal M, Le Geros RZ, Bonel G, Montel G (1971) Mise en evidence d'un phénomène d'order-désordre dans le réseau des carbonate-apatites strontique. Compt Rend Acad Sci 272:45-48

Naddari T, Feki HE, Savariault JM, Salles P, Salah AB (2003) Structure and ionic conductivity of the lacunary apatite $Pb_6Ca_2Na_2(PO_4)_6$. Solid State Ionics 158:157-166

Naddari T, Savariault JM, El Feki H, Salles P, Ben Salah A (2002) Conductivity and structural investigations in lacunary $Pb_6Ca_2Li_2(PO_4)_6$ apatite. J Solid State Chem 166:237-244

Nadezhina TN, Pushcharovskii DY, Khomyakov AP (1987) The refinement of crystal structure of belovite. Mineral Zh 9:45-48 (in Russian)

Nakayama S, Kagayama T, Aono H, Sadoaka Y (1995) Ionic conductivity of lanthanoid silicates, $Ln_{10}(SiO_4)_6O_3$ (Ln = La, Nd, Sm, Gd, Dy, Y, Ho, Er and Yb). J Mat Chem 5:1801-1806

Narasaraju TSB, Singh RP, Rao VLN (1972) A new method of preparation of solid solutions of calcium and lead hydroxlapatites. J Inorg Nucl Chem 34:2072-2074

Negas T, Roth RS (1968) High temperature dehydroxylation of apatitic phosphates. J Res Natn Bur Stand A 72A:783-787

Newkirk AE, Hughes VB (1969) Identification of the "lead(II) hydroxide" of Robin and Theólier. Inorg Chem 9:401-404

Newnham RE, Wolfe RW, Darlington CNW (1973) Prototype structure of $Pb_5Ge_3O_{11}$. J Solid State Chem 6: 378-383

Nishikawa H, Omamiuda K (2002) Photocatalytic activity of hydroxyapatite for methyl mercaptane. J Mol Catalysis 179:193-200

Noe DC, Hughes JM, Mariano AN, Drexler JW, Kato A (1993) The crystal structure of monoclinic britholite-(Ce) and britholite-(Y). Z Kristallogr 206:233-246

Nötzold D, Wulff H (1998) Determining the crystal structure of $Sr_5(PO_4)_3Br$, a new compound in the apatite series, by powder diffraction modeling. Powder Diffr 13:70-73

Nötzold D, Wulff H, Herzog G (1994) Differenzthermoanalyse der Bildung des Pentastrontiumchloridphospha ts und röntgenographische Untersuchung seiner Struktur. J Alloys Compd 215:281-288

Nötzold D, Wulff H (1996) Structural and optical properties of the system $(Sr, Ba, Eu)_5(PO_4)_3Cl$. Phys Stat Sol (a) 158:303-311

Nounah A, Lacout JL (1993) Thermal behaviour of cadmium-containing apatites. J Solid State Chem 107: 444-451

Nriagu JO, Moore PB (1984) Phosphate Minerals. Springer-Verlag, New York

Nyman H, Andersson S (1979) On the structure of Mn_5Si_3, Th_6Mn_{23} and γ-brass. Acta Crystallogr A35:580-583

O'Keeffe M, Hyde BG (1985) An alternative approach to crystal structures with emphasis on the arrangements of cations. Struct Bonding 61:79-144

Organova NI, Rastsvetaeva RK, Kuz'mina OV, Arapova GA, Litsarev MA, Fin'ko VI (1994) The crystal structure of low-symmetry ellestadite in comparison with other apatitelike structures. Crystallogr Rep 39:234-238

Owada H, Yamashita K, Umegaki T, Kanazawa T (1989) Humidity-sensitivity of yttrium substituted apatite ceramics. Solid State Ionics 35:401-404

Pan Y, Fleet ME (2002) Compositions of the apatite-group minerals: substitution mechanism and controlling factors. Rev Mineral Geochem 48:13-50

Parmentier J, Liddell K, Thompson DP, Lemercier H, Schneider N, Hampshire S, Bodart PR, Harris RK (2001) Influence of iron on the synthesis and stability of yttrium silicate apatite. Solid State Sci 3:495–502

Perret R, Bouillet AM (1975) The sulfate apatites $Na_3Cd_2(SO_4)_3Cl$ and $Na_3Pb_2(SO_4)_3Cl$. Bull Soc fr Minéral Cristallogr 98:254-255

Pekov IV, Kulikova IM, Kabalov YuK, Eletskaya OV, Chukanov NV, Menshikov YP, Khomyakov AP (1996) Belovite-(La), $Sr_3Na(La,Ce)[PO_4]_3(F,OH)$ - a new rare earth mineral in the apatite group. Zap Vser Mineral Obshch 125(2):101-109 (in Russian)

Piccoli MP, Candela PA (2002) Apatite in igneous systems. Rev Mineral Geochem 48:255-292

Piotrowski A, Kahlenberg V, Fischer RX (2002a) The solid solution series of the sulfate apatite system $Na_{6.45}Ca_{3.55}(SO_4)_6(F_xCl_{1-x})_{1.55}$. J Solid State Chem 163:398-405

Piotrowski A, Kahlenberg V, Fischer RX, Lee Y, Parise JB (2002b) The crystal structures of cesanite and its synthetic analogue - a comparison. Am Mineral 87:715-720

Piriou B, Fahmi D, Dexpert-Ghys J, Taïtaï A, Lacout JL (1987) Unusual fluorescent properties of Eu^{3+} in oxyapatites. J Lumines 39:97-103

Pletchov PY, Sinogeikin SV (1996) The origin of apatite-nepheline ores of the Khibina massive. Vestnik MGU (Herald of the Moscow State Univ.) (Russian) Ser. 4, Geology 1:77-80

Posner AS, Diorio AF (1958) Refinement of the hydroxylapatite structure. Acta Crystallogr 11:308-209

Povarennykh AS (1972) Crystal chemical classification of minerals. Plenum Press, New York/London

Prener JS (1967) The growth and crystallographic properties of calcium fluor- and chlorapatite crystals. J Electrochem Soc 114:77-83

Prener JS (1971) Nonstoichiometry in calcium chlorapatite. J Solid State Chem 3:49-55

Pritzkow W, Rentsch H (1985) Structure refinement with X-ray powder diffraction data for synthetic calcium hydroxyapatite by Rietveld method. Crystal Res Tech 20:957-960

Pushcharovskii DY, Dorokhova GI, Pobedimskaya EA, Belov NV (1978) Potassium-neodymium silicate $KNd_9[SiO_4]_6O_2$ with apatite structure. Sov Phys Dokl 23:694-696

Pushcharovskii DY, Nadezhina TN, Khomyakov AP (1987) Crystal structure of strontium apatite from Khibiny. Sov Phys Crystallogr 32:524-526

Rakovan JF, Hughes JM (2000) Strontium in the apatite structure: strontium fluorapatite and belovite-(Ce). Can Mineral 38:839-845

Redhammer GR, Roth G (2003) Lithium and sodium yttrium orthosilicate oxyapatite, $LiY_9(SiO_4)_6O_2$ and $NaY_9(SiO_4)_6O_2$ at both 100 K and near room temperature. Acta Crystallogr C59:i120-i124

Reichert J, Binner JGP (1996) An evaluation of hydroxyapatite-based filters for removal of heavy metal ions from aqueous solutions. J Mat Sci 31:1231-1241

Reinen D, Lachwa H, Allmann R (1986) EPR- und ligandenfeldspektroskopische Untersuchungen an Mn^V– haltigen Apatiten sowie die Struktur von $Ba_5(MnO_4)_3Cl$. Z Anorg Allg Chem 542:71-88

Robin J, Théolier A (1956) Preparation and properties of lead hydroxide. Bull Soc Chim France 680:9921

Rouse RC, Dunn PJ, Peacor DR (1984) Hedyphane from Franklin, New Jersey and Långban, Sweden: canon ordering in an arsenate apatite. Am Mineral 69:920-927

Roux P, Bonel G (1977) Sur la preparation de l'apatite carbonatée de type A, à haute température par evolution, sous pression de gaz carbonique, des arséniates tricalcique et tristrontique. Ann Chim 159-165

Roy R (1982) Radioactive waste disposal. Pergamon Press, New York

Saenger AT, Kuhs WF (1992) Structural disorder in hydoxyapatite. Z Kristallogr 199:123-148

Sansom JEH, Slater PR (2004) Oxide ion conductivity in the mixed Si/Ge apatite-type phases $La_{9.33}Si_{6-x}Ge_xO_{26}$. Solid State Ionics 167:23-27

Sansom JEH, Tolchard JR, Slater PR, Islam MS (2004) Synthesis and structural characterisation of the apatite-type phases $La_{10-x}Si_6O_{26+z}$ doped with Ga. Solid State Ionics 167:17-22

Schiff-Francois A, Savelsberg G, Schaefer H (1979) Preparation and crystal structure of $Ba_5S(AsO_4)_2SO_4$, $Sr_5S(AsO_4)_2SO_4$ und $Sr_5S(PO_4)_2SO_4$. Z Naturforsch Teil B34:764-765

Schneider W (1967) Caracolit, das $Na_3Pb_2(SO_4)_3Cl$ mit Apatitstruktur. N Jahrb Mineral Monatsh 1967:284-289

Schriewer MS, Jeitschko W (1993) Preparation and crystal structure of the isotypic orthorhombic strontium perrhenate halides $Sr_5(ReO_5)_3X$ (X = Cl, Br, I) and structure refinement of the related hexagonall apatite-like compound $Ba_3(ReO_5)_3Cl$. J Solid State Chem 107:1-11

Schroeder LW, Mathew M (1978) Cation ordering in $Ca_2La_8(SiO_4)_6O_2$. J Solid State Chem 26:383-387

Schwarz H (1967a) Apatite des Typs $Pb_6K_4(X^VO_4)_4(X^{VI}O_4)_2$ (X^V = P, As; X^{VI} = S, Se). Z Anorg Allgem Chem 356:29-35

Schwarz H (1967b) Apatite des Typs $M^{II}{}_{10}(X^{VI}O_4)_3(X^{IV}O_4)_3F_2$ (M^{II} = Sr, Pb; X^{VI} = S, Cr; X^{IV} = Si, Ge). Z Anorg Allgem Chem 356:36-45

Schwarz H (1967c) Vergindungen mit Aptitstruktur. I. Ungewohnliche Silicatapatite. Inorg Nucl Chem Lett 3:231-236

Schwarz H (1968) Strontiumapatite des Typs $Sr_{10}(PO_4)_4(X^{IV}O_4)_2$ (X^{IV}= Si, Ge). Z Anorg Allgem Chem 357: 43-53

Skakle JMS, Dickson CL, Glasser FP (2000) The crystal structures of $CeSiO_4$ and $Ca_2Ce_8(SiO_4)_6O_2$. Powder Diff 15:234-238

Sirotinkin SP, Oboznenko, Oboznenko Yu V, Nevskii NN (1989) Alkali metal lead double vanadates. Russ J Inorg Chem 34:1716-1718

Spear FS, Pyle JM (2002) Apatite, monazite, and xenotime in metamorphic rocks. Rev Mineral Geochem 48: 293-336

Steele IM, Pluth JJ, Livingstone A (2000) Crystal structure of mattheddleite: a Pb, S, Si phase with the apatite structure. Mineral Mag 64:915-921

Stemmermann P (1992) Silikatapatite: Struktur, Chemismus und Anwendung als Speichermineral zur Konditionierung von Rauchgasreinigungsrückständen. PhD Dissertation, Naturwissenschaftlichen Fakultaten, Friedrich-Alexander-Universität, Erlangen-Nurnberg, Germany

Sudarsanan K (1980) Structure of hydroxyellestadite. Acta Crystallogr B36:1636-1639

Sudarsanan K, Mackie PE, Young RA (1972) Comparison of synthetic and mineral fluorapatite, $Ca_5(PO_4)_3F$, in crystallographic detail. Mater Res Bull 7:1331-1338

Sudarsanan K, Young RA (1969) Significant precision in crystal structural details: holly springs hydroxyapatite. Acta Crystallogr 25:1534-1543

Sudarsanan K, Young RA (1972) Structure of strontium hydroxide phosphate, $Sr_5(PO_4)_3OH$. Acta Crystallogr B28:3668-3670

Sudarsanan K, Young RA (1974) Structure refinement and random error analysis for strontium 'chlorapatite', $Sr_5(PO_4)_3Cl$. Acta Crystallogr B30:1381-1386

Sudarsanan K, Young RA (1980) Structure of partially substituted chlorapatite (Ca, Sr)$_5(PO_4)_3Cl$. Acta Crystallogr B36:1525-1530

Sudarsanan K, Young RA, Wilson AJC (1977) The structures of some cadmium "apatites" $Cd_5(MO_4)_3X$. I. Determination of the structures of $Cd_5(VO_4)_3I$, $Cd_5(PO_4)_3Br$, $Cd_5(AsO_4)_3Br$ and $Cd_5(VO_4)_3Br$. Acta Crystallogr B33:3136-3142

Suetsugu Y, Takahashi Y, Okamura FP, Tanaka J (2000) Structure analysis of A-type carbonate apatite by a single-crystal X-ray diffraction method. J Solid State Chem 155:292-297

Sugiyama S, Ichii T, Fujisawa M, Kawashiro K, Tomida T, Shigemoto N, Hayashi H (2003) Heavy metal immobilization in aqueous solution using calcium phosphate and calcium hydrogen phosphates. J Colloid Interface Sci 259:408-410

Suitch PR, Taïtaï A, Lacout JL, Young RA (1986) Structural consequences of the coupled substitution of Eu, S, in calcium sulfoapatite. J Solid State Chem 63:267-277

Suwa Y, Naka S, Noda T (1968) Preparation and properties of yttrium magnesium silicate with apatite structure. Mat Res Bull 3:139-148

Tachihante M, Zambon D, Arbus A, Zahir M, Sadel A, Cousseins JC (1993) Optical properties of trivalent terbium doped calcium fluorapatite. Mat Res Bull 28:605-613

Takahashi M, Uematsu K, Ye ZG, Sato M (1998) Single-crystal growth and structure determination of a new oxide apatite, $NaLa_9(GeO_4)_6O_2$. J Solid State Chem 139:304-309

Takeda H, Ohgaki M, Kizuki T, Hashimoto K, Toda Y, Udagawa S, Yamashita K (2000) Formation mechanism and synthesis of apatite-type structure $Ba_{2+x}La_{8-x}(SiO_4)_6O_{2-\delta}$. J Am Cer Soc 83:2884-2886

Tao S, Irvine JTS (2001) Preparation and characterisation of apatite-type lanthanum silicates by a sol-gel process. Mat Res Bull 36:1245-1258

Tolchard JR, Saiful Islam M, Slater PR (2003) Defect chemistry and oxygen ion migration in the apatite-type materials $La_{9.33}Si_6O_{26}$ and $La_8Sr_2Si_6O_{26}$. J Mater Chem 13:1956-1961

Tomita K, Kawano M, Shiraki K, Otsuka H, (1996) Sulfatian apatite from the Katanoyama formation in Nishina-omote City, Kagoshima prefecture. J Mineral Petrol Econ Geol 91:11-20

Toumi M, Smiri-Dogguy L, Bulou A (2000) Crystal structure and polarized Raman spectra of $Ca_6Sm_2Na_2(PO_4)_6F_2$. J Solid State Chem 149:308-313

Trombe J-C, Montel G (1975) Sur les conditions de preparation d'une nouvelle apatite contenant des ions sulfure. Compt Rend Acad Sci 280:567-570

Trombe J-C, Montel G (1978) Some features of the incorporation of oxygen in different oxidation states in the apatite lattice-II. On the synthesis and properties of calcium and strontium peroxiapatites. J Inorg Nuclear Chem 40:23-26

Utsunomiya S, Yudintsev S, Wang LM, Ewing RC (2003) Ion-beam and electron-beam irradiation of synthetic britholite. J Nuclear Mater 322:180-188

Vance ER, Ball CJ, Begg BD, Carter ML, Day RA, Thorogood GJ (2003) Pu, U, and Hf incorporation in Gd silicate apatite. J Am Ceram Soc 86:1223-1225

Vegas A, Jansen M (2002) Structural relationships between cations and alloys; an equivalence between oxidation and pressure. Acta Crystallogr B58:38-51

Vegas A, Romero A, Martinez-Ripoll M (1991) A new approach to describing non-molecular crystal structures. Acta Crystallogr B47:17-23

Verbeeck RMH, Lassuyt CJ, Heijligers HJM, Driessens FCM, Vrolijk JWGA (1981) Lattice parameters and cation distribution of solid solutions of calcium and lead hydroxylapatite. Calcif Tissue Int 33:243-247

Wallaeys R (1952) Contribution à l'étude des apatites phosphocalciques. Ann Chim 7:808-848

Wang LM, Weber WJ (1999) Transmission electron microscopy study of ion-beam-induced amorphization of $Ca_2La_8(SiO_4)_6O_2$. Phil Mag A79:237-253

Wardojo TA, Hwu S-J (1996) Chlorapatite: $Ca_5(AsO_4)_3Cl$. Acta Crystallogr C52:2959-2960

Weiner S, Wagner HD (1998) The material bone: structure-mechanical function relations. Annu Rev Mat Sci 28:271-298

White TJ, Zhili D (2003) Structural derivation and crystal chemistry of apatites. Acta Crystallogr B59:1-16

Wilhemi KA, Jonsson O (1965) X-Ray studies on some alkali and alkaline-earth chromates. Acta Chem Scand 19:177-184

Wilke HJ (1997) Die mineralien und fundstellen von Schweden. Christian Weise Verlag, München

Wilson AJC, Sudarsanan K, Young RA (1977) The structures of some cadmium "apatites" $Cd_5(MO_4)_3X$. II. The distributions of the halogen atoms in $Cd_5(VO_4)_3I$, $Cd_5(PO_4)_3Br$, $Cd_5(AsO_4)_3Br$ and $Cd_5(VO_4)_3Br$ and $Cd_5(PO_4)_3Cl$. Acta Crystallogr B33:3142-3154

Wilson RM, Elliott JC, Dowker SEP (1999) Rietveld refinement of the crystallographic structure of human dental enamel apatites. Am Mineral 84:1406-1414

Wilson RM, Elliott JC, Dowker SEP (2003) Formate incorporation in the structure of Ca-deficient apatite: Rietveld structure refinement. J Solid State Chem 174:132-140

Wilson RM, Elliott JC, Dowker SEP, Smith RI (2004) Rietveld structure refinement of precipitated carbonate apatite using neutron diffraction data. Biomaterials 25:2205-2213

Wondratschek H (1963) Untersuchungen zur kristallchemie der blei-apatite (pyromorphite). N Jahrb Mineral Abh 99:113-160

Wondratschek H, Merker L, Schubert K (1964) Relations between the apatite structure and the structure of the compounds of the Mn_5Si_3 (D88)-type. Z Kristallogr 120:393-395

Wright J, Hansen B, Conca J (2005) PIMS: an apatite II permeable reactive barrier to remediate groundwater containing Zn, Pb and Cd. Submitted to J Contam Hydrol

Wyckoff RWG (1965) Inorganic compounds $R_x(MX_4)_y$, $R_x(MnX_p)_y$, Hydrates and Ammoniates. *In:* Crystal Structures. Vol 3. John Wiley and Sons, New York, p 228-234

Xu Y, Schwartz FW (1994) Lead immobilization by hydroxyapatite in aqueous solutions. J Contamin Hydr 15:187-206

Yasuda I, Hishinuma M (1995) Electrical conductivity and chemical stability of calcium chromate hydroxyl apatite, $Ca_5(CrO_4)_3OH$, and problems caused by the apatite formation at the electrode/separator interface in solid oxide fuel cells. Solid State Ionics 78:109-114

Yoshimura M, Suda H (1994) Hydrothermal processing of hydroxyapatite: past, present, and future. *In:* Hydroxyapatite and Related Materials. Brown P, Constantze B (eds), CRC Press, Inc., Boca Raton p 45-72

Young RA, Mackie PE, Von Dreele RB (1977) Application of the pattern-fitting structure-refinement method of X-ray powder diffractometer patterns. J Appl Crystallogr 10:262-269

Zahouily M, Abrouki Y, Bahlaouan B, Rayadh A, Sebti S (2003) Hydroxyapatite: new efficient catalyst for the Michael addition. Catalysis Com 4:521-52

APPENDIX A – APATITE LATTICE PARAMETERS

Table A1. Cell Parameters for apatites reported with $P6_3/m$ symmetry.

Table A2. Cell parameters for apatites reported with $P6_3$, $P\bar{3}$ or $P\bar{6}$ symmetry.

Table A3. Cell parameters for monoclinic apatites reported with $P2_1/m$, $P2_1/b$ or $P2_1$ symmetry.

Table A4. Cell parameters for apatites reported with $P6_3cm$ and $Pnma$ symmetry.

Table A1. Cell Parameters for apatites reported with $P6_3/m$ symmetry.

Composition	a (Å)	c (Å)	c/a	Volume (Å^3)	Reference
$Ba_{10}(AsO_4)_6F_2$	10.41	7.83	0.752	734.8	Kreidler and Hummel (1970)
$Ba_{10}(AsO_4)_6Cl_2$	10.54	7.73	0.733	743.7	Kreidler and Hummel (1970)
$Ba_{10}(AsO_4)_4(SO_4)_2S_2$	10.526(5)	7.737(1)	0.735	742.4	Schiff-Francois et al. (1979)
$Ba_{10}(CrO_4)_6Cl_2$	10.50	7.73	0.736	738.1	Banks and Jaunarajs (1965)
$Ba_{10}(CrO_4)_6Cl_2$	10.511	7.764	0.739	742.9	Wilhelmi and Jonsson (1965)
$Ba_{10}(CrO_4)_6(OH)_2$	10.428	7.89	0.757	743.0	Mattausch and Mueller-Buschbaum (1973)
$Ba_{10}(MnO_4)_6F_2$	10.3437	7.8639	0.760	728.7	Dardenne et al. (1999)
$Ba_{10}(MnO_4)_6Cl_2$	10.459	7.762	0.742	735.3	Grisafe and Hummel (1970a)
$Ba_{10}(PO_4)_6(Cu_{0.30}O_{0.86}H_y)_2$	10.2073(1)	7.7401(1)	0.758	698.4	Karpov et al. (2003)
$Ba_{10}(PO_4)_6Cl_2$	10.284(2)	7.651(3)	0.744	700.8	Hata et al. (1979)
$Ba_{10}(PO_4)_6CO_3$	10.20(1)	7.65(1)	0.750	689.3	Mohensi-Koutchesfehani (1961)
$Ba_{10}(PO_4)_6F_2$	10.153(2)	7.733(1)	0.762	690.3	Mathew et al. (1979)
$Ba_{10}(PO_4)_6(OH)_2$	10.177	7.731	0.760	693.4	Negas et al. (1968)
$Ba_{9.98}Eu_{0.02}(PO_4)_6Cl_2$	10.2717(1)	7.6500(1)	0.745	698.9	Nötzold and Wulff (1996)
$Ba_{10}(P_{0.20}Mn_{0.80}O_4)_6F_2$	10.300	7.825	0.760	718.9	Dardenne et al. (1999)
$Ba_{10}(P_{0.40}Mn_{0.60}O_4)_6F_2$	10.250	7.802	0.761	709.9	Dardenne et al. (1999)
$Ba_{10}(P_{0.50}Mn_{0.50}O_4)_6F_2$	10.247	7.784	0.760	707.8	Dardenne et al. (1999)
$Ba_{10}(P_{0.70}Mn_{0.30}O_4)_6F_2$	10.211	7.740	0.758	698.9	Dardenne et al. (1999)
$Ba_{10}(P_{0.80}Mn_{0.20}O_4)_6F_2$	10.194	7.738	0.759	696.4	Dardenne et al. (1999)
$Ba_{10}(P_{0.9}Mn_{0.1}O_4)_6F_2$	10.176	7.726	0.759	692.9	Dardenne et al. (1999)
$Ba_{10}(P_{0.95}Mn_{0.05}O_4)_6F_2$	10.163	7.720	0.760	690.5	Dardenne et al. (1999)
$Ba_{10}(P_{0.967}Mn_{0.033}O_4)_6F_2$	10.172	7.733	0.760	692.9	Dardenne et al. (1999)
$Ba_{10}(P_{0.98}Mn_{0.02}O_4)_6F_2$	10.150	7.712	0.760	688.1	Dardenne et al. (1999)
$Ba_{10}(P_{0.99}Mn_{0.01}O_4)_6F_2$	10.152	7.712	0.760	688.3	Dardenne et al. (1999)
$Ba_{10}(P_{0.999}Mn_{0.001}O_4)_6F_2$	10.155	7.720	0.760	689.5	Dardenne et al. (1999)
$Ba_2La_8(SiO_4)_6O_2$	9.77	7.32	0.749	605.1	Ito (1968)
$Ba_2Nd_8(SiO_4)_6O_2$	9.66	7.16	0.741	578.6	Ito (1968)
$Ba_2Sm_8(SiO_4)_6O_2$	9.62	7.06	0.734	565.8	Ito (1968)
$Ba_4Dy_6(SiO_4)_6(OH)_2$	9.66	6.95	0.719	561.7	Ito (1968)
$Ba_4Gd_6(SiO_4)_6(OH)_2$	9.72	7.03	0.723	575.2	Ito (1968)
$Ba_4Nd_6(SiO_4)_6(OH)_2$	9.76	7.18	0.736	592.3	Ito (1968)
$Ba_4Sm_6(SiO_4)_6(OH)_2$	9.73	7.10	0.730	582.1	Ito (1968)
$Ba_{10}(VO_4)_6Cl_2$	10.55	7.75	0.735	747.0	Kreidler and Hummel (1970)
$Ba_{10}(VO_4)_6F_2$	10.420	7.854	0.754	738.5	Grisafe and Hummel (1970a)

Table A1 continued.

Composition	a (Å)	c (Å)	c/a	Volume (Å^3)	Reference
Ba$_{10}$(VO$_4$)$_6$F$_2$	10.44	7.86	0.654	644.3	Kreidler and Hummel (1970)
Ba$_{10}$(V$_{0.60}$Mn$_{0.40}$O$_4$)$_6$F$_2$	10.368	7.851	0.757	730.9	Dardenne et al. (1999)
Ba$_{10}$(V$_{0.80}$Mn$_{0.20}$O$_4$)$_6$F$_2$	10.396	7.844	0.755	734.2	Dardenne et al. (1999)
Ba$_{10}$(V$_{0.98}$Mn$_{0.02}$O$_4$)$_6$F$_2$	10.39	7.86	0.756	734.8	Dardenne et al. (1999)
Bi$_2$Ca$_8$(PO$_4$)$_6$O$_2$	9.461(8)	6.95(1)	0.735	538.7	Buvaneswari and Varadaraju (2000)
Bi$_2$Ca$_6$Sr$_2$(PO$_4$)$_6$O$_2$	9.534(7)	7.032(7)	0.738	553.5	Buvaneswari and Varadaraju (2000)
Bi$_2$Ca$_4$Sr$_4$(PO$_4$)$_6$O$_2$	9.605(7)	7.121(8)	0.741	568.9	Buvaneswari and Varadaraju (2000)
Bi$_2$Ca$_2$Sr$_6$(PO$_4$)$_6$O$_2$	9.669(7)	7.209(8)	0.746	583.7	Buvaneswari and Varadaraju (2000)
Bi$_2$Sr$_8$(PO$_4$)$_6$O$_2$	9.725(8)	7.30(1)	0.751	597.9	Buvaneswari and Varadaraju (2000)
Ca$_{10}$(AsO$_4$)$_6$Cl$_2$	10.076(1)	6.807(1)	0.676	598.5	Wardojo and Hwu (1996)
Ca$_{10}$(AsO$_4$)$_6$F$_2$	9.63	6.99	0.726	561.4	Kreidler and Hummel (1970)
Ca$_{10}$(AsO$_4$)$_6$CO$_3$	9.858	7.010	0.711	590.0	Roux and Bonel (1977)
Ca$_4$Pb$_6$(AsO$_4$)$_6$Cl$_2$	10.140(3)	7.185(4)	0.709	639.8	Rouse et al. (1984)
Ca$_{10}$(CrO$_4$)$_6$Cl$_2$	10.03	6.78	0.676	590.7	Banks and Jaunarajs (1965)
Ca$_{10}$(CrO$_4$)$_6$(OH)$_2$	9.66	7.01	0.726	566.5	Banks and Jaunarajs (1965)
Ca$_{10}$(CrO$_4$)$_6$(OH)$_2$	9.683	7.01	0.724	569.2	Wilhelmi and Jonsson (1965)
Ca$_4$La$_6$(GeO$_4$)$_6$(OH)$_2$	9.850	7.200	0.731	605.0	Cockbain and Smith (1967)
Ca$_{10}$(PO$_4$)$_6$Br$_2$	9.761(1)	6.739(1)	0.690	556.1	Elliot et al. (1981)
Ca$_{10}$(PO$_4$)$_6$Br$_2$	9.7682(6)	6.7388(2)	0.690	643.0	Kim et al. (2000)
Ca$_9$Na$_{0.5}$(PO$_4$)$_{4.5}$(CO$_3$)$_{1.5}$(OH)$_2$	9.3892(4)	6.9019(3)	0.735	526.9	El Feki et al. (1999)
Ca$_{10}$(PO$_4$)$_6$F$_2$	9.3475(3)	6.8646(1)	0.743	519.4	Kim et al. (2000)
Ca$_{10}$(PO$_4$)$_6$F$_2$	9.367(1)	6.884(1)	0.735	523.1	Sudarsanan et al. (1972)
Ca$_{10}$(PO$_4$)$_6$F$_2$	9.374(3)	6.889(3)	0.735	524.2	Schwarz (1967b)
Ca$_{10}$(PO$_4$)$_6$F$_2$	9.374(2)	6.882(2)	0.734	523.7	Mayer and Semadja (1983)
Ca$_{10}$(PO$_4$)$_6$F$_2$	9.3842	6.8878	0.734	527.7	Boyer et al. (1997)
Ca$_{9.2}$(PO$_4$)$_{5.55}$F$_2$	9.3653(3)	6.8816(2)	0.735	522.7	Young et al. (1977)
Ca$_{8.80}$(PO$_4$)$_{5.34}$F$_2$	9.3661(4)	6.8826(3)	0.735	522.9	Young et al. (1977)
Ca$_{10}$(PO$_4$)$_6$Cl$_2$	9.5902(6)	6.7666(2)	0.720	539.0	Kim et al. (2000)
Ca$_{10}$(PO$_4$)$_6$Cl$_2$	9.52(3)	6.85(3)	0.720	537.6	Hendricks et al. (1932)
Ca$_{15}$(PO$_4$)$_9$IO (superstructure)	9.567(1)	20.754(2)	2.170	1645.1	Alberius-Henning et al. (1999a)
Ca$_{10}$(PO$_4$)$_6$(NCN)□	9.424	6.852	0.727	527.0	Habelitz et al. (1999)
Ca$_{10}$(PO$_4$)$_6$O□	9.416	6.8747	0.730	527.7	Boyer et al. (1997)
Ca$_{10}$(PO$_4$)$_6$(OH)$_2$	9.4302(5)	6.8911(2)	0.731	530.7	Kim et al. (2000)
Ca$_{10}$(PO$_4$)$_6$(OH)$_2$	9.432	6.881	0.730	530.1	Posner and Diorio (1958)

Table A1 continued.

Composition	a (Å)	c (Å)	c/a	Volume (Å^3)	Reference
$Ca_{10}(PO_4)_6(OH)_2$	9.4166	6.8745	0.730	527.9	Hughes et al. (1989)
$Ca_{10}(PO_4)_6(OH)_2$	9.424(4)	6.879(4)	0.730	529.1	Sudarsanan and Young (1969)
$Ca_{10}(PO_4)_6(OH)_2$	9.422(3)	6.885(3)	0.730	529.3	Schwarz (1967b)
$Ca_{10}(PO_4)_6[O_x(OH)_2]$	9.402	6.888	0.733	527.3	Engel et al. (1975a)
$Ca_{10}(PO_4)_6OH_{1.9}F_{0.1}$	9.418(2)	6.876(2)	0.730	528.2	Bertoni et al. (1998)
$Ca_{10}(PO_4)_6OH_{1.9}F_{0.1}$	9.404(5)	6.907(6)	0.734	529.0	Manjubala et al. (2001)
$Ca_{10}(PO_4)_6OH_{1.94}F_{0.06}$	9.412(3)	6.907(6)	0.734	529.9	Manjubala et al. (2001)
$Ca_{10}(PO_4)_6OH_{1.98}F_{0.02}$	9.423(2)	6.880(2)	0.730	529.1	Bertoni et al. (1998)
$Ca_{10}(PO_4)_6OH_{1.98}F_{0.02}$	9.427(5)	6.909(8)	0.733	531.7	Manjubala et al. (2001)
$Ca_{10}(PO_4)_6O_{0.75}(OH)_{0.5}\square_{0.75}$	9.402	6.888	0.733	527.3	Trombe and Montel (1978)
$Ca_{10}(PO_4)_6(Cu_{0.27}O_{0.86}H_y)_2$	9.4303(1)	6.9069(1)	0.732	531.9	Karpov (2003)
$Ca_{10}(P_{0.98}Mn_{0.02}O_4)_6F_2$	9.377	6.885	0.734	524.3	Dardenne et al. (1999)
$Ca_{10}(P_{0.999}Mn_{0.001}O_4)_6F_2$	9.376	6.885	0.734	524.2	Dardenne et al. (1999)
$Ca_{10}(P_{0.99}Mn_{0.01}O_4)_6F_2$	9.384	6.887	0.734	525.2	Dardenne et al. (1999)
$Ca_{9.45}Ag_{0.55}(PO_4)_6(OH)_{1.45}\square_{0.55}$	9.443	6.917	0.733	534.2	Badrour et al. (1998)
$Ca_{9.55}Ag_{0.45}(PO_4)_6(OH)_{1.55}\square_{0.45}$	9.439	6.911	0.732	533.2	Badrour et al. (1998)
$Ca_{9.6}Ag_{0.4}(PO_4)_6(OH)_{1.6}\square_{0.4}$	9.441	6.901	0.731	532.7	Badrour et al. (1998)
$Ca_{9.7}Ag_{0.3}(PO_4)_6(OH)_{1.7}\square_{0.3}$	9.436	6.897	0.731	531.8	Badrour et al. (1998)
$Ca_{9.8}Ag_{0.2}(PO_4)_6(OH)_{1.8}\square_{0.2}$	9.432	6.884	0.730	530.4	Badrour et al. (1998)
$Ca_{7.30}Co_{2.70}(PO_4)_6F_2$	9.347	6.833	0.731	517.0	Grisafe and Hummel (1970b)
$Ca_{8.20}Co_{1.80}(PO_4)_6F_2$	9.347	6.839	0.732	517.4	Grisafe and Hummel (1970b)
$Ca_{9.10}Co_{0.90}(PO_4)_6F_2$	9.358	6.849	0.732	519.4	Grisafe and Hummel (1970b)
$Ca_{9.55}Co_{0.45}(PO_4)_6F_2$	9.365	6.862	0.733	521.2	Grisafe and Hummel (1970b)
$Ca_{5.50}Co_{4.50}(PO_4)_6Cl_2$	9.644	6.749	0.700	543.6	Grisafe and Hummel (1970b)
$Ca_{6.40}Co_{3.60}(PO_4)_6Cl_2$	9.643	6.748	0.700	543.4	Grisafe and Hummel (1970b)
$Ca_{7.30}Co_{2.70}(PO_4)_6Cl_2$	9.643	6.751	0.700	543.7	Grisafe and Hummel (1970b)
$Ca_{8.20}Co_{1.80}(PO_4)_6Cl_2$	9.644	6.754	0.700	544.0	Grisafe and Hummel (1970b)
$Ca_{9.10}Co_{0.90}(PO_4)_6Cl_2$	9.643	6.755	0.701	544.0	Grisafe and Hummel (1970b)
$Ca_{9.55}Co_{0.45}(PO_4)_6Cl_2$	9.385	6.875	0.733	524.4	Piriou et al. (1987)
$Ca_8Eu_2(PO_4)_6O_2$	9.401	6.880	0.732	526.6	Piriou et al. (1987)
$Ca_{9.95}Eu_{0.05}(PO_4)_6O_{1.03}\square_{0.98}$	9.385(2)	6.893(3)	0.734	525.8	Mayer and Cohen (1983)
$Ca_6Eu_2Na_2(PO_4)_6F_2$	9.355	6.867	0.734	520.5	Kreidler and Hummel (1970)
$Ca_9Mg(PO_4)_6F_2$	9.3446(3)	6.9199(4)	0.741	523.3	Wilson et al. (2004)
$Ca_{8.11}Mg_{0.05}Na_{1.17}(PO_4)_{4.09}(CO_3)_{1.91}(H_2O)_{1.05}(OH)_{1.38}$					

Table A1 continued.

Composition	a (Å)	c (Å)	c/a	Volume (Å^3)	Reference
$Ca_{9.1}Na_{0.5}(PO_4)_{4.7}(HPO_4)_{1.3}\cdot 2.9H_2O$	9.436	6.880	0.729	530.5	Mandjiny et al. (1998)
$Ca_{9.1}Na_{0.5}(PO_4)_6(OH)_{0.7}\cdot 4.2H_2O$	9.436	6.880	0.729	530.5	Mandjiny et al. (1998)
$Ca_{9.8}Na_{0.06}(PO_4)_{5.6}(HPO_4)_{0.4}\cdot 0.84H_2O$	9.415	6.879	0.731	528.1	Mandjiny et al. (1998)
$Ca_{9.8}Na_{0.06}(PO_4)_6(OH)_{1.6}\cdot 1.2H_2O$	9.415	6.879	0.731	528.1	Mandjiny et al. (1998)
$Ca_{9.24}Cd_{0.70}Na_{0.06}(PO_4)_6(OH)_{1.6}\cdot 1.2H_2O$	9.410(5)	6.875(5)	0.731	527.2	Jeanjean et al (1996a)
$Ca_{9.50}(NH_4)_{0.10}(PO_4)_{5.05}(CO_3)_{0.95}(OH)_2$	9.437(1)	6.888(1)	0.730	531.2	Ivanova et al. (2001)
$(Ca_{9.02}Nd_{0.98}][(PO_4)_{5.1}(SiO_4)_{0.9}]F_{1.53}O_{0.27}$	9.3938(8)	6.9013(5)	0.735	527.4	Boyer et al. (1998)
$Ca_9Ni(PO_4)_6F_2$	9.364	6.870	0.734	521.7	Kreidler and Hummel (1970)
$Ca_{9.8}Sb_{0.2}(PO_4)_6F_2$	9.3750	6.8904	0.735	524.5	DeBoer et al. (1991)
$Ca_{9.7}Sb_{0.3}(PO_4)_6F_2$	9.3715	6.8867	0.735	523.8	DeBoer et al. (1991)
$Ca_8BiNa(PO_4)_6F_2$	9.396(2)	6.914(2)	0.736	528.6	Mayer and Semadja (1983)
$Ca_6Bi_2Na_2(PO_4)_6F_2$	9.449(2)	6.952(2)	0.736	537.5	Mayer and Semadja (1983)
$Ca_4Bi_3Na_3(PO_4)_6F_2$	9.465(2)	6.968(2)	0.736	540.6	Mayer and Semadja (1983)
$Ca_6Sm_2Na_2(PO_4)_6F_2$	9.3895(3)	6.8950(4)	0.734	526.4	Toumi et al. (2000)
$Ca_{3.2}Sr_{4.8}(PO_4)_6F_2$	9.859(1)	7.206(2)	0.731	606.6	Sudarsanan and Young (1980)
$Ca_{8.83}Sr_{1.17}(PO_4)_6F_2$	9.416(1)	6.924(1)	0.735	531.6	Rakovan and Hughes (2000)
$Ca_{9.37}Sr_{0.63}(PO_4)_6F_2$	9.3902(10)	6.9011(7)	0.735	527.0	Hughes et al. (1991)
$Ca_{5.15}Sr_{4.85}(PO_4)_6Cl_2$	9.737(2)	7.022(4)	0.721	576.6	Sudarsanan and Young (1980)
$Ca_{9.27}Sr_{0.73}(PO_4)_6Cl_2$	9.653(3)	6.777(1)	0.702	546.9	Sudarsanan and Young (1980)
$Ca_{9.77}Sr_{0.23}(PO_4)_6Cl_2$	9.643(1)	6.766(1)	0.702	544.9	Sudarsanan and Young (1980)
$Ca_{9.42}Sr_{0.18}H_{0.8}(PO_4)_6(OH)_2$	9.3920(7)	6.890(1)	0.734	526.3	Kikuchi et al. (1994)
$Ca_{10}(PO_4)_4(SO_4)(SiO_4)F_2$	9.45	6.96	0.737	538.0	Schwarz (1967b)
$Ca_{10}(PO_4)_2(SO_4)(SiO_4)(OH)_2$	9.44	6.96	0.737	537.0	Schwarz (1967b)
$Ca_6La_4(SiO_4)_4(PO_4)_2F_2$	9.5911	7.0717	0.737	563.4	Boyer et al. (1997)
$Ca_8La_2(SiO_4)_2(PO_4)_4F_2$	9.5027	6.9806	0.735	545.9	Boyer et al. (1997)
$Ca_6La_4(SiO_4)_4(PO_4)_2O\square$	9.5843	7.0444	0.735	561.9	Boyer et al. (1997)
$Ca_8La_2(SiO_4)_2(PO_4)_4O\square$	9.4802	6.9570	0.734	541.5	Boyer et al. (1997)
$Ca_{10}(PO_4)_{5.7}(VO_4)_{0.3}(OH)_2$	9.4267(5)	6.8944(5)	0.731	530.6	Boechat et al. (2000)
$Ca_{10}(PO_4)_{4.5}(VO_4)_{1.5}(OH)_2$	9.4904(6)	6.9166(5)	0.729	539.5	Boechat et al. (2000)
$Ca_{10}(PO_4)_3(VO_4)_3(OH)_2$	9.5624(10)	6.9400(8)	0.726	549.6	Boechat et al. (2000)
$Ca_{10}(PO_4)_{1.5}(VO_4)_{4.5}(OH)_2$	9.6713(7)	6.9732(7)	0.721	564.9	Boechat et al. (2000)
$Ca_{10}(SiO_4)_3(SO_4)_3F_2$	9.54	6.99	0.733	550.9	Klement and Dihn (1941)
$Ca_{10}(SO_4)_3(SiO_4)_3F_2$	9.447(3)	6.938(3)	0.734	536.2	Schwarz (1967b)
$Ca_2Dy_8(SiO_4)_6O_2$	9.37	6.81	0.727	517.8	Ito (1968)

Table A1 continued.

Composition	a (Å)	c (Å)	c/a	Volume (Å^3)	Reference
Ca$_4$La$_6$(SiO$_4$)$_6$F$_2$	9.6503	7.1412	0.740	576.0	Boyer et al (1997)
Ca$_2$Er$_8$(SiO$_4$)$_6$O$_2$	9.33	6.75	0.723	508.9	Ito (1968)
Ca$_2$Gd$_8$(SiO$_4$)$_6$O$_2$	9.39	6.87	0.732	524.6	Ito (1968)
Ca$_2$La$_8$(SiO$_4$)$_6$O$_2$	9.63	7.12	0.739	571.8	Ito (1968)
Ca$_4$La$_6$(SiO$_4$)$_6$O☐	9.6351	7.1341	0.740	573.6	Boyer et al (1997)
Ca$_6$La$_2$Ce$_2$(SiO$_4$)$_6$	9.52	7.00	0.735	549.4	Cockbain and Smith (1967)
Ca$_4$La$_3$Ce$_3$(SiO$_4$)$_6$(OH)$_2$	9.52	7.01	0.736	550.2	Cockbain and Smith (1967)
Ca$_2$Lu$_8$(SiO$_4$)$_6$O$_2$	9.28	6.68	0.720	498.2	Ito (1968)
Ca$_2$Na$_2$La$_6$Si$_4$P$_2$O$_{24}$(OH)$_2$	9.62	7.11	0.739	569.8	Ito (1968)
Ca$_2$Nd$_8$(SiO$_4$)$_6$O$_2$	9.52	7.00	0.735	549.4	Ito (1968)
Ca$_6$Nd$_4$(SiO$_4$)$_6$	9.52	7.00	0.735	549.4	Cockbain and Smith (1967)
Ca$_6$Nd$_4$(SiO$_4$)$_6$(OH)$_2$	9.52	7.02	0.737	550.9	Cockbain and Smith (1967)
Ca$_2$Pu$_8$(SiO$_4$)$_6$O$_2$	9.561	7.028	0.735	556.4	Ito (1968)
Ca$_2$Pu$_8$(SiO$_4$)$_6$O$_2$	9.561	7.028	0.735	556.4	Vance et al. (2003)
Ca$_2$Sm$_8$(SiO$_4$)$_6$O$_2$	9.44	6.93	0.734	534.8	Ito (1968)
Ca$_2$Y$_8$(SiO$_4$)$_6$O$_2$	9.34	6.77	0.725	511.5	Ito (1968)
Ca$_4$Ce$_6$(SiO$_4$)$_6$(OH)$_2$	9.61	7.09	0.738	567.1	Ito (1968)
(Ca$_{4.30}$Ce$_{5.70}$)(SiO$_4$)$_6$[F$_{1.0}$(OH)$_{1.0}$]	9.580(5)	6.980(3)	0.719	554.8	Noe et al. (1993)
Ca$_4$Dy$_6$(SiO$_4$)$_6$(OH)$_2$	9.40	6.83	0.727	522.6	Ito (1968)
Ca$_4$Er$_6$(SiO$_4$)$_6$(OH)$_2$	9.38	6.78	0.723	516.6	Ito (1968)
Ca$_4$Gd$_6$(SiO$_4$)$_6$(OH)$_2$	9.43	6.89	0.731	530.6	Ito (1968)
Ca$_4$La$_6$(SiO$_4$)$_6$(OH)$_2$	9.66	7.12	0.737	575.4	Ito (1968)
Ca$_4$La$_6$Si$_4$P$_2$O$_{26}$	9.62	7.07	0.735	566.6	Ito (1968)
Ca$_6$La$_4$Si$_2$P$_4$O$_{26}$	9.57	7.02	0.734	556.8	Ito (1968)
Ca$_8$La$_6$Ce$_6$(SiO$_4$)$_{12}$(OH)$_4$	9.52	7.01	0.736	550.2	Cockbain and smith (1967)
Ca$_4$Lu$_6$(SiO$_4$)$_6$(OH)$_2$	9.35	6.74	0.721	510.3	Ito (1968)
Ca$_4$Na$_2$La$_4$Si$_2$P$_4$O$_{24}$(OH)$_2$	9.51	7.00	0.736	548.3	Ito (1968)
Ca$_4$Nd$_6$(SiO$_4$)$_6$(OH)$_2$	9.56	7.00	0.732	554.0	Ito (1968)
Ca$_4$Sm$_6$(SiO$_4$)$_6$(OH)$_2$	9.50	6.94	0.731	542.4	Ito (1968)
Ca$_4$Y$_6$(SiO$_4$)$_6$(OH)$_2$	9.40	6.81	0.724	521.1	Ito (1968)
Ca$_4$Y$_6$Si$_4$P$_2$O$_{26}$	9.35	6.82	0.729	516.3	Ito (1968)
Ca$_6$Y$_4$Si$_2$P$_4$O$_{26}$	9.36	6.84	0.731	519.0	Ito (1968)
Ca$_6$Y$_4$Si$_4$P$_2$O$_{24}$(OH)$_2$	9.39	6.83	0.727	521.5	Ito (1968)
Ca$_8$Y$_2$Si$_2$P$_4$O$_{24}$(OH)$_2$	9.41	6.87	0.730	526.8	Ito (1968)

Table A1 continued.

Composition	a (Å)	c (Å)	c/a	Volume (Å^3)	Reference
CaY$_9$Si$_5$BO$_{26}$	9.28	6.78	0.731	505.7	Ito (1968)
Ca$_4$Na$_6$(SO$_4$)$_6$F$_2$	9.49	6.87	0.724	535.8	Klement and Dihn (1941)
Ca$_{10}$(VO$_4$)$_6$Cl$_2$	10.13	6.55	0.647	582.1	Kreidler and Hummel (1970)
Ca$_{10}$(VO$_4$)$_6$F$_2$	9.68	7.01	0.724	568.9	Kreidler and Hummel (1970)
Ca$_{10}$(VO$_4$)$_6$(OH)$_2$	9.7405(6)	7.0040(8)	0.719	575.5	Boechat et al. (2000)
Ca$_{10}$(VO$_4$)$_6$(OH)$_2$	9.818	6.981	0.711	582.8	Kutoglu (1974)
Ca$_{10}$(V$_{0.98}$Mn$_{0.02}$O$_4$)$_6$F$_2$	9.70	7.00	0.722	570.4	Dardenne et al. (1999)
Cd$_{10}$(AsO$_4$)$_6$Br$_2$	10.100(1)	6.519(1)	0.645	575.9	Sudarsanan et al. (1977)
Cd$_{10}$(AsO$_4$)$_6$Cl$_2$	10.04	6.51	0.648	568.3	Engel (1965)
Cd$_{10}$(PO$_4$)$_6$(OH)$_2$	9.335(2)	6.664(3)	0.714	502.9	Hata et al. (1978)
Cd$_{10}$(PO$_4$)$_6$Br$_2$	9.733(1)	6.468(1)	0.665	530.6	Sudarsanan et al. (1977)
Cd$_{9.64}$(PO$_4$)$_6$Br$_{3.04}$	9.733(1)	6.468(1)	0.665	530.6	Sudarsanan et al. (1977)
Cd$_{10}$(PO$_4$)$_6$F$_2$	9.30	6.63	0.713	496.6	Kreidler and Hummel (1970)
Cd$_{10}$(PO$_4$)$_6$Cl$_2$	9.633(4)	6.484(4)	0.673	521.1	Sudarsanan et al. (1977)
Cd$_{9.84}$(PO$_4$)$_6$Cl$_{1.814}$	9.633(4)	6.484(4)	0.673	521.1	Sudarsanan et al. (1977)
Cd$_{0.5}$Pb$_{9.5}$(PO$_4$)$_6$(OH)$_2$	9.861(1)	7.3604(1)	0.746	619.8	Badraoui et al. (2002a)
Cd$_{1.0}$Pb$_{9.0}$(PO$_4$)$_6$(OH)$_2$	9.810(1)	7.3315(6)	0.747	611.0	Badraoui et al. (2002a)
Cd$_{2.0}$Pb$_{8.0}$(PO$_4$)$_6$(OH)$_2$	9.741(1)	7.2522(7)	0.745	595.9	Badraoui et al. (2002a)
Cd$_{3.0}$Pb$_{7.0}$(PO$_4$)$_6$(OH)$_2$	9.692(1)	7.174(1)	0.740	583.6	Badraoui et al. (2002a)
Cd$_2$La$_8$(SiO$_4$)$_6$O$_2$	9.64	7.09	0.735	570.6	Ito (1968)
Cd$_{10}$(VO$_4$)$_6$Br$_2$	10.173(2)	6.532(1)	0.642	585.4	Sudarsanan et al. (1977)
Cd$_{9.72}$(VO$_4$)$_6$Br$_{2.82}$	10.173(2)	6.532(1)	0.642	585.4	Sudarsanan et al. (1977)
Cd$_{9.28}$(VO$_4$)$_6$I$_{2.78}$	10.307(1)	6.496(1)	0.630	597.6	Sudarsanan et al. (1977)
Cd$_{10}$(VO$_4$)$_6$F$_2$	10.11	6.52	0.645	577.1	Engel (1965)
Cd$_{10}$(VO$_4$)$_6$I$_2$	10.307(1)	6.496(1)	0.630	597.6	Sudarsanan et al. (1977)
Ce$_{9.33}$□$_{0.67}$(SiO$_4$)$_6$O$_2$	9.657	7.121	0.737	575.1	Felsche (1972)
(Ce$_{0.4}$Ca$_{0.35}$Sr$_{0.25}$)$_4$(Ce$_{0.86}$Ca$_{0.14}$)$_6$(SiO$_4$)$_6$(O$_{0.5}$F$_{0.38}$)$_2$	9.638(1)	7.081(1)	0.735	569.6	Genkina et al. (1991)
Dy$_{9.33}$□$_{0.67}$(SiO$_4$)$_6$O$_2$	9.373	6.784	0.724	516.2	Felsche (1972)
Er$_2$Sr$_6$Na$_2$(PO$_4$)$_6$F$_2$	9.605	7.112	0.740	568.2	Mayer et al. (1974)
Er$_{9.33}$□$_{0.67}$(SiO$_4$)$_6$O$_2$	9.324	6.686	0.717	503.4	Felsche (1972)
Eu$_{10}$(AsO$_4$)$_6$(OH)$_2$	9.995(2)	7.345(2)	0.735	635.5	Mayer et al. (1975)
Eu$_{10}$(PO$_4$)$_6$Cl$_2$	9.866(3)	7.187(2)	0.728	605.8	Mayer et al. (1975)
Eu$_{10}$(PO$_4$)$_6$F$_2$	9.726(3)	7.265(5)	0.747	595.2	Mayer et al. (1975)
Eu$_2$Ca$_6$Na$_2$(PO$_4$)$_6$F$_2$	9.374	6.882	0.734	523.7	Mayer et al. (1975)

Table A1 continued.

Composition	a (Å)	c (Å)	c/a	Volume (Å^3)	Reference
Eu$_3$Ca$_4$Na$_3$(PO$_4$)$_6$F$_2$	9.374	6.882	0.734	523.7	Mayer et al. (1975)
Eu$_4$Ca$_2$Na$_4$(PO$_4$)$_6$F$_2$	9.374	6.882	0.734	523.7	Mayer et al. (1975)
Eu$_2$Sr$_6$Na$_2$(PO$_4$)$_6$F$_2$	9.620	7.142	0.742	572.4	Mayer et al. (1975)
Eu$_3$Sr$_4$Na$_3$(PO$_4$)$_6$F$_2$	9.557	7.041	0.737	556.9	Mayer et al. (1975)
K$_8$Na$_2$(VO$_4$)$_6$	10.106(1)	7.451(1)	0.737	659.0	Sirotinkin et al. (1989)
Eu$_{0.33}$□$_{0.67}$(SiO$_4$)$_6$O$_2$	9.472	6.905	0.729	536.6	Felsche (1972)
Gd$_{0.33}$□$_{0.67}$(SiO$_4$)$_6$O$_2$	9.431	6.873	0.729	529.4	Felsche (1972)
Ho$_{0.33}$□$_{0.67}$(SiO$_4$)$_6$O$_2$	9.346	6.744	0.722	510.3	Felsche (1972)
KNd$_9$(SiO$_4$)$_6$O$_2$	9.576(2)	7.009(2)	0.732	556.6	Pushcharovskii et al. (1978)
La$_{9.33}$(GeO$_4$)$_6$O$_2$	9.9117(1)	7.2833(2)	0.735	619.7	Berastegui et al. (2002)
La$_3$Ba$_4$Na$_3$(PO$_4$)$_6$F$_2$	9.856	7.369	0.748	619.9	Mayer et al. (1975)
La$_2$Sr$_6$Na$_2$(PO$_4$)$_6$F$_2$	9.690	7.219	0.745	587.0	Mayer et al. (1974)
□$_{0.67}$La$_{9.33}$(SiO$_4$)$_6$□$_2$	9.55	7.14	0.748	563.9	Kuzmin and Belov (1965)
La$_{9.33}$□$_{0.67}$(SiO$_4$)$_6$O$_2$	9.713	7.194	0.741	587.8	Felsche (1972)
La$_4$Ca$_{3.5}$(SiO$_4$)$_6$(H$_2$O)F	9.664	7.090	0.734	573.4	Kalsbeek et al. (1990)
La$_3$Nd$_{11}$(SiO$_4$)$_9$O$_3$	9.638(2)	21.35(8)	2.215	1717.5	Malinovskii et al. (1990)
La$_{9.33}$Si$_6$O$_{26}$	9.719	7.183	0.739	587.6	Mazza et al. (2000)
La$_{9.66}$Si$_5$BO$_{26}$	9.630	7.196	0.747	577.9	Mazza et al. (2000)
La$_{10}$Si$_4$B$_2$O$_{26}$	9.5587(2)	7.2171(2)	0.755	571.1	Mazza et al. (2000)
LiY$_9$(SiO$_4$)$_6$O$_2$	9.34	6.72	0.719	507.7	Ito (1968)
LiY$_9$(SiO$_4$)$_6$O$_2$ (at 100K)	9.3108(14)	6.7088(10)	0.721	503.7	Redhammer et al. (2003)
LiY$_9$(SiO$_4$)$_6$O$_2$ (at 295K)	9.3376(14)	6.7321(10)	0.721	508.3	Redhammer et al. (2003)
LiLa$_9$(SiO$_4$)$_6$O$_2$	9.681	7.160	0.739	581.2	Felsche (1972)
LiCe$_9$(SiO$_4$)$_6$O$_2$	9.623	7.091	0.736	568.7	Felsche (1972)
LiPr$_9$(SiO$_4$)$_6$O$_2$	9.575	7.040	0.735	558.8	Felsche (1972)
LiNd$_9$(SiO$_4$)$_6$O$_2$	9.529	6.994	0.733	550.1	Felsche (1972)
LiSm$_9$(SiO$_4$)$_6$O$_2$	9.464	6.918	0.730	536.6	Felsche (1972)
LiEu$_9$(SiO$_4$)$_6$O$_2$	9.437	6.876	0.728	530.3	Felsche (1972)
LiGd$_9$(SiO$_4$)$_6$O$_2$	9.413	6.852	0.727	525.8	Felsche (1972)
LiTb$_9$(SiO$_4$)$_6$O$_2$	9.381	6.803	0.725	522.0	Felsche (1972)
LiDy$_9$(SiO$_4$)$_6$O$_2$	9.362	6.769	0.723	513.9	Felsche (1972)
LiHo$_9$(SiO$_4$)$_6$O$_2$	9.337	6.763	0.724	508.6	Felsche (1972)
LiEr$_9$(SiO$_4$)$_6$O$_2$	9.316	6.696	0.718	503.3	Felsche (1972)
LiTm$_9$(SiO$_4$)$_6$O$_2$	9.301	6.672	0.717	499.8	Felsche (1972)

Table A1 continued.

Composition	a (Å)	c (Å)	c/a	Volume (Å³)	Reference
LiYb$_9$(SiO$_4$)$_6$O$_2$	9.270	6.637	0.715	493.8	Felsche (1972)
LiLu$_9$(SiO$_4$)$_6$O$_2$	9.265	6.615	0.713	490.9	Felsche (1972)
Lu$_{9.33}$□$_{0.67}$(SiO$_4$)$_6$O$_2$	9.260	6.621	0.715	491.6	Felsche (1972)
Mg$_{1.0}$Ca$_{9.0}$(PO$_4$)$_6$(OH)$_2$	9.416(2)	6.858(3)	0.728	526.6	Bigi et al. (1996)
Mg$_{1.5}$Ca$_{8.5}$(PO$_4$)$_6$(OH)$_2$	9.418(2)	6.833(3)	0.726	524.9	Bigi et al. (1996)
Mg$_{2.5}$Ca$_{7.5}$(PO$_4$)$_6$(OH)$_2$	9.432(3)	6.806(6)	0.722	524.4	Bigi et al. (1996)
Mg$_{3.0}$Ca$_{7.0}$(PO$_4$)$_6$(OH)$_2$	9.459(5)	6.800(8)	0.719	526.9	Bigi et al. (1996)
Mg$_2$Dy$_8$(SiO$_4$)$_6$O$_2$	9.31	6.69	0.719	502.2	Ito (1968)
Mg$_2$Er$_8$(SiO$_4$)$_6$O$_2$	9.28	6.58	0.709	490.7	Ito (1968)
Mg$_2$Gd$_8$(SiO$_4$)$_6$O$_2$	9.33	6.75	0.723	508.9	Ito (1968)
Mg$_2$La$_8$(SiO$_4$)$_6$O$_2$	9.59	7.05	0.735	561.5	Ito (1968)
Mg$_2$Nd$_8$(SiO$_4$)$_6$O$_2$	9.45	6.86	0.726	530.5	Ito (1968)
Mg$_2$Sm$_8$(SiO$_4$)$_6$O$_2$	9.38	6.8	0.725	518.1	Ito (1968)
Mg$_2$Y$_8$(SiO$_4$)$_6$O$_2$	9.31	6.64	0.713	498.4	Ito (1968)
MgY$_9$Si$_5$BO$_{26}$	9.18	6.73	0.733	491.2	Ito (1968)
Mn$_{10}$(PO$_4$)$_6$Cl$_2$	9.30	6.20	0.667	464.4	Klement and Haselbeck (1965)
Mn$_{10}$(PO$_4$)$_6$Cl$_{1.8}$(OH)$_{0.2}$	9.532(1)	6.199(1)	0.650	487.8	Engel et al. (1975b)
Mn$_2$Dy$_8$(SiO$_4$)$_6$O$_2$	9.33	6.71	0.719	505.8	Ito (1968)
Mn$_2$Er$_8$(SiO$_4$)$_6$O$_2$	9.30	6.65	0.715	498.1	Ito (1968)
Mn$_2$Gd$_8$(SiO$_4$)$_6$O$_2$	9.34	6.79	0.727	513.0	Ito (1968)
Mn$_2$La$_8$(SiO$_4$)$_6$O$_2$	9.63	7.08	0.735	568.6	Ito (1968)
Mn$_2$Nd$_8$(SiO$_4$)$_6$O$_2$	9.47	6.91	0.730	536.7	Ito (1968)
Mn$_2$Sm$_8$(SiO$_4$)$_6$O$_2$	9.42	6.85	0.727	526.4	Ito (1968)
Mn$_2$Y$_8$(SiO$_4$)$_6$O$_2$	9.32	6.69	0.718	503.3	Ito (1968)
Mn$_4$Dy$_6$(SiO$_4$)$_6$(OH)$_2$	9.33	6.68	0.716	503.6	Ito (1968)
Mn$_4$Er$_6$(SiO$_4$)$_6$(OH)$_2$	9.28	6.63	0.714	494.5	Ito (1968)
Mn$_4$Gd$_6$(SiO$_4$)$_6$(OH)$_2$	9.38	6.78	0.723	516.6	Ito (1968)
Mn$_4$La$_6$(SiO$_4$)$_6$(OH)$_2$	9.66	7.05	0.730	569.7	Ito (1968)
Mn$_4$Nd$_6$(SiO$_4$)$_6$(OH)$_2$	9.50	6.90	0.726	539.3	Ito (1968)
Mn$_4$Sm$_6$(SiO$_4$)$_6$(OH)$_2$	9.43	6.82	0.723	525.2	Ito (1968)
Mn$_4$Y$_6$(SiO$_4$)$_6$(OH)$_2$	9.31	6.65	0.714	499.2	Ito (1968)
Na$_6$Ca$_4$(BeF$_4$)$_6$F$_2$	9.247	6.809	0.736	504.2	Engel (1978)
Na$_6$Pb$_4$(BeF$_4$)$_6$F$_2$	9.497	6.997	0.737	546.5	Engel (1978)
Na$_6$Pb$_4$(BeF$_4$)$_6$F$_2$	9.531(3)	7.028(2)	0.737	552.9	Engel and Fischer (1990)

Table A1 continued.

Composition	a (Å)	c (Å)	c/a	Volume (Å^3)	Reference
NaLa$_9$(GeO$_4$)$_6$O$_2$	9.8833(2)	7.267(3)	0.735	614.7	Takahashi et al. (1998)
Na$_2$Ca$_2$Sm$_2$(PO$_4$)$_6$F$_2$	9.3895(3)	6.895(4)	0.734	526.4	Toumi et al. (2000)
Na$_6$Cd$_4$(SO$_4$)$_6$Cl$_2$	9.574(4)	6.780(3)	0.708	538.2	Perret and Bouillet (1975)
Na$_6$Pb$_4$(SO$_4$)$_6$Cl$_2$	9.815(4)	7.105(3)	0.724	592.7	Perret and Bouillet (1975)
Na$_6$Pb$_4$(SO$_4$)$_6$Cl$_2$	9.810(20)	7.140(20)	0.728	595.1	Schneider (1967)
Na$_{6.35}$Ca$_{3.65}$(SO$_4$)$_6$F$_{1.65}$	9.4364(21)	6.9186(16)	0.733	533.5	Piotrowski et al. (2002a)
Na$_{6.39}$Ca$_{3.61}$(SO$_4$)$_6$Cl$_{1.61}$	9.5423(1)	6.8429(1)	0.717	539.6	Piotrowski et al. (2002a)
NaLa$_9$(SiO$_4$)$_6$O$_2$	9.69	7.18	0.741	583.9	Ito (1968)
NaLa$_9$(SiO$_4$)$_6$O$_2$	9.687	7.180	0.741	583.5	Felsche (1972)
NaCe$_9$(SiO$_4$)$_6$O$_2$	9.628	7.117	0.739	571.3	Felsche (1972)
NaPr$_9$(SiO$_4$)$_6$O$_2$	9.580	7.080	0.739	562.6	Felsche (1972)
NaNd$_9$(SiO$_4$)$_6$O$_2$	9.535	7.027	0.737	553.4	Felsche (1972)
NaSm$_9$(SiO$_4$)$_6$O$_2$	9.472	6.943	0.733	539.6	Felsche (1972)
NaEu$_9$(SiO$_4$)$_6$O$_2$	9.456	6.912	0.731	535.2	Felsche (1972)
NaGd$_9$(SiO$_4$)$_6$O$_2$	9.419	6.878	0.730	528.4	Felsche (1972)
NaTb$_9$(SiO$_4$)$_6$O$_2$	9.390	6.840	0.728	522.2	Felsche (1972)
NaDy$_9$(SiO$_4$)$_6$O$_2$	9.362	6.800	0.726	516.2	Felsche (1972)
NaHo$_9$(SiO$_4$)$_6$O$_2$	9.337	6.760	0.724	510.4	Felsche (1972)
NaEr$_9$(SiO$_4$)$_6$O$_2$	9.321	6.728	0.721	506.2	Felsche (1972)
NaTm$_9$(SiO$_4$)$_6$O$_2$	9.310	6.688	0.718	502.0	Felsche (1972)
NaYb$_9$(SiO$_4$)$_6$O$_2$	9.300	6.661	0.716	498.8	Felsche (1972)
NaLu$_9$(SiO$_4$)$_6$O$_2$	9.290	6.635	0.714	495.9	Felsche (1972)
Na$_2$La$_8$(SiO$_4$)$_6$(OH)$_2$	9.74	7.17	0.736	589.1	Ito (1968)
(Na$_{1.46}$La$_{8.55}$)(SiO$_4$)$_6$(F$_{0.9}$O$_{0.11}$)	9.678(1)	7.1363(3)	0.737	578.9	Hughes et al. (1992)
(Na,Th,La)$_{10}$(SiO$_4$)$_6$O$_2$	9.66	7.13	0.738	576.2	Ito (1968)
NaY$_9$(SiO$_4$)$_6$O$_2$	9.334(2)	6.759(1)	0.724	510.0	Gunawardane et al. (1982)
NaY$_9$(SiO$_4$)$_6$O$_2$	9.33	6.75	0.723	508.9	Ito (1968)
Na$_2$Y$_8$(SiO$_4$)$_6$(OH)$_2$	9.34	6.78	0.726	512.2	Ito (1968)
NaY$_9$(SiO$_4$)$_6$O$_2$ (at 100K)	9.3274(10)	6.7554(7)	0.724	508.9	Redhammer et al. (2003)
NaY$_9$(SiO$_4$)$_6$O$_2$ (at 270K)	9.3386(10)	6.7589(8)	0.724	510.5	Redhammer et al. (2003)
Nd$_2$Ca$_6$Na$_2$(PO$_4$)$_6$F$_2$	9.406	6.907	0.734	529.2	Mayer et al. (1975)
Nd$_3$Ca$_4$Na$_3$(PO$_4$)$_6$F$_2$	9.421	6.923	0.735	532.1	Mayer et al. (1975)
Nd$_4$Ca$_2$Na$_4$(PO$_4$)$_6$F$_2$	9.436	6.956	0.737	536.4	Mayer et al. (1975)
Nd$_5$Sr$_6$Na$_2$(PO$_4$)$_6$F$_2$	9.640	7.168	0.744	576.9	Mayer et al. (1975)

Table A1 continued.

Composition	a (Å)	c (Å)	c/a	Volume (Å^3)	Reference
$Nd_3Sr_4Na_3(PO_4)_6F_2$	9.638	7.168	0.744	576.6	Mayer et al. (1975)
$Nd_{0.33}\square_{0.67}(SiO_4)_6O_2$	9.563	7.029	0.735	556.9	Felsche (1972)
$Nd_8Mn_2(SiO_4)_6O_2$	9.4986(9)	6.944(2)	0.731	542.6	Kluver and Muller-Buschbaum (1995)
$Pb_{10}(AsO_4)_6F_2$	10.07	7.42	0.737	651.6	Merker and Wondratschek (1959)
$Pb_{10}(AsO_4)_6Cl_2$	10.250(2)	7.454(1)	0.727	678.2	Calos and Kennard (1990)
$Pb_{10}(AsO_4)_6Cl_2$	10.25	7.46	0.728	678.8	Merker and Wondratschek (1959)
$Pb_{10}(AsO_4)_6(OH)_2$	10.154(2)	7.515(2)	0.740	671.0	Engel (1970)
$Pb_6K_4(AsO_4)_4(SO_4)_2$	10.130	7.459	0.736	662.8	Schwarz (1967a)
$Pb_9K(AsO_4)_5(SiO_4)$	10.06	7.40	0.736	648.6	Wondratschek (1963)
$Pb_{10}(CrO_4)_3(GeO_4)_3F_2$	10.115(3)	7.466(3)	0.738	661.5	Schwarz (1967b)
$Pb_{10}(CrO_4)_3(SiO_4)_3F_2$	9.973(3)	7.401(3)	0.742	637.5	Schwarz (1967b)
$Pb_{10}(GeO_4)_4(CrO_4)_2$	10.105(3)	7.428(2)	0.735	656.9	Engel and Deppisch (1988)
$Pb_{10}(GeO_4)_4(SO_4)_2$	10.058(4)	7.416(1)	0.737	649.7	Engel and Deppisch (1988)
$Pb_{10}(GeO_4)_2(VO_4)_4$	10.099(3)	7.400(2)	0.733	652.9	Ivanov (1990)
$Pb_{10}(PO_4)_6(OH)_2$	9.8612(4)	7.4242(2)	0.753	625.2	Kim et al. (2000)
$Pb_{10}(PO_4)_6(OH)_2$	9.866(3)	7.426(2)	0.753	626.0	Brückner et al. (1995)
$Pb_{10}(PO_4)_6Br_2$	10.0618(3)	7.3592(1)	0.731	645.2	Kim et al. (2000)
$Pb_{10}(PO_4)_6Cl_2$	9.9767(4)	7.3255(1)	0.734	631.5	Kim et al. (2000)
$Pb_{10}(PO_4)_6Cl_2$	9.9764(10)	7.35511(20)	0.737	633.6	Dai and Hughes (1989)
$Pb_{10}(PO_4)_6Cl_2$	9.95(2)	7.31(2)	0.735	626.8	Hendricks et al. (1932)
$Pb_{10}(PO_4)_6Cl_2$	9.993(2)	7.334(6)	0.734	634.3	Hashimoto and Matsumoto (1998)
$Pb_{10}(PO_4)_6Cl_2$	9.9981(8)	7.344(1)	0.735	635.8	Akao et al. (1989)
$Pb_{10}(PO_4)_6F_2$	9.7547(4)	7.2832(2)	0.747	600.2	Kim et al. (2000)
$Pb_{10}(PO_4)_6F_2$	9.777	7.310	0.748	605.1	Grisafe and Hummel (1970a)
$Pb_{10}(PO_4)_6F_2$	9.760(8)	7.300(8)	0.748	602.2	Belokoneva et al. (1982)
$\square Pb_9(PO_4)_6\square_2$	9.826(4)	7.357(3)	0.749	615.2	Hata et al. (1980)
$Pb_{10}(PO_4)_6O$	9.826	2x7.431	2x0.756	2x621.3	Engel et al. (1975b)
$Pb_{10}(PO_4)_6O\square$	9.84	2x7.43	2x0.755	2x623.0	Wondratschek (1963)
$Pb_{10}(PO_4)_6S\square$	9.45(8)	6.84	0.724	529.0	Trombe and Montel (1975)
$Pb_{10}(PO_4)_4(GeO_4)_2$	9.891(3)	7.338(3)	0.742	621.7	Schwarz (1968)
$Pb_8Bi_2(PO_4)_4(SiO_4)_2$	9.76	7.26	0.744	598.9	Wondratschek (1963)
$Pb_8Bi_2Tl(PO_4)_4(SiO_4)$	9.78	7.34	0.750	608.0	Wondratschek (1963)
$Pb_6Bi_3Na_3(PO_4)_6F_2$	9.653(2)	7.149(2)	0.741	576.9	Mayer and Semadja (1983)
$Pb_6Bi_2Na_2(PO_4)_6F_2$	9.690(2)	7.183(2)	0.741	584.1	Mayer and Semadja (1983)

Table A1 continued.

Composition	a (Å)	c (Å)	c/a	Volume (Å^3)	Reference
$Pb_8BiNa(PO_4)_6F_2$	9.732(2)	7.246(2)	0.742	594.3	Mayer and Semadja (1983)
$Pb_6Bi_2Na_2(PO_4)_6Cl_2$	9.849(2)	7.195(2)	0.731	604.4	Mayer and Semadja (1983)
$Pb_8BiNa(PO_4)_6Cl_2$	9.888(2)	7.217(2)	0.730	611.1	Mayer and Semadja (1983)
$Pb_6Ca_2Li_2(PO_4)_6$	9.6790(15)	7.1130(7)	0.735	577.1	Naddari et al. (2002)
$Pb_8Na_2(PO_4)_6$	9.722(2)	7.193(2)	0.740	588.8	Mayer and Semadja (1983)
$Pb_9Na(PO_4)_6$	9.77	7.26	0.743	600.1	Wondratschek (1963)
$Pb_6Na_3Bi(PO_4)_6$	9.689(2)	7.154(2)	0.738	581.6	Mayer and Semadja (1983)
$Pb_4Na_4Bi_2(PO_4)_6$	9.657(2)	7.079(2)	0.733	571.7	Mayer and Semadja (1983)
$Pb_6KNa_2Bi(PO_4)_6$	9.725(2)	7.146(2)	0.735	585.3	Mayer and Semadja (1983)
$Pb_{10}(PO_4)_4(SiO_4)_2$	9.804	7.327	0.748	609.9	Schwarz (1967a)
$Pb_{10}(PO_4)_4(SiO_4)_2$	9.794	7.307	0.746	606.9	Engel et al. (1975b)
$Pb_8Ca_2(PO_4)_4(SiO_4)_2$	9.734	7.203	0.740	591.5	Engel et al. (1975b)
$Pb_6Ca_4(PO_4)_4(SiO_4)_2$	9.666	7.070	0.731	572.1	Engel et al. (1975b)
$Pb_4Ca_6(PO_4)_4(SiO_4)_2$	9.567	7.047	0.737	558.5	Engel et al. (1975b)
$Pb_6K_4(PO_4)_4(SeO_4)_2$	9.843	7.336	0.745	615.5	Schwarz (1967a)
$Pb_6K_4(PO_4)_4(SO_4)_2$	9.839	7.335	0.746	614.9	Schwarz (1967a)
$Pb_{0.57}Ca_{9.43}(PO_4)_6F_2$	9.7825(3)	7.0318(2)	0.719	582.8	Dong and White (2004b)
$Pb_{1.33}Ca_{8.67}(PO_4)_6F_2$	9.8208(6)	7.0521(5)	0.718	589.0	Dong and White (2004b)
$Pb_{1.76}Ca_{8.24}(PO_4)_6F_2$	9.8441(5)	7.0635(5)	0.718	592.8	Dong and White (2004b)
$Pb_{3.01}Ca_{6.99}(PO_4)_6F_2$	9.9175(7)	7.1154(6)	0.717	606.1	Dong and White (2004b)
$Pb_{3.06}Ca_{6.94}(PO_4)_6F_2$	9.9174(4)	7.1199(3)	0.718	606.5	Dong and White (2004b)
$Pb_{3.82}Ca_{6.18}(PO_4)_6F_2$	9.9559(2)	7.1508(2)	0.718	613.8	Dong and White (2004b)
$Pb_{4.42}Ca_{3.58}(PO_4)_6F_2$	9.9856(2)	7.1825(1)	0.719	620.2	Dong and White (2004b)
$Pb_{5.55}Ca_{4.45}(PO_4)_6F_2$	10.0233(2)	7.2410(2)	0.722	630.0	Dong and White (2004b)
$Pb_{7.00}Ca_{3.00}(PO_4)_6F_2$	10.0531(1)	7.3033(1)	0.726	639.2	Dong and White (2004b)
$PbCa_9(PO_4)_6(OH)_2$	9.468	6.925	0.731	537.6	Engel et al. (1975b)
$Pb_2Ca_8(PO_4)_6(OH)_2$	9.521	6.976	0.733	547.6	Engel et al. (1975b)
$Pb_4Ca_6(PO_4)_6(OH)_2$	9.619	7.056	0.734	565.4	Engel et al. (1975b)
$Pb_5Ca_5(PO_4)_6(OH)_2$	9.653	7.091	0.735	572.2	Engel et al. (1975b)
$Pb_6Ca_4(PO_4)_6(OH)_2$	9.691	7.130	0.736	579.9	Engel et al. (1975b)
$Pb_7Ca_3(PO_4)_6(OH)_2$	9.718	7.210	0.742	589.7	Engel et al. (1975b)
$Pb_8Ca_2(PO_4)_6(OH)_2$	9.780	7.298	0.746	604.5	Engel et al. (1975b)
$Pb_2Ca_8(PO_4)_6(O,(OH))_2$	9.496	6.978	0.735	544.9	Engel et al. (1975b)
$Pb_6Ca_4(PO_4)_6(O,(OH))_2$	9.668	7.116	0.736	576.0	Engel et al. (1975b)

Table A1 continued.

Composition	a (Å)	c (Å)	c/a	Volume (Å^3)	Reference
Pb$_8$K$_2$(PO$_4$)$_6$	9.827(1)	7.304(1)	0.743	610.8	Mathew et al. (1980)
Pb$_8$Na$_2$(PO$_4$)$_6$	9.7249(8)	7.190(1)	0.739	588.9	El Koumiri et al. (2000)
Pb$_8$Na$_2$(PO$_4$)$_6$	9.734	7.200	0.740	590.8	Engel et al. (1975a)
Pb$_8$Na$_2$(PO$_4$)$_2$(SiO$_4$)$_2$(SO$_4$)$_2$	9.79	7.29	0.744	605.1	Wondratschek (1963)
Pb$_4$Ca$_4$Na$_2$(PO$_4$)$_6$	9.595	7.025	0.732	560.1	Engel et al. (1975a)
Pb$_6$Ca$_2$Na$_2$(PO$_4$)$_6$	9.660	7.082	0.733	572.3	Engel et al. (1975a)
Pb$_6$Bi$_4$(SiO$_4$)$_6$	9.75	7.25	0.745	596.9	Wondratschek (1963)
Pb$_2$Dy$_8$(SiO$_4$)$_6$O$_2$	9.47	6.87	0.725	533.6	Ito (1968)
Pb$_{10}$(SiO$_4$)$_3$(SO$_4$)$_3$F$_2$	9.88	7.41	0.750	626.4	Kreidler and Hummel (1970)
Pb$_2$Gd$_8$(SiO$_4$)$_6$O$_2$	9.54	6.95	0.729	547.8	Ito (1968)
Pb$_2$La$_8$(SiO$_4$)$_6$O$_2$	9.71	7.20	0.742	587.9	Ito (1968)
Pb$_2$Nd$_8$(SiO$_4$)$_6$O$_2$	9.65	7.12	0.738	574.2	Ito (1968)
□Pb$_3$Pr$_6$(SiO$_4$)$_6$□$_2$	9.662(1)	7.162(1)	0.741	579.0	Grisafe and Hummel (1970b)
Pb$_2$Sm$_8$(SiO$_4$)$_6$O$_2$	9.58	7.06	0.737	561.1	Ito (1968)
Pb$_2$Y$_8$(SiO$_4$)$_6$O$_2$	9.42	6.80	0.722	522.6	Ito (1968)
Pb$_4$Ce$_6$(SiO$_4$)$_6$(OH)$_2$	9.77	7.19	0.736	594.4	Ito (1968)
Pb$_4$Dy$_6$(SiO$_4$)$_6$(OH)$_2$	9.70	6.90	0.711	562.2	Ito (1968)
Pb$_4$Er$_6$(SiO$_4$)$_6$(OH)$_2$	9.68	6.84	0.707	555.1	Ito (1968)
Pb$_4$Gd$_6$(SiO$_4$)$_6$(OH)$_2$	9.72	6.99	0.719	571.9	Ito (1968)
Pb$_4$La$_6$(SiO$_4$)$_6$(OH)$_2$	9.80	7.26	0.741	603.8	Ito (1968)
Pb$_4$Lu$_6$(SiO$_4$)$_6$(OH)$_2$	9.64	6.75	0.700	543.2	Ito (1968)
Pb$_4$Nd$_6$(SiO$_4$)$_6$(OH)$_2$	9.76	7.13	0.731	588.2	Ito (1968)
Pb$_4$Sm$_6$(SiO$_4$)$_6$(OH)$_2$	9.74	7.05	0.724	579.2	Ito (1968)
Pb$_4$Y$_6$(SiO$_4$)$_6$(OH)$_2$	9.68	6.86	0.709	556.7	Ito (1968)
Pb$_{10}$(SO$_4$)$_2$(GeO$_4$)$_4$	10.058(4)	7.416(1)	0.737	649.7	Engel and Deppisch (1988)
Pb$_{10}$(SO$_4$)$_3$(GeO$_4$)$_3$F$_2$	10.032(3)	7.450(3)	0.743	649.3	Schwarz (1967b)
Pb$_{10}$(SO$_4$)$_3$(SiO$_4$)$_3$F$_2$	9.890(3)	7.424(3)	0.751	628.9	Schwarz (1967b)
Pb$_4$Na$_6$(SO$_4$)$_6$F$_2$	9.63	7.11	0.738	571.0	Kreidler and Hummel (1970)
Pb$_8$Na$_2$(VO$_4$)$_6$	10.059	7.330	0.729	642.3	Sirotinkin et al. (1989)
Pb$_{10}$(VO$_4$)$_6$(OH)$_2$	10.165(2)	7.463(2)	0.734	667.8	Engel (1970)
Pb$_{10}$(VO$_4$)$_6$Cl$_2$	10.317(3)	7.338(3)	0.711	676.4	Dai and Hughes (1989)
Pr$_{10}$(VO$_4$)$_6$F$_2$	10.113	7.375	0.729	653.2	Grisafe and Hummel (1970a)
Pb$_{10}$(VO$_4$)$_6$F$_2$	10.11	7.34	0.726	649.7	Kreidler and Hummel (1970)
Pb$_{9.85}$(VO$_4$)$_6$I$_{1.7}$	10.442(5)	7.467(3)	0.715	705.1	Audubert et al. (1999)

Table A1 continued.

Composition	a (Å)	c (Å)	c/a	Volume (Å^3)	Reference
$Pr_{9.33}\square_{0.67}(SiO_4)_6O_2$	9.607	7.073	0.736	565.3	Felsche (1972)
$Pr_3Ba_4Na_3(PO_4)_6F_2$	9.813	7.300	0.744	608.8	Mayer et al. (1974)
$\square_2Pr_8(SiO_4)_6\square_2$	9.613(2)	7.068(2)	0.735	565.6	Grisafe and Hummel (1970b)
$Sm_3Ba_4Na_3(PO_4)_6F_2$	9.759	7.243	0.742	597.4	Mayer et al. (1974)
$Sm_{9.33\ 0.67}(SiO_4)_6O_2$	9.493	6.946	0.732	542.0	Felsche (1972)
$Sm_{10}(SiO_4)_6O_2$	9.4959(10)	7.0361(7)	0.741	549.5	Morgan et al. (2002)
$\square_{0.67}Sm_{9.33}(SiO_4)_6\square_2$	9.33	6.85	0.743	516.4	Kuzmin and Belov (1965)
$\square_2Sm_8(SiO_4)_6\square_2$	9.497(3)	6.949(2)	0.732	542.8	McCarthy et al. (1967)
$Sr_{10}(AsO_4)_6CO_3$	10.212	7.392	0.724	667.6	Roux and Bonel (1977)
$Sr_{10}(AsO_4)_6F_2$	9.99	7.40	0.741	639.6	Kreidler and Hummel (1970)
$Sr_{10}(AsO_4)_6Cl_2$	10.12	7.50	0.741	665.2	Klement and Harth (1961)
$Sr_{10}(AsO_4)_4(SO_4)_2S_2$	10.161	7.411	0.729	662.6	Schiff-Francois et al. (1979)
$Sr_{10}(CrO_4)_3(GeO_4)_3F_2$	10.002(3)	7.454(3)	0.745	645.8	Schwarz (1967a)
$Sr_{10}(CrO_4)_3(SiO_4)_3F_2$	9.900(3)	7.409(3)	0.748	628.9	Schwarz (1967a)
$Sr_{10}(CrO_4)_6(OH)_2$	9.98	7.40	0.741	638.3	Banks and Jaunarajs (1965)
$Sr_{10}(CrO_4)_6Cl_2$	10.149	7.333	0.723	654.1	Wilhemi and Jonsson (1965)
$Sr_{10}(CrO_4)_6F_2$	9.956(1)	7.437(1)	0.747	638.4	Herdtweck (1991)
$Sr_{10}(PO_4)_6(OH)_2$	9.745(1)	7.265(1)	0.746	597.5	Sudarsanan and Young (1972)
$Sr_{10}(PO_4)_6O_{0.9}(OH)_{0.2}\square_{0.9}$	9.710	7.279	0.750	594.3	Trombe and Montel (1978)
$Sr_{10}(PO_4)_6Br_2$	9.9641(1)	7.2070(1)	0.723	619.7	Nötzold and Wulff (1998)
$Sr_{10}(PO_4)_6Cl_2$	9.8777(1)	7.1892(1)	0.728	607.5	Nötzold et al. (1994)
$Sr_{10}(PO_4)_6Cl_2$	9.859(1)	7.206(2)	0.731	606.6	Sudarsanan and Young (1980)
$Sr_{10}(PO_4)_6F_2$	9.719	7.276	0.749	595.2	Akhavan-Niaki (1961)
$Sr_{10}(PO_4)_6F_2$	9.720(3)	7.289(3)	0.750	596.3	Schwarz (1967b)
$Sr_{10}(PO_4)_6F_2$	9.717	7.284	0.750	595.6	Grisafe and Hummel (1970a)
$Sr_{10}(PO_4)_6F_2$	9.776(2)	7.270(2)	0.744	601.7	Mayer and Semadja (1983)
$Sr_{10}(PO_4)_6CO_3$	9.882	7.239	0.733	612.2	Nadal et al. (1971)
$Sr_{5.08}Ba_{4.9}Eu_{0.02}(PO_4)_6Cl_2$	10.0265(2)	7.4099(2)	0.739	645.1	Nötzold and Wulff (1996)
$Sr_{3.98}Ba_{6.00}Eu_{0.02}(PO_4)_6Cl_2$	10.0699(1)	7.4599(1)	0.741	655.1	Nötzold and Wulff (1996)
$Sr_8BiNa(PO_4)_6F_2$	9.733(2)	7.290(2)	0.749	598.1	Mayer and Semadja (1983)
$Sr_6Bi_2Na_2(PO_4)_6F_2$	9.670(2)	7.246(2)	0.749	586.8	Mayer and Semadja (1983)
$Sr_{9.98}Eu_{0.02}(PO_4)_6F_2$	9.8775(3)	7.1893(1)	0.728	607.5	Nötzold and Wulff (1996)
$Sr_{9.92}Nd_{0.05}(PO_4)_6F_2$	9.7156(4)	7.2810(3)	0.749	595.2	Corker et al. (1995)
$Sr_{1.0}Pb_{9.0}(PO_4)_6F_2$	9.8725(8)	7.3497(5)	0.744	620.4	Badraoui et al. (2002b)

Table A1 continued.

Composition	a (Å)	c (Å)	c/a	Volume (Å^3)	Reference
$Sr_{2.0}Pb_{8.0}(PO_4)_6F_2$	9.8681(9)	7.3616(5)	0.746	620.8	Badraoui et al. (2002b)
$Sr_{4.0}Pb_{6.0}(PO_4)_6F_2$	9.8625(7)	7.3414(8)	0.744	618.4	Badraoui et al. (2002b)
$Sr_{4.9}Pb_{5.1}(PO_4)_6F_2$	9.8572(7)	7.33527(7)	0.744	617.2	Badraoui et al. (2002b)
$Sr_{6.9}Pb_{3.1}(PO_4)_6F_2$	9.8121(5)	7.3203(6)	0.746	610.4	Badraoui et al. (2002b)
$Sr_{8.9}Pb_{1.1}(PO_4)_6F_2$	9.7736(9)	7.2958(4)	0.746	603.5	Badraoui et al. (2002b)
$Sr_{10}(P_{0.70}Mn_{0.30}O_4)_6F_2$	9.746	7.308	0.750	601.1	Dardenne et al. (1999)
$Sr_{10}(P_{0.80}Mn_{0.20}O_4)_6F_2$	9.747	7.307	0.750	601.2	Dardenne et al. (1999)
$Sr_{10}(P_{0.90}Mn_{0.10}O_4)_6F_2$	9.736	7.301	0.750	599.3	Dardenne et al. (1999)
$Sr_{10}(P_{0.95}Mn_{0.05}O_4)_6F_2$	9.722	7.289	0.750	596.6	Dardenne et al. (1999)
$Sr_{10}(P_{0.98}Mn_{0.02}O_4)_6F_2$	9.731	7.294	0.750	598.2	Dardenne et al. (1999)
$Sr_{10}(P_{0.999}Mn_{0.001}O_4)_6F_2$	9.719	9.287	0.956	759.7	Dardenne et al. (1999)
$Sr_{10}(P_{0.99}Mn_{0.01}O_4)_6F_2$	9.733	7.290	0.749	598.1	Dardenne et al. (1999)
$Sr_{10}(PO_4)_4(GeO_4)_2$	9.827	7.340	0.747	613.9	Schwarz (1968)
$Sr_{10}(PO_4)_4(SiO_4)_2$	9.765	7.316	0.749	604.1	Schwarz (1968)
$Sr_{10}(PO_4)_4(SO_4)(SiO_4)F_2$	9.744	7.305	0.750	600.6	Schwarz (1967b)
$Sr_{10}(PO_4)_2(SO_4)_2(SiO_4)_2F_2$	9.787	7.330	0.749	608.0	Schwarz (1967b)
$Sr_{10}(SO_4)_3(GeO_4)_3F_2$	9.905	7.392	0.746	628.1	Schwarz (1967b)
$Sr_{10}(SO_4)_3(SiO_4)_3F_2$	9.796	7.343	0.750	610.2	Schwarz (1967b)
$Sr_2Dy_8(SiO_4)_6O_2$	9.42	6.90	0.732	530.3	Ito (1968)
$Sr_2Er_8(SiO_4)_6O_2$	9.36	6.81	0.728	516.7	Ito (1968)
$Sr_2Gd_8(SiO_4)_6O_2$	9.47	6.97	0.736	541.3	Ito (1968)
$Sr_2La_8(SiO_4)_6O_2$	9.69	7.22	0.745	587.1	Ito (1968)
$\square Sr_3La_6(SiO_4)_6\square_2$	9.710	7.244	0.746	591.5	Schwarz (1967c)
$Sr_2Nd_8(SiO_4)_6O_2$	9.57	7.09	0.741	562.3	Ito (1968)
$\square Sr_3Pr_6(SiO_4)_6\square_2$	9.611(2)	7.144(2)	0.743	571.5	Grisafe and Hummel (1970b)
$Sr_2Sm_8(SiO_4)_6O_2$	9.51	7.02	0.738	549.8	Ito (1968)
$Sr_2Y_8(SiO_4)_6O_2$	9.38	6.86	0.731	522.7	Ito (1968)
$Sr_4Ce_6(SiO_4)_6(OH)_2$	9.67	7.14	0.738	578.2	Ito (1968)
$Sr_4Dy_6(SiO_4)_6(OH)_2$	9.43	6.92	0.734	532.9	Ito (1968)
$Sr_4Er_6(SiO_4)_6(OH)_2$	9.42	6.83	0.725	524.9	Ito (1968)
$Sr_4Gd_6(SiO_4)_6(OH)_2$	9.53	7.00	0.735	550.6	Ito (1968)
$Sr_4Ho_6(SiO_4)_6(OH)_2$	9.42	6.91	0.734	531.0	Ito (1968)
$Sr_4La_6(SiO_4)_6(OH)_2$	9.71	7.23	0.745	590.3	Ito (1968)
$Sr_4Lu_6(SiO_4)_6(OH)_2$	9.50	6.79	0.715	530.7	Ito (1968)

Table A1 continued.

Composition	a (Å)	c (Å)	c/a	Volume (Å^3)	Reference
$Sr_4Nd_6(SiO_4)_6(OH)_2$	9.62	7.10	0.738	569.0	Ito (1968)
$Sr_4Sm_6(SiO_4)_6(OH)_2$	9.6	7.05	0.734	562.7	Ito (1968)
$Sr_4Y_6(SiO_4)_6(OH)_2$	9.52	6.91	0.726	542.4	Ito (1968)
$Sr_{10}(VO_4)_6Cl_2$	10.21	7.30	0.715	659.0	Kreidler and Hummel (1970)
$Sr_{10}(VO_4)_6F_2$	10.006	7.430	0.743	644.2	Grisafe and Hummel (1970a)
$Sr_{10}(VO_4)_6F_2$	10.01	7.43	0.742	644.7	Kreidler and Hummel (1970)
$Sr_{10}(V_{0.98}Mn_{0.02}O_4)_6F_2$	10.02	7.44	0.743	646.9	Dardenne et al. (1999)
$Sr_{10}(VO_4)_6(CuO_2)$	10.126(1)	7.415(1)	0.732	658.4	Carrillo-Cabrera and von Schnering (1999)
$Sr_{10}(VO_4)_6(CuO_2)_{2/3}$	9.7815(4)	7.3018(4)	0.746	605.0	Kazin et al. (2003)
$Sr_{9.82}Nd_{0.12}(VO_4)_6F_2$	10.0077(6)	7.4342(8)	0.743	644.8	Corker et al. (1995)
$Tb_{9.33}\square_{0.67}(SiO_4)_6O_2$	9.401	6.825	0.726	522.4	Felsche (1972)
$Tm_{9.33}\square_{0.67}(SiO_4)_6O_2$	9.300	6.666	0.717	499.3	Felsche (1972)
$Y_2Ca_6Na_2(PO_4)_6F_2$	9.358	6.866	0.734	520.7	Mayer et al. (1974)
$Y_3Ca_4Na_3(PO_4)_6F_2$	9.344	7.845	0.840	593.2	Mayer et al. (1974)
$Y_4Ca_2Na_4(PO_4)_6F_2$	9.313	7.81	0.839	586.6	Mayer et al. (1974)
$(Y_4Mg)Si_3O_{13}$	9.298(2)	6.635(1)	0.714	496.8	Suwa et al. (1968)
$(Y_{3.73}Mg_{1.31})Si_{3.05}O_{13}$	9.312	6.638	0.713	498.5	Suwa et al. (1968)
$(Y_{3.30}Mg_{1.33})Si_{3.36}O_{13}$	9.306	6.632	0.713	497.4	Suwa et al. (1968)
$Y_{10}Si_4B_2O_{26}$	9.15	6.75	0.738	489.4	Ito (1968)
$Yb_{9.33}\square_{0.67}(SiO_4)_6O_2$	9.275	6.636	0.715	494.4	Felsche (1972)

Table A2. Cell parameters for apatites reported with $P6_3$, $P\bar{3}$ or $P\bar{6}$ symmetry.

Composition	S.G.	a (Å)	c (Å)	c/a	Volume (Å^3)	Reference
$Ba_{10}(CrO_4)_6(OH)_2$	$P6_3$	10.428	7.89	0.757	743.0	Banks and Jaunarajs (1965)
$Ba_{10}((Ge,C)(O,OH)_4)_6(OH)_2$	$P6_3$	10.207(3)	7.734(2)	0.758	697.8	Malinovskii et al. (1975)
$Ba_{10}(PO_4)_6(OH)_2$	$P6_3$	10.1904(7)	7.721(2)	0.758	694.4	Bondareva and Malinovskii (1986)
$Bi_2Ca_8(VO_4)_6O_2$	$P6_3$	9.819(2)	7.033(2)	0.716	587.2	Huang and Sleight (1993)
$Ca_{10}(PO_4)_6S\square$	$P6_3$	9.455	8.84	0.935	684.4	Suitch et al. (1986)
$Ca_2Sr_2Ce_6(PO_4)_6F_2$	$P6_3$	9.485	7.000	0.738	545.4	Khomyakov et al. (1997)
$(Ca_{4.3}Ce_{5.7})(SiO_4)_6(F)(OH)$	$P6_3$	9.58	6.98	0.729	554.8	Noe et al. (1993)
$Ca_8Bi_2(VO_4)_6O_2$	$P6_3$	9.819(2)	7.033(2)	0.716	587.2	Huang and Sleight (1993)

Table A2 continued.

Composition	S.G.	a (Å)	c (Å)	c/a	Volume (Å^3)	Reference
$Cd_{10}(PO_4)_6(OH)_2$ (superstructure)	$P6_3$	16.199	6.648	0.820	1510.8	Hata and Marumo (1983)
$K_6Sn_4(SO_4)_6Cl_2$	$P6_3$	10.230(20)	7.560(20)	0.739	685.2	Howie et al. (1973)
$K_6Sn_4(SO_4)_6Cl_2$	$P6_3$	10.183(6)	7.540(2)	0.740	677.2	Donaldson and Grimes (1984)
$K_6Sn_4(SO_4)_6Br_2$	$P6_3$	10.280(20)	7.670(20)	0.746	701.9	Howie et al. (1973)
$K_6Sn_4(SO_4)_6Br_2$	$P6_3$	10.256(2)	7.582(4)	0.739	690.7	Donaldson and Grimes (1984)
$La_2Ca_8(PO_4)_6O_2$	$P6_3$	9.463(8)	6.92(1)	0.731	536.7	Buvaneswari and Varadaraju (2000)
$La_2Sr_8(PO_4)_6O_2$	$P6_3$	9.71(1)	7.30(1)	0.752	596.1	Buvaneswari and Varadaraju (2000)
$La_6Ca_{3.5}(SiO_4)_6(H_2O)F$	$P6_3$	9.664(3)	7.090(1)	0.734	573.4	Kalsbeek et al. (1990)
$La_{0.97}Ca_{1.40}La_{2.20}Ce_{3.69}Pr_{0.32}Nd_{0.80}[(Si_{5.69}P_{0.31})_6](OH,F)$	$P6_3$	9.664(3)	7.090(1)	0.734	573.4	Kalsbeek et al. (1990)
$Na_{6.39}Ca_{3.61}(SO_4)_6Cl_{1.61}$	$P6_3$	9.542	6.843	0.717	539.6	Piotrowskii et al. (2002a)
$Na_6Pb_4(SO_4)_6Cl_2$	$P6_3$	9.810	7.140	0.728	595.1	Schneider (1967)
$Pb_5GeV_{12}O_{12}$	$P6_3$	10.097	7.396	0.732	653.0	Ivanov and Zavodnik (1989)
$Sm_{10}(SiO_4)_6N_2$	$P6_3$	9.517(6)	6.981(4)	0.734	540.5	Gaude et al. (1975)
$(Sm_8Cr_2)(SiO_4)_6N_2$	$P6_3$	9.469(5)	6.890(4)	0.728	534.9	Maunaye et al. (1976)
$Sr_{10}(CrO_4)_6Cl_2$	$P6_3$	10.125	7.328	0.724	650.6	Banks and Jaunarajs (1965)
$Sr_6Ca_4(PO_4)_6F_2$	$P6_3$	9.63	7.22	0.750	579.9	Klevtsova (1964)
$Sr_{7.3}Ca_{2.7}(PO_4)_6F_2$	$P6_3$	9.565(8)	7.115(3)	0.744	563.7	Pushcharovskii et al. (1987)
$Sr_{10}(SiO_4)_3(CrO_4)_3F_2$	$P\bar{6}$	9.9	7.409	0.748	628.9	Schwarz (1967b)
$Ba_4Nd_3Na_3(PO_4)_6F_2$	$P\bar{3}$	9.786(2)	7.281(1)	0.744	603.9	Mathew et al. (1979), Mayer et al. (1974)
$Ca_{10}(PO_4)_6[(CO_3)_x(OH)_{2-2x}]$, $(x \geq 0.5)$	$P\bar{3}$	9.5211(3)	6.8725(2)	0.722	539.5	Fleet and Liu (2003)
$Na_2Ce_2Sr_6(PO_4)_6(OH)_2$	$P\bar{3}$	9.620	7.120	0.740	570.6	Klevtsova and Borisov (1964)
$Na_2Cr_4Sr_2Ce_2(PO_4)_6F_2$	$P\bar{3}$	9.51	7.01	0.737	549.0	Khomyakov et al. (1996)
$Na_{0.5}Ca_{0.3}Ce_{1.00}Sr_{2.95}(PO_4)_6(OH)_2$	$P\bar{3}$	9.692(3)	7.201(1)	0.743	585.8	Nadezhina et al. (1987)
$Na_{1.960}La_{1.998}Sr_{5.508}Ba_{0.24}Ca_{0.12}(PO_4)_6OH_2$	$P\bar{3}$	9.664	7.182	0.743	580.9	Kabalov et al. (1997)
$Nd_3Ba_4Na_3(PO_4)_6F_2$	$P\bar{3}$	9.786	7.281	0.744	603.9	Mayer et al. (1975)
$Sr_6Na_2Ce_2(PO_4)_6(OH)_2$	$P\bar{3}$	9.664	7.182	0.743	580.9	Borodin and Kazakova (1954)
$Sr_6Na_2La_2(PO_4)_6(OH)_2$	$P\bar{3}$	9.647	7.170	0.743	577.9	Pekov et al. (1996)
$Sr_{9.402}Na_{0.209}(PO_4)_6B_{0.996}O_2$	$P\bar{3}$	9.734(4)	7.279(2)	0.748	597.3	Calvo et al. (1975)
$Ba_4La_2Na_2(PO_4)_6F_2$	$P\bar{6}$	9.9392(4)	7.4419(5)	0.749	636.7	Mathew et al. (1979)
$Ca_{10}(PO_4)_6O$	$P\bar{6}$	9.432	6.881	0.730	530.1	Alberius-Henning et al. (1999b)
$Ca_{9.55}(PO_4)_{5.52}(CO_3)_{0.48})(CO_3)_{1.157}$	$P\bar{6}$	9.480(3)	6.898(1)	0.728	536.9	Suetsugu et al. (2000)
$Na_{6.9}Ca_{3.1}(SO_4)_6OH_{1.1}$	$P\bar{6}$	9.4434(13)	6.8855(14)	0.729	531.8	Piotrowski et al. (2002b)
$Na_6Pb_4(SO_4)_6Cl_2$	$P\bar{6}$	9.815	7.105	0.724	592.8	Perret and Bouillet (1975)
$Nd_2Ba_6Na_2(PO_4)_6F_2$	$P\bar{6}$	9.910	7.399	0.747	629.3	Mayer et al. (1975)

Table A3. Cell parameters for monoclinic apatites reported with $P2_1/m$, $P2_1/b$ or $P2_1$ symmetry.

Composition	S.G.	a (Å)	b (Å)	c (Å)	γ	Reference
$(Ca_{8.40}Sr_{1.61})(AsO_4)_{2.58}(PO_4)_{3.42}F_{0.69}(OH)_{1.31}$	$P2_1/m$	9.594(2)	9.597(2)	6.975(2)	120.0	Hughes and Drexler (1991)
$Ca_8Sr_2(PO_4)_6(OH)_2$	$P2_1/m$	9.594(2)	9.597(2)	6.975(2)	120.0	Hughes and Drexler (1991)
$Ca_{10}(SiO_4)_3(SO_4)_3(F_{0.16}Cl_{0.48}(OH)_{1.36})$	$P2_1/m$	9.476(2)	9.508(2)	6.919(1)	119.5	Sudarsaran (1980)
$Ca_{10}(VO_4)_6F_2$	$P2_1/m$	9.737	9.7358	7.00572	120.002	Dong and White (2004b)
$Na_6Pb_4(SO_4)_6Cl_2$	$P2_1/m$	19.62	9.81	7.14	120.0	Schneider (1967)
$Ca_{9.97}(PO_4)_6Cl_{1.94}$ (superstructure)	$P2_1/b$	9.632(7)	19.226(20)	6.776(5)	120.01	Bauer and Klee (1993)
$Ca_{9.983}(PO_4)_6Cl_{1.966}$	$P2_1/b$	9.643(5)	19.279(10)	6.766(3)	120.01	Bauer and Klee (1993)
$Ca_{9.9}(PO_4)_{5.98}Cl_{1.84}$	$P2_1/b$	9.426(3)	18.856(5)	6.887(1)	119.97	Ikoma et al. (1999)
$Ca_{10}(PO_4)_6Cl_2$	$P2_1/b$	9.628(5)	19.256(10)	6.764(5)	120.0	Mackie et al. (1972)
$Pb_{10}(AsO_4)_6Cl_2$	$P2_1/b$	10.189	20.372	7.456	119.0	Dai et al. (1991)
$(Ca_{4.30}Ce_{5.70})(SiO_4)_6(F_{1.0}(OH)_{1.0})$	$P2_1$	9.58	9.590	6.980	120.1	Noe et al. (1993)
$Ca_{10}(Si_{3.14}S_{2.94}C_{0.08}P_{0.02})O_{24}((OH)_{1.12}Cl_{0.316}F_{0.05})$	$P2_1$	9.526(2)	9.506(4)	6.922(1)	120.0	Organova et al. (1994)
$(Na_{1.46}La_{8.55})(SiO_4)_6(F_{0.9}O_{0.11})$	$P2_1$	9.678	9.682	7.1363	120.0	Hughes et al. (1992)

Table A4. Cell parameters for apatites reported with $P6_3cm$ and $Pnma$ symmetry.

Composition	S.G.	a (Å)	b (Å)	c (Å)	Reference
$Ba_{10}(ReO_5)_6Cl_2$	$P6_3cm$	10.935(7)	-	7.795(5)	Besse et al. (1979)
$Ba_{10}(ReO_5)_6Cl_2$	$P6_3cm$	10.926(1)	-	7.7816(8)	Schriewer and Jeitschko (1993)
$Ba_{10}(ReO_5)_6Br_2$	$P6_3cm$	10.967(7)	-	7.790(4)	Baud et al. (1979)
$Ba_{10}(ReO_5)_6I_2$	$P6_3cm$	10.932(7)	-	7.776(5)	Baud et al. (1979)
$Ba_{10}(ReO_5)_6F_2$	$P6_3cm$	10.830(2)	-	7.855(3)	Baud et al. (1980)
$Ba_{10}(ReO_5)_6CO_3$	$P6_3cm$	10.938(1)	-	7.788(3)	Baud et al. (1980)
$Ba_{10}(ReO_5)_6(O_2)_2$	$P6_3cm$	10.912(2)	-	7.774(3)	Besse et al. (1980)
$Na_6Pb_4(SO_4)_6F_2$	$P6_3cm$	9.630	-	7.110	Kreidler and Hummel (1970)
$Sr_{10}(ReO_5)_6Cl_2$	$Pnma$	7.4380(9)	18.434(2)	10.563(2)	Schriewer and Jeitschko (1993)
$Sr_{10}(ReO_5)_6Br_2$	$Pnma$	7.4341(9)	18.478(3)	10.576(2)	Schriewer and Jeitschko (1993)
$Sr_{10}(ReO_5)_6I_2$	$Pnma$	7.473(2)	18.646(5)	10.700(3)	Schriewer and Jeitschko (1993)

APPENDIX B – APATITE SYNTHESIS METHODS

Table B1. Solid state route
Table B2. Hydrothermal route
Table B3. Soft chemistry route
Table B4. Sol-gel route

Table B1. Solid state route for apatite synthesis.

Composition	Synthesis Procedure	Reference
$Ba_{10}(AsO_4)_4(SO_4)_2S_2$	Stoichiometric amounts of $BaSO_4$, As_2O_5 and BaS were heated at 1400°C under Ar-atmosphere for 3 h.	Schiff-Francois et al. (1979)
$Ba_{10}(CrO_4)_6(OH)_2$	$Ba_3(CrO_4)_2$ was mixed with $Ba(OH)_2$ and fired at 820°C in air, then at 980°C in N_2 atmosphere.	Banks and Jaunarajs (1965)
$Ba_{10}(CrO_4)_6Cl_2$	Stoichiometric mixture of $Ba(OH)_2$, Cr_2O_3, and $BaCl_2$ was fired at 820°C in air for 4 h.	Banks and Jaunarajs (1965)
$Ba_{10}(CrO_4)_6Cl_2$	$Ba_3(CrO_4)_2$ melts were mixed with an excess of $BaCl_2$ at 1000–1200°C in Ar atmosphere. The excess chlorides were extracted with H_2O.	Wilhelmi and Jonsson (1965)
$Ba_{10}(CrO_4)_6F_2$	Stoichiometric quantities of $Ba(OH)_2$, Cr_2O_3, and BaF_2 were mixed and fired at 800°C in air, followed by ignition at 890°C in a N_2 atmosphere containing H_2O vapor.	Banks and Jaunarajs (1965)
$Ba_{10}(MnO_4)_6Cl_2$	Stoichiometric amounts of K_2HPO_4, $BaCl_2$, BaO_2, Mn_2O_3 were ground and sintered at 850°C.	Reinen et al. (1986)
$Ba_{10}(MnO_4)_6F_2$	Stoichiometric mixtures of $BaCO_3$, Mn_2O_3 and NH_4F were pressed into pellets and repeatedly heated for 12 h at 1250°C in air.	Dardenne et al. (1999)
$Ba_{10}(PO_4)_6Cl_2$	Single crystals were grown by reaction of $BaCl_2.2H_2O$ and K_2HPO_4 in distilled water at 473 K for 2 weeks.	Hata et al. (1979)
$Ba_{10}(PO_4)_6F_2$	A mixture of $BaHPO_4$, BaO and BaF_2 were heated at 1100°C for 4–24 h.	Mathew et al. (1979); Mayer et al. (1974)
$Ba_{10}(PO_4)_6CO_3$	$Ba_{10}(PO_4)_6(OH)_2$ was fired at 900°C in a CO_2 atmosphere.	Mohensi-Koutchesfehani (1961)
$Ba_{10}(PO_4)_6(Cu_{0.30}O_{0.86}H_y)_2$	A mixture of $BaCO_3$, $NH_4H_2PO_4$ and CuO was calcined at 600–800°C for 0.5 h, and then further annealed at 1100°C for 24 h.	Karpov et al. (2003)
$Ba_{10}(PO_4)_{6-x}(CrO_4)_xF_2$ (x = 1.2, 2.4, 3.6, 4.8, 6.0)	Starting mixtures were fired in nitrogen atmosphere at 1000°C for 10 h.	Grisafe and Hummel (1970a)

Table B1 continued.

Composition	Synthesis Procedure	Reference
$Ba_{10}(PO_4)_{6-x}(MnO_4)_xF_2$ ($x = 0.6, 1.2, 2.4, 3.6, 4.8, 6.0$)	Respective starting materials were fired in oxygen atmosphere at 1000–1025°C for 10 h.	Grisafe and Hummel (1970a)
$Ba_{10}(PO_4)_{6-x}(MnO_4)_xCl_2$ ($x = 0.6, 1.2, 2.4, 3.6, 4.8, 6.0$)		Grisafe and Hummel (1970a)
$Ba_{10}(PO_4)_{6-x}(SbO_4)_xCl_2$ ($x \leq 1.2$)	Starting mixtures were fired in air at 950°C for 8 h.	Grisafe and Hummel (1970a)
$Ba_{10}(PO_4)_{6-x}(SbO_4)_xF_2$ ($x \leq 0.6$)	Starting mixtures were fired in air at 1000°C for 10 h.	Grisafe and Hummel (1970a)
$Ba_{10}(PO_4)_{6-x}(VO_4)_xF_2$ ($x = 1.2, 2.4, 3.6, 4.8, 6.0$)	Starting mixtures were fired in air at 1000°C for 8 h.	Grisafe and Hummel (1970a)
$Ba_{10}(P_{1-x}Mn_xO_4)_6F_2$ ($x = 0, 0.001, 0.01, 0.02, 0.033, 0.05, 0.1, 0.2, 0.3, 0.5, 0.6, 0.8, 1$)	Stoichiometric amounts of $BaCO_3$, NH_4HPO_4/Mn_2O_3, and NH_4F were pressed into pellets and heated for 12 h at 1250°C in air (for $x \leq 0.3$) and at 1000°C in an oxygen stream (for $x > 0.3$).	Dardenne et al. (1999)
$Ba_{10-x}Eu_x(PO_4)_6F_2$ ($0.05 \leq x \leq 0.5$) $Ba_{10-x}Eu_x(PO_4)_6Br_2$ ($0.05 \leq x \leq 0.5$) $Ba_{10-x}Eu_x(PO_4)_6Cl_2$ ($0.05 \leq x \leq 0.5$)	$BaHPO_4$, $BaCO_3$, $NH_4F/NH_4Br/BaCl_2 \cdot 2H_2O$ and Eu_2O_3 were mixed proportionately and heated at 1100°C for 2 h in a reducing atmosphere (either N_2 plus H_2 or CO gas produced by burning activated charcoal).	Kottaisamy et al. (1994)
$Ba_6La_2Na_2(PO_4)_6F_2$ $Ba_4Nd_3Na_3(PO_4)_6F_2$	$BaHPO_4$, BaF_2, La_2O_3 and Nd_3PO_4 were stoichiometrically mixed and heated at ~1100°C for 4–24 h.	Mathew et al. (1979); Mayer et al. (1974)
$Ba_{10}(ReO_5)_6Br_2$ $Ba_{10}(ReO_5)_6I_2$	Single crystal were obtained by heating Re and BaX_2 ($X = Br, I$) in air at 700–800°C.	Baud et al. (1979)
$Ba_{10}(ReO_5)_6Cl_2$	Annealed the powders of alkaline-earth metaperrhenates $Ba(ReO_4)_2 \cdot H_2O$ in vacuum-sealed tubes with an excess of $BaCl_2 \cdot H_2O$ at 800°C.	Besse et al. (1979); Schriewer and Jeitschko (1993)
$Ba_{10}(ReO_5)_6CO_3\square$ $Ba_5(ReO_5)_3F$	Single crystals were obtained by heating a mixture of $Re+BeF_2$ or $Re+BaCO_3$ at 700°C in air, followed by slow cooling to room temperature at 4°C/min.	Baud et al. (1980)
$Ba_{10}(ReO_5)_6(O_2)_2$	Single crystal was obtained by heating Re, BaO_2 at 900°C in an oxygen atmosphere.	Besse et al. (1980)

Table B1 continued.

Composition	Synthesis Procedure	Reference
$Ba_2La_8(SiO_4)_6O_2$	A stoichiometric mixtures of $BaCO_3$, amorphous SiO_2 and La_2O_3 was calcined at 1400°C for 2 h.	Takeda et al. (2000)
$\Box Ba_5Nd_4(SiO_4)_4(PO_4)_2\Box_2$	Stoichiometric amounts of metal carbonates or oxides, diammonium hydrogen phosphate and silicic acid were mortared in acetone. Dried sample was placed in Pt or silica crucible and heated to 600–650°C, remixed and reheated to 1350°C for 16 h.	Grisafe and Hummel (1970b)
$Ba_2Ln_8(SiO_4)_6O_2$ (Ln = La, Nd, Sm)	High temperature solid-state reaction of stoichiometric mixtures made from respective oxides of lanthanides, chlorides or hydroxides of alkaline earth metals, cadmium sulfate, manganese chloride, and potassium silicate at 1200°C in air.	Ito (1968)
$Ba_{10}(VO_4)_{6-x}(CrO_4)_xF_2$ ($x = 1.2, 2.4, 3.6, 4.8, 6.0$)	Starting mixtures were fired in N_2 atmosphere at 1000°C for 10 h.	Grisafe and Hummel (1970a)
$Ba_{10}(VO_4)_{6-x}(MnO_4)_xF_2$ ($x = 0.6, 1.2, 2.4, 3.6, 4.8, 6.0$) $Ba_{10}(VO_4)_{6-x}(MnO_4)_xCl_2$ ($x = 0.6, 1.2, 2.4, 3.6, 4.8, 6.0$)	Respective starting materials were fired in an O_2 atmosphere at 975–1000°C for 10 h.	Grisafe and Hummel (1970a)
$Ba_{10}(V_{1-x}P_xO_4)_6F_2$ ($x = 0, 0.02, 0.2, 0.6, 1$)	Stoichiometric amounts of $BaCO_3$, V_2O_5/NH_4HPO_4, and NH_4F were pressed into pellets and heated for 12 h at 1000°C in an oxygen stream.	Dardenne et al. (1999)
$Ba_{10}(VO_4)_{6-x}(SbO_4)_xCl_2$ ($x \leq 1.2$)	Starting mixtures were fired in air at 950°C for 8 h.	Grisafe and Hummel (1970a)
$Ba_{10}(VO_4)_{6-x}(SbO_4)_xF_2$ ($x \leq 0.6$)	Starting mixtures were fired in air at 1000°C for 10 h.	Grisafe and Hummel (1970a)
$Bi_2Ca_8(PO_4)_6O_2$ $Bi_2Ca_6Sr_2(PO_4)_6O_2$ $Bi_2Ca_4Sr_4(PO_4)_6O_2$ $Bi_2Ca_2Sr_6(PO_4)_6O_2$ $Bi_2Sr_8(PO_4)_6O_2$	Stoichiometric quantities of $NH_4H_2PO_4$, Bi_2O_3, $CaCO_3$ and/or $SrCO_3$ were initially heated at 300°C for 6 h to eliminate H_2O and NH_3, and then further heated at 700°C for 12 h and at 950°C for 24 h. Powders were then pelletized and sintered at 1100°C.	Buvaneswari and Varadaraju (2000)

Table B1 continued.

Composition	Synthesis Procedure	Reference
$Bi_2Ca_8(VO_4)_6O_2$	Single crystals were grown by mixing Bi_2O_3, CaO, and V_2O_5 at 700–1240°C for a total of 22 h.	Huang and Sleight (1993)
$Ca_{10}(AsO_4)_6Cl_2$	Crystals were grown via a eutectic flux of 31% $CaCl_2$ and 69% NaCl (m.p. 773 K) from a reaction mixture of As_2O_3, CuO and Cu_2O in a fused-silica ampoule. The reactants were heated at 973 K for 6 days before being slowly cooled to room temperature.	Wardojo and Hwu (1996)
$Ca_{10}(AsO_4)_6CO_3$	Sintered stoichiometric mixture of $Ca_3(PO_4)_2$ and $CaCO_3$ at 900°C in CO_2 atmosphere.	Roux and Bonel (1977)
$Ca_{10}(CrO_4)_6(OH)_2$	Partial oxidation of a mixture of CaO and Cr_2O_3 in humid air at 900°C.	Wilhelmi and Jonsson (1965)
$Ca_{10}(CrO_4)_6(OH)_2$	$CaCrO_4·2H_2O$ and $Ca(OH)_2$ were mixed stoichiometrically and fired at 900°C for 12 h in air humidified at room temperature.	Yasuda and Hishinuma (1995)
$Ca_{10}(CrO_4)_6Cl_2$ $Ca_{10}(CrO_4)_6F_2$	$CaCO_3$, CrO_3 and $CaCl_2/CaF_2$ were stoichiometrically mixed and calcined at 900°C in air, followed by ignition at 950°C in a N_2 atmosphere.	Banks and Jaunarajs (1965)
$Ca_4La_6(GeO_4)_6(OH)_2$	A stoichiometric mixture of CaO, La_2O_3 and GeO_2 was fired at 1350°C for 1 h in air.	Cockbain and Smith (1967)
$Ca_{10}(PO_4)_6Br_2$	Crystals were grown by slowly cooling a mixture of powdered bromapatite and $CaBr_2$ in an HBr atmosphere at 1073 K.	Elliot et al. (1981)
$Ca_{10}(PO_4)_6Br_2$	$Ca_{10}(PO_4)_6(OH)_2$ was reacted with CH_2Br_2 under oxygen flushing at 800°C for 2 h.	Kim et al. (2000)
$Ca_{10}(PO_4)_6Cl_2$	$CaCl_2$ and $Ca_3(PO_4)_2$ were mixed in a stoichiometric ratio and fused in a platinum crucible.	Hendricks et al. (1932)
$Ca_{10}(PO_4)_6Cl_2$	Single crystals were prepared by annealing Ca_2PO_4Cl in argon at 1200°C for several hours, followed by rapid cooling. Rapid cooling is required to prevent reconversion to Ca_2PO_4Cl.	Prener (1971)
$Ca_{10}(PO_4)_6Cl_2$	Crystals were grown from solutions of the apatite in fused $CaCl_2$.	Mackie et al. (1972)

Table B1 continued.

Composition	Synthesis Procedure	Reference
$Ca_{10}(PO_4)_6CO_3$	$Ca_{10}(PO_4)_6(OH)_2$ was treated with dry CO_2 at 1170 K for several days.	Elliot et al. (1980)
$Ca_{10}(PO_4)_6F_2$	$Ca_3(PO_4)_2$ and CaF_2 were mixed and calcined in the range 350-1250°C for 0.5–24 h.	Kreidler and Hummel (1970)
$Ca_{10}(PO_4)_6F_2$	$Ca_4(PO_4)_2O$ and CaF_2 were mixed and calcined at 900°C for several hours.	Elliot (1994); Wallaeys (1952)
$Ca_{10}(PO_4)_6F_2$	Crystals were grown by the Czochralski method in an Ar atmosphere from a seed crystal pulled from an induction heated Ir-crucible at rates of 3–5 mm h^{-1} and rotations of 30–100 rpm.	Mazelsky et al. (1968)
$Ca_{10-x}(PO_4)_6Cl_{2-2x}\square_{2x}$ ($x = 0.03$)	Crystals were grown from a flux of molten $CaCl_2$.	Bauer and Klee (1993)
$Ca_{10}(PO_4)_6[(CO_3)(OH)]_{2-2x}$ ($x \geq 0.5$)	Single crystals were prepared by direct reaction of stoichiometric amounts of $Ca_2P_2O_7$, CaO and $CaCO_3$ in the ratio 3:3:1 at a pressure of 2 Gpa and at 1400–1500°C.	Fleet and Liu (2003)
$Ca_{10}(PO_4)_6Cu_{0.54}O_{1.72}H_{2y}$	$CaCO_3$, $NH_4H_2PO_4$ and CuO were mixed and calcined from 600–800°C for 0.5 h, then further annealed at 1100°C for 24 h.	Karpov et al. (2003)
$Ca_{10}(PO_4)_{6-x}(CrO_4)_xF_2$ ($x \leq 4.8$)	Respective starting materials were fired in a nitrogen atmosphere at 1000°C for 10 h.	Grisafe and Hummel (1970a)
$Ca_{10-x}(PO_4)_{6-x}(HPO_4)_x(OH)_{2-x}$ ($0 \leq x \leq 1$)	H_3PO_4 was added to $(CH_3COO)_2Ca \cdot xH_2O$ with Ca/P=1.5, 1.55, 1.6 and 1.67 for 2 h. NH_3 was used to keep the solution at pH 10. After filtering, the mixture was dried at 100°C overnight.	Liou et al. (2004)
$Ca_{10}(PO_4)_6(NCN)$	Hydroxyapatite was subjected to a nitrogen atmosphere. At 600°C, nitrogen flow was replaced by ammonia and the temperature raised from 800 to 1300°C over 16 h.	Habelitz et al. (1999)
$Ca_{10}(PO_4)_6O_{0.75}(OH)_{0.5}\square_{0.75}$	Peroxiapatites are formed by thermal treatment of calcium hydroxyapatite at 900°C under pure O_2. In order to eliminate traces of water, the stream of oxygen has been allowed to run through a coil cooled by carbon dioxide-ice before reaching the product.	Trombe and Montel (1978)
$Ca_{10}(PO_4)_6S\square$	$Ca_3(PO_4)_2$ was mixed with an excess of CaS and heated at 950°C under vacuum.	Trombe and Montel (1975)

Table B1 continued.

Composition	Synthesis Procedure	Reference
$Ca_{10}(P_{1-x}Mn_xO_4)_6F_2$ $(0.001 \leq x \leq 0.02)$	$CaCO_3$, NH_4HPO_4/Mn_2O_3, and NH_4F were mixed in stoichiometric amounts, pressed into pellets and heated for 12 h at 1250°C in air (for $x \leq 0.3$) and 1000°C in an oxygen stream (for $x > 0.3$).	Dardenne et al. (1999)
$Ca_{10-2x}Bi_xNa_x(PO_4)_6F_2$ $(x = 1, 2, 3)$	Stoichiometric amounts of $Ca_3(PO_4)_2$, CaF_2, $NH_4H_2PO_4$, NaF, $Na_3PO_4 \cdot 12H_2O$ and $Bi(NO_3)_3 \cdot 5H_2O$ were mixed together and heated at 1100°C for 24–48 hrs in a closed gold or Pt-crucible.	Mayer and Semadja (1983)
$Ca_{10-x}Co_x(PO_4)_6F_2$ $(x = 0.45, 0.9, 1.8$ and $2.7)$ $Ca_{10-x}Co_x(PO_4)_6Cl_2$ $(x = 0.45, 0.9, 1.8$ and $2.7)$	Stoichiometric amounts of metal carbonates or oxides and diammonium hydrogen phosphate and/or silicic acid were mixed and dried in sealed platinum crucibles and heated at 975°C for 8 h.	Grisafe and Hummel (1970b)
$Ca_{10-x}Eu_x(PO_4)_6F_2$ $(0.05 \leq x \leq 0.5)$ $Ca_{10-x}Eu_x(PO_4)_6Br_2$ $(0.05 \leq x \leq 0.5)$ $Ca_{10-x}Eu_x(PO_4)_6Cl_2$ $(0.05 \leq x \leq 0.5)$	$CaHPO_4$, $CaCO_3$, $CaF_2/NH_4Br/NH_4Cl$ and Eu_2O_3 were mixed proportionately and heated at 1100°C for 2 h in a reducing atmosphere (either N_2 plus H_2 or CO gas produced by burning activated charcoal).	Kottaisamy et al. (1994)
$Ca_{10-x}Eu_x(PO_4)_6O_{1+x/2}\square_{1-x/2}$ $(0.05 \leq x \leq 2)$	A mixture of dry β-calcium pyrophosphate, Ca-carbonate and Eu-oxide were fired at 1350°C in air. The samples were annealed in vacuum at 900°C until a single-phase oxyapatite was obtained.	Piriou et al. (1987)
$Ca_{10-2x}Na_xTb_x(PO_4)_6F_2$ $(0 \leq x \leq 2)$	Solid-state reactions between a stoichiometric mixture of $Ca_3(PO_4)_2$, NaF, $TbPO_4$ and CaF_2 at 1060°C for 15 h under argon.	Tachihante et al. (1993)
$Ca_{10-x}Ni_x(PO_4)_6Cl_2$ $(x = 0.2, 0.5$ and $1.0)$	Stoichiometric amounts of metal carbonates or oxides and diammonium hydrogen phosphate and/or silicic acid were mixed and dried in covered crucibles and heated at 800°C for 4 h.	Grisafe and Hummel (1970b)
$Ca_8RE_2(PO_4)_6O_2$ $(RE = La, Pr, Nd)$	Stoichiometric amounts of metal carbonates or oxides and diammonium hydrogen phosphate were mortared in acetone. Dried sample was placed in Pt or silica crucible and heated to 600–650°C, remixed and reheated at 1320°C for 72 h.	Grisafe and Hummel (1970b)
$Ca_{10-x}Sb_x(PO_4)_6F_2$ $(x = 0.2, 0.3)$	CaO, CaF_2, $Ca_2P_2O_7$, and $CaSb_2O_6$ were stoichiometrically mixed and fired at 1473 K for 2.5 h in air.	De Boer et al. (1991)

Table B1 continued.

Composition	Synthesis Procedure	Reference
$Ca_6Sm_2Na_2(PO_4)_6F_2$	Crystals were obtained by mixing $CaCO_3$, $(NH_4)_2HPO_4$, Sm_2O_3, and NaF and heating at 400°C for 4 h, then at 1200°C for 1 h.	Toumi et al. (2000)
$Ca_{10-x}Sr_x(PO_4)_6Cl_2$ ($0.23 \leq x \leq 4.85$)	Single crystals were prepared by standard flux-growth techniques. The starting material was calcium chlorapatite mixed with $CaCl_2$ and $SrCl_2$. The mixture was placed in a lightly covered Pt-crucible and heated in a N_2 atmosphere at 1553 K for 24 h. The furnace was cooled at a rate of 30 K/day to 1333 K and then shut off.	Sudarsanan and Young (1980)
$Ca_9Nd(PO_4)_5(SiO_4)F_{1.5}O_{0.25}$	Stoichiometric mixtures of CaF_2, P_2O_5, $CaCO_3$, Nd_2O_3, and SiO_2 were heated at 1973 K over a period of 2 h followed by cooling at 50 K min^{-1}.	Boyer et al. (1998)
$(Ca_{9.02}Nd_{0.98})[(PO_4)_{5.1}(SiO_4)_{0.9}]F_{1.53}O_{0.27}$	Single crystals were obtained from a stoichiometric mixture of CaF_2, P_2O_5, $CaCO_3$, Nd_2O_3 and SiO_2 with rapid heating to 1973 K and kept there for 2 h.	Boyer et al. (1998)
$Ca_{10}(PO_4)_{6-x}(VO_4)_xF_2$ ($x \leq 4.8$)	Respective starting materials were fired in air at 1000°C for 10 h.	Grisafe and Hummel (1970a)
$Ca_{3.05}Ce_{2.38}Fe_{0.25}Gd_{5.37}Si_{4.88}O_{26}$ $Ce_{1.45}Zr_{0.78}Fe_{0.14}Nd_{2.15}Eu_{0.50}Si_{6.02}O_{26}$	Synthesized by inductive melting of an oxide mixture in a cold crucible.	Utsunomiya et al. (2003)
$Ca_2La_8(SiO_4)_6O_2$	Single crystals grown from stoichiometric melt.	Wang and Weber (1999)
$Ca_2Ln_8(SiO_4)_6O_2$ (Ln = La, Nd, Sm, Gd, Dy, Y, Er, Lu)	Solid-state reaction of stoichiometric mixtures made from respective oxides of lanthanides, chlorides or hydroxides of alkaline earth metals, cadmium sulfate, manganese chloride, and potassium silicate at 1200°C in air.	Ito (1968)
$Ca_{10-x}La_x(SiO_4)_x(PO_4)_{6-x}F_2$ ($0 \leq x \leq 6$) $Ca_{10-x}La_x(SiO_4)_x(PO_4)_{6-x}O\square$ ($0 \leq x \leq 6$)	Stoichiometric amounts of La_2O_3, CaF_2, SiO_2, $Ca_2P_2O_7$ and $CaCO_3$ were mixed and sintered at 1100–1400°C.	Boyer et al. (1997)
$Ca_2Nd_8(SiO_4)_6O_2$	Appropriate amounts of Nd_2O_3, CaO were dissolved in 35% nitric acid and the resulting solution was added to an ammonia stabilized silicate solution containing the desired amount of SiO_2. The solution was evaporated to dryness and heated to 500°C for 2 hr. Pressed into pellets and fired at 1250°C for 24 h.	Fahey et al. (1985)

Table B1 continued.

Composition	Synthesis Procedure	Reference
$Ca_2Pu_8(SiO_4)_6O_2$ $Ca_2Gd_{7-A}A(SiO_4)_6O_2$ (A = U, Hf) $Ca_3Gd_6Hf(SiO_4)_6O_2$	Aqueous metal nitrate solutions were mixed with colloidal silica and were calcined at 1350°C for 20 h.	Vance et al. (2003)
$Ca_{5.43}La_{3.22}(SiO_4)_{1.68}(PO_4)_{4.32}O_{0.411}$ $Ca_5La_{4.49}(SiO_4)_{2.8}(PO_4)_{3.2}O_{1.23}$ $Ca_{4.15}La_{5.38}(SiO_4)_{3.74}(PO_4)_{2.26}O_{1.31}$ $CaLa_{8.66}(SiO_4)_6O_2$	Stoichiometric mixtures of $CaCO_3$, La_2O_3, SiO_2, and $(NH_4)_2HPO_4$ were fired at 200–1400°C for 12 h.	El Ouenzerfi (2003)
$Ca_6Th_4(SiO_4)_6O_2$	Stoichiometric mixtures of ThO_2, $CaCO_3$ and SiO_2 were calcined at 1400°C.	Engel (1978)
$Ca_{10}(VO_4)_6F_2$ $Ca_{10}(VO_4)_6Cl_2$	$Ca_3(VO_4)_2$ was mixed with CaF_2 or $CaCl_2$ and fired at 970°C for 1 h.	Kreidler and Hummel (1970)
$Ca_{10-x}La_x(VO_4)_6O_{1+x/2}\square_{1-x/2}$ ($0 \leq x \leq 2$)	Single crystals were grown by the Czochralski method.	Benmoussa et al. (2000)
$Ca_{10}(V_{1-x}Mn_xO_4)_6F_2$ ($x = 0.02$)	$CaCO_3$, V_2O_5/Mn_2O_3, and NH_4F were stoichiometrically mixed, pressed into pellets and heated for 12 h at 1000°C in an O_2 stream.	Dardenne et al. (1999)
$Ca_8Bi_2(VO_4)_6O_2$	A stoichiometric mixture of Bi_2O_3, CaO, and V_2O_5 was first heated at 700°C for 12 h and then at 950°C for 20 h. It was further heated to 1240°C and held at this temperature for 5 min, followed by cooling to 1200°C at a rate of 15 °C/h, held there for 1 h, cooled to 1150°C at a rate of 3°C/h, then cooled to 600°C at 15 °C/h, and finally cooled to room temperature.	Huang and Sleight (1993)
$Cd_{10}(AsO_4)_6Br_2$ $Cd_{10}(AsO_4)_6I_2$ $Cd_{9.84}(AsO_4)_6Br_{3.04}$	Single crystals were grown from $Cd_3(MO_4)_2$ and an excess of CdX_2 (M = As, V, or P and X = Br or I).	Sudarsanan et al. (1977) Wilson et al. (1977)
$Cd_{10}(PO_4)_6Br_2$ $Cd_{10}(PO_4)_6Cl_2$	Crystals were grown from the melt in Pt crucibles charged with $Cd_3(PO_4)_2$ and an excess of $CdBr_2$ or $CdCl_2$. Crucibles were not perfectly closed against air.	Sudarsanan et al. (1977)

Table B1 continued.

Composition	Synthesis Procedure	Reference
$Cd_2La_8(SiO_4)_6O_2$	Solid-state reaction of stoichiometric mixtures made from respective oxides of lanthanides, chlorides or hydroxides of alkaline earth metals, cadmium sulfate, manganese chloride, and potassium silicate at 1200°C in air.	Ito (1968)
$Cd_9Nd(PO_4)_5(SiO_4)F_2$	Stoichiometric amounts of Nd_2O_3, CaF_2, SiO_2, CaP_2O_7 and $CaCO_3$ were mixed and sintered at 1400°C for 6 h in air.	Bregiroux et al. (2003)
$Cd_{10}(VO_4)_6Br_2$ $Cd_{10}(VO_4)_6I_2$	Crystals were grown from melt in platinum crucibles charged with $Cd_3(VO_4)_2$ and an excess of $CdBr_2$ or CdI_2. The crucibles were not perfectly closed against air.	Sudarsanan et al. (1977)
$Eu_{10}(PO_4)_6Cl_2$ $Eu_{10}(PO_4)_6F_2$	$Eu_3(PO_4)_2$, which was prepared by the reduction of $EuPO_4$ by metallic Eu, was mixed with $EuCl_2$ or EuF_2 in vitreosil tubes at 1000°C for 24 h.	Mayer et al. (1975)
$Eu_{10-x}Ba_x(PO_4)_6F_2$ ($0 \le x \le 10$) $Eu_{10-x}Ba_x(PO_4)_6Cl_2$ ($0 \le x \le 10$)	Solid solutions were prepared by reacting the fluoride and chloride apatites of Ba with that of Eu(II) at 1000°C for 24 h.	Mayer et al. (1975)
$Eu_{10-x}Ca_x(PO_4)_6F_2$ ($0 \le x \le 10$)	$Eu_3(PO_4)_2$, which was prepared by the reduction of $EuPO_4$ by metallic Eu, was mixed with $Ca(PO_4)_2$ and EuF_2, and the reactions performed in vitreosil tubes, evacuated and sealed off. This was fired at 1000°C for 24 h.	Mayer et al. (1975)
$Eu_xCa_{10-2x}Na_x(PO_4)_6F_2$ ($2 \le x \le 4$)	$Eu_3(PO_4)_2$ was prepared by heating $(NH_4)_2HPO_4$ and Eu_2O_3 at 500°C for 2–3 h, grinding the mixture, and reheating at 1100°C for 4-12 h. $Eu_3(PO_4)_2$, Na_3PO_4, $Ca_3(PO_4)_2$, CaF_2 were then mixed in stoichiometric ratios and fired at 1100°C for 4-24 h.	Mayer et al. (1974)
$K_2La_2Ba_6(PO_4)_6Z_2$ ($Z = F, Cl$)	Stoichiometric amounts of metal carbonates or oxides and diammonium hydrogen phosphate were mortared in acetone. The dried sample was placed in Pt or silica crucible and heated to 600–650°C, remixed and reheated to 950°C for 8 h.	Grisafe and Hummel (1970b)
$K_6Ca_4(SO_4)_6F_2$	Single crystals were obtained by firing stoichiometric mixtures of $CaSO_4$, K_2SO_4 and CaF_2 together with 3% excess CaF_2 and K_2SO_4 at 960°C for 17 h followed by quenching with water.	Fayos et al. (1987)
$K_8Na_2(VO_4)_6$	Stoichiometric quantities of sodium and potassium carbonates, and NH_4VO_3 were mixed and annealed at 600°C for 48 h.	Sirotinkin et al. (1989)

Table B1 continued.

Composition	Synthesis Procedure	Reference
$K_2La_2Ba_6(VO_4)_6Z_2$ (Z = F, Cl)	Stoichiometric amounts of metal carbonates or oxides and ammonium metavanadate were mortared in acetone. Dried sample was placed in Pt or silica crucible and heated to 600–650°C, remixed and reheated to 950°C for 8 h.	Grisafe and Hummel (1970b)
$La_{9.33}(GeO_4)_6O_2$	Stoichiometric amounts of La_2O_3 and GeO_2 were ground and sintered below 1100°C in air for 24 h followed by sintering at 1000–1250°C for 2 days.	Berastegui et al. (2002)
$La_2Ca_8(PO_4)_6O_2$ $La_2Sr_8(PO_4)_6O_2$	$CaCO_3$/ $SrCO_3$, La_2O_3 and $NH_4H_2PO_4$ mixed stoichiometrically and heaed at 300–1100°C for a total of 42 h.	Buvaneswari and Varadaraju (2000)
$La_{9.33}(SiO_4)_6O_2$	Stoichiometric mixtures of La_2O_3 and SiO_2 were fired at 1200–1400°C for 12 h.	El Ouenzerfi (2003)
$La_{9.33+x/3}Si_{6-x}Ga_xO_2$ ($0 \leq x \leq 2$)	Stoichiometric amounts of La_2O_3, Ga_2O_3 and SiO_2 were ground and heated to 1300°C for 16 h. It was then reground and reheated to 1350°C for 16 h.	Sansom et al. (2004)
$Li_2La_2R_6(PO_4)_6F_2$ (R = Ca, Sr, Ba)	Stoichiometric amounts of metal carbonates or oxides and diammonium hydrogen phosphate were mortared in acetone. Dried sample was placed in Pt or silica crucible and heated to 600–650°C, remixed and reheated to 800 and 900°C each for 6 h.	Grisafe and Hummel (1970b)
$Li_2La_2Pb_6(PO_4)_6F_2$	Stoichiometric amounts of metal carbonates or oxides and diammonium hydrogen phosphate were mortared in acetone. Dried sample was placed in Pt or silica crucible and heated to 600–650°C, remixed and reheated to 850°C for 3 h.	Grisafe and Hummel (1970b)
$LiY_9(SiO_4)_6O_2$	Solid-state reaction of the respective oxides of lanthanides, Mg-nitrate, Li-carbonate, K-silicate and Na-silicate at 1050°C in air.	Ito (1968)
$LiY_9(SiO_4)_6O_2$	Single crystals were obtained by mixing Li_2CO_3, Y_2O_3 and SiO_2 (in proportions corresponding to the chemical composition of $LiYSi_2O_6$) and Li_2MoO_4 (high-temperature solvent 1:10 ratio). The mixture was placed in a Pt-crucible heated slowly to 1573 K, maintained at this temperature for 24 h and then cooled slowly (2 K/h) to 673 K.	Redhammer et al. (2003)

Table B1 continued.

Composition	Synthesis Procedure	Reference
LiRE$_9$[SiO$_4$]$_6$O$_2$ (RE = La→Lu)	Pellets containing the corresponding oxide/salt mixtures 7RE$_2$O$_3$·9SiO$_2$, Li$_2$CO$_3$·9RE$_2$O$_3$·12SiO$_2$ were annealed at different T in the range 1000–1900°C for 24 h after pre-sintering at 900°C.	Felsche (1972)
Ln$_x$M$_{10-2x}$Na$_x$(PO$_4$)$_6$F$_2$ (Ln = La, Pr, Nd, Sm, Eu, Dy, Er, Lu, and Y; M = Sr, and Ba)	Starting materials used were SrHPO$_4$/BaHPO$_4$, the fluorides of Sr and Ba, oxides and phosphates of the rare earths (Ln = La, Pr, Nd, Sm, Eu, Dy, Er, Lu, and Y) and Na$_3$PO$_4$. LnPO$_4$ were prepared by heating (NH$_4$)$_2$HPO$_4$ and Ln$_2$O$_3$ at 500°C for 2–3 h, and reheating at 1100°C for 4–12 h. Stoichiometric mixtures were sintered at 1100°C for 4–24 h.	Mayer et al. (1974)
Mg$_2$La$_8$(SiO$_4$)$_6$O$_2$ (Ln = La, Nd, Sm, Gd, Dy, Y, Er)	Solid-state reaction of the respective oxides of lanthanides, Mg-nitrate, Li-carbonate, K-silicate and Na-silicate at 1050°C in air.	Ito (1968)
Mn$_{10}$(PO$_4$)$_6$Cl$_2$	Mn$_3$(PO$_4$)$_2$ and MnCl$_2$ were mixed and put in a sealed silica tube at 1000°C for several hours.	Klement and Haselbeck (1965)
Mn$_2$Ln$_8$(SiO$_4$)$_6$O$_2$ (Ln = La, Nd, Sm, Gd, Dy, Y, Er)	High temperature solid-state reaction of stoichiometric mixtures made from respective oxides of lanthanides, chlorides or hydroxides of alkaline earth metals, Cd-sulfate, Mn-chloride, and K-silicate at 1200°C in air.	Ito (1968)
Na$_6$Pb$_4$(BeF$_4$)$_6$F$_2$ Na$_6$Ca$_4$(BeF$_4$)$_6$F$_2$	Stoichiometric amounts of Na$_2$BeF$_4$, (NH$_4$)$_2$BeF$_4$, PbF$_2$ and CaCO$_3$ in Pt-crucible were mixed and fired at 400–600°C in N atmosphere.	Engel (1978)
NaLa$_9$(GeO$_4$)$_6$O$_2$	Sintering Na$_2$CO$_3$, La$_2$O$_3$ and GeO$_2$ at 1173 K for 6 h and then at 1373 K for 1 h. Crystals grown by both high temperature flux and melt system.	Takahashi et al. (1998)
Na$_2$RE$_2$Ca$_6$(PO$_4$)$_6$Cl$_2$ (RE = La, Pr, Nd) Na$_2$RE$_2$Ca$_6$(PO$_4$)$_6$F$_2$ (RE = La, Pr, Nd) Na$_2$La$_2$Sr$_6$(PO$_4$)$_6$F$_2$ Na$_2$La$_2$Sr$_6$(PO$_4$)$_6$Cl$_2$ Na$_2$La$_2$Ba$_6$(PO$_4$)$_6$Cl$_2$ Na$_x$La$_x$Ba$_{10-x}$(PO$_4$)$_6$F$_2$ (x = 1, 2, 3)	Stoichiometric amounts of metal carbonates or oxides and diammonium hydrogen phosphate were mortared in acetone. Dried sample was placed in Pt or silica crucible and heated to 600–650°C, remixed and reheated to 1000°C for 8 h.	Grisafe and Hummel (1970b)

Table B1 continued.

Composition	Synthesis Procedure	Reference
$NaRE_9[SiO_4]$ (RE = La→Lu)	Pellets containing the corresponding oxide/salt mixtures $7RE_2O_3 \cdot 9SiO_2$ and $Na_2CO_3 \cdot 9RE_2O_3 \cdot 12SiO_2$ were annealed at different temperatures in the range of 1000–1900°C for 24 h after pre-sintering at 900°C.	Felsche (1972)
$Na_2RE_2Pb_6(PO_4)_6Z_2$ (RE = La, Pr, Nd; Z = F, Cl)	Stoichiometric amounts of metal carbonates or oxides and diammonium hydrogen phosphate were mortared in acetone. Dried sample was placed in Pt or silica crucible and heated to 600–650°C, remixed and reheated to 850°C for 6 h.	Grisafe and Hummel (1970b)
$(Na_{1.46}La_{8.55})(SiO_4)_6(F_{0.9}O_{0.11})$	Single crystals were grown by slow cooling and evaporation of a NaF flux at 1350°C to 900°C.	Hughes et al. (1992)
$NaLn_9(SiO_4)_6O_2$ (Ln = La, Y)	Solid-state reaction of the respective oxides of lanthanides, Mg-nitrate, Li-carbonate, K-silicate and Na-silicate at 1050°C in air.	Ito (1968)
$(Na,Th,La)_{10}(SiO_4)_6O_2$	Solid-state reaction of the respective oxides of lanthanides, Mg-nitrate, Li-carbonate, K-silicate and Na-silicate at 1050°C in air.	Ito (1968)
$NaY_9(SiO_4)_6O_2$	Crystals were obtained by mixing Na_2CO_3, Y_2O_3 and SiO_2 (in proportions to the chemical composition of $NaYSi_2O_6$) and Na_2MoO_4 (high-temperature solvent 1:10 ratio). The mixture was placed in a Pt-crucible heated slowly to 1473 K, maintained at this temperature for 24 h and then cooled slowly (2 K/h) to 673 K.	Redhammer et al. (2003)
$NaY_9(SiO_4)_6O_2$	Single crystals were grown by melting respective starting materials at 1748 K and quenched to yield a glass which was subsequently annealed in air at 1538 K for 12 h.	Gunawardane et al.(1982)
$Na_{6.35}Ca_{3.65}(SO_4)_6F_{1.65}$ $Na_{6.39}Ca_{3.61}(SO_4)_6Cl_{1.61}$	Stoichiometric mixtures of $CaSO_4 \cdot 2H_2O$, Na_2SO_4, CaF_2 or $CaCl_2$ were heated in covered corundum crucibles at 740°C for 10 h and subsequently cooled to 100°C with 4 °C/h.	Piotrowski et al. (2002a)
$Na_6Cd_4(SO_4)Cl_2$ $Na_6Pb_4(SO_4)_6Cl_2$	Solid state reaction of NaCl, Na_2SO_4, $CdSO_4/PbSO_4$ at 500°C.	Perret and Bouillet (1975)
$Na_6Pb_4(SO_4)_6F_2$	Stoichiometric mixtures of NaF_2 were heated with a pre-reacted intermediate at 570°C for 2 h. The intermediate was formed by sintering the starting metal carbonates/sulfates at 500–1000°C for 15 h.	Kreidler and Hummel (1970)

Table B1 continued.

Composition	Synthesis Procedure	Reference
$Nd_xCa_{10-2x}Na_x(PO_4)_6F_2$ ($2 \leq x \leq 4$)	$Nd_3(PO_4)_2$ was prepared by heating $(NH_4)_2HPO_4$ and Nd_2O_3 at 500°C for 2–3 h, grinding the mixture, and reheating at 1100°C for 4–12 h. $Nd_3(PO_4)_2$, Na_3PO_4, $Ca_3(PO_4)_2$ and CaF_2 were then mixed in stoichiometric ratios and fired at 1100°C for 4–24 h.	Mayer et al. (1974)
$Nd_{9.33}(SiO_4)_6O_2$	Stoichiometric mixture of Pr_6O_{11}, Sm_2O_3, and SiO_2 was calcined at 1200–1600°C for 10–20 h in air.	Higuchi et al. (2000)
$Nd_8Mn_2(SiO_4)_6O_2$	Single crystals prepared from stoichiometric sintering of Nd_2O_3, $MnCO_3$ and SiO_2 at 950°C for 48 h using a Bi_2O_3 flux in closed Cu-tubes.	Kluver and Mueller-Busch-baum (1995)
$Pb_6K_4(AsO_4)_4(SO_4)_2$	Solid-state reaction of stoichiometric mixture of $Pb_3(X^VO_4)_2$, $K_2X^{VI}O_4$ (X^V = As; X^{VI} = S) at 550–1000°C.	Schwarz (1967b)
$Pb_6K_2(AsO_4)_5(SiO_4)$	Solid-state reaction of a stoichiometric mixture of PbO, $NaNO_3$, $NH_4H_2AsO_4$ and SiO_2 was fired at 1000–1400°C.	Wondratschek (1963)
$Pb_{10}(CrO_4)_2(GeO_4)_4$	Solid-state reaction of PbO, GeO_2 and $PbCrO_4$ at 730–830°C.	Engel and Deppisch (1988)
$Pb_{10}(GeO_4)_2(VO_4)_2$	Yellow monocrystals were obtained from the melt of the starting materials.	Ivanov and Zavodnik (1989); Ivanov (1990)
$Pb_8Bi_2(PO_4)_4(SiO_4)_2$ $Pb_6Bi_2Tl(PO_4)_4(SiO_4)$	Solid-state reaction of a stoichiometric mixture of PbO, $NH_4H_2PO_4$, Bi_2O_3, SiO_2 (and Tl_2CO_3) was fired at 1000–1400°C.	Wondratschek (1963)
$Pb_6Ca_2Li_2(PO_4)_6$	Single crystals were grown by mixing Li_2CO_3, $(NH_4)_2HPO_4$, $CaCO_3$ and PbO between 800–900°C in air for 24 h.	Naddari et al. (2002)
$Pb_6Ca_2Na_2(PO_4)_6$	Na_2CO_3, $(NH_4)_2HPO_4$, $CaCO_3$ and PbO powders were stoichiometrically mixed and heated at 1073 K in air for 12 h and at 1173 K for 12 h.	Naddari et al. (2003)
$Pb_8K_2(PO_4)_6$	Powders and single crystals of $Pb_8K_2(PO_4)_6$ were prepared by mixing $Pb_3(PO_4)_2$, KPO_3, and PbO at 600–700°C for 48 h.	Mathew et al. (1980)

Table B1 continued.

Composition	Synthesis Procedure	Reference
$Pb_8La_2(PO_4)_6O_2$	Stoichiometric amounts of metal carbonates or oxides and diammonium hydrogen phosphate were mortared in acetone. Dried sample was placed in Pt or silica crucible and heated to 600–650°C, remixed and reheated at 1000°C for 10 h.	Grisafe and Hummel (1970b)
$Pb_{10}(PO_4)_{6-x}(VO_4)_xF_2$ ($x = 1.2, 2.4, 3.6, 4.8, 6.0$)	Starting mixtures were fired in air at 800°C for 4 h.	Grisafe and Hummel (1970a)
$Pb_{10-x}Ni_x(PO_4)_6F_2$ ($x = 0.2, 0.5$ and 1.0)	Stoichiometric amounts of metal carbonates or oxides and diammonium hydrogen phosphate and/or silicic acid were mixed and dried in covered crucibles and heated at 800°C for 4 h.	Grisafe and Hummel (1970b)
$Pb_6K_4(PO_4)_4(SO_4)_2$ $Pb_6K_4(PO_4)_4(SeO_4)_2$	Solid-state reaction of stoichiometric mixtures of $Pb_3(X^VO_4)_2$, $K_2X^{VI}O_4$ ($X^V = P$; $X^{VI} = S$, Se) at 550–1000°C.	Schwarz (1967b)
$Pb_9Na(PO_4)_5(SiO_4)$ $Pb_8Na_2(PO_4)_2(SiO_4)(SO_4)$	Solid-state reaction of a stoichiometric mixture of PbO, $NaNO_3$, $NH_4H_2PO_4$, SiO_2 (and $(NH_4)_2SO_4$) at 1000–1400°C.	Wondratschek (1963)
$Pb_{10-2x}Bi_xNa_x(PO_4)_6F_2$ ($x = 1, 2, 3$)	Stoichiometric amounts of $Pb_3(PO_4)_2$, $NH_4H_2PO_4$, NaF, $Na_3PO_4 \cdot 12H_2O$ and $Bi(NO_3)_3 \cdot 5H_2O$ were mixed together and heated in a closed Au- or Pt-tube at 900°C for 24–24 h.	Mayer and Semadja (1983)
$Pb_{10-2x}Bi_xNa_x(PO_4)_6Cl_2$ ($x = 1, 2$)	Stoichiometric amounts of $Pb_3(PO_4)_2$, $NH_4H_2PO_4$, NaCl, $Na_3PO_4 \cdot 12H_2O$ and $Bi(NO_3)_3 \cdot 5H_2O$ were mixed together and heated in closed Au or Pt tube at 600°C for 24–48 h.	Mayer and Semadja (1983)
$Pb_{10-(2x+2)}Na_{x+2}Bi_x(PO_4)_6$ ($x = 0, 1, 2$)	Stoichiometric amounts of $Pb_3(PO_4)_2$, $NH_4H_2PO_4$, $Na_3PO_4 \cdot 12H_2O$ and $Bi(NO_3)_3 \cdot 5H_2O$ were mixed together and heated in closed Au or Pt-tube at 900°C for 24–48 h.	Mayer and Semadja (1983)
$Pb_8La_2(SiO_4)_2(PO_4)_4Z_2$ ($Z = F, Cl$)	Stoichiometric amounts of metal carbonates or oxides and diammonium hydrogen phosphate and silicic acid were mortared in acetone. Dried sample was placed in Pt or silica crucible and heated to 600–650°C, remixed and reheated to 875°C for 12 h.	Grisafe and Hummel (1970b)

Table B1 continued.

Composition	Synthesis Procedure	Reference
$\square Pb_5Nd_4(SiO_4)_4(PO_4)_2\square_2$	Stoichiometric amounts of metal carbonates or oxides, diammonium hydrogen phosphate and silicic acid were mortared in acetone. Dried sample was placed in Pt or silica crucible and heated to 600–650°C, remixed and reheated at 1000°C for 12 h.	Grisafe and Hummel (1970b)
$Pb_6Bi_4(SiO_4)_6$	Solid-state reaction of a stoichiometric mixture of PbO, Bi_2O_3 and SiO_2 at 1000–1400°C.	Wondratschek (1963)
$Pb_4La_6(SiO_4)_6F_2$	Stoichiometric amounts of metal carbonates or oxides and silicic acid were mortared in acetone. Dried sample was placed in Pt or silica crucible and heated to 600–650°C, remixed and reheated to 950°C for 12 h.	Grisafe and Hummel (1970b)
$Pb_2Ln_8(SiO_4)_6O_2$ (Ln = La, Nd, Sm, Gd, Dy, Y)	Solid-state reaction of respective oxides of lanthanides, Pb-nitrate and K-silicate at 900°C in air.	Ito (1968)
$\square PbLa_8(SiO_4)_6F_2$	Stoichiometric amounts of metal carbonates or oxides and silicic acid were mortared in acetone. Dried sample was placed in Pt or silica crucible and heated to 600–650°C, remixed and reheated to 1000°C for 16 h.	Grisafe and Hummel (1970b)
$Pb_4Nd_6(SiO_4)_6O\square$	Stoichiometric amounts of metal carbonates or oxides and silicic acid were mixed. Dried sample was placed in Pt or silica crucible and heated to 600–650°C, remixed and reheated at 1000°C for 10 h.	Grisafe and Hummel (1970b)
$\square Pb_3RE_6(SiO_4)_6\square_2$ (RE = La, Pr, Nd)	Stoichiometric amounts of metal carbonates or oxides and silicic acid were mixed. Dried sample was placed in Pt or silica crucible and heated to 600–650°C, remixed and reheated at 1000°C for 10 h.	Grisafe and Hummel (1970b)
$Pb_{10}(SO_4)_2(GeO_4)_4$	Solid-state reaction of PbO, GeO_2 and $PbSO_4$ at 730–830°C.	Engel and Deppisch (1988)
$Pb_{10}(VO_4)_6Cl_2$	$Pb_3(VO_4)_2$ and $PbCl_2$ were mixed and fired at 550°C for 1 h.	Kreidler and Hummel (1970)
$Pb_{9.85}(VO_4)_6I_{1.7}$	Single crystal was grown from a melt of $Pb_5(VO_4)_2$ and PbI_2 in stoichiometric amounts at 773–1073 K.	Aubert et al. (1999)

Table B1 continued.

Composition	Synthesis Procedure	Reference
$Pb_8Na_2(VO_4)_6$	Stoichiometric quantities of Na_2CO_3, PbO and NH_4VO_3 were mixed and annealed at 600°C for 48 h.	Sirotinkin et al. (1989)
$(Pb_xCa_{10-x})(VO_4)_6F_{2\delta}$ $(0 < x < 9)$	CaO, PbO, V_2O_5 and CaF_2 were mixed in a stoichiometric ratio and fired at 1073 K for 4 weeks.	Dong and White (2004a)
$Pr_{9.33}(SiO_4)_6O_2$	Stoichiometric mixture of Pr_6O_{11} and SiO_2 was calcined at 1200–1600°C for 10–20 h in air.	Higuchi et al. (2000)
$R_4Nd_6(SiO_4)_6O\square$ (R = Ca, Sr, Ba)	Stoichiometric amounts of metal carbonates or oxides and silicic acid were mixed. Dried sample was placed in Pt or silica crucible and heated to 600–650°C, remixed and reheated to 1350°C for 16 h.	Grisafe and Hummel (1970b)
$\square R_5Nd_4(SiO_4)_4(PO_4)_2\square_2$ (R = Ca, Sr)	Stoichiometric amounts of metal carbonates or oxides, diammonium hydrogen phosphate and silicic acid were mortared in acetone. Dried sample was placed in Pt or silica crucible and heated to 600–650°C, remixed and reheated to 1350°C for 16 h.	Grisafe and Hummel (1970b)
$R_8La_2(SiO_4)_2(PO_4)_4Z_2$ (R = Ca, Sr, Ba ; Z = F, Cl)	Stoichiometric amounts of metal carbonates or oxides and diammonium hydrogen phosphate and silicic acid were mortared in acetone. Dried sample was placed in Pt or silica crucible and heated to 600–650°C, remixed and reheated to 1000°C for 18 h.	Grisafe and Hummel (1970b)
$R_4La_6(SiO_4)_6(PO_4)_4F_2$ (R = Ca, Sr, Ba)	Stoichiometric amounts of metal carbonates or oxides and diammonium hydrogen phosphate and silicic acid were mortared in acetone. Dried sample was placed in Pt or silica crucible and heated to 600–650°C, remixed and reheated to 1100°C for 12 h.	Grisafe and Hummel (1970b)
$\square RLa_8(SiO_4)_6F_2$ (R = Ca, Sr, Ba)	Stoichiometric amounts of metal carbonates or oxides and silicic acid were mixed. Dried sample was placed in Pt or silica crucible and heated to 600–650°C, remixed and reheated to 1000°C for 16h.	Grisafe and Hummel (1970b)
$\square R_3RE_6(SiO_4)_6\square_2$ (R= Ca, Sr, Ba ;RE = La, Pr, Nd)	Stoichiometric amounts of metal carbonates or oxides and silicic acid were mixed. Dried sample was placed in Pt crucible, heated to 600–650°C, remixed and reheated to 1350–1400°C for 36 h.	Grisafe and Hummel (1970b)

Table B1 continued.

Composition	Synthesis Procedure	Reference
$RE_{9.33}\square_{0.67}[SiO_4]_6O_2$ (RE = La→Lu)	Pellets containing the corresponding oxide/salt mixtures $7RE_2O_3 \cdot 9SiO_2$, were annealed at different temperatures in the range 1000–1900°C for 24 h after pre-sintering at 900°C.	Felsche (1972)
$\square_2RE_8(SiO_4)_6\square_2$ (RE = La, Pr, Nd, Sm, Eu)	Stoichiometric amounts of metal carbonates or oxides and silicic acid were mixed. Dried sample was placed in Pt or silica crucible and heated to 600–650°C, remixed and reheated to 1350–1400°C for 72 h.	Grisafe and Hummel (1970b)
$Sm_{10}Si_6N_2O_{24}$	Stoichiometric amounts of Sm_2O_3, Si_3N_4 and SiO_2 were heated at 1250°C for 36 h in a sealed Ni-tube sealed under a N_2 atmosphere.	Gaude et al. (1975)
$Sm_{9.33}(SiO_4)_6O_2$	Stoichiometric mixture of Sm_2O_3 and SiO_2 was calcined at 1200–1600°C for 10–20 h in air.	Higuchi et al. (2000)
$Sm_{10}(SiO_4)_6O_2$	Elemental Samarium was heated in an evacuated fused-silica tube at 1223 K for 4 days, cooled to 773 K over 4 days, and then cooled to room temperature over 1.5 days. Partial attack was observed on the walls of the silica tube, on which were deposited dark-red needle-shaped crystals.	Morgan et al. (2002)
$(Sm_8Cr_2)(SiO_4)_6N_2$	Powders were prepared by heating $8Sm_2O_3 - 2Cr_2O_3 - 1Si_3N_4 - 9SiO_2$ at 1250°C in a sealed nickel tube.	Maunaye et al. (1976)
$Sr_{10}(AsO_4)_4(SO_4)_2S_2$	Stoichiometric amounts of $SrSO_4$, As_2O_5 and SrS were heated at 1400°C under argon atmosphere for 3 h.	Schiff-Francois et al. (1979)
$Sr_{10}(CrO_4)_6Cl_2$	Stoichiometric mixture of $Sr(OH)_2 \cdot 8H_2O$, Cr_2O_3 and $SrCl_2 \cdot 6H_2O$ was heated in air at 845°C for 5 h.	Mueller-Buschbaum and Sander (1978)
$Sr_{10}(CrO_4)_6Cl_2$	Stoichiometric mixture of $Sr(OH)_2 \cdot 8H_2O$, Cr_2O_3 and $SrCl_2 \cdot 6H_2O$ was heated in air at 845°C for 5 h, then in N_2 atmosphere at 920°C for 5 h.	Banks and Jaunarajs (1965)
$Sr_{10}(CrO_4)_6Cl_2$	$Sr_3(CrO_4)_2$ melts were put in an excess of $SrCl_2$ at 1000–1200°C in argon. The excess chlorides were extracted with H_2O.	Wilhelmi and Jonsson (1965)

Table B1 continued.

Composition	Synthesis Procedure	Reference
$Sr_{10}(CrO_4)_6F_2$	Stoichiometric mixture of $SrCO_3$, SrF_2 and Cr_2O_3 was heated in a corundum crucible at 1400 K in a dry N_2 atmosphere.	Herdtweck (1991)
$Sr_{10}(CrO_4)_6(OH)_2$	$Sr(OH)_2 \cdot 8H_2O$ was mixed with Cr_2O_3 and heated in air at 845°C for 5 h, then in N_2 atmosphere at 920°C for a further 5 h.	Banks and Jaunarajs (1965)
$Sr_{10}(CrO_4)_3(SiO_4)_3F_2$	$PbCrO_4$, Pb_2SiO_4 and PbF_2 were stoichiometrically mixed and heated at 500–600°C for 24–48 h in a N_2 atmosphere.	Schwarz (1967a)
$Sr_{10}(PO_4)_6Br_2$	After a mixture of $SrHPO_4$ and $SrCO_3$ was fired for 0.5 h at 800°C, dehydrated $SrBr_2$ was added and the combined mixture fired for 90 min at 1180°C.	Nötzold and Wulff (1998)
$Sr_{10}(PO_4)_6Cl_2$	$SrHPO_4$, $SrCO_3$ and $SrCl_2 \cdot 6H_2O$ were mixed and heated at temperatures ranging from 25 to 1200°C for a few hours.	Nötzold et al. (1994)
$Sr_{10}(PO_4)_6Cl_2$	Single crystals were prepared by mixing $SrCl_2$ with $Sr_{10}(PO_4)_6OH_2$ and heating to 1280°C for 24 h.	Sudarsanan and Young (1974)
$Sr_{10}(PO_4)_6F_2$	$Sr_3(PO_4)_2$ and SrF_2 were mixed and fired for 0.5 h at 1230°C.	Kreidler and Hummel (1970)
$Sr_{10}(PO_4)_6CO_3$	A stream of CO_2 was added to strontium hydroxyapatite.	Nadal et al. (1971)
$Sr_{10}(PO_4)_6(CuO)_2$	$SrCO_3$, $NH_4H_2PO_4$ and CuO were mixed stoichiometrically and heated stepwise at 400, 600 and 850°C for 32 h, then pressed into pellets and annealed in air at 1100°C for 24 h.	Kazin et al. (2003)
$Sr_{10}(PO_4)_{6-x}(CrO_4)_xF_2$ ($x = 1.2, 2.4, 3.6, 4.8, 6.0$)	Starting mixtures were fired in a nitrogen atmosphere at 1000°C for 10 h.	Grisafe and Hummel (1970a)
$Sr_{10}(PO_4)_4(GeO_4)_2\square_2$	Solid-state reaction of stoichiometric amounts of $Sr_3(PO_4)_2$, Sr_2GeO_4 at 1200°C in air.	Schwarz (1968)
$Sr_{10}(PO_4)_6O_{0.9}(OH)_{0.2}\square_{0.9}$	Peroxiapatites are formed by thermal treatment of strontium hydroxyapatite at 900°C, under pure O_2. In order to eliminate traces of water, the stream of oxygen was allowed to run through a coil cooled by carbon dioxide-ice, before reaching the product.	Trombe and Montel (1978)

Table B1 continued.

Composition	Synthesis Procedure	Reference
$Sr_{10}(PO_4)_6F_2$ $Sr_8BiNa(PO_4)_6F_2$ $Sr_6Bi_2Na_2(PO_4)_6F_2$	Stoichiometric amounts of $SrCO_3$, $NH_4H_2PO_4$, SrF_2, (and $NaF/Na_3PO_4 \cdot 12H_2O$/ $Bi(NO_3)_3 \cdot 5H_2O$) were mixed together and heated at 1100°C for 24–48 h in a closed Au- or Pt-crucible.	Mayer and Semadja (1983)
$Sr_{5-x-y}Ba_xEu_y(PO_4)_3Cl$ $(0 \le x \le 4.99, 0 \le y \le 0.05)$	Starting mixtures were fired at 700°C and 1050°C in a N_2/H_2 atmosphere. H_3BO_3 with an excess of $SrCl_2$ and/or $BaCl_2$ as flux.	Nötzold and Wulff (1996)
$Sr_{10-x}Co_x(PO_4)_6F_2$ ($x = 0.45, 0.9, 1.8,$ and 2.7)	Stoichiometric amounts of metal carbonates or oxides and diammonium hydrogen phosphate and/or silicic acid were mixed and dried in sealed platinum crucibles and heated at 975°C for 8 h.	Grisafe and Hummel (1970b)
$Sr_{10-x}Co_x(PO_4)_6F_2$ ($x = 0.45, 0.9, 1.8,$ and 2.7)		
$Sr_{10-x}Eu_x(PO_4)_6F_2$ $(0.05 \le x \le 0.5)$ $Sr_{10-x}Eu_x(PO_4)_6Br_2$ $(0.05 \le x \le 0.5)$ $Sr_{10-x}Eu_x(PO_4)_6Cl_2$ $(0.05 \le x \le 0.5)$	$SrHPO_4$, $SrCO_3$, $NH_4F/NH_4Br/SrCl_2 \cdot 6H_2O$ and Eu_2O_3 were mixed proportionately and heated at 1100°C for 2 h in a reducing atmosphere (either N_2 plus H_2 or CO gas).	Kottaisamy et al. (1994)
$Sr_{9.402}Na_{0.209}(PO_4)_6B_{1.992}$	Crystals were grown in a Pt-crucible from a mixture of $(NH_4)_2HPO_4$, $SrCO_3$ and $Na_2B_4O_7 \cdot 10H_2O$ (2:3:6) at 1450°C.	Calvo et al. (1975)
$Sr_{10-x}Ni_x(PO_4)_6F_2$ ($x = 0.2, 0.5$ and 1.0)	Stoichiometric amounts of metal carbonates or oxides and diammonium hydrogen phosphate and/or silicic acid were mixed and heated at 800°C for 4 h in covered crucibles.	Grisafe and Hummel (1970b)
$Sr_{10}(P_{1-x}Mn_xO_4)_6F_2$ ($x = 0.01, 0.1, 0.2, 0.3$)	$SrCO_3$, NH_4HPO_4/Mn_2O_3, and NH_4F were stoichiometrically mixed, pressed into pellets and repeatedly heated for 12 h at 1250°C in air (for $x \le 0.3$) and at 1000°C in an O_2 stream (for $x > 0.3$).	Dardenne et al. (1999)
$Sr_{10}(PO_4)_6O$	$Sr_{10}(PO_4)_6(OH)_2$ was heated at 1000°C for 2 days under pure O_2. To eliminate traces of H_2O, the stream of oxygen was run through a coil cooled by CO_2-ice before reaching the product.	Trombe and Montel (1978)
$Sr_{10}(PO_4)_{6-x}(VO_4)_xF_2$ ($x = 1.2, 2.4, 3.6, 4.8, 6.0$)	Starting mixtures were fired in air at 1000°C for 8 h.	Grisafe and Hummel (1970a)
$Sr_{9.92}Nd_{0.05}(PO_4)_6F_2$	Crystals were grown by the Czochralski method.	Corker (1995)

Table B1 continued.

Composition	Synthesis Procedure	Reference
$Sr_8RE_2(PO_4)_6O_2$ (RE = La, Pr, Nd)	Stoichiometric amounts of metal carbonates or oxides and diammonium hydrogen phosphate were mortared in acetone. Dried sample was placed in Pt or silica crucible and heated to 600–650°C, remixed and reheated at 1350°C for 72 h.	Grisafe and Hummel (1970b)
$Sr_{10}(ReO_5)_6X_2$ (X= Cl, Br, I)	Powders of respective alkaline-earth metaperrhenates $M(ReO_4)_2 \cdot H_2O$ were annealed with an excess of corresponding alkaline-earth halides at 500–800°C.	Schriewer and Jeitschko (1993)
$Sr_{10}(SiO_4)_3(CrO_4)_3F_2$	Solid-state reaction of $SrCrO_4$, Sr_2SiO_4, SrF_2 (3:3:1) at 1000°C for 18 h in O_2.	Schwarz (1967a)
$Sr_2Ln_8(SiO_4)_6O_2$ (Ln =La, Nd, Sm, Gd, Dy, Y, Er)	Solid-state reaction of stoichiometric mixtures made from respective oxides of lanthanides, chlorides or hydroxides of alkaline earth metals, cadmium sulfate, manganese chloride, and potassium silicate at 1200°C in air.	Ito (1968)
$Sr_{10}(SO_4)_3(GeO_4)_3F_2$	Solid-state reaction of $SrSO_4$, Sr_2GeO_4, SrF_2 (3:3:1) at 1000°C for 23 h in air.	Schwarz (1967a)
$Sr_{10}(VO_4)_6F_2$	Crystals were grown by the Czochralski method.	Corker et al. (1995)
$Sr_{10}(VO_4)_6(CuO_2)$	Solid-state reactions of $SrCO_3$, V_2O_5 and CuO were carried out in air at 1173–1740 K for a total of 59 h.	Carrillo-Cabrera and von Schnering (1999)
$Sr_{10}(VO_4)_{6-x}(CrO_4)_xF_2$ (x = 1.2, 2.4, 3.6, 4.8, 6.0)	Starting mixtures were fired in nitrogen atmosphere at 1000°C for 10 h.	Grisafe and Hummel (1970a)
$Sr_{9.818}Nd_{0.122}(V_{0.972}O_4)_6F_{1.96}$	Crystals were grown by the Czochralski method.	Corker et al. (1995)
$Sr_{10}(V_{1-x}Mn_xO_4)_6F_2$ (x = 0.02)	$SrCO_3$, V_2O_5/Mn_2O_3, and NH_4F were mixed in stoichiometric amounts, pressed into pellets and heated for 12 h at 1000°C in an oxygen stream.	Dardenne et al. (1999)
$Y_xCa_{10-2x}Na_x(PO_4)_6F_2$ ($2 \leq x \leq 4$)	$Y_3(PO_4)_2$ was prepared by heating $(NH_4)_2HPO_4$ and Y_2O_3 at 500°C for 2–3 h, grinding the mixture, and reheating at 1100°C for 4–12 h. $Y_3(PO_4)_2$, Na_3PO_4, $Ca_3(PO_4)_2$, CaF_2 were mixed in stoichiometric ratios and fired at 1100°C for 4–24 h.	Mayer et al. (1974)
$(Y_4Mg)Si_3O_{13}$ $(Y_{3.73}Mg_{1.31})Si_{3.05}O_{13}$ $(Y_{3.30}Mg_{1.33})Si_{3.36}O_{13}$	Stoichiometric amounts of yttria, silica and magnesia were mixed in ethanol using an agate mortar pestle and heated at 1200°C for 2 h. It was then reground, pelletized and heated at 1450–1550°C for several hours.	Suwa et al. (1968)

Table B2. Hydrothermal route for apatite synthesis.

Composition	Synthesis Procedure	Reference
$Ba_{10}(PO_4)_6(OH)_2$	Crystals were obtained from hydrothermal treatment of a Na_2O, $BaO-Lu_2O_3$, P_2O_5, H_2O mixture (500°C and P of ~1 kbar).	Bondareva and Malinovskii (1986)
$Ba_4Ln_6(SiO_4)_6(OH)_2$ (Ln = Nd, Sm, Gd, Dy)	Hydrothermal treatment at 650°C and 2 kbars of gels made from respective oxides of lanthanides, chlorides or hydroxides of alkaline earth metals, Cd-sulfate, Mn-chloride, K-silicate and Na-silicate.	Ito (1968)
$Ca_{10}(CrO_4)_6(OH)_2$	Hydrothermal treatment of $Ca_3(CrO_4)_2$ and CaO saturated steam pressures at 200–300°C for 2–5 days.	Banks and Jaunarajs (1965)
$Ca_{10}(PO_4)_6(OH)_2$	Hydrothermal reaction of $Ca(NO_3)_2$ and $(NH_4)_2HPO_4$ at 100–200°C, 1–2 Mpa.	Yoshimura and Suda (1994)
$Ca_{10}(PO_4)_6O$	$Ca_{10}(PO_4)_6(OH)_2$ was prepared under hydrothermal conditions at 523 K for 7 days using stoichiometric amounts of $Ca(NO_3)_2$ and $(NH_4)_2HPO_4$. The dehydration product under high vacuum and beam irradiation is $Ca_{10}(PO_4)_6O$.	Alberius-Henning et al. (1999b)
$Ca_2Na_2La_6Si_4P_2O_{24}(OH)_2$	Hydrothermal reaction of Ca-nitrate, Y-oxide, La-oxide, ammonium dihydrogen phosphate, K-silicate and Na-silicate at 550°C at 2 kbar.	Ito (1968)
$Ca_{9.50}(NH_4)_{0.10}(PO_4)_{5.05}(CO_3)_{0.95}(OH)_2$	$CaCO_3$ was reacted with $NH_4H_2PO_4$ solution under hydrothermal conditions at 250°C and 1 kbar for 10 days. A small amount of NH_4OH was added to the starting solution to bring the pH value to 9.	Ivanova et al. (2001)
$Ca_{9.42}Sr_{0.18}H_{0.8}(PO_4)_6(OH)_2$	Crystals were grown hydrothermally from Ca and Sr hydrogenphoshates by gradual heating from 493–603 K with 40.2 MPa at pH 2.	Kikuchi et al. (1994)
$CaY_9Si_5BO_{26}$	Hydrothermal reaction of Ca/Mn-nitrate, Y-oxide, K-silicate and boric acid at 1150°C, 1 bar.	Ito (1968)
$Ca_4La_6Si_4P_2O_{26}$	Hydrothermal reaction of Ca-nitrate, Y-oxide, ammonium dihydrogen phosphate and K-silicate at 1200°C, 1 bar.	Ito (1968)

Table B2 continued.

Composition	Synthesis Procedure	Reference
$Ca_4Ln_6(SiO_4)_6(OH)_2$ (Ln = La, Ce, Nd, Sm, Gd, Dy, Y, Er, Lu)	Hydrothermal treatment at 650°C and 2 kbars of gels made from respective oxides of lanthanides, chlorides or hydroxides of alkaline earth metals, Cd-sulfate, Mn-chloride, K-silicate and Na-silicate.	Ito (1968)
$Ca_4Na_2La_4Si_2P_4O_{24}(OH)_2$	Hydrothermal reaction of Ca-nitrate, Y-oxide, La-oxide, ammonium dihydrogen phosphate, potassium silicate and sodium silicate at 550°C, 2 kbar.	Ito (1968)
$Ca_4Y_6Si_4P_2O_{26}$ $Ca_6La_4Si_2P_4O_{26}$ $Ca_6Y_4Si_2P_4O_{26}$	Hydrothermal reaction of Ca-nitrate, Y-oxide, ammonium dihydrogen phosphate and potassium silicate at 1200°C, 1bar.	Ito (1968)
$Ca_6Y_4Si_4P_2O_{24}(OH)_2$	Hycrothermal reaction of Ca-nitrate, Y-oxide, La-oxide, ammonium dihydrogen phosphate, K-silicate and Na-silicate at 550°C, 2 kbar.	Ito (1968)
$Ca_8Y_2Si_2P_4O_{24}(OH)_2$	Hycrothermal reaction of calcium nitrate, yttrium oxide, ammonium dihydrogen phosphate and potassium silicate at 500°C, 2 kbar.	Ito (1968)
$Ca_{10}(VO_4)_6(OH)_2$	Oxide phases of vanadium underwent hydrothermal treatment.	Kutoglu (1974); Kutoglu and Schulien (1972)
$Cd_{10}(PO_4)_6(OH)_2$ ($P6_3$ - superstructure)	Hycrothermal treatment of $Cd_5H_2(PO_4)_4 \cdot 4H_2O$ at 200°C for 2 weeks in the pH range 2.5–7.0.	Hata and Marumo (1983)
$Cd_{10}(PO_4)_6(OH)_2$ ($P6_3/m$)	Hycrothermal reaction of $Cd_3H_2(PO_4)_4 \cdot 4H_2O$, controlled by H_3PO_4 solution, at 200°C for 2 weeks in pH range 2.8–3.3.	Hata et al. (1978)
$Cd_4La_6(SiO_4)_6(OH)_2$	Hydrothermal treatment at 650°C and 2 kbars of gels made from respective oxides of lanthanides, chlorides or hydroxides of alkaline earth metals, Cd-sulfate, Mn-chloride, K-silicate and Na-silicate.	Ito (1968)
$KNd_9(SiO_4)_6O_2$	Hydrothermal synthesis from stoichiometric amounts of K_2O, Nd_2O_3, SiO_2 and H_2O.	Pushcharovskii et al. (1978)
$Mg_4La_6(SiO_4)_6(OH)_2$	Hyd-othermal treatment at 650°C and 2 kbars of gels made from respective oxides of lanthanides, chlorides or hydroxides of alkaline earth metals, Cd-sulfate, Mn-chloride, K-silicate and Na-silicate.	Ito (1968)

Table B2 continued.

Composition	Synthesis Procedure	Reference
$MgY_9Si_5BO_{26}$	Hydrothermal reaction of Ca/Mg-nitrate, Y-oxide, K-silicate and boric acid at 1150°C, 1 bar.	Ito (1968)
$Mn_{10}(PO_4)_6Cl_{1.8}(OH)_{0.2}$	Hydrothermal treatment of $Mn_3(PO_4)_2$, $MnCl_2 \cdot 4H_2O$ and H_2O mixture at 425°C.	Engel et al. (1975a)
$Mn_4Ln_6(SiO_4)_6(OH)_2$ (Ln = La, Nd, Sm, Gd, Dy, Y, Er)	Hydrothermal treatment at 650°C and 2 kbars of gels made from respective oxides of lanthanides, chlorides or hydroxides of alkaline earth metals, Cd-sulfate, Mn-chloride, K-silicate and Na-silicate.	Ito (1968)
$Na_2Ln_8(SiO_4)_6(OH)_2$ (Ln = La, Y)	Hydrothermal treatment at 650°C and 2 kbars of gels made from respective oxides of lanthanides, chlorides or hydroxides of alkaline earth metals, Cd-sulfate, Mn-chloride, K-silicate and Na-silicate.	Ito (1968)
$Na_{6.9}Ca_{3.1}(SO_4)_6OH_{1.1}$	Stoichiometric mixtures of Na_2SO_4, $CaSO_4$, and $Ca(OH)_2$ were heated under hydrothermal conditions and carried out at 250°C for 14 days.	Piotrowski et al. (2002b)
$Pb_{10}(AsO_4)_6(OH)_2$	Hydrothermal synthesis using stoichiometric powders of PbO and As_2O_5 with water at 425°C.	Engel (1970)
$\square Pb_6(PO_4)_6 \square_2$	Hydrothermal reaction of $Pb(CH_3COO)_2 \cdot 3H_2O$ and K_2HPO_4 at 473 K for 1 week. This hydroxyapatite was fused at 1363 K and cooled to room temperature, yielding single crystals of $Pb_3(PO_4)_2$.	Hata et al. (1980)
$Pb_{10}(PO_4)_6F_2$	A mixture of PbO, P_2O_5, KF, H_2O was heated in a hydrothermal system at 1200°C.	Belokoneva et al. (1982)
$Pb_2^{4+}Pb_4^{2+}Y_4Si_6O_{26}$ $Pb_2^{4+}Pb_8^{2+}P_4Si_2O_{26}$ $Pb_3^{4+}Pb_5^{2+}Y_2Si_6O_{26}$ $Pb_4^{4+}Pb_3^{2+}Y_6Si_6O_{26}$	Hydrothermal reaction of Pb-nitrate, Y-oxide, K-silicate at 900°C, 1 bar.	Ito (1968)
$Pb_4Ln_6(SiO_4)_6(OH)_2$ (Ln = La, Ce, Nd, Sm, Gd, Dy, Y, Er, Lu)	Hydrothermal treatment at 450°C and 2 kbars of gels made from respective oxides of lanthanides, Pb-nitrate and K-silicate.	Ito (1968)

Table B2 continued.

Composition	Synthesis Procedure	Reference
$Pb_{10}(VO_4)_6(OH)_2$	PbO was firstly reacted with V_2O_5. Resulting $Pb_{10}(VO_4)_6O$ was treated hydrothermally with H_2O at 425°C.	Engel (1970)
$Sr_4Ln_6(SiO_4)_6(OH)_2$ (Ln = La, Ce, Nd, Sm, Gd, Dy, Y, Ho, Er, Lu)	Hydrothermal treatment at 650°C and 2 kbars of gels made from respective oxides of lanthanides, chlorides or hydroxides of alkaline earth metals, Cd-sulfate, Mn-chloride, K-silicate and Na-silicate.	Ito (1968)
$Y_{10}Si_4B_2O_{26}$	Hydrothermal reaction of Ca/Mg-nitrate, Y-oxide, K-silicate and boric acid at 1150°C, 1 bar.	Ito (1968)

Table B3. Soft chemistry route for apatite synthesis.

Composition	Synthesis Procedure	Reference
$Ca_{10}(PO_4)_6(OH)_2$	$Ca(OH)_2$ mixed with H_3PO_4 at 100°C/2 h then calcined at 850°C/24 h.	Bigi et al. (1991)
$Ca_{10}(PO_4)_6(OH)_2$	$Ca(NO_3)_2 \cdot 4H_2O$ and $(NH_4)_2HPO_4$ were dissolved in H_2O. 25 vol% NH_4OH solution was added and heated to 65°C for 1.5 h, then further heated to boiling for 2 h. Solution was then cooled and allowed to precipitate overnight.	Engin and Tas (2000)
$Ca_{10}(PO_4)_6(OH)_2$	$Ca(OH)_2$ and H_3PO_4 were mixed and aged for 3 days, with the final pH 7.5. Resulting hydroxyapatite suspension was freeze-dried and heated in air at 1473 K for 1 h.	Ikoma et al. (1999)
$Ca_{10}(PO_4)_6OH_{2-x}F_x$ $(0.02 \leq x \leq 2)$	Varying amounts of NaF was added to $Ca(NO_3)_2 \cdot 4H_2O$ and $(NH_4)_2HPO_4$ with stirring in N_2 atmosphere. pH was adjusted using NH_4OH.	Bertoni et al. (1998)
$Ca_{10}(PO_4)_6(OH)_{2-x}F_x$ $(0 < x < 1)$	Ca-acetate and orthophosphoric acid were used as starting materials for Ca and P precursors. Na-fluoride (1, 3, 5 mol%) was added to the Ca precursor solution and precipitates filtered after 24 h of hydrolysis. Calcination was carried out at 900°C for 12 h.	Manjubala et al. (2001)

Table B3 continued.

Composition	Synthesis Procedure	Reference
$Ca_{10-x}Ag_x(PO_4)_6(OH)_{2-x}\square_x$ ($0 < x < 0.55$)	Apatite was precipitated from a solution of Na_2HPO_4 and a mixture of $Ca(NO_3)_2 \cdot 4H_2O$ and $AgNO_3$ in stoichiometric proportions.	Badrour et al. (1998)
$Ca_{10-x}Cd_x(PO_4)_6(OH)_2$ ($0 \leq x \leq 10$)	A double decomposition method is used in a boiling aqueous medium.	Nounah and Lacout (1993)
$Ca_{10-x}Cd_x(PO_4)_6(OH)_2$ ($x = 0.70, 0.73, 0.76$)	Ca-hydroxyapatite was introduced into Cd-nitrate solution of known concentrations. The solution was shaken at 20°C for 2 days.	JeanJean et al. (1996a,b)
$Ca_{10-x}Cd_x(PO_4)_6F_2$ ($0 \leq x \leq 6$)	A double decomposition method was used in a boiling aqueous medium in the presence of a large excess of fluoride ions.	Nounah and Lacout (1993)
$Ca_9Na_{0.5}(PO_4)_{4.7}(HPO_4)_{1.3} \cdot 2.9H_2O$ $Ca_9Na_{0.5}(PO_4)_6(OH)_{0.7} \cdot 4.2H_2O$ $Ca_{9.8}Na_{0.06}(PO_4)_{5.6}(HPO_4)_{0.4} \cdot 0.84H_2O$ $Ca_{9.8}Na_{0.06}(PO_4)_6(OH)_{1.6} \cdot 1.2H_2O$	Cd-nitrate was dissolved in H_2O and pH adjusted to 5 with HNO_3. Cal-hydroxyapatite was introduced and suspensions shaken at 20°C from 1 min to 10 days.	Mandjiny et al. (1998)
$Ca_{10-x}Na_{2x/3}(PO_4)_{6-x}(CO_3)_x(H_2O)_x$ $(OH)_{2-x/3}$ ($0 \leq x \leq 3$)	Stoichiometric amounts of $Ca(CH_3COO)_2$ solution was added to Na_2HPO_4 and $NaHCO_3$ solutions at 95°C over 3 h. The solution was maintained at 91-95°C for 5 days, filtered, washed and dried at 80°C overnight.	Wilson et al. (2004)
$Ca_9Na_{0.5}(PO_4)_{4.5}(CO_3)_{1.5}(OH)_2$	An aqueous solution of $Ca(NO_3)_2 \cdot 4H_2O$ containing 0.027mol of Ca^{2+} ions was added dropwise (187.5 µl s^{-1}) using a peristaltic pump into a solution of ammonium phosphate and ammonium carbonate (0.018mol of PO_4^{3-} and 0.036 mol CO_3^{2-}) and kept at 80°C. The precipitate was filtered when hot, washed with distilled water and dried at 70°C for 15 h in oven and then heated at 400°C in air for 24 h.	El Feki et al. (1999)
$Ca_{10-x}Pb_x(PO_4)_6(OH)_2$ ($0 < x < 10$)	Solutions of $Ca(NO_3)_2 \cdot 4H_2O$ and $Pb(NO_3)_2$ were added separately to $(NH_4)_2HPO_4$ and mixed in stoichiometric amounts. The pH was adjusted by addition of NH_3.	Andres-Verges et al. (1983)
$[Ca_{10-x}Y_x](PO_4)_6[(OH)_{2-x}O_2$ ($x \leq 2$)	Apatites were synthesized by aqueous solutions and sintered at 1000°C in a mulitte tube under steam flow.	Owada et al. (1989)

Table B3 continued.

Composition	Synthesis Procedure	Reference
$Ca_{10}(PO_4)_{6-x}(VO_4)_x(OH)_2$ ($x = 0.3, 1.5, 3.0, 4.5, 6.0$)	A double precipitation of phosphate and vanadate anions through stoichiometric addition of ammoniacal solutions of phosphate and vanadate to Ca-nitrate, refluxing near boiling point. The precipitate was dried at 100°C for 24 h and then calcined at 900°C for 3 h under inert atmosphere.	Boechat et al. (2000)
$Cd_{10-x}Pb_x(PO_4)_6(OH)_2$ ($0 \leq x \leq 10$)	A double decomposition method was used, where stoichiometric Cd-nitrate and Pb-acetate, were added to disodium hydrogen phosphate and heated at 100°C under nitrogen for 2 h. Ammonia was then added to maintain pH at 11 and the mixture filtered and dried at 110°C for 24 h.	Badroui et al. (2002a)
$Eu_{10}(AsO_4)_6(OH)_2$	Boiling $EuSO_4$ under Ar atmosphere with a solution containing an excess of Na_2HAsO_4 and NaOH. It was then filtered, washed with acetone and stored in dessicator to avoid oxidation. $EuSO_4$ was prepared by dissolving Eu_2O_3 in HCl and passing the solution through Jones reductor in to 8 N H_2SO_4.	Mayer et al. (1975)
$K_6Sn_4(SO_4)_6Cl_2$ $K_6Sn_4(SO_4)_6Br_2$	Acid solution of Zn(II) sulphate (33%) was mixed with H_2SO_4 (0.5 M) and an appropriate volume of a saturated solution of potassium salt in distilled water, and set aside at 5°C overnight. The crystals were recovered by filtration and vacuum drying.	Howie et al. (1973)
$K_6Sn_4(SO_4)_6Cl_2$ $K_6Sn_4(SO_4)_6Br_2$	Colorless acicular crystals were formed from a mixture of K-acetate and the appropriate Zn(II) halide, in the molar ratio 5:1, heated in the minimum amount of H_2SO_4 (2 mol dm^{-3}). The crystals were recovered by filtration and vacuum drying.	Donaldson and Grimes (1984)
$Mg_xCa_{10-x}(PO_4)_6(OH)_2$ ($1 \leq x \leq 3$)	$Ca(NO_3)_2 \cdot 4H_2O$ and $Mg(NO_3)_2 \cdot 6H_2O$ were added in appropriate amounts to $(NH_4)_2HPO_4$ with stirring in N_2 atmosphere. pH was adjusted with NH_4OH.	Bigi et al. (1996)
$Pb_{10}(PO_4)_6(OH)_2$	Lead oxynitrate, $Pb_6O_5(NO_3)_2$, was treated with phosphoric acid at 100°C, and the pH raised with NH_4OH under heating and stirring for 2 h. Product was dried at 100°C for 15 h.	Kim et al. (2000); Robin and Théolier (1956); Newkirk and Hughes (1969); Bigi et al. (1991)
$Pb_{10}(PO_4)_6(OH)_2$	Hot aqueous solutions of Pb(II) acetate trihydrate was mixed with disodium hydrogenphosphate and dried at 100°C for 24 h.	Brückner et al. (1995)

Table B3 continued.

Composition	Synthesis Procedure	Reference
$Pb_{10}(PO_4)_6Br_2$ $Pb_{10}(PO_4)_6Cl_2$ $Pb_{10}(PO_4)_6F_2$	Pb-oxynitrate, was treated with phosphoric acid at 100°C, and pH raised with $NH_3Br/NH_3Cl/NH_3F$ under heating and stirring for 2 h. Product was dried at 100°C for 15 h.	Kim et al. (2000); Robin and Theólier (1956); Newkirk and Hughes (1969); Bigi et al. (1991)
$Sr_{10}(AsO_4)_6CO_3$	The arsenate was prepared by double decomposition between aqueous solutions of ammonium arsenate and Sr-nitrate. The precipitate tri Sr-arsenate was dried and calcined at 900°C in CO_2 at 60 bars pressure.	Hitmi et al. (1986)
$Sr_{10}(PO_4)_6(OH)_2$	Hot $(NH_4)_2HPO_4$ was added to $Sr(NO_3)_2 \cdot 4H_2O$ and mixed. 25 vol% NH_4OH solution was added and heated to 95°C for 6 h. Solution was then cooled and allowed to precipitate overnight. The precipitate was filtered and dried at 110°C, then calcined at 950°C for 4 h.	Collin (1959)
$Sr_{10-x}Ca_x(PO_4)_6(OH)_2$ $(0 \le x \le 10)$	Ethylenediamine was added to required amounts of Ca- and Sr-nitrate. Hot ammonium phosphate solution was added dropwise. The precipitate and solution was maintained at 95°C for 6 h, and the precipitate allowed settling overnight. The supernatant was pulled off by an aspirator and washed until pH was 7. The precipitate was filtered, dried at 110°C, and heated to 950°C for 4 h.	Collin (1959)
$Sr_{9.5}Eu_{0.5}(PO_4)_6(OH)_2$	Sr-nitrate is mixed with Eu-nitrate solution (7.3 ml, 5 mol% Eu) and heated at 90°C. Ammonia was added until pH reaches 8. $(NH_4)_2HPO_4$ (2.04 g in 200 ml H_2O) is added drop-wise under stirring and the precipitate digested at 90°C for 6 h and dried at 110°C. It is then sintered at 950°C for 2 h in a reducing atmosphere.	Kottaisamy et al. (1994)
$Sr_{10-x}Pb_x(PO_4)_6(OH)_2$ $(0 < x \le 10)$	A double decomposition method was used, where $(NH_4)H_2(PO_4)$ was added dropwise to stoichiometric mixture of $Pb(CH_3COO)_2 \cdot 3H_2O$, $Sr(NO_3)_2$ and $(NH_4)H_2(PO_4)$ solutions and heated at boiling under N_2 stream. It was dried at 100°C for 12 h and calcined at 600°C for 4 h.	Badraoui et al. (2002b)

Table B4. Sol-gel route for apatite synthesis.

Composition	Synthesis Procedure	Reference
$Ca_{10}(PO_4)_6(OH)_2$	0.03 mol of triethylphosphite was hydrolyzed for 24 h with distilled water (molar ratio of water to phosphite is 8) under vigorous stirring. 0.05 mol of Ca-nitrate was dissolved first in 25 ml of distilled water and added dropwise into the hydrolyzed phosphite sol. The mixed sol solution was then continuously agitated for additional 3 min and kept static (aging) at 45°C and subjected to thermal treatment at 80°C for 16 h until a white, dried gel was obtained. The dried gels were further calcined at 200°C, 300°C for 2h and at 400°C, 500°C for 10 min (10°C/min).	Liu et al. (2002)
$La_{10}(Si_{3.96}B_{1.98}O_4)_6O_2$	A sol was prepared from $La(NO_3)_3 \cdot 6H_2O$, H_3BO_3, tetraethylorthosilicate, ethanol and glycerol (pH~0.5–1.0, adjusted by dil.HNO_3), converted to gel and heated in air at 1100°C for 3 h.	Mazza et al. (2000)
$La10(SiO4)6O3$ $La9.33(SiO4)6O2$	Sol-gel prepared from tetraethoxysilane (TEOS) and La_2O_3 and sintered at 800°C for 6 h.	Tao and Irvine (2001)
$La_{10}Si_4B_2O_{26}$	A sol was prepared from $La(NO_3)_3 \cdot 6H_2O$, H_3BO_3, tetraethylorthosilicate and glycerol; converted to gel and heated in air at 1100°C for 3 h.	Mazza et al. (2000)
$La_{9.33}Si_6O_{26}$	A sol was prepared from $La(NO_3)_3 \cdot 6H_2O$, H_3BO_3, tetraethylorthosilicate and glycerol; converted to gel and heated in air at 1100°C for 3 h.	Mazza et al. (2000)
$La_{9.66}Si_5BO_{26}$	A sol was prepared from $La(NO_3)_3 \cdot 6H_2O$, H_3BO_3, tetraethylorthosilicate and glycerol; converted to gel and heated in air at 1100°C for 3 h.	Mazza et al. (2000)
$Y_4(SiO_4)_3$ $Y_4Fe_{0.2}(SiO_4)_3O_{0.2}$	A gel was prepared from a sol prepared from an ethanol-water mixture of $Y(NO_3)_3 \cdot 5H_2O$, $Fe(NO_3)_3 \cdot 9H_2O$, tetraethylorthosilicate at 80°C and calcined at 1650–1700°C.	Parmentier et al. (2001)

Micro- and Mesoporous Sulfide and Selenide Structures

Emil Makovicky

*Geological Institute
University of Copenhagen
Oester Voldgade 10
DK-1350 Copenhagen, Denmark
e-mail: emilm@geol.ku.dk*

INTRODUCTION

Investigations and syntheses of meso- and microporous sulfides and selenides received in general much less attention than those of silicate and oxide families. Moreover, data on such compounds are scattered over sizable literature and it is very difficult to give a qualified overview of the subject. Scanning potential candidates in crystal structure databases leads in most cases to chain and layer structures, usually with only one or two microporous and a rare mesoporous structure in the entire list of compounds with the given promising three-element combination involving a large cation, an octahedral, tetrahedral or linear cation and an anion. In this situation, the present review presents a set of outstanding examples, preferably of smaller or larger structural families, rather than a systematic list of all cases published, synthesized or investigated. Compounds and families based on inorganic framework modifiers and those related to naturally occurring compounds have preference. For crystal-chemical reasons, they are divided into a category of sulfosalts and that of complex sulfides/selenides/ arsenides without the lone-electron pair metalloids participating in the structure.

SULFOSALTS: GENERAL FEATURES

Sulfosalts are complex sulfides/selenides in which formally trivalent lone-electron pair metalloids As^{3+}, Sb^{3+} and/or Bi^{3+} are combined with different cations: in a half of natural sulfosalt species it is Pb^{2+}, furthermore Cu, Ag, Fe, Tl, Mn, Sn and other, less frequently occurring elements. Rare natural finds and a plethora of synthetic products by a number of research groups enlarge this spectrum of compositions with sulfosalts of Li, Na, K, Rb, Cs, as well as with those of the entire spectrum of alkaline earths and lanthanoids, and with a number of sulfosalts that contain $(NH_4)^+$ and organic cations of various sizes. Li emulates Cu in sulfosalts (ionic conductivity), Na is a fairly small cation as also is Ca. However, potassium appears to be a 'channel-building' element in sulfosalts, besides playing the role of replacement for Pb^{2+} in the structures by means of coupled substitution, with Sb^{3+} (Bi^{3+}) or a lanthanoid replacing another Pb atom. In its channel- and interlayer-building role, K^+ also replaces Tl^+ whereas the heavy alkalis and Ba^{2+} are large-size structure modifiers.

There is no space for an outline of the crystal chemistry of sulfosalts here; this is to be found in Makovicky (1989, 1997) and Ferraris et al. (2004). Still, we have to mention the difference between the stereochemical activity of the lone electron pair (LEP) in As^{3+} and Sb^{3+} (pronounced activity resulting in capped trigonal coordination prisms of these elements and an SnS-like archetypal arrangement of structural portions) and Bi^{3+} (limited stereochemical activity of LEP, yielding octahedral, distorted octahedral as well as capped trigonal prismatic coordinations, and PbS-like archetypal structure arrangement in structural blocks/layers). Other configurations occur in structures with Cu or Ag as principal cations. Pb^{2+} and Tl^+ have close to insignificant LEP activity and can therefore be replaced by alkali metals or alkaline earths.

The family of cetineites

Cetineites are compounds of Sb^{3+} with a general formula $A_6[Sb_{12}O_{18}][SbX_3]_2[D_xY_{6-y}]$ where A = Na^+, K^+, Rb^+, Sr^{2+}, and Ba^{2+}; X = S^{2-}, Se^{2-}; D = Na^+, Sb^{3+}, CO_3 groups (C^{4+}), and Y = H_2O, OH^- or O^{2-}.

The story of cetineites starts with the preparation of the synthetic potassium compound $K_3SbS_3 \cdot 3Sb_2O_3$ from the alkaline aqueous solution of K_2S and Sb_2S_3 by Graf and Schäfer (1975) who also determined its crystal structure. In 1982 Nakai (in Miyawaki et al. 1993) prepared, and determined the structure of the Na analogue, giving its formula as $Na_3Sb_7O_9S_3$ with the absorbed OH groups present in the channel. Kluger and Pertlik (1985) synthesized and structurally analyzed the analogous compound, given by them as $Na_3SbSe_3 \cdot 3Sb_2O_3 \cdot 0.5Sb(OH)_3$. Sabelli and Vezzalini (1987) described natural cetineite as $(K,Na)_{3+x}(Sb_2O_3)_3(SbS_3)(OH)_x \cdot (2.8-x)H_2O$ and Sabelli et al. (1988) described its structure together with, once more, the structure of a synthetic Na compound, $Na_{3.6}(Sb_2O_3)_3(SbS_3)(OH)_{0.6} \cdot 2.4H_2O$.

Extensive investigation of the cetineite family was undertaken by Wang (1995a) as well as Wang and Liebau (1998, 1999). Parallel to the structure investigations, electronic properties of cetineites were investigated because some of them were found to be photo-semiconducting (refs. in Wang and Liebau 1999). According to these latest investigations, the above outlined compositional volume of cetineites contains eight structurally investigated members (Table 1).

The description of synthetic potassium compound by Graf and Schäfer (1975) was perhaps incomplete. The cetineite structure type was completely described by Sabelli et al. (1988) and their results were perfected by Wang and Liebau (1998). They define the following structure units (Figs. 1, 2):

(a) Tubes of a composition $[Sb_{12}O_{18}]$, composed of pyramidal, covalently bonded $[SbO_3]^{3-}$ groups. These groups share all three corners and form a tubular net (a rolled-up hexagonal net) with Sb vertices oriented to its exterior whereas the interior

Table 1. Structurally characterized members of the cetineite group.

Compound	Code[①]	Lattice Parameters (Å)		Space group	Ref.
		a	c		
$K_6[Sb_{12}O_{18}][SbS_3]_2[Sb_{0.12}(H_2O)_{5.64}(OH)_{0.36}]$	K; Se	14.318(3)	5.633(1)	$P6_3/m$	(1)
$Sr_6[Sb_{12}O_{18}][SbSe_3]_2[OH]_6$	Sr; Se	14.320(3)	5.515(1)	$P6_3/m$	(1)
$Ba_6[Sb_{12}O_{18}][SbSe_3]_2[(CO_3)_{1.5}O_{1.5}]$	Ba; Se	14.504(3)	5.616(1)	$P6_3/m$	(1)
$K_6[Sb_{12}O_{18}][SbSe_3]_2[(H_2O)_6]$	K; Se	29.260(7)	5.616(1)	$P6_3$	(1)
$(K,Na)_{6+x}[Sb_{12}O_{18}][SbS_3]_2[(OH)_x \cdot 2.8-x(H_2O)]$	(K,Na); S	14.251(3)	5.590(1)	$P6_3$	(2) (3)
$K_6[Sb_{12}O_{18}][SbS_3]_2$	K; S	14.256(5)	5.621(2)	$P6_3$	(4)
$Na_6[Sb_{12}O_{18}][SbSe_3]_2[Na_{1.86}Sb_{0.14}(OH)_{2.28}(H_2O)_{4.02}]$	Na; Se	14.423(3)	5.565(1)	$P6_3/m$	(5)
$Rb_6[Sb_{12}O_{18}][SbSe_3]_2[Sb_{0.22}(OH)_{0.66}(H_2O)_{3.48}]$	Rb; Se	14.715(3)	5.653(2)	$P6_3/m$	(5)
$Na_6[Sb_{12}O_{18}][SbS_3]_2[Na_{1.2}(OH)_{1.2}(H_2O)_{4.8}]$	Na; S	14.152(3)	5.576(1)	$P6_3$	(3)

① Shows the shorthand devised by Wang and Liebau (1999).

References: (1) Wang and Liebau 1999; (2) Sabelli and Vezzalini 1987; (3) Sabelli et al. 1988; (4) Graf and Schäfer 1975; (5) Wang 1995a

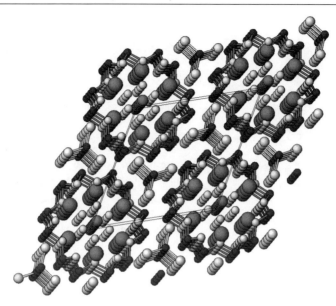

Figure 1. The crystal structure of natural cetineite $(K,Na)_{6+x}[Sb_{12}O_{18}][SbS_3][(OH)_x(H_2O)_{2.8-x}]$. Tubes of SbO_3 pyramids parallel to [001] with intervening stacks of SbS_3 pyramids contain octahedral columns of Na and Sb coordinated by OH and H_2O and large K ions situated along the walls of the tube. Projection along [001]; black spheres: Sb, light small spheres: O, light larger spheres: S, grey: K (large), and Na (small).

is lined by O atoms. Triangular pyramidal groups form hexagonal rings; through these "windows" the cations from the tube interior can get into contact with intertubular S atoms.

(b) Isolated [SbS₃] or [SbSe₃] pyramids, stapled in columns parallel to [001]. Their lone electron pairs may be all directed in the same way (space group $P6_3$) or in the opposite way, the two orientations being distributed at random over columns.

(c) The A cations from the above quoted formula line the tube walls from inside (Fig. 2), being coordinated to six oxygen atoms of the tube wall, one extra tubular S or Se atom through the wall "window" and to 1–4 ligands from the central octahedral column situated along the tube axes.

(d) The interior of the $[Sb_{12}O_{18}]$ tubes is filled by Y ligands (see the formula) which alternatively can be water molecules, hydroxyl groups or O^{2-} anions. They form trigonal antiprisms (distorted octahedra) sharing basal faces and forming a chain (Fig. 2). These octahedra may either be empty or occupied by additional, D cations to various extents. The latter elements were not recognized by Graf and Schäfer (1975) but they were defined by Sabelli et al. (1988).

All cetineite phases with the exception of K; Se have simple hexagonal unit cells a, c which were refined in the space group $P6_3/m$ by Wang (1995a) and Wang and Liebau (1999) for the oxyselenides of Na, Sr, Ba and Rb as well as for K; S. This is contrary to the earlier refinements of Na and K oxysulfides and Na; Se (Table 1) which were refined in $P6_3$.

The reason for differences is primarily the orientation of SbX_3 groups (X = S, Se). In the (a, c) $P6_3$ structure these pyramidal groups should have parallel orientation in all columns (the earlier $P6_3$ refinements). Wang and Liebau (1999) favor $P6_3/m$ for *all* the structures with a simple a, c—a random parallel and antiparallel orientation of SbX_3 stacks. This will be favored

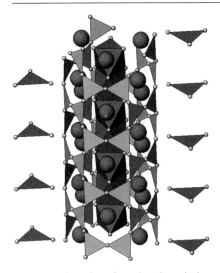

Figure 2. Channel configurations in cetineite. A tube with a hexagonal net of SbO_3 pyramids, a central column of DY_6 octahedra (see text), K sites (large spheres) inside the channel walls and adjacent horizontal SbS_3 groups. Only pyramidal bases of the SbX_3 groups are shown, omitting the Sb positions. Oblique projection upon ($11\bar{2}0$).

by the presence of mirror planes (001) in the surrounding [$Sb_{12}O_{18}$] tubes. An ordered (a, c) $P\bar{3}$ structure was not found (Wang and Liebau 1999). The exceptional K; Se structure ($2a$, c) appears to have two SbX_3 stacks on the three-fold rotation axes pointing in the [001] direction whereas the remaining six stacks (in a unit cell), in general positions, point in the opposite, [00$\bar{1}$] direction. The ($2a$, c) superstructure reflections are extremely weak although sharp, leaving the question of potential superstructures for other chemical compositions fairly open.

The A cations have very irregular coordinations which are freely adjustable in order to obtain the usual A-O distances (2.68 Å in Na; Se to 3.06 Å in Ba; Se). Conspicuous are the short A-Y distances, from 2.44 Å (Na; Se) to 2.72 Å (Ba; Se). Y is the (statistically occupied) corner site of the central octahedral column. The A-X distance varies from 2.89 Å in Na; S to 3.52 Å in Sr; Se (Wang and Liebau 1999).

The central octahedral columns are split into two subsets rotated by 30° against one another in the sodium cetineites whereas a single column orientation exists in all other cetineites. Only one short Y-A bond exists for Na^+, four bonds for other A cations. Wang and Liebau (1999) devoted special attention to the occupancy of DY_6 octahedra. They conclude that only Na^+ and Sb^{3+} fit these coordinations. As a consequence, Sb^{3+} has been used to explain residual maxima in these octahedra (antiprisms) in the K, Rb, and also Na cetineites.

For the mesoporous character of cetineite is important that the tubes appear rigid, rod group $P6_3/m$ close to $P6_3/mcm$; tube diameter [defined by Wang and Liebau (1999) as the projection of interatomic distances on (001)] ranges from 10.037 Å to 10.265 Å for Na; S and K; Se, respectively, and from 10.016 Å (Na; Se) (even 9.912 Å for Sr; Se) to 10.400 Å Rb; Se (i.e., a difference of ~5%). The entire structure expands with the size of A cations (2.5% change in volume) and with the Se-for-S exchange (< 1.5% in volume). A contraction of at least 2% is observed when A^+ is replaced by the A^{2+} element adjacent to A^+ in the periodic system.

In the tube packing, tubes are rotated in such a way that the lone electron pairs on the tube exteriors are in the most favorable situation; the resulting packing is $P6_3/m$, fairly independent of the type of A cations in the tubes. The SbX_3 groups rotate much more, accommodating perhaps the A-S bonds. Detailed data for all these correlations are in the summary work by Wang and Liebau (1999).

Halogen sulfides of Bi with anions in the channels

Several complex sulfides of Bi and Pb-Sb form a family of hexagonal structures in which double walls composed of tetragonal pyramids meet in a triangular fashion and form triangular channels. In the lowest homologue, $Bi_4Cl_2S_5$ (Krämer 1979) these channels are too small, empty, and situated between three interconnected trigonal prisms (one in each channel corner). Chlorine atoms are distributed statistically in the S sites of the framework, i.e. they are parts of the framework rather than channel-hosted ions.

In the next higher homologue, $Bi_{0.67}Bi_{12}S_{18}Hal_2$, where *Hal* is either iodine or bromine (Table 2), the trigonal channels accommodate individual halogen anions and the lone electron pairs (Fig. 3). The hexagonal channels, typical for the entire homologous series (Makovicky 1985) are blocked by partly occupied trigonal pyramidal Bi positions (or by Sb positions in the higher, natural Pb-Sb homologue zinkenite). Higher homologues do not exhibit channels suitable for the accommodation of anions.

Table 2. Selected halogen sulfides of bismuth.

Compound	Lattice Parameters (Å)		Space group	Ref.
	a	*c*		
$Bi_{0.67}Bi_6S_8Cl_4$	19.80[1]	12.36[1]	$R\bar{3}$	(1)
$Bi(Bi_2S_3)_9I_3$	15.63	4.02	$P6_3$ (or $P6_3/m$)	(2)
$Bi(Bi_2S_3)_9Br_3$	15.55	4.02	$P6_3$	(3)

[1] 11.43 × 2cos 30° and 3 × 4.12 Å, a supercell produced by ordering of Bi in hexagonal channels and framework distortion.

References: (1) Krämer 1979; (2) Miehe and Kupčík 1971; (3) Mariolacos 1976

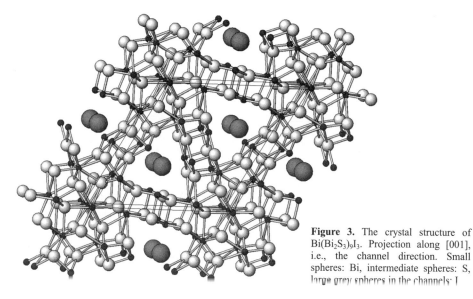

Figure 3. The crystal structure of $Bi(Bi_2S_3)_9I_3$. Projection along [001], i.e., the channel direction. Small spheres: Bi, intermediate spheres: S, large grey spheres in the channels: I

KBi_3S_5 and related channel structures

The crystal structure of KBi_3S_5 (McCarthy et al. 1995) consists of large lozenge-shaped channels able to accommodate four strings of K atoms (Fig. 4). Their walls are ribbons of octahedra derived from single layers of somewhat irregular coordination octahedra of Bi. Columns of octahedra in these walls are coordinated to two, three or four other such columns; the wall surfaces are alternatively 2 and 3 octahedra wide. The minimum diameter of the tunnel is 9.45 Å, the maximum value is 14 Å.

The five-coordinated K1 site (K-S = 3.19–3.43 Å) is 78% occupied, whereas the less advantageous four-coordinated K2 site in the shallow corner of the channel is only 22% occupied. The distorted octahedral coordinations of Bi have a short Bi-S bond equal to 2.64–2.75 Å opposed by a long distance 2.92–3.12 Å, as well as pairs of opposing bonds 2.78–2.92 Å.

This open structure formed from K_2S, Bi and S in a sealed silica glass tube at 300°C, followed by slow cooling. The phase decomposes endothermally at 520°C. KBi_3S_5 does not exhibit cation exchange with alkali and ammonium ions, with the exception of RbCl which (via solid state reaction) produces isostructural β-$RbBi_3S_5$ below 400°C. At or above 400°C another, denser α-$RbBi_3S_5$ structure forms instead.

$RbBi_3S_5$ (Schmitz and Bronger 1974) has a related structure (Table 3), with the walls of channels 1×3 octahedra wide (Fig. 5). Rb atoms occupy two halves of the channel with elongated cross-section which also accommodates an additional S atom in the channel center. Thus, the Rb atoms have bicapped (tricapped) prismatic coordination.

Walls again are formed by octahedral layers but they meet in a way topologically different from that observed in KBi_3S_5, that is to say, $RbBi_3S_5$ is not just a lower homologue of KBi_3S_5. The range of bonds in the irregular Bi octahedra is 2.72–2.99 Å, nine-coordinated Rb has Rb-S distances equal to 3.23–3.93 Å.

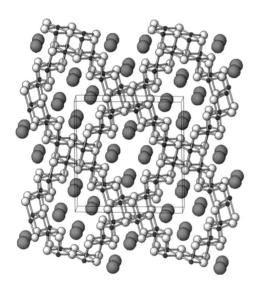 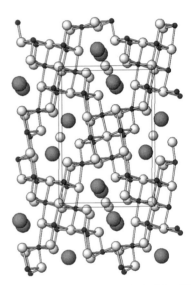

Figure 4. The crystal structure of KBi_3S_5. Projection on (010), a axis horizontal. Octahedral Bi-S framework (small spheres: Bi) enclosing large channels parallel to [010] and occupied by potassium (large grey spheres).

Figure 5. The crystal structure of $RbBi_3S_5$ with two strings of Rb ions (large grey circles) and additional S (light circles) in the [100] channels. Projection along a, b axis horizontal.

Table 3. Crystallographic data for $Me Bi_3S_5$ and related structures.

Compound	a (Å)	b (Å)	c (Å)	β	Space group	Ref.
KBi_3S_5	17.013(5)	4.076(2)	17.365(4)		$Pnma$; $Z = 4$	(1)
$CsBi_3S_5$	4.064(1)	12.098(3)	21.098(4)		$Pmnb$; $Z = 4$	(2)
β-$RbBi_3S_5$	4.16(1)	12.90(2)	18.47(8)		$Pmnn$; $Z = 4$	(3)
α-$CsPbBi_3Se_6$	23.564(6)	4.210(2)	13.798(3)		$Pnma$; $Z = 4$	(4)
$[(CH_3NH_3)_{0.5}(NH_4)_{1.5}]Sb_8S_{13}\cdot 2.8H_2O$	7.193(0)	25.770(1)	16.000(1)	96.856(1)	$P2_1/m$; $Z = 4$	(5)
$Rb_2Sb_8S_{13}\cdot 3.28H_2O$	7.190(1)	25.760(3)	15.973(2)	96.541(2)	$P2_1/m$; $Z = 4$	(5)

References: (1) McCarthy et al. 1995; (2) Kanishcheva et al. 1980; (3) Schmitz and Bronger 1974; (4); Chung et al. 1999 (5) Wang et al. 2000

Yet another structure of this family is α-CsPbBi$_3$Se$_6$ (Chung et al. 1999). Channels in this structure (Fig. 6) are constricted in the middle by the protruding octahedra of M2 that also supply the S atoms needed to create the tricapped prismatic coordination of Cs. Removing these octahedra makes a 3 × 2 cavity as in KBi$_3$S$_5$ and M2 actually stand for the K1 positions in the latter compound. Meeting of octahedral walls is again compound-specific, different from the other two compounds, although it can be interpreted as that in RbBi$_3$S$_5$ (minus the constricting octahedra).

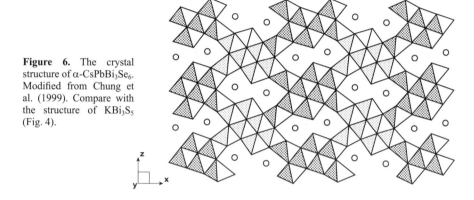

Figure 6. The crystal structure of α-CsPbBi$_3$Se$_6$. Modified from Chung et al. (1999). Compare with the structure of KBi$_3$S$_5$ (Fig. 4).

The intersecting portions of walls are thinnest in KBi$_3$S$_5$; they form four joint columns in RbBi$_3$S$_5$ and blocks of 6 octahedral columns in CsPbBi$_3$Se$_6$, this trend being concurrent with the increasing constriction of the channels. A sort of an end-member to this trend is the structure of CsBi$_3$S$_5$ (Kanishcheva et al. 1980). The channels are again the 3 × 2 channels (Fig. 7) with octahedral walls but these walls are the walls of condensed, block-like 2 × 3 octahedral columns and not of single-octahedral ribbons as in the previous cases. The blocks (columns) are joined only via common S atoms. Cesium occurs as two columns of bicapped trigonal coordination prisms (Cs-S = 3.46–3.85 Å). Coordination properties of bismuth—present as irregular coordination octahedra—are very similar to those in previous compounds.

A somewhat distinct representative of this broad family is Rb$_2$Sb$_8$S$_{13}$·3.3H$_2$O (Wang et al. 2000). The framework is based on a SnS archetype structure (Makovicky 1993), that is to say, a structure with tightly-bonded double-layers separated by lone electron pair interspaces with a lateral shift of double-layers when compared to the simple PbS-like motif of the Bi compounds. The framework can be divided into polyhedral ribbons separating large triangular channels and 2 × 3 blocks of coordination polyhedra (Fig. 8).

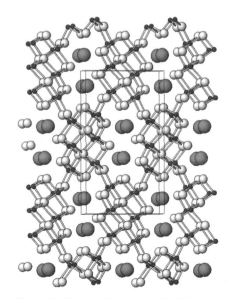

Figure 7. The crystal structure of CsBi$_3$S$_5$ with rods of Bi-S octahedra and Cs in [010] channels. Projection along b, c axis horizontal. In the order of increasing size circles indicate Bi, S and Cs, respectively.

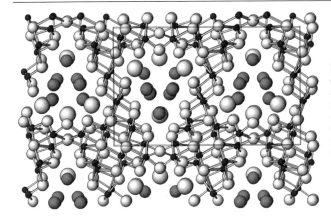

Figure 8. The crystal structure of $Rb_2Sb_8S_{13}\cdot 3H_2O$. Projection along a, b axis horizontal. Sb-S framework based on the SnS archetype with large channels enclosing H_2O (large dark spheres) and Rb (large light spheres).

The bulk of channel volume is occupied by H_2O molecules whereas the Rb ions reside in the corners and at one face of the channel, only partly coordinated by sulfur. Six Rb-S bonds lie between 3.38 and 3.63 Å, Rb-O_{H_2O} bonds range from 2.97–3.42 Å. There are up to three Rb-H_2O bonds (distances) per one Rb atom.

The version with organic cations in the channels has identical framework. It is a methyl ammonium tridecathiooctaantimonate hydrate, $[(CH_3NH_3)_{0.5}(NH_4)_{1.5}]Sb_8S_{13}\cdot 2.75(H_2O)$ (Wang et al. 2000). Hydrogen positions were not determined. Nitrogen atoms of the NH_4 and NH_3CH_3 groups assume the same positions as K^+ in the previous structure (Fig. 9), the shortest van der Waals distances occur to four S atoms (3.33–3.54 Å) and two to three O_{H_2O} sites (2.96–3.30 Å). H_2O positions are partially occupied (0.26–0.59 H_2O per site) in the case of ammonium but only one H_2O site (occupancy 0.78) is partially occupied in the case of Rb.

Hutchinsonite merotypes

Hutchinsonite merotypes (Makovicky 1997) is a group of complex sulfides of As or Sb and large uni- and divalent cations (Tl^+, Pb^{2+}, Na^+, Cs^+, NH_4^+, and others, including organic cations (Table 4). The structures of these sulfosalts are regular 1:1 intergrowths of slabs A which can

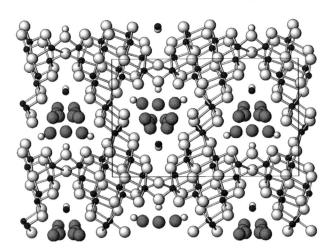

Figure 9. The crystal structure of $[(CH_3NH_3)_{0.5}(NH_4)_{1.5}]Sb_8S_{13}\cdot 2.8H_2O$. Projection along a, b axis horizontal. The SnS-like framework of Sb-S coordination polyhedra hosts water molecules (dark large spheres) and the organic (C - black, N - white) and ammonia (N - small white spheres) component.

Table 4. Selected hutchinsonite merotypes.

Compound/Mineral	HN[1]	Chemical formula	Lattice parameters (Å)			Angles (°)		Space group	Ref.
Hutchinsonite	2;3	$TlPb_{\approx 8.5}S_9$	a 10.79	b 35.39	c 8.14			$Pbca$	(1)
Bernardite	2;2	$TlAs_{\approx 8}$	c 10.75	a 15.65	b 8.04	β 91.27		$P2_1/c$	(2)
Imhofite[2]	1;2	$Tl_{5.8}A_{\approx 15.4}S_{26}$	c 5.74	b 24.43	a 8.76	β 108.28		$P2_1/n$	(3)
Gillulyite (subcell)[3]	1;2	$Tl_2(A_{-},Sb)_8S_{13}$	b 5.68	c 21.50	a 9.58	β 100.07		$P2/n$	(4), (5)
Synthetic	1;2	$Rb_2A_{\approx}S_{13} \cdot H_2O$	b 11.52	c 21.76	a 9.60	β 98.8		$P2_1/n$	(6)
Edenharterite[3]	2;3	$PbTl_{\approx 3}S_6$	c 5.85	a 47.45	b 15.48			$Fdd2$	(7)
Jentschite[3]	2;3	$TlPb_{\approx 2}SbS_6$	c 5.89	b 23.92	a 8.10	β 108.06		$P2_1/n$	(8), (9)
Synthetic[3]	1;2	$(NH_4)_2Sb_4S_7$	a 11.33	b 26.25	c 9.94			$Pbca$	(10)
Synthetic[3]	1;2	$Rb_2SL_{\approx}S_{12}(S_2) \cdot 2H_2O$	a 7.08	b 25.40	a 8.05	β 97.84		$P2_1/n$	(11)
Synthetic	2;2	$Cs_2Sb_8S_{13}$	b 11.48	a 15.43	c 8.29[4]	α 71.89 β 102.45 γ 95.16		$P\bar{1}$	(12)
Synthetic	2;2	$[CH_3 H_3]_2Sb_8S_{13}$	b 11.58	a 15.87	c 8.30[4]	α 71.46 β 75.71 γ 82.25		$P\bar{1}$	(13)
Synthetic	2;2	$[H_3N(CH_2)_3NH_3]Sb_{10}S_{16}$	b 10.93	a 18.36	c 17.39	β 111.44		$P2_1/n$	(14)
Kermesite	1;1	$Sb_2S_2\bullet$	c 5.79	b 10.71	a 8.15	α 102.78 β 110.63 γ 101.00		$P\bar{1}$	(15), (16)
Pääkkönenite	2;2	$Sb_2(A_{\approx 1.84}Sb_{0.16})S_2$	a 10.75	b 12.49	b 3.60	β 115.25		$C2/m$	(17)
Synthetic[3]		$(C_2H_8)_2Sb_8S_{12}(S_2)$	b 11.65	c 25.98	a 9.97			$Cmca$	(19)
Gerstleyite[3]	1;2	$Na_2(SL_{\approx}As)_9S_{13} \cdot 2H_2O$	c 7.10	b 23.05	a 9.91	β 127.85		Cm	(20)
Synthetic		$(NH_3CH_2CH_2NH_3)Sb_8S_{13}$ ethylen-diammonium sulfide	c 11.34	a 22.87	b 10.06			$Cmc2_1$	(18)

Notes: [1] Homologue order (HN) is determined by numbers of square coordination pyramids (SnS subcells) across the width of the tightly bonded SnS-like ribbon *in this slab*. [2] OD character described by Balić-Žunić and Makovicky (1993). [3] B slabs are reduced out (see text). Structures of edenharterite and jentschite respectively contain 2-fold axes and $\bar{1}$ as operators of unit-cell twinning. [4] Tightly-bonded SnS-like ribbons are parallel to [011]. [5] PbS archetype.

References: (1) Matsushita and Takéuchi 1994; (2) Pašava et al. 1989; (3) Divjakovič and Nowacki 1976; (4) Foit et al. 1995; (5) Makovicky and Balič-Žunič 1999; (6) Sheldrick and Kaub 1985; (7) Balič-Žunič and Engel 1983; (8) Berlepsch et al. 2000; (9) Berlepsch et al. 2001; (11) Berlepsch et al. 2001; (12) Volk and Schäfer 1979; (13) Wang and Liebau 1994; (14) Wang 1995b; (15) Bonazzi et al. 1987; (16) Kupčík 1967 ; (17) Bonazzi et al. 1995; (18) Tan et al. 1994; (19) Tan et al. 1996; (20) Nakai and Appleman 1981 be described as $(010)_{SnS}$ cut-outs of different widths from the SnS-archetype or $(110)_{PbS}$ cut-outs from the PbS-archetype, with layers B of variable thickness and configuration. The B layers contain primarily MeS_3 pyramids (Me = As, Sb) with active lone electron pairs which are combined with coordination polyhedra of large (eventually organic) cations. Slabs A and B share certain S atoms in common.

The definition of the hutchinsonite-related sulfosalts as a family of merotypes means that the slabs A are built according to the common principles stated above in all these structures, whereas the B slabs may differ, being always adapted to the requirements of different large cations (Makovicky 1997). In rare cases, B slabs are reduced out as in edenharterite $TlPbAs_3S_6$ and jentschite $TlPbAs_2SbS_6$ (Balić-Žunić and Engel 1983; Berlepsch 1996; Berlepsch et al. 2000). The width of SnS-like layers has been quantified by giving N values independently to the two opposing sides of the SnS-like tightly-bonded strips (Makovicky 1989). This scheme can equally well be applied to PbS-like tightly-bonded strips, for example, $Rb_2Sb_8S_{12}(S_2)\cdot 2H_2O$ is $N_{1,2}$ = 1;2 (Berlepsch et al. 2001) (Fig. 10).

In the layers based on PbS-archetype structural allowances are made for a more active role of lone electron pairs by inflating the appropriate structure portions. In a number of these structures, the pattern of short Sb-S bonds forms chains composed of corner-sharing SbS_3 coordination pyramids. This type of chains is found in $(NH_4)_2Sb_4S_7$ (Dittmar and Schäfer 1977), gerstleyite $Na_2(Sb,As)_8S_{12}\cdot 2H_2O$ (Nakai and Appleman 1981), gillulyite $Tl_2(As,Sb)_8S_{13}$ (Foit et al. 1995; Makovicky and Balić-Žunić 1999), $Rb_2Sb_8S_{12}(S_2)\cdot 2H_2O$ (Berlepsch et al. 2001), $[C_4H_8N_2][Sb_4S_7]$ (Parise and Ko 1992), and $[C_2H_8N]_2(Sb_8S_{12}(S_2))$ (Tan et al. 1996). In all these cases, via weak Sb(As)-S interactions, the same A slab configuration arises, that of distorted PbS-like motif N = 1,2 with the N = 2 layers containing additional, periodically sideways inserted SbS_3 groups and with the coordination polyhedra of trapezoidal cross section.

Presence or absence of Sb, S based interconnections of adjacent A slabs is of importance for the microporous character of these structures. Tan et al. (1996) classified interconnections of A slabs across the B slabs as follows: the lateral Sb-S chains of A slabs can be (1) isolated as in $[C_4H_8N_2]Sb_4S_7$, (2) linked via common sulfurs $[C_2H_{10}N_2]Sb_8S_{13}$ (Tan et al. 1994), (3) linked via disulfide groups $[C_2H_8N]_2Sb_8S_{12}(S_2)$, and $Rb_2Sb_8S_{12}(S_2)\cdot 2H_2O$, or (4) via additional SbS_3 groups $[C_3H_{12}N_2]Sb_{10}S_{16}$ (Wang 1995b) (Fig. 11).

Imhofite $Tl_3As_{7.66}S_{13}$, hutchinsonite $TlPbAs_5S_9$, bernardite $TlAs_5S_8$, edenharterite $PbTlAs_3S_6$, jentschite $PbTlAs_2SbS_6$, pääkkönenite $Sb_2(As,Sb)S_2$, and kermesite Sb_2S_2O belong to the branch based on SnS-archetype; the configuration of A slabs are designed to accommodate active lone electron pairs, often with large cations (Tl$^+$) alternating with As polyhedra along the slab margins. These margins can sometimes be strongly modified relative

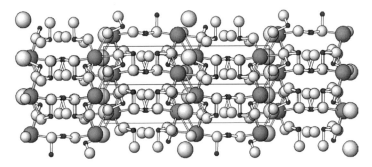

Figure 10. Crystal structure of $Rb_2Sb_2S_{12}(S_2)\cdot 2H_2O$, a member of the hutchinsonite merotype series. Projection along [101], b axis horizontal. SnS-like slabs $N_{1,2}$ = 1; 2 house Sb polyhedra, complex interlayers (010) contain Rb (black) and H_2O molecules (white).

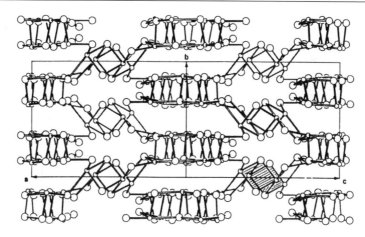

Figure 11. The crystal structure of [$H_3N(CH_2)_3NH_3$]$Sb_{10}S_{16}$ with columns of paired SbS_5 pyramids interconnecting PbS-based layers of Sb-S polyhedra. The organic component (not shown) is housed in the channels. Projection along [101], small circles: Sb; large circles: S. Reproduced with permission from Wang (1995b).

to the archetype. In the case of $(NH_4)_2Sb_4S_7$ (Dittmar and Schäfer 1977), the distribution of bonds in A slabs is transitional between the PbS and SnS principles.

Rod-based sulfosalts

This group of microporous (predominantly) bismuth sulphosalts (Table 5) differs in several important aspects from the KBi_3S_5 family. They belong to the broad family of so-called rod-based sulfosalts; the fundamental element of these is a rod of PbS or SnS-like topology, $\underline{n}$ coordination polyhedra broad and $\underline{m}$ atomic levels thick. These rods are interconnected by various polyhedral configurations into layers which are periodically widened (rods) and constricted (interconnections). Cyclic or chess-board arrangement of rods is less common. The rods have two sets of surfaces:

(a) (100) surfaces of the PbS or SnS-like arrangement. These represent square- or trapezoidal nets of S (or Se) atoms, with cations (in most minerals Pb^{2+}) positioned

Table 5. Microporous sulfosalts of bismuth with rod-layers.

Compound	Lattice parameters (Å)			Angles (°)	Space group	Channel*	Ref.
	a	b	c				
$KLa_{1.28}Bi_{3.72}S_8$	16.652(1)	4.071(0)	21.589(1)	—	Pnma	SC	(1)
$RbCe_{0.84}Bi_{4.16}S_8$	16.747(0)	4.053(0)	21.753(1)	—	Pnma	SC	(1)
$KBi_{6.33}S_{10}$	24.05(1)	4.10(2)	19.44(1)	—	Pnma	SC	(2)
$K_{1.4}Sn_{2.2}Bi_{7.4}Se_{14}$	17.402(2)	4.205(1)	21.227(3)	109.52(0)	$P2_1/m$	PSC	(3)
$Sr_4Bi_6Se_{13}$	18.398(7)	4.241(2)	17.037(7)	90.58(2)	$P2_1/m$	PSC	(4)
α-$K_2Bi_8Se_{13}$	13.768	12.096	4.166	α 89.98 β 98.64 γ 87.96	$P\bar{1}$	TC	(5)

* Channel type: SC = single cation; PSC = paired single-cation; TC = two-cation

References: (1) Iordanidis et al. 1999; (2) Kanatzidis et al. 1996; (3) Mrotzek et al. 2001; (4) Cordier et al. 1985; (5) McCarthy et al. 1993;

inside this mesh and moved slightly out of the plane of the S net. Lone electron pairs of these cations point outwards, into the interlayer space. These are so-called pseudotetragonal, Q, surfaces.

(b) (111) surfaces of the PbS-like arrangement; these are hexagonal anion nets without cations situated in them (the pseudohexagonal, H, surfaces). They always face the pseudotetragonal surfaces in a non-commensurate fashion; long cation-S bonds/distances interconnect the Q layer with the H layer. In the case of SnS archetype, modified H surfaces occur.

As illustrated here using the structure of jamesonite, $FePb_4Sb_6S_{14}$ (Fig. 12), in the natural sulfosalts the surface cations are mostly Pb^{2+} whereas the interior polyhedra are occupied especially by Sb^{3+} or Bi^{3+}. Introduction of large cations such as Ba^{2+} or K^+ into what can originally be considered as Pb^{2+}-like positions in pseudotetragonal surfaces, complicates the simple linear sequence of cations, introducing various offsets in them. Introduction of M^+ also asks for a parallel replacement of Pb^{2+}-like sites by trivalent elements (e.g., Bi^{3+} or REE^{3+}).

The first example is the structure of $K_{1.4}Sn_{2.2}Bi_{7.4}Se_{14}$ (Mrotzek et al. 2001). In this structure, layers with rods 3 square pyramids (which are completed to octahedra across the interlayer space) wide and four layers thick alternate with rod-layers in which extended rods are only 2 pyramids (octahedra) wide but 6 atomic layers thick (Fig. 13). Positions of K can be interpreted as completing the surface layers of these lozenges. In the first layer type this results in a pair of K^+ channels (K-Se distances from 3.42 to 3.57 Å), separated by a zig-zag of "paired" Se atoms with the (possibly average) Se-Se distance equal to 3.33 Å. All the other K^+ sites alternate with Sn along the 4 Å axis, such that they cannot be interpreted as channels.

The structure of $KLa_{1.28}Bi_{3.72}S_8$ (Iordanidis et al. 1999) has single-cation channels with K^+. These can be counted as a somewhat protruding square-pyramidal coordination belonging to a lozenge-shaped rod of PbS-like structure, 3 square pyramids wide and four atomic layers thick (Fig. 14). The edges of rods in one (100) layer are interconnected via 7-coordinated Bi (partly substituted by La), combined with the La-Bi sequences at the blunt edges of the rods situated in adjacent (100) layers. Iordanidis et al. (1999) noticed the similarity of these portions to the structure of Gd_2S_3 (Schleid 1990).

Potassium in the channels forms bicapped trigonal coordination prisms, with bond lengths 3.25–3.29 Å but with one of the caps extended to 3.55 Å. In the isostructural $RbCe_{0.84}Bi_{4.16}S_8$

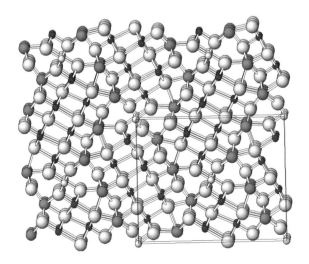

Figure 12. The crystal structure of jamesonite $FePb_4Sb_6S_{14}$, a natural rod-layer sulfosalt with Pb (large grey spheres) situated on rod surfaces and Sb (black) in the rod interior. Fe (white, small) forms octahedral columns interconnecting rods into layers. Projection along c, b axis horizontal.

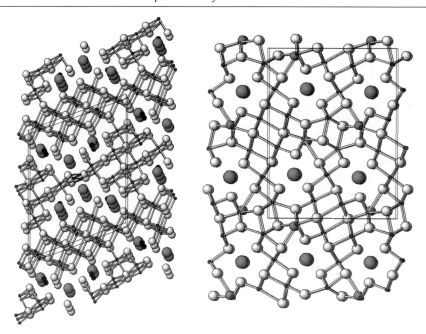

Figure 13 (left). The crystal structure of $K_{1.4}Sn_{2.2}Bi_{7.4}Se_{14}$. Two types of (001) layers composed of rods of distorted Bi coordination octahedra with K and Sn in some positions of their pseudotetragonal surfaces. Note the paired columns of K^+ parallel to [010] and adjacent to the Se-Se columns in the centre of the unit cell. Projection along b, c axis vertical. In the order of increasing size spheres indicate Bi, Sn, Se, and K.

Figure 14 (right). The crystal structure of $KLa_{1.3}Bi_{3.7}S_8$. Projection upon (010), a axis horizontal. In the order of increasing size spheres indicate Bi, La (grey), Se (light) and K. Octahedral rod-like Bi arrays are interconnected by trigonal prismatic portions into "rod-layers" (100). K and La form columns that can be described as belonging to pseudotetragonal surfaces of the rods.

(Iordanidis et al. 1999), again two Bi-Ce sites were detected; the Rb-S distances are longer than in the previous case, 3.30–3.40 Å, with the additional 8th distance of 3.56 Å.

The same structure type has been obtained for $RbLa_{1\pm x}Bi_{4\pm x}S_8$, $KCe_{1\pm x}Bi_{4\pm x}S_8$, $KPr_{1\pm x}Bi_{4\pm x}S_8$ and $KNd_{1\pm x}Bi_{4\pm x}S_8$ (Iordanidis et al. 1999). They are fairly close to typical rod-layer structures of Pb-(Sb,Bi) sulfosalts (Makovicky 1993). Cesium forms a structure type different from these sulfosalts.

All these compounds were synthesized in a dry way, reacting the corresponding alkali metal sulfide, Bi_2S_3, La_2S_3 (or Ce and S) in the appropriate ratios above 800°C. They melt incongruently between 770 and 880°C. They are semiconductors with a relatively large bandgap. Ion exchange has not been attempted.

The next example, α-$K_2Bi_8Se_{13}$ contains two-cation channels filled by potassium. Layers are modified, the more extensive ones correspond to stepped (sheared) layers observed in junoite $Cu_2Pb_3Bi_8(S,Se)_{16}$ (Mumme 1975) but with a more extensive overlap (Fig. 15). The other type is reduced to columns Bi_2S_4 of paired square coordination pyramids, interconnecting the former layers and closing in this way the K-containing channels.

The semiconducting $KBi_{6.33}S_{10}$ (Kanatzidis et al. 1996) has the crystal structure that is analogous to cosalite, with approximate composition $Pb_4Bi_4S_{10}$ (Srikrishnan and Nowacki 1974). In this structure (Fig. 16), bismuth assumes an irregular to regular octahedral coordination, forming four octahedra wide and four atomic layers thick rods of PbS-like

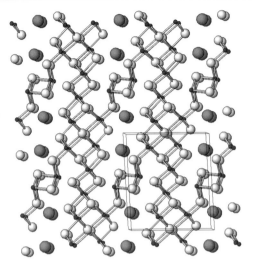

Figure 15. The crystal structure of α-$K_2Bi_8Se_{13}$. Projection along c, b axis horizontal. Stepped octahedral layers composed of Bi octahedra (small black spheres: Bi) and paired columns of BiS_5 coordination pyramids enclose double columns of K atoms (large grey spheres) parallel to [001].

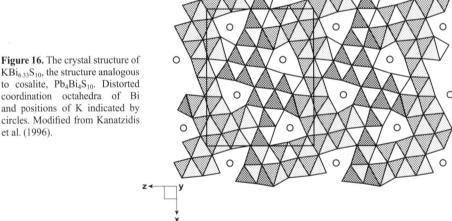

Figure 16. The crystal structure of $KBi_{6.33}S_{10}$, the structure analogous to cosalite, $Pb_4Bi_4S_{10}$. Distorted coordination octahedra of Bi and positions of K indicated by circles. Modified from Kanatzidis et al. (1996).

arrangement. These are interconnected into wavy layers (001) by short single-octahedral layer intervals which merge imperceptibly into the rods at both ends. Columns of bicapped trigonal coordination prisms of K replace one of the Pb positions observed on the pseudotetragonal surfaces of PbS-like rods in cosalite, the other becomes Bi. The K-S distances range from 3.15 Å to 3.44 Å. This compound was synthesized by reacting K_2S and Bi_2S_3 in an evacuated, carbon-coated silica glass tube at 800°C, with a slow subsequent cooling. No ion exchange was attempted. The potassium channels in $KBi_{6.33}S_{10}$ resemble, by their position and cross section, the thallium channels in the lillianite homologue no. 3, $TlSb_3S_5$ (Gostojic et al. 1982).

The crystal structure of $Sr_4Bi_6Se_{13}$ (Cordier et al. 1985) consists of two types of layers. The first type has rods two pyramids broad and 4 layers thick, the other type 3 pyramids broad. Rod interconnections in the two layer types proceed via H and Q fragments, respectively. It is in these Q-type interconnections (details in Makovicky 1993) that two Sr atoms with close-to-regular tricapped trigonal prismatic coordination occur, forming two close, parallel channels (Makovicky 1997). The Sr-Se distances range from 3.190 Å to 3.499 Å.

Galkhaite: a cage-like sulphide

Galkhaite was described by Gruzdev et al. (1972) as a cubic mineral (a = 10.41 Å, space group $I\bar{4}3m$) from Gal Khaya (Yakutia) and Khaydarkan (Kirgizia). They reported the composition range $Hg_{0.74-0.80}Cu_{0.17-0.15}Zn_{0.14-0.03}Tl_{0.01-0.05}As_{0.98-0.85}Sb_{0.02-0.15}S_{2.01-1.97}$, missing out—as we shall see—some of the principal and most interesting cations. Botinelly et al. (1973) found galkhaite in the Getchell Mine (Nevada), with the microprobe composition $Hg_{4.26}Cu_{0.91}Zn_{0.31}Tl_{0.29}As_{3.60}S_{12}$. In the first structure determination on galkhaite, Divjaković and Nowacki (1975) found a heavy atom positioned centrally in the cavity of a tetrahedrite-like framework which they described as thallium. Only 0.48 Tl was found in the cavity and these authors described galkhaite as $[Hg_{0.76}(Cu,Zn)_{0.24}]_6(Tl_{0.48}\square_{0.52})As_4S_{12}$. Another model, by Kaplunnik et al. (1975), with much worse agreement factors, placed As_4 groups into the cavities instead of a single atom. Chen and Szymanski (1981) explain this model by Kaplunnik's attempt to fit the structure to the chemistry reported by Gruzdev et al. (1972). The puzzle of galkhaite was finally solved by Chen and Szymanski (1981). They found that galkhaite from the Getchell Mine (Nevada) varies between $(Hg_{4.74}Cu_{0.46}Ag_{0.27}Zn_{0.08})_{\Sigma5.55}Cs_{0.98}(As_{3.56}Sb_{0.47})_{4.03}S_{12}$ and $(Hg_{4.23}Cu_{0.95}Zn_{0.61})_{\Sigma5.76}Tl_{0.57}Cs_{0.33}(As_{3.56}Sb_{0.04})_{\Sigma3.60}S_{12}$, the average being $(Hg_{4.42}Cu_{0.88}Zn_{0.48})_{\Sigma5.78}Tl_{0.21}Cs_{0.67}(As_{3.55}Sb_{0.04})_{\Sigma3.59}S_{12}$.

In their structure determination, the cubic crystal, with a = 10.365(3) Å and the space group $I\bar{4}3m$ has a tetrahedrite-like skeleton, with nearly regular (Hg, Cu, Zn)S_4 tetrahedra (Me-S = 2.496 Å) and As in flat trigonal pyramids AsS_3 (As-S = 2.265 Å).

(Cs,Tl) in the cavity centre (Fig. 17) is 12-coordinated, with Cs-S distances equal to 3.863 Å and with very variable S-Cs-S angles, ranging from 50.2° to 145.1°. The coordination polyhedron is a Laves polyhedron, i.e. a truncated tetrahedron. The equivalent isotropic displacement factor B_{iso} of Cs is 3.8, whereas that of S in the framework is only 2.1. This indicates large movements ("rattling") of (Cs,Tl) in the cavity. The structure refinement indicated 19% vacancies in the position of the large cation. Contrary to Chen and Szymanski (1981) who suggest the 2 Tl for 3 Cs exchange in the cavity, connected with a rather complex explanation, we can read their plot as the 0.8 Tl for 0.8 Cs exchange, with the exception of the Cs ≈ 1.0 case.

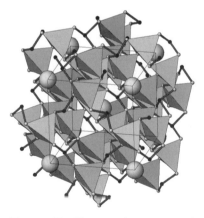

Figure 17. The crystal structure of galkhaite. Oblique projection upon (100). Tetrahedral openwork, with (Hg,Cu,Zn)S_4 tetrahedra and (AsS_3) pyramids (As: black small spheres) forming large cages which contain (Cs,Tl) (large spheres).

The formation of galkhaite has to be ascribed to the presence of Cs in the ore-forming environment; Tl can only play a minor role in the mineral. There are only constricted triangular "windows" in the walls between two adjacent cavities of the galkhaite structure. They do not allow passage of larger cations. Only in Cu-rich synthetic tetrahedrite $Cu_{14}Sb_4S_{13}$ (Makovicky and Skinner 1979; Pfitzner 1997) they serve as passageways for mobile Cu atoms. In this connection it should be stressed that in the tetrahedrite-tennantite family the cavities host a central S atom and 6 copper atoms disposed octahedrally around it. These Cu atoms are triangularly coordinated, by the central sulfur and two sulfur atoms of the framework. They show extreme displacement factors perpendicular to the plane of the triangle.

SULFIDES AND SELENIDES: GENERAL FEATURES

Micro- and mesoporous sulfides and selenides contain typically a framework-building cation and a large cation forcing the formation of channels, cages or interlayers. Among the prominent framework builders belong chromium and indium, primarily forming octahedral frameworks, furthermore copper, iron, as well as silver and mercury. These build frameworks based on combined tetragonal/triangular/linear configurations. In the latter category, metal-metal interactions are often very important, extending sometimes to the contacts with the large channel cations. Higher polarizability of anions and bond covalency contribute to the specificity of these structures. Coordinations of large cations differ in principle from those observed in oxysalts and oxides because of the altered cation: anion radius ratios. A short excursion is made into cage-like arsenides, exemplified by skutterudite and kutinaite.

Djerfisherite and bartonite: derivatives of pentlandite

Djerfisherite is a complex sulfide of iron and copper, containing large cations, K and lesser amounts of Na, as well chlorine in an independent site. Its structure was determined by Dmitrieva et al. (1979) and refined on a synthetic analogue $K_6LiFe_{23}S_{26}Cl$ by Tani et al. (1986). Natural djerfisherite, $K_6Na_{0.81}(Fe_{0.84}Cu_{0.16})_{24}S_{26}Cl$, is an iron-copper sulfide but its synthetic analogue and the related bartonite $K_{5.68}Fe_{20.37}S_{26.93}$ (Evans and Clark 1981) indicate that copper can be fully exchanged by Fe and, in bartonite and owensite, even Cl is replaced by S. A nickel-barium analogue of djerfisherite, $Ba_6Ni_{25}S_{27}$, with S instead of Cl, has been synthesized (Gelabert et al. 1997). The full compositional variation of this group can be seen in Table 6.

The crystal structure of the parent compound, pentlandite or Co_9S_8 (Rajamani and Prewitt 1975) contains cubic clusters of eight metal-sulfur tetrahedra. Tetrahedra in a cluster share edges, yielding 3 direct short metal-metal interactions, not extending beyond the cluster. Clusters are hinged via common apical S atoms and the cubic space between 6 adjacent clusters is assumed by a metal atom (Co, Fe and Ni, Ag or PGE in different pentlandites) in octahedral coordination (Fig. 18). Cages are closed in pentlandite structure, preventing movement of atoms between cages.

In cubic djerfisherite such units—cubic spaces with 6 surrounding clusters—are interconnected into a loose framework with channels [100], and equivalent, i.e. along all edges of the cubic unit cell when the cubic cage, occupied by Na(Li), is at ½ ½ ½ (Fig. 19).

Table 6. Djerfisherite and related phases.

Compound	Lattice parameters (Å)			Space group	Ref.
	a	b	c		
Bartonite; $K_{5.68}Fe_{20.368}S_{26.925}$	10.424(1)	—	20.626(2)	$I4/mmm$	(1)
Chlorbartonite; $K_6(Fe,Cu)_{24}S_{26}(Cl,S)$	10.381(8)	—	20.614(2)		(2)
Djerfisherite; $K_6LiFe_{23}S_{26}Cl$	10.353(1)			$Pm\bar{3}m$	(3)
Synthetic; $Ba_6Ni_{25}S_{27}$	10.057(1)			$Pm\bar{3}m$	(4)
Thalfenisite; $Tl_6Fe(Fe,Ni)_{24}S_{26}Cl$	10.92			$Pm\bar{3}m$	(5)
Owensite; $(Ba,Pb)_6(Cu,Fe,Ni)_{25}S_{27}$	10.349(1)			$Pm\bar{3}m$	(6)
Argentopentlandite; $AgFe_8S_8$	10.521(3)			$Fm\bar{3}m$	(7)
Co-pentlandite; Co_9S_8	9.923(1)			$Fm\bar{3}m$	(8)

References: (1) Evans and Clark 1981; (2) Yakovenchuk et al. 2003; (3) Tani et al. 1986; (4) Gelabert et al. 1997; (5) Rudashevski et al. 1979; (6) Szymanski 1995; (7) Hall and Stewart 1973; (8) Rajamani and Prewitt 1975

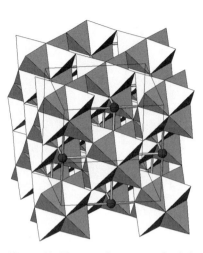
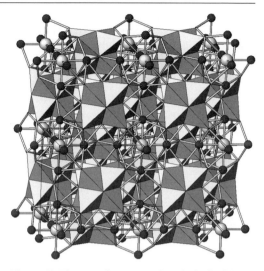

Figure 18. The crystal structure of cobalt pentlandite Co_9S_8. Oblique projection upon (001). Corner-sharing tetrahedral clusters Co_8S_{14} and octahedrally coordinated cobalt atoms (black) in the intervening cages.

Figure 19. The crystal structure of synthetic djerfisherite $K_6LiFe_{23}S_{26}Cl$. Oblique projection upon (100). The framework of corner-sharing Fe_8S_{14} tetrahedral clusters contains cubic cages with Li (hidden from view, at ½ ½ ½) and intersecting channels with K (dark spheres) and Cl (light spheres).

Chlorine atoms lie at intersections of the channel systems and are octahedrally surrounded by K atoms, occurring as two cations per [100] channel interval (Fig. 19). No constrictions occur in the channels.

In thalfenisite (Table 6), K is replaced by Tl and Na (Li) by a divalent octahedrally coordinated metal (as in pentlandite). In owensite, the univalent metals in the channels are replaced by Ba and, in the ratio 1:9, also by Pb with an active lone electron pair. Chlorine is replaced by S and Na by octahedrally coordinated metal with the *Me*-S distance equal to 2.50 Å.

Tetragonal bartonite (Fig. 20) contains intersecting channels [100] and [010] but lacks continuous channels in the [001] direction. The "nuclei" of vertical channels reach only one cage up and one cage down from each intersection, with appropriate K atoms again completing an octahedron around the non-framework S atoms in the channel intersection. Lacking enclosed cages, bartonite lacks the appropriate smaller cation (Na or Fe, Ni, Cu).

Djerfisherite and bartonite can be considered two polytypes, where the channel-containing (in terms of cluster interconnection "loose") (001) layers are stacked in the AAAA sequence in djerfisherite whereas in bartonite

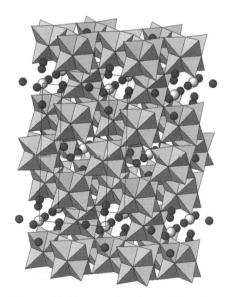

Figure 20. The crystal structure of bartonite $K_{5.7}Fe_{20.4}S_{27}$. Oblique projection upon (100). Framework of corner-sharing tetrahedral clusters Fe_8S_{14} and intersecting channels with K (dark) and additional S (light).

they follow the ABAB stacking sequence. Thus, any occurrence of the "bartonite sequence" in djerfisherite will block the [001] channels. Eight K-S bonds in djerfisherite (square antiprism) are 4 × 3.31 and 4 × 3.44 Å, the single K-Cl bond is 3.10 Å long. The corresponding Ba-S bonds in owensite are 4 × 3.24 and 4 × 3.30 Å, plus a single Ba-S bond equal to 3.15 Å.

Microporous sulfides with octahedral walls

A simple microporous example of this category is the structure of $K_{0.3}Ti_3S_4$ (Schöllhorn et al. 1980). This hexagonal structure (Fig. 21) has $a = 9.505$ Å and $c = 3.414$ Å; the space group is $P6_3/m$. The hexagonal channels have walls one octahedron wide and the octahedra form a dense octahedral framework in the nodes of which three octahedra meet, sharing faces and the common S atoms in the centre of every such triplet. Fig. 21 also suggests that every wall separating two adjacent channels could be split, and additional octahedra inserted, leading to a potential homologous series. Potassium in the channels (partly occupied positions forming a quasi-continuous column of K sites) has two sets of distances: 3 × 3.18 Å and 6 × 3.61 Å (Schöllhorn et al. 1980).

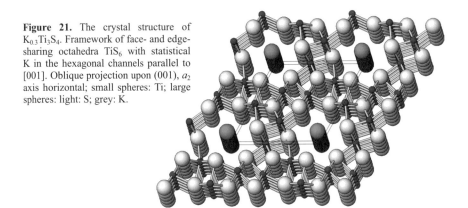

Figure 21. The crystal structure of $K_{0.3}Ti_3S_4$. Framework of face- and edge-sharing octahedra TiS_6 with statistical K in the hexagonal channels parallel to [001]. Oblique projection upon (001), a_2 axis horizontal; small spheres: Ti; large spheres: light: S; grey: K.

Another simple and typical channel structure of this category is that of KCr_5Se_8 (Dung et al. 1987). In this monoclinic structure with $a = 18.66$ Å, $b = 8.98$ Å, $c = 3.59$ Å and $\gamma = 104.53°$ (Fig. 22), infinite straight octahedral walls are interconnected by two octahedra wide fragments forming lozenge-shaped, 2 × 2 octahedra broad channels with a single row of potassium atoms. In the point of branching, the octahedra share faces, such that these points exhibit short Cr-Cr distances. Potassium in the channels is disordered at two positions 0.4 Å apart, the other K-K distance is 3.18 Å. Ten K-Se distances range from 3.34 to 3.74 Å.

The crystal structure of KIn_5S_8 (Carré and Pardo 1983; Deiseroth 1986) (monoclinic, $B2/m$, $a = 19.05$ Å, $b = 9.21$ Å, $c = 3.85$ Å, $\gamma = 103.27°$) is similar although not described as such. The reason is that the In atoms avoid short In-In contacts by moving, in the appropriate octahedra of the infinite walls, away from the attached octahedra (Fig. 23). This results in distorted tetrahedral coordinations for these polyhedra. The K atoms in the channels are disordered as well (K-S 4 × 3.26 Å and 4 × 3.74 Å).

One of the most spectacular channel structures among sulfides/selenides is the family of mesoporous $A_{1-p}Cr_2X_{4-p}$ compounds (Brouwer and Jellinek 1979) where A = Ba, Sr, Eu, Pb; X = S or Se; $p \approx 0.29$. These are hexagonal structures with $a \approx 21.5$ Å and $c_o \approx 3.45$ Å for S, whereas $a \approx 22.5$ Å and $c_o \approx 3.62$ Å for Se. At least for the chromium sulfide framework the space group is $P6/m$. The structure (Fig. 24) consists of seven octahedra wide fragments of single-octahedral walls, interconnected via octahedral faces, building Cr-Cr pairs in the

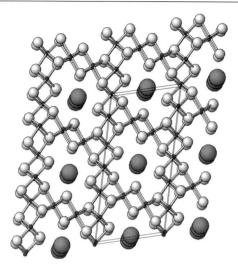

Figure 22. The crystal structure of KCr_5Se_8. Projection along c, a axis vertical. In the order of increasing size spheres indicate Cr, Se and K. Layers of $CrSe_6$ octahedra are interconnected by octahedral partitions, with short Cr-Cr contacts in the points of joining. Pseudohexagonal [001] channels host columns of potassium ions.

Figure 23. The crystal structure of KIn_5S_8. Oblique projection along b, c axis vertical. Horizontal layers of In octahedra (small spheres: In) and flattened tetrahedra are interconnected by octahedral partitions; K (large spheres) is in the channels parallel to [010].

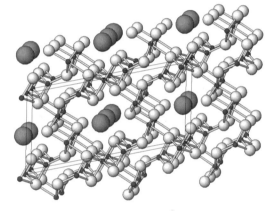

process. The unit cell content of the hexagonal "honey-comb" is $Cr_{21}X_{36}$. It contains hexagonal channels with walls that are two octahedra wide and trigonal channels with walls that are 4 octahedra wide. Channels surrounding the six-fold axes contain a central row of Cr atoms alternating with horizontal triangles of Se atoms. They form a stack of octahedra $CrSe_6$ that share faces. Large atoms, such as Ba, cling to this column from outside, squeezed between it and the octahedral walls of the channel; composition is $A_6Cr_2X_6$ per a period of the column infill. Channels about the three-fold axes contain a central row of X atoms surrounded by three A sites; composition is $A_{2.7}X$ per a repeat distance of this column.

These remarkable compounds have three individual (sub)periodicities, (a) for the chromium sulfide framework, (b) for the contents of the hexagonal channels and (c) for the fill of the trigonal channels. For $Ba_{1-p}Cr_2Se_{4-p}$, there is a common superstructure with $c = 5c_o = 4c_3 = 3c_6$, but for other compounds Brouwer and Jellinek (1979) found incommensurate infillings with $c_6/c_o = 1.64$–1.66 instead of 5/3 and $c_3/c_o = 1.21$–1.29 instead of exactly 5/4.

Detailed sequences in the trigonal channels are ordered and they, and the octahedral walls, modulate each other, giving rise to sharp satellite reflections (except for $Pb_{1-p}Cr_2S_{4-p}$). The chains about the six-fold axes are ordered only in the Ba selenide but disordered in other compounds, yielding diffuse reflections.

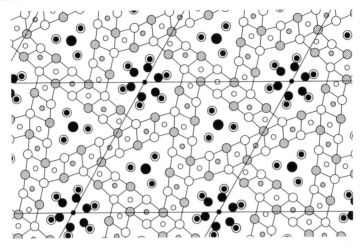

Figure 24. The crystal structure of $Ba_{1-p}Cr_2Se_{4-p}$. Hexagonally arranged octahedral walls form trigonal and hexagonal channels parallel to [0001], with non-commensurate channel cations and triangles of Se atoms. Projection along c, a_2 axis horizontal. Small circles: Cr, larger circles: Se, double circles: Sr, Ba. Framework elements are light, channel elements black. Modified from Brouwer and Jellinek (1979).

Cage-like arsenide and antimonide structures

Skutterudites. Skutterudite, $CoAs_3$ or, for natural samples, $(Co,Fe,Ni)As_3$ has a cubic structure composed of corner-sharing octahedra $CoAs_6$ (space group $Im\bar{3}$). The structure is essentially a collapsed perovskite structure (Chakhmouradian, written comm.) with four octahedral vertices close enough to build square anions As_4 (As-As is 2×2.48 and 2×2.56 Å). Icosahedral cavities limited by the faces of 8 octahedra and 6 edges of As_4 groups are substantial element of the structure (Fig. 25). The As-As radius of the cage can be estimated as 6.17 Å. Octahedral Co-As bonds all are 2.34 Å. Icosahedra do not share any structural element, they are interconnected via the octahedra and As_4 groups and they have no interconnecting pathways. It is in synthetic derivatives that the icosahedral cavity becomes occupied.

Sales et al. (1997) synthesized filled skutterudite antimonides $R_{1-y}Fe_{4-x}Co_xSb_{12}$ where R = lanthanides or Th, $0 < y < 1$, and $0 < x < 4$. They used premelted charges, annealed and pressure-sintered at 700°C, as well as the Bridgmann technique for growing single crystals. A common feature of these products is a strong Fe-Co zoning and partial occupancy of the large cavity, negatively correlated with the contents of cobalt.

Single crystal data (Jeitschko and Braun 1977; Braun and Jeitschko 1980) confirm a considerable "rattling" of the encapsulated lanthanide atom. Sales et al. (1997) demonstrate this effect for $La_{0.75}Fe_3CoSb_{12}$, showing the strong temperature dependence of the rattling effect and giving an average rattling amplitude as 0.15 Å.

Bauer et al. (2000) describe skutterudite $Yb_xM_4Sb_{12}$ (M = Fe, Co, FeCo, Rh and Ir) in

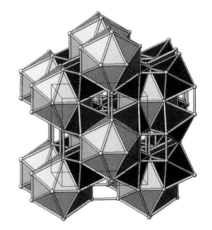

Figure 25. Icosahedral cavities (walls light), coordination octahedra of Co (walls black), and As_4 groups (As: small spheres) in the crystal structure of skutterudite, $CoAs_3$.

which the Yb content decreases in the sequence $x = 0.8$ (M = Fe), $x = 0.5$ (FeCo), $x = 0.2$ (Co), $x = 0.1$ (Rh), and $x \approx 0.02$ (Ir). They stress that Fe does not form a stable skutterudite without the filling cations whereas the remaining ones do. Symmetry is cubic, $P\bar{6}2c$ with the unit cell parameter a equal to 9.150, 9.086, 9.049, 9.229 and 9.243 Å, in the above sequence.

Shirotani et al. (1997) describe superconductors $LaRu_4As_{12}$ and $PrRu_4As_{12}$ prepared at high pressures and Korenstein et al. 1977 synthesized rhombohedral sulphur(selenium)-germanium skutterudites $CoGe_{1.5}Y_{1.5}$ where Y = S or Se.

Kutinaite. Another structure that can be interpreted as a cage-like structure is the copper-silver arsenide, kutinaite $Cu_{14}Ag_6As_7$ (Karanović et al. 2002). One of the possible interpretations of this structure is its description as a Cu-As framework of composite "supertetrahedra" with large cavities hosting octahedral Ag_6 clusters (Fig. 26).

The "supertetrahedron" consists of four $CuAs_4$ tetrahedra sharing corners and the interspace is enclosed by (not fully occupied) $CuAs_3$ coordination triangles (Fig. 26). "Supertetrahedra" share edges and the octahedral cavities (edge length = 8.33 Å) do not communicate. Cu-Cu distances suggest metal-metal interactions of different intensity (2.58–2.78 Å), some Ag-Cu contacts are shorter as well (2.78–2.82 Å) whereas the Ag-Ag contacts are 2.95–3.04 Å. Ag-As contacts are the longest of all (3.03–3.10 Å).

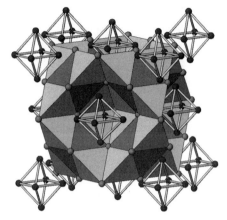

Figure 26. The crystal structure of kutinaite $Cu_{14}Ag_6As_7$. Oblique projection upon (100). "Supertetrahedra" of Cu, composed of $CuAs_4$ tetrahedra (light) and $CuAs_3$ triangles (dark) form octahedral cavities with Ag_6 groups (skeletal).

Kutinaite was synthesized in a "dry" system Cu-Ag-As at 350 and 450°C from pure elements and hydrothermally at 230°C. It forms in hydrothermal copper-silver arsenide deposits in nature.

Crookesite and the related thallium-copper chalcogenides

The crystal structure of $TlCu_7Se_4$ (Eriksson et al. 1991), a synthetic analogue of the mineral crookesite which was first described by Nordenskjöld in 1866 (Berger 1987) but has had a long story of uncertain chemical composition, is a typical channel structure. $(NH_4)Cu_7S_4$ (Gattow 1957) is isostructural, as also is $TlCu_7S_4$ (Berger and Sobott 1987) (Table 7).

The copper-selenium framework (Fig. 27) consists of a combination of $CuSe_4$ tetrahedra (Cu-Se = 2.37–2.63 Å) and non-planar $CuSe_3$ groups (Cu-Se = 2.43–2.49 Å). It contains larger square channels in 1:1 combination with small square channels. The large channels are occupied by thallium and are lined by selenium (Tl-Se = 3.39 Å) and tetrahedral copper (Tl-Cu = 4.01 Å), as well as by triangular copper in the wall (Tl-Cu 3.84 and 4.08 Å, respectively). The small channels are "filled" by the Cu-Cu interactions at the distances of 2.82 Å. Triangular coordinated Cu interacts also with tetrahedral Cu (2.56 and 2.72 Å) whereas adjacent tetrahedral copper atoms are only 2.57 Å apart. Tl-Tl distances in the channels are equal to the overall periodicity of the structure (3.97 Å). The Cu-Se and Cu-Cu interactions form a dense net producing the presumably stiff channel walls; the square nets (120) of Cu-Cu interactions are staggered giving the overall (110) net system (Fig. 28).

In the isostructural $TlCu_7S_4$ (Berger and Sobott 1987), the tetrahedral Cu-S distances again are slightly asymmetric (2.23–2.85 Å); triangular Cu-S distances are 2.28–2.36 Å. The

Table 7. Thallium-copper selenides and related structures.

Compound	Lattice parameters (Å; °)				Space group	Ref.
	a	b	c	β		
$TlCu_2Se_2$	3.857(0)	—	14.038(1)	—	$I4/mmm$	(1)
$TlCu_4Se_3$	3.975(1)	—	9.834(1)	—	$P4/mmm$	(2)
KCu_4Se_3	4.019(3)	—	9.720(1)	—	$P4/mmm$	(3)
$TlCu_{3.99}Se_3$	12.431(0)	12.800(0)	3.935(1)	—	$Pnnm$	(4)
$TlCu_5Se_3$	12.900	—	3.968	—	$P4_2/mnm$	(5)
$TlCu_3Se_2$	15.213(1)	4.012(0)	8.394(0)	111.70(1)	$C2/m$	(2)
$Tl_5Cu_{14}Se_{10}$	18.097(2)	3.958(0)	18.118(2)	116.09(1)	$C2/m$	(6)
$TlCu_7Se_4$	10.448(1)	—	3.968(0)	—	$I4/m$	(7)
$TlCu_7S_4$	10.180(0)	—	3.859(0)	—	$I\bar{4}$	(8)
$NH_4Cu_7S_4$	10.25(2)	—	3.84(1)	—	$I\bar{4}$	(9)
$Rb_3Cu_8Se_6$	18.458(6)	4.010(1)	10.212(3)	104.44(2)	$C2/m$	(10)
$Cs_3Cu_8Se_6$	19.076(4)	4.078(1)	10.449(3)	106.04(3)	$C2/m$	(10)
$K_2Hg_6S_7$	13.805(8)	-	4.080(3)	-	$P42_1m$	(11)

References: (1) Berger and van Bruggen 1984; (2) Berger 1987; (3) Stoll et al. 1999; (4) Berger et al. 1995; (5) Berger et al. 1990; (6) Berger and Meerschaut 1988; (7) Eriksson et al. 1991; (8) Berger and Sobott 1987; (9) Gattow 1957; (10) Schils and Bronger 1979; (11) Kanatzidis 1990

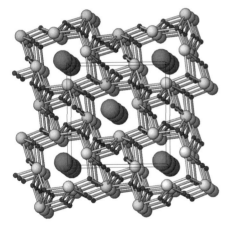

Figure 27. The crystal structure of Tl Cu_7Se_4 with $CuSe_4$ tetrahedra and $CuSe_3$ triangles (in the walls of octahedral columns) forming a skeleton with square channels parallel to [001] and hosting Tl. Projection along c, a_2 axis horizontal. In order of increasing size spheres indicate Cu, Se and Tl.

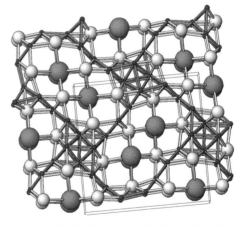

Figure 28. Cu-Cu contacts (dark lines) in the structure of $TlCu_7Se_4$. Conventions from Figure 27 apply.

Cu-Cu distances are 2.82–2.92 Å for triangular copper, 2.50–2.65 Å between trigonal and tetrahedral Cu. The structure proposed indicates an erroneously short Cu-Cu distance of only 2.25 Å between adjacent tetrahedrally coordinated copper atoms, bringing the existence of an ordered structure in doubt. The Tl-S distances are 8 × 3.37 Å.

Two-cation channels are present in $TlCu_5Se_3$ (Berger et al. 1990). Ignoring slight distortion, the channels of this tetragonal compound can be understood as amalgamation of two adjacent channels of $TlCu_7Se_4$ leaving a constriction along the median plane of the channel (Fig. 29). The Tl-Se distances are 4×3.49 Å and 4×3.34 Å, spacing of tetrahedral Cu atoms is 2.74 Å whereas their distances to triangular Cu are 2.60 and 2.72 Å, and triangular coordinated Cu atoms have contacts of 2.80 and 2.81 Å. Tl-Tl distances are 3.91 Å across the channels and 3.99 Å along the channels.

The structure of $TlCu_5Se_3$ can be obtained from that of $TlCu_7Se_4$ by unit cell twinning on the {110} set of mirror planes. The small square columns in $TlCu_5Se_3$ and $TlCu_7Se_4$ are essentially the columns of broad, flattened octahedra Se_6, with triangular Cu in alternative triangular surfaces of the column.

In $TlCu_4Se_3$ (or $TlCu_{3.99}Se_3$, Berger et al. 1995) the octahedral character of these columns is accentuated, octahedra are more elongated and the pattern of Cu sites, exclusively tetrahedral in this case, is altered (Fig. 30). The tetrahedral partitions and the double channels are essentially unchanged (Tl-Se distances range from 3.32 to 3.53 Å). Cu symmetrically fills both tetrahedral interspaces in the octahedral columns and is accompanied by copper sites outside the column, one above a face of Cu tetrahedron and another one above a face of the empty octahedron; these sites already belong to tetrahedral walls. Cu-Cu contacts in $TlCu_4Se_3$ (Berger et al. 1995) enclose the two-cation channels by two wavy nets of interactions, Cu-Cu distances vary in them from 2.44 Å between two adjacent tetrahedral sites which correspond to the tetrahedral-tetrahedral contacts in the previously mentioned structures, to 2.74 Å. The rich system of interactions observed in the little square columns of the previous structures is reduced to a single, 2.60 Å contact, interconnecting the above-mentioned undulating but otherwise separated nets of Cu-Cu interactions across the octahedral column. This phase has the symmetry reduced to orthorhombic. The crystal structure of KCu_4Se_3 (Stoll et al. 1999) follows a different pattern: it is a pure layered structure with double tetrahedral layers with all tetrahedra occupied and K in 8-fold cube-like coordination, although $TlCu_4Se_3$ with the KCu_4Se_3 structure has also been repeatedly synthesized (data in Berger 1987).

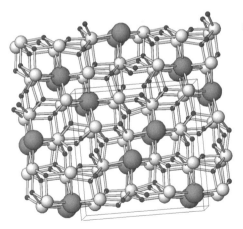

Figure 29. The crystal structure of $TlCu_5Se_3$ with tetrahedral and triangular Cu, and with Tl (large spheres) in paired columns of rectangular [001] channels. Projection along the c axis, a_2 axis horizontal.

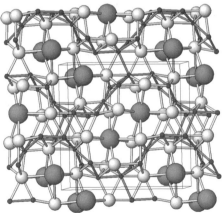

Figure 30. The crystal structure of $TlCu_4Se_3$ with paired columns of Tl parallel to [001]. Projection along c, b axis horizontal. In order of increasing size, spheres indicate Cu, Se and Tl. Cu-Cu contacts of tetrahedrally coordinated Cu are expressed as dark connections. Compare with the structures of $TlCu_5Se_3$ and $TlCu_7S_4$.

In spite of a number of features common with the previous structures, TlCu$_3$Se$_2$ (Berger and Eriksson 1990) is a layer structure (Fig. 31). Still, "copies" of double-columns with surrounding flat-octahedral and tetrahedral portions in the right positions, with trigonal and tetrahedral copper sites, can be traced in it. They repeat along the [100] direction. TlCu$_2$Se$_2$ and Tl(Cu,Fe)$_2$Se$_2$ (Berger and van Bruggen 1984; Makovicky et al. 1980) is a layered structure with cubic coordination of Tl and unmodified, single tetrahedral layers with all tetrahedra occupied.

In the TlCu$_3$Se$_2$ structure, there are 2 tetrahedral columns between the adjacent octahedral columns in each copper-based layer. In the similar structure of Tl$_5$Cu$_{14}$Se$_{10}$ (Berger and Meerschaut 1988) intervals of two tetrahedra and of four tetrahedra alternate regularly along the layers. A structure based on four-tetrahedra intervals exists for Rb$_3$Cu$_8$Se$_6$ (Schils and Bronger 1979).

The only analogous silver-sulfides are the thallium-silver sulfide, TlAg$_3$S$_2$ (Klepp 1985) and KAg$_5$S$_3$ (Emirdag et al. 1998). The crystal structure of TlAg$_3$S$_2$ is orthorhombic, *Pbcn*, a = 8.15 Å, b = 8.79 Å and c = 7.03 Å (Fig. 32). It contains elliptical (flattened-octagonal) channels enclosed by the silver-sulfur framework. This framework consists of square "chimneys," formed by a combination of AgS$_4$ tetrahedra (Ag-S = 2.47–2.64 Å) in the corners and two walls of these "chimneys" and of linear S-Ag-S coordinations (2.47 Å) in the other two walls. Both coordinations result in Ag lining the walls of the channels, unhindered by

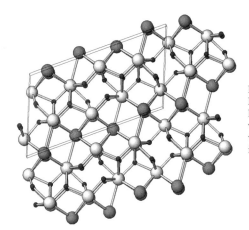

Figure 31. The layer structure of TlCu$_3$Se$_2$. Projection axis *b*; *c* axis vertical. Small spheres: Cu; light spheres: Se; grey large spheres: Tl. Slightly warped (001) layers of triangular (in the walls of void octahedra) and tetrahedral Cu-Se layers are interleaved by paired Tl columns.

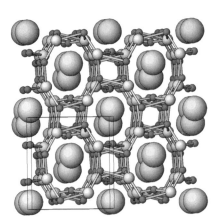
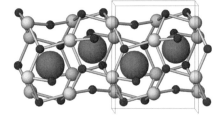

Figure 32. The crystal structure of TlAg$_3$S$_5$. (*top*) Elliptical Tl (large spheres) channels in a silver-sulfur framework of AgS$_4$ tetrahedra and S-Ag-S linear groups. Projection along the channel direction *c*, *a* axis horizontal. (*bottom*) A single [001] channel. Ag: dark small spheres.

sulfur. Tl forms zig-zag rows parallel to the c-axis. Its coordination number is only six (pairs of bonds 3.12, 3.31 Å and opposing bonds 3.72 Å), with addition of several Tl-Ag contacts (2 × 3.30 Å, 3 × 3.32 Å and 2 × 2.79 Å). The Tl-Tl contacts are 3.76 Å. Thus, the structure is full of cation-cation contacts.

The structure of KAg_5S_3 (Emirdag et al. 1998) is hexagonal, a = 13.20 Å, c = 7.9 Å, space group $P\bar{6}2c$. The hexagonal channels (Fig. 33) contain potassium in octahedral coordination (3.17–3.19 Å); spacing of K in the channels is 3.97 ± 0.03 Å. The skeleton consists of a combination of linear-coordinated Ag (Ag-S 2.40 Å) and triangular planar Ag (Ag-S ranges from 2.50 to 2.75 Å), the former positioned on 2_1 axes, the latter between the $\bar{6}$ axes. Horizontal AgS_3 triangles are interconnected by diagonal S-Ag-S groups, as three edges of an imaginary (empty) octahedron. These octahedra share the AgS_3 triangles, forming six triangular columns around each K column. The crystal structure of $CsAg_5Te_3$ (Li et al. 1995) with a = 14.67 Å, c = 4.60 Å, space group $P4_2/mnm$ is virtually identical to that of $TlCu_5Se_3$.

Crookesite $TlCu_7Se_4$ and sabatierite $TlCu_6Se_4$ are the natural phases of this family (Strunz and Nickel 2001), alongside bukovite $Tl(Cu,Fe)_2Se_2$ which has the layer structure described above. All the phases mentioned were prepared by annealing of appropriate mixtures of TlSe, Cu and Se in evacuated silica glass tubes at 400°C. Additionally, some phases were obtained also by leaching Cu out of these products by means of aerated ammonia solution (Berger 1987). $TlCu_{3.99}Se_3$ is such a low-temperature product whereas $Tl_5Cu_{14}Se_{10}$ appears to be only a high-temperature product. No attempt of ion exchange in the channels was described. Attempts to synthesize KCu_7S_4 and $NaCu_7S_4$ instead of $NH_4Cu_7S_4$ by means of the wet-chemical synthesis (Gattow 1957) were unsuccessful.

Related to these structure types is the structure of $K_2Hg_6S_7$ (Kanatzidis 1990). The framework contains narrow square channels limited by four distorted Hg tetrahedra and pairs of large channels limited by these Hg tetrahedra combined with linear S-Hg-S coordinations. The K-S distances in the large channels range from 3.30 to 3.62 Å. This structure does not tolerate exchange of S by Se while maintaining K as the large cation. A nearly isostructural compound $Cs_2Hg_6Se_7$ was synthesized, in which larger Cs plays the role of a channel cation (Kanatzidis 1990).

Attempts to combine copper/silver coordinations with coordinations of a lone-electron-pair element result in quite different structures. Many are complicated layer structures (e.g., α-$KAgTeS_3$ and $RbCuTeS_3$ (Zhang and Kanatzidis 1994) but a very interesting openwork

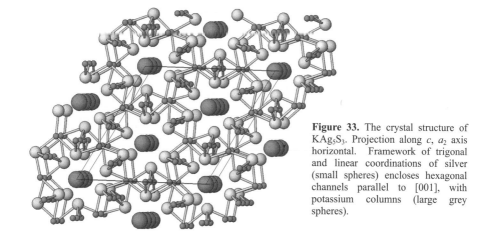

Figure 33. The crystal structure of KAg_5S_3. Projection along c, a_2 axis horizontal. Framework of trigonal and linear coordinations of silver (small spheres) encloses hexagonal channels parallel to [001], with potassium columns (large grey spheres).

cubic structure resulted for CsCuTeS$_3$ (Zhang and Kanatzidis 1994). The anionic [CuTeS$_3$]$^-$ framework is composed of trigonal planar Cu$^+$ centers and TeS$_3$$^{2-}$ pyramids. Large tunnels run along [100] and equivalent directions. Cesium is close to the channel wall, with six Cs-S distances 3.65–3.70 Å and it is not interacting with Te which maintains its stereochemically active lone electron pair.

There are surprisingly few mesoporous structures with frameworks based on coordination squares of platinum or palladium with S or Se as ligands. The monoclinic structure of Tl$_2$Pt$_5$S$_6$ (Klepp 1993) with $a = 6.97$ Å, $b = 6.94$ Å, $c = 11.09$ Å and $\beta = 97.83°$, space group $P2_1/n$, may serve as an example. In this structure (Fig. 34), gently folded ribbons of edge-sharing PtS$_4$ squares parallel to b are interconnected (via corners) by folded triplets [101] of partly warped, edge-sharing coordination squares. This 'house of cards' framework leads to intersecting channels parallel, respectively, to [010] and [100]. The access along [001] is blocked, however. The Tl atoms which fill the channels as zig-zag double-columns avoid direct metal-metal bonding with the exposed Pt atoms in the channel walls by shifts along [010]. Still, they have two short Pt-Tl distances each, equal to 3.10 and 3.11 Å, respectively. The shortest Tl-Tl distance is 3.54 Å.

The other Pt, Pd sulfides or selenides with large cations form mostly layered structures although they often possess openwork layers with various accommodation possibilities for the large cations. The interesting cubic structure of LaPd$_3$S$_4$ (Wakeshima et al. 1997), with $a = 6.74$ Å and space group $Pm3n$, has PdS$_4$ coordination squares stacked in non-intersecting columns parallel to the unit cell axes (Fig. 35). Eight-coordinated La sites are in the intersections of alternative [100, etc.] square channels which are lined on two walls by PdS$_4$ groups. This arrangement will certainly be a hindrance to any movement of La ions along these channels. Similar reasons can be evoked for the inability of the structural arrangements of PdS and PtS to accommodate additional components in the channels.

Structures based on supertetrahedra

Large mesoporous structures are based on supertetrahedra of especially In, Sn or Ge. Supertetrahedra are large, higher-order structural elements. They are blocks of tetrahedral archetype limited by tetrahedral faces parallel to those of unit coordination tetrahedra of In or other metals. The smallest supertetrahedron contains four coordination tetrahedra and a central

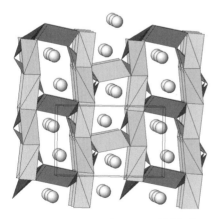

Figure 34. The crystal structure of Tl$_2$Pt$_5$S$_6$. Large spheres: Tl; rectangles and broken rectangles: PtS$_4$ polyhedra.

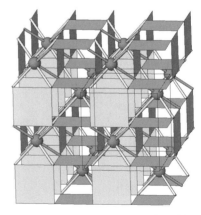

Figure 35. The crystal structure of LaPd$_3$S$_4$. Spheres indicate sites of La, squares and broken squares the PdS$_4$ polyhedra.

octahedral cavity open to the outside. Besides these supertetrahedra, with two tetrahedra along each edge, those with up to four tetrahedra per edge length (20 component tetrahedra) were observed.

Some structures with supertetrahedra are layer structures, in which the supertetrahedra share corners, forming an openwork layer. Large cations occur between these corrugated layers as well as in the layer windows. As an example, the structure of $Cs_2MnS_2S_6$, with tetrahedrally coordinated Sn and Mn (or Zn with analogous role, with K as the large cation) (Liao 1993). This is a microporous structure, with channels of about 3.2 Å in diameter, both between and across the layers.

True mesoporous structures are formed by metalloorganic sulfides with large supertetrahedra. These compounds lie rather far from the domain of mineralogy; two impressive examples will be illustrated here. Parallel account and listing of these compounds are given by Krivovichev (2005) in this volume.

The tetragonal structure of M-In sulfide (Wang et al. 2001) with M = Cd and tri-methylenedipiperidine as the organic radical (a = 42.12 Å, c = 16.70 Å, space group $I\overline{4}2d$) is a 3D network of supertetrahedra with edges equal to 4 coordination tetrahedra. The $\overline{4}$ axes of supertetrahedra are parallel to the two-fold axes of the space group. The structure contains large [001] channels, 4 × 4 in terms of component coordination tetrahedra, interconnected laterally. The openwork is characterized by only two supertetrahedra joining at each corner (Fig. 36).

The crystal structure of the indium-cadmium amino-propylpiperazine hydrate (Li et al. 2001; Fig. 37) is composed of corner sharing "N = 4" supertetrahedra as well. In this tetragonal structure, with a = 23.58 Å and c = 43.92 Å, space group $I4_1/acd$, the supertetrahedra have their $\overline{4}$ axes oriented along the c and (approximately) the a axes of the unit cell, respectively. Again, only two supertetrahedra share corners, resulting in two interpenetrating tetrahedral frameworks of cristobalite type which leave a complex system of cavities and 2 × 2 channels parallel to [110].

Figure 36. The sulfide framework of Cd-In sulfide (4-4'-trimethylenedipiperidine). Cd positions indicated by darkened tetrahedra.

Figure 37. The sulfide framework of Cd-In sulfide 1,4-bis (3-aminopropyl) piperazine hydrate. The Cd and In tetrahedra are not distinguished.

Large supertetrahedra require divalent metals as charge-moderating constituents. Thus, Wang et al. (2001) used Mn, Co, Zn and Cd for the syntheses of $(M_4In_{16}S_{33})^{10-}$ clusters whereas Li et al. (2001) used cadmium for the $(Cd_4In_{16}S_{35})^{14-}$ skeleton. In the first study, the divalent metals formed a small tetrahedral group in the central portions of the supertetrahedra.

EPILOGUE

The review presented shows the broad spectrum of microporous sulfides and the paucity of mesoporous cases. Most probably, an equal number of further examples was overlooked and some were omitted for the reason of space. They are rare in nature, principally because the majority of elements involved—large cations and some of the framework cations—have a pronounced lithophile character and only special conditions, for example, hyperagpaicity or occurrence in residual Cu-rich sulfide magmas (with Cl) force them into sulfides. This is the preferred mode of occurrence for djerfisherite whereas sulfides of chromium are only known from meteorites. Cetineite, galkhaite and gerstleyite as well as the Tl sulfides and sulfosalts occur in hydrothermal environments. Synthetic products were prepared both at dry, higher-temperature conditions and in wet, hydrothermal runs. Ion-exchange experiments are almost universally lacking.

In general, it is the larger uni- and divalent cations (K, Tl, Ba, Cs, Rb, NH_4 ...) which form the microporous structures among sulfides and selenides. The structures based on small cations, such as Li, may display ionic conduction along narrower and more complicated paths.

Mobility of cations in the rigid channels of microporous sulfides (such as KCr_5Se_8) is expected to be fair; this is also indicated by the observed disorder in these channels. Lesser mobility might be displayed by some of the single columns of K, Ba or Sr in the bismuth sulphosalts, due to different cation coordinations. Increase in mobility is expected when the large cations occur as double- or multiple columns in larger channels (e.g., in $CsBi_3S_5$) or they assume fairly regular trigonal prismatic coordination capped on all three faces of the prism.

A different situation occurs in the hutchinsonite merotypes, in which "porous interlayers" occur, with a different degree of expected cation mobility along complicated paths (e.g., in $Cs_2Sb_8S_{13}$). A clathrate-like topology is present in galkhaite (encapsulating Cs), skutterudites (encapsulating REE) and kutinaite (Ag_6 groups) although from the point of view of atom valencies these are not true clathrates. The rare instance of pure anion-hosting channels is represented by $Bi(Bi_2S_3)_9I_3$ whereas in the case of cetineite family and the Brouwer-Jellinek phases a combination of channel cations and anions occurs, these anions compensating for the excess of the positive valence and completing the coordination spheres of channel cations. There are many striking similarities between the two latter families in the configurations of channel constituents.

The absence of structures with large cages/channels filled with cations and associated dipoles (water molecules) is striking, although we should stress again that this is not an exhaustive review. As an exception, the structure of $Rb_2Sb_8S_{13} \cdot 3.3H_2O$ can be quoted, with cations along the walls and corners of the channels and H_2O filling the channel volume in a semi-ordered fashion, a situation similar to zeolites. We speculate that, besides experimental conditions (activity of H_2O in the mixture/solution during the synthesis), the different character of the hydrogen bond to sulfur (selenium) rather than to oxygen of the silicate (oxysalt) framework may be the reason of the paucity of large zeolite-like structures. The success of the cetineite structure type, in which the bulk of interactions takes place in the oxygen-lined and (H_2O, OH) filled tubes, supports this assumption.

A number of sulfide/selenide structures with large channels/cavities were synthesized as sulfosalts/sulfides of organic radicals (the hutchinsonite merotypes and the "supertetrahedral" structures with Sn, Ge and In as the principal metal cations). Although treated here to a degree, they lie outside the scope of a basically mineralogical text. The stability of these structures after the organic component has been exchanged for inorganic cations (+ water) is not known.

ACKNOWLEDGMENTS

The encouragement and interest of the volume editor, Prof. G. Ferraris, the qualified assistance of Mrs. Camilla Sarantaris and Mrs. Britta Munch as well as the careful reviews by Profs. G. Ferraris, M. Mellini, S. Krivovichev and S. Merlino are gratefully acknowledged.

REFERENCES

Balić-Žunić T, Engel P (1983) Crystal structure of synthetic $PbTlAs_3S_6$. Z Kristallogr 165:261-269
Balić-Žunić T, Makovicky E (1993) Contributions to the crystal chemistry of thallium sulphosalts. I. The O-D nature of imhofite. N Jb Mineral Abh 165:317-330
Bauer ED, Galatanu A, Michor H, Holscher G, Rogl P, Boulet P, Noel H (2000) Physical properties of skutterudites $Yb_xM_4Sb_{12}$, M = Fe, Co, Rh, Ir. Eur Phys J B14:483-493
Berger R (1987) A phase-analytical study of the Tl-Cu-Se system. J Solid State Chem 70:65-70
Berger R, Eriksson L (1990) Crystal structure refinement of monoclinic $TlCu_3Se_2$. J Less-Common Metals 161:101-108
Berger R, Eriksson L, Meerschaut A (1990) The crystal structure of $TlCu_5Se_3$. J Solid State Chem 87:283-288
Berger R, Meerschaut A (1988) The crystal structure of $Tl_5Cu_{14}Se_{10}$. Eur J Solid State Inorg Chem 25:279-288
Berger R, Tergenius L-E, Noren L, Eriksson L (1995) The crystal structure of room-temperature synthesized orthorhombic $TlCu_4Se_3$ from direct methods on X-ray powder data. J Alloys Compd 224:171-176
Berger R, van Bruggen CF (1984) $TlCu_2Se_2$: A p-type metal with a layer structure. J Less-Common Metals 99:113-123
Berger RA, Sobott RJ (1987) Characterization of $TlCu_7S_4$, a crookesite analogue. Monat Chemi Wiss 118:967-972
Berlepsch D (1996) Crystal structure and crystal chemistry of the homeotypes edenharterite ($TlPbAs_3S_6$) and jentschite ($TlPbAs_2SbS_6$) from Lengenbach, Binntal (Switzerland). Schweiz Mineral Petrogr Mitteilungen 76:147-157
Berlepsch P, Makovicky E, Balić-Žunić T (2000) Contribution to the crystal chemistry of Tl-sulfosalts. VI. Modular-level structure relationship between edenharterite $TlPbAs_3S_6$ and jentschite $TlPbAs_2SbS_6$. N Jb Mineral Mh 2000:315-332
Berlepsch P, Miletich R, Makovicky E, Balić-Žunić T, Topa D (2001) The crystal structure of synthetic $Rb_2Sb_8S_{12}(S_2)\cdot 2H_2O$, a new member of the hutchinsonite family of merotypes. Z Kristallogr 216:272-277
Bonazzi P, Borrini D, Mazzi F, Olmi F (1995) Crystal structure and twinning of Sb_2AsS_2, the synthetic analogue of pääkkönenite. Am Mineral 80:1054-1058
Bonazzi P, Menchetti S, Sabelli C (1987) Structure refinement of kermesite: symmetry, twinning, and comparison with stibnite. N Jb Mineral Mh 1987:557-567
Botinelly T, Neuerburg GJ, Conklin NM (1973) Galkhaite (Hg, Cu, Tl, Zn) (As, Sb)S_2 from the Getchell mine, Humboldt County, Nevada. J Res US Geol Surv 1:515-517
Braun DJ, Jeitschko W (1980) Thorium containing pnictides with the lanthanum phosphide ($LaFe_4P_{12}$) structure. J Less-Common Metals 76:33-40
Brouwer R, Jellinek F (1979) Modulation of the intergrowth structures of $A_{1-p}Cr_2X_{4-p}$ (A = Ba, Sr, Eu, Pb; X = S, Se; p ≈ 0.29). Am Inst Phys Conf Proc 53:114-116
Carré D, Pardo MP (1983) Structure de l'octasulfure de pentaindium et de potassium, In_5KS_8. Acta Crystallogr C39:822-824
Chen TT, Szymanski JT (1981) The structure and chemistry of galkhaite, a mercury sulfosalt containing Cs and Tl. Can Mineral 19:571-581
Chung D-Y, Iordanidis L, Rangan KK, Brazis PW, Kannewurf CR, Kanatzidis MG (1999) First quaternary A-Pb-Bi-Q (A = K, Rb, Cs; Q = S, Se) compounds. Synthesis, structure, and properties of α-, β-$CsPbBi_3S_6$, $APbBi_3Se_6$, (A = K, Rb), and $APbBi_3S_6$ (A = Rb, Cs). Chem Mater 11:1352-1362
Cordier G, Schaefer H, Schwidetzky C (1985) Darstellung und Struktur der Verbindung $Sr_4Bi_6Se_{13}$. Rev Chim Minér 22:631-638

Deiseroth H-J (1986) Splitpositionen für Alkalimetallkationen in den Thioindaten MIn_5S_8 (M=K,Rb,Cs)? Z Kristallogr 177:307-314
Dittmar G, Schäfer M (1977) Darstellung und Kristallstruktur von $(NH_4)_2Sb_4S_7$. Z Anorg Allg Chem 437: 183-187
Divjaković V, Nowacki W (1975) Die Kristallstruktur von Galchait $[Hg_{0.76}(Cu,Zn)_{0.24}]_{12}Tl_{0.96}(AsS_3)_8$. Z Kristallogr 142:262-270
Divjaković V, Nowacki W (1976) Die Kristallstruktur von Imhofit, $Tl_{5.6}As_{15}S_{25.3}$. Z Kristallogr 144:323-333
Dmitrieva MT, Ilyukhin VV, Bokii GB (1979) Close packing and cation arrangement in the djerfisherite structure. Sov Phys Crystallogr 24:683-685
Dung N-H, Tien V-V, Behm HJ, Beurskens PT (1987) Superstructure with pseudotranslation. II. Monopotassium pentachromium octaselenide: a tunnel structure. Acta Crystallogr C43:2258-2260
Emirdag M, Schimek GL, Kolis JW (1998) KAg_5S_3. Acta Crystallogr C54:1376-1378
Eriksson L, Werner P-E, Berger R, Meerschaut A (1991) Structure refinement of $TlCu_7Se_4$ from X-ray powder profile data. J Solid State Chem 90:61-68
Evans HT Jr, Clark JR (1981) The crystal structure of bartonite, a potassium iron sulfide, and its relationship to pentlandite and djerfisherite. Am Mineral 66:376-384
Ferraris G, Makovicky E, Merlino S (2004) Crystallography of Modular Materials. IUCr Monographs in Crystallography. Oxford University Press, Oxford
Foit FF, Robinson PD, Wilson JR (1995) The crystal structure of gillulyite, $Tl_2(As,Sb)_8S_{13}$, from the Mercur gold deposit, Tooele County, Utah, U.S.A. Am Mineral 80:394-399
Gattow G (1957) Die Kristallstruktur von $NH_4Cu_7S_4$. Acta Crystallogr 10:549-553
Gelabert MC, Ho MH, Malik A-S, Di Salvo FJ, Deniard P (1997) Structure and properties of $Ba_6Ni_{25}S_{27}$. Chem Eur J A3:1884-1889
Gostojić M, Nowacki W, Engel P (1982) The crystal structure of synthetic $TlSb_3S_5$. Z Kristallogr 159:217-224
Graf HA, Schäfer H (1975) $K_3SbS_3 \cdot 3Sb_2O_3$, ein Oxothioantimonit mit Röhrenstruktur. Z Anorg Allg Chem 414:220-230
Gruzdev VS, Stepanov VI, Shumkova NG, Chernistova NM, Yudin RN, Bryzgalov IA (1972) Galkhaite $(HgAsS_2)$, a new mineral from arsenic-antimony-mercury deposits of the U.S.S.R. Dokl Akad Nauk SSSR 205:1194-1197 (in Russian)
Hall SR, Stewart JM (1973) The crystal structure of argentian pentlandite, $(Fe,Ni)_8AgS_8$, compared with the refined structure of pentlandite $(Fe,Ni)_9S_8$. Can Mineral 12:169-177
Iordanidis L, Schindler JL, Kannewurf CR, Kanatzidis MG (1999) $ALn_{1\pm x}Bi_{4\pm x}S_8$ (A = K, Rb; Ln = La, Ce, Pr, Nd): new semiconducting quaternary bismuth sulfides. J Solid State Chem 143:151-162
Jeitschko W, Braun DJ (1977) $LaFe_4P_{12}$ with filled $CoAs_3$-type structure and isotypic lanthanoid-transition metal polyphosphides. Acta Crystallogr B33:3401-3406
Kanatzidis MG (1990) Molten alkali-metal polychalcogenides as reagents and solvents for the synthesis of new chalcogenide materials. Chem Mater 2:353-363
Kanatzidis MG, McCarthy TJ, Tanzer TA, Chen L-H, Iordanidis L, Hogan T, Kannewurf CR, Uher C, Chen B (1996) Synthesis and thermoelectric properties of the new ternary bismuth sulfides $KBi_{6.33}S_{10}$ and $K_2Bi_8S_{13}$. Chem Mater 8:1465-1474
Kanishcheva AS, Mikhailov YuN, Lazarev BV, Trippel AF (1980) A new cesium sulfobismuthite $CsBi_3S_5$: synthesis and structure containing extensive channels. Sov Phys Dokl 25:319-320
Kaplunnik LN, Pobedimskaya EA, Belov NV (1975) The crystal structure of galkhaite $HgAsS_2$. Dokl Akad Nauk SSSR 225:561-563 (in Russian)
Karanović L, Poleti D, Makovicky E, Balić-Žunić T, Makovicky M (2002) The crystal structure of synthetic kutinaite, $Cu_{14}Ag_6As_7$. Can Mineral 40:1437-1449
Klepp KO (1985) The crystal structure of $TlAg_3S_2$. J Less-Common Metals 107:139-146.
Klepp KO (1993) $Tl_2Pt_5S_6$ - a new thioplatinate with a channel-type structure. J Alloys Compd 196:25-28
Kluger F, Pertlik F (1985) Synthese und Strukturanalyse von $Na_3SbSe_3 \cdot 3Sb_2O_3 \cdot 0.5Sb(OH)_3$. Monat Chem 116:149-156
Korenstein R, Soled S, Wold A, Collin G (1977) Preparation and characterization of skutterudite-related phases $CoGe_{1.5}S_{1.5}$ and $CoGe_{1.5}Se_{1.5}$. Inorg Chem 16:2344-2346
Krämer V (1979) Structure of the bismuth chloride sulphide $Bi_4Cl_2S_5$. Acta Crystallogr B35:139-140
Krivovichev S (2005) Topology of microporous structures. Rev Mineral Geochem 57:17-68
Kupčík V (1967) Die Kristallstruktur des Kermesits, Sb_2S_2O. Naturwiss 54:114-115
Li H-L, Kim J, Groy TL, O'Keeffe M, Yaghy OM (2001) 20 Å $(Cd_4In_{16}S_{35})^{14-}$ supertetrahedral T4 clusters as building units in decorated cristobalite frameworks. J Am Chem Soc 123:4867-4868
Li J, Guo H-Y, Zhang X, Kanatzidis MG (1995) $CsAg_5Te_3$: a new metal rich telluride with a unique tunnel structure. J Alloys Compd 218:1-4
Liao J-H (1993) Ph.D. Thesis, Michigan State University, East Lansing, Michigan

Makovicky E (1985) Cyclically twinned sulphosalt structures and their approximate analogues. Z Kristallogr 173:1-23

Makovicky E (1989) Modular classification of sulfosalts – current status. Definition and application of homologous series. N Jb Mineral Abh 160:269-297

Makovicky E (1993) Rod-based sulphosalt structures derived from the SnS and PbS archetype. Eur J Mineral 5:545-591

Makovicky E (1997) Modular crystal chemistry of sulfosalts and other complex sulphides. EMU Notes Mineral 1:237-271

Makovicky E, Balič-Žunič T (1999) Gillulyite $Tl_2(As, Sb)_8S_{13}$: reinterpretation of the crystal structure and order-disorder phenomena. Am Mineral 84:400-406

Makovicky E, Johan Z, Karup-Møller S (1980) New data on bukovite, thalcusite, chalcothallite and rohaite. N Jb Miner Abh 138:122-146

Makovicky E, Skinner BJ (1979) Studies of the sulfosalts of copper VII. Crystal structures of the exsolution products $Cu_{12.3}Sb_4S_{13}$ and $Cu_{13.8}Sb_4S_{13}$ of unsubstituted synthetic tetrahedrite. Can Mineral 17:619-634

Mariolacos K (1976) The crystal structure of $Bi(Bi_2S_3)_9Br_3$. Acta Crystallogr B32:1947-1949

Matsushita Y, Takéuchi Y (1994) Refinement of the crystal structure of hutchinsonite, $TlPbAs_5S_9$. Z Kristallogr 209:475-478

McCarthy TJ, Ngeyi S-P, Liao J-H, De Groot DC, Hogan T, Kannewurf CR, Kanatzidis MG (1993) Molten salt synthesis and properties of three new solid state ternary bismuth chalcogenides. Chem Mater 5:331-340

McCarthy TJ, Tanzer TA, Kanatzidis MG (1995) A new metastable three-dimensional bismuth sulfide with large tunnels: synthesis, structural characterization, ion-exchange properties, and reactivity of KBi_3S_5. J Am Chem Soc 117:1294-1301

Miehe G, Kupčík V (1971) Die Kristallstruktur des $Bi(Bi_2S_3)_9J_3$. Naturwiss 58:219-220

Miyawaki R, Nakai I, Nagashima K (1993) Reexamination of the crystal structure of $Na_3Sb_7O_9Se_3 \cdot Sb_{0.319}(H_2O, OH)_3$. Bull Chem Soc Japan 66:3671-3675

Mrotzek A, Iordanidis L, Kanatzidis MG (2001) New members of the homologous series $A_m(M_6Se_8)_m(M_{5+n}Se_{9+n})$: the quaternary phases $A_{1-x}(M')_{3-x}Bi_{11+x}Se_{20}$ and $A_{1+x}(M')_{3-2x}Bi_{7+x}Se_{14}$ (A = K, Rb, Cs; M' = Sn, Pb). Inorg Chem 40:6204-6211

Mumme WG (1975) Junoite, $Cu_2Pb_3Bi_8(S,Se)_{16}$, a new sulfosalt from Tennant Creek, Australia: its crystal structure and relationship with other bismuth sulfosalts. Am Mineral 60:548-558

Nakai I, Appleman DE (1981) The crystal structure of gerstleyite $Na_2(Sb, As)_8S_{13} \cdot 2H_2O$: The first sulfosalt mineral of sodium. Chem Letters (Japan) 1981:1327-1330

Parise JB, Ko YH (1992) Novel antimony sulfides-synthesis and X-ray structural characterization of $Sb_3S_5N(C_3H_7)_4$ and $Sb_4S_7N_2C_4H_8$. Chem Mater 4:1446-1450

Pašava J, Pertlik F, Stumpfl EF, Zemann J (1989) Bernardite, a new thallium arsenic sulphosalt from Allchar, Macedonia, with a determination of the crystal structure. Mineral Mag 53:531-538

Pfitzner A (1997) Die Präparative Anvendung der Kupfer (I)-halogenid-Matrix zur Synthese neuer Materialen. Habilitationschrift Universität Siegen.

Rajamani V, Prewitt CT (1975) Refinement of the structure of Co_9S_8. Can Mineral 13:75-78

Rudashevski NS, Karpenkov AM, Shipova GS, Shishkin NN, Ryabkin VA (1979) Thalfenisite, the thallium analogue of djerfisherite. Zap Vses Mineral Obs 108:696-701 (in Russian)

Sabelli C, Nakai I, Katsura S (1988) Crystal structures of cetineite and its synthetic analogue $Na_{3.6}(Sb_2O_3)_3(SbS_3)(OH)_{0.6} \cdot 2.4H_2O$. Am Mineral 73:398-404

Sabelli C, Vezzalini G (1987) Cetineite, a new antimony oxide-sulfide mineral from Cetine mine, Tuscany, Italy. N Jb Mineral Mh 1987:419-425

Sales BC, Mandrus D, Chakoumakos BC, Keppens V, Thompson JR (1997) Filled skutterudite antimonides. Electron crystals and phonon glasses. Phys Rev B56:15081-15089

Schils H, Bronger W (1979) Ternäre Selenide des Kupfers. Z Anorg Allg Chem 456:187-193

Schleid T (1990) Das System Na_2GdClH_x/S. II Einkristalle von Gd_2S_3 im U_2S_3-Typ. Z Anorg Allg Chem 590:111-119

Schmitz D, Bronger W (1974) Die Kristallstruktur von $RbBi_3S_5$. Z Naturforsch 29b:438-439

Schöllhorn R, Schramm W, Fenske D (1980) Nichtstöchiometrische Alkalimetalltitansulfide mit Kanalstruktur. Angew Chem 92:477-478

Sheldrick WS, Kaub J (1985) Darstellung und Struktur von $Rb_2As_8S_{13} \cdot H_2O$ und $(NH_4)_2As_8S_{13} \cdot H_2O$. Z Naturforsch 40B:1130-1133

Shirotani I, Uchiumi T, Ohno K, Sekine C, Nakazawa Y, Kanoda K, Todo S, Takehiko Y (1997) Superconductivity of filled skutterudites $LaRu_4As_{12}$ and $PrRu_4As_{12}$. Phys Rev B56:7866-7869

Srikrishnan T, Nowacki W (1974) A redetermination of the crystal structure of cosalite, $Pb_2Bi_2S_5$. Z Kristallogr 140:114-136

Stoll P, Näther C, Jess I, Bensch W (1999) KCu_4Se_3. Acta Crystallogr C55:286-288

Strunz H, Nickel EH (2001) Strunz Mineralogical Tables. Stuttgart: E. Schweizerbart'sche Verlagsbuchhandlung

Szymanski JT (1995) The crystal structure of owensite $(Ba,Pb)_6(Cu,Fe,Ni)_{25}S_{27}$, a new member of the djerfisherite group. Can Mineral 33:671-677

Tan K, Ko Y, Parise JB (1994) A novel antimony sulfide templated by ethylenediammonium. Acta Crystallogr C50:1439-1442

Tan K, Parise JB, Ko Y, Dorovsky A, Norby P, Hanson JC (1996) Applications of synchrotron imaging plate system to elucidate the structure and synthetic pathways to open framework antimony sulfides. XVII IUCr Congress C-402

Tani B, Mrazek F, Faber J Jr, Hitterman RL (1986) Neutron diffraction study of electro-chemically synthesized djerfisherite. J Electrochem Soc 133:2644-2649

Volk K, Schäfer H (1979) $Cs_2Sb_8S_{13}$, ein neuer Formel- und Strukturtyp bei Thioantimoniten. Z Naturforsch 34b:1637-1640

Wakeshima M, Fujiro T, Sato N, Yamada K, Masuda H (1997) Crystal structure and electrical conductivity of palladium sulfide bronzes MPd_3S_4 (M = La, Nd, and Eu). J Solid State Chem 129:1-6

Wang C, Li Y-Q, Bu X-H, Zheng N-F, Zivkovic O, Yang C-S, Feng P-Y (2001) Three-dimensional superlattices built from $(M_4In_{16}S_{33})^{10-}$ (M = Mn, Co, Zn, Cd) supertetrahedral clusters. J Am Chem Soc 123:11506-11507

Wang X (1995a) Crystal structures of oxoselenoantimonates(III) of sodium and rubidium, $Na_6(Sb_{12}O_{18})$ $(SbSe_3)_2(Na_{1.86}Sb_{0.14})((OH)_{2.28}(H_2O)_{4.02}$ and $Rb_6(Sb_{12}O_{18})(SbSe_3)_2Sb_{0.22}((OH_{0.66}(H_2O)_{3.48})$. Z Kristallogr 210:693-694

Wang X (1995b) Synthesis and structure of a new microporous thioantimonate (III) $[H_3N(CH_2)_3 NH_3] Sb_{10}S_{16}$. Eur J Solid State Inorg Chem 32:303-312

Wang X, Liebau F (1994) Synthesis and structure of $[CH_3NH_3]_2Sb_8S_{13}$: A nanoporous thioantimonate (III) with a two-dimensional channel system. J Solid State Chem 111:385-389

Wang X, Liebau F (1998) An investigation of microporous cetineite-type phases $A_6[B_{12}O_{18}][CX_3]_2[D_x(H_2O, OH,O)_{6-y}]$. I. The cetineite structure field. Eur J Solid State Inorg Chem 35:27-37

Wang X, Liebau F (1999) Crystal structures of cetineites, $A_6[Sb_{12}O_{18}][SbX_3]_2[D_xY_{6-y}]$, and their changes with chemical composition. Z Kristallogr 214:820-834

Wang X-Q, Liu L-M, Jacobson AJ (2000) Hydrothermal synthesis and crystal structures of $(CH_3NH_3)_{0.5}$ $(NH_4)_{1.5}S_8S_{13}·2.8(H_2O)$ and $Rb_2Sb_8S_{13}·3.3(H_2O)$. J Solid State Chem 155:409-416

Yakovenchuk, VN, Pakhomovsky, YA, Men'shikov YP, Ivanyuk GYu, Krivovichev SV, Burns PC (2003) Chlorbartonite, $K_6Fe_{24}S_{26}(Cl,S)$, a new mineral species from a hydrothermal vein in the Khikina massif, Kola Peninsula, Russia: description and crystal structure. Can Mineral 41:503-511.

Zhang X, Kanatzidis MG (1994) $AMTeS_3$ (A = K,Rb,Cs; M = Cu,Ag): A new class of compounds based on a new polychalcogenide anion, TeS_3^{2-}. J Am Chem Soc 116:1890-1898

Micro- and Mesoporous Carbon Forms, Chrysotile, and Clathrates

Marcello Mellini

Dipartimento di Scienze della Terra
Via Laterina 8
53100 Siena, Italy
mellini@unisi.it

INTRODUCTION

This chapter offers a not-exhaustive overview of structural mesoporosity in selected, natural mineral phases (carbon forms, chrysotile, gas hydrates). After a short introduction devoted to the introduction of the most general features of porosity, attention will be paid mostly to natural mesoporous mineral phases, leaving microporosity to other chapters of this volume. However, a few peculiar microporous structures will be also considered, because of their close resemblance with the mesoporous substances.

Conversely, no attempt will be made to report properties and structures of the many examples of man-made mesoporous materials including silica xerogels, mesoporous synthetic silica, calcium phosphate and the fantastic arrangements of shapes, surface patterns and channels that can occur (e.g., Yang et al. 1997; Maschmeyer 1998; Sayari 2003; Xia et al. 2003; Zheng et al. 2003; White et al. 2005). Similarly, the overview will not deal with crystal engineering processes, such as the fabrication of hollow porous shells of calcium carbonate from self-organizing media (oil-water-surfactant microemulsions supersaturated with calcium carbonate; Walsh and Mann 1995). Finally, one more aspect outside the scope of this article will be micro-to-mesoporosity conversion in synthetic materials, such as activated palygorskite and sepiolite, or the so-called pillared clays (e.g., Mass et al. 1997; Dékany et al. 1999; Salerno and Mendioroz 2002; Ferraris and Gula 2005). Evidently, these issues represent extremely important technological targets, often achieved following synthetic routes that may be totally different from those used by nature to form minerals.

Microporosity, mesoporosity and macroporosity

Pore size. Porous structures are derived from a framework of linked atoms ("host"), that create volumetrically important voids ("pores"), possibly capable of including several different "guest" species. Three groups (micro-, meso- and macropores) are discriminated based upon pore size. According to International Union for Pure and Applied Chemistry (IUPAC) recommendations, pores with free diameters less than 2 nm should be called micropores (McCusker et al. 2001, 2003; McCusker 2005). Pores in the 2–50 nm range are mesopores while pores larger than 50 nm are characterized as macroporous materials. Ideally, pores repeat in a regular manner, forming long-range ordered structures; otherwise, the material may be non-crystalline, even if other kinds of short-range or long-range order are present (e.g., in the internal structure of the pores, or the manner in which adjacent pores come together, respectively). A systematic nomenclature of porous structures, as well as criteria for obtaining comprehensive crystal-chemical formulae that reflect the chemical composition of host and guest, structures of host and pores, symmetry, atom connectivity and pore dimensionality may be found in McCusker et al. (2001, 2003).

Synthetic micro- to macroporous structures. Interest in micro- and mesoporous materials arises from the important technological properties that these materials possess. For instance, Zheng et al. (2003) reported the synthesis of a series of microporous, three-dimensional open-framework sulfides and selenides, containing highly mobile alkali metal cations, exhibiting fast-ion conduction with potential applications in devices such as batteries, fuel cells, photocatalysts and electrochemicals sensors.

As regards mesoporous materials, their possible uses are simply extraordinary, ranging over catalysis, oxidation, hydrogenation, halogenation, polymerization, and membranes and offering unique electronic, magnetic and optical properties (Maschmeyer 1998 and references therein). One more important feature derives from the development of synthetic routes leading to derivatized products tailored for particular chemical aims, through fine-tuning of structure, composition and physical properties (e.g., Sayari 2003). Consequently, mesoporous materials are actively studied within the framework of so-called "green chemistry," that focuses on replacement of resource- and energy-demanding chemical routes by environmentally friendly catalytic methods. Also from the point of view of basic science, the mesoporous world offers exciting perspectives, being intimately linked with topics such as self-assembled structures and spontaneous patterning (e.g., Aizenberg et al. 1999). Last but not least, the need to deal with the nanoworld of mesoporous phases is leading to advanced investigation techniques, such as electron microscopy imaging (e.g., Sakamoto et al. 2003) and non-trivial applications of X-ray powder diffraction and spectroscopies (electron-spin resonance, Raman, NMR, EXAFS, UV-Vis).

In some cases, technological exploitation has been inspired by the study of natural materials. For instance, opal is a natural silica form, characterized by mesoporous arrangements (Graetsch 1994). In particular, precious opal is a non-crystalline material, with $SiO_2 \cdot nH_2O$ chemical composition, consisting of close-packed homometric spheres of amorphous silica, usually 150–350 nm in diameter (Rossman 1994). Diffraction effects originate because each sphere has diameter close to the visible light wavelength; depending upon the actual size of the amorphous spheres and stone orientation, with iridescence leading to the well-known play of color in opal. Packing of spheres necessarily leaves some open space as interstices and voids with sizes that depend on sphere diameter. However, detailed determination of water content and specific surface area reveals values lower than expected, consistent with the presence of intersphere cement that partially clogs the ideal close-packing of rigid spheres. This feature is immediately evident in scanning electron microscope images as sphere coalescence or partial polygonalization. Distinct from monodispersed precious opals, non precious opals (*"potch opals"*) fail to show optical diffraction effects, because they are formed by irregular packing of polydispersed heterometric spheres (Gauthier et al. 1995). Both diffracting and non-diffracting opals may be stained to different colors by mineral pigments (hematite, red; copper minerals, blue) as pore-fillers.

After 1992, an explosion in the study of mesoporous silica led to over 3000 papers (Sayari 2003), dealing with three main mesophases (MCM-41, hexagonal; MCM-48, cubic; MCM-50, lamellar). These are often synthesized by supramolecular templating, using long chain alkyltrimethilammonium surfactants whose electrostatic interaction triggers a self-assembly process that controls the nature of the final mesophase; other surfactants lead to different silica mesophases including SBA-1 to SBA-16, MSU-n, MSU-V, MSU-G, HMS (Firouzi et al. 1995). Varying the synthesis conditions produces silica mesophases of varying morphologies. Thin films, spheres, fibers and monoliths are the simplest shapes; otherwise, particles with spiral, discoidal, toroidal, pinwheel, doughnut and other exotic shapes have been reported. Mesostructured chalcogenides and mixed oxides further increase the chemical, structural and morphological diversity of synthetic materials. According to Sayari (2003), the amount of information is now so large as to frustrate any attempt to write a book-sized, comprehensive review on periodic mesoporous materials.

Spherical colloids have been used as versatile templates, capable of generating macroporous phases (Xia et al. 2003). In this method, the voids in the colloidal crystal resulting from regular sphere packing are infiltrated with a precursor solution or gel. Finally, chemical removal of the sacrificial template produces three-dimensional macroporous materials with interesting bandgap properties ("photonic" crystals), and potential application in fields such as filtration, separation and purification. Sometimes, these materials are referred to as inverted, or inverse, opals.

Finally, one more stimulating research topic, at least in part connected with porosity, can be drawn from the nanowire world. Nanowires (often named as "quantum wires" by physicists) are anisotropic nanocrystals of extreme length/diameter ratio (Yan and Yang 2003). Their technological importance derives from exceptional properties, such as mechanic toughness, photoluminescent efficiency, thermoelectric behavior, nonlinear optical behavior, and low lasing threshold.

Structural constraints. More than a mere classification issue, the distinction among micro-, meso- and macropores perhaps masks a very important issue, associated with different types of particle interactions. Namely, it is crucial to understand what forces actually control porosity development. I tentatively propose here some schematic guesses, with the aim to verify their validity by the details reported in the following paragraphs.

Possibly, void development may be controlled i) by strong, first-neighbor atom-atom bonds in microporous systems; ii) by weak intraparticle non-bonding interactions in mesoporous systems; or iii) by particle coalescence in macroporous systems.

Furthermore, voids may arise from a combination of intrinsic and extrinsic factors. One intrinsic reason for porosity development would be a strongly anisotropic shape of the building particle; namely, internally strong bonded units pack together leaving open-space, as observed in the partially empty rod-close packing of chrysotile (e.g., Mellini 1986). Examples for extrinsic control may be found in the reaction environment, for instance in terms of outer anisotropic fields (*e.g.*, tectonic shear strain) that impose orienting forces capable of controlling void shape.

Finally, a still open problem deserving consideration is connected with the appraisal of the self-assembly mechanisms operative in micropatterned mesoporous systems. In particular, an extremely exciting perspective are biotic-abiotic, organic-inorganic interactions in biomineralization processes (Banfield and Nealson 1997; Dove et al. 2003).

Mesoscale and Earth Sciences. From the point of view of Earth Sciences (Baronnet and Belluso 2002), synthetic nanostructured meso- and macroporous materials are worthy of study as they offer insights into the origin of shape complexity in the mineral kingdom (e.g., in terms of mineralization processes, their intensive P and T parameters, the role of catalytic species, and the effects of dilute chemical components that may not immediately appear within the crystallization reaction). As already anticipated, even more complex and interesting effects are found in moving from totally inorganic mineral to biomineral systems (McLean et al. 2001; Dove et al. 2003), or to the interactions between minerals and microbes (Banfield and Nealson 1997). Here, templating effects are required to quantitatively explain the crystallization of phases far from their stability field (e.g., deposition of calcium carbonate as the high-pressure polymorph, aragonite, within a shell living at low-pressure, in a shallow level of the coral reef).

CARBON FORMS

Basically, the low-pressure forms of carbon are based upon the graphene sheet structure, namely upon a six-membered carbon sheet, that can be folded and rolled to yield different carbon forms (Buseck 2002).

Anthracite

Anthracite offers several examples of externally-controlled porous arrangements of carbon, with the final pore size determined by metamorphic and tectonic shear stresses. This mineral has been long known to be microporous, with pores flattened parallel to the bedding by pressure yielding a long-range statistical preferred orientation (Bonijoly et al. 1982), that results in anisotropic texture and biperiodic turbostratic crystallization. Mesopores, and even macropores, can evolve with the changing conditions. For instance, meta-anthracite differs from anthracite only by the increasing coalescence of adjacent pores. Semi-graphites are formed by single macropores, occurring as hollow distorted polyhedral shells (Bonijoly et al. 1982). Finally, increasing P, T and shear stress destroys porosity, with lamellar graphite representing the limit of a flattened macropore. More recently, evolution of carbonaceous material to graphite in a high-pressure, low-temperature metamorphic environment has been studied in detail by Beyssac et al. (1994), who describe micropores and onion-ring-like phases, preferentially transforming to graphite in the outer ring, namely in the areas with the highest radius of curvature.

While the carbon structure is difficult to study, even using accurate local probes such as HRTEM, detailed information is important for understanding the behavior of coal char during combustion or gasification. Therefore, the need to deal quantitatively with the char structure has led to the development of HRTEM filtering techniques, that generate numerical information concerning graphene layer size, interlayer spacing, the number of layers per stack, and their textural distribution (Sharma et al. 1999). From such studies (e.g., Rouzaud and Clinard 2002) it has been shown that carbon materials offer multiscale organization from the subnanometric to the millimetric scale, whose formation is determined by synthesis conditions, that results in a variety of industrially important chemical and physical properties.

In the recent years, carbon research has focused on two specific forms, nanotubes and fullerenes.

Carbon nanotubes

Carbon nanotubes have been extensively investigated, beginning with the first images reported by Iijima (1991). In fact, their hollow cores, coupled with large aspect ratio, make them candidates for a host of nanotechnological applications. Synthetic carbon nanotubes have interesting porosity, that may be modulated according to the specific need. For instance, it is possible to grow carbon nanotubes with controlled inner diameter, from 4.3 ± 2.3 nm, or definitely larger. These tubes may adopt a strongly oriented pattern, forming membranes with tailored transport properties. By filling nanotubes with metal, composite nanodevices are created consisting of a metal nanowire surrounded by a dielectric shield (Ajayan and Iijima 1993). Finally, carbon nanotubes are expected to behave as tough nanomanipulators, suitable as nanoscale mass conveyors (e.g., Regan et al. 2004). Notwithstanding their apparent resemblance with chrysotile nanotubes (see later), carbon nanotubes of natural origin have not yet been reported.

Carbon nanotubes may be considered as modifications of the basic sp^2 carbon graphite sheet, that due to the absence of strong three-dimensional chemical bonding may be rolled into tubes while conserving the six-membered carbon rings.

Fullerenes

Alternatively, the carbon sheet may be further modified, possibly by introduction of sp^3 defects (Hiura et al. 1994), to achieve a different connectivities. In particular, a few of the six-membered rings may become five-membered (pentagons), obliging the sheet to form large molecules arranged in closed surfaces ("soccer balls") with truncated icosahedron shape and C_{60} (or C_{70}, C_{76} and C_{84}) chemical compositions. These cage-like molecules, named fullerenes, may be synthesized in large amounts, by passing an electric current between two graphite electrodes, under helium (e.g., Ball 2001). Also in the case of fullerenes, several potential applications have been proposed and investigated, with variable success (superlubricants, superconductors, exceptionally stiff materials).

Whatever the future applications of synthetic fullerenes and nanotubes, they have already activated a huge amount of research. Therefore, it seems appropriate to spend a few words about the possible natural occurrences of these exotic carbon forms.

Following the synthesis of fullerenes by Kroto et al. (1985), it was not until 1992 that their first natural occurrences were reported as thin films within fractures in a carbon-rich rock from Karelia, named shungite (Buseck et al. 1992). In the same year, C_{60} and C_{70} fullerenes were also extracted directly from coal by chromatography (Wilson et al. 1992) and from a fulgurite (Daly et al. 1993). A few years later, fullerene-like carbon nanostructures were found also in the Allende carbonaceous-chondritic meteorite (Harris et al. 2000), as closed nanoparticles 2–10 nm in diameter, supposedly capable behaving as carriers of primordial planetary gases in the extraterrestrial space.

Notwithstanding the large excitement arising after those discoveries natural occurrences may be rare, and many unresolved questions remain about the identity and formation of fullerenes in geological environments (Buseck 2002, and references therein).

SERPENTINE

Chrysotile mesopores

Among the different serpentine minerals—the flat-layer lizardite, the polygonally-shaped polygonal serpentine, the corrugated-layer antigorite, and the curled-layer chrysotile; (Wicks and O'Hanley 1989)—it is the last that offers the most interesting mesoporosity properties. Historically, one of the first successful mineralogical applications of electron microscopy was the demonstration that "both natural and synthetic chrysotile crystallize in the form of hollow cylindrical tubes" (Bates et al. 1950; Noll and Kirchner 1950). Although quite short, that early report captured other features common for chrysotile, namely the "uniform diameter" of several hundreds Å, the presence of multiple tubes and "the interesting conical development of the tubes." The hollow-tube model was further investigated by Pundsack (1955, 1956, 1961), who investigated in detail the colloidal behavior of the chrysotile suspensions, including its surface chemistry, density and structure. Basically, Pundsack questioned the hollow-tube model, by postulating that only a limited volume was actually void. In particular, density measurements suggested a maximum void volume of 6%, rather than the 20–30% expected for the fiber bundle. Therefore, Pundsack (1955 and 1956) was obliged to postulate that "distorted strips or ribbons of fibers rather than …hollow tubes" were present, and to conclude that "the sample viewed in the electron microscope no longer bears a one to one relationship with the native fiber." In any case, not all the porosity of the hollow-tube model was actually available for nitrogen adsorption, for reasons then unknown. Finally, Pundsack (1961) concluded on the basis of adsorption-desorption isotherms that 80% of the observed void volume existed in pores less than 60 Å in diameter, that the cylindrical chrysotile fibers have average outer diameters of 200–250 Å and average inner diameters of 20–50 Å, and that interfibril space was irregular but with an effective pore size similar to the intrafibril pores. In a contemporaneous electron microscope study, Bates and Comer (1959) proposed that the tubes contain amorphous material with chrysotile-like composition. Using X-ray diffraction, Whittaker (1957) suggested that the hollow tubes were filled with curved laths of the same composition as the tubes themselves, and that this material should be amorphous, as indicated by contrast analysis (Whittaker 1966). Similarly, Martinez and Comer (1964) supported the presence of amorphous interfiber material, after extracting the material by ultrasonic treatment of chrysotile crudes.

The partially-filled model of Pundsack was however, at least in part, refuted by Huggins and Shell (1965). These authors also reported extensive data in favor of the existence of totally hollow tubes. More recently, the pioneer high resolution TEM work of Yada (1967 and 1971) apparently supported the predominantly hollow-tube model, concluding that the central hollow was most frequently 70–80 Å in diameter, that the outer diameter of the fiber was 220–270 Å and that the wall had constant thickness of up to 100 Å (Fig. 1). Comparison of subsequent TEM investigations indicates that the actual values are specimen-dependent, as different occurrences lead to slightly different sizes. In any case, it is now confirmed, that from 350 Å diameter, further growth of chrysotile leads to non-porous polygonal shapes (Mellini 1986; Baronnet et al. 1994).

Pore-dependent properties

The variable porosity of the chrysotile fiber bundles suggests that this property may be specimen-dependent, as well as treatment-dependent. This point was first addressed by Naumann and Dresher (1966) in a study dedicated to the influence of sample texture on chrysotile dehydroxylation. The higher the surface area of chrysotile, the faster its thermal transition to an intermediate amorphous phase, and subsequent crystallization as olivine. Therefore, at that time it was already clear that differently textured chrysotiles might exist, that they might vary in the amount of interstitial material, that the actual amount of void space had an important bearing on dehydroxylation kinetics, and that chrysotile samples from different sources might display different porosity-related properties.

Subsequently, the issue of porosity and reaction kinetics became important in view of the technological utilization of chrysotile (e.g., Monkman 1971; Choi and Smith 1972; Atkinson 1973; Bleiman and Mercier 1975), in addition to the health hazard possibly arising from chrysotile inhalation (as a result of interaction with physiological fluids) and its influence on water dissolution reactions within groundwater systems at landfill sites (Gronow 1987; Tartaj et al. 2000). Basically, these studies demonstrated acidic decomposition of chrysotile, through the breakdown of fiber-bundles into individual fibrils (simultaneous with the release and/or reaction of the interfibrillar material), and by migration of ions along the tube. Magnesium cations appeared continuously released from the fibers, in a manner not unlike magnesium hydroxide (Pundsack and Reimschussel 1956) leaving a silica skeleton that preserved the primitive fiber shape. As the rate limiting factor was the acidic removal of the basic brucite layer from the fiber surface, the smaller the particle size the faster the reaction.

Increasing interest in the adsorption properties of chrysotile, and of leached chrysotile, led to comparative analyses of several specimens. For example, Suquet (1989) reported sharply different adsorption properties in different samples of chrysotile. For instance, one of them adsorbed twelve times less phenantrene and 45% less CO_2 than the other. Furthermore, chrysotile adsorption properties were dependent on preparation; in fact, dry grinding converts the long fibers into shapeless, non crystalline material, with strongly active basic sites, that can adsorb CO_2 and H_2O directly from the atmosphere (Suquet 1989). Titulaer et al (1993)

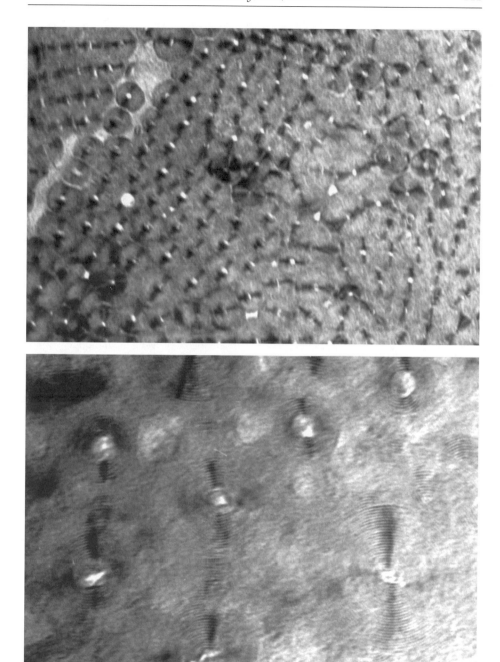

Figure 1. (*top*) Rod close-packing of chrysotile fibers, as seen along the fiber axis (ion-thinned section). Average diameter of the fiber close to 300 Å. Pores appear as low contrast regions, in the center of the individual fiber and at the junctions among three adjacent fibers. (*bottom*) Enlarged view of the chrysotile rod-close packing, showing bent (001) layers, deformed central holes and poorly crystalline interfiber filling.

applied diffuse reflectance infrared spectroscopy (DRIFTS), extended sorption studies using nitrogen, and performed thermoporometric characterization to determine values of 50-80 Å for the hollow-tube inner diameter, and of 36–46 Å for the diameter of the channel running among clustered tubes. These results were in good agreement with pre-existing and their own TEM determinations.

An interesting application of adsorption in mesoporous chrysotile was presented by Soma et al. (1993), with the aim to enhance optical chrysotile recognition during environmental monitoring. They used fluorescent xanthene dyes with carboxyl or hydroxyl groups to mark chrysotile, to exploit the preferential adsorption of basic dyes in chrysotile, with respect to kaolinite, gypsum, sulfur and glass wool. In particular, calcein and umbelliferone appeared to be suitable dyes for selective detection of chrysotile by fluorescence microscopy. The same monitoring issue was also theoretically approached by Balan et al. (2002); they observed multiple ionic-plasmon resonance in the IR-spectra of chrysotile nanotubes, and modeled these IR effects to calculate the geometrical parameters of nanotube based materials.

The origin of pores in chrysotile

The porosity of chrysotile therefore depends upon the configuration of the tetrahedral-octahedral 1:1 layer, the organization into cylindrical or spirally wrapped tubes, the rod-close packing of strongly textured chrysotile fiber-bundles, and the empty space resulting from poorly textured arrangements. Following Yada and Iishi (1977) we know that the morphological and structural features of synthetic chrysotile are controlled by reaction parameters such as pH, time and temperature. To summarize, early formed nuclei consist of membranes a few unit cells thick, laterally extending for several hundred Angstroms. Within 30 minutes at 300–400°C, these quickly curl into cylindrical or conical shapes. So, in absence of strong three-dimensional chemical bonds, internal tensions lead to wrapping of the serpentine membrane, and the resultant curled shapes.

The chrysotile growth process has been further studied in detail by Baronnet and Devouard (1996), and Amelinckx et al. (1996). In both cases the growth of chrysotile was analyzed in terms of elastic strain and departure from equilibrium, and ultimately the formation of 15- and 30-sector fan-like arrangements of flat lizardite sectors.

In the assessment of the present author, it is evident that chrysotile is an open structure with 5–30% void space. As nature will circumvent void formation if possible (*"horridum vacui"*), transition from open-space chrysotile to fan-shaped polygonal serpentine represents an effective space-saving system. In other words, it is to be expected that chrysotile will be favored under extensional regimes (e.g., Andreani et al. 2004), and polygonal serpentine under compressive regimes.

More water than space?

Retrograde serpentinites often consist of meshes, formed by replacement of olivine. Whereas the mesh rim is mostly formed by oriented, non-topotactic lizardite (Rumori et al. 2004), the mesh core consists of a chaotic mixture of lizardite flakes, polygonal serpentine sectors, chrysotile fibers and more-or-less amorphous material (Viti and Mellini 1998). The variable amount and distribution of magnetite leads to the dark color of serpentinites; in particular, the darker varieties have been extensively used as decorative stones for at least twenty centuries, as in the Romanic and Gothic churches of Tuscany that are often adorned by alternating strips of white marbles and dark-green serpentinites.

During a study on serpentinite alteration, Bralia et al. (1995) systematically measured anomalous water saturation indices. This parameter measures the percentage of pores accessible to water diffusion, with respect to the whole porosity determined using helium diffusion. The values were 131 and 161%, for weathered stones extracted from the Siena Cathedral and from altered quarry surfaces, respectively. Therefore, unlike other stones, the quantity of pores accessible to water was >100%, indicating that water diffuses through serpentinites more easily than the helium gas used for rock porosity determination (close to 4–5%). In other words, the amount of water entering the serpentinite was greater than the available space. This anomaly was interpreted as due to the presence of inaccessible, sealed pores within the complex serpentinite texture; excess water uptake was due to permeability through pre-existing soluble barriers, that prevented gas flow into the sealed pore-space. The barrier material was thought to be akin to the brucite-depleted $MgSi_2O_5$ residue envisaged by Pundsack (1956) and Suquet (1989), possibly formed by weathering processes acting on very short time scales, as demonstrated by the normal water saturation indices of very fresh specimens.

Serpentine mesopores: from gas carriers to nanowires templates

Serpentine porosity may offer exciting perspectives also under certain extreme scenarios in the primitive Solar System and the deep ocean floor. In the first case, primitive meteorites like the Cold Bokkeveld or the Mighei CM carbonaceous chondrites are rich in serpentine nanotubes, formed by aqueous alteration prior to the arrival on Earth (Zega and Buseck 2003; Zega et al. 2004). As noted by these authors, an intriguing possibility is that the silicate nanotubes could have served as carriers of primordial fluids. The other scenario is connected with present-day ocean floors, and the continuous release of hydrocarbons abiotically formed during the peridotite serpentinization; we may wonder whether the serpentine micropores could behave as methane traps to control methane transfer from the lithosphere to the ocean, possibly interacting with gas hydrates in the global carbon budget.

Chrysotile nanotubes have also been studied as nanowires templates (Grobéty et al. 2004). By filling the hollow-tube of chrysotile by appropriate metals, it would be possible to obtain a composite material having a tiny conducting core surrounded by a dielectric silicate wall. Apparently, this property was recognized as early as in 1960 by Cosslett and Horne (1960). They modified the TEM contrast of chrysotile from light to dark fibril core, by filling it by potassium phosphotungstate; the experiment was repeated by Clifton et al. (1966), introducing lead nitrate and decomposing *in situ* to lead oxide.

Finally, mesopores with average diameter of 20 Å diameter have been obtained by acid treatment of antigorite (Kosuge et al. 1995). If successful, this approach might lead to substitutes that would comply with current regulations limiting the use of chrysotile fibers.

CLATHRATES

Enclosure compounds

Clathrates are non-stoichiometric enclosure compounds, formed by combining two stable compounds, with no chemical bond between the two components (Sloan 1998). One of them (the host) forms a cage-defining three-dimensional frame; the other component (the guest) is physically trapped within the cages and stabilizes the clathrate, that would not exist without the trapped guest. In Earth Sciences gas hydrates represent the most significant class of clathrates. The most important is methane hydrate, with end-member composition of $CH_4 \cdot 5.76H_2O$ in the case of completely occupied cages. Although quite paradoxical, no gas hydrate has ever been recognized as an accepted mineral species.

Physically, gas hydrates look like ordinary water ice, from which they differ in the larger density value of 0.95 g/cm^3 as compared to ~0.92 g/cm^3. From the structural point of view, gas hydrates (known as clathrate hydrates as well) present three different crystal structures. The most common ones are cubic (structure I, $Pm3n$, $a = 12$ Å; structure II, $Fd3m$, $a = 17.3$ Å); they are stabilized by guest gases ranging in size from argon to p-dioxane (i.e., diameters 3.8–6.8 Å). The third phase (hexagonal $P6/mmm$, $a = 12.26$, $c = 10.17$ Å) can host still larger molecules (e.g., methylciclohexane or hexacloroethane; Ripmeester et al. 1987). As the diameter of the cavities range from 7.8 to 9.2 Å, the upper limit for the stability of gas hydrates is assured by molecular diameters up to 9 Å and therefore these compounds are more properly included in the microporous group. The basic cage of gas hydrates is a pentagonal dodecahedron, consisting of twelve pentagonal faces, often indicated as 5^{12}. As dodecahedra alone cannot fill space, polyhedral packing is completed with tetrakaidecahedra $5^{12}6^2$, namely polyhedra with twelve pentagonal and two hexagonal faces. Two 5^{12} and six $5^{12}6^2$ cages, linked together through vertices, form the so-called structure I. Structure II still contains sixteen $5^{12}6^2$ cages, linked to eight hexakaidecahedra (12 pentagonal and 6 hexagonal faces), that meet by face sharing (Englezos 1993, and references therein). Finally, structure III is based upon two 5^{12} cages, one $4^35^66^3$ cage, and a very large $5^{12}6^8$ cage (Ripmeester et al. 1987). Comprehensive reviews of structure and properties of gas hydrates may be found in Sloan (2003a,b), and more recently, almost a whole issue of American Mineralogist has been dedicated to this topic (Sloan 2004, and companion papers).

Quite interesting, natural gas hydrates develop a complex porosity, because they contain both the micropores deriving from their constituent cages, as well as larger eso- and macropores, usually ranging from 100 to 500 nm and occasionally reaching 1 μm (Kuhs et al. 2000, 2004).

Owing to their peculiar nature, the study of gas hydrates requires an approach based upon specific techniques (Sloan 2003b; Genov et al. 2004). The most frequently used have been X-ray powder diffraction (e.g., used to test the possible structural models); solid-state nuclear magnetic resonance (to support the structure models and to determine the absolute occupancy cages by the gas); Raman spectroscopy (to determine the environment of hydrate guests); electron microscopies and finally, theoretical modeling.

Energy resource and/or geological hazard

Negative aspects of gas hydrates are well known, as they may lead to the obstruction of natural gas pipelines in cold regions unless expensive nucleation inhibitors are used. However, starting from 1990 (e.g., Englezos 1993), gas hydrates attracted interest, because of their possible economic and environmental importance (Haq 1998; Holder and Bishnoi 2000).

Stable at accessible pressures and temperatures (e.g., $P > 50$ bar, $T < 7°C$), they occur extensively in permafrost. Furthermore, gas hydrates are found widely on the ocean floors, from the Arctic to the Antarctic, including some tropical seas (but not in the Mediterranean Sea). Their stratigraphy may be easily mapped, due to sharp contrast in acoustic velocities. Sonar investigations show the existence of a layer of gas hydrates and sediments, at depths below 500 m, and several hundred meters thick. This cemented layer contains large reserves of gas hydrate; for instance, the sediments of the Gulf of Cadiz contain 3–16 vol% of gas hydrate.

As a given volume of gas hydrate contains approximately 164 volumes of STP (Standard Temperature and Pressure) methane gas, the methane hydrate behaves as a storage medium, capable of fixing methane in a way similar to compression at 164 bar. Worldwide, clathrates have therefore accumulated an immense amount of methane, estimated to total twice the quantity of carbon present in all the known fossils fuels of the Earth. This raises the possibility of developing technologies to store and transport methane as methane-hydrates, with no necessity of cryogenic refrigeration or high pressure.

As long as they remain stable, gas-hydrates seal the underlying free gas. Consequently, methane hydrates represent a hazard to drillers demanding careful consideration during exploitation (Grauls 2001). Also temperature and/or pressure variations may well destabilize the methane hydrate layer, triggering landslides on the continental slope and even tsunamis (Maslin et al. 2004). Furthermore, the rapid release of methane from frozen gas hydrates has been invoked to explain abrupt climatic inversions from cold to warm periods (Kennett et al. 2002), as the greenhouse gas capacity of methane is 3 to 10 times more effective than carbon dioxide. One more aspect related to destabilization of gas hydrates is the generation of mud volcanoes; in fact, destabilization of gas hydrates would produce mud upwelling and hydrocarbon release.

CONCLUSIONS

Using the IUPAC definition of pore size, the number of examples of natural mesoporous phases does not seem large. For instance, gas hydrates have been included within this review, although they should be more properly included among the microporous ones. Therefore, it is useful to explore briefly the geological, structural and thermodynamic factors determining their occurrence.

Geologically, the limited incidence of mesoporous arrangements reflects the high pressure conditions dominating the solid Earth; namely, the occurrence of any phase with density lower than the average crustal density of 2.6–2.7 g/cm^3 is unlikely. Therefore, this limits the presence of mesoporous phases to the outermost crust, or to very specific environments, such as extensional veins (chrysotile tubes), water-sediment interfaces (clathrates), abrupt high-temperature events (fullerenes), low-pressure extraterrestrial environments.

Structurally, mesoporous arrangements seem to be connected with interactions other than direct chemical bonding among nearest neighbors. For instance, mesoporosity of chrysotile derives from folding and wrapping of a strongly internally connected layer, with only limited layer-to-layer interactions. Similarly, the submesoporous arrangement of gas hydrates is connected with a physical interaction between host and guest species.

Finally, the presence of a stability field for any of the previous mesoporous phases seems dubious, indicating that the formation of these phases may be kinetically controlled rather than thermodynamically stabilized. However, as mineralogical, thermodynamical and geochemical constraints may not hold for man-made materials (at least in the very short time-scale), we may anticipate further future developments in the design, engineering and use of synthetic mesoporous materials.

ACKNOWLEDGMENTS

The author is indebted with Giovanni Ferraris, Emil Makovicky and Tim White, for their careful revisions of an earlier draft.

REFERENCES

Aizenberg J, Black AJ, Whitesides GM (1999) Control of crystal nucleation by patterned self-assembled monolayers. Nature 398:495-498

Ajayan PM, Iijima S (1993) Capillarity-induced filling of carbon nanotubes. Nature 361:333-334

Amelinckx S, Devouard B, Baronnet A (1996) Geometrical aspects of the diffraction space of serpentine rolled microstructures: their study by means of electron diffraction and microscopy. Acta Crystallogr A52: 850-878

Andreani M, Baronnet A, Boullier AM, Gratier JP (2004) A microstructural study of a "crack-seal" type serpentine vein using SEM and TEM techniques. Eur J Mineral 16:585-595
Atkinson RJ (1973) Chrysotile asbestos: colloidal silica surfaces in acidified suspensions. J Colloid Interface Sci 42:624-628
Balan E, Mauri F, Lemaire C, Brouder C, Guyot F, Saitta AM, Devouard B (2002) Multiple-ionic plasmon resonances in naturally occurring multiwall nanotubes: infrared spectra of chrysotile asbestos. Phys Rev Lett 89:177401_1- 177401_4
Ball P (2001) Roll up for the revolution. Nature 414:142-144
Banfield JF, Nealson KH (1997) Geomicrobiology: interactions between microbes and minerals. Rev Mineral 35:1-448
Baronnet A, Belluso E (2002) Microstructures of the silicates: key information about mineral reactions and a link with the Earth and materials sciences. Mineral Mag 66:709-732
Baronnet A, Devouard B (1996) Topology and crystal growth of natural chrysotile and polygonal serpentine. J Cryst Growth 166:952-960
Baronnet A, Mellini M, Devouard B (1994) Sectors in polygonal serpentine, a model based on dislocations. Phys Chem Mineral 21:330-343
Bates TF, Sand LB, Mink JF (1950) Tubular crystals of chrysotile asbestos. Science 111:512-513
Bates TF, Comer JJ (1959) Further observations on the morphology of chrysotile and halloysite. Proceedings of the Sixth National Conference on Clays and Clay Minerals. Pergamon Press, New York p. 237-248
Beyssac O, Rouzaud JN, Goffé B, Chopin C (1994) Graphitization in a high-pressure, low-temperature metamorphic gradient: a Raman microspectroscopy and HRTEM study. Contrib Mineral Petrol 143:19-31
Bleiman C, Mercier JP (1975) Attaque acide et chloration de l' asbeste chrysotile. Bull Soc Chimie France 3-4:529-534
Bonijoly M, Oberlin M, Oberlin A (1982) A possible mechanism for natural graphite formation. Int J Coal Geol 1:283-312
Bralia A, Ceccherini S, Fratini F, Manganelli Del Fa' C, Mellini M, Sabatini G (1995) Anomalous water absorption in low-grade serpentinites: more water than space? Eur J Mineral 7:205-215
Buseck PR (2002) Geological fullerenes: review and analysis. Earth Planet Sci Lett 203:781-792
Buseck PR, Tsipursky SJ, Hettich R (1992) Fullerenes from the geological environment. Science 257:215-217
Choi I, Smith RW (1972) Kinetic study of dissolution of asbestos fibers in water. J Colloid Interface Sci 40:253-262
Clifton RA, Huggins JW, Shell HR (1966) Hollow chrysotile fibers. Am Mineral 51:508-511
Cosslett VE, Horne RW (1960) Private communication to Whittaker EJW and Zussman (1971). *In:* Gard JA (1971) The Electron-optical Investigation of Clays. Mineralogical Society Monograph no. 3, p 159-191
Daly TK, Buseck PR, Williams P, Lewis CF (1993) Fullerenese from a fulgurite. Science 259:1599-1601
Dékany I, Turi L, Fonseca A, Nagy JB (1999) The structure of acid treated sepiolites: small angle X-ray scattering and multi MAS-NMR investigations. Appl Clay Sci 14:141-160
Dove PM, De Yoero JJ, Weiner S (2003) Biomineralization. Rev Mineral Geochem 54:1-381
Englezos P (1993) Clathrate hydrates. Ind Eng Chem Res 32:1251-1274
Firouzi A, Monnier A, Bull LM, Besier T, Sieger T, Huo Q, Walker SA, Zasadzinski JA, Glinka C, Nicol J, Margolese D, Stucky GD, Chmelka BF (1995) Cooperative organization of inorganic-surfactant and biomimetic assemblies. Science 267:1138-1143
Ferraris G, Gula A (2005) Polysomatic aspects of microporous minerals –heterophyllosilicates, palysepioles and rhodesite-related structures. Rev Mineral Geochem 57:69-104
Gauthier JP, Caseiro J, Rantsordas S, Bittencourt Rosa D (1995) Nouvelle structure d' empilement compact dans de l' opale noble du Bréil. Compt Rendus Acad Science Paris 320:373-379
Genov G, Kuhs WF, Staykova DK, Goreshnik E, Salamatin AN (2004) Experimental studies on the formation of porous gas hydrates. Am Mineral 89:1228-1239
Graetsch H (1994) Structural characteritics of opaline and microcrystalline silica minerals. Rev Mineral 29:209-232
Grauls D (2001) Gas hydrates: importance and applications in petroleum exploration. Mar Pet Geol 18:519-523
Grobéty BH, Metraux C, Ulmer P (2004) Chrysotile, a template for metal nanowires. Abstracts volume of the 32nd International Geological Congress, Florence, August 20-28, p 308
Gronow JR (1987) The dissolution of asbestos fibres in water. Clay Minerals 22:21-35
Haq BU (1998) Gas hydrates: greenhouse nightmare? Energy panacea or pipe dream? GSA Today 8:1-6
Harris PJF, Vis RD, Heymann D (2000) Fullerene-like carbon nanostructures in the Allende meteorite. Earth Planet Sci Lett 183:355-359

Hiura H, Ebbesen TW, Fujita J, Tanigaki K, Takada T (1994) Role of sp^3 defects structures in graphite and carbon nanotubes. Nature 367:148-151

Holder GD, Bishnoi PR (2000) Gas hydrates: challenges for the future. Proceedings National Academy of Science, vol. 912, New York, p 1044

Huggins CW, Shell HR (1965) Density of bulk chrysotile and massive serpentine. Am Mineral 50:1058-1067

Iijima S (1991) Helical microtubules of graphitic carbon. Nature 354:56-58

Kennett JP, Cannariato KG, Hendy IL, Behl RJ (2002) Methane hydrates in Quaternary climate change: the clathrate hypothesis. American Geophysical Union, Washington D.C. p 216

Kosuge K, Shimada K, Tsunashima A (1995) Micropore formation by acid treatment of antigorite. Chem Mater 7:2241-2246

Kroto HW, Heath JR, O'Brien SC, Curl RF, Smalley RE (1985) C_{60} backminsterfullerene. Nature 318:162-163

Kuhs WF, Klapproth A, Gotthardt F, Techmer KS, Heinrichs T (2000) The formation of meso- and macroporous gas hydrates. Geophys Res Lett 27:2929-2932

Kuhs WF, Genov G, Staykova D, Zeller A, Techmer KS, Heinrichs T, Bohrmann G (2004) Porous microstructures of gas hydrates. Abstracts Meeting "Micro- and mesoporous mineral phases," Rome 6-7 December 2004, p 103-106

Martinez E, Comer JJ (1964) The concentration and study of the interstitial material in chrysotile asbestos. Am Mineral 49:153-157

Maschmeyer T (1998) Derivatised mesoporous solids. Curr Opin Solid State Mat Sci 3:71-78

Maslin M, Owen M, Say S, Long D (2004) Linking continental-slope failures and climate change: testing the clathrate gun hypothesis. Geology 32:53-56

Mass N, Heylen I, Cool P, Vansant EF (1997) The relation between the synthesis of pillared clays and their resulting porosity. Appl Clay Sci 12:43-60

McCusker LB (2005) IUPAC nomenclature for ordered microporous and mesoporous materials and its application to non-zeolite microporous mineral phases. Rev Mineral Geochem 57:1-16

McCusker LB, Liebau F, Engelhardt G (2001) Nomenclature of structural and compositional characteristics of ordered microporous and mesoporous materials with inorganic hist (IUPAC recommendations 2001). Pure Appl Chem 73:381-394

McCusker LB, Liebau F, Engelhardt G (2001) Nomenclature of structural and compositional characteristics of ordered microporous and mesoporous materials with inorganic hist (IUPAC recommendations 2001). Microporous Mesoporous Mater 58:3-13

McLean RG, Schofield MA, Kean WF, Sommer CV, Robertson DP, Toth D (2001) Botanical iron minerals: correlation between nanocrystal structure and modes of biological self-assembly. Eur J Mineral 13:1235-1242

Mellini M (1986) Chrysotile and polygonal serpentine from the Balangero serpentinite. Mineral Mag 50:301-306

Monkman LJ (1971) Some chemical and mineralogical aspects of the acid decomposition of chrysotile. Proceedings 2nd International Conference Physics Chemistry Asbestos Minerals, Louvain (Belgium), paper 3-2, p 1-9

Naumann AW, Dresher WH (1996) The influence of sample texture on chrysotile dehydroxylation. Am Mineral 51:1200-1211

Pundsack FL (1955) The properties of asbestos. I. The colloidal and surface chemistry of chrysotile. J Phys Chem 59:892-895

Pundsack FL (1956) The properties of asbestos. II. The density and structure of chrysotile. J Phys Chem 60:361-364

Pundsack FL (1961) The pore structure of chrysotile asbestos. J Phys Chem 65:30-33

Pundsack FL, Reimschussel G (1956) The properties of asbestos. III. Basicity of chrysotile suspensions. J Phys Chem 60:1218-1222

Regan BC, Aloni S, Ritchie RO, Dahmen U, Zetti A (2004) Carbon nanotubes as nanoscale mass conveyors. Nature 428:924-926

Ripmeester JA, Tse JS, Ratcliffe CI, Powell BM (1987) A new clathrate hydrate structure. Nature 325:135-136

Rossman G (1994) Colored varieties of the silica minerals. Rev Mineral 29:433-468

Rouzaud JN, Clinard C (2002) Quantitative high-resolution transmission electron microscopy: a promising tool for carbon materials characterization. Fuel Proc Technol 77-78:229-235

Rumori C, Mellini M, Viti C (2004) Oriented, not-topotactic olivine > serpentine replacement in mesh-textured, serpentinized peridotites. E J Mineral 16:731-741

Sakamoto Y, Diaz I, Terasaki O, Zhao D, Pérez-Pariente J, Kim JM, Stucky GD (2002). Three-dimensional cubic mesoporous structures of SBA-12 and related materials by electron crystallography. J Phys Chem B 106:3118-3123

Salerno P, Mendioroz S (2002) Preparation of Al-pillared montmorillonite from concentrated dispersions. Appl Clay Science 22:115-123
Sayari A (2003) Mesoporous materials. *In:* The Chemistry of Nanostructured Materials. Yong P (ed) World Scientific Publishing Co., Singapore, p. 39-68
Sharma A, Kyotani T, Tomita A (1999) A new quantitative approach for microstructural analysis of coal char using HRTEM images. Fuel 78:1203-1211
Sloan ED Jr (1998) Clathrate hydrates of natural gases. Marcel Dekker Inc., New York
Sloan ED Jr (2003a) Fundamental principles and applications of natural gas hydrates. Nature 426:353-359
Sloan ED Jr (2003b) Clathrate hydrate measurements: microscopic, mesoscopic, and macroscopic. J Chem Thermodyn 35:41-53
Sloan ED Jr (2004) Introductory overview: hydrate knowledge development. Am Mineral 89:1155-1161.
Soma Y, Seyama H, Soma M (1993) Adsorption of fluorescent dyes to chrysotile asbestos. Clay Science 9: 9-20
Suquet (1989) Effects of dry grinding and leaching on the crystal structure of chrysotile. Clays Clay Mineral 37:439-445
Tartaj P, Cerpa A, Garcia-Gonzalez MT, Serna CJ (2000) Surface instability of serpentine in aqueous suspensions. J Colloid Interface Sci 231:176-181
Titulaer MK, van Miltenburg JC, Jansen JBH, Geus, JW (1993) Characterization of tubular chrysotile by thermoporometry, nitrogen sorption, DRIFTS, and TEM. Clays Clay Mineral 41:496-513
Viti C, Mellini M (1998) Mesh textures and bastites in the Elba retrograde serpentinites. Eur J Mineral 10: 1341-1359
Walsh D, Mann S (1995) Fabrication of hollow porous shells of calcium carbonate from self-organizing media. Nature 377:320-323
White T, Ferraris C, Kim J, Madhavi S (2005) Apatite – an adaptive framework structure. Rev Mineral Geochem 57:307-401
Whittaker EJW (1957) The structure of chrysotile. V. Diffuse reflexions and fibre texture. Acta Cryst 10:149-156
Whittaker EJW (1966) Diffraction contrast in electron microscopy of chrysotile. Acta Cryst 21:461-466
Wicks FJ, O'Hanley DS (1989) Serpentine minerals: structure and petrology. Rev Mineral 19:91-167
Wilson MA, Pang LSK, Vassallo AM (1992) C_{60} separation on coal. Nature 355:117-118
Xia Y, Lu Y, Kamata K, Gates B, Yin Y (2003) Macroporous materials containing three-dimensionally periodic structures. *In:* The Chemistry of Nanostructured Materials. Yong P (ed) World Scientific Publishing Co., Singapore, p 69-100
Yada K (1967) Study of chrysotile asbestos by a high resolution electron microscope. Acta Crystallogr 23: 704-707
Yada K (1971) Study of microstructure of chrysotile asbestos by high resolution electron microscopy. Acta Crystallogr A27:659-664
Yada K, Iishi K (1977) Growth and microstructure of synthetic chrysotile. Am Mineral 62:958-965
Yan H, Yang P (2003) Semiconductor nanowires: functional building blocks for nanotechnology. *In* The Chemistry of nanostructured materials. Yong P (ed) World Scientific Publishing Co., Singapore, p 182-226
Yang H, Coombs N, Ozin GA (1997) Morphogenesis of shapes and surface patterns in mesoporous silica. Nature 386:692-695
Zega TJ, Buseck PR (2003) Fine-grained mineralogy of the Cold Bokkeveld CM chondrite. Geochim Cosmochim Acta 67:1711-1721
Zega TJ, Garvie LAJ, Dodony I, Buseck PR (2004) Serpentine nanotubes in the Mighei CM chondrite. Earth Planet Sci Lett 223:141-146
Zheng N, Bu X, Feng P (2003) Synthetic design of crystalline inorganic chalcogenides exhibiting fast-ion conductivities. Nature 426:428-432